D1786493

VIIIth INTERNATIONAL CONGRESS ON

MATHEMATICAL PHYSICS

VIIIth INTERNATIONAL CONGRESS ON

MATHEMATICAL PHYSICS

July 16-25, 1986
Marseille, France

Editors:
M. Mebkhout
R. Sénéor

Published by

World Scientific Publishing Co Pte Ltd.
P. O. Box 128, Farrer Road, Singapore 9128

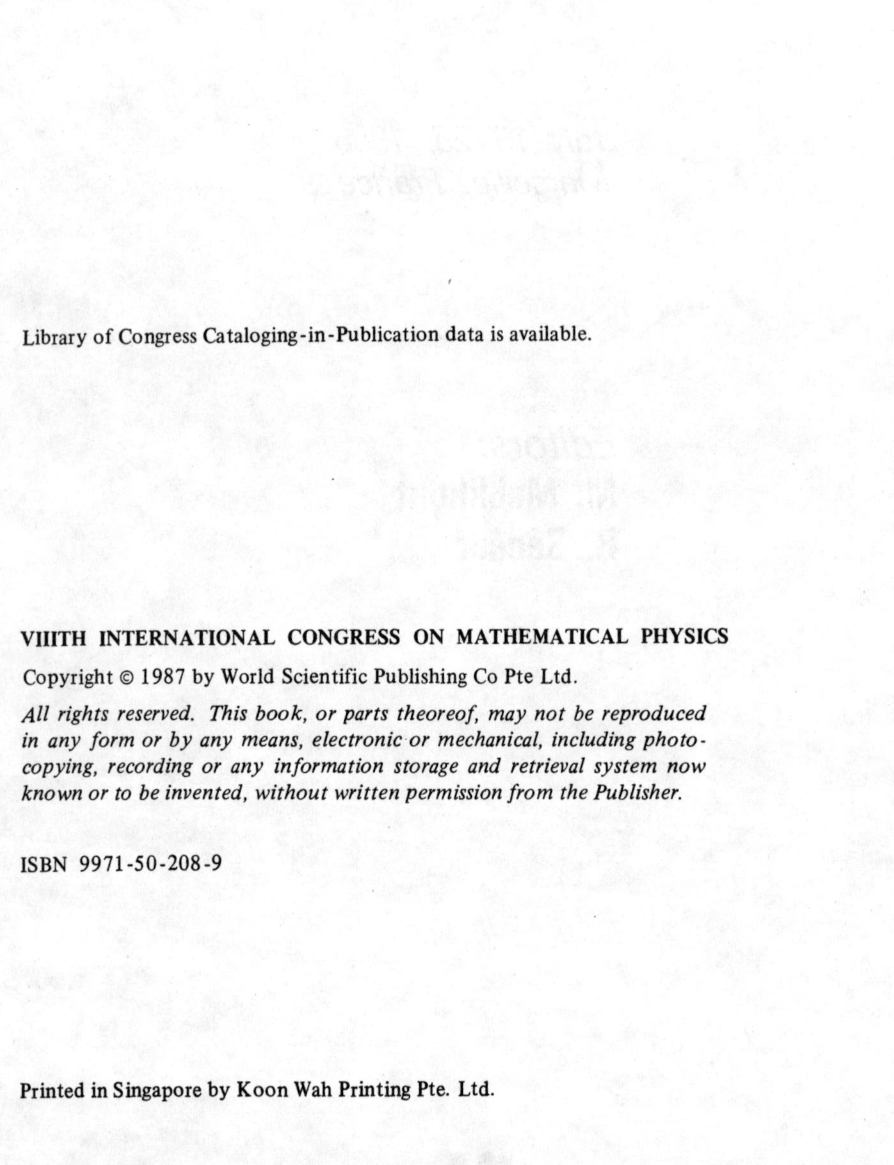

Library of Congress Cataloging-in-Publication data is available.

VIIITH INTERNATIONAL CONGRESS ON MATHEMATICAL PHYSICS

ISBN 9971-50-208-9

Printed in Singapore by Koon Wah Printing Pte. Ltd.

Avant-propos

Ce volume contient les notes écrites de la plupart des conférences prononcées lors de la VIIIème Conférence Internationale de Physique Mathématique qui s'est tenue du 16 au 25 juillet 1986 à Marseille, France, sur le campus de l'Université d'Aix-Marseille II, Luminy. Cette conférence a réuni plus de 500 participants venant de 40 pays différents. Elle fait partie d'une série, patronnée par l'Union Internationale de Physique Mathématique, qui a débuté en 1972 à Moscou et qui s'est poursuivie en 1974 à Varsovie, en 1975 à Kyoto, en 1977 à Rome, en 1979 à Lausanne, en 1981 à Berlin et en 1983 à Boulder.

La prochaine conférence aura lieu en 1988 à Swansea en Grande Bretagne.

Cette conférence est le résultat de plusieurs choix : choix de la durée, 10 jours, qui en a fait une conférence relativement longue; choix d'une ouverture vers certains domaines sortant du cadre traditionnel soit par leur nouveauté, soit par leur marginalité. Le premier de ces choix était une conséquence de l'espacement, 3 ans, plus long que d'habitude, décidé entre la réunion de Boulder et celle de Marseille; il en était espéré néanmoins une plus grande aération. Le second choix amenait à ne pas recouvrir tous les domaines traditionnels d'activité des physiciens mathématiciens et à proposer soit en session soit en conférence pleinière des thèmes dans lesquels notre communauté n'avait d'évidence que peu investi. Il en a résulté une conférence peut être plus dense que ce que nous aurions souhaité mais d'une grande richesse. Elle a suscité une importante activité spontanée sous forme de séminaires et de sessions nocturnes portant sur : la mécanique statistique à l'équilibre, la géométrie non-commutative, les systèmes intégrables, les probabilités et l'analyse non-standard, les fractals, les approches à une théorie quantique de la gravitation, la limite semi-classique, les théories de Kaluza-Klein et les cordes, la théorie des opérateurs de Schrödinger.

Les conférences principales ont soit traité de travaux originaux, soit présenté une revue sur un thème développé ensuite en session. Il y a eu 22 conférences pleinières et 15 sessions spécialisées. MM. les Professeurs S.P. Novikov et A.S. Wightman ont ouvert le congrès. Il a été clos par MM. les Professeurs A. Jaffe et K. Ostervalder.

Le Comité de la Conférence était composé de MM. M.Atiyah (GB), Ph.Combe (F), G.F.Dell'Antonio (I), L.Faddeev (USSR), A.Jaffe (USA), M.Mebkhout (F), K.Osterwalder (CH), R.Seneor (F), J.M.Souriau (F), L.Streit (Président) (RFA).

Le Comité d'organisation était composé de Ph.Combe (F), M.Mebkhout (Président) (F), J.M.Souriau (F).

Le Comité a bénéficié des conseils du Comité Scientifique composé de MM. S.Albeverio (Bochum), J.P.Antoine (Louvain-la-Neuve), H.Araki (Kyoto), J.E.Avron (Haïfa), A.Böhm (Austin), D.Buchholz (Hamburg), A.Connes (Bures-sur-Yvette), H.Ezawa (Tokyo), M.Flato (Dijon), J.Fröhlich (Zurich), G.Gallavotti (Rome), KGawedzki (Bures-sur-Yvette), J.Glimm (New York), S.Graffi (Bologna), S.W.Hawking (Cambridge), Ch.A.Hurst (Adelaïde), C.Itzykson (Gif-sur-Yvette), G.Jona-Lasinio (Rome), D.Kastler (Marseille), J.R.Klauder (Murray Hill), H.Kunz (Lausanne), O.E.Lanford (Bures-sur-Yvette), J.L.Lebowitz (New Brunswick), E.Lieb (Princeton), G.Mack (Hamburg), A.Martin (Geneva), M.Moshinsky (Mexico), D.Ruelle (Bures-sur-Yvette), E.Seiler (München), B.Simon (Pasadena), Y.G.Sinai (Moscow),

H.E.Stanley (Cambridge), W.Thirring (Wien), I.T.Todorov (Sofia), A.Uhlmann (Leipzig), S.R.S.Varadhan (New York), G.Velo Bologna), R.Vilela Mendes (Lisbon), S.Weinberg (Austin), A.S.Wightman (Princeton), E.Witten (Princeton), W.Wyss (Boulder).

Le présent volume comprend, outre les contributions des conférenciers qui nous sont parvenues, des textes provenant de conférenciers invités qui n'ont pu participer à la conférence. Comme nombre de contributions correspondent à des travaux non encore publiés un effort particulier a été fait afin d'accélérer la parution de ce livre.

Remerciements

Ce congrès n'aurait pu être réalisé sans les aides nombreuses et efficaces dont a bénéficié le Comité d'Organisation.

Nous avons eu, dès le début, de la part de M. Gaston Defferre, le regretté Maire de Marseille, un soutien décisif auprès des organismes nationaux et régionaux.

Nous remercions également M. Alain Devaquet, Ministre délégué auprés du Ministre de l'Education Nationale chargé de la Recherche et l'Enseignement Supérieur, pour l'intérêt qu'il a porté à ce congrès.

La collaboration et la disponibilité des collègues du Comité International a été très apprécié.

Le soutien financier des organismes cités ci-dessous a été décisif:

Centre National de la Recherche Scientifique, Commissariat à l'Energie Atomique, Conseil Général des Bouches-du Rhône, Conseil régional PACA, Direction de la Coopération et des Relations Internationales, Direction des Recherches Etudes et Techniques, European Physical Society, Institut National de Physique Nucléaire et de Physique des Particules, International Association of Mathematical Physics, International Union of Pure and Applied Physics, Mairie de Marseille, Ministère de l'Education Nationale, Ministère de la Recherche et de la Technologie, Ministère des Relations Extérieures, Société Mathématique de France, UNESCO, Universités d'Aix-Marseille I, II, Avignon, Nice, Toulon, DIGITAL, IBM, OCE, Société Ricard.

Pour terminer nous tenons à remercier tout le personnel qui a eu en charge l'organisation matérielle considérable du colloque:

Tout le secrétariat du Centre de Physique Théorique, notamment MMes P. Bourgeois, M. Cohen-solal, G. Escalon, N. Lambert, A. Sueur.

Lc personnel de l'Université et de la Faculté des Sciences et leurs responsables : M. Douchy, Mlle Lagier, MM. Sharffe et Vanni.

Le personnel du CROUS (Centre Régional des Oeuvres Universitaires et Scolaires) et leurs responsables : M. Moulin et Mme Phillipe.

M. Mebkhout et R. Sénéor
Marseille-Luminy et Ecole Polytechnique Paris
Décembre 1986

Lettre du Ministre

Message à l'occasion du VIIIème Congrès International de Physique Mathématique
Marseille 16-25 Juillet 1986

J'ai appris avec satisfaction que, pour le première fois, le congrès international de physique mathématique se tient en France, et je regrette vivement que mes obligations ne me permettent pas d'accueillir à Marseille les spécialistes de cette discipline venus du monde entier. Le choix de la ville de Marseille pour ce huitième congrès me paraît très judicieux : il témoigne de la réputation internationale acquise par le centre de physique théorique de Marseille, laboratoire propre du CNRS, et de la qualité des recherches qui y sont menées.

Je souhaite que les multiples sessions qui se dérouleront pendant les dix jours de la conférence permettent de faire un bilan solide de l'évolution des idées depuis le congrès de Boulder, dans des domaines de la physique qui peuvent apparaître fort éloignés les uns des autres, mais qui sont tous caractérisés par le très haut niveau mathématique des méthodes employées. Il est remarquable qu'un même congrès puisse réunir des chercheurs s'intéressant à la théorie des champs, aux systèmes turbulents, aux milieux aléatoires ou à des problèmes cristallographiques. Il est vrai que l'expérience a déjà montré à quel point des méthodes développées dans certains domaines pouvaient se révéler fructureuses dans d'autres disciplines, et il s'agit le plus souvent d'une même approche mathématique des questions posées.

J'espère que de nouvelles idées naîtront de cette rencontre exceptionnelle entre les meilleurs spécialistes de la physique mathématique, et que le congrès de Marseille marquera une importante étape dans le développement de cette discipline.

Alain DEVAQUET

Ministre délégué auprès du
Ministre de l'Education Nationale
Chargé de la Recherche et de
l'Enseignement Supérieur

FOREWORD

This volume contains written notes of most of the communications delivered at the VIIIth Congress of Mathematical Physics held in Marseille, France, on the Luminy campus of the University of Aix-Marseille II, from July 16th to July 25th 1986.

This Congress was attended by over 500 participants from 40 different countries. It is part of a series of conferences (sponsored by International Union of Mathematical Physics) which began in 1972 in Moscow, and was held subsequently in Warsaw 1974, Kyoto 1975, Rome1977, Lausanne 1979, Berlin 1981 and Boulder 1983. The next conference of the series will take place in 1988 in Swansea (G.B).

Two features characterized this conference : the decision to have a rather long conference (10 days) and the introduction of new or unconventional domains. This choice of duration for the congress was a consequence of the rather long time interval between it and the previous one, it was nevertheless hoped to be not too heavy. As for the introduction of new domains, it led us to omit some of the traditional domains of mathematical physics and to treat, in plenary or parallel sessions, subjects which had not yet been explored. This led to a conference more dense than we would have wished but very rich. It also brought about a large amount of spontaneous activity in the form of seminars or night sessions on the following subjects: statistical mechanics at equilibrium, non-commutative geometry, integrable systems, probability and non-standard analysis, fractals, approaches to quantum theory of gravitation, semi-classical limit, Kaluza-Klein theories, strings and Schrödinger operators.

The main talks either presented original work, or were reviews on a theme which was subsequently developped during a session. There were 22 plenary talks and 15 topical sessions. The opening addresses were done by Professor S.P. Novikov and A.S. Wightman. The congress was closed by Professor A. Jaffe and K. Ostervalder.

The Conference Committee included : MM. M.Atiyah (GB), Ph.Combe (F), G.F.Dell'Antonio (I), L.Faddeev (USSR), A.Jaffe (USA), M.Mebkhout (F), K.Osterwalder (CH), R.Seneor (F), J.M.Souriau (F), L.Streit (Président) (RFA).

The Organizing Committee was composed by : Ph.Combe (F), M.Mebkhout (Chairman) (F), J.M.Souriau (F).

The Conference Committee benefited from the suggestions of the Advisory Committee including : MM. S.Albeverio (Bochum), J.P.Antoine (Louvain-la-Neuve), H.Araki (Kyoto), J.E.Avron (Haïfa), A.Böhm (Austin), D.Buchholz (Hamburg), A.Connes (Bures-sur-Yvette), H.Ezawa (Tokyo), M.Flato (Dijon), J.Fröhlich (Zurich), G.Gallavotti (Rome), KGawedzki (Bures-sur-Yvette), J.Glimm (New York), S.Graffi (Bologna), S.W.Hawking (Cambridge), Ch.A.Hurst (Adelaïde), C.Itzykson (Gif-sur-Yvette), G.Jona-Lasinio (Rome), D.Kastler (Marseille), J.R.Klauder (Murray Hill), H.Kunz (Lausanne), O.E.Lanford (Bures-sur-Yvette), J.L.Lebowitz (New Brunswick), E.Lieb (Princeton), G.Mack (Hamburg), A.Martin (Geneva), M.Moshinsky (Mexico), D.Ruelle (Bures-sur-Yvette), E.Seiler (München), B.Simon (Pasadena), Y.G.Sinai (Moscow), H.E.Stanley (Cambridge), W.Thirring (Wien), I.T.Todorov (Sofia), A.Uhlmann (Leipzig), S.R.S.Varadhan (New York), G.Velo Bologna), R.Vilela Mendes (Lisbon), S.Weinberg (Austin), A.S.Wightman (Princeton), E.Witten (Princeton), W.Wyss (Boulder).

This book collects the contributions that we received from the speakers and also some invited papers from speakers who were not able to attend the congress.

As many of the contributions refer to unpublished work, we made an effort to speed up the publication of this book.

ACKNOWLEDGMENTS

This Congress could not have been held without the extensive and efficient help we received.

From the very beginning, we obtained from Mr.Gaston Defferre, the late Mayor of Marseille, a decisive support from the national and regional authorities. We also wish to thank Mr. Alain Devaquet, Minister of Education for the interest he showed in this Congress.

The participation of our Conference Committee Colleagues and the time they devoted to the Congress were very much appreciated.

From the financial point of view we could not have succeded without the sponsors listed below:

Centre National de la Recherche Scientifique, Commissariat à l'Energie Atomique, Conseil Général des Bouches-du Rhône, Conseil régional PACA, Direction de la Coopération et des Relations Internationales, Direction des Recherches Etudes et Techniques, European Physical Society, Institut National de Physique Nucléaire et de Physique des Particules, International Association of Mathematical Physics, International Union of Pure and Applied Physics, Mairie de Marseille, Ministère de l'Education Nationale, Ministère de la Recherche et de la Technologie, Ministère des Relations Extérieures, Société Mathématique de France, UNESCO, Universités d'Aix-Marseille I, II, Avignon, Nice, Toulon, DIGITAL, IBM, OCE, Société Ricard.

Last but not least, we want to thank all the staff who took charge of the organization of the Congress: all the secretariat of CPT (Centre de Physique theorique de Marseille) and in particular, P. Bourgeois, M. Cohen-solal, G. Escalon, N. Lambert, A. Sueur, the staff of the University and of the Faculty of Sciences and their heads : M. Douchy, Mlle Lagier, MM. Sharffe et Vanni, the staff of C R O U S (Centre Régional des Oeuvres Universitaires et Scolaires) especially M.Moulin et Mme Phillipe.

M. Mebkhout and R. Sénéor
Marseille-Luminy and Ecole Polytechnique Paris
December 1986

Table of Contents

Avant-propos. v

Foreword ix

Main Talks

V. I. ARNOLD* : "First steps of symplectic topology". 1

J. M. BISMUT : "Filtering equation, equivariant cohomology and the Chern character" 17

T. DAMOUR : "New problems and new approximation methods in general relativity". 57

R. L. DOBRUSHIN** : "Induction on volume and no cluster expansion" 73

E. S. FRADKIN, A. A. TSEYTLIN : "Effective action in superstring theory". 92

V. GUILLEMIN : "Quasi-classical aspects of reduction". 106

M. HAZEWINKEL : "Lie algebraic methods in filtering and identification" 120

M. R. HERMAN : "Recent results and some open questions on Siegel's linearization theorem of germs of complex analytic diffeomorphisms of C^n near a fixed point". 138

E. H. LIEB : "A model for crystallization: a variation on the Hubbard model". . . . 185

V. P. MASLOV : "Deterministic theory of turbulence in hydrodynamics (coherent structures)". 197

A. NEVEU : "Superstrings". 217

S. P. NOVIKOV : "Two dimensional Schrödinger operator and solitons. 3-dimensional integrable systems" 226

D. OLIVE : "Infinite dimensional algebras in modern theoretical physics" 242

V. RIVASSEAU : "Constructive renormalization" 257

D. RUELLE: "Theory and experiment in the ergodic study of chaos and strange attractors" 273

---

* Presented by V. Guillemin

** Presented by M. Aizenman

Ya. G. SINAI : "Anderson localization for one-dimensional difference Schrödinger operators with quasi-periodic potentials" 283

I. M. SIGAL, A. SOFFER : "Asymptotic completeness for N-body quantum mechanics" 292

S. T. YAU : "A survey on the interaction between mathematical physics and geometry" 305

Topical Talks

Equilibrium statistical mechanics
Session organizers: J. BRICMONT and YU. SUKHOV

D. C. BRYDGES, T. KENNEDY : "Convergence of multiscale Mayer series" 311

K. GAWĘDZKI : "Many scales picture of first order phase transitions"...... 318

R. KOTECKÝ, S. MIRACLE-SOLE : "A roughening transition indicated by the behaviour of ground states" 331

Ch. E. PFISTER : "The wetting transition. New rigorous results for the semi-infinite Ising model" 338

M. ZAHRADNÍK : "Stable and unstable phases in the Pirogov Sinai theory" 347

Fields, quanta, entropy, understanding heritage and perspectives
Session organizer : R. HAAG

H. ARAKI : "Recent progress on entropy and relative entropy" 354

L. BAULIEU : "Non commutative diffeomorphisms" 366

D. BUCHHOLZ : "On particles, infraparticles, and the problem of asymptotic completeness" 381

Integrable systems and Kac-Moody algebras
Session organizers: M. JIMBO and M. C. POLIVANOV

P. GODDARD : "Kac-Moody and Virasoro algebras in mathematical Physics" 390

E. K. SKLYANIN : "Boundary conditions for integrable systems" 402

Stochastic methods
Session organizer: G. JONA-LASINIO

S. ALBEVERIO : "Some personal remarks on nonstandard analysis in probability theory and mathematical physics" 409

PH. COMBE, M. SIRUGUE, M. SIRUGUE-COLLIN : "Point processes and quantum physics : some recent developments and results" 421

E. GETZLER : "The Degree theory of the Nicolai map"................. 431

M. YOR : "Some recent studies of Brownian paths intersections".......... 439

Non equilibrium statistical mechanics
Session organizer: J. L. LEBOWITZ

S. J. AL'BER, M. S. AL'BER* : "Hamiltonian formalism for finite-zone solutions of non-linear integrable equations"........................... 447

G. CASATI : "Relevance of classical chaos in quantum mechanics" 463

P. COLLET, J.-P. ECKMANN : "Pattern selection in hydrodynamical equations" 467

M. FREIDLIN* : "Some general properties of evolution processes quasi-deterministic approximation".................... 470

S. GOLDSTEIN : "Diffusive and subdiffusive behavior in mechanical and stochastic models" 482

E. PRESUTTI : "Collective phenomena in stochastic interacting particle systems" 486

General quantum field theory
Session organizer: R. HAAG

S. DOPLICHER, J. E. ROBERTS : " C*-algebras and duality for compact groups: why there is a compact group of internal gauge symmetries in particle physics" 489

K. FREDENHAGEN : "Criteria for quark confinement"................. 499

O. PIGUET, K. SIBOLD : "Models with asymptotic scale invariance"....... 505

---

* Presented by J. L. Lebowitz

Dynamical systems
Session organizer : O. E. LANFORD III

V. G. BARYAKHTAR : "The relaxation terms of the Landau-Lifschitz equation and the problem of soliton damping 512

R. DE LA LLAVE, D. RANA : "Accurate bounds in K.A.M. theory"....... 516

J.-P. ECKMANN, H. EPSTEIN : "Fixed points of composition operators" 517

J. FRÖHLICH, T. SPENCER, E. WAYNE : "Nonlinear localization and an infinite dimensional K.A.M. theorem" 531

O. E. LANFORD III : "Renormalization group methods for critical circle mappings with general rotation number......... 532

D. RAND : "Universality for the breakdown of dissipative golden invariant tori" 537

Quantum gravity
Session organizer : N. SANCHEZ

I. A. BATALIN, E. S. FRADKIN : "Operatorial quantization of dynamical systems with irreducible first and second class constraints"...................................... 548

P. D. D'EATH, J. J. HALLIWELL : "The inclusion of fermions in quantum cosmological models"........ 549

S. DESER : "String corrections to Einstein gravity" 557

J. B. HARTLE : "Initial conditions and quantum cosmology" 566

H. RUMPF : "Stochastic quantization of gravity and string fields" 582

Constructive quantum field theory
Session organizers : G. GALLAVOTTI and R. SENEOR

P. FEDERBUSH : "Towards a four dimensional Yang-Mills theory"........ 597

G. FELDER : "Non-trivial renormalization group fixed points"........... 605

J. FELDMAN, T. HURD, L. ROSEN, J. WRIGHT : "Renormalizability of QED_4" 614

D. IAGOLNITZER : "Asymptotic completeness and multiparticle analysis in field theories" 622

A. KUPIAINEN : "Construction of renormalizable and non-renormalizable quantum field theories"............... 628

Supersymmetry
Session organizer : J. WESS

A. BOHM : "Superconformal algebra and relativistic supersymmetric quantum mechanics" . . . 638

M. FORGER : "Higher conservation laws for ten-dimensional supersymmetric Yang-Mills theories" . . . 641

J. HARNAD, S. SHNIDER : "Superconnection integrability conditions and supersymmetric Yang-Mills equations in ten dimensions" . . . 644

G. GIRARDI, R. GRIMM : "Chern Weil forms in superspace" . . . 651

M. MULLER : "Minimal N = 2 supergravity" . . . 658

H. ROEMER : "Classification of homogeneous Kähler manifolds" . . . 662

Schrödinger operators and semi-classical methods
Session organizer : J. M. COMBES

B. BAUMGARTNER, H. GROSSE, A. MARTIN : "Level comparison theorems and supersymmetric quantum mechanics" . . . 667

I. DAUBECHIES, T. PAUL : "Wavelets and applications" . . . 675

B. HELFFER : "Resonances in semi-classical analysis" . . . 687

Computational physics
Session organizers : J. GLIMM and A. JAFFE

J. JONES, B. BUKIET : "The competition between curvilinear geometry and chemistry in reactive fluid flow" . . . 696

H. KOCH, P. WITTWER : "Rigorous computer-assisted renormalization group analysis" . . . 702

B. J. PLOHR, D. H. SHARP : "Riemann problems and their application to ultra-relativistic heavy ion collisions" . . . 708

A. D. SOKAL : "New Monte Carlo algorithms for quantum field theory and critical phenomena, or how to beat critical slowing-down" . . . 714

Disordered systems
Session organizers : J. BELLISSARD and J. IMBRIE

M. DUNEAU, A. KATZ : "Models of quasiperiodic structures" . . . 724

J. Z. IMBRIE : "The Ising model in a random magnetic field" . . . 729

F. MARTINELLI, E. SCOPPOLA : "Rigorous results on Anderson localization" 731

R. RAMMAL : "Physics of disordered media : some new recent results" 736

B. SOUILLARD : "Waves in non-linear and non-homogeneous media" 738

Anomalies
Session organizer : L. ALVAREZ-GAUME

L. BAULIEU : "Local BRS symmetry, and higher order cocycles" 743

M. J. DUFF : "Superstrings from the bosonic string on the group manifold" 746

R. STORA : "Differential algebras in field theory and their anomalies : two examples" 757

Strings and superstrings
Session organizer : L. ALVAREZ-GAUME

J.-B. BOST : "Conformal and holomorphic anomalies on Riemann surfaces and determinant line bundles" 768

G. MOORE : "Modular forms and multiloop string physics" 776

Contributed papers from invited speakers who were not able to attend the congress

P. M. BLEHER : "The Thouless effect in the hierarchical model" 786

P. M. BLEHER : "Symmetry breaking in the classical N-Vector hierarchical model" 789

E. I. DINABURG, A. E. MAZEL : "Low-temperature phase transitions in ANNNI model" 796

S. N. ISAKOV : "Nonanalytic features of the first order phase transition in classical lattice gas models" 816

S. B. SHLOSMAN : "Graph colouring : a way to variety of new correlation inequalities" 839

Poster Sessions 848

Talks not published 856

Authors Index 858

List of participants 859

FIRST STEPS OF SYMPLECTIC TOPOLOGY

V. ARNOLD
Department of Mathematics,
Moscow University

ABSTRACT

Conjectures and theorems on fixed points of canonical transformations, on intersections of Lagrangian manifolds, on Legendrian knots and on the singularities of optical caustics are discussed.

Symplectic geometry is the geometry of phase space, endowed with its symplectic structure - the closed nondegenerate 2-form "$dp \wedge dq$". One may consider it a particular case of ordinary geometry in the presence of an additional structure. I believe, however, that it is more suggestive to view symplectic geometry as being rather a variant of ordinary geometry, having equal rights, like the complex variant.

Thus, the symplectic group is a particular subgroup of the matrix group, but it is better to view it as the simple Lie group C_k, having its own roots system, Cartan and Weyl groups and so on, which are as nice as those of the general matrix group. Similarly, symplectic topology seems to be rather a sister than a descendant of ordinary topology.

Rather than make this analogy formal, I have used it to guess symplectic results by a "symplectisation" of the notions and results of ordinary geometry. Such symplectisations may look very strange. Thus, the "symplectic boundary"* of a symplectic manifold has codimension 2, and the "symplectohomology" should be very different from all the existant homology theories, including the "perverse" ones.

----

* for details on the symplectic boundary, see the Lagrangian and Legendrian cobordisms theories in [1), 2), 3)]. By the way, the complex analogue for "boundary" is "the ramification divisor", for "Z_2"-"Z", for "O"-"U" (see [4)]).

Conjectures, arising from these unformal ideas, were published between 1965 and 1976 (see [5) 8)]). Some of them have recently been proved by Conley, Zehnder, Chaperon, Sikorav, Gromov and others, and these powerful new methods provide some hope for the whole symplectisation program.

I. GENERALIZED POINCARE THEOREM

The simplest results in symplectic topology are the following two theorems (see 7), 9))

Poincare-Birkhoff theorem. An orientation - and area - preserving transformation of the annulus, turning the boundary circles in opposite directions, has at least two fixed points.

Shnirelman-Nikishin theorem. An orientation - and area - preserving mapping of a 2-sphere has at least two fixed points.

In both cases the first author discovered the result and the second proved it.

To formulate a generalization * of the Poincare theorem we need :

Definition. A symplectic, orientation - and area - preserving, mapping of the 2-torus preserves the gravity center, if it can be lifted to a plane transformation $x \to x + f(x)$, where the Z^2 - periodic mapping f has mean value 0 on the unit square.

Theorem [10)]. Such a mapping has at least 4 fixed points if they are nondegenerate, and at least three of them are geometrically different in any case.

The Poincaré-Birkhoff theorem follows, since one can construct such a transformation of the torus from two copies of the given transformation of the annulus.

For an arbitrary symplectic manifold, the gravity center condition reads as follows :

Definition [7)]. A mapping is homologous to the identity, if it is connected to the identity mapping by a smooth path, whose derivative is a (time-depending) Hamiltonian vector field (with a single-valued Hamilton function).

---

* a generalization to transformations of cotangent ball bundles satisfying a linking condition at the boundary is formulated in [6)] but seems to be still unproved (even for the unit bundle over the 2-torus with its standard symplectic structure).

Remark. These mappings form the subgroup of commutators of the component of the identity in the group of all symplectic diffeomorphisms, see [11].

Conjecture [7), 8)]. Such a mapping on a compact manifold M has at least as many fixed points as a smooth function on M has critical points.

This conjecture is now proved i) for the 2-surfaces (see [12), 13), 17)]), ii) for the standard 2n-torus (see [10), 14)]), iii) for the standard CP^n (see [15)]), iv) for C_0-small mappings (see [16)]), v) for some negatively curved Kahler manifolds (see [13), 17)]).

To obtain a mapping with no more fixed points than the minimal number, take the small time value of the flow generated by a hamiltonian vector field.The minimal numbers of critical points on simple manifolds are given by the table

manifold	c_a	c_g
circle	2	2
2-torus	4	3
sphere with g handles	2g+2	3
2n-torus	2^{2n}	2n+1
CP^n	n+1	n+1

If the critical points are all nondegenerate their number equals at least the sum c_a, of the Betti numbers. In any case the number of geometrical distinct criticall points is not less than the Ljusternik-Shnirelmann category c_g - which is greater than the cuplength.

Remark. For symplectic mappings which are not homologous to the identity, the "function" in the conjecture should be replaced by "closed 1-form" and the Morse inequalities by those of Novikov [18), 19)].

The Conley-Zehnder method is a "hyperbolic" generalization of Morse theory (both inertia indîces being infinite) in the same fashion as the Anosov systems are "hyperbolic" generalizations of the Ljapounov asymptotically stable attractors. In spite of the important progress of last years, it is still unknown whether every transformation of the 4-torus , homologous to the identity, has a fixed point (assuming the symplectic structure nonstandard).

2. INTERSECTIONS OF LAGRANGIAN MANIFOLDS

According to a general principle of symplectic geometry, "everything" is a Lagrangian manifold [20].

Definition. A Lagrangian submanifold of a symplectic manifold is a submanifold of maximal dimension on which the symplectic structure vanishes (this maximal dimension is n in a phase space of dimension 2n).

Typical examples are the graph of the differential of a function and the Liouville or Kolmogorov invariant tori of integrable or almost integrable Hamiltonian systems.

A canonical transformation, that is a symplectomorphism $f : M \to M$, may be viewed as a Lagrangian manifold * : its graph is a Lagrangian submanifold of the product space MxM, endowed with the symplectic form "$dP \wedge dQ - dp \wedge dq$", i.e. the difference of the pull-backs of the symplectic structures of the two factors..

The fixed points of f are the intersections of two Lagrangian manifolds, the graph and the diagonal.

Definition. Two Lagrangian manifolds are Hamilton homologous if one of them is transformed into the other by the time one mapping of a Hamilton flow with single -valued (but in general time dependent) Hamilton function.

Conjecture (see [5), 8)]). The number of intersection points of two such compact manifolds (L and L') has the same lower bound as the number of critical points of a

----

* this leads to an interesting generalization of the canonical transformations : the symplectic correspondences which are arbitrary Lagrangian submanifolds of the product.

function on one of them, provided the integral of the symplectic form is 0 on every disk whose boundary lies in L (or L').

Theorem (see [14), 17a)]). The number of intersection points of the zero section (p=0) of the cotangent bundle ** T* X with any Lagrangian compact manifold, Hamilton homologous to it, is greater than the cuplength of X , and at least equal to the sum of Betti numbers of X if all the intersections are transversal.

Example. For T^n we obtain at least n+1 intersection points, and at least 2^n if all of them are transversal.

Remark. A neigbourhood of any Lagrangian submanifold is symplectomorphic to a neighbourhood of the zero section of the cotangent bundle [20)]. Thus, the theorem provides a local version of the above conjecture.

The simplest case of the theorem says that a circle, embedded into a cylinder, homotopic to the equator and bounding with the equator a surface of oriented area zero must intersect the equator at least at two points.

Such an intersection result fails for immersions (as was noted in [5)]), see fig. 1, even if they are regularly homotopic to the equator. An immersion variant of the above theorem, due to Yu. V. Chekanov, is formulated in the next section.

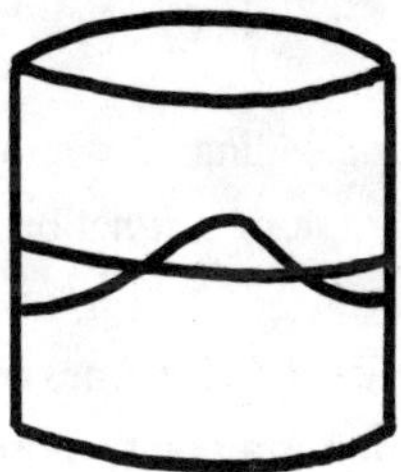

Fig.1

** the cotangent bundle of a configuration manifold X is its phase space together with its natural projection on t and the linear space structure of the fibers (of the momentum spaces at every configuration space point). It carries the natural symplectic structure - the Poincaré integral invariant "dp∧dq" in classical notations - and is denoted by T* X.

3. LEGENDRIAN SUBMANIFOLDS OF CONTACT MANIFOLDS

Contact geometry is the odd-dimensional sister of symplectic geometry.

Definition. A contact structure on an odd-dimensional manifold is a field of tangent hyperplanes, which is generic at each point.

Example. The complex lines, tangent to S^3, the boundary of a ball in C^2, define a contact structure on S^3-its standard contact structure. Similarly for S^{2n+1}.

A local model for all the 2n+1-contact manifolds is the space with coordinates x_i, p_i, y (i=1,...,n) endowed with the contact structure dy=pdx. This space is called the space of 1-jets* of functions $J^1(R^n, R)$ (x=argument, y=value, p=first derivative).

Forgetting the value of functions, we project this 1-jets space onto the cotangent bundle defining the line fibration $J^1(X,R) \to T^* X$.

Definition. The Legendrian submanifolds of a contact manifold are those integral manifolds of the contact structure of the greatest possible dimension (this equals n in a 2n+1-manifold).

Example. The 1-graph of a function f $(y=f(x), p=\partial f/\partial x)$.

Definition. A quasifunction on X is a Legendrian embedding of X into its space of 1-jets of functions, which is regular homotopic to a 1-graph of a function (among the Legendrian embeddings).

The projection of a quasifunction to the phase space T* X defines an immersed Lagrangian submanifold. A critical point of a quasifunction is a point whose projection belongs to the zero section of T* X.

Theorem (Ju. V. Chekanov). The number of critical points of a quasifunction on a compact manifold X is greater than the cuplength of X, and at least equal to the sum of the Betti numbers of X, if all of these critical points are non degenerate.

----

* k-jet = Taylor polynomial of degree k (at a given point).

Example. The homotopy condition is essential. One sees at the left side of fig. 2 the projection of an immersion $S^1 \; J^1(S^1,R)$ to $T\,S^1$, which does not intersect the zero section. Hence it is not homotopic to a quasifunction, whose projection is represented at the right side, in the class of Legendrian embeddings (while they are obviously homotopic in the class of Legendrian immersions).

fig. 2

4. LAGRANGIAN AND LEGENDRIAN KNOTS

These are defined as the connected components of the spaces of Lagrangian and Legendrian embeddings ($C^r, r \geq 1$).

Example 1. Let us consider the Legendrian knots in the standard 3-sphere. Every knot in the usual sense has a Legendrian representative (this follows from the Rashevski-Chow generalization of the Caratheodory's theorem in thermodynamics).

Still, the classification of Legendrian knots is very different from that of ordinary knots. For instance, there exist "purely Legendrian" knots, nonhomotopic Legendrian embeddings of the circle, unknotted in the usual sense .

To give an example, we may define the "Maslov index"[21)] of a Legendrian curve as the degree of the composed mappings

$S^1 \to U(2) \to S^1$ (left arrow : the frame, formed by the velocity vector of the curve and the exterior normal to the sphere in C^2, right arrow : the determinant).

Theorem (A.V. Alekseev)*. The index is the unique homotopy invariant of Legendrian immersions into the standard 3-sphere : all Legendrian immersions having the same index are regular homotopic among the Legendrian immersions and all the values of the index are attainable by Legendrian embeddings arbitrarily C^0-close to any given one.

It seems that the ordinary knot type and the index value do not define the Legendrian knot type (fig. 2).

The multidimensional version of the above construction provides a Gauss type mapping of a Legendrian manifold

$L^{n-1} \hookrightarrow S^{2n-1}$ to the Lagrangian Grassmannian Λ_n=U(n)/0(n).

Example 2. Let us consider the Lagrangian embeddings $R^2 \hookrightarrow R^4$, equal to the embedding of the plane p=0 outside a ball in the standard R^4. The Lagrangian knots problem reads :

(i) do there exist such embeddings, knotted in the usual sense (nonhomotopic to the standard embedding $\mathbb{R}^2 \to \mathbb{C}^2$ in the space of embeddings);

(ii) if the answer is positive, find whether every usual knot admits a Lagrangian representative ;

(iii) do there exist purely Lagrangian knots (Lagrangian embeddings, homotopic to the plane among the usual embeddings, but not among the Lagrangian ones).

Remark. A Lagrangian knot $R^n \to R^{2n}$ defines an element of the Lagrangian Grassmannian homotopy group, $\pi_n(\Lambda_n)$. For n=2 this Grassmannian is the nontrivial** 2-spheres bundle over a circle [22)]. It is unknown whether this element of the homotopy group may be nontrivial. In any case this cannot be detected from the characteristic numbers, since they vanish on these elements (A.B. Givental).

----

*this is a particular case of the general Alekseev variant of Smale immersion theory.

** the two proofs of the opposite in 23) are wrong.

5. GIENTAL'S THEOREMS ON LAGRANGIAN EMBEDDINGS

Let us consider the projection of a Lagrangian knot, embedded in R^{2n}, to the configuration space (fig. 3), left :

$$R^n \rightarrow T R^n \rightarrow R^n.$$

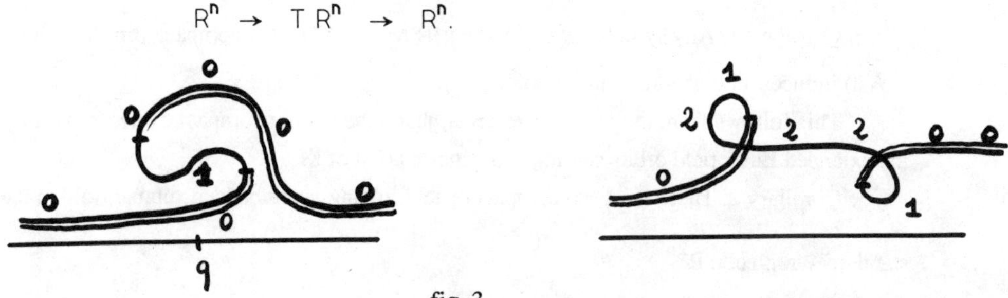

fig. 3

For every point of the knot above any regular value q of the projection, one defines the Maslov index of the point as the intersection number of any curve, arriving at this point from infinity along the knot with the hypersurface consisting of the singularities of the restricted projection [21]. One sees the values of the index for some points of the knot on fig. 3, left.

Theorem. Above every regular value there exists a point of index zero.

This theorem has been conjectured by V. Kolokolzov, who proved it for plane curves.

The second theorem results from an attempt to symplectise the following fact : an embedding of a compact manifold into a manifold of the same dimension induces an embedding of the fundamental class in homology.

Definition. A Lagrangian submanifold of the cotangent bundle space is exact, if it is the projection of a Legendrian submanifold of the space of 1-jets of functions, (that is if the "generating function" $\int pdq$ is single-valued).

Theorem. Any exact embedding $M \rightarrow T^*N$ of a compact Lagrangian submanifold admits an extension to a symplectic embedding $T^*M \hookrightarrow T^*N$ (the Lagrangian manifold being the image of the zero section).

The proof depends on an extension of the image of the Euler field of T^*M from a tubular neighbourhood of the embedded manifold M to the whole of T* N.

Conjecture (Givental). For almost every point n of M, the orbits of the extended Euler field starting at n go to infinity in T* N.

Admitting this conjecture, Givental proves :

Corollary 1. Any symplectic embedding $T^*M \rightarrow T^*N$ (M compact, dim M = dim N = n) induces an embedding in homology.

This follows from the fact M intersects, just once the noncompact cycle, formed by the extended Euler field orbits starting at a generic point of M.

Corollary 2. There exists no compact exact Lagrangian embedded submanifold in the standard symplectic R^{2n}.

This old conjecture (see [24]) has been recently proved by M. Gromov.

Corollary 3. The image of the zero section of T* M under a symplectomorphism homologous to the identity mapping intersects the zero section.

Otherwise there would exist an exact Lagrangian embedding of the union of two disjoint copies of M into T* M, whose fundamental class maps to zero.

6. ODD-DIMENSIONAL FORMULATIONS

Let a contact structure have a transverse orientation. Then, one may choose a contact 1-form, that is a form α, whose field of kernels is the contact structure. The 2-form $d\alpha$ defines a line field (that of its kernels), and its integral curves are called the characteristics. They are transversal to the contact hyperplanes and depend on the choice of a particular contact form defining the given contact structure.

If the characteristics form a fibration, its base space is a symplectic manifold. The Legendrian embeddings project to Lagrangian immersions along the characteristics. The theorems and conjectures on the symplectic topology of the base space admit formulations in terms of the initial contact space and of the characteristic fibration. Avoiding references to the global fibration and to the base space, we obtain new conjectures in contact geometry.

Example. Let us consider a Legendrian S^1 in the standard 3-sphere. For the standard contact form, the characteristic fibration is the Hopf fibration $S^3 \rightarrow S^2$. The Lagrangian projection of the Legendrian curve is a curve on the 2-sphere, bounding zero area mod 4π. Such a curve has a self intersection point, hence :

<u>Conjecture</u>. Any Legendrian S^1 in the standard contact S^3 has a characteristic cord for every choice of a contact form.

For the same reason, we may expect a lower bound on the number of characteristic paths from any Legendrian submanifold of the standard R^{2n+1} or S^{2n+1} to any nearby Legendrian immersion in terms of the topology of the submanifold for every choice of the contact form .

In this direction, only the following "magnetic field" variant of the Poincare theorem is proved.

Let $p:E^{2n+1} \to M^{2n}$ be a compact fibration into oriented circles, ω a closed 2-form on E, which is nondegenerate at every point (has 1-dimensional kernels). We impose the following "gravity center preservation" condition : the cohomology class of ω is lifted (belongs to $p^* H^2(M)$), in other words the integrals of ω along the vertical cycles are all zero.

<u>Theorem</u> (V.L. Ginsburg). Suppose that the field of characteristic directions (kernels) of ω is C^1-close to the vertical (fibers directions) field, and that the form preserves the gravity center.
Then, the number of closed characteristics (close to the fibers as embeddings) is greater than the cuplength of M , and at least equal to the sum of the Betti numbers of M if all of these characteristics are nondegenerate.

One hopes this is still true for C^0 neighbourhoods, and for more general cases.

If the gravity center restriction is dropped, the Morse type estimates should be replaced by Novikov's.

<u>Example</u>. Let us consider the motion of a "charged particle" along a Riemannian torus in a magnetic field B which is normal to the surface and nowhere zero [30)].
The above theorem implies that there exist at least 3 closed contractible orbits for every sufficiently small initial velocity value v_o (at least 4, if nondegenerate).

Geometrically these orbits are curves, whose geodesic curvature has at every point the prescribed value $B(x)/v_o$. It is conjectured that such curves exist for all $v_o \neq 0$ (and there are $2g+2$ of them on a sphere with g handles if they are nondegenerate). See (19) for another approach to this problem by Novikov and Taimanov.

7. OPTICAL LAGRANGIAN MANIFOLDS

Solutions of the eikonal equation $(\partial S/\partial q)^2 = 1$ define Lagrangian manifolds $p = (\partial S/\partial q)$ lying on the hypersurface $p^2=1$ of the phase space, this hypersurface has a convex (spherical) intersection with the fiber above any point q.

Definition. A Lagrangian submanifold of the cotangent bundle space is optical if it lies in a hypersurface, which is transversal to the fibers and whose intersections with the fibers are convex quadrics.

Every Lagrangian singularity [4] has a local optical realization [25]. However, being globally optical is a topological restriction on the coexistence of caustic singularities and hence on their local metamorphoses. This effect was first observed by J.Nye [26] for the eikonal case when attempting a laser realization of the typical caustic metamorphoses. The general theory is due to Yu.V. Chekanov [27].

Definition. The characteristic field of a hypersurface in a symplectic manifold is the line field whose value at each point is the symplectic orthogonal of the tangent hyperplane (it is the direction of these Hamilton vector fields whose Hamiltonian is constant on the hypersurface).

Theorem (Chekanov). The characteristic field of a fiberwise convex hypersurface containing a Lagrangian optical submanifold is nowhere tangent to the set of critical points of the restriction of the canonical projection to the Lagrangian manifold (even at singular points of the critical set).

Corollary. The regular part of the critical set of an optical Lagrangian manifold carries a smooth field of tangent directions coinciding with the kernel projection at those points where the latter is 1-dimensional, and tangent to the critical manifold (at points A_k, $k \geq 3$).

Corollary. The Euler characteristic of a smooth compact critical manifold of an optical Lagrangian projection is equal to zero.

This implies the impossibility of optical "flying saucers" ("lips" in Thom's terminology, "pancakes" in that of Zeldovich [28),29)]), since the Euler characteristic of the 2-sphere is 2 (fig. 4).

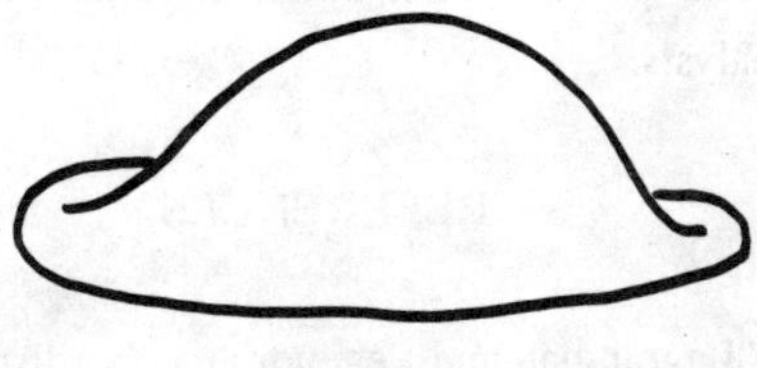

fig. 4

The "lips" are impossible for the metamorphoses of optical caustics for all the dimensions and signature values (these metamorphoses are generic among nonoptical caustics, see 4)). In 3 dimensions the metamorphosis D_4^-(birth of two pyramides) is also optically impossible, while all the others (A_4, A_5, D_4^+, D_5) have optical realizations.

The Lagrangian knot problem, cobordisms theories and so on have optical variants. An optical Lagrangian manifold is invariant under the corresponding Hamilton flow : therefore global optical Lagrangian manifolds are generically rigid, as Kolmogorov tori on a fixed energy level are. Thus, to define homotopies in the "optical knots" theory, we must allow deformations of the ambiant hypersurface (leaving it fiberwise convex).

Let us consider an invariant (optical, Lagrangian, Kolmogorov) torus of a geodesic flow on a Riemannian 2-torus. A geodesic is called minimal, if its lift to the covering plane is minimal between any two of its points.

Theorem (M. Bjalyi, L. Polterovich). An invariant torus of the geodesic flow is filled by the lifts of the minimal geodesics to the phase space if and only if it is a section of the cotangent bundle.

This is probably true for other optical Lagrangian manifolds, not just for those of the geodesic flow. The minimality condition is an obstruction to certain metamorphoses of optical caustics (like in Chekanov's theory). Thus, this conjecture is related with the optical unknottedness conjecture for optical Lagrangian manifolds, homotopic to a section of the cotangent bundle.

Remark. A Lagrangian torus embedded in $T^* T^n$ can't project to the base space with multiplicity greater than one. Otherwise, after a q-sheeted covering, we obtain two Hamilton homologous tori with no intersections, contradicting the theorems of sections 2 and 3.

It is interesting to note, that such natural problems and theorems in symplectic topology as the Lagrangian knots problem and Chekanov's theorem on optical critical sets were only discovered as a result of laser optics experimentation and Percival-Aubry... variational principles analysis.

REFERENCES

1) Arnold, V.I., "Lagrangian and Legendrian cobordisms", Funct. Anal. 14 : 3, 1-13; 14 : 4, 8-17 (1980).

2) Vasiliev, V.A.,"Characteristical classes of Lagrangian and Legendrian manifolds, dual to the singularities of caustics and of wave fronts. Selfintersections of wavefronts and Legendrian (Lagrangian) characteristic numbers". Funct. Anal. 15 : 3, 10-22 (1981), 16 : 3, 68-69 (1982).

3) Audin, M. "Cobordismes d'immersions lagrangiennes et legendriennes", These, Universite Paris-Sud, Orsay, 198 p., 1986.

4) Arnold, V.I., Varchenko, A.N., Gusein-Zade, S.M., "Singularities of differentiable maps", Vol. I , Birkhauser, 1985.

5) Arnold, V.I., "Sur une propriété topologique des applications globalement canoniques de la mécanique classique", Comptes Rendus Ac. Sc. Paris, 261, 3719-3722 (1965).

6) Arnold, V.I., "Stability problem and ergodic properties of classical dynamical systems", Proc. Int. Congr. Math. (Moscow 1966), Moscow. : MIR (1968), p. 387-392.

7) Arnold, V.I., Comments to "On a geometrical theorem"", in : Poincare, H., Collected works, Moscow. : NAUKA, 1972, vol. 2, pp. 987-989.

8) Arnold, V.I., "Fixed points of symplectic diffeomorphisms", in : "Mathematical development, arising from Hilbert problems", Proc. Symp. Pure Math. 28, Amer. Math. Soc. 1976, p. 66.

9) Nikishin, N.A., "On fixed points of diffeomorphisms of 2-sphere preserving areas", Funct. Anal. 8 :I,84-85 (1974).

10) Conley, C.C., Zehnder, E., "The Birkhoff-Lewis fixed point theorem and a conjecture of V.I. Arnold", Invent. Math. 73, 33-49 (1983).

11) Banyaga, A., "Sur le groupe des diffeomorphismes symplectiques", Lect. Notes in Math., Springer, 484, 50-56 (1975).

12) Eliashberg, Ya.M., "Estimation of the number of fixed points of area presering mappings, Syktyvkar 1978, 105 p.

13) Floer, A., "Proof of the Arnold conjecture for surfaces and generalizations to certain Kahler manifolds", Duke Math. Journ., 53 : I, I-32 (1986).

14) Chaperon, M., "Quelques questions de geometrie symplectique", Asterisque, 105-106, 231-249 (1983) ; C.R. 298 (1984).

15) Fortune, B., Weinstein, A., "A symplectic fixed point theorem for complex projective spaces", Berkeley, 1984.

16) Weinstein, A., "Co perturbation theorems for symplectic fixed points and Lagrangian intersections, "Lecture Notes Amer. Math. Soc. Summer Inst. on nonlinear funct. anal. and appl., Berkeley, 1983.

17) Sikorav, J.-C., "Points fixes d'une application symplectique homologue à l'identité", J. Diff. Geom. 22, 49-79 (1985).

17a) Hofer, H., "Lagrangian embeddings and critical point theory", Ann. Inst. H. Poincaré, Anal. non lin., Vol. 2 n° 6 (1985), 407-462.

Laudenbach, F., Sikorav, J.C., "Persistance d'intersection avec la section nulle...", Invent. Math. 82 (1985), 349-357.

18) Novikov, S.P., "Hamiltonian formalism and multivaled variant of Morse theory", Uspehi Math. Nauk, 37 : 5, 3-49 (1982).

19) Novikov, S.P., Taimanov, I.A., Doklady AN SSSR 274 : I, 26-29 (1984).

20) Weinstein, A., "Lectures on symplectic geometry", CBMS, Reg. Conf. Series, Amer. Math. Soc. 29 (1977).

21) Arnold, V.I., "On the characteristical class entering the quantization conditions", Funct. Anal. I : I, I-14 (1967).

22) Arnold, V.I., "Sturm theorems and symplectic geometry", Funct. Anal. 19 : 4, I-10 (1985).

23) Guillemin, V., Sternberg, S., "Geometric asymptotics", Amer. Math. Soc. Survey 14 (1977).

24) Arnold, V.I., "Singularities of ray systems", Uspehi Math. Nauk 38 : 2, 77-147 (1983).

25) Guckenheimer, J., "Caustics and nondegenerate Hamiltonians", Topology 13, 127-133 (1974).

26) Nye, J.F., Hannay, J.H., "The orientations and distortions of caustics in geometrical optics", Optica Acta, 31, 115-130 (1984).

27) Chekanov, Yu.V., "Caustics of geometrical optics", Funct. Anal. 20 : 3 (1986).

28) Zeldovich, Ya.B., "Gravitational instability : an approximate theory for large density perturbations", Astron. Astrophys. 5, 84-89 (1970).

29) Arnold, V.I., Shandarin, S.F., Zeldovich, Ya.B., "The large scale of the Universe, 1", Geophys. Astrophys. Fluid Dyn. 20, 111-130 (1982).

30) Koslov, V.V., Uspehi Math. Nauk 40 (1985).

FILTERING EQUATION, EQUIVARIANT COHOMOLOGY AND THE CHERN CHARACTER

Jean-Michel Bismut

Département de Mathématique de l'Université Paris-Sud
Bâtiment 425 91405 Orsay FRANCE

Introduction.

I - Index Theorem and equivariant cohomology on the loop space.

a) The Dirac operator.
b) The loop space of a Riemannian manifold.
c) The index of the Dirac operator and equivariant path integrals.
d) Localization formulas and the Index Theorem.

II - Filtering equation, superconnections and the Chern character.

a) Description of the fibered manifold.
b) The loop spaces of M and B.
c) Integration along the fiber in the loop space.
d) A connection on an infinite dimensional vector bundle.
e) The filtering equation.
f) Integration along the fiber and the filtering equation.
g) Superconnections and the filtering equation.
h) The Levi-Civita superconnection and the filtering equation.
i) The filtering equation and the Chern character.

In [A] , Atiyah and Witten have done a formal link between Index Theory [AS1] for the Dirac operator on the spin complex of a manifold M and the equivariant cohomology of the loop space LM associated with the vector field X generating the natural action of S_1 on LM by rotations. Namely they observed that the heat equation formula of [ABP] for the index could be expressed as the integral of a differential form ω on LM, which is X equivariantly closed, i.e. $(d+i_X)\omega = 0$. Noting that M embeds in LM as the set of zeros of X, and applying in an infinite dimensional situation a formula of Duistermaat-Heckman [DH], Berline-Vergne [BV], which localizes formally the integral of on the fixed point set M, they found in [A] the correct formula for the Index.

In [B4], we extended the observation of [A] to the case of a Dirac operator on a twisted spin complex. Namely, we showed that the heat equation formula for the index can be written as the pairing of the Atiyah-Witten form ω with an X equivariantly closed form β, which is a natural lift to LM of the Chern character forms of the twisting finite dimensional bundle. β is obtained by integrating a linear differential equation along each loop. It was shown in [B4] that β could be viewed as a X equivariant characteristic class over LM, associated with an infinite dimensional principal fiber bundle, whose structure group is a Kac-Moody loop group associated with the finite dimensional group $U(k)$.

In [B3, B5], we gave two heat equations proofs of the Index Theorem of Atiyah-Singer for families of Dirac operators [AS2], this by using the superconnection formalism of Quillen [Q] in an infinite dimensional situation. The two facts which motivated our proofs were :

• The observation in [B5, Section 1] that the probabilistic proof of the Index Theorem [B2] has a finite dimensional analogue, which proves the formulas of [BV], [DH].

• The formal analogy between Quillen's superconnections and integration along the fiber in the loop space, developed in [B5, Section 3].

These two observations permitted us to give in [B3] two proofs of the Index Theorem for families, which as shown in [B5, Section 1], are the infinite dimensional analogues of well-defined finite dimensional results.

The purpose of this paper is to show how the superconnection formalism of Quillen [Q] is, in some sense, forced upon us by the mere existence of Index Theory. Namely let $M \xrightarrow[\pi]{} B$ be a compact fibering of manifolds with compact fibers Z. Let D' be the Dirac operator on M. The heat equation formula for the Index of D'_+ can be expressed as the integral of a well-defined signed measure R^t [B4, Definition 1.8] on LM or of the formal differential form $\omega \wedge \beta$ on LM [A], [B4]. Now LM fibers on LB. The integral along the fiber of the signed measure R^t on LM is then possible, and corresponds formally to integration along the fiber of the corresponding differential form.

On the other hand, filtering theory is a well known probabilistic technique, whose purpose is to calculate the conditional law of a process $x.$ given the trajectory of a process y_s $(0 \leq s \leq t)$ (see Liptser-Shiryayev [LS, Chapter 8]). In particular the filtering equation [LS, Chapter 8.5], [BM] is a partial differential equation with random coefficients depending on $y.$ which at each time t calculates the law of x_t conditional on $(y_s / s \leq t)$.

We here use the filtering equation technique to calculate the integral along the fiber $LM \longrightarrow LB$ of the measure R^t. Here $x.$ is the Brownian motion in M, and $y.$ the Brownian motion in B. Of course the results of [LS] have to be adequately modified since we are dealing ultimately with matrix valued kernels instead of scalar kernels.

By transforming the obtained result into a differential form on LB, we ultimately show that the integral along the fiber $LM \longrightarrow LB$ of $\omega \wedge \beta$ will be obtained by integrating along each loop $y.$ in LB a partial differential equation, which is itself well-defined.

Note here that the equivalence between

- integrating a linear partial differential equation, which is an analytic procedure.
- integrating along the fiber a differential form, which is a geo-

metric procedure

will later be viewed as the equality between an analytic index and a topological index. Let Y be the generator of the action of S_1 on LB.

On a priori grounds, we know that the differential form β' on LB, which is obtained by integration along the fiber $LM \longrightarrow LB$ of an X equivariantly closed form, should be Y equivariantly closed. This is of course a non rigorous argument.

On the other hand, the filtering equation is a well-defined and rigorous object. We prove in Theorem 2.10 that the differential form β' associated with the filtering equation is indeed Y equivariantly closed. β' appears to be an extension of the form β constructed in [B4] in two ways :

• We need the superconnection formalism of Quillen [Q] instead of the usual connections of [B4].

• We are in an infinite dimensional situation, i.e. we consider a partial differential equation over each loop, instead of a differential equation.

β' should be viewed as an infinite dimensional characteristic class over LB associated with an ill-defined infinite dimensional principal fiber bundle, whose structure group is a Kac-Moody loop group of an infinite dimensional gauge group. Since this object is not really well-defined, the proof that β' is Y equivariantly closed is less pleasant than in [B4].

By using a slightly different expression for the index of the Dirac operator on M, we obtain in Section 2h) a much more complicate filtering equation, in which the Levi-Civita superconnection of [B3, Section 3], [B5] appears naturally. The fact that ultimately, the Levi-Civita superconnection is the right simple object to consider is proved in[B3, Section 4] and [B5, Section 3].

In Section 2i), the filtering equation looses its mystery. In fact using our results in [B3, Section 2], we find that β' restricts on B to a closed form which represents the Chern character associated with the family of vertical Dirac operators indexed by B. Again, β' appear to be the natural lift to LB of the differential forms appearing in [B3]. In some sense, this paper is a new version of [B3, B5], which is written upside down.

This paper is organized in the following way. In Section 1, we briefly summarize the results of [A], [B4], [B5] on the relations of equivariant cohomology to Index Theory. In Section 2, we study the case of a Dirac operator on a fibered manifold $M \longrightarrow B$, introduce the filtering equation, and exhibit its relation with Quillen's superconnections [Q], equivariant cohomology and the Index Theorem for families [B3, B5].

For a general introduction to Brownian motion, we refer to [B1]. For the theory of filtering, we refer to [LS] and [BM]. The operator $d+i_X$ was introduced in explicit form in Witten [W], Berline-Vergne [BV], and extensively studied in Atiyah-Bott [AB]. Finally note that in [BF], the filtering equation is used as a tool to study the asymptotics of certain êta invariants.

I - Index Theorem and equivariant cohomology on the loop space.

In this section, we briefly describe the results given in Atiyah [A] and ourselves [B4] concerning the relation of the Index Theory for Dirac operators to the equivariant cohomology of the loop space.

In a), we introduce the Dirac operator. In b), and along the lines of [A], we describe some properties of the loop space of a Riemannian manifold. In c) and d), we summarize the results of [A] and [B4] and their relation to the localization formulas of [BV], [DH].

a) The Dirac operator.

Let M be a compact connected Riemannian manifold of dimension n. We assume that M is orientable and spin.

Let F be the Hermitian bundle of spinors over M.

The Levi Civita connection ∇ on TM lifts into a unitary connection on F, which we still note ∇.

Let ξ be a complex Hermitian bundle, endowed with a unitary connection ∇^{ξ}.

Then $F \otimes \xi$ is naturally endowed with a unitary connection, which we still note ∇.

$F \otimes \xi$ is a TM Clifford module i.e. TM acts by Clifford multiplication on $F \otimes \xi$.

$\Gamma(F \otimes \xi)$ is the set of C^{∞} sections of $F \otimes \xi$.

$e_1, \dots, e_n$ is an orthonormal base of TM.

Definition 1.1. D denotes the Dirac operator acting on $\Gamma(F \otimes \xi)$

$$D = \sum_1^n e_i \nabla_{e_i} \tag{1.1}$$

b) The loop space of a Riemannian manifold

Let LM be the set of C^{∞} loops $s \in R/Z \to x_s \in M$. If $x. \in LM$, $T_x LM$ identifies naturally with the set of C^{∞} loops $s \in R/Z \to (x_s, Y_s)$ in TM, with $Y_s \in T_{x_s} M$. $T_x LM$ is endowed with the scalar product

$$Y, Y' \to \int_0^1 \langle Y, Y' \rangle_s \, ds . \tag{1.2}$$

LM is then a Riemannian manifold.

We now follow Atiyah [A]. S_1 acts naturally on LM, by

$$s \in S_1 = R/Z \quad k_s x_. = x_{.+s}$$

$(k_s)_{s \in S_1}$ is a group of isometries of LM. The generating Killing vector field X on LM is given by

$$[X(x)]_s = \frac{dx}{ds} .$$

Since LM is Riemannian we can define a one form X' by the relation

(1.3) $$Y \in TLM \to X'(Y) = < X,Y >$$

If $x \in LM$, let $\frac{D}{Ds}$ be the covariant differentiation operator along x. with respect to the Levi-Civita connection.

dX' is the 2 form

(1.4) $$(Y,Z) \in TLM \to dX'(Y,Z) = 2 \int_0^1 < \frac{DY}{Ds} , Z > ds$$

If i_X is the interior multiplication operator then

(1.5) $$i_X X' = \int_0^1 |\frac{dx}{ds}|^2 ds .$$

Also since X is Killing, X' is also X invariant and so

(1.6) $$L_X X' = 0 .$$

Now when acting on forms

(1.7) $$L_X = (d + i_X)^2$$

From (1.6), (1.7), we find that

(1.8) $$(d + i_X)[(d + i_X)X'] = 0$$

In other words $(d + i_X)X' \in \Lambda^\circ(T^*LM) \oplus \Lambda^2(T^*LM)$ is in the kernel of the operator $d + i_X$.

The operator $d + i_X$ was introduced in Witten [W] and Berline-Vergne [BV]. Acording to the terminology of [BV], we will say that $(d + i_X)X'$ is X equivariantly closed. For the relations of the operator $d + i_X$ to equivariant cohomology, we refer to Atiyah-Bott [AB] and Berline Vergne [BV].

Set

(1.9) $\Omega_X = \{\mu \in \Gamma(\Lambda(T^*LM))\ ;\ L_X\mu = 0\}$.

By (1.7), we find that on Ω_X, $(d+i_X)^2 = 0$.

We can then define in Ω_X cohomology classes with respect to $d+i_X$.

c) The index of the Dirac operator and equivariant path integrals.

For $t>0$, $\exp\{-\frac{(d+i_X)X'}{2t}\}$ is a formal power series in $\Lambda^{even}(T^*LM)$. By (1.8), it is clear that

$$(1.10)\qquad (d+i_X)[\exp - \frac{(d+i_X)X'}{2t}] = 0 .$$

Following [B4], we now show how we can construct natural X equivariantly closed forms associated with ξ .

Let L be the curvature tensor of ξ .

If $x \in LM$, $s \in R/Z$, L_{x_s} is the 2 form with values in $\mathrm{End}\ \xi_{x_s}$

$$(1.11)\qquad Y,Z \in T_x(LM) \to L_{x_s}(Y_s, Z_s) .$$

τ_0^s denotes the parallel transport operator from fibers at x_s into fibers at x_0 .

If $Y,Z \in T_x(LM)$, $\tau_0^s L_{x_s}(Y_s,Z_s)$ is an element of $\mathrm{End}\ (\xi_{x_0})$.

$\tau_0^s L_{x_s}$ can be then be viewed as an element of the algebra $\Lambda^{even}(T_x^*(LM)) \otimes \mathrm{End}\ \xi_{x_0}$. The computations which follow will be done in this algebra.

Definition 1.2. H_s is the process taking its values in $\Lambda^{even}(T_x^*(LM)) \otimes \mathrm{End}\ \xi_{x_0}$ defined by the differential equation

$$(1.12)\qquad \frac{dH}{ds} = H_s(\tau_0^s L_{x_s})$$

$$H(0) = I_{\xi_{x_0}}$$

(1.12) can be solved by a formal power series

$$H_s = I + \int_0^s \tau_0^u L_{X_u} du + \int_{0 \leq u \leq u' \leq s} \tau_0^u L_{X_u} \wedge \tau_0^{u'} L_{X_{u'}} du\, du' + \ldots$$

Let i be the embedding $M \to LM$.

We now have the result proved in [B4, Theorem 3.9].

Theorem 1.3. The even form on LM

(1.13) $\quad \beta = \mathrm{Tr}[H_1 \tau_0^1]$

is X equivariantly closed. Moreover the $d + i_X$ equivariant cohomology class of β does not depend on the connection on ξ. $i^* \beta$ is the normalized Chern character form $\mathrm{Tr}[\exp L]$.

Assume now that n is even so that $n = 2\ell$. F splits into $F = F_+ \oplus F_-$

D interchanges $\Gamma(F_+ \otimes \xi)$ and $\Gamma(F_- \otimes \xi)$. Let $D_\pm$ be the restriction of D to $\Gamma(F_\pm \otimes \xi)$.

$F \otimes \xi = (F_+ \otimes \xi) \oplus (F_- \otimes \xi)$ is a Z_2 graded vector bundle on M. Let τ be the involution defining the grading i.e. $\tau = \pm 1$ on $F_\pm \otimes \xi$.

Similarly $\Gamma(F \otimes \xi) = \Gamma(F_+ \otimes \xi) \oplus \Gamma(F_- \otimes \xi)$ is a Z_2 graded vector space. We still allow τ to act on $\Gamma(F \otimes \xi)$ with $\tau = \pm 1$ on $\Gamma(F_\pm \otimes \xi)$.

If $A \in \mathrm{End}(F \otimes \xi)$, we write A in matrix form as

(1.14) $$A = \begin{bmatrix} E & F \\ G & H \end{bmatrix}$$

where E maps $\Gamma(F_+ \otimes \xi)$ into itself etc...

If A is trace class, set

(1.15) $\quad \mathrm{Tr}_s A = \mathrm{Tr}\, \tau A$.

Recall that the index $\mathrm{Ind}\, D_+$ of D_+ is defined by

(1.16) $\quad \mathrm{Ind}\, D_+ = \dim \ker D_+ - \dim \ker D_-$.

The well-known heat equation formula for the index $\mathrm{Ind}\, D_+$ in [ABP] asserts that for any $t > 0$

$$(1.17) \qquad \text{Ind } D_+ = \text{Tr}_s \exp\{-\frac{t\,D^2}{2}\} .$$

By the classical theory of Brownian motion on M [B4, Theorem 1.9], we know that there is a signed measure $dR^t(x)$ on the set L^oM of continuous loops $s \in R/Z \to x_s \in M$ such that

$$(1.18) \qquad \text{Tr}_s \exp(-\frac{t\,D^2}{2}) = \int_{L^oM} dR^t(x) .$$

By some non trivial algebraic transformations, it follows from Atiyah [A] and ourselves [B4, Section 2] that $\int_{L^oM} dR^t(x)$ can be written formally as

$$(1.19) \qquad \int_{L^oM} dR^t(x) = \frac{\prod_1^{+\infty} (m^2)^\ell \, i^\ell}{(2\pi)^\ell} \int_{L^oM} \exp\{-\frac{(d+i_X)X'}{2t}\} \wedge \beta$$

Note in the r.h.s. of (1.19) the infinite normalizing constant $\prod_1^{+\infty} (m^2)^\ell_-$. Also if N is an oriented manifold and $\mu \in \Gamma(\bigoplus_o^{\dim N} \Lambda(T^*N))$, by definition $\int_N \mu$ is the integral over N of the top order form in μ , whose degree is equal to $\dim N$.

Here $\dim L^oM = +\infty$. It turns out that the well-defined $dR^t(x)$ should be throught of as the top order form in the product of forms written in the r.h.s. of (1.19).

Finally observe that β was constructed for C^∞ loops while dR^t is carried by continous loops. This is largely irrelevant since parallel transport still makes sense for Brownian loops [B6, Section 2].

d) <u>Localization formulas and Index Theorem.</u>

From (1.17) - (1.19), we get

$$(1.20) \qquad \text{Ind } D_+ = \frac{\prod_1^{+\infty} (m^2)^\ell}{(2\pi)^\ell} \int_{L^oM} \exp\{-\frac{(d+i_X)X'}{2t}\} \wedge \beta$$

Note that the differential form in the r.h.s. of (1.20) is X equivariantly closed.

However the form $\exp - \frac{(d+i_X)X'}{2t}$ and β do not live in the same class of forms.

In fact β lives in the ordinary X equivariant cohomology $H^*_{eq}(LM)$.

$\exp\{-\frac{(d+i_X)X'}{2t}\}$ should be thought as living in the dual space of $H^*_{eq}(LM)$, which we note $H^{*c}_{eq}(LM)$. In fact $\exp - \frac{(d+i_X)X'}{2t}$ decays quickly as $|X|^2 \to +\infty$ and so is formally a representative of compactly supported cohomology on LM.

None of these statements are to be taken too literally. In fact it is known by Goodwillie [G] that the cohomology of the operator $d+i_X$ on all the X invariant forms of LM only depends on $\pi_1(M)$. So only a subcomplex of the complex of differential forms on LM - which is made of iterated path integrals - should be considered to define $H^*_{eq}(LM)$. (See Petrack [P])

Now by formulas of Berline-Vergne [BV] and Duistermaat-Heckman [DH], we know that if L^oM was finite dimensional, we could equal on a priori grounds the r.h.s. of (1.20) to an integral over M, where M should be thought of as the set of zeros of X.

The formal application of the formula of [BV], [DH] rather stunningly produces in [A] and [B4] the right formulas for $\mathrm{Ind}\, D_+$.

To explain this fact, and instead of trying to obtain an infinite dimensional version of [BV], [DH], we showed in [B5, Section 1] that <u>even in finite dimensions</u>, introducing the cocycle

$\exp\{-\frac{(d+i_X)X'}{2t}\}$ is a perfectly reasonable method of proving the formulas of [BV], [DH]. In retrospect the heat equation method just does this in infinite dimensions.

II. FILTERING EQUATION, SUPERCONNECTIONS AND THE CHERN CHARACTER.

In this Section , we specialize the results of Section 1 to the case of a fibering of compact manifolds $M \to B$.

In a), we describe the fibering $M \to B$ and the corresponding Dirac operators on M D' and D'^{L}. We here follow closely [B3, Section 1]. In b), we consider the loop spaces LM and LB. In c), we introduce formally the integration of differential forms on LM along the fibers of $LM \to LB$. In d), and along the lines of [B3, Section 2], we introduce a Z_2 graded infinite dimensional bundle $H^{\infty} = H_{+}^{\infty} \oplus H_{-}^{\infty}$ on B and a connection $\tilde{\nabla}$ on H^{∞}.

In e), we describe the filtering equation. In f), we show that formally, the differential form β' on LB which is constructed by using the filtering equation is formally obtained by integration along the fiber of an equivariantly closed differential form on LM. In g), we prove rigorously that β' is Y equivariantly closed. In h), we introduce naturally the Levi-Civita superconnection , the corresponding filtering equation, and the associated differential form $\overline{\beta}'$ on LB. Finally in i), we relate the filtering equation to the Chern character of a family Dirac operators along the lines of [B3, Sections 2, 3, 4].

a) Description of the fibered manifold.

$n = 2\ell$, $m = 2\ell'$ are even integers. M, B are compact connected manifolds of dimension $n+m$ and m.

We assume that $M \xrightarrow[\pi]{} B$ is a fibration of M over B with fiber Z, where Z is compact and connected.

TZ is a subbundle of TM. We assume that TZ is oriented and spin.

We also assume that B is oriented and spin.

Let $T^H M$ a smooth subbundle of TM such that

$$TM = T^H M \oplus TZ .$$

Clearly $T^H M \simeq \pi^* TB$. Let P_H, P_Z be the projection operators on $T^H M, TZ$.

We now assume that TB and TZ are endowed with smooth metrics g_B and g_Z. g_B lifts to T^HM. $g_B \oplus g_Z$ denotes the metric on TM which coincides with g_B on T^HM, with g_Z on TZ, and is such that T^HM and TZ are orthogonal. $<,>$ is the corresponding scalar product.

Let F be the Hermitian bundle of spinors over TZ, F' the Hermitian bundle of spinors over TB. We will write F' instead of π^*F'. F,F' split into $F = F_+ \oplus F_-$, $F' = F'_+ \oplus F'_-$. The Z_2 graded Hermitian bundle of spinors over TM is given by $F' \hat{\otimes} F$.

Let ∇^B, ∇^L be the Levi-Civita connections on TB, TM for the metric g_B and $g_B \oplus g_Z$.

Let ∇^Z be the connection on T^Z

$$\nabla^Z = P_Z \nabla^L \quad .$$

∇^B lifts into a Euclidean connection on T^HM. Let ∇ be the connection $\nabla = \nabla^B \oplus \nabla^Z$ on $TM = T^HM \oplus TZ$. ∇ still preserves the metric $g_B \oplus g_Z$. Let T be the torsion of ∇. Set

$$S = \nabla^L - \nabla \ .$$

In [B3, Theorem 1.9], we proved that

- ∇^Z does not depend g_B.
- T takes its values in TZ.
- The (3,0) tensor $<S(.).,.>$ does not depend on g_B.
- For $U \in TM$, S(U) maps TZ into T^HM.
- For $U,V,W \in T^HM$ $<S(U)V,W> = 0$.

Also in [BF, Section 1], we proved that if $U \in T^HM$, $S(U)U = 0$.

ξ is a complex Hermitian bundle on M, endowed with a unitary connection ∇^ξ. ∇^L still denotes the connection on $(F' \hat{\otimes} F) \otimes \xi$

$$\nabla^L = \nabla^L \otimes 1 + 1 \otimes \nabla^\xi \quad .$$

∇ denotes the connection on $(F' \hat{\otimes} F) \otimes \xi$

$$\nabla = \nabla \otimes 1 + 1 \otimes \nabla^{\xi}$$

If $Y \in TB$, Y^H denotes the horizontal lift of Y in $T^H M$, i.e. $Y^H \in T^H M$, $\pi_* Y^H = Y$.

$e_1, \dots, e_i, \dots e_n$ is an orthonormal base of TZ, $f_1, \dots, f_\alpha \dots f_m$ an orthonormal base of TB. We identify f_α and $(f_\alpha)^H$.

e_i, f_α act on $(F' \hat{\otimes} F) \otimes \xi$ by Clifford multiplication.

Definition 2.1. D', D'^L denote the Dirac operators acting on $\Gamma((F' \hat{\otimes} F) \otimes \xi)$

$$(2.1) \qquad D' = f_\alpha \nabla_{f_\alpha} + e_i \nabla_{e_i}$$

$$D'^L = f_\alpha \nabla^L_{f_\alpha} + e_i \nabla^L_{e_i}$$

By [B3,(3.11)], D'^L is given by the formula

$$(2.2) \qquad D'^L = e_i(\nabla_{e_i} + \tfrac{1}{2}\langle S(e_i)e_j, f_\alpha\rangle e_j f_\alpha + \tfrac{1}{4}\langle S(e_i)f_\alpha, f_\beta\rangle f_\alpha f_\beta)$$

$$+ f_\alpha(\nabla_{f_\alpha} + \tfrac{1}{2}\langle S(f_\alpha)e_j, f_\beta\rangle e_j f_\beta) .$$

We know that, $S(f_\alpha)f_\alpha = 0$. We can use [BF, Proposition 1.18] to get

$$(2.3) \qquad D'^L = f_\alpha(\nabla_{f_\alpha} - \tfrac{1}{2}\langle S(e_i)e_i, f_\alpha\rangle) + D - \tfrac{1}{8}\langle T(f_\alpha, f_\beta), e_i\rangle e_i f_\alpha f_\beta$$

Let $D'_\pm$, $D'^L_\pm$ be the restriction of D', D'^L to $\Gamma((F' \otimes F)_\pm \otimes \xi)$. We still write

$$D' = \begin{bmatrix} 0 & D'_- \\ D'_+ & 0 \end{bmatrix} \qquad D'^L = \begin{bmatrix} 0 & D'^L_- \\ D'^L_+ & 0 \end{bmatrix} .$$

Note that the operator $e_i \nabla_{e_i}$ acts fiberwise, so that D' is made of an horizontal part $f_\alpha \nabla_{f_\alpha}$ and a vertical part $e_i \nabla_{e_i}$. On the contrary, D'^L mixes horizontal and vertical variables in a more intricate way.

We now have

Theorem 2.2. For any $t>0$.

$$(2.4)\qquad \operatorname{Ind} D_+^{'L} = \operatorname{Ind} D_+' = \operatorname{Tr}_s[\exp - \frac{t(D^{'L})^2}{2}] = \operatorname{Tr}_s[\exp - \frac{tD^{'2}}{2}]$$

Proof : D_+^L and D_+' have the same symbol and so have the same index. Since $D^{'L}$ is self-adjoint, by [ABP, Theorem EIII]

$$\operatorname{Ind} D_+^{'L} = \operatorname{Tr}_s[\exp - \frac{t(D^{'L})^2}{2}]$$

Since D' is not self-adjoint, the last equality is less trivial than the reader likes to think. However it easily follows from Quillen's formalism [Q], and is proved in [B7, Theorem 1.2].

b) The loop spaces of M and B .

Let X, Y be the generators of the action of S_1 on LM, LB and X',Y' the corresponding one forms. Clearly π extends into a smooth map π':$LM \to LB$. Obviously

$$(2.5)\qquad \pi'_* X = Y .$$

Let L be the curvature of ξ . We still define β as the X equivariantly closed form on LM of Theorem 1.3.

Using (1.20), we still have

$$(2.6)\qquad \operatorname{Ind} D_+^{'L} = \operatorname{Tr}_s \exp - \frac{t(D^{'L})^2}{2} = i^{\ell+\ell'} \frac{\left(\prod_1^{+\infty} m^2\right)^{\ell+\ell'}}{(2\pi)^{\ell+\ell'}} \int_{L^oM} \exp\{-\frac{(d+i_X)X'}{2t}\}\ \beta$$

We note $<T(.,.)_\wedge\theta>$ the antisymmetrization of the tensor on TM $(U,V,W) \to <T(U,V),W>$. $<T(.,.)_\wedge\theta>$ extends as a 3 form on LM

$$(2.7)\qquad (U,V,W)\in T_{x_\cdot}LM \to \int_0^1 T_{x_s}(U_s,V_s)_\wedge,W_s>ds .$$

which is X invariant.

In [B5, Section 2], we briefly indicated that

$$\text{Ind } D'_+ = \mathrm{Tr}_s \exp\left(-\frac{tD'^2}{2}\right) = \frac{i^{\ell+\ell'}}{(2\pi)^{\ell+\ell'}}\left(\prod_1^{+\infty} m^2\right)^{\ell+\ell'} \tag{2.8}$$

$$\int_{L^oM} \exp\left\{-\frac{(d+i_X)X'}{2t} - \frac{(d+i_X)<T(.,.)_\wedge\, \Theta>}{2t}\right\}\beta$$

Of course ultimately (2.6) and (2.8) coincide . What we mean by the final equality of (2.8) is that the natural X equivariantly closed form on LM associated with $\mathrm{Tr}_s[\exp - \frac{tD'^2}{2}]$ appears in the r.h.s. of (2.8).

Of course, on purely formal grounds, since $\exp\left\{-\frac{(d+i_X)<T(.,.)_\wedge\Theta>}{2t}\right\}$ differs from 1 by a $d+i_X$ exact X invariant form, the fact that the integrals in (2.6) and (2.8) coincide can be "seen" by differential geometric arguments on LM .

c) <u>Integration along the fiber in the loop space</u>.

Take $x. \in LM$ If $U \in T_xLM$, we have

$$X'(U) = \int_0^1 <U_s, \frac{dx}{ds}> = \int_0^1 <U, P_H \frac{dx}{ds}> ds + \int_0^1 <P_Z U , \frac{dx}{ds}> ds. \tag{2.9}$$

Set $y. = \pi x..$ Since T^HM inherits the metric g_B ,

$$< U_s, P_H \frac{dx}{ds} > = < \pi_* U_s, \frac{dy}{ds} > \tag{2.10}$$

It follows from (2.10) and (2.11) that

(2.11) $$X' = \pi'^{*} Y' + P_Z^{*} X'$$

Set
$$C = \frac{i^{\ell+\ell'} \left(\overset{+\infty}{\Pi} m^2\right)^{\ell+\ell''} \quad 1}{(2\pi)^{\ell+\ell'}}$$

π' defines a "fibration" of LM on LB . One verifies easily that since the fibers Z_y are connected, the fibers Z' of $LM \underset{\pi'}{\to} LB$ have the homotopy type of the free loop space in the fibers Z .

If ω is a C^∞ differential form on M , since TZ is oriented, the integral along the fiber $\pi_*\omega$ is a well defined C^∞ form on B . Moreover d and π_* commute .

In the sequel, we will proceed as if the same principle applies to the fibration $LM \underset{\pi}{\to} LB$. We will of course forget about the orientability of TZ' . It may help to think of Z' as an even (and still infinite) dimensional manifold .

Then using (2.11), we find

(2.12) $$C\int_{L^oM} \exp - \frac{(d+i_X)X'}{2t} \quad \beta = C\int_{L^oB} \exp - \frac{(d+i_Y)Y'}{2t}$$

$$\pi'_*[\exp - \frac{(d+i_X)P_Z^*X'}{2t} \beta]$$

$$C\int_{L^oM} \exp\{- \frac{(d+i_X)X'}{2t} - \frac{(d+i_X)<T(.,.)_\wedge\theta>}{2t}\}\beta =$$

$$C\int_{L^oB} \exp - \frac{(d+i_Y)Y'}{2t} \quad \pi'_*[\exp\{- \frac{(d+i_X)P_Z^*X'}{2t}$$

$$- \frac{(d+i_X)<T(.,.)_\wedge\theta>}{2t}\}\beta]$$

Clearly $L_X P_Z^* X' = 0$. It follows that

$$(2.13) \qquad (d+i_X)[\exp - \frac{(d+i_X)P_Z^* X'}{2t} \beta \;] = 0 \; .$$

Since $\pi_* X = Y$ and $d\pi'_* = \pi'_* d$, we find that

$$(2.14) \qquad (d+i_Y)\pi'_* \; [\exp - \frac{(d+i_X)P_Z^* X'}{2t} \beta] = 0 \; .$$

and also

$$(2.15) \qquad (d+i_Y)\pi'_*[\exp\{- \frac{(d+i_X)P_Z^* X'}{2t} - \frac{(d+i_X)<T(.,.)_\wedge \Theta>}{2t}\}\beta] = 0 \; .$$

So the forms on LB appearing in the r.h.s. of (2.12) are Y equivariantly closed.

It is now our purpose to show that a well-known probabilistic technique, the filtering equation, constructs explicitly the integrals along the fiber $\pi'_*[\exp\{ \quad \}...]$ appearing in (2.12), by a construction which is very similar to what we did in Theorem 1.3, but using instead infinite dimensional bundles.

d) <u>A connection on an infinite dimensional vector bundle</u>.

For $y \in B$, $H^\infty_{\pm,y}$ is the set of C^∞ sections of $F_\pm \otimes \xi$ over the fiber Z_y .

$H^\infty = H^\infty_+ \oplus H^\infty_-$ is a Z_2 graded infinite dimensional vector bundle over B .

As in [B3, Section 1], we define a connection on H^∞ . Namely if $h \in H^\infty$, $U \in TB$, set

$$(2.16) \qquad \tilde{\nabla}_U h = \nabla_{U^H} h$$

One immediately verifies that $\tilde{\nabla}$ is a connection on H^∞ .

Also if $y \in B$, if $e_1,\ldots,e_n$ is an orthonormal base of TZ_y, set

(2.17) $\qquad D_y = e_i \nabla_{e_i}$

D_y obviously acts on H_y^∞. Let $D_{\pm,y}$ be the restriction of D_y to $H_{\pm,y}$. Then

(2.18) $$D_y = \begin{bmatrix} 0 & D_{-,y} \\ D_{+,y} & 0 \end{bmatrix}$$

Of course, D will also be considered as acting on the C^∞ sections of H^∞. In particular

(2.19) $$D' = f_\alpha \nabla_{f_\alpha} + D$$

Let R^Z be the curvature of ∇ on TZ.

<u>Theorem 2.3</u>. The curvature $\tilde{\nabla}^2$ of the connection $\tilde{\nabla}$ is such that if $U,V \in TB$

(2.20) $$\tilde{\nabla}^2(U,V) = R^Z(U^H,V^H) \otimes 1 + 1 \otimes L(U^H,V^H) - \nabla_{T(U^H,V^H)}.$$

Also if $U \in TB$

(2.21) $$\tilde{\nabla}_U D = - e_i \nabla_{T(U^H,e_i)} + e_i [R^Z(U^H,e_i) \otimes 1 + 1 \otimes L(U^H,e_i)]$$

<u>Proof</u> (2.20) is proved in [B3, Proposition 1.11]. (2.21) follows from [B3, Theorem 2.5]. □

Remember that (f_α) is an orthonormal base of TB, and that we identify f_α and f_α^H

<u>Definition 2.4</u>. Δ^B denotes the operator acting on the C^∞ sections of H^∞

(2.22) $$\Delta^B = (\nabla_{f_\alpha})^2 - \nabla_{\nabla_{f_\beta} f_\beta}.$$

Let K^B be the scalar curvature of B.

We now have

Proposition 2.5. The following formula holds

$$(2.23)\qquad D'^2 = -\Delta^B + \frac{K^B}{4} + D^2 + \frac{1}{2} f_\alpha f_\beta \tilde{\nabla}^2(f_\alpha, f_\beta) + f_\alpha \tilde{\nabla}_{f_\alpha} D \,.$$

Proof: By (2.20), (2.21) and [B3, Proposition 5.2] (2.23) follows. An alternative proof is simply to use the identity (2.19) and Theorem 2.3.

Observe that in (2.23), D^2, $\frac{1}{2} f_\alpha f_\beta \tilde{\nabla}^2 (f_\alpha, f_\beta)$ and $f_\alpha \tilde{\nabla}_{f_\alpha} D$ act fiberwise.

e) The filtering equation.

Let us briefly recall what is filtering. Take a Markov diffusion $x_t = (y_t, y'_t)$. The purpose of filtering is to calculate at each time t the conditional law of x_t given $(y_s | s \leq t)$. It is a basic result of filtering theory that is certains situations, this can be done recursively : there is a partial differential equation with non anticipating random coefficients depending on $(y.)$ whose solution gives at each time t the conditional law of x_t given $(y_s | s \leq t)$

We will use here the ideas of filtering theory, but now we are interested in kernels acting on vector bundles.

Let D^B be the Dirac operator on B acting on $\Gamma(F')$.

Take $y_0 \in B$. Let y_s be the Brownian motion on B starting at y_0 [B6, Chapter 2]. Let τ_0^s be the parallel transport operator from F'_{y_s} into F'_{y_0} along y .

Recall that by [B2, Theorem 2.5], for any $s > 0$, $k \in \Gamma(F')$

$$(2.24)\qquad \exp\left(-\frac{s(D^B)^2}{2}\right)k(y_0) = E\left[\exp\left\{-\int_0^s \frac{K^B(y_h)}{8}\, dh\right\}\tau_0^s k(y_s)\right] \quad .$$

We now calculate $\exp - \frac{s(D')^2}{2}$ by only using the Brownian motion y. and a subordinated partial differential equation.

First note that as shown in [B3, Section 5], by using the theory of stochastic flows, we can lift horizontally $s \to y_s$ into horizontal curves in M. Namely if $x_o \in Z_{y_o}$, consider the Stratonovitch differential equation

$$(2.25) \qquad d\,x_s = (dy_s)^H \quad ; \quad x(0) = x_o$$

By [B3, Section 5], (2.25) has a unique solution which depends smoothly on x_o. For a.e. $y.$, we have a diffeomorphism τ_s^o of Z_{y_o} into Z_{y_s} depending continuously on s. In (2.25), $x_s = \tau_s^o(x_o)$. Set $\tau_o^s = (\tau_s^o)^{-1}$. Similarly, we can parallel transport (with respect to the connection ∇) $h \in (F \otimes \xi)_x$ along $s' \to \tau_{s'}^o(x_o)$ into $\tau_s^o h \in (F \otimes \xi)_{\tau_s^o(x_o)}$. If $h \in H_{y_o}^\infty$, let $\tau_s^o h \in H_{y_s}^\infty$ be defined

$$(2.26) \qquad x' \in Z_{y_s} \quad ; \quad \tau_s^o h(x') = \tau_s^o[h(\tau_o^s x')]$$

Of course in the r.h.s. of (2.26), τ_s^o is a linear map, and τ_o^s a diffeomorphism. It is still legitimate to keep the same notation for both objects. $\tau_s^o h$ is the parallel transport of h for the connection $\tilde{\nabla}$.

Similarly if $h \in H_{y_s}^\infty$, set

$$(2.27) \qquad x \in Z_{y_o} \quad ; \quad (\tau_o^s h)(x) = \tau_o^s[h(\tau_s^o x)] \quad .$$

We now define

$$(2.28) \qquad \tau_o^s\, D_{y_s}^2 = \tau_o^s\, D_{y_s}^2\, \tau_s^o \quad .$$

$\tau_o^s D_{y_s}^2$ is an operator acting on $H_{y_o}^\infty$. We will use a similar notation for other operators than D^2.

$f_1, \dots, f_\alpha, \dots f_m$ is now an orthonormal base of $T_{y_o} B$.

<u>Definition 2.6</u>. Take $x_o \in Z_{y_o}$. Let $Q_s(x_o, .)$ be the operator acting on $(F' \otimes H^\infty)_{y_o}$ defined by the parabolic equation

$$(2.29) \qquad \frac{\partial Q}{\partial s} = Q[-\tau_0^s \frac{D^2}{2} y_s - \frac{1}{4} f_\alpha f_\beta \otimes \tau_0^s \tilde{\nabla}^2_{y_s} (\tau_s^0 f_\alpha, \tau_s^0 f_\beta)$$

$$- \frac{f_\alpha}{2} \tau_0^s (\nabla_{\tau_s^0 f_\alpha} D_{y_s})]$$

$$Q_0 = \delta_{\{x_0\}}$$

Observe that $\tau_0^s D^2_{y_s}$ is elliptic of order 2, while the other operators are of order 1. By elliptic regularity, (2.29) has a unique solution $Q_s(x_0,.)$ given by a C^∞ kernel $Q_s(x_0,x)$ on Z_{y_0}, which is C^∞ in (x_0,x), and continuous in (s,x_0,x).

$Q_s \tau_0^s$ is an operator from $F'_{y_s} \otimes H^\infty_{y_s}$ into $F'_{y_0} \otimes H^\infty_{y_0}$ given by a smooth kernel on $Z_{y_0} \times Z_{y_s}$. If $h' \in \Gamma((F' \hat{\otimes} F) \otimes \xi)$, we regard h' as a C^∞ section of $F' \otimes H^\infty$, which we note $h'(y,.)$ where . is a dummy variable in Z_y.

We now have the basic result

<u>Theorem 2.7</u>. For any $h' \in \Gamma((F' \hat{\otimes} F) \otimes \xi)$, any $x_0 \in Z_{y_0}$, $s > 0$,

$$(2.30) \qquad \exp(-\frac{sD^2}{2})\, h'(x_0) = E[\exp\{-\int_0^s \frac{K^B(y_h)dh}{8}\}(Q_s \tau_0^s) h'(y_s,\cdot)]$$

<u>Proof</u> : A formal proof can be easily obtained by using Itô's formula, Proposition 2.5, and equation (2.29). A rigorous proof is obtained by the methods of filtering theory [LS, Chapter 8],[BM]. Note here that because of τ_0^s, the coefficients of the parabolic equation are not uniformly bounded. Hewever by using the methods of the Malliavin calculus, as in [BF, Section 3g)], one can prove that for any $p \geq 1$

$$E[\sup_{(x_0,x') \in Z_{y_0} \times Z_{y_s}} |(Q_s(x_0,.)\tau_0^s)(x')|^p] < +\infty$$

Similar estimates hold for the derivatives of $Q_s(x_0,.)\tau_0^s(x')$. In particular the r.h.s. of (2.30) is integrable. □

<u>Remark 1</u>. At this stage, we have not used the main property of equation (2.29), which is that $Q_s \tau_0^s$ disintegrates $\exp(-\frac{sD'^2}{2})$

conditionally on $(y_h | h \leq s)$. This property, which is fundamental in filtering theory will be used in Section 2f).

Recall for $y \in B$, dx^y is the volume element of Z_y. The volume element of M is given by $dy\, dx^y$.

Let P_s be the kernel of $\exp(-\frac{s D'^2}{2})$.

Let q_s be the scalar heat kernel on B . Finally let $E^B_{y_o,y}$ be the law of the Brownian bridge in B starting at y_o and ending at y at time 1 .

From Theorem 2.7, we deduce :

Theorem 2.8. For any $x_o \in Z_{y_o}$

$$P_1(x_o,x_o) = q_1(y_o,y_o)\, E^B_{y_o,y_o}\Big[\exp\Big\{-\int_o^1 \frac{K^B(y_s)}{8}\, ds\Big\} (Q_1\, \tau_o^1)(x_o,x_o)\Big] \tag{2.31}$$

In particular

$$\int_{Z_{y_o}} \mathrm{Tr}_s[P_1(x,x)]\, dx^{y_o} = q_1(y_o,y_o)\, E^B_{y_o,y_o}\Big\{\exp\Big\{-\int_o^1 \frac{K^B(y_s)}{8}\, ds\Big\} (\mathrm{Tr}_s \otimes \mathrm{Tr}_s)[Q_1\, \tau_o^1]\Big\} \tag{2.32}$$

Proof : (2.31) is obvious by disintegration of (2.30). Using (2.31), (2.32) immediately follows. □

f) Integration along the fiber and the filtering equation.

We regard $\tilde{\nabla}^2_{y_s}$ as the 2 form on LB

$$(U,V) \in T_y LB \to \tilde{\nabla}^2_{y_s}(U_s,V_s) .$$

Similarly $(\tilde{\nabla}.D)_{y_s}$ is identified with the 1 form on LB

$$U \in T_y LB \to \tilde{\nabla}_{U_s} D_{y_s} .$$

Note that $\tilde{\nabla} - \frac{iD}{\sqrt{2}}$ is a superconnection on H^∞ in the sense of Quillen [Q] .

We will do our computations in $\mathrm{End}\, H^\infty \hat{\otimes} \Lambda(T^*B)$.
In particular

$$(2.33) \qquad (\tilde{\nabla} - \frac{iD}{\sqrt{2}})^2 = \tilde{\nabla}^2 - \frac{D^2}{2} - \frac{i}{\sqrt{2}} \tilde{\nabla}.D$$

Of course $(\tilde{\nabla} - \frac{iD}{\sqrt{2}})^2_{y_s}$ extends into an element of

$\mathrm{End}\, H^\infty_{y_s} \hat{\otimes} \Lambda(T^*_y L B)$ in the obvious way.

<u>Definition 2.9.</u> Take $y \in L B$. Q'_s denotes the element of $\mathrm{End}\, H^\infty_{y_o} \hat{\otimes} \Lambda(T^*_y L B)$ defined by the parabolic equation

$$(2.34) \qquad \frac{\partial Q'}{\partial s} = Q' \, \tau_o^s (\tilde{\nabla} - \frac{iD}{\sqrt{2}})^2_{y_s}$$

$$Q'_o = \mathrm{Id} \, .$$

Of course (2.34) still makes sense along a Brownian loop y which lives in L^oB .

For $s > 0$, Q'_s is given by a C^∞ kernel.

Again (2.34) can be solved by a power series expansion in the Grassmann variables over $L B$.

Remember that

$$(2.35) \qquad \mathrm{Ind}\, D'_+ = \int_B dy_o \int_{Z_{y_o}} \mathrm{Tr}_s[P_1(x_o,x_o)] \, dx_o^{y_o} \, .$$

Using (2.35), we claim that by proceeding formally as in Atiyah [A] and ourselves [B4], we can write formally

$$(2.36) \qquad \mathrm{Ind}\, D'_+ = \frac{1}{(2\pi)^{\ell'}} \left(\prod^{+\infty} m^2\right)^{\ell'} i^{\ell'} \int_{L^oB} \exp\{\frac{-(d + i_Y) Y'}{2}\} \wedge \mathrm{Tr}_s[Q'_1 \, \tau_o^1]$$

The only point is to note that in [B4, Section 2], $f_\alpha f_\beta$ is changed into $-2\,dy^\alpha dy^\beta$, and so f_α should be changed into $\sqrt{2}\,i\,dy_\alpha$

We now claim that at least formally

$$\text{(2.37)} \qquad \mathrm{Tr}_s[Q_1' \tau_o^1] = i^\ell \frac{(\Pi m^2)^\ell}{(2\pi)^\ell} \pi_*' [\exp\{\frac{-(d+i_X)P_Z^* X'}{2} - \frac{(d+i_X)<T(.,.)_\wedge \theta>}{2}\beta]$$

The idea will be to prove that for sufficiently many differential even forms η on LB, at least formally

$$\text{(2.38)} \qquad \int_{LB} \exp \frac{-(d+i_Y)Y'}{2} \mathrm{Tr}_s[Q_1' \tau_o^1] \wedge \eta =$$

$$\frac{i^\ell (\Pi m^2)^\ell}{(2\pi)^\ell} \int_{LB}^{+\infty} \exp\{-\frac{(d+i_Y)Y'}{2}\} \pi_*' [\exp\{-\frac{(d+i_X)P_Z^* X'}{2} - \frac{(d+i_X)<T(.,.)_\wedge \theta>}{2}\}\beta] \wedge \eta \quad .$$

Let $\eta_1, \ldots \eta_p$ be C^∞ differential form on B, such that $\Sigma \deg(\eta_j)$ is even.

Let $g(s_1, \ldots s_p)$ be a bounded C^∞ function defined on R^p with values in R.

We will consider even differential forms η on LB of the type

$$\text{(2.39)} \qquad \eta(y) = \int_{0 \leq s_1 \leq s_2, \ldots \leq s_p \leq 1} g(s_1, \ldots s_p) \eta_{1,y_{s_1}} \wedge \eta_{2,y_{s_2}} \wedge \ldots \wedge \eta_{p,y_{s_p}} \, ds_1 \ldots ds_p$$

In (2.24) and in Theorem 2.7, we saw how to construct the semi-groups $\exp(-\frac{s(D^B)^2}{2})$ and $\exp(-\frac{s(D')^2}{2})$ using the Brownian motion

y. in B. We briefly show how to construct $\exp(-\frac{sD'^2}{2})$ using the Brownian motion x. in M.

In fact take $x_o \in M$. Let x. be the Brownian motion starting at x_o, τ_o^s temporarily denotes the parallel transport operator from $(F'\hat{\otimes}F\otimes\xi)_{x_s}$ into $(F'\otimes F\otimes\xi)_{x_o}$ along x. Set $y.=\pi x.$ $e_1 \dots e_n$ is an orthonormal base of $T_{x_o}Z$ and $f_1 \dots f_m$ an orthonormal base of $T_{\pi x_o}B$.

Using Proposition 2.5 and a modified version of [B2, Theorem 2.5] it is not difficult to prove that if U_s is the solution of an equation which we only write in part

$$dU_s = U_s[(-\frac{1}{4}e_i\, e_j \otimes \tau_o^s L_{x_s}(\tau_s^o e_i, \tau_s^o e_j)$$

$$-\frac{1}{4} f_\alpha\, f_\beta \otimes \tau_o^s L_x(\tau_s^o f_\alpha, \tau_s^o f_\beta) - \frac{1}{2} f_\alpha\, e_i \otimes \tau_o^s L_{x_s}(\tau_s^o f_\alpha, \tau_s^o e_i)$$

$$-\frac{1}{2} f_\alpha\, e_i\, \tau_o^s R^Z(\tau_s^o f_\alpha, \tau_s^o e_i))ds + \frac{1}{2} f_\alpha\, e_j < T_{x_s}(\tau_s^o f_\alpha, \tau_s^o e_j), dx_s >$$

$$+\frac{1}{4} f_\alpha\, f_\beta < T_{x_o}(\tau_s^o f_\alpha, \tau_s^o f_\beta), dx_s > + \dots]$$

$$U_o = I_{(F'\hat{\otimes}F\otimes\xi)_{x_o}}$$

(where ... involves quadratic expressions in T) then for any $h' \in \Gamma(F'\hat{\otimes}F\otimes\xi)$

$$\exp - \frac{s(D')^2}{2} h'(x_o) = E[\exp\{-\int_o^s \frac{K^B(y_h)dh}{8}\} U_s\, \tau_o^s h'(x_s)] \tag{2.40}$$

If $k_t x_. = x_{.+t}$, clearly $k_t U_s$ (which is U_s calculated on the path $k_t x_.$) is given by

$$k_t U_s = \tau_t^o U_t^{-1} U_{s+t} \tau_o^t$$

For $y \in B$, $1 \leq j \leq p$, let $\eta'_{j,y} \in c\,(T_y B)$ be given by

$$\eta'_{j,y} = (\sqrt{2}i)^{\deg \eta_j} \sum_{\alpha_1 < \ldots < \alpha_{\deg \eta_j}} \eta_j(f_{\alpha_1}, \ldots f_{\alpha_{\deg \eta_j}}) f_{\alpha_1} \ldots f_{\alpha_{\deg \eta_j}}$$

Let $\Phi(x_.)$ be the random variable in $\mathrm{End}((F'\hat{\otimes} F \otimes \xi)_{x_1}, (F'\hat{\otimes} F \otimes \xi)_{x_o})$

$$(2.41) \qquad \Phi(x_.) = \int_{0 \leq s_1 \leq s_2 \ldots \leq s_p} g(s_1 \ldots s_p)\, U_{s_1} \tau_o^{s_1} \eta'_{1,y_{s_1}} k_{s_1}$$

$$[U_{s_2-s_1} \tau_o^{s_2-s_1} \eta'_{2,y_{s_2-s_1}} k_{s_2-s_1}$$

$$[\cdots \eta'_{p,y_{s_p-s_{p-1}}} U_{1-s_p} \tau_o^{1-s_p}]] ds_1 \ldots ds_p \; .$$

Let $E^M_{x_o,x}$ be the law of the Brownian motion $x_.$ conditioned on $x_1 = x$.

By proceeding formally as in [A] and in [B4, Section 2], we find that

$$(2.42) \qquad \int_B dy \int_{Z_y} p_1(x_o,x_o)\, E^M_{x_o,x_o}[\exp\{-\int_o^1 \frac{K^B(y_h)dh}{8}\} \mathrm{Tr}_s \Phi(x_.)] dx_o^y$$

$$= \frac{i^{\ell+\ell'}}{(2\pi)^{\ell+\ell'}} (\prod_1^{+\infty} m^2)^{\ell+\ell'} \int_{LM} \exp\{-\frac{(d+i_X)X'}{2} \cdot (d+i_X)\frac{<T(.,.)_\wedge \theta>}{2}\}$$

$$\wedge \beta \wedge \pi'^* \eta$$

If $y.$ is a continuous path in B , set $k_t y_. = y_{.+t}$.

If $y.$ is the Brownian motion in B starting at $y_o = \pi x_o$, $k_t Q_s$ (which is Q_s calculated on k_t $y.$) is well defined.

Let $\Psi(y) \in \mathrm{End}((F' \otimes H^\infty)_{y_1}, (F' \otimes H^\infty)_{y_o})$ be defined by

$$(2.43) \qquad \Psi(y) = \int_{0 \leq s_1 \leq \ldots \leq s_p \leq 1} g(s_1 \ldots s_p)(Q_1 \tau_o^{s_1}) \eta'_{1,y_{s_1}} k_{s_1}$$

$$[Q_{s_2-s_1} \tau_o^{s_2-s_1} \eta'_{2,y_{s_2-s_1}} k_{s_2-s_1} [\ldots Q_{1-s_p} \tau_o^{1-s_p} \ldots]] ds_1 \ldots ds_p$$

We claim that if $h' \in \Gamma(F' \hat{\otimes} F \otimes \xi)$

(2.44) $$E[\exp\{-\int_o^1 \frac{K^B(y_s)ds}{8}\}\Phi(x_.)h'(x_1)] = E[\exp\{-\int_o^1 \frac{K^B(y_s)ds}{8}\}(\Psi(y_.))_{x_o} h'(y_1,.)]$$

(2.44) is in fact a consequence of the basic property of $Q_s \tau_o^s$, which is a disintegration of the kernel P_s conditional on $(y_h(h \leq s)$. From (2.44), we find that

(2.45) $$\int_B dy \int_{Z_y} p_1(x_o,x) E^M_{x_o,x} [\exp\{-\int_o^1 \frac{K^B(y_s)ds}{8}\} \Phi(x_.) h'(x_1)]dx^y$$
$$= \int_B dy\, q_1(y_o,y) E^B_{y_o,y} [\exp\{-\int_o^1 \frac{K^B(y_s)ds}{8}\} (\Psi(y_.))_{x_o} h'(y_1,.)]$$

By disintegrating (2.45) with respect to y_1 we find in particular that

(2.46) $$\int_{Z_{y_o}} p_1(x_o,x_o) E^M_{x_o,x_o} [\exp\{-\int_o^1 \frac{K^B(y_s)ds}{8}\} \mathrm{Tr}_s\, \Phi(x_.)]dx_o^{y_o}$$
$$= q_1(y_o,y_o) E^B_{y_o,y_o} [\exp\{-\int_o^1 \frac{K^B(y_s)ds}{8}\} \mathrm{Tr}_s[\Psi(y)]] .$$

By proceeding again formally as in [A] and [B4, Section 2], we also find that formally

(2.47) $$\int_B q_1(y_o,y_o) E^B_{y_o,y_o} [\exp\{-\int_o^1 \frac{K^B(y_h)dh}{8}\} \mathrm{Tr}_s[\Psi(y)]]dy_o$$
$$= \frac{i^{\ell'}}{(2\pi)^{\ell'}} (\prod_1^{+\infty} m^2)^{\ell'} \int_{LB} \exp\{-\frac{(d+i_\gamma)\gamma'}{2}\} \wedge \mathrm{Tr}_s[Q_1' \tau_o^1] \wedge n$$

By (2.42), (2.46), (2.47), we find that (2.38) has been proved for the differential form η given by (2.39).
Since such η generate formally all the differential forms in LB, we find that (2.37) follows.

Remark 2. We certainly do not claim to have "proved" (2.37). Our idea is that although the arguments leading to (2.37) are formal, the consequences one can possibly draw from (2.37) can be proved rigorously. This was already the case in [B3], [B4], [B5].

From (2.15), (2.37), we find that

$$(d+i_\gamma)\mathrm{Tr}_s(Q_1' \tau_0^1) = 0 .$$

As we shall see in the next section, this can be proved rigorously.

g) Superconnections and the filtering equation.

Remember that $\mathrm{Tr}_s[Q_1' \tau_0^1]$ is a well defined series of smooth even forms on LB.

We now prove

Theorem 2.10. If $\beta' = \mathrm{Tr}_s[Q_1' \tau_0^1]$, then

$$(d+i_\gamma)\beta' = 0 \tag{2.48}$$

In particular β' restricts on B to the closed form $\mathrm{Tr}_s \exp(\tilde{\nabla} - \frac{iD}{\sqrt{2}})^2$.

Proof. Take a loop $y.$ in LB. Let $y'_.$ be another loop in LB which is close enough to y. Then y'_0 and y_0 are connected by a unique geodesic in B. We can then identify $H^\infty_{y'_0}$ and $H^\infty_{y_0}$ by parallel transport along the geodesic, which consists of a diffeomorphism of $Z_{y'_0}$ into Z_{y_0} and of a linear parallel transport operator along the horizontal lift of the geodesic.

Similarly, we identify $H^\infty_{y'_s}$ and $H^\infty_{y'_0}$ by parallel transport along y'. Ultimately for y' close enough to y, $H^\infty_{y'_s}$ is identified with $H^\infty_{y_0}$ through two successive parallel transports. It is here essential to observe that now s should considered as an element of R, since parallel transport along a loop is in general not trivial.

For y' close enough to y, for $s \in R$, we thus have a well defined isomorphism u'_s from $H^\infty_{y_0}$ into $H^\infty_{y'_s}$.

For y' close enough to y, we define a connection 1 form $\bar{\lambda}$ on $T_{y'}LB$ with values in $C^\infty(R\ ;\ \mathrm{End}\ H^\infty_{y_0})$. In fact if $Y' \in T_{y'}LB$, set

$$\bar{\lambda}_s\ (Y') = u_s'^{-1}(\tilde{\nabla}_{Y'_s}\ u'_s)$$

One readily verifies that $\bar{\lambda}_s(Y')$ is a first order differential operator acting on $H^\infty_{y_0}$. Moreover the curvature of the connection $\bar{\lambda}$ is such that if $Y', Z' \in T_{y'}LB$

$$\bar{\Lambda}_s(Y',Z') = u_s'^{-1}\ (\tilde{\nabla}^2_{y'_s}(Y'_s,Z'_s))u'_s \tag{2.49}$$

$\bar{\Lambda}_s$ also takes its values in first order differential operators acting on $H^\infty_{y_0}$.

Incidently note that the action of S_1 on LB lifts into an action of R, since the rotation of a loop by 1 is no longer trivial on the u'_s. Let K be the vector field generating this action of R.

Let $\bar{D}_s$ be the equivariant representation of D, i.e.

$$\bar{D}_s = u_s'^{-1}\ D_{y'_s}\ u'_s \tag{2.50}$$

We now use the superconnection formalism of Quillen [Q]. The computations which follow are done in the graded algebra

$A = \mathrm{End}\, H^{\infty}_{y_0} \hat{\otimes} \Lambda(T^*LB)$. Recall that if $a,b \in A$, the supercommutator $[a,b]_s$ is defined by

$$(2.51) \qquad [a,b]_s = ab - (-1)^{\deg a \deg b}\, ba$$

Clearly

$$(2.52) \qquad u_s'^{-1}\, \tilde{\nabla} D\, u_s' = d\bar{D} + [\bar{\lambda},\bar{D}]_s$$

$$= [d + \bar{\lambda},\bar{D}]_s$$

Set

$$(2.53) \qquad M_s = \bar{\Lambda}_s - \frac{i}{\sqrt{2}}[d + \bar{\lambda},\bar{D}]_s - \frac{\bar{D}^2}{2}$$

We have

$$(2.54) \qquad \left(d + i_K + \bar{\lambda} - \frac{i\bar{D}}{\sqrt{2}}\right)^2 = L_K + i_K\, \bar{\lambda} + M .$$

By (2.54), we find that

$$(2.55) \qquad [d + i_K, M] = \left[M + i_K\bar{\lambda}, \bar{\lambda} - \frac{i\bar{D}}{\sqrt{2}}\right] + L_K\left(\bar{\lambda} - \frac{i\bar{D}}{\sqrt{2}}\right) - d(i_K\bar{\lambda}) .$$

Also the equation of the connection $\bar{\lambda}$ is given by

$$(2.56) \qquad d\bar{\lambda} = -\bar{\lambda}^2 + \bar{\Lambda}$$

We can rewrite (2.55) in the form

$$(2.57) \qquad [d + i_K, M] = \left[M, \bar{\lambda} - \frac{i\bar{D}}{\sqrt{2}}\right] + i_K\, \bar{\Lambda} - \left[i_K\bar{\lambda}, \frac{i\bar{D}}{\sqrt{2}}\right] - L_K\left(\frac{i\bar{D}}{\sqrt{2}}\right)$$

Consider now the partial differential equation

$$(2.58) \qquad \frac{\partial H}{\partial s} = H_s M_s \quad ; \quad H(0) = I_{\mathrm{End}(H^{\infty}_{y_0})}$$

Since M_s is elliptic, (2.58) has a unique solution given by a C^{∞} kernel on Z_{y_0} . Using (2.52) one verifies that

$$H_s = u_0'^{-1}\, Q_s'\, u_0' .$$

and so

$$(2.59) \qquad H_1\, u_1'^{-1}\, u_0' = u_0'^{-1}\, Q_1'\, u_0'\, u_1'^{-1}\, u_0'$$

Now by definition

$$(2.60) \qquad \tau^1_0(y') = u_0'\, u_1'^{-1}$$

We get

$$(2.61) \qquad H_1\, u_1'^{-1}\, u_0' = u_0'^{-1}(Q_1'\, \tau^1_0)(y')\, u_0'$$

We thus find that

$$(2.62) \qquad \beta' = \mathrm{Tr}_s[H_1\, u_1'^{-1}\, u_o'\,].$$

Since (2.59) has uniformly elliptic coefficients, as y' varies, it is standard that we can differentiate the kernel of H with respect to y', and moreover that

$$(2.63) \qquad (d+i_K)H_t = \int_o^t (d+i_K)H_s\, M\, ds + \int_o^1 H_s[d+i_K, M]ds\ .$$

If $y' \in LB$ is close enough to y, for h small enough, let $\overline{\tau}_h$ be the parallel transport operator from $H^\infty_{y_o}$ into $H^\infty_{y'_h}$ along the unique geodesic connecting y_o and y'_h. Clearly since $u'_{s+h}\, u_h'^{-1}$ is the parallel transport operator from $H^\infty_{y'_h}$ into $H^\infty_{y'_{s+h}}$

$$(2.64) \qquad (i_K\overline{\lambda})_s = u_s'^{-1} \frac{D}{Dh}\, [u'_{s+h}\, u_h'^{-1}\overline{\tau}_h]_{h=o} = \frac{d}{dh}\, [u_h'^{-1}\overline{\tau}_h]_{h=o}.$$

Also

$$(2.65) \qquad (e^{hK}\overline{D})_s = (\overline{\tau}_h^{-1}\, u'_h)\, \overline{D}_{s+h}\, (u_h'^{-1}\, \overline{\tau}_h)\ .$$

From (2.64), (2.65), we get

$$(2.66) \qquad \frac{d\overline{D}}{ds} = [i_K\overline{\lambda}, \overline{D}] + L_K\overline{D}$$

Similarly, we find that since u'_s is the parallel transport of u'_o along y', then

$$(2.67) \qquad \frac{d\overline{\lambda}}{ds} = i_K\, \overline{\Lambda}$$

Using (2.57), (2.65), (2.67), we find that (2.63) can be rewritten in the form

$$(2.68) \qquad (d+i_K)H_t = \int_o^1 (d+i_K)H_s\ M\ ds$$
$$+ \int_o^1 H_s[M, \overline{\lambda} - \frac{i\overline{D}}{\sqrt{2}}]ds + \int_o^t H\frac{d}{ds}(\overline{\lambda} - \frac{i\overline{D}}{\sqrt{2}})ds\ .$$

Now using (2.58) again, we have

$$(2.69)\quad H_t(\bar\lambda_t - \frac{i}{\sqrt{2}}\bar D_t) - (\bar\lambda_o - \frac{i}{\sqrt{2}}\bar D_o) H_t = \int_o^t HM(\bar\lambda - \frac{i}{\sqrt{2}}\bar D)ds + \int_o^t H\frac{d}{ds}(\bar\lambda - \frac{i}{\sqrt{2}}D)ds - \int_o^t (\bar\lambda_o - \frac{i}{\sqrt{2}}\bar D_o) HM\,ds$$

or equivalently

$$(2.70)\quad H_t(\bar\lambda_t - \frac{i}{\sqrt{2}}\bar D_t) - (\bar\lambda_o - \frac{i}{\sqrt{2}}\bar D_o)H_t = \int_o^t \left(H(\bar\lambda - \frac{i\bar D}{\sqrt{2}}) - (\bar\lambda_o - \frac{i\bar D_o}{\sqrt{2}})H \right) M\,ds + \int_o^t H[M,\bar\lambda - \frac{i\bar D}{2}]ds + \int_o^t H\frac{d}{ds}(\bar\lambda - \frac{i\bar D}{\sqrt{2}})ds \quad .$$

Now both sides of (2.68) use (2.70) are given by C^∞ kernels on Z_{y_o}. Since linear elliptic equations have unique solutions, we find that

$$(2.71)\quad (d + i_K)H_t = H_t(\lambda_t - \frac{iD_t}{\sqrt{2}}) - (\bar\lambda_o - \frac{i}{\sqrt{2}}\bar D_o)H_t$$

Now

$$(2.72)\quad (d + i_K)\beta' = Tr_s[(d + i_K)H_1 u_1'^{-1} u_o'] + Tr_s[H_1(-\bar\lambda_1 u_1'^{-1} u_o' + u_1'^{-1} u_o' \bar\lambda_o)]$$

Since Tr_s vanishes on supercommutators, using (2.71), (2.72), we find that

$$(2.73)\quad (d + i_K)\beta' = -\frac{i}{\sqrt{2}} Tr_s\{(H_1\bar D_1 - \bar D_o H_1)u_1'^{-1} u_o'\}$$
$$= -\frac{i}{\sqrt{2}} Tr_s\{(H_1 u_1'^{-1} D_{y_o'} u_1' - u_o'^{-1} D_{y_o'} u_o' H_1)u_1'^{-1} u_o'\}$$
$$= 0 \quad .$$

The Theorem is proved. □

Remark 3 : The proof is much easier if $H^\infty = H_+^\infty \oplus H_-^\infty$ is a finite dimensional Z_2 graded bundle. If $n_\pm = \dim H_\pm^\infty$, let O be the set of graded orthonormal frames in H^∞. O is a $U(n_+)xU(n_-)$ fiber bundle over B. Since $U(n_+)xU(n_-)$ is connected, as shown in [B4, Theorem 3.2], the loop space LO is a $LU(n_+)xLU(n_-)$ principal fiber bundle. As in [B4, Remark 3.2], $LOxS_1$ becomes a principal bundle over LB, whose structure group is a Mac-Moody group i.e. the semi direct product of $LU(n_+)xLU(n_-)$ and of S_1. S_1 then acts on $LOxS_1$ by bundle homomorphisms which preserve a natural connection $\overline{\mu}$ on $LOxS_1$. From the family D, we construct a family $\overline{D}'$ of S_1 invariant operators acting on $C^{n+}xC^{n-}$. Imitating [B4, Remark 3.2]

$$\frac{\partial}{\partial s} + \overline{\lambda}(K) + \overline{\Lambda} - \frac{i}{\sqrt{2}}[d + \overline{\lambda}, D]_s - \frac{\overline{D}^2}{2} \tag{2.74}$$

becomes a natural moment map. (2.47) easily follows.

h) The Levi-Civita superconnection and the filtering equation.

Instead of calculating r.h.s. of (2.37), we can try to explicitly calculate the form appearing in (2.14).

Note that as we saw in Section 2a) $S(f_\alpha)f_\alpha = 0$, and $\langle S(f_\alpha)f_\beta, f_\gamma\rangle = 0$. Also $S(e_i)e_i \in T^H M$. Using [B3, (3.14)] and cancelling the unnecessary terms, we find that if K^M is the scalar curvature of M and if

$$C = \frac{K^M}{4} - \frac{\langle \nabla_{f_\alpha}(S(e_j)e_j), f_\alpha\rangle}{2} + \frac{1}{4}\langle S(e_j)e_j, S(e_k)e_k\rangle \tag{2.75}$$

then

$$\begin{aligned}(D^L)^2 = {} & -(\nabla_{e_i} + \tfrac{1}{2}\langle S(e_i)e_j, f_\alpha\rangle e_j\, f_\alpha + \tfrac{1}{4}\langle S(e_i)f_\alpha, f_\beta\rangle f_\alpha\, f_\beta)^2 \\ & - (\nabla_{f_\alpha} + \tfrac{1}{2}\langle S(f_\alpha)e_i, f_\beta\rangle e_i\, f_\beta - \frac{\langle S(e_j)e_j, f_\alpha\rangle}{2})^2 \\ & + \tfrac{1}{2}\, e_i\, e_j \otimes L(e_i, e_j) + \tfrac{1}{2} f_\alpha\, f_\beta \otimes L(f_\alpha, f_\beta) + e_i\, f_\alpha \otimes L(e_i, f_\alpha) + C\end{aligned} \tag{2.76}$$

Remember that by [B3, (1.28)], since T takes its values in TZ

(2.77) $$<S(f_\alpha)e_i, f_\beta> = \frac{1}{2}<T(f_\alpha, f_\beta), e_i>.$$

In Theorem 2.8, we replace $Q_s(x_0,.)$ by $\overline{Q}_s(x_0,.)$ which is now an element of $\mathrm{End}(F'\otimes H^\infty)_{y_0}$.
$\overline{Q}_s(x_0,.)$ is the solution of

(2.78) $$\frac{\partial\overline{Q}}{\partial s} = \overline{Q}_s[\frac{1}{2}\tau_0^s(\nabla_{e_i} + \frac{1}{2}<S(e_i)e_j, f_\alpha> e_j\, f_\alpha + \frac{1}{2}<S(e_i)f_\alpha, f_\beta>)f_\alpha\, f_\beta)^2$$
$$+\frac{1}{4}<T(\frac{dy^H}{ds}, f_\beta), e_i>\tau_0^s(e_i\, f_\beta)$$
$$-\frac{1}{2}<S(e_j)e_j, \frac{dy^H}{ds}> - \tau_0^s(\frac{e_i\, e_j\otimes L(e_i, e_j)}{4}$$
$$+\frac{1}{4}f_\alpha\, f_\beta\otimes L(f_\alpha, f_\beta) + \frac{e_i\, f_\alpha\otimes L(e_i, f_\alpha)}{2})]$$

$$Q_0 = \delta_{\{x_0\}}$$

To describe the analogue $\overline{Q}'$ of Q', we must introduce the Levi-Civita superconnection $\tilde{\nabla}^{LC}$ on H^∞ of Bismut [B4, Section 2], which by [BF, Proposition 1.18] is given by

(2.79) $$\tilde{\nabla}^{LC} = \tilde{\nabla} - \frac{1}{2}<S(e_i)e_i, f_\alpha>dy^\alpha - \frac{1}{8}<T(f_\alpha, f_\beta), e_i>e_i\, dy^\alpha\, dy^\beta$$
$$+ D$$

We now use the formula of [B4, Theorem 3.5] which calculate $(\tilde{\nabla}^{LC})^2$.

After adequate rescaling of the f_α, the analogue of Q' is now $\overline{Q}'\in \mathrm{End}\, H_{y_0}^\infty \hat{\otimes}\Lambda(T_y^* LB)$ given by

(2.80) $$\frac{\partial\overline{Q}'}{\partial s} = \overline{Q}'\,\tau_0^s[(\tilde{\nabla} - \frac{1}{2}<S(e_i)e_i, f_\alpha>dy^\alpha - \frac{i}{4\sqrt{2}}<T(f_\alpha, f_\beta), e_i>$$
$$e_i\, dy^\alpha\, dy^\beta - \frac{iD}{\sqrt{2}})^2 - \frac{1}{2}<S(e_i)e_i, \frac{dy^H}{ds}>$$
$$+\frac{i}{2\sqrt{2}}<T(\frac{dy^H}{ds}, f_\beta), e_i>e_i\, dy^\beta]$$

$$\overline{Q}'_0 = \mathrm{Id}_{H^\infty_{y_0}}$$

The analogue of Theorem 2.10 is now :

Theorem 2.11. If $\overline{\beta}' = \mathrm{Tr}_s[Q_1' \, \tau_0^1]$, then

$$(2.81) \qquad (d + i_\gamma)\, \overline{\beta}' = 0 \,.$$

$\overline{\beta}' - \beta'$ is $d + i_\gamma$ exact .

$\overline{\beta}'$ restricts on B to the closed form

$$(2.82) \qquad \mathrm{Tr}_s[\exp(\widetilde{\nabla} - \frac{1}{2} < S(e_i)e_i, f_\alpha > dy^\alpha - \frac{i}{4\sqrt{2}} < T(f_\alpha, f_\beta), e_i > e_i \, dy^\alpha \, dy^\beta - \frac{i\,D}{\sqrt{2}})^2 \,]$$

Proof : Since the analysis of the proof is the same as the proof of Theorem 2.10, we concentrate on the algebra, while keeping the same notations.

The key fact is that

$$(2.83) \qquad (d + i_K + \overline{\lambda} - \frac{i\overline{D}}{\sqrt{2}} - \frac{1}{2} < S(e_i)e_i, f_\alpha > dy^\alpha - \frac{i}{4\sqrt{2}} < T(f_\alpha, f_\beta), e_i > e_i \, dy^\alpha \, dy^\beta)^2 = L_K + i_K\overline{\lambda} + (d + \overline{\lambda} - i\,\overline{D}/\sqrt{2} - \frac{1}{2} < S(e_i)e_i, f_\alpha > dy^\alpha - \frac{i}{4\sqrt{2}} < T(f_\alpha, f_\beta), e_i >)^2 - \frac{1}{2} < S(e_i)e_i, \frac{dy^H}{ds} > + \frac{i}{2\sqrt{2}} < T(\frac{dy^H}{ds}, f_\beta), e_i > e_i \, dy^\beta \,.$$

Note that the final terms in (2.83) exactly appear in the r.h.s. of (2.80). The proof is then identical to the proof of Theorem 2.10.

The fact that $\beta' - \beta$ is $d + i_\gamma$ exact is a straightforward application of the techniques of [B3, Section 2] and of Theorem 2.10. □

Remark 4. Using the main properties of the filtering equation, we have shown that formally

$$(2.84) \qquad \overline{\beta}' = \frac{\overset{+\infty}{(\Pi} m^2)^\ell}{(2\pi)^\ell} \, i^\ell \, \pi_*'[\exp\{-\frac{(d + i_X) P_Z^* X'}{2}\}\beta]$$

i) The filtering equation and the Chern character.

If E is a finite dimensional complex bundle with connection ∇^E, set

$$\overline{ch}\, E = Tr_s[\exp(\nabla^E)^2]$$

$\overline{ch}\, E$ is a scaled representative of the Chern character of E.

Let i be the embedding $B \to LB$.

In Theorems 2.10 and 2.11, we have calculated $i^*\beta'$ and $i^*\overline{\beta}'$.

Recall that by Atiyah-Singer [AS2, Proposition 2.2], $\ker D_+ - \ker D_-$ is a well defined element of $K(B)$.

In [B4, Theorems 2.6 and 3.4], we proved that $i^*\beta'$ and $i^*\overline{\beta}'$ are closed forms which represent in cohomology $\overline{ch}(\ker D_+ - \ker D_-)$.

So the similarity of equation (1.12) with equations (2.34) and (2.80) is not purely formal : like β, β' and $\overline{\beta}'$ represent lifts as Y equivariantly closed forms on LB of Chern character forms. The fact that $\ker D_+ - \ker D_-$ is only a virtual bundle is reflected in the infinite dimensional character of equations (2.34) and (2.80).

On the other hand, we have seen how the Levi-Civita superconnection is natural, in a problem where a priori no family index problem considered.

The basic explanation for this is that, as we have shown in [B5, Section 3], the explicit calculation of $\overline{ch}(\ker D_+ - \ker D_-)$ is directly by related to integration along the fiber on the loop space.

Incidently note that equalities (2.37) and (2.84) exactly say that a "topological" quantity - an integral along the fiber in the loop space - can be calculated "analytically", i.e. by solving a partial differential equation. This is very much index Theory.

As shown in [B3, Section 5], the proof of the families Index Theorem in cohomological form can be immediately derived from such considerations.

As to our first proof [B3, Section 4], which gives a local version of the families index, it is based on the use of the Levi-Civita superconnection .

[A] ATIYAH M.F. : Circular symmetry and stationary phase approximation. In proceedings of the Conference in honor of L. Schwartz pp. 43-59. Astérisque n° 131 (1985).

[AB] ATIYAH M.F., BOTT R. : The moment map and equivariant cohomology. Topology 23, 1-28 (1984).

[ABP] ATIYAH M.F., BOTT R., PATODI V.K. : On the heat equation and the Index Theorem. Invent. Math. 19(1973), 279-330.

[AS1] ATIYAH M.F., SINGER I.M. : The Index of elliptic operators. I. Ann. of Math. 87 (1968), 484-530. III. Ann.of Math. 87 (1968), 546-604.

[AS2] ATIYAH M.F., SINGER I.M. : The Index of elliptic operators. IV Ann. of Math. 93 (1971), 119-138.

[B1] BISMUT J.M. : Transformations différentiables du mouvement Brownien. In Proceedings of the Conference in honor of L. Schwartz, pp. 61-87, Astérisque n° 131 (1985).

[B2] BISMUT J.M. : The Atiyah-Singer Theorems: a probabilistic approach I. J. Funct. Ann. 57 (1984), 56-99.

[B3] BISMUT J.M. : The Atiyah-Singer Index Theorem for families of Dirac operators : two heat equation proofs. Invent. Math. 83 (1986), 91-151.

[B4] BISMUT J.M. : Index Theorem and equivariant cohomology on the loop space. Comm. Math. Phys. 98 (1985), 213-237.

[B5] BISMUT J.M. : Localization formulas, Superconnections and the Index Theorem for families. Comm. Math. Phys. 103 (1976), 127-166.

[B6] BISMUT J.M. : Large deviations and the Malliavin calculus. Progress in Math. n° 45. Boston : Birkaüser 1984.

[B7] BISMUT J.M. : Formules de Lichnerowicz et théorème de l'indice. Proc. of the Conference in honor of A. Lichnerowicz. Paris - Hermann (to appear).

[BF] BISMUT J.M., FREED D.S. : The analysis of elliptic families. II. Dirac operators, êta invariants and the holonomy theorem. Comm. Math. Phys. (to appear 1986).

[BM] BISMUT J.M., MICHEL D. : Diffusions conditionnelles. I. Hypoellepticité partielle. J. Funct. Anal. 44 (1981), 174-211. II. Générateur conditionnel. Application au filtrage. J. Funct. Anal. 45(1982), 274-292.

[BV] BERLINE N., VERGNE M. : Zéros d'un champ de vecteurs et classes caractéristiques équivariantes. Duke Math. J. 50 (1983), 539-549.

[DH] DUISTERMAAT J.J., HECKMAN C. : On the variation of the cohomology of the reduced phase space. Invent. Math. 259-268 (1982). Addendum 72, 153-158 (1983).

[G] GOODWILLIE T.G. : Cyclic homology, derivations and the free loop space. Topology 24 (1985), 187-215.

[IMK] ITÔ K., Mc KEAN H. : Diffusion processes and their sample paths. Grundl. Math. Wiss. Band 125. Berlin, Springer 1974.

[LS] LIPTSER R.S., SHIRYAYEV A.T. : Statistics of Random processes. Vol. 1, Berlin, Springer 1977.

[P] PETRAK S. : Analytic and algebraic cohomology and the free loop space (1985).

[Q] QUILLEN D. : Superconnections and the Chern character. Topology 24 (1985), 89-95.

[W] WITTEN E. : Supersymmetry and Morse theory. J. Diff. Geom. 17, 661-692 (1982).

NEW PROBLEMS AND NEW APPROXIMATION METHODS IN GENERAL RELATIVITY

Thibault DAMOUR
Groupe d'Astrophysique Relativiste
CNRS - Observatoire de Paris
92195 Meudon Principal Cedex
FRANCE

ABSTRACT

The mathematical and physical status of approximation methods in General Relativity is briefly discussed. It is stressed that some new problems of relativistic astrophysics (motion of condensed bodies, clumpy cosmology, ...) require the use of new approximation methods. These methods, instead of being global, use several local charts and several local asymptotic expansions. Some recent progresses in the problem of the motion of two strongly self-gravitating bodies, obtained by means of such a "patchwork" method, are presented. The comparison of the (approximate) theoretical predictions with observations leads to a strong confirmation of General Relativity, because this is the first "non perturbative" test of gravity theories.

1. INTRODUCTION

The aim of Physics is to provide a link between Mathematics and "Reality". For this programme to be successful, many conditions have to be fulfilled. Any physical theory must base itself on precise mathematical axioms. The mathematical consequences of the axioms must be drawn using rigourous methods, up to a point where sufficiently detailed results concerning well defined theoretical models can be directly compared with the observations. It is essential for the meaning of the scientific enterprise that there be no weak link in the long chain going from the axioms to the ultimate comparison with experimental results. In real life, this is rarely so. Modern physical theories are so complex (both mathematically and physically) that there are always weak links, and sometimes even real gaps, in the chain of deductions. One of the roles of Mathematical Physics is to work towards bridging these gaps. A type of gap, present in many physical theories, is that one usually compares observations with theoretical results obtained not via a rigourous deduction, but by using some approximation method. Now, most of the time, the approximation methods used in Physics constitute a kind of "black art" which is not rigourously tied with mathematical axioms defining physical theories.

The aim of this article is to draw mathematical physicists' attention to the presence of gaps in the theoretical deductions of General Relativity, and especially with regards to the use of approximation methods. Furthermore there are still several fundamental mathematical problems of General Relativity which need better solutions. In particular, the Cauchy problem for General Relativity, first solved by (Choquet) Fourès-Bruhat [1)], still needs, in spite of recent advances [2)], to be improved upon, both to give more global results, and to give results (apparently lacking, and well needed in applications) about the evolution of separated bodies (with stress-energy distributions having spatially compact supports). For a review of such fundamental mathematical problems, see the recent work by Choquet-Bruhat [3)].

In this article, we wish to emphasize the need for mathematical justifications of the approximation methods used in General Relativity. This problem is in fact becoming more and more complex for the following reasons. When the theory of General Relativity was created (circa 1915) the concept of differentiable manifold had not yet been clearly defined by mathematicians. As a consequence, Riemannian spaces were more or less thought (at least by physicists) as being globally diffeomorphic to $\mathbb{R}^n$.

This semi-implicit impression was strengthened by the intuitive idea that the space-time curvature in the real world (i.e. mainly the solar system, in 1915) was so small that an approximation starting with a more familiar (global) Newtonian or Minkowskian description of the space-time should be adequate. This led to the introduction by physicists of some approximation methods (the so-called post-Newtonian and post-Minkowskian methods) using only <u>one global coordinate system</u>,

$$x^a \in \mathbb{R}^4, \tag{1}$$

and based on a <u>global weak-field ansatz</u> of the type :

$$g_{ab}(x^c) = f_{ab} + \varepsilon\, h^{(1)}_{ab}(x^c) + \varepsilon^2\, h^{(2)}_{ab}(x^c) + \ldots , \tag{2}$$

where f_{ab} denotes the flat Minkowski metric (diag (-1,+1,+1,+1)) and where ε denotes a suitable "small" parameter (or a formal expansion parameter). For a recent review of this type of methods, that for the sake of exposition we shall call the "old" approximation methods, see e.g. reference [4].

For a long time the only contact between the theory of General Relativity and the "real world" consisted of observations of phenomena taking place in the solar system. These observations have been fitted to theoretical formulae obtained by means of "old" approximation methods. This led to good quantitative confirmations of General Relativity (down to the level of one part per thousand). For a review see Will [5]. However this result is not fully satisfactory for at least two kinds of reasons. First, in spite of the existence "on the market" of the "old" approximation methods for seventy years, they have not yet been mathematically justified starting from the axioms of General Relativity. For a review of the state of the art see e.g. Ehlers [6], reference [4] and references therein. The second reason which makes the comparison between General Relativity and the solar system observations somewhat dissatisfying is that, as shown notably by Will [5], there are at least ten other relativistic theories of gravity which explain all gravitational phenomena in the solar system as well as the Einstein theory. The reason for this being that the latter phenomena concern only the quasi-stationary weak-field limit of any relativistic theory of gravity, so that two theories of gravity which are in fact very different in the dynamic and/ or strong-field regime can still lead to identical predictions in the solar system case. This means that the solar system tests may be of good quantitative value, but that they are poor qualitative tests of the theories of gravity.

This situation has changed recently by the (observational or conceptual) discovery of new astrophysical objects, and also by the willingness to describe more precisely previously known astrophysical objects. Indeed, on the one hand the discovery of binary systems containing "gravitationally condensed" objects (i.e. neutron stars or black holes) has posed to the theorist the new problem of the motion of strongly self-gravitating bodies, a problem where, contrarily to the solar system case, the strong-field regime of General Relativity comes into play. On the other hand, the necessity of a more detailed description of the Universe at large has renewed the cosmological problem by compelling theorists to take into account, right from the start, the "clumpiness" of the matter distribution. The preceding "new problems" (together with other problems concerning gravitational radiation phenomena) require the use of "new approximation methods".

By contrast with the "old" approximation methods based on the idea of a global asymptotic expansion (2), described in one global coordinate system (1), the "new" approximation methods can be characterised by the use of several coordinate systems,

$$x^a \in U_\varepsilon \subseteq \mathbb{R}^4 , \tag{3a}$$

$$\hat{x}^a \in \hat{U}_\varepsilon \subseteq \mathbb{R}^4 , \text{ etc } \ldots \tag{3b}$$

together with several asymptotic expansions :

$$g_{ab}(x^c) = g_{ab}^{(0)}(x) + \delta_1(\varepsilon) \; g_{ab}^{(1)}(x) + \ldots \quad , \tag{4a}$$

$$\hat{g}_{ab}(\hat{x}^c) = \hat{g}^{(0)}_{ab}(\hat{x}) + \hat{\delta}_1(\varepsilon)\, \hat{g}^{(1)}_{ab}(\hat{x}) + \ldots \quad , \text{etc} \qquad (4b)$$

As a side remark, we can notice that although the use of several coordinate charts (3) is motivated by physical reasons (in general a necessity to describe several different physical regimes taking place at different length or time scales) the formal setting of these "new" approximation methods is somewhat parallel to the mathematical definition of differentiable manifolds by an atlas of charts (brought out only in 1936 by Whitney [7]). There is no relation of cause to effect behind this parallelism, but only an a-posteriori confirmation of the fittingness of the mathematical definition.

In the following section we shall describe some results on the motion of two condensed bodies which have been obtained by means of such a "new" approximation method, and how it has been possible to push the theoretical calculations up to a point where a direct complete comparison with observational data was possible. However, before entering into this detailed application, we would like to emphasize that the mathematical status of these "new" approximation methods is unsatisfactory for several reasons.

First, as in the case of the "old" approximation methods, equation (2), there exists no theorem guaranteeing that the expansions (4) are asymptotic to a family of solutions of Einstein's equations (see Ehlers [6] for a review and references to this problem). But the problem is worse than that in the case of the "new" methods. Indeed, even at the formal level, where one is not asking for convergence or asymptotic character, it has not yet been proved that the expansions (4) can be algorithmically defined in a consistent way to all orders. Moreover, the problem of checking the <u>formal completeness</u> and the <u>formal compatibility</u> of the expansions (4) is not yet clearly understood. Indeed, in the case of the mathematical definition of a manifold, an essential role is played by the completeness (union of domains of charts = manifold) and the compatibility conditions (diffeomorphisms connecting two charts). In the case of a "patchwork" of asymptotic expansions like (4), one expects that there should exist corresponding completeness and compatibility conditions.

However, this issue is not trivial, because, in many cases, the natural domains of uniform validity of the expansions (4) are disconnected (when $\varepsilon \to 0$),

$$U_\varepsilon \cap \hat{U}_\varepsilon = \phi \,, \tag{5}$$

and ε appears in a non trivial way in the transformation between charts:

$$\hat{x}^a = F^a(x^b \,;\, \varepsilon\,) \,. \tag{6}$$

For preliminary formal discussions of these issues, see Kates [8] and Damour [9]. Up to now these problems have been somewhat by-passed by using some formal "matching rules" borrowed from perturbation techniques used in fluid dynamics. Work is in progress towards clarifying these formal problems, so we hope that soon the only mathematical gap which will remain in the applications of these methods will be the lack of a proof that the expansions (4) are asymptotic to a family of solutions of Einstein's equations.

2. THE BINARY PULSAR.

In 1974 Hulse and Taylor discovered a pulsar member of a binary system : the "binary pulsar" PSR 1913 + 16. This discovery of a very precise clock (the pulsar) moving in a highly elliptic orbit around the centre of mass of the binary system opened up a new testing field for General Relativity. This system has been monitored, since its discovery, by Taylor and his collaborators [10] with an ever increasing precision. The raw data consist of measurements of the (atomic) time of arrival on Earth of some of the electromagnetic pulses emitted by the pulsar. Mathematically this means that one has a finite sequence of real numbers which is thought of as a <u>sub-sequence</u> of the hypothetical complete sequence of measurements of the arrival times of all the pulses emitted by the pulsar :

$$t_N^{arriv} = F^{theor}(N) + \xi_N \quad , \tag{7}$$

where N runs over all the set of integers. In equation (7) $F^{theor}(N)$ is an explicit function ("timing formula") calculable, in principle, from the theories of gravity and of electromagnetism (used to describe the motions of the pulsar and of the Earth, and the propagation of electromagnetic pulses from the pulsar to the Earth) and from some theoretical models (used for the emission mechanism of pulsars, the structure of the binary system, and the effect of interstellar dispersion). Finally ξ_N, in equation (7), is assumed to be a random variable representing all the effects not included in the theoretical models (experimental errors, flickering of the pulsar emission, fluctuation of the interstellar index, etc ...).

Now the role of Physics is to give an explicit expression for the "timing formula" $F^{theor}(N)$. This problem can be split into three subproblems :

1) to derive the equations of motion of two condensed bodies ;

2) to solve the latter equations of motion ; and

3) to derive the explicit timing formula for the propagation of electromagnetic pulses between one condensed body and the Earth.

We shall discuss successively these problems in the following three sub-sections. In the third sub-section we shall also discuss the fourth problem of how

4) to analyze and to interpret the observational data.

3. THE TWO-CONDENSED-BODY PROBLEM IN GENERAL RELATIVITY.

One calls a body "gravitationally condensed", or simply "condensed", if its physical radius, a, is of the order of its gravitational radius, $G(\text{mass})/c^2$. In other words this means that the relativistic self-gravitating potential,

$$\frac{Gm}{c^2 a} \sim 1. \tag{8}$$

Relativistic astrophysics has predicted the existence of two families of condensed objects : the neutron stars and the black holes. It has been now convincingly shown that neutron stars exist in the "real world", and, in particular, that pulsars are rotating neutron stars emitting pulses by an electromagnetic "lighthouse effect".

Now the gravitational interaction of several condensed bodies poses a new problem in General Relativity. Indeed each body being strongly self-gravitating the space-time curvature generated by a system of condensed bodies will not be everywhere small, as in the simpler case of the solar system. In fact there is a mixing of several strong-gravitational-field regions (near and in each body) and of a weak-gravitational-field region (obtained by cutting out several balls, each enclosing one body). Because of the presence of strong fields, one cannot use the "old" approximation methods, applied in the solar system, and, based on the uniform weakness of the gravitational field (expansion of the type (2), with ε being the small self-gravity potential Gm/c^2a). New methods for dealing with the problem of the motion of n condensed bodies have been devised by Manasse [11)], Demianski and Grishchuk [12)], D'Eath [13)], Kates [14)], Damour [15)] and Thorne and Hartle [16)]. The common feature of all these approaches is to use a "new" approximation method, in the sense of §1 above, i.e. to employ simultaneously several asymptotic expansions, of the type of equation (4). One such expansion being appropriate to the common weak field domain ("external region") and the others being appropriate to each strong field domain associated with each body ("internal regions"). Then, one compares these expansions, by some matching procedure, so as to propagate the information between the domains of validity of the approximation schemes. In the following we shall briefly outline the most complete results obtained by such a method (for a general review of the problem of motion in General Relativity, including a discussion of other less complete or less satisfactory results, see Damour [4)]).

3.1 The Internal Problems

Intuitively, the physical fact which permits to develop an approximation scheme in each strong-field internal region is that there should exist a frame of reference, linked to each condensed body, in which the influence of the other bodies will only cause small corrections to the (supposedly known) equilibrium configuration of one isolated condensed body. Mathematically this leads to assuming, around each body, the existence of a coordinate chart $\hat{x}^a \in \mathbb{R}^4$ (first latin indices : a,b,c=0,1,2,3; last latin indices : i,j,k=1,2,3) in which

$$\hat{g}_{ab}(\hat{x}^c) = g^{isol}_{ab}(\hat{x}^i) + \sum_n \varepsilon^n \hat{g}^{(n)}_{ab}(\hat{x}^i,\hat{x}^o) \quad , \tag{9}$$

where

$$\varepsilon = \frac{Gm}{c^2R} \tag{10}$$

is the small gravitational coupling parameter between the bodies (m, being a characteristic mass of the bodies, and R, a characteristic separation between the bodies).

A detailed investigation of the structure of the hierarchy of equations satisfied by the $\hat{g}^{(n)}_{ab}$ (see Damour [15], §5) shows that one can choose each internal chart $\hat{x}^a$ so as to "efface" the influence of the other bodies up to the order ε^3 (i.e. $n \geqq 3$ in equation (9)).

3.2 The Matching : Internal → External

The information contained in the structure of the expansion (9) can be propagated to the external region by assuming that the transformation between each "internal" coordinate chart $\hat{x}^a$ and the common "external" coordinate chart x^a can be expanded, near each body, as :

$$x^a = z^a(\hat{x}^o) + e^a_i(\hat{x}^o)\,\hat{x}^i - \frac{1}{2} f^a_{ij}(\hat{x}^o)\hat{x}^i\hat{x}^j + O((\hat{x}^i)^3). \tag{11}$$

This leads to a partial knowledge of the structure of the external gravitational field near each body, which says essentially that it looks like a boosted Schwarzschild solution associated with the "centre of field" world-line $z^a(\hat{x}^o)$, and containing the "Schwarzschild mass" of each isolated body.

3.3 The External Problem

In the external region one assumes a usual weak field expansion,

$$g_{ab}(x) = f_{ab} + \varepsilon\, g^{(1)}_{ab}(x) + \varepsilon^2 g^{(2)}_{ab}(x) + \varepsilon^3 g^{(3)}_{ab}(x) + \ldots \quad , \tag{12}$$

where f_{ab} is the metric of flat space.

Plugging the expansion (12) into the vacuum Einstein equations,

$$R_{ab}(g) = 0 \quad , \tag{13}$$

leads to a hierarchy of partial differential equations. We must complete these local equations by boundary conditions : 1) the Fock "no incoming radiation" conditions at past null infinity, and, 2) the third order "Dominant Schwarzschild" conditions in the vicinity of each body. It is then possible to show that these boundary conditions are sufficient, given several world-lines, $z^a(s)$, $z'^a(s')$, ..., and several constant parameters, m,m',..., to characterize, up to order ε^3 included, at most one formally expanded metric (12). Moreover the world-lines must necessarily fulfill some integro-differential constraints ("equations of motion") for a solution to exist. Then these equations of motion will contain only the parameters, m,m',..., which represent the "Schwarzschild masses" of the isolated bodies. This property of "effacing of the internal structure" of the bodies holds to a very high accuracy (ε^6) for initially non rotating condensed objects. This is a kind of classical analogue of a "renormalization" property of the masses : indeed all the strong field effects associated with each strongly self-gravitating body can be renormalized away in its "Schwarzschild mass".

3.4 A Convenient Auxiliary Mathematical Technique.

The above quoted uniqueness result allows one to use any convenient technique to compute the third order gravitational field and equations of motion. It has been found convenient to introduce, in the external problem, a fictitious energy-momentum tensor whose role is to incorporate the Schwarzschild-like boundary conditions. A priori one would think of representing each body, as seen in the external region, by a Dirac distribution supported on the centre-of-field world-line. However this is meaningless because of the non-linearity of the Einstein equations. However, following Riesz [17)], we can replace each Dirac distribution,

$$\delta^4(x-z), \tag{14}$$

by the following function,

$$A^2(-f_{ab}(x^a-z^a)(x^b-z^b))^{\frac{A-4}{2}}\ Y(-f_{ab}(x^a-z^a)(x^b-z^b))\ Y(x^o-z^o)\ , \quad (15)$$

where A is a complex number, and Y denotes Heaviside's step function. Then it makes sense to use the functions (15) in the right-hand side of the Einstein equations (13). Using complex analytic continuation in A, one can then define the first three terms of an expansion of the type (12) by an hierarchy of integral equations containing the functions (15) in their right-hand sides. At the end of the calculation one makes the analytic continuation of A down to zero, so that the functions (15) tend, pointwise, to zero (although, in the sense of distributions, they tend to (14)). This procedure generates a solution of the (harmonically relaxed) vacuum Einstein equations (13), outside a set of world-lines, which satisfies the third order "Dominant Schwarzschild" conditions. A refinement of the method also gives the third order equations of motion of the world-lines (see ref. [15]).

Note that the preceding technique of analytic continuation plays no privileged role. It is used only because it is well-defined, and it permits to explicitly construct a solution of the Einstein equations satisfying the required boundary conditions. It would be more convenient to work directly with the "point mass" limits (14), but I know of no well-defined mathematical theory allowing one to use distributional sources in non-linear equations, and leading unambiguously to a solution belonging to the set of real functions of real variables. The recent theory of "generalized functions" of Colombeau [18] is a-priori interesting in this respect, because it permits a general multiplication of distributions. However, it seems to me that it does not really help, because one has still to prove that the final "generalized" solution is an ordinary real function.

4. SOLVING THE EQUATIONS OF MOTION.

The method outlined in §3 leads to equations of motion for n condensed bodies which have the structure of retarded integro-differential equations : the hyperbolic structure of Einstein's equations, i.e. the finite velocity of propagation of gravity, causes the acceleration of (the centre of field) of each body to be a functional of the past history of the motions of the n bodies. In order to solve explicitly these equations it is convenient to make a further expansion in the small "retardation parameter",

$$\rho = \frac{\text{retardation}}{\text{orbital period}} = \frac{R/c}{P} \sim \frac{v}{c} , \tag{16}$$

where v is a characteristic orbital velocity. Using also the virial theorem (which implies $\varepsilon \sim \rho^2$), we finally get equations of motion in the form of a quasi-Newtonian system of ordinary differential equations expanded in powers of ρ :

$$\frac{d^2 z^i}{dt^2} = \sum_{n=0}^{5} \rho^n A^i_n (z(t),v(t),z'(t),v'(t),\ldots)+O(\rho^6). \tag{17}$$

The explicit values of the A^i_n s ($n \leq 5$) have been computed by Damour and Deruelle [19] and Damour [20] for the case of two bodies (see also references in ref. [4]). The system (17) truncated up to order ρ^4 has been shown to be deducible from an acceleration-dependent Lagrangian [21,20,15]. This allows one to use the method of variation of constants to approximately solve the more exact system (17). The most interesting result so found [22-23] is a small secular acceleration of the mean orbital motion of a binary system (of order ρ^5), which is a direct consequence of the retardation effects caused by the finite velocity of propagation of the gravitational interaction ("relativistic Laplace effect", see ref. [4]). This result agrees with previous heuristic predictions [24-25] based on the assumption of a balance between an outgoing flux of "gravitational wave energy" and some "quasi-Newtonian energy" of the binary system (for a critical discussion, see ref. [4], and references therein).

5. COMPARISON WITH THE OBSERVATIONS.

In order to compare the theory with the observations, or, more precisely, the theoretical model of a clean general relativistic two-condensed-body system, treated by means of the previous approximation methods, with the observational model (7), we still need to compute explicitly the function F^{theor} appearing in the "timing equation" (7). This has been done in an increasingly accurate manner in references [26-29]. This "timing" problem poses both theoretical and pragmatical problems. The theoretical problems are again due to the "mixing" of strong-field and weak-field effects : indeed the electromagnetic pulse originates from the vicinity of the pulsar (strong-field region) and travels afterwards in weak-field regions (between the condensed objects, and in the solar system). The pragmatical problems come from the fact that the theoretical "timing formula", F^{theor}, must be, at the same time, very accurate, fully explicit, and simple enough for allowing one to understand analytically the influence of each physical effect in the final result. These aims have been met in the latest derivation of the timing formula [29], which has been recently used for the comparison with the observational data [30].

For a detailed discussion of the results of this comparison see references [10,30]. Here we wish only to quote the fact that the secular acceleration of the mean orbital motion of the binary system has been measured with a value in excellent agreement (better than 4 %) with the theoretical value deduced from the explicit integration of the equations of motion (17). On the other hand semi-heuristic calculations [31,32,33,5], within the framework of other relativistic theories of gravity, have led to predictions for this effect which differ by orders of magnitude from the general relativistic prediction. The reason of this great difference in magnitude (and, sometimes, also in sign) lies in the fact that, contrarily to what happens in General Relativity, in many other theories of gravity the strong-field effects are not "effaced", and "renormalized" in a total mass. Rather they show up at a low order of approximation, and contribute, in general, a big term to the secular acceleration of the orbital motion (or order ρ^3, instead of the order ρ^5 of the general relativistic effect).

6. CONCLUSIONS

The binary pulsar PSR 1913+16 appears to be a marvellous laboratory for testing the relativistic theories of gravity. The present state of the art, in the development of approximation methods in General Relativity, leads us to believe that the binary pulsar provides us with a very profound confirmation of General Relativity. For the first time we have a confirmation both of the strong-field effects and of the propagation effects in General Relativity. In other words, the nonlinear hyperbolic structure of Einstein's equations has been checked. Many other theories of gravity, which were confirmed, like General Relativity, by the observations in the solar system, seem to be ruled out by the observations of the binary pulsar.

We started by recalling that the aim of physical theories is to connect mathematics and the "real world" (as it reveals itself in observations). For this connection to be fully satisfactory an extreme care must be brought both in the mathematical derivations, used to draw conclusions from the axioms of physical theories, and in the obtention, and interpretation, of observations. This means that some further work needs to be done with respect to the binary pulsar. From the mathematical point of view, one needs, first, a better formalization of the "new" approximation methods, and, secondly, to bridge the gap which still exists between the axioms of General Relativity and the use of approximation methods. From the observational point of view, one needs more refined measurements, or a finer analysis of existing measurements, to justify the hypotheses entering into the model of the binary system (in particular the hypotheses of "cleanliness" of the system, concerning this point see ref.[29]). In the long term, one would very much like to observe several other clean relativistic binary pulsars, so as to rule out any fortuitous "fine tuning" of parameters.

REFERENCES

1. Fourès-Bruhat, Y., Acta Mathematica 88, 141 (1952).
2. Christodoulou, D. and O'Murchada, N., Commun.Math.Phys. 80,271(1981).
3. Choquet-Bruhat, Y., "Mathematical problems in General Relativity", expanded text of lectures delivered at the International School "Cosmology and Gravitation", organized by M. Novello (Rio de Janeiro, 1984).
4. Damour, T., "The problem of motion in Newtonian and Einsteinian gravity", in "300 Years of Gravitation", S.W. Hawking and W. Israel editors, Cambridge University Press, in press.

5. Will, C.M., "Theory and experiment in gravitational physics", Cambridge University Press, Cambridge (1981).
6. Ehlers, J., talk given at GR11 (Stockholm, July 1986), to be published in the proceedings of the conference.
7. Whitney, H., Annals of Mathematics 37, 645 (1936).
8. Kates, R.E., Ann.Phys. (N.Y.) 132, 1 (1981).
9. Damour, T., "Sur les nouvelles méthodes d'approximation en relativité générale", in "Géométrie, Groupes et Relativité", eds Y. Choquet, B. Coll, R. Kerner and A. Lichnerowicz, Hermann, Paris (série "Travaux en cours"), in press.
10. Taylor, J.H. and Weisberg, J.M., Astrophys.J. 253, 908(1982). Weisberg, J.M. and Taylor, J.H., Phys.Rev.Lett. 52, 1348 (1984).
11. Manasse, F.K., J.Math.Phys. 4, 746 (1963).
12. Demianski, M. and Grishchuk, L.P., Gen.Rel.Grav. 5, 673 (1974).
13. D'Eath, P.D., Phys.Rev. D 11, 1387 and 2183 (1975).
14. Kates, R.E., Phys.Rev. D 22, 1853 and 1871 (1980).
15. Damour, T., in "Gravitational Radiation", eds N. Deruelle and T. Piran (Les Houches 1982), pp 59-144, North Holland, Amsterdam (1983).
16. Thorne, K.S. and Hartle, J.B., Phys.Rev. D 31, 1815 (1985).
17. Riesz, M., Acta Mathematica 81, 1-223 (1949).
18. Colombeau, J.F., "New Generalized Functions and Multiplication of Distributions", North Holland Math. Studies 84 (1984).
19. Damour, T. and Deruelle, N., Phys.Lett. 87A, 81 (1981).
20. Damour, T., C.R.Acad.Sci.Paris 294, série II, 1355 (1982).
21. Damour, T. and Deruelle, N., C.R.Acad.Sci.Paris 293, série II, 537, (1981).
22. Damour, T., Phys.Rev.Lett. 51, 1019 (1983).
23. Damour, T., "Un nouveau test de la relativité générale", in "Proceedings of Journées Relativistes 1983", eds S. Benenti, M. Ferraris, and M. Francaviglia, pp 89-110, Pitagora Editrice, Bologna (1985).
24. Peters, P.C. and Mathews, J., Phys.Rev. 131, 435 (1963).
25. Peters, P.C., Phys.Rev. 136, B 1224 (1964).
26. Blandford, R. and Teukolsky, S.A., Astrophys.J. 205, 580 (1976).
27. Epstein, R., Astrophys.J. 216, 92(1977) ; 231, 644 (1979).

28. Haugan, M.P., Astrophys.J. 296, 1 (1985).

29. Damour, T. and Deruelle, N., Ann.Inst.H.Poincaré (Physique Théorique) 43, 107 (1985), and, 44, 263 (1986).

30. Taylor, J.H., talk given at GR 11 (Stockholm, July 1986), to be published in the proceedings of the conference.

31. Will, C.M., and Eardley, D.M., Astrophys.J. 212, L91 (1977).

32. Will, C.M., Astrophys.J. 214, 826 (1977).

33. Weisberg, J.M. and Taylor, J.H., Gen.Rel.Grav. 13, 1 (1981).

INDUCTION ON VOLUME AND NO CLUSTER EXPANSION

R.L. DOBRUSHIN

Institute of Information Transmission Problems,
Academy of Sciences, Moscow, U.S.S.R.

1. THE METHOD

The main body of rigorous results in statistical mechanics and constructive quantum field theory is obtained by applying perturbation methods in situations close to degenerate ones, which, in turn, can be studied straightforwardly. The conceptual novelty on the corresponding mathematical problems is that one is only interested here in studying systems of very large size. Mathematically, it is the same as studying the infinite limit systems, or, equivalently, to have expansions for finite size systems, uniform in their size. Standard analytic methods do not meet these requirements, while their formal application leads often to qualitatively wrong results.

The main method to solve this problem is the cluster expansion method, introduced by Glimm, Jaffe and Spencer [1] over decade ago. It was developed then further by many peoples (see [2], [3] for references). This wonderful method enables one to control completely the infinite dimensional case. In fact, almost all recent results are based on its applications. But, in spite of many recent improvements, the cluster expansion machinery is still very cumbersome. Further dissatisfaction (not commonly accepted, may be) comes from the fact, that while both the problems treated and the results obtained have a clear physical meaning (or probabilistic, if one prefer), the intermediate steps have not.

The aim of the present report is to present a more elementary and more transparent way to handle these problems. To some degree it can be considered as an alternative to the cluster expansion method. The method under discussion is based on some fundamental bounds for finite volume systems, which can be proven by induction on the volume with uniformly bounded constants. This is possible due to the Markov property of the random fields, which enables one to express the characteristics of the system through those of its subsystems. The induction method was developed by R. Dobrushin together with L. Bassalygo, M. Martirosyan, S. Shlosman. The list of problems solved by this method is up to now not very long. But in all the considered cases it was possible to find easy and probabilistically-appealing proofs in situations of natural generality. We hasten to add, however, that if induction proofs are easier for the presentation, their invention is more involved : to find the natural self-generating induction statement requires usually to look over many alternatives.

In the next paragraph we present a detailed application of our method to the simplest case of high temperature discrete spin systems. Other sections of our report are less detailed ; they contain only the results and the ideas of the proofs. For some of them it seems that they cannot be obtained by the usual cluster expansion method, at least without considerable modifications.

2. HIGH-TEMPERATURE LATTICE SYSTEMS

Let $\mathbb{Z}^r$ be the r-dimensional lattice, X - any finite set (single spin space). Suppose that the distributions $P_t(x_t)$, $x_t \in X$, $t \in \mathbb{Z}^r$ are fixed and, for any finite $B \subset \mathbb{Z}^r$ let

$$P_B(x_B) = \prod_{t\in B} P_t(x_t), \quad x_B = (x_t,\ t\in B) \in X^B \tag{2.1}$$

By a finite interaction system we mean a system $\mathfrak{U} = \{U_\alpha\ ;\ A_\alpha\ ,\ \alpha\in\mathbb{A}\}$ where $\mathbb{A}$ is some finite set of indices α, to each of them corresponding a finite set $A_\alpha \subset \mathbb{Z}^r$ (not necessarily distinct for different α) and a complex function $U_\alpha(x)$ of $x \in X^{A_\alpha}$. Let

$$\|U_\alpha\| = \sup_{x \in X^{A_\alpha}} |e^{-U_\alpha(x)} - 1| \ . \tag{2.2}$$

For any $\gamma > 0$ define the norm

$$\|\mathcal{U}\|_\gamma = \sup_{t \in \mathbb{Z}^r} \sum_{\alpha \in \mathbb{A} \,:\, t \in A_\alpha} \|U_\alpha\| \, e^{\gamma |A_\alpha|} \tag{2.3}$$

where $|A|$ stands for the cardinality of A .
Let

$$\text{supp}\, \mathcal{U} = \bigcup_{\alpha \in \mathbb{A}} A_\alpha \tag{2.4}$$

Define the partition function as

$$Z(\mathcal{U}) = \sum_{x \in X^{\text{supp}\, \mathcal{U}}} \exp\left\{ - \sum_{\alpha \in \mathbb{A}} U_\alpha(x_{A_\alpha}) \right\} P_{\text{supp}\, \mathcal{U}}(x) \tag{2.5}$$

where for $A \subset B$ we denote by x_A the restriction of $x \in X^B$ to A.

The following result will lead us to many corollaries :

<u>Proposition 2.1</u> : For any finite interaction $\mathcal{U}$, such that

$$\|\mathcal{U}\|_\gamma \leq 1 - e^{-\gamma/2} \tag{2.6}$$

the partition function $Z(\mathcal{U}) \neq 0$

and for $\mathcal{U}' = \{U'_\alpha \ , \ A'_\alpha \ , \ \alpha \in \mathbb{A}'\}$

such that $\mathbb{A}' \subset \mathbb{A}$ and

$$A'_\alpha = A_\alpha \quad , \quad U'_\alpha = U_\alpha, \quad \alpha\in\mathbb{A}' \tag{2.7}$$

the following estimate holds :

$$\left|\ln\left|\frac{Z(\mathfrak{U})}{Z(\mathfrak{U}')}\right|\right| \le K \sum_{\alpha\in \mathbb{A}\setminus \mathbb{A}'} \|U_\alpha\| \, e^{\gamma|A_\alpha|} \tag{2.8}$$

where $K = \gamma(2(1-e^{-\gamma/2}))^{-1}$.

<u>Proof</u> : We shall obtain (2.8) by induction on $|\mathbb{A}|$. We put by definition $Z(\mathfrak{U}) = 1$ for $\mathbb{A} = \emptyset$ thus having the initial step of induction. This definition is justified by the forthcoming formula (2.9). Suppose now that (2.8) holds for any $\mathbb{A}$ with $|\mathbb{A}| < N$ where N is fixed. Because one can obtain $\mathbb{A}'$ by deleting elements of $\mathbb{A}$ one by one, it is enough to establish (2.8) when $|\mathbb{A}| = N$ and $\mathbb{A} = \mathbb{A}' \cup \{\alpha_o\}$, $\alpha_o\in \mathbb{A}$. Let $A_o = A_{\alpha_o}$, $U_o = U_{\alpha_o}$.

It is easy to see, that

$$\frac{Z(\mathfrak{U})}{Z(\mathfrak{U}')} - 1 = \frac{Z(\mathfrak{U})/Z(\mathfrak{U}^c) - Z(\mathfrak{U}')/Z(\mathfrak{U}^c)}{Z(\mathfrak{U}')/Z(\mathfrak{U}^c)} \tag{2.9}$$

where $\mathfrak{U}^c = \{U_\alpha, A_\alpha, \ \alpha\in \mathbb{A}^c\}$, $\mathbb{A}^c = \{\alpha\in \mathbb{A} : A_\alpha \cap A_o = \emptyset\}$. Having $\mathbb{A}^c \subset \mathbb{A}'$, $|\mathbb{A}'| = N-1$, we infer from the induction hypothesis (2.8), that

$$\left|\frac{Z(\mathfrak{U}')}{Z(\mathfrak{U}^c)}\right| \ge \exp\left\{-K \sum_{\alpha\in \mathbb{A}':A_\alpha\cap A_o \neq \emptyset} \|U_\alpha\| \, e^{\gamma|A_\alpha|}\right. \tag{2.10}$$

Consider now the set $\bar{\mathbb{A}} = \{\alpha\in \mathbb{A} : A_\alpha \cap A_o^c \neq \emptyset\}$ where $A_o^c = \mathbb{Z}^r\setminus A_o$, and let the configuration $\bar{x} \in X^{A_o}$ be fixed. We define the potential

$\mathfrak{U}^{\bar{x}} = \{ U_\alpha^{\bar{x}}, \bar{A}_\alpha, \alpha \in \bar{\mathbb{A}} \}$ by

$$\bar{A}_\alpha = A_\alpha \cap A_o^c \ , \ U_\alpha^{\bar{x}}(x) = U_\alpha(x_{\bar{A}_\alpha} \ , \ \bar{x}_{A_o \cap A_\alpha}) \tag{2.11}$$

From definition (2.5) it follows that

$$Z(\mathfrak{U}) = \sum_{\bar{x} \in X^{A_o}} Z(\mathfrak{U}^{\bar{x}}) \exp\left\{- \sum_{\alpha \in \mathbb{A} : A_\alpha \subset A_o} U_\alpha(\bar{x}_{A_\alpha})\right\} P_{A_o}(\bar{x}), \tag{2.12}$$

$$Z(\mathfrak{U}') = \sum_{\bar{x} \in X^{A_o}} Z(\mathfrak{U}^{\bar{x}}) \exp\left\{- \sum_{\alpha \in \mathbb{A}' : A_\alpha \subset A_o} U_\alpha(\bar{x}_{A_\alpha})\right\} P_{A_o}(\bar{x}),$$

Hence

$$Z(\mathfrak{U})/Z(\mathfrak{U}^c) - Z(\mathfrak{U}')/Z(\mathfrak{U}^c) =$$

$$= \sum_{\bar{x} \in X^{A_o}} Z(\mathfrak{U}^{\bar{x}})(Z(\mathfrak{U}^c))^{-1} \exp\left\{- \sum_{\alpha \in \mathbb{A}' : A_\alpha \subset A_o} U_\alpha(\bar{x}_{A_\alpha})\right\}(1-e^{-U_o(\bar{x})}) \ P_{A_o}(\bar{x}) \tag{2.13}$$

Because

$$\mathbb{A}^c \subset \bar{\mathbb{A}} \quad , \quad |\bar{\mathbb{A}}| < N \quad , \quad U_\alpha^{\bar{x}} = U_\alpha$$

for $\alpha \in \mathbb{A}^c$, we can use the induction hypothesis (2.8), from which it follows that,
for all $\bar{x} \in X^{A_o}$

$$|Z(\mathfrak{U}^{\bar{x}})(Z(\mathfrak{U}^c))^{-1}| \le \exp\left\{K \sum_{\alpha \in \mathbb{A} : A_\alpha \cap A_o \neq \emptyset, A_\alpha \cap A_o^c \neq \emptyset} \| U_\alpha \| \, e^{\gamma |A_\alpha|}\right\} \tag{2.14}$$

Furthermore

$$|1-e^{-U_\circ(\bar{x})}| \le \|U_\circ\| \quad , \quad \bar{x} \in X^{A_\circ} \quad , \tag{2.15}$$

Finally

$$\sum_{\bar{x}\in X^{A_\circ}} \left| \exp\left\{- \sum_{\alpha\in \mathbb{A}' : A_\alpha \subset A_\circ} U_\alpha(\bar{x}_{A_\alpha})\right\}\right| P_{A_\circ}(\bar{x}) = Z(\mathfrak{U}), \tag{2.16}$$

where $\mathfrak{U} = \{\tilde{U}_\alpha, \tilde{A}_\alpha, \alpha\in\mathbb{A}\}$ is a new finite interaction, given by

$$\mathbb{A} = \{\alpha\in\mathbb{A}' : A_\alpha \subset A_\circ\} \ , \ \tilde{A}_\alpha = A_\alpha \ , \ \alpha\in\mathbb{A},$$
$$\tilde{U}_\alpha(x_{A_\alpha}) = \text{Re } U_\alpha(x_{A_\alpha}) \ . \tag{2.17}$$

It follows from definition (2.2) that

$$\|\tilde{U}_\alpha\| \le \|U_\alpha\| \ , \ \alpha\in\mathbb{A} \ , \tag{2.18}$$

hence the inductive bound (2.8) is also valid for the case $\mathfrak{U} = \mathfrak{U}$. Using it with the potential $\mathfrak{U}'$ corresponding to the empty index set, one infers that

$$Z(\mathfrak{U}) \le \exp\left\{K \sum_{\alpha\in\mathbb{A}' : A_\alpha \subset A_\circ} \|U_\alpha\| \, e^{\gamma|A_\alpha|}\right\} \tag{2.19}$$

Putting together the bounds (2.10), (2.14), (2.15), (2.19) and the identities (2.9), (2.13), (2.16), using the definition (2.3) and the definition of K, and finally applying the condition (2.6), we arrive to the bound

$$|Z(\mathfrak{U})(Z(\mathfrak{U}'))^{-1} - 1| \le \|U_{\alpha_o}\| \exp\{2K \sum_{\alpha\in\mathbb{A}' : A_\alpha \cap A_o \neq \emptyset} \|U_\alpha\| e^{\gamma|A_\alpha|}\}$$

$$\le \|U_{\alpha_o}\| \exp\{2K \|U\|_\gamma |A_o|\} \le \|U_{\alpha_o}\| e^{\gamma|A_o|} \tag{2.20}$$

Note, that for any complex z and real a, such that 0<a<1 one has

$$|\ln|z|| \le -\frac{\ln(1-a)}{a} |z-1| \quad \text{if} \quad |z-1| < a \tag{2.21}$$

Applying it to (2.20) with $a = 1-e^{-\gamma/2}$, one has

$$\left|\ln\left|\frac{Z(\mathfrak{U})}{Z(\mathfrak{U}')}\right|\right| \le K \|U_o\| e^{\gamma|A_o|} . \tag{2.22}$$

Really it follows from (2.20) because of (2.3) and (2.6) that

$$\left|\frac{Z(\mathfrak{U})}{Z(\mathfrak{U}')} - 1\right| \le \|\mathfrak{U}\|_\gamma \le 1-e^{-\gamma/2} \tag{2.23}$$

The bound (2.22) is the same as (2.8) for the case we consider, hence the proposition is proven.

We are going on to applications. Suppose $\mathfrak{U}^o = \{U^o_B(x_B),\ x_B\in X^B,\ B\subset\mathbb{Z}^r\}$ is some real translation-invariant interaction, such that for some $\gamma>0$

$$\sum_{B:0\in B} (\sup_{x\in X^B} |U^o_B(x_B)|)\ e^{\gamma|B|} < \infty . \tag{2.24}$$

The free energy is defined by the usual formula

$$F(\beta) = \lim_{\Lambda\to\infty} \left[|\Lambda|^{-1} \ln(\sum_{x\in X^\Lambda} \exp\{-\beta \sum_{B\subset\Lambda} U_B^\circ(x_B)\})\right] \tag{2.25}$$

<u>Corollary 1</u> : There exist a value $\beta_\circ$ such that for complex $|\beta| \leq \beta_\circ$ the limit (2.25) exists and is an analytic function of β.

<u>Proof</u> : Applying the bound (2.8) to the case $\mathfrak{U} = \{\beta\, U^\circ_B\,(x_B),\ B\subset\Lambda\}$ with empty $\mathfrak{U}'$, one has that for $\beta_\circ$ small the bracketed expression in (2.25) is uniformly bounded in Λ and $|\beta| \leq \beta_\circ$. By van Hove theorem, the limit (2.25) exists for real β, so our statement follows from the well-known compactness criterion for analytic functions.

Let $\Lambda \subset \mathbb{Z}^r$ be finite, $A_1,\ldots,A_k \subset \Lambda$ and the functions $\psi_i(x_{A_i}),\ x_{A_i} \in X^{A_i}$, be given together with some boundary condition $\bar{x} \in X^{\Lambda^c}$. Consider a truncated correlation function

$$<\psi_1^{'n_1},\ \psi_2^{'n_2},\ldots,\psi_k^{'n_k}>_{\bar{x},\beta}$$

of order $n_1,\ldots,n_k$, calculated with respect to the Gibbs state in Λ specified by the interaction $\mathfrak{U}^\circ$, the boundary condition $\bar{x}$ and the reciprocal temperature β. In what follows we assume for simplicity that $\mathfrak{U}^\circ$ has finite range : $U^\circ_A \equiv 0$ for diam $A > r$, where r is the range of interaction.

<u>Corollary 2</u> : For any interaction $\mathfrak{U}^\circ$ there exist constant $\beta_\circ$ and C, c such that for any $\Lambda,k,n_1,\ldots,n_k,\psi_1,\ldots,\psi_k,\bar{x}$ and $\beta < \beta_\circ$

$$|<\psi_1^{'n_1},\ldots,\psi_k^{'n_k}>_{\bar{x},\beta}| \leq$$

$$\leq C^{\sum_{i=1}^{k} |A_i|} (n_1!\dots n_k!) \exp\{-cd(A_1,\dots,A_k)\} \prod_{i=1}^{k} \sup_{x_{A_i}} |\psi(x_{A_i})|,$$

Here $d(A_1,\dots,A_k)$ is the smallest volume of sets $D \subset \mathbb{Z}^r$, such that the union $D \cup A_1 \cup\dots\cup A_k$ is connected (if $\mathbb{Z}^r$ is viewed as a graph with bonds between nearest neighbours).

<u>Proof</u> : Because the correlations are linear in all variables, it is enough to consider the case

$$\sup_{x_{A_i}} |\psi_i(x_{A_i})| \leq e^{-\alpha|A_i|}, \quad i=1,\dots,k \tag{2.26}$$

where $\alpha > 0$. By definition (2.5) and that of the generating function for truncated correlations

$$\langle \psi'^{n_1}_1,\dots,\psi'^{n_k}_k \rangle_{\bar{x},\beta} = \frac{\partial^{n_1+\dots+n_k}}{\partial\lambda_1^{n_1}\dots\partial\lambda_k^{n_k}} \ln \frac{Z(\mathfrak{U}_{\beta,\lambda_1,\dots,\lambda_n})}{Z(\mathfrak{U}_\beta)}\Bigg|_{\lambda_1=\dots=\lambda_k=0} \tag{2.27}$$

where the interaction $\mathfrak{U}_{\beta,\lambda_1,\dots,\lambda_k}$ corresponds to the family $\mathbb{A} = \{B: B\cap\Lambda \neq \emptyset, A_1,\dots A_k\}$ of finite volumes and is given by the functions $U_B(x_{B\cap\Lambda}) = \beta\, U^\circ_B(x_{B\cap\Lambda}, \bar{x}_{B\cap\Lambda^c})$, $U_{A_i}(x_{A_i}) = \lambda_i\, \psi_i(x_{A_i})$ while $\mathfrak{U}_\beta$ is the restriction of $\mathfrak{U}_{\beta,\lambda_1,\dots,\lambda_k}$ to the family $\{B : B\cap\Lambda \neq \emptyset\}$. It is easy to see that for $\alpha>\gamma$, $|\lambda_i| \leq \lambda_o$, $|\beta| \leq \beta_o$ and λ_o, β_o small enough the interactions $\mathfrak{U}_{\beta,\lambda_1,\dots,\lambda_k}$, $\mathfrak{U}$ obeys to the condition (2.6) of the proposition 2.1. Hence from (2.8) and Cauchy formula for the derivatives in λ_i at $\lambda_1=\dots=\lambda_k=0$ one has

$$|\langle \psi_1'^{n_1}, \ldots, \psi_k'^{n_k} \rangle_{\bar{x},\beta}| \le n_1! \ldots n_k! \, C^k \tag{2.28}$$

for some $C > 0$. Let us denote this last correlation by $f(\beta)$. This function is analytic for $|\beta| \le \beta_o$, and its derivatives at $\beta=0$ are nothing else than the correlations with respect to the measure P_Λ, defined by (2.1). From it, it is easy to see that for

$$l < r^{-1} \, d(A_1, \ldots, A_k)$$

all the derivatives $f^{(l)}(0) = 0$. The rest can be estimated by the a priori bound (2.28) and Cauchy formula. Summing up the Taylor series at the point $\beta=0$, we have for some $C'>0$, $c>0$

$$|\langle \psi_1'^{n_1}, \ldots, \psi_k'^{n_k} \rangle_{\bar{x},\beta}| \le n_1! \ldots n_k! (C')^k \exp\{-c \, d(A_1, \ldots, A_k)\} \tag{2.29}$$

Dropping the restriction (2.26), we arrive (because of linearity) to the desired bound

<u>Corollary 3</u> : Under the conditions of the corollary 2, for any boundary conditions $\bar{x}^1, \bar{x}^2$, one has

$$|\langle \psi_1'^{n_1}, \ldots, \psi_k'^{n_k} \rangle_{\bar{x}^1,\beta} - \langle \psi_1'^{n_1}, \ldots, \psi_k'^{n_k} \rangle_{\bar{x}^2,\beta}| \le$$

$$\le C^{\sum_{i=1}^{k} |A_i|} (n_1! \ldots n_k!) \exp\{-C[d(A_1, \ldots, A_k) + \mathrm{dist}(\Lambda^c, \bigcup_{i=1}^{k} A_i)]\} \tag{2.30}$$

The proof is the same. From (2.30) the uniqueness of the infinite volume state and the bounds for its correlations follows as in corollary 2.

The derivation of analogous bounds by the cluster expansion method can be found in [3]. The above statements follows closely a recent paper by the author and M. Martirosyan.

Finally we want to emphasize that the condition of exponential decay of the interaction with $|B|\to\infty$, included in (2.24), is crucial. Suppose that a function $\varphi(n)$ is given, such that $\varphi(n)e^{-\alpha n}\to 0$ for all $\alpha>0$. Then for all $\epsilon>0$, there exists a potential $\mathfrak{U}^{\circ}$, such that

$$\sum_{B:0\in B} (\sup_{x\in X^B} |U_B^{\circ}(x_B)|)\ \varphi(|B|) < \epsilon \tag{2.31}$$

while

$$Z(\{U_B\ ,\ B\subset \Lambda_n\}) \equiv 0$$

for some Λ_n of volumes, $|\Lambda_n| \to \infty$ (Dobrushin, M. Martirosyan).

3. GENERALIZATIONS

The range of applications of the above method is not restricted to the high temperature region. It works as well in the case of perturbations of any "sufficiently nice" interaction. For example, Dobrushin and Martirosyan have considered recently the case of interactions of type $\mathfrak{U}^{\circ} + \bar{\mathfrak{U}}$, where $\bar{\mathfrak{U}}$ is a many body interaction with norm (2.3) small enough, while $\mathfrak{U}^{\circ}$ is a finite range, translation invariant, real and completely analytical potential (this last property has been introduced by Dobrushin and Shlosman [4]) and have proven results of the type of those of section 2. The class of completely analytical interactions can be defined in many ways, and they turn out to be equivalent. The following are several conditions for the interaction $\mathfrak{U}^{\circ}$ to be completely analytical.

1. For some $\epsilon>0$, any finite Λ, any boundary condition $\bar{x}\in X^{\Lambda^c}$ and any complex translation-invariant interaction $\bar{\mathfrak{U}}$ with finite range r

and norm $|\overline{U}|_\gamma < \varepsilon$

$$Z(\{U^{\circ}_B(x_{B\cap\Lambda}, \overline{x}_{B\cap\Lambda^c}) + \overline{U}_B(x_{B\cap\Lambda}, \overline{x}_{B\cap\Lambda^c}), B\cap\Lambda \neq \emptyset\} \neq 0$$

2. For some $\delta<1$, $\rho>0$, any finite Λ, any pair $\overline{x}^1, \overline{x}^2 \in X$, coinciding outside a point $t_0 \in \Lambda^c$

$$\sum_{x_B \in X_B} |P_\Lambda(x_B|\overline{x}^1) - P_\Lambda(x_B|\overline{x}^2)| \leq \delta|B|^{-1} \tag{3.1}$$

where

$$B = \{s\in\Lambda:\ \rho < |s-t| \leq \rho+r\} ,$$

r being the range of the interaction U°, and where $P_\Lambda(\,.\,|\overline{x}^i)$ is the conditional Gibbs state in Λ with interaction U° and boundary condition $\overline{x}^i$ restricted to $B \subset \Lambda$.

3. The truncated correlations of Gibbs states with completely analytical interaction satisfy the tree-type estimates of corollary 2.

These and other properties imply that the class of completely-analytical interactions is physically the same as the class of interactions without phase transitions. To prove the equivalence of the different definitions one again uses the method of sect. 2. In particular, the condition (2.1) of independence has to be exchanged with condition (3.1). Recently this method was also applied by the author to the case of perturbations of Gaussian fields with exponential decay of correlations (= with a positive mass). One can treat, in particular, $\mathbb{R}^k$-valued Gibbs fields, given by the following translation-invariant Hamiltonian

$$\sum_{s,t:|s-t|\leq r} \Phi_{s-t} x_s x_t + \lambda \sum_{A:\operatorname{diam} A\leq r} \Phi_A(x_A) \qquad (3.2)$$

Here the interactions Φ_A are bounded from below, λ is small enough, while the quadratic part is positively defined : for some C>0

$$\sum_{s,t\in\Lambda:|s-t|\leq r} \Phi_{s-t} x_s x_t \geq C \sum_{t\in\Lambda} (x_t)^2$$

for any $(x_s, s\in\Lambda) \in (\mathbb{R}^k)^\Lambda$ and any Λ. The results thus obtained are stronger and more general than the corresponding cluster expansion results (see [3], [5] and compare with [6]). In particular, together with analyticity of the free energy one has also analyticity of the infinite volume densities ; the above tree-type estimates (see corollaries 2,3 of sect. 2) are also valide in the case of unbounded observables. These results were further generalized by Martirosyan, who considered the case of infinite range interactions Φ_A.

One can apply here the induction method, because for the perturbation a condition analogous to (3.1) is valid, however, under the restriction that the boundary conditions $\bar{x}^1, \bar{x}^2$ are uniformly bounded. This is the main source of additional complications : one has to estimate separately the case of large boundary conditions. As a result, the proof is somehow cumbursome ; it seems, nevertheless, that the proofs using cluster expansions ([3], [5], [6]) in the case of small mass, are even more cumbursome.

The following innovation is useful for the study of correlations of unbounded observables. The usual definition of correlations by a generating function is of no use in the case where the corresponding integrals are divergent. However, if $n_i \leq n$

$$\langle \psi_1'^{\,n_1},\ldots,\psi_1'^{\,n_k}\rangle = \frac{\partial^{n_1+\ldots+n_k}}{\partial\lambda_1^{n_1}\ldots\partial\lambda_k^{n_k}} \ln \langle \prod_{i=1}^{k} \left(\sum_{j=1}^{n} \frac{(\lambda_i\psi_i)^j}{j!}\right)\rangle \Big|_{\lambda_1=\ldots=\lambda_k=0}$$

where one only needs the existence of moments of order n . This last definition can be used in the situations of corollary 2 of sect. 2.

4. HIGH-TEMPERATURE RANDOM FIELDS ON ARBITRARY GRAPHS AND RANDOM INTERACTIONS

Let Γ be any countable graph. Suppose that the order of any site (i.e. the number of adjacent bonds) is finite. Suppose that are given for any site $t\in\Gamma$ a finite set X_t together with a probability distribution p_t on it. Consider a nearest-neighbour interaction

$$U_{s,t}(x_s,x_t)\ ,\quad (s,t)\in\Gamma\ ,\ x_s\in X_s\ ,\ x_t\in X_t$$

<u>Definition</u> : A random field on $\prod_{t\in\Gamma} X_t$ is a Gibbs field, if for any finite $V\subset\Gamma$ the conditional distribution on $\prod_{t\in V} X_t$ under a fixed condition $x\in\prod_{t\in V^c} X_t$ is given by the Gibbs formula

$$p_V(x_V|x_{V^c}) = Z^{-1}\exp\left\{-\sum_{(s,t)\in\Gamma,s\in V} U_{s,t}(x_s,x_t)\right\}\prod_{t\in V} p_t(x_t) \tag{4.1}$$

where Z is the corresponding partition function.

Consider the following natural question : does the uniqueness of Gibbs fields follow from the fact that

$$|U_{s,t}(x_s,x_t)| \le \epsilon\ ,\ (s,t)\in\Gamma\ ,\ x_s\in X_s\ ,\ x_t\in X_t \tag{4.2}$$

with ε small ? In the case where the orders of the sites are uniformly bounded, the answer is, of course, positive, and the generalization of the case $\Gamma = \mathbb{Z}^r$ is immediate. But without any restrictions the result certainly fails. The simplest example is provided by the graph $\Gamma = \bigcup_{r=1}^{\infty} \mathbb{Z}^r$, where lattices with different r are disjoint, and with on each lattice the usual Ising model.

The following result was obtained recently by the author and L. Bassalygo [9].

<u>Theorem 4.1</u> : Suppose there exist some > 0, such that for all $s,t \in \Gamma$ the length of any path between s and t is not smaller than $c\ \ln(\min(m_s,m_t))^2$, where m_s and m_t are the orders of the sites s and t. Then for $\varepsilon = \varepsilon(c)$ small enough, with an interaction satisfying (4.2) the Gibbs field is unique.

It seems that the known methods (in particular, the cluster expansion) can not be applied here (at least immediately). It is possible, however, to prove the above theorem by elementary induction arguments. We shall present the main ideas of the proof.

The main estimate is : for any finite $\Lambda \subset V \subset \mathbb{Z}^r$ and $x^1_{V^c}, x^2_{V^c} \in \prod_{t \in V^c} X_t$ coinciding outside $t = t_0 \in V^c$

$$\left| \frac{p_V(x_\Lambda | x^1_{V^c})}{p_V(x_\Lambda | x^2_{V^c})} - 1 \right| \le \sum_{L: t \to \Lambda} \alpha^{|L|}, \quad x_\Lambda \in \prod_{t \in \Lambda} X_t \tag{4.3}$$

if the right-hand part of (4.3) is less than 1. Here $P_V(x_\Lambda | x^i_{V^c})$ is the restriction of the conditional Gibbs measure (4.1) to $\prod_{t \in \Lambda} X_t \subset \prod_{t \in V} X_t$, and the summation goes over all allowed paths in V, which start at t and terminate in Λ. We call a path to be allowed, if it does not leave V, and if between any two visits to the same site it necessarily visits also some other site with at least the same order. Finally, by $|L|$ we denote the length of the path L, and α, $0 < \alpha < 1$, is some constant.

Given (4.3), the theorem follows immediately by passing to the limit $\Lambda \to \infty$ and by estimating the number of allowed paths. In turn, the bound (4.3) is obtained by induction on the maximal order of sites in V and on the number of sites of the such maximal order. Each induction step results from surgery one of the maximal order sites being replaced by a corresponding number of sites of order 1 (see Fig. 1).

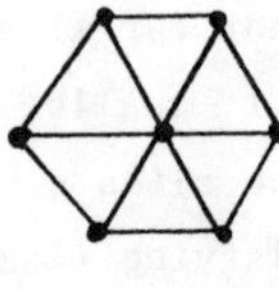

Initial

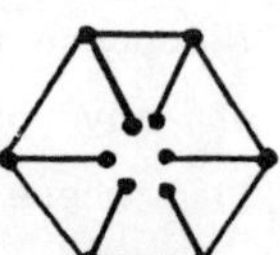

Surgery graph

Fig. 1

The single point sets, the distributions p_t and the interactions are the same before and after surgery. The induction is based on the observation that Gibbs probabilities on the initial graph are the same as conditional Gibbs probabilities on the surgery graph under the condition that the field values in all the newly created sites are the same.

The above theorem can be applied straightforwardly to the case of random interactions (or to spin glasses). The following theorem is due to the author and Bassalygo [9].

<u>Theorem 4.2</u> : Let $\Gamma = \mathbb{Z}^r$, the sets X_t are all the same, and the interactions $U_{s,t}(x_s,x_t)$, $s,t \in \mathbb{Z}^r$, $|s-t| = 1$, are for fixed x_s, x_t and all $s,t \in \mathbb{Z}^r$, $|s-t| = 1$ independent and identically distributed random variables. Then for some $\beta > 0$ the Gibbs field with interaction $\beta U_{s,t}(x_s,x_t)$ is unique for almost all realizations of the interaction.

This result generalizes earlier results by Fröhlich and Imbrie [7] and Beretti [8], obtained under some additional restrictions on the distribution of the interaction. They used some highly involved versions of cluster expansions, together with a renormalization group approach. (In case the interaction variable are uniformly bounded, the problem is trivial). Now we shall describe how to deduce theorem 4.2 from theorem 4.1. Let be given a realization of the interactions. The site $t \in \mathbb{Z}^r$ is called "malignant" (or m-site), if for some nearest-neighbour site s and for some values $x_s, x_t \in X$, $\beta U_{s,t}(x_t, x_t) > \epsilon$. By Borel-Cantelli and easy combinatoric one sees that for almost all realizations the set $G \subset \mathbb{Z}^r$ of m-sites can be decomposed as $G = \bigcup_i G_i$, in such a way that $|G_i| < \infty$ and

$$\mathrm{dist}(G_i, G_j) \geq C \ln(\min(|G_i|, |G_j|))^3 + 2 \ , \ i \neq j \qquad (4.4)$$

Consider now the factor-graph, which sites are all the "benign" sites $t \in G^c$ (b-sites) of $\mathbb{Z}^r$ together with the sets G_i. The bonds between b-sites are unchanged, while a b-site t is connected with a newly created site G_i iff $\mathrm{dist}(t, G_i) = 1$ (see Fig. 2)

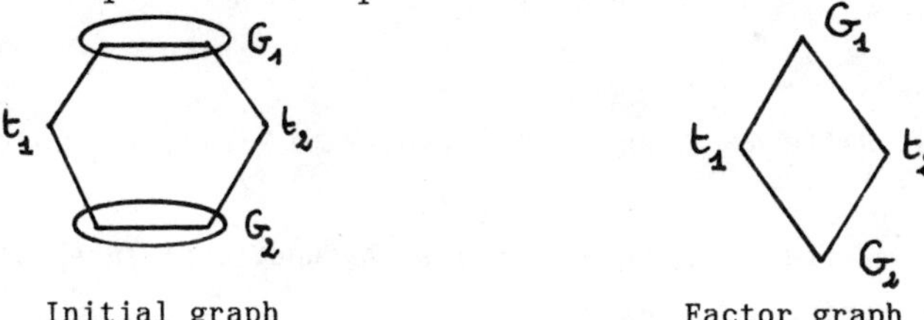

Initial graph Factor graph

Fig. 2

The sets X_t for t benign are unchanged, as well as the one-point distributions and interactions.

The one-point set of a site G_i is the product X^{G_i}, the free measure P_{G_i} on G_i is the Gibbs measure on X with free boundary condition, while the interaction

$$U_{G_i,t}(x_{G_i},x_t) = \sum_{s\in G_i:|s-t|=1} U_{s,t}(x_s,x_t),\quad x_{G_i} = (x_s, s\in G_i)$$

In this way we obtain a one-to-one correspondence between Gibbs states on the initial graph and on the factor-graph, while for the latter the conditions of the theorem 4.1 are satisfied.

With the help of estimate (4.3) of weak dependence (analogous to bound (3.1)), one also proves analyticity of the free energy as a function of complex perturbations, and one obtains the tree-like bounds on the correlations. The main feature of this problem ("Griffiths singularity") is manifested in the fact that such perturbations can be applied only far from sites of large order (in theorem 4.1) or from large component G_i of sites (in theorem 4.2).

REFERENCES

[1] Glimm J., Jaffe A., Spencer T., Comm. Math. Phys. 45, 203 (1975).

[2] Glimm J., Jaffe A., Quantum Physics. A functional Integral Point of View, Springer-Verlag (1981).

[3] Malyshev V.A., Minlos R.A., Gibbs random fields. Moscow, "Nauka" (1985).

[4] Dobrushin R.L., Shlosman S.B., in "Statistical Physics and Dynamical Systems Rigorous Results", 347, Boston, Birkhäuser (1985),

[5] Dobrushin R.L., Zahradnik M., in Mathematical Problems of Statistical Mechanics and Dynamics", 1, D. Reidel Publ. C., (1986).

[6] Imbrie J., Comm. Math. Phys., 82, 261 and 305 (1981).

[7] Fröhlich J., Imbrie J., Comm. Math. Phys., 96, 145 (1984).

[8] Beretti A., J. Stat. Phys., 38, 3/4, 483 (1985).

[9] Bassalygo L.A., Dobrushin R.L., Theor. Prob. and Appl., 31, n°4, 651 (1986).

EFFECTIVE ACTION IN SUPERSTRING THEORY

E.S.Fradkin and A.A.Tseytlin

P.N.Lebedev Physical Institute, Moscow 117924 USSR

1. INTRODUCTION

Superstring theories are presently considered as prime candidates for a unified theory of all interactions in nature [1]. They are generalizations of field theories of supergravity and super Yang-Mills interactions. For example a closed Bose string theory can be thought as a theory of a finite number of massless (graviton $h_{\mu\nu} = G_{\mu\nu} - \delta_{\mu\nu}$, dilaton ϕ, antisymmetric tensor $A_{\mu\nu}$) and an infinite number of massive particles (tachyon φ, spin 4 $B_{\mu\nu\lambda\rho}$, etc.). Curiously, it appears that fundamental consistent superstring theories only exist in ten spacetime dimensions, thus making difficult straightforward comparisons with observed reality. It is possible to describe effective string theories in lower dimensions through the mechanism of compactification [2], but at present there is a lack of understanding of how a superstring theory could dynamically single out vacua unambiguously corresponding to the known phenomenology.

A useful framework for addressing such issues is a gauge-covariant second-quantized field theory of strings [3]. Within it one can study quantization about classical backgrounds of string fields, eventually trying to make contact with phenomenology. Suppose we expand a string field in terms of components, i.e. local fields on spacetime. If we are interested in low-energy dynamics we are to avoid asymptotic massive fields and so are simply to integrate them out in a string field theory path integral. At the classical level (i.e. ignoring string theory loop corrections) this amounts to elimination of the massive fields from the classical string field theory action using the corresponding equations of motion. The result is what is usually called effective action (EA) for the massless string modes [4,5].

Let φ denote all massless fields ($G_{\mu\nu}$, ϕ, $B_{\mu\nu}$, ...) and S be the EA. Assuming that the effective scale of the massless vacuum fields is much larger than the string scale $\sqrt{\alpha'}$ (and that all massive field have vanishing vacuum values) it is possible to study the ground-state problem using a low-energy EA (expanded in α'), i.e. by solving $\delta S/\delta\varphi = 0$. The effective action is also the basic tool in establishing connection of string theory with a low-energy theory of particles as well as for studing string-theory predictions for gravitational physics (e.g., modifications of cosmological and black hole solutions).

In what follows we shall discuss a number of approaches to computation of EA (at the tree level) and summarize some explicit results found for EA in various string theories.

2. S-MATRIX APPROACH

Since a string field theory must reproduce the first quantized string S-matrix, the EA (defined as a result of "integration over" all massive fields) is a field theory action which generates an S-matrix which coincides with the massless sector of the string S-matrix. In other words, the EA represents an "off-shell extension" of string S-matrix with all massless exchanges subtracted out. Thus defined it is non-unique since we can make field redefinitions $\varphi \to \varphi' = \varphi + \ldots$ which does not influence the S-matrix because of the equivalence theorem. Different EA's correspond to different gauge-invariant off-shell extensions of "subtracted" string amplitudes. To get a local expression for EA we are to expand all massive propagators in inverse powers of masses, i.e. in powers of α'. Then the massive exchanges produce contact vertices with arbitrary number of massless "legs".

A construction of EA proceeds in perturbative fashion. One first writes an effective action S_o that describes the free massless particles and their three point couplings. One then considers the four point scattering amplitude. The unitarity of the theory guarantees that the massless poles will be produced by the tree graphs constructed using S_o. What remains has no singularities for vanishing external momenta and can thus be expanded in a power series in $\alpha'(\text{momentum})^2$. One then adds four-point local vertices to S_o. This procedure goes over for n=4,5,...-particle amplitudes resulting in the full EA as a power series in α'.

Let us now review the known perturbative results for EA depending on bosonic massless fields in several (super)string theories. We shall start with the open string theory. Here the massless bosons are the gauge vectors A_μ. The scattering amplitudes of the vectors in the open Bose string theory are given by expectation values of the corresponding vertex operators (we use Polyakov's covariant path integral approach [6,7])

$$\mathcal{A}_N = \sum_{\text{non-cyclic}} \bar{\mathcal{A}}_N \, tr(\lambda_1 \ldots \lambda_N) \,, \quad \bar{\mathcal{A}}_N = \langle V(1) \ldots V(N) \rangle \,, \tag{1}$$

$$\langle \ldots \rangle = \int [\mathcal{D}x^\mu \mathcal{D}g_{ab}] \, e^{-I_o} \,, \quad I_o = \frac{1}{4\pi\alpha'} \int d^2z \sqrt{g} \, g^{ab} \partial_a x^\mu \partial_b x^\mu \,, \tag{2}$$

$$V(n) = \oint d\tau \, \dot{x}^\mu e^{ik_n x} \zeta_\mu^{(n)}(k_n) \,, \quad k_n^2 = 0, \quad k_\mu \zeta_\mu = 0 \,. \tag{3}$$

Here the path integration goes over disc-like 2-surfaces and τ parametrizes the boundary. For example,

$$\bar{\mathcal{A}}_3 = (\zeta_1 \cdot \zeta_2 \, \zeta_3 \cdot k_1 + \text{cyclic permutations}) + 2\alpha' \, \zeta_1 \cdot k_2 \, \zeta_2 \cdot k_3 \, \zeta_3 \cdot k_1 \,. \tag{4}$$

The 3-point amplitude in the open superstring theory is (4) without the last ζ^3-term [8]. From (4) we conclude that in addition to the Yang-Mills $AA\partial A$-vertex $S_0^{(B)}$ contains also the $\alpha'\,\partial A\,\partial A\,\partial A$ correction [4]. The 4-point amplitude is given by [8,9]

$$\bar{A}_4 = \frac{\Gamma(-\alpha' s)\,\Gamma(-\alpha' t)}{\Gamma(1+\alpha' u)} \Big\{ (ut\, b_s\, \zeta_1\cdot\zeta_2\, \zeta_3\cdot\zeta_4 + \text{cyclic permutation})$$

$$+ \text{"}\zeta\cdot\zeta\,\zeta\cdot k\,\zeta\cdot k\text{"} + \text{"}\zeta\cdot k\,\zeta\cdot k\,\zeta\cdot k\,\zeta\cdot k\text{"} - \text{structures} \Big\} \; . \tag{5}$$

$b_s = (1+\alpha' s)^{-1}$ in the Bose string case and 1 in the superstring case.

The most general gauge-invariant EA has the following expansion

$$S = \int d^Dx\, \mathcal{L}\,, \quad \mathcal{L} = -\frac{1}{8g^2}\, tr \Big\{ F_{\mu\nu}^2 + 2\alpha' \big(a_1 F_{\mu\lambda} F_{\lambda\nu} F_{\nu\mu}$$

$$+ a_2 \mathcal{D}_\lambda F_{\mu\nu} \mathcal{D}_\lambda F_{\mu\nu} \big) + (2\alpha')^2 \big(a_3 F_{\mu\lambda} F_{\lambda\nu} F_{\mu\rho} F_{\rho\nu} +$$

$$+ a_4 F_{\mu\lambda} F_{\lambda\nu} F_{\nu\rho} F_{\rho\mu} + a_5 F_{\mu\nu} F_{\mu\nu} F_{\lambda\rho} F_{\lambda\rho} + a_6 (F_{\mu\nu} F_{\lambda\rho})^2$$

$$+ a_7 F_{\mu\nu} \mathcal{D}_\lambda F_{\mu\nu} \mathcal{D}_\rho F_{\rho\lambda} + a_8 \mathcal{D}_\lambda F_{\lambda\mu} \mathcal{D}_\rho F_{\rho\nu} F_{\mu\nu} + a_9 (\mathcal{D}_\rho \mathcal{D}_\lambda F_{\lambda\mu})^2 \tag{6}$$

$$+ O(\alpha'^3) \Big\}$$

The coefficients a_2, a_7, a_8 and a_9 are not fixed by the S-matrix method, i.e. are ambiguous since they can be changed by a suitable field redefinition [10]: $A_\mu \to A_\mu + \alpha' b_1 \mathcal{D}_\rho F_{\rho\mu} + \dots$. Computing the 3-point and 4-point amplitudes corresponding to (6) and comparing them with (4) and (5) we find that [10]

$$a_1 = 4/3\,, \quad a_3 = 2a_4 = \pi^2/3\,, \quad a_5 = -\frac{\pi^2}{12} - \frac{1}{2}\,, \quad a_6 = -\frac{\pi^2}{24} + \frac{1}{2} \tag{7}$$

in the Bose string case and

$$a_1 = 0\,, \quad a_3 = 2a_4 = \pi^2/3\,, \quad a_5 = 2a_6 = -\pi^2/12 \tag{8}$$

in the superstring case. In the case of the abelian background (6) reduces to (both in the Bose string and superstring theories)

$$\mathcal{L}_{abelian} = \frac{1}{4g^2} \Big\{ F_{\mu\nu}^2 - \frac{1}{2}(2\pi\alpha')^2 \big[F^4 - \frac{1}{4}(F^2)^2 \big] + O(\alpha'^3) \Big\} \tag{9}$$

(here we have put all ambiguous coefficients in (6) equal to zero).

Next let us consider gravitational EA's in closed string theories. In general (for a moment we set ϕ = const and $B_{\mu\nu} = 0$)

$$S = -\frac{1}{\kappa^2} \int d^Dx \sqrt{G}\, \Big\{ R + \alpha' \big(a_1 R_{\lambda\mu\nu\rho}^2 + a_2 R_{\mu\nu}^2 + a_3 R^2 \big)$$

$$+ \alpha'^2 \big(b_1 R^{\mu\nu}_{\ \ \alpha\beta} R^{\alpha\beta}_{\ \ \lambda\rho} R^{\lambda\rho}_{\ \ \mu\nu} + b_2 R^{\mu\alpha}_{\ \ \nu\beta} R^{\beta\gamma}_{\ \ \alpha\delta} R^{\delta\mu}_{\ \ \gamma\nu} \big) + \tag{10}$$

$+$ terms with $R_{\mu\nu}$) $+ \alpha'^3 \left(R^{\alpha\beta}_{\lambda\rho} R^{\lambda\rho}_{\mu\nu} R^{\mu\nu}_{\gamma\delta} R^{\gamma\delta}_{\alpha\beta} + \ldots \right) + O(\alpha'^4) \}$.

The closed string 3- and 4-point amplitudes can be represented as products of two corresponding open string amplitudes (one for the left- and one for the right-moving modes) [9]. Symbolically, if $A_3^{(B)}(\xi) \sim \xi^3 k + \alpha' \xi^3 k^3$, $A_3^{(S)} \sim \xi^3 k$ then the 3-graviton amplitudes in the Bose string, heterotic string and type II superstring theories are given by: $A_3^{(B)} \sim A_3^{(B)}(\xi) A_3^{(B)}(\bar{\xi})$, $A_3^{(H)} \sim A_3^{(B)}(\xi) A_3^{(S)}(\bar{\xi})$, $A_3^{(S)} \sim A_3^{(S)}(\xi) A_3^{(S)}(\bar{\xi})$ (the polarization tensor of the graviton is $\xi_{(\mu} \bar{\xi}_{\nu)}$). Substituting $G_{\mu\nu} = \delta_{\mu\nu} + h_{\mu\nu}$, expanding (10) to the third order in $h_{\mu\nu}$ and comparing with the string 3-point amplitudes we find that: (i) the Einstein 3-vertex is correctly reproduced [4]; (ii) $a_1 = 1/4$, 1/8 and 0 in the Bose, heterotic and type II theories [11,12]; (iii) only the combination $\tilde{b}_1 = b_1 + 1/2\, b_2$ of b_1 and b_2 is fixed by the 3-point amplitude: $\tilde{b}_1 = 1/48$ in the Bose string theory and 0 in the heterotic and type II superstring theory [13].

The coefficients of the terms in (10) which contain "reducible" contractions of the curvature tensor representable as a products of two non-trivial (not equal to the metric) second rank tensors (in particular,the terms, which contain the Ricci tensor) are, as a rule, ambiguous, i.e. cannot be fixed from the S-matrix [10]. In fact, consider the field redefinition $G_{\mu\nu} \to G_{\mu\nu} + T_{\mu\nu}$, $T_{\mu\nu} = \alpha'(c_1 R_{\mu\nu} + c_2 G_{\mu\nu} R) + O(\alpha'^2)$, under which $R\sqrt{G} \to R\sqrt{G} - R_{\mu\nu} T_{\mu\nu} + 1/2\, G_{\mu\nu} T_{\mu\nu} R + O(T^2)$. Any term in (10) which can be written as $R_{\mu\nu} \Omega^{\mu\nu}$ where $\Omega_{\mu\nu}$ has a definite power in α' can thus be transformed away (without influencing perturbative S-matrix). As for the "reducible terms" of the type $M_{\mu\nu} L^{\mu\nu}$ (where M and L are constructed of the curvature tensor, its covariant derivatives and the metric) they in general have the "analogs" $M_{\mu\nu} G^{\mu\nu}$ and $L_{\mu\nu} G^{\mu\nu}$ at lower orders in α' and hence can be transformed away by a redefinition $G_{\mu\nu} \to G_{\mu\nu} + K_{\mu\nu}$ where $K_{\mu\nu}$ is a properly chosen tensor. For example, the coefficients of the terms $R^2_{\mu\nu}$, R^2 , $R_{\lambda\mu\alpha\beta} R_{\lambda\mu\alpha\gamma} R^{\beta\gamma}$, etc. in (10) are arbitrary in the S-matrix approach. It is clear that the possibility to transform away the term $M_{\mu\nu} L^{\mu\nu}$ depends on the actual presence of the lower-order irreducible terms $M_{\mu\mu}$ or $L_{\mu\mu}$. For example, if in $\mathcal{L} \sim \sqrt{G} \{ R + \alpha' a_1 R^2_{\lambda\mu\nu\rho} + \alpha'^3 c_1 (R^2_{\lambda\mu\nu\rho})^2 + \ldots \}$ $a_1 \neq 0$ (as is the case for the Bose and heterotic string theories) then c_1 is ambiguous because it changes under $G_{\mu\nu} \to G_{\mu\nu} + \alpha'^2 d_1 G_{\mu\nu} R^2_{\lambda\mu\nu\rho}$. However, if $c_1 = 0$ (as is true in the type II theory) then c_1 is unambiguously determined by the S-matrix (in fact, $c_1 \sim \zeta(3)$ according to ref. [21]). Thus the $M_{\mu\nu} L^{\mu\nu}$-representation is only a necessary condition for a term to be ambiguous.

Let us now discuss in more detail the α'^2-terms in (10) [13]. In writing (10) we have used the following facts about the R^3-invariants (see, e.g., Appendix of ref.[14]): if $R_{\mu\nu} = 0$ (i.e. if we ignore the "Ricci terms") then the five possible "R^3" curvature contractions

$$I_1 = R^{\mu\nu}_{\alpha\beta} R^{\alpha\beta}_{\lambda\rho} R^{\lambda\rho}_{\mu\nu} , \quad I_2 = R^{\mu\nu}_{\alpha\beta} R^{\beta\gamma}_{\nu\lambda} R^{\lambda\mu}_{\gamma\alpha} , \quad I_3 = R^{\alpha\nu}_{\mu\beta} R^{\beta\gamma}_{\nu\lambda} R^{\lambda\mu}_{\gamma\alpha} ,$$

$$I_4 = R_{\mu\nu\alpha\beta} R^{\mu\alpha}_{\gamma\delta} R^{\nu\beta\gamma\delta} , \quad I_5 = R_{\lambda\mu\nu\rho} \mathcal{D}^2 R^{\lambda\mu\nu\rho} \tag{11}$$

satisfy the following identities

$$I_4 = \tfrac{1}{2} I_1 , \quad I_5 = -I_1 - 4I_2 , \quad I_3 = I_2 - \tfrac{1}{4} I_1 \tag{12}$$

Hence only two invariants are independent. In (10) we have chosen them to be I_1 and I_2. It is useful, however, to change a basis to I_1 and $\tilde{I}_2 = I_1 - 2I_2$ so that $b_1 I_1 + b_2 I_2 = \tilde{b}_1 I_1 + \tilde{b}_2 \tilde{I}_2$, $\tilde{b}_1 = b_1 + 1/2\, b_2$, $\tilde{b}_2 = - 1/2\, b_2$. To understand why $\tilde{b}_2$ is not fixed by the 3-point amplitude let us consider the following geometrical invariants [11,15]

$$\int d^D x \sqrt{G}\, \Omega_n = \int \underbrace{R^{ab} \wedge R^{cd} \wedge \ldots \wedge R^{mn}}_{n} \wedge \underbrace{e^p \wedge \ldots \wedge e^q}_{m} \underbrace{\varepsilon_{ab\ldots q}}_{2n+m = D} \tag{13}$$

$\Omega_{D/2}$ are the Euler number integrands. It is possible to check that (see, e.g. [16]): $\Omega_1 = R$, $\Omega_2 = R^2_{\lambda\mu\nu\rho} - 4R^2_{\mu\nu} + R^2$, $\Omega_3 = 32\, \tilde{I}_2 +$ "Ricci terms". The expansion of Ω_n near a flat background ($e = e_0 + e_1$, $e^a_0 = \delta^a_\mu dx^\mu$, $e^a_1 = h^a_\mu dx^\mu$) starts with $O(h^{n+1})$ - terms($\omega = \omega_1 \sim O(h)$, $R = d\omega_1 + O(h^2)$, $\int d\omega_1 \wedge \ldots \wedge d\omega_1 \wedge e_0 \wedge \ldots \wedge e_0\, \varepsilon_{\ldots} = 0$ because $de_0 = 0$). Thus $\tilde{I}_2$ may contribute only to four- and higher-point amplitudes (see also [17]).

To compute $\tilde{b}_2$ and to determine the $O(\alpha'^3)$ - terms in (10) one is to consider the four-point amplitude. It is useful to fix the ambiguous coefficients in (10) so that to avoid the propagator corrections ($a_2 = -4\, a_1$, $a_3 = a_1$, etc.). Then there are the following contributions to the 4-point amplitude corresponding to (10): (i) contact 4-vertices coming from R, "R^2", "R^3" and "R^4"-terms; (ii) one graviton exchange diagrams "$V_3\, \Box^{-1} V_3$", where V_3 is a 3-point vertex coming from the R or "R^2" or I_1-terms. At this point it is necessary to recall that we must account for all the massless exchanges in a field theory in order to be able to compare with the string-theory S-matrix. The relevant terms in EA which we have not yet written in (10) correspond to the dilaton couplings [4]. The structure of these couplings can be most easily established in the path integral approach [5]. It turns out to be the following [5]

$$S = -\frac{1}{k^2} \int d^D x \sqrt{G} \Big\{ R - \frac{4}{D-2} (\partial_\mu \phi)^2 + \alpha' e^{-\frac{4\phi}{D-2}} \big(a_1 R^2_{\lambda\mu\nu\rho} + \ldots \big) + \alpha'^2 e^{-\frac{8\phi}{D-2}} \big(\tilde{b}_1 I_1 + \tilde{b}_2 \tilde{I}_2 + \ldots \big) + O\big(\alpha'^3 e^{-\frac{12\phi}{D-2}}\big) \Big\} \tag{14}$$

Because of the $\phi R^2_{\lambda\mu\nu\rho}$ -coupling there is the dilaton exchange contribution $\sim a_1^2 R^2_{\lambda\mu\nu\rho}\, \Box^{-1} R^2_{\lambda'\mu'\nu'\rho'}$ to the 4-graviton amplitude (note that $\phi R^2_{\mu\nu}$ - and ϕR^2 -couplings give vanishing contributions when the graviton legs are put on shell). Note that this contribution is absent in the type II superstring theory where $a_1=0$. In the heterotic string theory (where $\tilde{b}_1=0$ and hence there is no "R^3"- 3-vertex) there are just two $O(\alpha'^2)$ exchange contributions to the 4-graviton amplitude: "$R^2\ldots\Box^{-1}R^2\ldots$" with the graviton exchange and "$R^2\ldots\Box^{-1}R^2\ldots$" with the dilaton exchange. Expanding the 4-graviton amplitude [8,12,9] in powers of momenta p we find that the $O(p^2)$ - piece is reproduced by the Einstein term in (10), (14) [18] (contact 4-vertex plus exchange diagram). $O(p^4)$ - piece is reproduced by the "$R+R^2$"-terms in (10) [19]. Direct inspection shows that there is no $O(p^6)$-term in the type II superstring amplitude. Hence $\tilde{b}_2=0$ and there is no (unambiguous) $O(\alpha'^2)$ terms in the EA in this theory. $O(p^8)$-terms in the type II theory amplitude are reproduced by the contact "R^4"-vertices. The corresponding terms in EA (which do not involve derivatives of curvature - these are probably the only "irreducible" $O(\alpha'^3)$ terms) were recently found in [20,21,22](cf. [23]). As for the heterotic string theory, there <u>are</u> $O(p^6)$-terms in the corresponding 4-graviton amplitude [12]. To find $\tilde{b}_2$ one is to compare the $O(p^6)$-part of the amplitude with the 4-vertex from the $\tilde{I}_2$-term plus the graviton and dilaton "$R^2\ldots\Box^{-1}R^2\ldots$"-exchange contributions. If $\tilde{b}_2$ will turn out to be non-vanishing this will not invalidate the R^4-results of refs.[21,22] (which ignored the $\tilde{I}_2$-term in (10)) since $\tilde{I}_2$ does not lead to exchanges contributing to the $O(p^8)$-part of the amplitude. The presence of $\tilde{I}_2$ in the heterotic string EA may have interesting implications for compactification on a <u>six</u>-dimensional compact space ($\int d^6x\, \tilde{I}_2$ is proportional to its Euler number if $R_{\mu\nu}=0$). $\tilde{b}_2$ should certainly be non-vanishing in the Bose string case.

3. PATH INTEGRAL APPROACH

There are certain shortcomings of the S-matrix approach to computation of the EA. First, the method is indirect: one computes on-shell amplitudes, expands them in $\alpha' p^2$, "extends" off-shell, etc. Second, it is not manifestly covariant being based on flat-space expansions. Third, it is perturbative in number of fields (one first compares 3-point amplitudes, then 4-point ones, etc.). Thus this method is not well suited for establishing the expression for EA to all orders in α' and the fields. It would be desirable to have more efficient, manifestly covariant off-shell procedure for computation of EA.

The path integral approach [5] is based on the observation that the generating functional for string theory S-matrix has a simple representation in terms of Polyakov's path integral with a free string action replaced by a sigma-model action. The generating functional for the S-matrix is given by

$$\mathcal{S}[\varphi_{(in)}] = \langle e^{V\cdot\varphi_{(in)}} \rangle = \sum_N \int dp_1 \ldots dp_N\, A_N(p_1,\ldots,p_N)\,\varphi_{(in)}(p_1)\ldots\varphi_{(in)}(p_N) \quad (15)$$

where $\varphi_{(in)}$ is the on-shell value of a massless field and A_N are the on-shell amplitudes (cf.(1)-(3)). For example, for the closed string gravitons: $V\cdot\varphi_{(in)} = \int d^2z \sqrt{g}\, \partial_a x^\mu \partial^a x^\nu\, h_{(in)\mu\nu}(x)$, $\Box h_{(in)}=0$,

$\partial_\mu h_{\mu\nu(in)}=0,\ h^\mu_{\mu(in)}=0$, so that (15) can be rewritten as follows

$$\mathcal{S}[h_{(in)}] = \int [\mathcal{D}x^\mu \mathcal{D}g_{ab}]\, exp\left(-I[x,g,G_{(in)}]\right), \tag{16}$$

$$I = \frac{1}{4\pi\alpha'}\int d^2z\sqrt{g}\,\partial_a x^\mu \partial^a x^\nu G_{\mu\nu(in)}(x),\quad G_{\mu\nu(in)} = \delta_{\mu\nu} + h_{\mu\nu}\ .$$

Here the path integral goes over sphere-like surfaces (we consider the tree approximation). Effective action is a field theory action $S[\varphi]$ with the corresponding generating functional for the S-matrix coinciding with (15) or (16). As is well-known, in a tree approximation

$$\mathcal{S}_{field\ theory}[\varphi_{(in)}] = S[\varphi_{cl}] - \int \varphi_{(in)}\Delta\varphi_{cl}\ , \tag{17}$$

where $\varphi_{cl} = \varphi_{in} + O(\square^{-1}\varphi_{in}^2)$ is a classical solution of $\delta S/\delta\varphi = 0$ and Δ is a free kinetic operator. Since $\mathcal{S}[\varphi_{(in)}]$ has a path integral representation a natural conjecture is that $S[\varphi]$ also has a path integral representation analogous to (16) [5],

$$S[G] = \int [\mathcal{D}x^\mu \mathcal{D}g_{ab}]\, exp\left(-I[x,g,G]\right) \tag{18}$$

We have replaced $G_{(in)}$ in (16) by an arbitrary metric $G_{\mu\nu}$ thus having gone off-shell. Our prescription for off-shell continuation is to impose a Weyl gauge on g_{ab} in computing (18). While a conformal factor ρ of 2-metric was decoupled in (16) due to on-shell restrictions on $G_{(in)}$ it does not automatically decouple in (18). To make the transition on the mass shell smooth we are to impose (as usual in gauge theories) a gauge on ρ, i.e. not to integrate over it. The resulting S (18) in general depends on a Weyl gauge used (it depends on a choice of a 2-metric on a sphere - like surface). This dependence is expected to be irrelevant on the effective mass shell ($\delta S/\delta G = 0$).

We have defined EA in sect.2 as being an "off-shell extension" of the string theory S-matrix with the massless exchanges subtracted out. However, (18) seems to be an off-shell extension of the full string S-matrix (16). A resolution to this puzzle is provided by an observation that a renormalization of the 2d σ-model (18) corresponds to subtraction of the massless exchanges [24,25,10]. If $G_{\mu\nu}$ in I in (18) is a bare field then the effective action is a renormalized path integral $S[G_R]$. A transformation from a bare to a renormalized field corresponds to the transformation from $\varphi_{(in)}$ to φ_{cl} in (17). Thus the basic assumption of the path integral approach is that EA can be represented as a renormalized partition function (18) of a generalized 2d σ-model on a sphere - like surface, i.e. of a string moving in the massless background fields. Though the above discussion suggests plausibility of this assumption the equivalence of the S-matrix and path integral approaches is at present proved only on some particular examples.

Eq.(18) can be generalized in a number of ways [5]. First, we may include the other massless fields of the Bose string theory, replacing $I[x,g,G]$ by

$$I = \frac{1}{4\pi\alpha'}\int d^2z\sqrt{g}\left\{\partial_a x^\mu \partial^a x^\nu G_{\mu\nu}(x) + \alpha' R^{(2)}\phi(x) + \right.$$

$$+ i \frac{\varepsilon^{ab}}{\sqrt{g}} \partial_a x^\mu \partial_b x^\nu B_{\mu\nu}(x) \Big\} \tag{19}$$

Here $R^{(2)}$ is a curvature of g_{ab}. Next, we may include string theory loop corrections by path integrating over different world sheet topologies. Also, we may consider a superstring generalization of (18). In a covariant formalism this is possible using a Fermi-string (Neveu-Schwarz -Ramond) formalism. The corresponding string action is the N=1 2d supersymmetric extension of the σ-model action (19). In the light-cone gauge we may use the Green-Schwarz formalism. (see [5], [26]).

In the latter case we have the following action for the string interaction with a background metric which is non-trivial only in transverse dimensions [26]

$$I = \frac{1}{4\pi\alpha'} \int d^2z \sqrt{g} \Big\{ \partial_a x^i \partial^a x^j G_{ij}(x) - i\bar{\theta}^m \rho^a \mathcal{D}_a \theta^m \tag{20}$$

$$+ \frac{\alpha'^2}{256} R_{ijkl}\, \bar{\theta}\gamma^{ij}\rho^a(1+\rho_3)\theta\, \bar{\theta}\gamma^{kl}\rho_a(1-\rho_3)\theta + \dots \Big\}$$

$\{\theta^m\}$, $m = 1,\dots,8$ is a 2d spinor and a Weyl spinor of O(8). Finally, we may consider the open-string generalization of (18). In the Bose string case [26,27]

$$S[A_\mu] = \int [\mathcal{D}x^\mu \mathcal{D}g_{ab}]\, e^{-I_0}\, \mathrm{tr}\, P \exp\Big(i \oint d\tau\, \dot{x}^\mu A_\mu(x)\Big), \tag{21}$$

where we integrate over disc-like surfaces (cf. (1)-(3)), A_μ (belonging to the algebra of a gauge group) couples to the boundary circle, P is the ordering along the boundary and the trace is taken over the group indices. This path integral can be easily computed if the background is abelian and has constant field strength, $A_\mu = -1/2\, F_{\mu\nu} x^\nu$. This corresponds to computing all the terms in EA which depend only on a field strength itself and not on its derivatives. Splitting x^μ on a constant and non-constant parts, $x^\mu = y^\mu + \xi^\mu(z)$, and computing the gaussian integral over ξ one finds (D=26) [27]

$$S[A]_{F_{\mu\nu}=const} = c_0 \int d^D y \sqrt{\det(\delta_{\mu\nu} + 2\pi\alpha' F_{\mu\nu})} \tag{22}$$

Note that this Born-Infeld-type action contains terms of all orders in α'. This demonstrates how the path integral approach makes possible to compute EA exactly in α'. The result (22) is in correspondence with the S-matrix approach. In fact, expanding (22) in powers of α' we find that the F^2 and F^4-terms in (22) coincide with (9).

It is possible also to find the type I open superstring generalization of the Born-Infeld action (22) [10]

$$S'[A]_{F_{\mu\nu}=const} = c_0' \int d^D y \Big\{ \frac{1}{4} F_{ij}^2 + \frac{2}{3} \sqrt{\frac{\det(\delta_{ij} + 2\pi\alpha' F_{ij})}{\det(\delta_{mn} + \frac{\pi\alpha'}{2}\gamma^{ij}_{mn} F_{ij})}} \Big\} \tag{23}$$

Here it is assumed that the background field has only transverse components being non-vanishing, i,j = 1,...8. The indices m,n = 1,...8 are in the spinor representation of SO(8) and γ^{ij} is a product of two Dirac matrices.

Let us now dicuss the perturbative results for the closed string EA within the path integral approach. Assuming the fields to be weak we compute EA expanding it in number of derivatives, i.e. in α'. This corresponds to the loop expansion in the σ -model (18). The measure of integration in (18) is $[\mathcal{D}x\,\mathcal{D}g] = \prod_z dx(z)\sqrt{G(x(z))}$ (ghost factor). Let $x^\mu = y^\mu + \xi^\mu(z)$, $y^\mu = x^\mu(z_0)$, $\xi(z_0) = 0$ and replace ξ^μ by the normal coordinates η^μ so that

$$G_{\mu\nu}(y+\xi(\eta)) = G_{\mu\nu}(y) - \frac{1}{3}R_{\mu\alpha\nu\beta}(y)\eta^\alpha\eta^\beta + \dots ,$$

$$\phi(y+\xi(\eta)) = \phi(y) + \mathcal{D}_\mu\phi(y)\,\eta^\mu + \dots \quad . \tag{24}$$

Then using that $\int d^2z\sqrt{g}\,R^{(2)} = 8\pi$ on a sphere - like surface we find [5]

$$S = \int d^D y\sqrt{G(y)}\,e^{-2\phi(y)}\,\mathcal{L}(R_{\dots},\ \mathcal{D}\phi\ ,\ \mathcal{D}R_{\dots},\dots), \tag{25}$$

where $\mathcal{L}$ is a manifestly covariant function of $G_{\mu\nu}$ and ϕ (we set $B_{\mu\nu} = 0$) which depends on ϕ only through its derivatives. Note that the expansion near one point y^μ may fail non-perturbatively in α' (in the case of large values of derivatives of fields) when one is also to include the contribution of other σ-model classical solutions (instantons). $\mathcal{L}$ is equal to $\exp(-W)-1$ (we assume that D=26 and normalize $\mathcal{L}$ so that to have zero cosmological constant), where $W[y,g]$ is a σ-model 2d effective action depending on a 2d background metric (g_{ab} is fixed according to a Weyl gauge). W (and hence $\mathcal{L}$) is in general renormalization scheme and Weyl gauge dependent. This ambiguity corresponds to field redefinition ambiguity ($G_{\mu\nu} \to G_{\mu\nu}$ + $T_{\mu\nu}(G\ ,\phi)$, $\phi \to \phi$ + $N(G,\phi)$) present in the S-matrix approach. W has complicated expansion in powers of 2d curvature starting with the characteristic "central charge" term (other terms in W are likely to have renormalization scheme dependent coefficients)

$$W = c(y)\int R^{(2)}\Delta^{-1}R^{(2)} + \dots , \tag{26}$$

$$C(y) = \frac{26-D}{96\pi} + \frac{\alpha'}{64\pi}\left(R + 4\mathcal{D}^2\phi - 4\partial_\mu\phi\,\partial^\mu\phi\right) + \frac{\alpha'^2}{64\pi}\times$$

$$\left(a_1 R^2_{\lambda\mu\nu\rho} + \text{RS - dependent terms}\right) + O(\alpha'^3), \qquad a_1 = \frac{1}{4} \quad . \tag{27}$$

Here $O(\alpha'^0)$ term is the Polyakov's result [6] and $O(\alpha')$-terms were computed in [5] (see also [28]). $O(\alpha'^2)$-terms were discussed in [10]. As a result,

$$S[G,\phi] = \text{const}\int d^D y\sqrt{G}\,e^{-2\phi}\{R + 4(\partial_\mu\phi)^2 + \alpha'(a_1 R^2_{\lambda\mu\nu\rho} +$$

$$+ a_2 R_{\mu\nu}^2 + a_3 R^2) + O(\alpha'^2) \} \tag{28}$$

where a_2 and a_3 are ambiguous (RS and (or) Weyl gauge dependent [10]). EA (28) is equivalent to that found in the S-matrix approach (14) after the rescaling of the metric $G_{\mu\nu} \to G_{\mu\nu} \exp\left(\frac{4\phi}{D-2}\right)$. Thus at the 3-loop order of the σ-model perturbation theory and up to ambiguous terms [5,10,28,29]

$$S = \text{const} \int d^D y \sqrt{G}\, e^{-2\phi} c(y) \tag{29}$$

The above discussion can be generalized to the closed heterotic string case [12] in which there is the additional coupling of the string to a gauge field background,

$$I_{int} = \int d^2 z \{ i \bar{\xi}^I \not{\partial} \xi^I +$$

$$+ i \bar{\xi}^I \rho^a \xi^J \left(\partial_a x^\mu A_\mu^{IJ}(x) - \frac{i}{2} \bar{\psi}^\mu \rho_a \psi^\nu F_{\mu\nu}^{IJ}(x) \right) \} . \tag{30}$$

Here I, J = 1,...,32 are, e.g., the SO(32) indices, ξ^I (ψ^μ) are 2d Majorana-Weyl left (right) -handed spinors. Intergating over ψ and ξ in the leading approximation we get the standard $\int d^D x \sqrt{G}\, e^{-2\phi}\, tr F_{\mu\nu}^2$-term in addition to (28). The total leading order result for EA in the heterotic string theory is

$$S[G,\phi,B,A] = - \frac{1}{K^2} \int d^D y \sqrt{G}\, e^{-2\phi} \{ R + 4(\partial_\mu \phi)^2$$

$$- \frac{1}{12} F_{\mu\nu\rho}^2(B) + \frac{\alpha'}{8} tr F_{\mu\nu}^2(A) + \frac{\alpha'}{8} R_{\lambda\mu\nu\rho}^2 + \dots \} , \tag{31}$$

where we included also the antisymmetric tensor contribution [5]. Eq. (31) coincides with the bosonic part of the N=1, D=10 supergravity plus N=1, D=10 super Yang-Mills action (up to Chern-Simons term which is of subleading order, cf. [2]).

4. SIGMA MODEL APPROACH

The basic observation of this approach is that the σ-model describing string propagating in true vacuum must be conformal (Weyl) invariant in order to remain "critical" (and hence to preserve unitarity). Thus the necessary condition that the vacuum fields (G, ϕ, B) must satisfy is the condition of conformal invariance of the σ-model, i.e. the vanishing of the trace T_a^a of its quantum energy-momentum tensor [30,2,28,24]. In other words, the vanishing of "ß-functions" should provide equations of motion for the corresponding massless fields. It was checked [28,29] that this approach is in agreement with the perturbative results (28), (29), (31) of the path integral approach.

In general, all solutions of the equations of motion of the σ-model approach (" $\langle T^a_a \rangle = 0$") are also the solutions of the equations of motion of the path integral approach. This follows from the fact that expectation values of vertex operators $\langle V \rangle$ in a conformal invariant theory on a sphere vanish [2,28] and that, e.g., $(\delta S/\delta G_{\mu\nu})_{G=G_{vac}}$ with S defined by (18) is just such an expectation value. The converse statement which is not yet rigorously proved (see, however, ref. [24] and below) should certainly be true (i.e. all solutions of the path integral approach should correspond to conformal invariant σ-models). In fact, only in this case the whole path integral approach is consistent (for example, on-shell conformal invariance is necessary for Weyl gauge independence of the vacuum configuration). The proof of this statement should be based on a justification of the representation (29) to all loop orders and the use of the theorem [31] (proved only for a flat world sheet) that the conditions of conformal invariance of a renormalizable 2d σ-model are equivalent to the conditions of stationarity of some scalar function equal to the central charge at a conformal point. In any case, the theorem of Zamolodchikov proves the consistency of the σ-model approach, i.e. that the conformal invariance conditions can be derived from a single EA. Let us also remark that the proof of the equivalence of the S-matrix and σ-model approaches should be based on demonstration of their equivalence with the path integral approach.

In the above we assumed that the space-time dimension is critical, D = 26 or 10. However, it seems that this restriction is not necessary in the path integral and σ-model approaches and hence these approaches are more general than the S-matrix one. In fact, in the S-matrix approach we start with computation of the on-shell amplitudes on the flat background and hence must fix $D = D_{critical}$. As for the path integral approach, here we in any case do not have conformal invariance off-shell and impose a Weyl gauge. So in principle we may keep D to be arbitrary. Analogously, in the σ-model approach we may study the conditions of conformal invariance for arbitrary D. The only place where D appears explicitly in equations of motion is the constant term in the dilaton equation, or C=0, see (27). Varying (29) with respect to ϕ and putting ϕ=const we get

$$C = -\frac{1}{96\pi}\left[D - 26 - \frac{3}{2}\alpha'\left(R + \alpha' a_1 R^2_{\lambda\mu\nu\rho} + \dots\right)\right] = 0 \quad (32)$$

It is the total "central charge" (32) that should vanish for conformal invariance. Thus EA in arbitrary D should start with a "cosmological constant" (in fact, dilaton potential) proportional to D-26 (this was first suggested in [32])

$$S = \text{const} \int d^D y \sqrt{G}\, e^{-2\phi} \left\{ D - 26 - \frac{3}{2}\alpha' R + \dots \right\}, \quad (33)$$

This action may be considered as a consistent extension of the EA in the S-matrix approach for D ≠ "flat" critical dimension= 26 .

5. CONCLUDING REMARKS

We have seen that the path integral (and σ-model) approaches provide a manifestly covariant derivation of string-theory EA. It may even turn out to be possible to derive exact (in α') expressions for EA. However, it should be kept in mind that the EA is a low-energy concept (note, e.g., that the string-theory S-matrix is unitary in the massless sector only below the threshold of production of massive particles). Thus results derived from a EA may be trustable only if the effective scale of the vacuum fields is much larger than the string scale $\sqrt{\alpha'}$ (e.g., if $|\alpha' R| \ll 1$).

If (29) is correct to all orders in α' and if $e^{-2\phi}\,\delta C/\delta\phi$ = = total derivative (as is true to the order α'^2) then the ϕ-equation of motion is $C=0$. This implies that the vacuum value of the cosmological term vanishes (the vanishing of the action at the vacuum point follows trivially from (25)). C=0 gives a stringent condition on ϕ and G which is not always possible to satisfy with ϕ = =const. For example, if we set D=26, ϕ =const. and truncate the α'-series in (28) at the $\alpha' R$ - order we easily find that the only solution of G - and ϕ-equations of motion is the flat space (cf. [16, 33], see also [34]). The only case in which it is clear how to prove that C=0 for D=10 and ϕ=const. (and $B_{\mu\nu}=0$) is that of the superstring compactified on a hyperkählez space. Since the N=4 supersymmetric σ-models are known to be ultraviolet finite [35] for an <u>arbitrary</u> (in particular very small) coupling and since the Zamolodchikov's theorem [31] implies <u>stationarity</u> of C, C must be simply equal to its free value $\sim$ D - 10.

There is well-known problem with a massless scalar particle in a gravitational theory like (14) or (31): ϕ couples to gauge (and spinor) fields and hence produces additional long-range interactions. To avoid contradictions with observations the dilaton should get a mass $m \sim 1/\sqrt{\alpha'}$ due to non-perturbative effects (in fact, dilaton may get mass due to the potential in (33)). Suppose that this happened but all other terms in the effective lagrangian (14) remained the same (we consider the heterotic string case and take D=10)

$$\mathcal{L} \sim -R + \frac{1}{2}(\partial_\mu \phi)^2 + \frac{m^2}{2}\phi^2 - \frac{\alpha'}{8} R^2_{\lambda\mu\nu\rho} e^{-\phi/2} + \dots \qquad (34)$$

Then the ϕ-equation of motion looks like

$$-\Box\phi + m^2\phi = -\frac{\alpha'}{16} R^2_{\lambda\mu\nu\rho} e^{-\phi/2} + \dots \qquad (35)$$

If $\phi \approx$ const then the growth of the curvature produces the growth in $|\phi|$ because of $m \neq 0$. But $e^{\phi} = g$ plays a role of a string-theory coupling constant. Therefore it has tendency to decrease with an increase of the curvature. This suggests that it may be possible to avoid singularities of black holes in string theory (see also [20]).

If the cosmological term vanishes due to the ϕ-equation of motion then the $G_{\mu\nu}$-equation takes the form (we start with (31) and set ϕ =const)

$$R_{\mu\nu} = \frac{1}{4} F_{\mu\alpha\beta} F_{\nu\alpha\beta} + \frac{\alpha'}{4} \operatorname{tr}(F_{\mu\alpha} F_{\nu\beta}) + \frac{\alpha'}{4} R_{\mu\alpha\beta\gamma} R_{\nu\alpha\beta\gamma} + \dots \qquad (36)$$

Suppose we are looking for a compactification of the type $M^4 \times M^6$ and assume that the vacuum values of $F_{\mu\nu\lambda}$ and $F_{\mu\nu}$ vanish on M^4. Then there is always a solution of (36) with <u>flat</u> M^4, i.e. $R^{(4)}_{\mu\nu\alpha\beta} = 0$ (see also [2]). Another observation that follows from (36) is that a scale of compactification $(R^{(6)})^{-1/2}$ is typically of order of $\sqrt{\alpha'}$, i.e. $\alpha' R^{(6)} \sim 1$. This implies that all terms in the α'-expansion of EA may be important and in fact, that we cannot trust the effective action approach. This raises a problem of how to generate a hierarchy ($\alpha' R^{(6)} \ll 1$ in order to be able to treat the compactification problem in the framework of a low-energy effective action approach. One possibility is that hierarchy is generated by the same non-perturbative mechanism that gives mass to the dilaton. Another suggestion is that string loop corrections may contribute terms which are non-analytic in α' (e.g., $\mathcal{L} \sim R + \hbar g^2 R \ln(\alpha' R) + \ldots$, $R^{-1} \sim \alpha' \exp(-\mathrm{const}/g^2)$).

REFERENCES

1. "Superstrings: The first 15 years of Superstring Theory", ed. J.H. Schwarz (World Scientific, Singapore, 1985).
2. Candelas, P., Horowitz, G., Strominger, A. and Witten, E., Nucl. Phys. B258, 46 (1986).
3. Kaku, M. and Kikkawa, K., Phys. Rev. D10, 1110 (1974).
   Siegel, W., Phys. Lett. 151B, 391, 396 (1985).
   Banks, T. and Peskin, M., Nucl. Phys. B264, 573 (1986).
   Witten E., Nucl. Phys. B268, 253 (1986).
4. Scherk, J. and Schwarz, J.H., Nucl. Phys. B81, 118 (1974);
   Phys. Lett. 52B, 347 (1974).
   Yoneya, T., Progr. Theor. Phys. 51, 1907 (1974).
5. Fradkin, E.S. and Tseytlin A.A., Phys. Lett. 158B, 316 (1985);
   Nucl. Phys. B261, 1 (1985).
   Tseytlin, A.A., Yad. Fiz. 43, 1012 (1986).
6. Polyakov, A.M., Phys. Lett. 103B, 207 (1981).
7. Aoyama, H., Dhar, A. and Namazie, M.A., Nucl. Phys. B267, 605 (1986).
8. Schwarz, J.H., Phys. Reports. 89, 223 (1982).
9. Kawai, H., Lewellen, D.C. and Tye, S.-H.H., Nucl. Phys. B269, 1 (1986).
10. Tseytlin, A.A., JETP Lett. 43, 263 (1986);
    Nucl. Phys. B276, 391 (1986).
11. Zwiebach, B., Phys. Lett. 156B, 315 (1985).
    Nepomechie, R., Phys. Rev. D32, 3201 (1985).
12. Gross, D.J., Harvey, J.A., Martinec, E. and Rohm, R., Nucl. Phys. B256, 253 (1985); B267, 75 (1986).
13. Tseytlin, A.A., unpublished.
14. Fradkin, E.S. and Tseytlin, A.A., Nucl. Phys. B227, 252 (1983).
15. Zumino, B. , Phys. Rep. 137, 109 (1986).
16. Wheeler, J.T., Nucl. Phys. B268, 737 (1986).
17. Ferrara, S., preprint CERN-TH 4497/1986.
18. Sannan, S., Santa Barbara preprint UCSB-86-0424.
19. Deser, S. and Redlich, A.N., Brandeis preprint BRX-TH-200 (1986).
20. Gross, D.J. and Witten, E. Princeton preprint (1986).
21. Kikuchi, Y., Marzban, C. and Ng., Y.J., Univ. North Carolina preprint IFP-272-UNC (1986).
22. Cai, Y. and Nunez C.A., Texas preprint UTTG-13-86.

23. Grisaru, M. and Zanon, D., Harvard preprint HUTP-86/A046.
Freeman, M.D., Pope, C.N., Sohnius, M.F. and Stelle, K.S., Imperial College preprint TP/85-86/27.
24. Lovelace, C., Nucl. Phys., B273, 413 (1986).
25. Fridling, B.E. and Jevicki, A., Phys. Lett. 174B, 75 (1986).
26. Fradkin, E.S. and Tseytlin, A.A., Phys. Lett. 160B, 69 (1985).
27. Fradkin, E.S. and Tseytlin, A.A., Phys. Lett. 163B, 123 (1985).
28. Callan, C., Friedan, D., Martinec, E. and Perry M., Nucl. Phys. B262, 593 (1985).
29. Callan, C., Klebanov, I. and Perry M., Princeton preprint (1986).
30. Lovelace, C., Phys. Lett., 135B, 75 (1984).
Sen, A., Phys. Rev. D32, 2102 (1985); Phys. Rev. Lett. 55, 1846 (1985).
31. Zamolodchikov, A.B., Pisma Zh. Eksp. Teor. Fiz. 43, 565 (1986).
32. Duff, M., Nilsson, B. and Pope, C., Phys. Lett. 163B, 343 (1985).
33. Boulware, D.G. and Deser, S., Phys. Rev. Lett. 55, 2656 (1986).
34. Boulware, D.G. and Deser, S., Santa Barbara preprint NSF-ITP-86-57.
35. Alvarez-Gaumé, L. and Freedman, D.Z., Commun. Math. Phys. 80, 443 (1981).
Alvarez-Gaumé, L. and Ginsparg, P., Commun. Math. Phys. 102, 311 (1985).

QUASI-CLASSICAL ASPECTS OF REDUCTION

Victor Guillemin

M.I.T.
Department of Mathematics
Cambridge, MA 02139

ABSTRACT

By exploiting its symmetries one can reduce a physical problem to a problem with fewer degrees of freedom. We will discuss some mathematical formulations of this (more or less self-evident) fact and describe some theorems on quasi-classical limits which the "theory of reduction" seems to illuminate. The results described here are joint with Alejandro Uribe and are closely related to recent results of Hogreve-Potthoff-Schräder on the quantum partition function.

QUASI-CLASSICAL ASPECTS OF REDUCTION

VICTOR GUILLEMIN
MASSACHUSETTS INSTITUTE OF TECHNOLOGY

Introduction

In a moment I will give a technical definition of what I mean by "reduction"; however, let me illustrate the idea with a simple example: the symmetric top. As you all know from highschool physics, the motion of a top can be thought of as a composite motion consisting of: a) rotation of the top about its axis of symmetry, and b) precession of the axis of symmetry about the top's fulcrum (the point, F, in the figure below). By Noether's principle (symmetries $\longleftrightarrow$ conserved quantities) the angular momentum of the top about its axis of symmetry is constant; so one can ignore the motion a) and concentrate on how some fixed reference point, say P, on the axis of symmetry moves about F. That is, one can regard the points P and F and the rod, PF, joining them as the components of a new system (which has one degree of freedom less than the original system).

The theory of reduction says that this new system is isomorphic to a system consisting of an electrically charged point mass, P, attached by a rigid rod to the fulcrum, F, and acted on by a uniform gravitational force

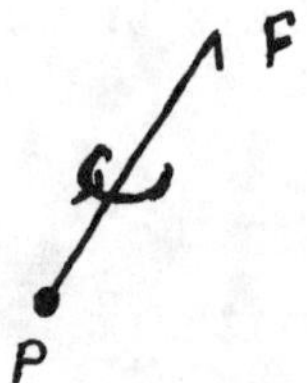

We will call this model of the spinning top the <u>reduced</u> top and the process by which one equates the motion of a simple mechanical system of three degrees of freedom and a one-dimensional symmetry group (top) with that of a slightly more complicated mechano-electro-magnetic system of two degrees of freedom (electrically charged spherical pendulum) <u>reduction.</u>

This process is apparently well understood by physicists both at the classical and at the quantum mechnanical level. I will report on how mathematicians look at it in §2 and §3. Alejandro Uribe and I have been attempting to see how quantum reduction and classical reduction are related. The relation appears to involve a type of asymptotics familiar to physicists from the work of Lieb, Berezin, Simon, Hogreve, Potthoff, Schrader and other but, as far as I can tell, relatively unfamiliar to mathematicians. I will discuss this asymptotics in §4.

§2. QUANTUM REDUCTION

I will begin by discussing quantum reduction, which is a little easier to understand than classical reduction. Let $\mathscr{H}$ be a Hilbert space, G a compact Lie group and ρ: G $\longrightarrow$ U($\mathscr{H}$) a unitary representation of G on $\mathscr{H}$. Let A be a self-adjoint operator on $\mathscr{H}$ which is G invariant. It

is clear that if we decompose $\mathcal{H}$ into irreducibles, A permutes irreducibles of the same type among themselves. More explicitly let V be a finite dimensional Hilbert space and $\alpha: G \longrightarrow U(V)$ an irreducible unitary representation.
Let

$$\mathcal{H}_\alpha = \mathrm{Hom}_G(V, \mathcal{H})$$

and let

$$\text{ev}: V \otimes \mathcal{H}_\alpha \longrightarrow \mathcal{H} \tag{2.1}$$

be the evaluation map, $(v,T) \longrightarrow T(v)$. The image of the map (2.1) is the subspace of $\mathcal{H}$ spanned by all copies of α. Moreover, if one lets G act on $V \otimes \mathcal{H}_\alpha$ by acting trivially on $\mathcal{H}_\alpha$ and acting by the representation, α, on V, then (2.1) is an isomorphism of $V \otimes \mathcal{H}_\alpha$ onto this subspace. In particular thee exists an operator A^o on $V \otimes \mathcal{H}_\alpha$ such that $A \circ (\text{ev}) = (\text{ev}) \circ A^o$. Finally since A^o is G invariant and the representation of G on V is irreducible, there exists an operator, A_α, on $\mathcal{H}_\alpha$ such that $A^o = I \otimes A_\alpha$. One calls A_α the <u>reduction of A with respect to the representation</u> α.

Lets see what reduction looks like in a typical concrete case. Let $M \subset \mathbb{R}^{3N}$ be the configuration space of an N-particle system of the type encountered in elementary quantum mechanics. For simplicity we will assume this system has no exterior forces acting on it, so that its quantum hamiltonian, $H = H_{K.E.}$, is just the usual Laplace operator, Δ_g, g_M being the metric on M coming from the imbedding of M into $\mathbb{R}^{3N}$.

Suppose now that G acts on M as a group of isometries. We will assume this action is as simple as possible, namely that the stabilizer group at every point is the identity. Since G is compact, this implies that the orbit space is a smooth manifold, B, and that the point $\longrightarrow$ orbit mapping

$$\kappa: M \longrightarrow B \tag{2.2}$$

makes M into a principal bundle over B with structure group G. We will show that k and g_M give rise to the following data on B:

(2.3)

i. A metric, g_B

ii. A connection, Φ.

iii. A section, θ, of $S^2(\mathfrak{g}_{ad})$,

$\mathfrak{g}_{ad}$ being the adjoint bundle of the fibration (2.2).

Indeed, imbed κ^*T^*B into T^*M by the imbedding $(p,\eta) \longrightarrow (p, dk^t_p(\eta))$. The metric, g_M, defines an inner product structure on T^*M hence, by restriction, an inner product structure on κ^*T^*B and hence by G-invariance an inner product structure on T^*B, which we can think of, dually, as an inner product structure on TB or, in other words, a metric on B. This is the g_B in (2.3), i.

Next because of the inner product structure on T^*M we can split T^*M into a direct sum of subbundles

$$T^*M \cong V \oplus \kappa^*T^*B$$

in a G-invariant fashion. The corresponding splitting of TM into horizontal and vertical components defines a connection, Φ, for the fibration (2.2).

Finally consider the restriction of the inner product on T^*M to V. V is a trivial bundle with fiber $\mathfrak{g}^*$ so this inner product can be thought of as a map

$$\theta: M \longrightarrow S^2(\mathfrak{g}) \tag{2.4}$$

which transforms under the action of G on M as Ad(G). Therefore it can also be regarded as a section of the adjoint bundle $S^2(\mathfrak{g}_{ad}) \longrightarrow B$. We will call this section the Higgs data of the fibration 2.2.

Now let α be an irreducible unitary representation of G and let E_α be the vector bundle over B associated, via (2.2), with α. Under the associated bundle construction the connection, Φ, gets converted into a connection on E_α. Therefore, the usual Laplace operator can be extended to a Laplace operator

(2.5) $$\Delta_\alpha : \underline{E}_\alpha \longrightarrow \underline{E}_\alpha$$

on sections of E_α. Notice also that the Higgs field, θ, defines a bundle map

(2.6) $$\theta_\alpha : E_\alpha \longrightarrow E_\alpha$$

which is positive-definite on each fiber.

<u>Theorem I</u> If $\mathcal{H} = L^2(M)$ and $A = \Delta_g$ the reduced Hamiltonian, A_α, is the self-adjoint operator $\Delta_g + \theta_\alpha$.

<u>Remark</u> Δ_α is the quantum observable associated with the <u>external</u> Kinetic energy and θ_α the quantum observable associated with the <u>internal</u> Kinetic energy of the reduced system. However, for a quantum observer positioned on B, θ_α looks like a potential energy contribution to the reduced Hamiltonian. It is therefore sometimes called the <u>effective</u> potential energy of the reduced system.

§3. CLASSICAL REDUCTION

Let (X,ω_X) be a symplectic manifold and $H: X \longrightarrow \mathbb{R}$ the energy function of a given dynamical system on X. We will assume that H is invariant with respect to a group, G, of symplectomorphisms of X. We will also assume that G is a compact Lie group and that it acts on X in a

Hamiltonian fashion with moment mapping

$$\Phi: X \longrightarrow \mathfrak{g}^*.$$

Finally to avoid problems with stabilizer groups we will assume as in § 2 that the stabilizer group of every point is trivial. Now let O be a co-adjoint orit in $\mathfrak{g}^*$ and let

$$Z = \{x \in X, \Phi(x) \in O\}.$$

It is known (see [KKS]) that Z is a G-invariant co-isotropic submanifold of X, and that its null-foliation is fibrating i.e., there exists a symplectic manifold (B,ω_B) and a fiber mapping

$$k: Z \longrightarrow B$$

such that $\iota^*\omega_X = k^*\omega_B$, $\iota: Z \longrightarrow X$ being the inclusion map. Moreover, B can be factored into a product of symplectic manifolds

$$B = O \times X_o \tag{3.1}$$

in such a way that the action of G on B is the product of its standard action on O and the trivial action on X_O.

Since the Hamiltonian, H, is G invariant there exists a G-invariant function, H', on B such that $\iota^*H = k^*H'$. Moreover since G acts transitively on O and B is the product (3.1), H' is of the form

$$H' = 1_O \otimes H_O.$$

H_O being a smooth function on X_O and 1_O the constant function, 1, on O. We will call the pair (X_O, H_O) the <u>reduction of the system</u> (X,H) <u>with respect to</u> O.

<u>Remark</u> It is worth pondering the similarities between (3.1) etc. and the corresponding quantum relations, (2.1) etc., described in the first paragraph of §2.

Lets now go through the details of this reduction construction in a typical concrete case. Suppose as in §2 that $M \hookrightarrow \mathbb{R}^{3N}$ is the configuration space of a classical

N-particle system. As in §2 we will assume there are no external forces acting on this system, so that its Hamiltonian

(3.2) $$H: T^*M \longrightarrow \mathbb{R}$$

involves only the Kinetic energy contribution coming from the Riemann metric. Let the compact Lie group G act freely on M as a group of isometries and, as in §2, let

$$\kappa: M \longrightarrow B$$

be the principal G-fibration describing this G-action. Recall that B acquires from M a metric g_B, a connection Φ, and a Higgs field Θ. (See 2.3, items i, ii and iii.) Let B_O be the associated bundle

$$M \times_G O.$$

We will show that the data (2.3) can be converted into the following data:

(3.3)
i)' a pre-symplectic form ω_O, on B_O,
ii)' a Hamiltonian function, H_B, on T^*B,
iii)' a function, V_{Higgs}, on B_O.

To define ω_B, let Φ be the connection form (2.3), ii. This form is a $\mathfrak{g}$-valued one form on M; so one can define from it an $\mathbb{R}$-valued one-form, μ, on $M \times O$ by setting

$$\mu_{p,\alpha} = (pr_1)^* \langle \alpha, \Phi_p \rangle .$$

Now set

(3.4) $$\nu = d\mu + (pr_2)^* \omega_O .$$

It is easy to check that for every $\xi \in \mathfrak{g}$

$$\iota(\xi^\#)\nu = D_{\xi^\#}\nu = 0 ;$$

so ν is the pull-back to $M \times O$ of a closed two-form, ω_B, on B_O. This is our definition of ω_B.

To define V_{Higgs} recall that, by (2.4), θ can be regarded as a function on M with values in $S^2(\mathfrak{g})$, so we can define a scalar function, $V = V_{Higgs}$, on $M \times O$ by setting

$$V_{Higgs}(p,\alpha) = \langle \theta_p, \alpha \otimes \alpha \rangle.$$

This function is G invariant, so it can actually be thought of as a function on B_O.

Finally H_B is just the Hamiltonian function on T^*B whose Kinetic energy component is that associated with the metric, g_B, and whose potential energy component is zero.

Now lets come back to the N-particle system which we started with. What happens when we reduce the Hamiltonian system (3.2) with respect to the co-adjoint orbit O in $\mathfrak{g}^*$? The answer is given by the following proposition:

<u>Theorem II</u> Let X_O be the fiber product

$$\begin{array}{ccccc} & & X_O & & \\ & \overset{pr_1}{\swarrow} & & \overset{pr_2}{\searrow} & \\ T^*B & & & & B_O \\ & \searrow & & \swarrow & \\ & & B & & \end{array} \tag{3.5}$$

Let ω_O be the closed two-form

$$(pr_1)^* \, d\alpha_B \; + \; (pr_2)^* \omega_B, \tag{3.6}$$

where α_B is the standard one form on T^*B, and let H_O be the function

$$H_O = (pr_1)^* H_B + (pr_2)^* \, V_{Higgs}. \tag{3.7}$$

Then (X_O, ω_O, H_O) is the reduction with respect to O of the system (3.2).

HISTORICAL COMMENTS

As far as I can tell the general idea of quantum reduction is due to Wigner. (The main ideas are clearly spelled out in his paper on the quantum theory of molecular vibrations, c. 1930.) Versions of Theorem I have been known to mathematicians since the early fifties as the Weitzenbock formalism. However, physicists were probably familiar with this material a good deal earlier in connection with Kaluza-Klein models of quantum relativity. Concerning classical reduction, the formulation at the beginning of Section 3 is due basically to Marsden and Weinstein [M-W]. However, the analogies which we pointed out above between quantum reduction and classical reduction were observed for the first time by Kazhdan, Kostant and Steinberg [KKS]. The idea of basing reduction on the co-isotropicity of $\Phi^{-1}(0)$ is also due to them. For the tie-in between the version of reduction described here and the version found in the physics literature, Richard Montgomery's article [M] is an excellent source. Finally most of the material in the second half of §3 is due to Sternberg [S].

§4 QUASI-CLASSICAL BEHAVIOR OF THE REDUCED SYSTEM IN THE HIGH ENERGY LIMIT.

Consider the dynamical system (3.2). The Hamiltonian $H\colon T^*M \longrightarrow \mathbb{R}$ and the moment map $\Phi\colon T^*M \longrightarrow \mathfrak{g}^*$ satisfy the homogeneity conditions

$$H(m,r\xi) = r^2 H(m,\xi)$$

(4.1)

$$\Phi(m,r\xi) = r\,\Phi(m,\xi)$$

for $r \in \mathbb{R}^+$, and from these conditions one can show that certain of the reduced systems are isomorphic. Explicitly let (X_0,H_0) be the reduction of the system (3.2) with respect to the co-adjoint orbit O and let $\Xi_{O,E}$ be the restriction of the Hamiltonian vector field associated with H_O to the energy hypersurface, $H_O = E$. Let us denote by exp t $\Xi_{O,E}$ the one-parameter group of diffeomorphisms generated by this vector field. From (4.1) one easily deduces

<u>Proposition</u> Let $O' = rO$ and $E' = r^2E$. Then the dynamical systems exp rt $\Xi_{O,E}$ and exp t $\Xi_{O',E'}$ are isomorphic.

In other words if we set $\Xi_{e,E} = \Xi_{eO,E}$ the dynamical systems, exp t $\Xi_{e,E}$, are essentially the same for all e and E for which the ratio of $\sqrt{E}$ to e is equal to a fixed number r. It is unreasonable to expect a similar result at the quantum level; however what one might hope to be true is that the quantum reduced system exhibits interesting quasi-classical behavior as the quantum analogues of the quantities E and e tend to infinity providing the ratio of $\sqrt{E}$ to e is kept fixed. It is not, however, as easy to do this as it sounds, for now the quantities, E and e, take on discrete values; so a prescribed value of $\sqrt{E}$: e may not be realized by <u>any</u> physical state of the reduced system. We will show how to get around this problem by "smearing out" the ratio of $\sqrt{E}$ to e; however, lets first give a more careful description of e and E for the quantum reduced systems. We recall that by the Bott-Borel-Weil-Kostant theorem there is a one-one correspondence between irreducible unitary representations of G and integral co-adjoint orbits. Now fix an integral co-adjoint orbit $O = O_1$ and consider the sequence of integral co-adjoint orbits

(4.2) $\qquad O_e = eO, \; e = 1,2,3,\cdots$

Corresponding to (4.2) is a sequence of irreducible representations

(4.3) $\qquad \alpha_e, \; e = 1,2,3,\cdots$

For each e consider the reduced system associated with α_e. To insure that the eigenvalues of this reduced system are discrete, we will assume that the manifold, M, in (3.2) is compact. (One could, of course, achieve the same effect by assuming that the effective potential energy grows sufficiently fast at infinity.) Let us denote these eigenvalues by

(4.4) $\qquad \lambda_{i,e}, \; i = 1,2,3,\cdots$

and display them in increasing order $(0 \leq \lambda_{i,e} \leq \lambda_{2,e} \leq \cdots)$ For any eigenstate of the e-th reduced system, $E = \lambda_{i,e}$ is the total energy of the system in that eigenstate and e itself can be interpreted as an internal quantum number such as charge, spin, isospin etc. Now fix $r \in \mathbb{R}^+$ and consider the Weyl counting function

(4.5) $\qquad N_r(\lambda) = \#\{\sqrt{\lambda_{i,e}} < \lambda, \; \sqrt{\lambda_{i,e}}: \; e = r\}.$

For certain simple dynamical systems (to be explicit, those associated with Hopf fibrations) Uribe and I have been able to show this has interesting quasi-classical behavior for λ large. (See [G-U, 1].) However, in general this won't be the case because of the problem alluded to above. There is, however, a way of making the counting function (4.5) more effective by smearing it out with respect to r. More explicitly choose a function, d on the real line which is smooth, positive, rapidly decreasing at infinity and has total integral one. For each r consider the counting function

$$(4.6) \qquad N_{d,r}(\lambda) = \sum_{e} \sum_{\lambda_{i,e} < \lambda^2} d(\sqrt{\lambda_{i,e}} - re).$$

This function counts the number of eigenstates of <u>all</u> the reduced systems having energy $<\lambda^2$; however the eigenstates for which the ratio of $\sqrt{E}$ to e is approximately r are heavily favored in this count.

Uribe and I have been able to show that $N_{d,r}(\lambda)$ has very nice asymptotic behavior as $\lambda \longrightarrow \infty$ providing d isn't too sharply blipped at 0. (For instance it suffices to assume that $\hat{d}$ is compactly supported, which forces d itself to be rather spread out.) In particular under this assumption we establish a Weyl estimate

$$(4.7) \qquad N_{d,r}(\lambda) \sim \text{volume}(X_{O,E})\lambda^N.$$

Here E = r, $X_{O,E}$ is the hypersurface, $H_O = E$, and N = dim B + (dim O)/2. We have also been able to obtain a generalized Selberg trace formula for $N_{d,r}$ analogous to the "wave trace" formulas in [C] and [DG]. Unfortunately I won't have time to describe this formula here but roughly speaking it is an identity whose right hand side is

$$\int e^{i\lambda s} \, dN_{d,r}(\lambda)$$

and whose left hand side is a generating function involving symplectic invariants of the system exp t $\Xi_{O,E}$. For details see [G-U, 2].

BIBLIOGRAPHY

[C] J. Chazarain, "Formule of Poisson pour les variétés riemanniennes" Invent. Math. 24(1974) 65-82.

[D-G] J.J. Duistermaat and V. Guillemin, "The spectrum of positive elliptic operators and periodic geodesics" Invent. Math. 29(1975) 184-269.

G U,1] V. Guillemin and A Uribe, "Clustering theorems with twisted spectra". Math. Ann. 1-28(1986) 212-239.

G U,2] ______________________________, "The trace formula for vector bundles" (to appear).

KKS] D. Kazhdan, B. Kostant, and S. Sternberg, "Hamiltonian group actions and dynamical systems of Calogero type", Comm. Pure Appl. Math. 31(1978) 481-507.

MW] J. Marsden and A. Weinstein "Reduction of symplectic manifolds with symmetry", J. Math. Phys. 5(1974) 121-130.

M] R. Montgomery, "Canonical formulations of a classical particle in a Yang-Mills field and Wong's equations", Lett. Math. Phys. 8(1984) 59-67.

S] S. Sternberg, "Minimal coupling and the symplectic mechanics of a classical particle in the presence of a Yang-Mills field", Proc. Nat. Acad. Sci. USA 74(1977) 5253-5254.

Lie algebraic methods in filtering and identification

Michiel Hazewinkel
CWI, Amsterdam
P.O.Box 4079
1009 AB Amsterdam
The Netherlands

Abstract and Preface.

These lectures concern (nonlinear) filtering. Very roughly the art of obtaining best estimates for some stochastic time-varying variable x on the basis of observations of another process y. The more concrete object under consideration being a stochastic dynamical system $dx = f(x)dt + G(x)dw$, where w is Wiener noise, with observations $dy = h(x)dt + dv$, corrupted by further noise. The subject as presented here involves ideas and techniques from Lie algebra theory, stochastics, differential topology, approximation theory and partial differential equations and has relations with quantum theory and stochastic physics.

The lectures are adressed to practitioners in any one of these areas assuming that as a rule they are not experts in the other ones.

AMS 1980 Classification: 93E11, 93B30, 93E10, 60H15, 93B15, 17B65, 17B99, 57R25.
Keywords and phrases: nonlinear filtering, estimation Lie algebra, asymptotic expansion, Weyl algebra, Heisenberg algebra, Kalman-Bucy filter, conditional density, Duncan-Mortensen-Zakai equation, BC-principle, identification, Lie algebra of vectorfields, finite dimensional filter, robustness.

1. Introduction

Filtering is concerned with making estimates of quantities associated with a stochastic process $\{x_t\}$ on the basis of information gleaned from a related process $\{y_t\}$. The process $\{x_t\}$ is called the *signal* or *state* process and $\{y_t\}$ is the *observation* process. In this paper the following more concrete realization will be considered

$$dx_t = f(x_t)dt + G(x_t)dw_t, \quad x_t \in \mathbb{R}^n, \ w_t \in \mathbb{R}^m \tag{1.1}$$

$$dy_t = h(x_t)dt + dv_t, y_t \in \mathbb{R}^p, \ v_t \in \mathbb{R}^p \tag{1.2}$$

Here f is a function $\mathbb{R}^n \to \mathbb{R}^n$; G is an $n \times m$ matrix valued function on $\mathbb{R}^n$, h is a function $\mathbb{R}^n \to \mathbb{R}^p$ and w_t and v_t are Wiener processes, assumed independent of each other and also independent of the initial random variable x_0. More precisely these equations can be written

$$x_t = x_0 + \int_0^t f(x_s)ds + \int_0^t G(x_s)dw_s \tag{1.3}$$

$$y_t = \int_0^t h(x_s)ds + v_t \tag{1.4}$$

where the last term of (1.3) is a stochastic integral in the sense of *Ito*.

Much more loosely one can look at equations (1.1) and (1.2) as

$$\dot{x} = f(x) + G(x)\dot{w} \tag{1.5}$$

$$\dot{y} = h(x) + \dot{v} \tag{1.6}$$

with $\dot{w}$ and $\dot{v}$ white noise. Thus we have a differential equation $\dot{x} = f(x)$ on $\mathbb{R}^n$ which is subject to continuous random shocks whose intensity and direction (distribution) is state dependant and as observations we have an integral of some function of x and these observations are corrupted by more noise.

The general filtering problem for the state process $\{x_t\}$ with observation process $\{y_t\}$ is now to calculate for (interesting) functions ϕ of the state the conditional expectation

$$E[\phi(x_t)\,|\,y_s,\ 0 \leqslant s \leqslant t] = \widehat{\phi(x_t)}, \tag{1.7}$$

i.e. the best (least squares) estimate of $\phi(x_t)$ given the observations y_s up to time t. That is we are interested in calculation procedures for $\widehat{\phi(x_t)}$. In many (engineering) applications the data come in sequentially and one does not really want a calculating procedure which needs all the data y_s, $0 \leqslant s \leqslant t$, every time t that it is desired to find $\widehat{\phi(x_t)}$; rather we would like to have a procedure which uses a statistic m_t which can be updated using only the new observations y_s, $t \leqslant s \leqslant t'$ to its value $m_{t'}$ i.e.

$$m_{t'} = a(m_t, t', t, \{y_s : t \leqslant s \leqslant t'\}) \tag{1.8}$$

and from which the desired conditional expectation can be calculated directly, i.e.

$$\widehat{\phi(x_t)} = E[\phi(x_t)\,|\,y_s, 0 \leqslant s \leqslant t] = b(t, y_t, m_t). \tag{1.9}$$

Finally to actually implement the filter it would be nice if m_t were a finite dimensional quantity. All this leads to the (ideal) notion of a *finite dimensional recursive filter*. By definition such a filter is a system

$$d\xi_t = \alpha(\xi_t)dt + \sum_{i=1}^{p} \beta_i(\xi_t)dy_{it} \tag{1.10}$$

driven by the observations y_{it}; y_{it} is the i-th component of y_t, $i = 1,\dots,p$; together with an output map

$$\widehat{\phi(x_t)} = \gamma(\xi_t) \tag{1.11}$$

More precisely formulated our problem is now the following: given a system (1.1)-(1.2) and a function ϕ on $\mathbb{R}^n$, how can we decide whether for these data there exists a finite dimensional recursive filter (1.10)-(1.11) which calculates $\widehat{\phi(x_t)}$, the best least squares estimate, and how do we find the functions (vectorfields) $\alpha, \beta_1, \cdots, \beta_p, \gamma$ of (1.10)-(1.11).

Now this may of course be a totally unreasonable question to ask. It could be that such nice filters virtually never exist. That is not the case though. In the case of linear systems

$$dx_t = Ax_t dt + Bdw_t \tag{1.12}$$

$$dy_t = Cx_t dt + dv_t \tag{1.13}$$

where now A,B,C are matrices of the appropriate sizes (which may be time varying), the well known Kalman-Bucy filter is precisely such a filter as (1.10)-(1.11). The equations are as follows. The statistic ξ_t is a pair (m_t, P_t) consisting of an n-vector and a symmetric $n \times n$ matrix P_t. These evolve according to

$$dP_t = (AP_t + P_t A^T + BB^T - P_t C^T C P_t)dt \tag{1.14}$$

$$dm_t = Am_t dt + P_t C^T(dy_t - Cm_t dt). \tag{1.15}$$

Here X^T denotes the transpose of a matrix X. This filter was discovered in 1961 and it is hard to overestimate its importance: whole books are devoted to its applications into single specialized fields and substantial companies can make a good living doing little more than Kalman-Bucy filtering. Naturally, efforts immediately started to find similar filters for more general systems than (1.12)-(1.13). This turned out to be unexpectedly difficult and this is still the case though there exists hosts of approximate filters of various kinds which (seem to) work well in a variety of situations; there is very little systematically known about how to construct approximate filters or about how to predict that a given one or class will work well when applied to a given collection of systems.

The approach based on Lie-algebraic considerations which I will try to discuss and explain below seems to hold great promise both in understanding the difficulties involved and in providing some kind of systematic foothold in the area of constructing approximate filters. For, as will become clear below, the existence of finite dimensional recursive filters for a nontrivial statistic will be a rare event.

Let me pause at this point to point out that identification problems can easily be construed as filtering problems. By way of illustrating this point consider again a linear system

$$dx_t = Ax_t + Bdw_t, \quad dy_t = Cx_t dt + dv_t \tag{1.16}$$

where now the matrices A,B,C are (partially) unknown. By adding to (1.16) the stochastic equations

$$da_{ij} = 0, \quad db_{kl} = 0, \quad dc_{qx} = 0 \tag{1.17}$$

for all unknown a_{ij}, b_{kl}, c_{qr}, one obtains a system (1.16)-(1.17) (of much larger state space dimension). And solving the filtering problem for the functions which project the vector $(x, (a_{ij}), (b_{kt}), (c_{qr}))$ onto a suitable component means identifying that particular coefficient.

2. The DMZ-equation and the estimation algebra

Let $\{x_t\}$ be a diffusion process as in (1.1)-(1.2) above. Given sufficient regularity of f,G,h the conditional expectation $\hat{x}_t$ will have a density $\pi(x,t)$.

2.1. Theorem. Under appropriate regularity conditions there exists an unnormalized version $\rho(x,t)$ of $\pi(x,t)$ which satisfies an equation

$$d\rho = \mathfrak{L}\rho dt + \sum_{i=1}^{p} h_i(x)\rho dy_{it} \tag{2.2}$$

where $\mathfrak{L}$ is the second order differential operator given by

$$(\mathfrak{L}\psi) = \frac{1}{2}\sum_{i,j=1}^{n} \frac{\partial^2}{\partial x_i \partial x_j}(GG^T)_{ij}\psi) - \sum_{i=1}^{n} \frac{\partial}{\partial x_i}(f_i\psi) - \frac{1}{2}\sum_{j=1}^{p} h_j^2\psi \tag{2.3}$$

Here $(GG^T)_{ij}$ is the (i,j)-th component of the $n \times n$ matrix $G(x)G(x)^T$ and f_i, h_j are the i-th and j-th component respectively of f and h.

Several comments are in order. First of all equation (2.2) is in Fisk-Stratonovic form. The corresponding Ito equation looks the same with $\mathfrak{L}$ changed by removing the $-\frac{1}{2}\sum h_j^2\psi$ term. The word "unnormalized" means that $\rho(x,t) = \sigma(t)\pi(x,t)$ where $\sigma(t)$ is an unknown function of time. Under appropriate reachability conditions on (1.1) $\rho(x,t)$ is a positive function. That ρ is unnormalized does not hurt much as $\rho(x,t)$ still suffices to calculate such things as $\widehat{\phi(x_t)}$. Indeed

$$\widehat{\phi(x_t)} = (\int \rho(x,t)dx)^{-1} \int \rho(x,t)\phi(x)dx \tag{2.4}$$

Theorem 2.1 was proved by Duncan [13], Mortensen [28] and Zakai [36] and the corresponding equation 2.1 is often refered to as the Duncan-Mortensen-Zakai or DMZ equation.

It is a stochastic partial differential equation being driven by the stochastic processes $y_1, \ldots, y_\rho$.

It is important to note (Brockett [5]), that equations (2.2), (2.4) together constitute in fact a recursive filter in the sense of (1.10)-(1.11). The role of ξ_t is played by $\rho(x,t)$ so that instead of a point ξ evolving on a finite dimensional M we have an evolving density, i.e. a point ρ in an infinite dimensional space of positive functions evolving with time.

The simplest nontrivial example of a system (1.1)-(1.2) is

$$dx = dw, \quad dy = xdt + dv \tag{2.5}$$

i.e. one dimensional Wiener noise linearly observed corrupted by further noise. In this case the DMZ-equation becomes

$$d\rho = (\frac{1}{2}\frac{\partial^2}{\partial x^2} - \frac{1}{2}x^2)\rho dt + x\rho dy_t, \quad (\frac{\partial \rho}{\partial t} = \frac{1}{2}\frac{\partial^2 \rho}{\partial x^2} - \frac{1}{2}x^2\rho + x\rho\dot{y}) \tag{2.6}$$

i.e. we are dealing with the Euclidean Schrödinger equation with an extra forcing term. This is not an accident but part of a general pattern of which we shall see a further manifestation below in section 8, cf. also Mitter [26,27] for other remarks on this theme. I do not know wether the use Bismut makes of the filtering equations when dealing with a stochastic approach to index theorems and the Dirac operator can also be fitted into this framework.

3. Robustness and numerical matters

As it stands equation (2.2) is not a very useful object for applications. It is a stochastic partial differential equation (with as probability space a space of paths $\{y\}$) and as such a solution is in principle only defined apart from a set of measure zero. On the other hand actual observations will always consist of piecewise smooth $y(t)$ and the class of all such is of measure zero. Thus there arises the question whether there exist a version of (2.2) which can be interpreted pathwise for all $y(t)$ and for which the solutions of (2.2) for piecewise smooth $(y(t)$ carry (approximative) information, cf. Clark [9] and Davis [11]. Fortunatedly the time dependent gauge transformation

$$\tilde{\rho}(x,t) = \exp(-h_1(x)y_1(t) - \cdots - h_p(x)y_p(t))\rho(t,x) \tag{3.1}$$

transforms (2.2) into an equation

$$\frac{\partial \tilde{\rho}}{\partial t} = \mathfrak{L}\tilde{\rho} - \sum_{i=1}^{p} y_i(t)\mathfrak{L}_i\tilde{\rho} + \sum_{i,j=1}^{p} y_i(t)y_j(t)\mathfrak{L}_{ij}\tilde{\rho} \tag{3.2}$$

where $\mathfrak{L}_i = [h_i,\mathfrak{L}] := h_i\mathfrak{L}-\mathfrak{L}h_i$ and $\mathfrak{L}_{ij} = \mathfrak{L}_{ji} = \frac{1}{2}[h_i,[h_j,\mathfrak{L}]]$, and this equation, which does not anymore involve derivatives of y, can simply be interpreted as a family of partial differential equations parametrized by the possible observation paths $y(t)$.

Equation (3.2) can of course be verified directly (remembering that (2.2) is a Fisk-Stratonovic integral so that the ordinary rules of calculus apply; removing the term $-\frac{1}{2}\sum h_i^2$ from $\mathfrak{L}$ gives the corresponding Ito equation and then Ito calculus of course also gives (3.2). An easier way of obtaining (3.2) is to observe that (3.1) in (2.2) gives $d\tilde{\rho} = \exp(-\sum h_i y_i)\mathfrak{L}\exp(\sum h_i y_i)\tilde{\rho}$ and to use the version of the Baker-Campbell-Hausdorff formula which says

$$\exp(-rA)B\exp(rA) = \sum_{k=0}^{\infty}(-1)^k\frac{r^k}{k!}ad_A^k(B) \tag{3.3}$$

where $ad_A(B) = [A,B] = AB-BA$, $ad_A^k(B) = ad_A(ad_A^{k-1}(B))$ for linear operators A,B. In our case the contributions of (3.3) for $k\geqslant 2$ disappear because then A is a function, $B=\mathfrak{L}$ is a second order differential operator, so $[A,B]$ is first order, $[A,[A,B]]$ is a function and $[A,[A,[A,B]]] = 0$.

Also of course there still remains the question of how to use equation (3.2) or (2.2) effectively to calculate certain desired conditional expectations. A direct numerical discretization approach is out of the question. Typically x is a fairly large dimensional object; for example around 27 for certain problems involving helicopters. Taking three data points per coordinate axis (which is ridiculous) then gives $3^{27}\approx 2.10^{14}$ space grid points! So other methods must be tried. It seems likely that the Lie-algebraic considerations to be discussed below will help. Other promising work into the numerics of the nonlinear filtering equations has been started by Pardoux-Talay [29].

4. Wei-Norman theory

It is important to note that the filtering equation (3.2) (or (2.2)) is of the general form

$$\dot{x} = (A_1x)u_1 + \cdots + (A_kx)u_k \tag{4.1}$$

where the A_i are linear operators and the u_i known functions of time. Of course in (3.2) the role of x is played by ρ, an infinite dimensional object. Here for the moment lets consider (4.1) as a finite dimensional object. Let us also assume that the $A_1,...,A_k$ who are now, say, $n\times n$ matrices, form the basis of a Lie algebra. (By adding a few more terms with corresponding u_i equal to zero this can of course always be assured.) Let us look for solutions of the form (Wei-Norman [35]).

$$x(t) = e^{g_1A_1}e^{g_2A_2}\dots e^{g_kA_k}x(0) \tag{4.2}$$

Differentiating this gives

$$\dot{x} = \dot{g}_1A_1e^{g_1A_1}e^{g_2A_2}\dots e^{g_kA_k}x(0) + e^{g_1A_1}\dot{g}_2A_2e^{g_2A_2}\dots e^{g_kA_k}x(0) + \dots \tag{4.3}$$

and inserting

$$e^{-g_1A_1}e^{-g_2A_2}\dots e^{-g_iA_i}e^{g_iA_i}\dots e^{g_1A_1}$$

just after $\dot{g}_iA_i$ in the i-th term equation (1.1) can be rewritten

$$\dot{x} = \sum_{i=1}^{k}\dot{g}_i(A_i+\sum_{r=1}^{i-1}\sum_{\substack{j_1,\dots,j_{i-1}\\ j_1+\dots+j_{i-1}>0}}\frac{g_1^{j_1}\dots g_{i-1}^{j_{i-1}}}{j_1!\dots j_{i-1}!}ad_{A_1}^{j_1}\dots ad_{A_{i-1}}^{j_{i-1}}(A_i))x \tag{4.4}$$

$$= \sum_{i=1}^{k}\dot{g}_i(A_i+h_{ij}(g_1,\dots,g_k)A_j)$$

with $h_{ij}(0,...,0) = 0$, where, again, the Campbell-Baker-Hausdorff formula (3.3) has been used. Note that h_{ij} are universal functions which only depend on the Lie algebra and the chosen basis. Thus it remains to solve (equating the coefficients of the basic elements A_i in (4.4) and (4.1))

$$\dot g_1 + \dot g_2 h_{11}(g_1,...,g_k) + \dot g_2 h_{21}(g_1,...,g_k) + ... + \dot g_k h_{k1}(g_1,...,g_k) = u_1$$
$$\dot g_2 + \dot g_1 h_{12}(g_1,...,g_k) + \dot g_2 h_{22}(g_1,...,g_k) + ... + \dot g_k h_{k2}(g_1,...,g_k) = u_2 \tag{4.5}$$
$$...$$
$$\dot g_k + \dot g_1 h_{1k}(g_1,...,g_k) + \dot g_2 h_{2k}(g_1,...,g_k) + ... + \dot g_k h_{kk}(g_1,...,g_k) = u_k$$

which can be done for small t and $g_1(0) = ... = g_k(0) = 0$ because $h_{ij}(0,...,0) = 0$. In general a representation (4.2) for the solution is only possible for small t. However things change if the Lie-algebra in question is solvable, then ([35]) there is such a representation for all t. More precisely there is a suitable basis such that there is such a representation for all t. How this comes about is easy to see in the case that the Lie algebra L is nilpotent. Indeed let

$$L \underset{\neq}{\supset} L^{(1)} = [L,L] \underset{\neq}{\supset} L^{(2)} = [L,L^{(1)}] \underset{\neq}{\supset} ... \underset{\neq}{\supset} L^{(m)} = [L,L^{(m-1)}] = 0 \tag{4.6}$$

be a basis such that $A_1,...,A_{k_1}$, $A_{k_1+1},...,$ $A_{k_2},...,A_{k_{m-1}+1},...,$ $A_{k_m} = A_k, | \, k_1<k_2<...<k_m$ such that $A_{k_i+1},...,A_{k_m}$ is a basis for $L^{(i)}$, $i=0,...,m-1$ $(k_0=1, k_m=k)$. Then it immediately follows from (4.4) that $h_{ij}=0$ for $j \leqslant i$ and the set of equations (4.5) gets a nice triangular structure. Moreover $h_{ij}(g_1,...,g_k)$ involves only $g_1,...,g_{i-1}$ (this is always the case, cf.(4.5), so the h_{1j} in (4.5) are always all zero) and the resulting equations (4.5) for the nilpotent case are therefore of the form

$$\dot g_1 = u_1,...,\dot g_{k_1} = u_{k_1}$$
$$\dot g_{k_1+1} = u_{k_1+1} + \alpha_{k_1+1}(u_1,...,u_{k_1};g_1,...,g_{k_1}),...,\dot g_{k_2} = u_{k_2} + \alpha_{k_2}(u_1,...,u_{k_1};g_1,...,g_{k_1}) \tag{4.7}$$
$$\dot g_{k_2+1} = u_{k_2+1} + \alpha_{k_2+1}(u_1,...,u_{k_2};g_1,...,g_{k_2}),...,\dot g_{k_3} = u_{k_3} + \alpha_{k_3}(u_1,...,u_{k_2};g_1,...,g_{k_2})$$
$$...$$

where the α_j are known (universal) functions of the u's and g's.

These considerations are not limited to Lie-algebras of matrices. Indeed the left hand sides of equations (4.5) only depend on the abstract structure of the Lie algebra in question and the choice of basis. Thus all this equally applies to Lie-algebras of say differential operators (given suitable definitions of $\exp(tA)$), though in order to have a finite set of equations (4.5) one needs of course a finite dimensional algebra. It also follows from (4.4) that the Wei-Norman equations are compatible with homomorphisms of Lie algebras, more precisely quotients. Indeed if $\mathfrak{A} \subset L$ is an ideal and $A_1,...,A_{k_1},A_{k_{1+1}},...,A_k$ is a basis of L such that $A_{k_1+1},...,A_k$ is a basis of $\mathfrak{A}$ then the h_{ij} are zero for $j \in \{1,...,k_1\}$ and $i \in \{k_1+1,...,k\}$. So in the case of a topologically nilpotent algebra L, or more generally one with a chain of ideals $\mathfrak{A}_1 \supset \mathfrak{A}_2 \supset \mathfrak{A}_3 \supset ...$ such that $\cap \mathfrak{A}_i = 0$ and $L / \mathfrak{A}_i$ finite dimensional for all i one can in principle still do Wei-Norman theory with now infinite ordered product expressions $x = e^{g_1 A_1} e^{g_2 A_2} ... e^{g_k A_k} ... x_0$ in the sense that the equations for the g_i belonging to a quotient $L / \mathfrak{A}_j$ involve only those same g_i. Of course now questions of convergence arise.

5. The estimation Lie algebra

The considerations of the previous section already make it clear that the Lie algebra generated by the operators which occur in equation (2.2) or (3.2) contains important information concerning the filtering problem. One therefore defines the estimation Lie algebra $EL(\Sigma)$ of a system Σ given by (1.1)-(1.2) as the Lie algebra of differential operators generated by the 2-nd order differential operator $\mathfrak{L}$ and the multiplication operators $h_1,...,h_p$.

$$EL(\Sigma) = Lie(\mathfrak{L}, h_1,...,h_p). \tag{5.1}$$

Note that the Lie algebra generated by the operators which occur in (3.2) is in any case a subalgebra of $EL(\Sigma)$. Often it is equal.

5.2. Example. Consider again the simplest nonzero linear system (2.5). Then $p=1$ and $\mathfrak{L}=\frac{1}{2}d^2/dx^2-\frac{1}{2}x^2$. So we have in this case the Lie algebra $Lie(\frac{1}{2}d^2/dx^2-\frac{1}{2}x^2,x)$. Now $[\frac{1}{2}d^2/dx^2-\frac{1}{2}x^2,x]=d/dx$ (as operators on functions), $[\frac{1}{2}d^2/dx^2-\frac{1}{2}x^2,\frac{d}{dx}]=x$, $[\frac{d}{dx},x]=1$ and $[?,1]=0$. So in this case we obtain the well-known oscillator Lie algebra, which is four dimensional with basis $\frac{1}{2}d^2/dx^2-\frac{1}{2}x^2$, x, d/dx, 1. It is solvable (but not nilpotent) with as derived algebra the nilpotent Heisenberg algebra with basis x, d/dx, 1.
$EL(\Sigma)$ is (of course) an invariant of Σ meaning that a change of coordinates in Σ (a diffeomorphism $x\to x'$ taking Σ to Σ') will yield isomorphic estimation Lie algebras. The algebra also has a gauge transformation invariance. A gauge transformation $\rho(x,t)\to\psi(x)\rho(x,t)$, where $\psi(x)\neq 0$ for all x, transforms the DMZ-equation in such a way that the operators in the new equation generate an isomorphic Lie algebra.

The new equation may again have the form of a DMZ-equation, and in this way systems which are definitely not equivalent as systems may have equivalent filtering problems associated to them. An example are the 1-dimensional Benes systems (cf. various contributions in [19]).

In a way which will (hopefully) become clearer below the estimation Lie algebra $EL(\Sigma)$ encodes information about how difficult the filtering problem for Σ is. For example if it is finite dimensional (a very rare case) Wei-Norman theory does the job for small time; if it is also solvable one thus gets a filter. If it is infinite dimensional but solvable things become more difficult but asymptotic expansions are possible, cf. below; etc.

6. The BC principle

Let me now describe a second reason why the Lie algebra $EL(\Sigma)$ of a system Σ is important for filtering problems. I like to call it the *BC principle,* not because it is very old, though it could have been maybe, nor is it named after Johny Hart's chartoon character; the BC stand for Brockett and Clark [6] who first enunciated it.

Suppose we have a filter (1.10)-(1.11) on a finite dimensional manifold M for a statistic $\widehat{\phi(x_t)}$. We may as well assume that it is minimal, i.e. has minimal dim(M). The α and $\beta_1,\ldots,\beta_p$ in (1.10) are vectorfields on M. Let $V(M)$ denote the Lie algebra of smooth vectorfields on M. Then the BC principle states the following

6.1. BC principle. If (1.10)-(1.11) is a minimal filter for a statistic then $\mathfrak{L}\mapsto\alpha$, $h_1\mapsto\beta_1,\ldots,h_p\mapsto\beta_p$ defines an antihomomorphism of Lie algebras from $EL(\Sigma)$ into $V(M)$.

Here "anti" means the following: if $\phi: L_1\to L_2$ is a map of vectorspaces from the Lie-algebra L_1 to the Lie-algebra L_2, it is called an antihomomorphism of Lie-algebras if $\phi([A,B]) = -[\phi(A),\phi(B)]$ for all $A,B\in L_1$.

6.2. Example. Consider again the simplest nonzero linear system (2.5). It is linear so there is the Kalman-Bucy filter for the conditional state $\hat{x}$. This filter is

$$dP_t = (1-P_t^2)dt, \quad dm_t = P_t(dy_t-m_t dt). \tag{6.3}$$

So the two vectorfields α and β of the filter are respectively

$$\alpha = (1-P^2)\frac{\partial}{\partial P} - Pm\frac{\partial}{\partial m}, \quad \beta = P\frac{\partial}{\partial m}. \tag{6.4}$$

A simple calculation shows $[\alpha,\beta]=\frac{\partial}{\partial m}$, and it is now indeed a simple exercise to show that $\frac{1}{2}\frac{d^2}{dx^2}-\frac{1}{2}x^2\mapsto\alpha$, $x\mapsto\beta$, induces an antimorphism of Lie-algebras. (It also induces a homomorphism, but that is an accident which happens for linear systems (1.12)-(1.13) if the drift term Ax is absent).

A feeling of why the BC principle should be true can be generated as follows. Think for the moment of two automata with given initial state and with outputs (Moore automata), which, when fed the same string of input data, produce exactly the same string of output data. Suppose the second automaton is minimal. Then it is wellknown (and easy to prove by constructing the minimal automaton from the input-output data) that there is a homomorphism of the subautomaton of the first consisting of the states reachable from the initial state to the second automaton; this homomorphism so to speak makes visible that the two machines do the same job. A similar theorem holds for initialized finite dimensional systems [Sussmann [34]), in particular for systems of the form

$$\dot{x} = \alpha(x) + \sum_{i=1}^{m} \beta_i(x)u_i, \; y = \gamma(x) \tag{6.5}$$

Here the picture produced by theorem is the following commutative diagram

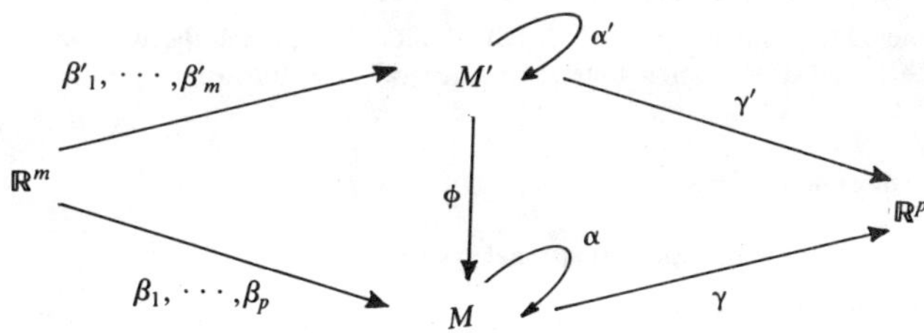

(The theorem asserts the existence of a differentiable map ϕ defined on the reachable from x'_0 subset of M' which makes the diagram commutative. This in particular implies that $d\phi$ takes the vectorfields $\alpha', \beta'_1, ..., \beta'_m$ into $\alpha, \beta_1, ..., \beta_m$ respectively, and, ϕ being a differentiable map, $d\phi$ induces a homomorphism from the Lie algebra generated by $\alpha', \beta'_1, ..., \beta'_m$ to $V(M)$.

In the case of the BC principle we also have two "machines" which do the same job: one is the postulated minimal filter, the other is the infinite dimensional machinge given by the DMZ-equation (2.2) and the ouput map (2.4). So we are in a similar situation as above but with M' infinite dimensional. A proof in this case follows from considerations of Hijab [20].

The fact that in the case of the BC-principle we get an antihomomorphism arises from the following. Given a linear space V and an operator A on it we can define a (linear) vectorfield on V by assigning to $v \in V$ the tangent vector Av. (So we are considering the equation $\dot{v} = Av$.) This defines an anti-isomorphism of the Lie algebra of operators on V to the Lie algebra of linear vectorfields on V.

What about a converse to the BC principle? I.e. suppose that we have given an antihomomorphism of Lie-algebras $EL(\Sigma) \rightarrow V(M)$ into the vectorfields of some finite dimensional manifold. Does there correspond a filter for some statistic of Σ. Just having the homomorphism is clearly insufficient. There are also explicit counterexamples. This is understandable for in any case we completely ignored the output aspect when making the BC-principle plausible. This is not trivial contrary to what the diagram above may suggest. It is not true that given ϕ and any γ one can take $\gamma' = \gamma \circ \phi$. The problem is that γ' as a function on M' = space of unnormalized densities is of a very specific type cf. (2.4).

Even apart from that things are not guaranteed. What we need of course is a ϕ making the left half of the diagram above commutative. Then, if $m' \in M'$ is going to the mapped on $m \in M$, obviously the isotropy subalgebra of $EL(\Sigma)$ at m' will go into the isotropy subalgebra of $V(M)$ at m.

For the case of finite dynamical systems there are positive results of Krener [21] stating that in such a case this extra condition is also sufficient to guarantee the existence of ϕ locally.

The whole cleariy relates to seeing to what extend a manifold can be recovered from its Lie algebra of vectorfields (via its maximal subalgebras of finite codimension) and whether differentiable maps can be recovered from the map between Lie-algebras they induce. This question has been examined by Pursell-Shanks [30].

A more representation theoretic way of looking at things is as follows. Both $EL(\Sigma)$ and $V(M)$ come with a natural representation on the space of functionals on M' and the space of functions on M respectively. If there were a ϕ as in the diagram above ϕ would also induce a map between these representation spaces compatible with the homomorphism of Lie algebras. That therefore is clearly a necessary condition. This way of looking at things contains the isotropy subalgebra condition and also contains output function aspects. Thus the total picture regarding a converse to the BC-principle is not unpromising but nothing is established.

Except for one quite positive aspect. If $EL(\Sigma)$ is finite dimensional, the Wei-Norman equations practically define the filter, for small time in the general case, for arbitrary time in the solvable case.

7. Examples of estimation algebras

7.1. The cubic sensor. This is the one dimensional system

$$dx_t = dw_t, \quad dy_t = x_t^3 dt + dv_t \tag{7.2}$$

and it is about the simplext nonlinear system imaginable. Its estimation Lie algebra is generated by $\frac{1}{2}\frac{d^2}{dx^2} - \frac{1}{2}x^6$, x^3.

7.3. Theorem. (Hazewinkel-Marcus [17]). EL(cubic sensor) $= W_1$, where $W_1 = \mathbb{R}\langle x, \frac{d}{dx}\rangle$ is the Lie algebra of the differential operators (any order, zero included) with polynomial coefficients.

7.4. Example. $dx_1 = dw$, $dx_2 = x_1^2 dt$; $dy_1 = x_1 dt + dv_1$, $dy_2 = x_2 dt + dv_2$. In this case the estimation Lie algebra is $W_2 = \mathbb{R}\langle x_1, x_2, \frac{\partial}{\partial x_1}, \frac{\partial}{\partial x_2}\rangle$, the Lie algebra of all differential operators in two variables with polynomial coefficients.

7.5. Example. $dx_1 = dw$, $dx_2 = x_1^2 dt$, $dy = x_1 dt + dv_1$. In this case the estimation Lie algebra has a basis A, B_i, C_i, D_i $i = 1,2,\ldots$ with the commutation relations $[A, B_i] = C_i$, $[A, C_i] = B_1 + 2B_{i+1}$, $[B_i, C_j] = -D_{i+j}$ and all other commutation relations between basis elements are zero. Note that in this case the Lie algebra is infinite dimensional but has many ideals $\mathfrak{A}_i$ such that $L / \mathfrak{A}_i$ is finite dimensional.

7.6. Example. $dx = dw_1 + \epsilon x dw_2$, $dy = x dt + dv$. Here again $EL = W_1$.

It has become clear that as a rule estimation algebras tend to be infinite dimensional (except in the linear case: then EL(linear system) has dimension $2n+2$ if the linear system is completely reachable and observable); it has also become noticeable that the Weyl-Heisenberg algebras or Weyl algebras W_n have a tendency to appear very often.

7.7. Conjecture. Consider systems (1.1)-(1.2) with polynomial f,G,h. Then generically, i.e. for almost all f,G,h, the estimation algebra will be W_n.

8. The Segal-Shale-Weil representation and all Kalman-Bucy filters

8.1. The linear systems Lie-algebra ls_n. Consider all differential operators in n indeterminates with polynomial coefficients

$$D = \Sigma c_{\alpha\beta} x^{\alpha} \frac{\partial^{\beta}}{\partial x^{\beta}} \tag{8.2}$$

where $\alpha = (\alpha_1,...,\alpha_n)$, $\beta = (\beta_1,...,\beta_n)$ are multiindices $\alpha_i,\beta_j \in \mathbb{N} \cup \{0\}$. Consider those D which are of total degree $\leqslant 2$; i.e. such that $|\alpha|+|\beta|>2 \Rightarrow c_{\alpha\beta}=0$ where $|\alpha| = \alpha_1 + \cdots + \alpha_n$. As is readily verified these form a finite dimensional Lie algebra (under the commutator product $[D_1,D_2] = D_1D_2 - D_2D_1$) of dimension $2n^2+3n+1$. A basis is

$$1; x_1,...,x_n; \frac{\partial}{\partial x_1},...,\frac{\partial}{\partial x_n}; \frac{\partial}{\partial x_i \partial x_j}, i,j = 1,...,n; x_ix_j, i,j = 1,...,n; x_i\frac{\partial}{\partial x_j}, i,j = 1,...,n. \tag{8.3}$$

The operators of total degree $\leqslant 1$ form a subalgebra h_n (basis: $1; x_1,...,x_n; \partial/\partial x_1,...,\partial/\partial x_n$) which is in fact an ideal. The quotient is isomorphic to the symplectic algebra sp_n of all real $2n \times 2n$ matrices M such that

$$MJ + JM^T = 0, \ J = \begin{pmatrix} 0 & I_n \\ -I_n & 0 \end{pmatrix}. \tag{8.4}$$

The isomorphism is given by

$$E_{i,n+j} + E_{j,n+i} \mapsto x_ix_j; \ E_{n+i,j} - E_{n+j,i} \mapsto \frac{\partial^2}{\partial x_i \partial x_j}; \tag{8.5}$$

$$E_{i,j} - E_{n+j,n+i} \mapsto x_i \frac{\partial}{\partial x_j} + \tfrac{1}{2}\delta_{ij}; \ i,j = 1,...,n.$$

Here $E_{i,j}$ is the matrix with a 1 at spot (i,j) and 0 everywhere else; these linear combinations of the $E_{i,j}$ form a basis of sp_n; this isomorphism exhibits sp_n as a subalgebra complementary to h_n; i.e. as a Levi-factor for the short exact sequence $0 \to h_n \to ls_n \to sp_n \to 0$).

8.6. The oscillator representation. There is a famous representation of sp_n which occurs in the framework of symmetries of boson fields (Shale, Segal), in algebraic number theory (Weil), and a multitude of other places, known variously as the Segal-Shale-Weil representation or the oscillator representation. One way to obtain it is as follows. Let H_n denote the Heisenberg group, $H_n = \mathbb{R}^n \times \mathbb{R}^n \times S^1$ with multiplication

$$(x,y,z)(x',y',z') = (x+x', y+y', e^{-2\pi i <x,y'>} zz') \tag{8.7}$$

where $< \, , \, >$ denotes the standard scalar product on $\mathbb{R}^n$. The Lie algebra of H_n is of course $h_n = \mathbb{R}^n \times \mathbb{R}^n \times \mathbb{R}$. And the Lie-bracket of h_n can be interpreted as giving (and given by) a bilinear form $\mathbb{R}^{2n} \times \mathbb{R}^{2n} \to \mathbb{R}$ defined by the matrix J, cf. (8.4) above. Thus the Lie group Sp_n of sp_n can be seen as a group of automorphism of h_n and H_n which is moreover the identity on the centre $S^1 \subset H_n$. Let ρ be the standard Schrödinger representation of H_n in $L^2(\mathbb{R}^n)$

$$\begin{aligned} &(x,0,0) \to M_x, \ (M_xf)(x') = e^{2\pi i <x,x'>} f(x'), \ f \in L^2(\mathbb{R}^n) \\ &(0,y,0) \to T_y, \ (T_yf(x') = f(x'-y), \ f \in L^2(\mathbb{R}^n) \\ &(0,0,z) \to S_z, \ (S_zf)(x') = z f(x'), \ f \in L^2(\mathbb{R}^n) \end{aligned} \tag{8.8}$$

Now let $g \in Sp_n$ be seen as a group of automorphisms of H_n. Then $h \mapsto \rho(g(h))$ is another irreducible representation of H_n with the same central character. So by the Stone-von Neumann theorem there is an $\omega(g)$ intertwining them, i.e. such that

$$\omega(g)\rho(h)\omega(g)^{-1} = \rho(g(h)). \tag{8.9}$$

These $\omega(g)$ are unique up to scalar factors and therefore define a projective representation of Sp_n. The factors can be fixed up to define a representation of the two-fold covering $\tilde{Sp}_n$ of Sp_n. This is the Segal-Shale-Weil representation.

8.10. All Kalman-Bucy filters. Now consider something apparently totally unrelated, namely the DMZ-filtering-equation (2.2) for a linear dynamical system

$$dx = Axdt + Bdw,\ dy = Cxdt + dv,\ x \in \mathbb{R}^n,\ w \in \mathbb{R}^m,\ y, v \in \mathbb{R}^p. \tag{8.11}$$

It is a trivial remark that the operators occurring in (2.2) are all in ls_n in this case. And in fact the Lie algebra generated by them will consist of the second order operator $\mathfrak{L}$ and a subalgebra of h_n stable under $\mathfrak{L}$. In most cases, to be precise in the case that the system (A,B,C) is completely reachable and completely observable, this will be all of h_n, giving us generically an estimation algebra of dimension $2n+2$ which is a subalgebra of ls_n, which has dimension $2n^2+3n+1$.

The Kalman-Bucy filter defines by BC principle an antihomomorphism of this Estimation Lie algebra $EL(A,B,C)$ into the vector Lie algebra of vector fields $V(\mathbb{R}^N)$, $N = n + \frac{1}{2}n(n+1)$.

8.12. Theorem [16]. For varying (A,B,C) these anti-homomorphisms fit together to define a antihomomorphism of all of ls_n into $V(\mathbb{R}^N)$ with as kernel the centre $\mathbb{R}1$. This representation can be lifted to one on $V(\mathbb{R}^{N+1})$ which is faithful.

The explicit formulas are as follows. Interpret a point $x \in \mathbb{R}^{N+1}$ as a triple $x = (c,m,P)$ consisting of a scalar c, an n-vector m, and a symmetric $n \times n$ matrix P. The antihomomorphism is then given by

$$1 \to \frac{\partial}{\partial c} \tag{8.13}$$

$$x \to m_i \frac{\partial}{\partial c} + \sum_t P_{it} \frac{\partial}{\partial m_t} \tag{8.14}$$

$$\frac{\partial}{\partial x_i} \to -\frac{\partial}{\partial x_i} \tag{8.15}$$

$$\begin{aligned} x_i x_j \to & (m_i m_j + P_{ij}) \frac{\partial}{\partial c} + \sum_t (m_i P_{jt} + m_j P_{it}) \frac{\partial}{\partial m_t} \\ & + \sum_{s,t} P_{is} P_{jt} \frac{\partial}{\partial P_{st}} + \sum_t P_{it} P_{jt} \frac{\partial}{\partial P_{tt}} \end{aligned} \tag{8.16}$$

$$x_i \frac{\partial}{\partial x_j} \to -m_i \frac{\partial}{\partial m_j} - \delta_{ij} \frac{\partial}{\partial c} - P_{ij} \frac{\partial}{\partial P_{jj}} - \sum_t P_{it} \frac{\partial}{\partial P_{jt}} \tag{8.17}$$

$$\frac{\partial^2}{\partial x_i \partial x_j} \to \frac{\partial}{\partial P_{ij}} \text{ if } i \neq j,\quad \frac{\partial^2}{\partial x_i^2} \to 2\frac{\partial}{\partial P_{ii}}. \tag{8.18}$$

Inversely these formulas can be checked directly to give an antihomomorphism of Lie algebras and this thus verifies the BC principle for linear dynamical systems and also for families of such depending on a parameter.

Changing all the minus signs in (8.17) and (8.15) into plus signs gives a faithful homomorphism of ls_n into $V(\mathbb{R}^{N+1})$.

Restricting this homomorphism to sp_n as given by (8.5) then defines a homomorphism of sp_n into $V(\mathbb{R}^{N+1})$.

The final remark is that this realization of sp_n as a Lie algebra of vector fields on $\mathbb{R}^{N+1}$ has much to do with the Segal-Shale-Weil representation. The precise statement is that the mapping

$$(c,m,p) \rightarrow \exp(c + <2\pi i m, x> - 2\pi^2 P(x)) \in L^2(\mathbb{R}^n) \tag{8.19}$$

(where $P(x)$ is the quadratic form defined by the symmetric matrix P; note that apart from a scaling factor this is the normal distribution with mean m and covariance P) linearizes the vectorfields in the image of) ls_n in $V(\mathbb{R}^{N+1})$ and switching from a linear vectorfield to the operator which defines it then defines a representation of $ls_n \supset sp_n$. This is another real form of the Segal-Shale-Weil representation meaning that after tensoring with $\mathbb{C}$ (=extending scalars to the complexes), they become isomorphic.

9. W_n and $V(M)$.

We have seen that the Weyl-Heisenberg algebra $W_n = \mathbb{R}<x_1,...,x_n; \partial/\partial x_1,...,\partial/\partial x_n>$ of all differential operators with polynomial coefficients often occurs in filtering problems, i.e. as an Estimation Lie algebra. Given the BC-principle it is therefore of interest to know something about its relations with another class of infinite dimensional Lie algebras, viz the Lie algebras $V(M)$ of smooth vectorfields on a finite dimensional manifold. The algebra W_n has a one-dimensional centre $\mathbb{R}.1$ consisting of the scalar multiples of the identity operator.

9.1. Theorem (Hazewinkel-Marcus [17]). Let α: $W_n \rightarrow V(M)$ or $W_n/\mathbb{R}.1 \rightarrow V(M)$ be a homomorphism or antihomomorphism of Lie algebras, where M is a finite dimensional manifold. Then $\alpha = 0$.

The original proof of this result ([17]) was long and computational. Another much shorter proof based on the nonexistence of finite dimensional representations of h_n for which 1 gets mapped onto the unit operator has more recently been given by Toby Stafford.

10. The cubic sensor.

Consider again the cubic sensor, i.e. the one-dimensional system

$$dx = dw, \quad dy = x^3 dt + dv \tag{10.1}$$

consisting of Wiener noise, cubically observed with further independent noise corrupting the observations. As noted before (theorem 7.3)

$$EL(\text{cubic sensor}) = W_1. \tag{10.2}$$

Now suppose that we have a finite dimensional filter for some conditional statistic $\widehat{\phi(x_t)}$ of the cubic sensor. By the BC-principle (6.1) if follows that there is an antihomomorphism of Lie-algebras $W_1 = EL$ (cubic sensor) $\overset{\alpha}{\rightarrow} V(M)$. By theorem 9.1 it follows that $\alpha = 0$ and from this it is not hard to see that the only statistics of the cubic sensor for which there exists a finite dimensional exact recursive filter are the constants.

A direct proof of this, which sort of proves the BC-principle in this particular case along the way, is contained in Hazewinkel-Marcus-Sussmann [18].

11. Perturbations and approximations

Let us start with an example. Consider the weak cubic sensor

$$dx = dw, \quad dy = xdt + \epsilon x^3 dt + dv \quad (\Sigma_\epsilon) \tag{11.1}$$

If $\epsilon = 0$, this the simplest nontrivial linear system for which there is the Kalman filter. For $\epsilon \neq 0$ one can prove that $EL(\Sigma_\epsilon) = W_1$ again, [15]. So for all $\epsilon \neq 0$ there is no recursive exact filter for any nonconstant statistic. Yet it is hard to believe that for small ϵ the Kalman-Bucy filter would not do a good job of first approximation. A question thus arises whether the estimation Lie algebra also has things to say about approximate filters. In this section and the following ones I shall argue that it does.

The first observation is as follows. If one actually commutes the two generators $\frac{1}{2}d^2/dx^2 - \frac{1}{2}(x+\epsilon x^3)^2$, $(x+\epsilon x^3)$ repeatedly of course eventually all the basis elements of W_1 appear. But they appear with higher and higher powers of ϵ and the ϵ-degree grows faster than the degree in (α) and (β) of the $x^\alpha \partial^\beta / \partial x^\beta$.

A precise version of this is as follows. Consider the two generators just listed as operators over the ring $\mathbb{R}[\epsilon]$ (or $\mathbb{R}[[\epsilon]]$), i.e. consider ϵ as an extra variable. Then it makes sense to consider

$$EL(\Sigma_\epsilon)\otimes_{\mathbb{R}[\epsilon]}\mathbb{R}[\epsilon]/\epsilon^n =: EL(\Sigma_\epsilon)\mathrm{mod}(\epsilon^n). \tag{11.2}$$

This simply amounts to setting $\epsilon^m = 0$ for $m \geqslant n$ whenever it appears. The set of all $\epsilon^i x^j d^k / dx^k$ with $\epsilon \geqslant n$ form an ideal in $\mathbb{R}[\epsilon]<x, d/dx>$, so this makes sense. Now observe

11.3. Proposition [15]. The Lie-algebras $EL(\Sigma_\epsilon)$ mod ϵ^n are finite-dimensional for all n.

As an example $EL(\Sigma_\epsilon)$ mod ϵ^2 turns out to be 14 dimensional with basis

$$\frac{1}{2}\frac{d^2}{dx^2} - \frac{1}{2}x^2 - \epsilon x^4,\ x,\ \epsilon x^3,\ \frac{d}{dx},\ 1,\ \epsilon,\ \epsilon x^2\frac{d}{dx},\ \epsilon x,$$

$$ex\frac{d}{dx},\ \epsilon\frac{d^2}{dx^2},\ \epsilon\frac{d}{dx},\ \epsilon\frac{d^3}{dx^3},\ \epsilon x\frac{d^2}{dx^2},\ \epsilon x^2.$$

This is a general phenomenon.

11.4. Theorem, [15]. Let Σ_ϵ be a system of the form

$$dx = (Ax + \epsilon P_A(x))dt + (B + \epsilon P_B(x))dw,\ dy = (Cx + \epsilon P_C(x))dt + dv \tag{11.5}$$

where P_A, P_B, P_C are polynomial vector and matrix valued functions of the approximate dimensions. Then $EL(\Sigma_\epsilon)$ mod (ϵ^n) is finite dimensional for all n. It is also solvable.

In [15] this is proved for the case $P_B = P_A = 0$. The proof generalizes immediately (simply give ϵ a negative enough degree to make degree decreasing all terms in the generators of $EL(\Sigma_\epsilon)$ in which ϵ appears (both $x_i, \partial/\partial x_j$ are given degree 1 in this argument).)

The next obvious question is: do these "finite dimensional quotients of $EL(\Sigma_\epsilon)$" actually compute anything, do they correspond to filters for some statistic? In the case of the weak cubic sensor this is (11.1) easy to answer. Consider the unnormalized conditional density $\rho(x,t,\epsilon)$ and (formally) expand it as a power series in ϵ

$$\rho(x,t,\epsilon) = \rho_0(x,t) + \epsilon\rho_1(x,t) + \epsilon^2\rho_2(x,t) + \ldots \tag{11.6}$$

Then $EL(\Sigma_\epsilon)$ mod (ϵ^n) corresponds to the first n coefficients $\rho_0(x,t), \ldots, \rho_{n-1}(x,t)$, and via Wei-Norman theory actually computes them. This is generally true, also in the setting of theorem 11.4. In the case of the weak cubic sensor (11.6) actually converges (for small ϵ). That, it appears, is not generally true. But it is still true that (11.6) gives an asymptotic expansion (Blankenschip-Liu-Marcus [4]). The Lie algebras being solvable one can of course implement these approximate filters, using the Wei-Norman technique. This was done in [4] and also the results were compared with the extended Kalman filter (EKF). The zero-th order approximation (of course) performed worse than EKF but the first order approximation performed better!

These Lie algebras $EL(\Sigma_\epsilon)$ tend to become large rapidly and to actually produce the, say FORTRAN, code is a long, but mechanical, job, prone to errors. Even the simplest nontrivial case needs several pages of densely written code. It is thus natural to try to let the computer do the job itself and in this way these ideas and techniques are being implemented in an expert system which is a joint effort of INRIA and the Department of Electrical Engineering of the University of Maryland (cf. Blankenschip [3]; the system also contains many other facets of stochastic control, filtering and optimization).

From the point of view developed in section 4 above the fact that calculating $\rho_0(x,t),...,\rho_{n-1}(x,t)$ corresponds to $EL(\Sigma_\epsilon)$ mod (ϵ^n) can be understood as follows. Choosing a basis suitably the remarks made in section 4 about the compatibility of Wei-Norman theory with quotients say that $\rho(x,t,\epsilon)$ admits an "expansion"

$$\rho(x,t,\epsilon) = e^{g_1(t)A_1}e^{g_2(t)A_2}...e^{g_n(t)A_n}...\rho_0(x) \tag{11.7}$$

with $g_1,...,g_{m(n)}$ where $m(n)=\dim(EL(\Sigma_\epsilon) \bmod (\epsilon^n))$ depending only on $EL(\Sigma_\epsilon)$ mod (ϵ^n). The operators A_i in (11.7) involve higher and higher powers of ϵ. Writing out the exponentials one recovers (11.6). (And this point of view also strongly suggests (because also higher derivatives appear in the A_i) that the best one can hope for in general is an asymptotic expansion.)

12. The profinite dimensional case.

A Lie-algebra L is said to be profinite dimensional if there is a sequence of ideals $L \supset \mathfrak{A}_1 \supset \mathfrak{A}_2 \supset \mathfrak{A}_3 \supset ...$ such that

$$\dim L / \mathfrak{A}_i < \infty \text{ for all } i \tag{12.1}$$

$$\bigcap_i \mathfrak{A}_i = \{0\}. \tag{12.2}$$

Suppose the estimation Lie algebra $EL(\Sigma)$ has this property. Then again, as in the previous section, one can write an expansion

$$\rho(x,t) = e^{g_1A_1}e^{g_2A_2}...e^{g_nA_n}...\rho_0(x) \tag{12.3}$$

and consider the possible approximants

$$\rho^{(n)}(x,t) = e^{g_1A_1}e^{g_2A_2}...e^{g_nA_n}\rho_0(x). \tag{12.4}$$

Using, again, that the equations for $g_1,...,g_n$; $n=n(m)=\dim L / \mathfrak{A}_m$ do not depend on $g_{n+1},...$. Abstractly, there is no immediate reason to expect the higher $e^{g_kA_k}$ to be small, though one would expect this to the case in the majority of the interesting cases, even in more general cases than this, as I shall argue below in section 14.

Profinite dimensional estimation Lie algebras occur frequently. Consider systems

$$dx = f(x)dt + G(x)dw_t, \quad dy = h(x)dt + dv \tag{12.5}$$

with the additional assumptions that f,G and h are analytic (totally around zero) and that $f(0)=G(0)=0$.

12.6. Theorem [17]. Under the assumptions made immediately above $EL(\Sigma)$ is profinite dimensional.

If one adds the condition that $h(0)\neq 0$ (which surely does no harm; removing a known constant from the observation equation is a triviality) the resulting estimation Lie algebra is even solvable (meaning that all the quotients $L / \mathfrak{A}_i$ are solvable).

Another case of a profinite dimensional estimation Lie algebra (different from the class of theorem 12.6, the identification case to be treated below, and the perturbation case of section 11 above) is example 7.5. As a rule one should probably not expect that "the statistic calculated by $L / \mathfrak{A}_i$" of a system whose estimation Lie algebra happens to be profinite dimensional, is easily interpretable (recognizable) as the statistic of an interesting quantity. In the case of example 7.5 this is however the case, (Liu-Marcus [23]).

13. Identification of linear dynamical systems

Suppose now that we are faced with a somewhat different problem. Namely suppose one has reason to believe, or simply does not know anything better to do, that a given phenomenon, say a time series, is modeled by a linear dynamical system

$$dx = Axdt + Bdw, \quad dy = Cxdt + dv \tag{13.1}$$

Now, however, the coefficients in A,B,C are unknown and also have to be estimated from the observation $y(t)$. That is the system (13.1) has to be identified. It is easy to turn this into a filtering problm by adding the (stochastic) equations

$$dA = 0, \quad dB = 0, \quad dC = 0 \tag{13.2}$$

(or just $dr_{ij} = 0$ whether the r_{ij} run through the coefficients which are unknown, if A,B,C are partly known; for example because of structural considerations). The resulting filtering problem is nonlinear.

13.3 Observation. The estimation Lie algebra of the system (13.1)-(13.2) is a sub-Lie-algebra of the current Lie algebra $ls_n \otimes \mathbb{R}[A,B,C]$ where $\mathbb{R}[A,B,C]$ stands for the ring of polynomials in the indeterminates a_{ij}, b_{kl}, c_{rs}.

A corollary is that these estimation algebras are profinite dimensional. And looking a bit more closely at them, they are solvable [37]. Thus the ideas and considerations of the previous two sections can be brought into play and one can try to do infinite dimensional Wei-Norman theory etc. This is attempted in Krishnaprasad-Marcus-Hazewinkel [37]. In this rather special case it turns out that the higher approximations (the zero-th approximation is simply the family of Kalman-Bucy filters parametrized by A,B,C also discussed in section 8 above) have to do with sensitivity equations: sensitivities of the ouput $y(t)$ with respect to changes in the parameters A,B,C.

As stated above, though, the problem is degenerate and likely to cause all kind of difficulties. The problem is that the conditional density $\rho(x,A,B,C,t)$ will be degenerate because the A,B,C are not uniquely determined by the observations. Indeed if S is an invertible $n \times n$ matrix then the system (13.1) given by the matrices SAS^{-1}, SB, CS^{-1} instead of A,B,C gives exactly the same input-output behaviour. Thus we should really be considering this problem on a suitable quotient space $\{(A,B,C)\} / GL_n$. These quotient spaces as a rule are not diffeomorphic to open sets in some $\mathbb{R}^n$. This is one way in which stochastic systems like (1.1)-(1.2) on nontrivial manifolds naturally arise and it leads to the necessity of finding a DMZ-equation in this more general context. Work in this direction has been done by Ji Dunmu and T.E. Duncan.

Let me add one observation. For the filters giving $\hat{x}, \hat{A}, \hat{B}, \hat{C}$ for problem (13.1)-(13.2) one expects $\hat{x}$ to move fast relative $\hat{A}, \hat{B}, \hat{C}$. Thus it would make sense to consider a system

$$dx = (A_0 + \epsilon A_1)xdt + (B_0 + \epsilon B_1)dw, \quad dy = (C_0 + \epsilon C_1)dt + dv \tag{13.4}$$

$$dA_1 = 0, \quad dB_1 = 0, \quad dC_1 = 0$$

(where A_0, B_0, C_0 are assumed known) and apply the ideas of section 11 above to find optimal directions of change (i.e. the A_1, B_1, C_1).

14. Asymptotic expansions and approximate homomorphisms.

The ideas to be outlined below in this section are still speculative but there are quite a number of positive signs.

First however let me point out that the procedures based on Wei-Norman techniques as described in sections 11 and 12 above clearly indicate that existence, uniqueness and regularity results for solutions of the DMZ-equation have a lot to do with the existence of asymptotic expansions ([2,4]). For regularity results etc. cf. e.g. work of D. Michel, J.-M. Bismut, E. Pardoux, M.

Chaleyat-Maurel, D. Ocone, Th. Kurtz, W.E. Hopkins Jr., H.J. Sussmann a.o. ([25,8,22,2] and references in these papers).

Let us consider a control system of the form

$$\dot{x} = f(x) + \Sigma u_i g_i(x) \tag{14.1}$$

where the f and g_i are vectorfields. To make thinking easier assume that 0 is a stable and asymptotically stable equilibrium for the unforced equation. A system like (14.1) is intended as a model of something and as such one can argue that say the values of $f(x), g_i(x)$ are relatively well known, the values of their (partial) derivatives (w.r.t. the x_i) will be less well known, the second partial derivations are still less well determined etc..

Thus, intuitively, for systems which represent or model real (stable) things one would expect that in many cases the behaviour of (14.1) will depend primarily on the first few terms which appear in the Lie algebra generated by f and the g_i. The higher brackets should matter less and less.

That means that instead of looking at $Lie\{f, g_1, ..., g_m\}$, the Lie algebra generated by $f, g_1, ..., g_m$ as a Lie algebra without further structure, we should look at it as a Lie algebra with a given set of generators and sort of keep track of how often these generators are used to generate further elements of the algebra. (For each time a bracket is taken a differentiation is applied, and thus the higher brackets of the $f, g_1, ..., g_m$ depend only on the deeper parts of the Taylor expansions of $f, g_1, ..., g_m$.)

Personally I would also say that having noises rather than precise deterministic controls u_i would enhance this type of (structural?) stability.

A precise way to keep track of how often the generators are used is to introduce one extra counting indeterminate z and to consider instead of $L = Lie\{f, g_1, ..., g_m\}$ the Lie algebra generated by the vectorfields $\{zf, zg_1, ..., zg_m\}$. This Lie algebra L_z is topologically nilpotent, i.e. if $L_z^{(n)} = [L_z, L_z^{(n-1)}]$, $L_z^{(0)} = L_z$, then $\cap L_z^{(m)} = \{0\}$. And a homomorphism $L_z \to V(M)$ into the vectorfields on M with kernel $L_z^{(n)}$ precisely means "respecting the structure of the Lie algebra L up to brackets of order n". All this is very much related to the ideas of nilpotent approximation as introduced by Stein, Rothschild, Goodman and Rockland, [32,14,31] in the study of hypoellipticity and taken up by Crouch in system theory [10].

Thus in filtering theory it would seem natural to look at the Lie algebra of operators $EL_z(\Sigma)$ generated by the operators

$$z_0\mathfrak{L},\ z_1h_1, ..., z_ph_p$$

where the $z_0, z_1, ..., z_p$ are additional variables (so as to give, if desired, certain observations more weight than others and to be able to set certain of them, especially z_0, equal to 1). The idea would be then to study the filters produced by Wei-Norman type techniques for the various finite dimensional quotients and to see whether this produces viable expansions.

15. Removing outliers

A final idea in much the same spirit as before is the following. Suppose we are again dealing with a system

$$dx = f(x)dt + G(x)dw, \quad dy = h(x)dt + dv. \tag{15.1}$$

Suppose also to make thinking easier that the thing is more or less stable so that x tends to remain in some bounded partion of $\mathbb{R}^n$ (f asymptotically stable) and maybe suppose also that h is proper, so that large y observations are exceedingly rare and should probably be discounted. Suppose that $e^{-\|x\|^2}$ is differential algebraically independent of f, G, h. This is for example this case if f, G, h are polynomial and also if they are of compact support. In other cases other functions with similar properties can presumably be found. Now instead of (15.1) consider the modified system

$$dx = f(x)dt + G(x)dw, \quad dy = e^{-a\|x\|^2} h(x)dt + dv \tag{15.2}$$

where $a > 0$ is a small parameter. Note that the only thing which (15.2) does with respect to (15.1) is to discount large y observations.

Now consider the estimation Lie algebra of the sytem (15.2).

15.3. Theorem. If $e^{-a\|x\|^2}$ is differentially algebraically independent of f,G,h then the estimation Lie algebra of (15.2) is pro-finite dimensional and solvable. To be more precise it is finite dimensional and solvable mod $(a^i e^{-ja\|x\|^2}, i+j \geqslant n)$ for all n.

Thus the yoga of the previous sections can again be applied and the behaviour of the resulting filters as a goes to zero could be studied.

References

[1] J.S. Baras, Group invariance methods in nonlinear filtering of diffusion processes, In: [19], 565-572.

[2] J.S. Baras, G.L. Blankenschip, W.E. Hopkins Jr., Existence, uniqueness and asymptotic behaviour of solutions to a class of Zakai equations with unbounded coefficients, IEEE Trans. AC-28 (1983), 203-214.

[3] G.L. Blankenschip, Lecture at MTNS'85, Stockholm, To appear North Holland Publ. Cy.

[4] G.L. Blankenschip, C.-H. Liu, S.I. Marcus, Asymptotic expansions and Lie algebras for some nonlinear filter problems, IEEE Trans. AC-28 (1983), 787-797.

[5] R.W. Brockett, Remarks on finite dimensional nonlinear estimation, In: C. Lobry (ed.), Analyse des systèmes, Astérisque 76 (1980), Soc. Math. de France.

[6] R.W. Brockett, J.M.C. Clark, The geometry of the conditional density equation In: O.L.R. Jacobs et al. (eds), Analysis and optimization of stochastic systems, New York, 1980, 299-309.

[7] R.S. Bucy, J.M.F. Moura (eds), Nonlinear stochastic problems, Reidel, 1983.

[8] J.M.C. Clark, The design of robust approximations to the stochastic differential equations of nonlinear filtering, In: J.K. Skwirzynski (ed.), Communication systems and random process theory, Sijthoff & Noordhoff, 1978.

[9] M. Chaleyat-Maurel, D. Michel, Hypoellipticity theorems and conditional laws, Z. Wahrsch. und verw. Geb. 65 (1984), 573-597.

[10] P.E. Crouch, Solvable approximations to control systems, SIAM J. Control and Opt. 32 (1984), 40-54.

[11] M.H.A. Davis, Pathwise nonlinear filtering, In [19], 505-528.

[12] M.H.A. Dsavis, S.I. Marcus, An introduction to nonlinear filtering, In [19], 565-572.

[13] T.E. Duncan, Probability densities for diffusion processes with applications to nonllinear filtering theory, Ph. D. thesis, Stanford, 1967.

[14] R.W. Goodman, Nilpotent Lie groups, structure and applications to analysis, LNM 562, Springer, 1976.

[15] M. Hazewinkel, On deformations, approximations and nonlinear filtering, Systems Control Lett. 1 (1982), 29-62.

[16] M. Hazewinkel, The linear systems Lie algebra, the Segal-Shale-Weil representation and all Kalman-Bucy filters, J. Syst. Sci. & Math. Sci. 5 (1985), 94-106.

[17] M. Hazewinkel, S.I. Marcus, On Lie algebras and finite dimensional filtering, Stochastics 7 (1982), 29-62.

[18] M.Hazewinkel, S.I. Marcus, H.J. Sussmann, Nonexistence of finite dimensional filters

for conditional statistics of the cubic sensor problem, Systems Control Lett. 3 (1983), 331-340.

[19] M. Hazewinkel, J.C. Willems (eds), Stochastic systems: the mathematics of filtering and identification and applications, Reidel, 1981.

[20] O.B. Hijab, Finite dimensional causal functionals of brownian motion, In [6], 425-436.

[21] A.J. Krener, On the equivalence of control systems and the linearization of nonlinear systems, SIAM J. Control 11 (1973), 670-676.

[22] Th.G. Kurtz, D. Ocone, A martingale problem for conditional distributions and uniqueness for the nonlinear filtering equations, Lect. Notes Control and Inf. Sci. 69 (1985), 224-235.

[23] C.-H. Liu, S.I. Marcus, The Lie algebraic structure of a class of finite dimensional nonlinear filters, In: C.I. Byrnes, C.F. Martin (eds), Algebraic and geometric methods in linear systems theory, Amer. Math. Soc., 1980, 277-297.

[24] S.I. Marcus, Algebraic and geometric methods in nonlinear filtering, SIAM J. Control Opt. 22 (1984), 817-844.

[25] D. Michel, Régularité des lois conditionelles en théorie du filtrage non-linéaire et calcul des variations stochastiques, J. Funct. Anal. 14 (1981), 8-36.

[26] S.K. Mitter, On the analogy between mathematical problems of non-linear filtering and quantum phisics, Ric. di Automatica 10 (1980), 163-216.

[27] S.K. Mitter, Nonlinear filtering and stochastic mechanics, In: [19], 479-504.

[28] R.E. Mortensen, Optimal control of continuous time stochastic systems, Ph.D. thesis, Berkeley, 1966.

[29] E. Pardoux, D. Talay, Discretization and simulation of stochastic differential equations, Acta Appl. Math. 3 (1982), 182-203.

[30] L.E. Pursell, M.E. Shanks, The Lie algebra of a smooth manifold, Proc. Amer. Math. Soc. 5 (1954), 468-472.

[31] Ch. Rockland, Intrinsic nilpotent approximation, preprint MIT, LIDS-R-1482, 1985, to appear Acta Appl. Math.

[32] L.P. Rothschild, E.M. Stein, Hypoelliptic differential operators and nilpotent groups, Acta Math. 137 (1976), 247-320.

[33] H.J. Sussmann, Approximate finite dimensional filters for some nonlinear problems, Stochastics 7 (1982), 183-203.

[34] H.J. Sussmann, Existence and uniqueness of minimal realizations of nonlinear systems, Math. Syst. Theory 10 (1977), 349-356.

[35] J. Wei, E. Norman, On the global representation of the solutions of linear differential equations as a product of exponentials, Proc. Amer. Math. Soc. 15 (1964), 327-334.

[36] M. Zakai, On the optimal filtering of diffusion processes, Z. Wahrsch. verw. Geb. 11 (1969), 230-243.

[37] P.S. Krishnaprasad, S.I. Marcus, M. Hazewinkel, Current algebras and the identification problem, Stochastics 11 (1983), 65-101.

RECENT RESULTS AND SOME OPEN QUESTIONS ON SIEGEL'S LINEARIZATION THEOREM OF GERMS OF COMPLEX ANALYTIC DIFFEOMORPHISMS OF C^n NEAR A FIXED POINT

Michael R. HERMAN

Centre de Mathématiques de l'Ecole Polytechnique
F-91128 Palaiseau cedex - France
"U.A. 169 du C.N.R.S."

INTRODUCTION.

We propose to survey some of the classical and recent results on the linearization of germs of analytic diffeomorphisms.

The main point will be the analytic difficulties due to small divisors, and we will concentrate on the case where all eigenvalues have modulus 1. The results are illustrated by many open questions and numerous examples which, for the sake of simplicity, will not be studied in the most general setting. Some new results are stated without complete proofs ; the details will appear elsewhere.

Typeset by $\mathcal{A}_{\mathcal{M}}\mathcal{S}$-TEX

TABLE OF CONTENTS.

Chapter I (§ 1 to § 7) : the problem and the statement of the results ; Siegel's and Brjuno's theorems.

Chapter II (§ 8 to § 12) : the idea of the proof, using majorant series, of Siegel's and Brjuno's theorems ; various remarks.

Chapter III (§ 13 to § 22) : Boundaries of Siegel singular disks.

Chapter IV (§ 23 to § 27) : Some examples in $\mathbf{C}^n$, $n \geq 2$; for some partial genral results, see § 23 and § 27.

Chapter V (§ 28 to § 34) : Center manifolds and their relation to invariant circles of twist maps ; examples.

I would like to thank Raphaël Douady and Jean-Christophe Yoccoz for many discussions and suggestions. I would also like to thank Jean-Christophe Yoccoz for helping me to improve considerably the manuscript, Marie-Jo Lécuyer for typing it marvellously and with great care, and Thomas Ehrhard for his "TEX-nical" assistance.

Chapter I

THE PROBLEM AND THE STATEMENT OF THE RESULTS ; SIEGEL'S AND BRJUNO'S THEOREM

1. Let $f \in (\mathbf{C}[[z_1, \cdots, z_n]])^n$ a germ of formal diffeomorphism of $(\mathbf{C}^n, 0)$. With $z = (z_1, \cdots, z_n)$, we write :

$$f(z) = Az + O(z^2) \,, \tag{1}$$

with $A \in GL(n, \mathbf{C})$, and denote by $\lambda_1, \cdots, \lambda_n$ the eigenvalues of A.

For $k = (k_1, \cdots, k_n) \in \mathbf{N}^n$, we write λ^k for $\lambda_1^{k_1} \cdots \lambda_n^{k_n}$ and $|k|$ for $\sum k_i$. We say that the matrix A satisfies condition (*) if we have :

$$\lambda^k - \lambda_j \neq 0 \tag{*}$$

for any $1 \leq j \leq n$, any $k \in \mathbf{N}^n$ with $|k| \geq 2$.

The following proposition is elementary.

PROPOSITION 1. *If A satisfies condition (*), there exists* a unique *germ of formal diffeomorphism h of $(\mathbf{C}^n, 0)$ of the form :*

$$h(z) = z + O(z^2) \tag{2}$$

which satisfies (formally) :

$$f \circ h(z) = h(Az) \,. \tag{3}$$

2. Example : For $n = 1$, $Az = \lambda z$, with $\lambda \in \mathbf{C}^\star$; then condition (*) means that λ is not a root of unity.

3. Let $g: \mathbf{C} \to \mathbf{C}$ be a holomorphic function (entire function) of the form $g(z) = z + a_2 z^2 + \cdots$ such that :

$$g(z) \not\equiv z \ . \tag{4}$$

For $\lambda \in \mathbf{C}$, let $f_\lambda = \lambda g$. If $\lambda^q = 1$, one has :

$$f_\lambda^q(z) \not\equiv z, \tag{5}$$

so one cannot find h satisfying (3) if (*) is violated.

When one has :

$$\lambda = \lambda_\alpha = \exp(2\pi i \alpha), \quad \alpha \in \mathbf{T}^1 - (\mathbf{Q}/\mathbf{Z}) \ , \quad \mathbf{T}^1 = \mathbf{R}/\mathbf{Z} \ , \tag{6}$$

Proposition 1 applies and there exists $h_\alpha(z) = z + b_2 z^2 + \cdots$ such that (formally) we have $f_{\lambda_\alpha} \circ h_\alpha(\lambda_\alpha z) = h_\alpha(\lambda_\alpha z)$.

The following proposition is easy ([H 7], [L 4], see also [C 7]), [C 10], [C 11]).

PROPOSITION 2. : *There is a G_δ-dense set of $\alpha \in \mathbf{T}^1 - (\mathbf{Q}/\mathbf{Z})$ for which the radius of convergence of h_α is 0.*

In [H 4] we constructed, for a G_δ-dense set of $\alpha \in \mathbf{T}^1 - (\mathbf{Q}/\mathbf{Z})$, rational functions f_α of degree $d \geq 2$ such that $f_\alpha(z) = \lambda_\alpha z + O(z^2)$ at 0 and f_α has a dense orbit on the Riemann sphere. I.N. Baker and P.J. Rippon ([B 1]) showed that, with $g(z) = e^z - 1$, f_{λ_α} has a dense orbit in $\mathbf{C}$ for a G_δ-dense subset of $\alpha \in \mathbf{T}^1 - (\mathbf{Q}/\mathbf{Z})$.

4. STATEMENT OF THE PROBLEM OF CONVERGENCE.

4.1 We suppose in the following that f is a germ of $\mathbf{C}$-analytic diffeomorphism of $\mathbf{C}^n$ at 0, with $f(0) = 0$ and

$$A = Df(0) \quad \text{satisfies (*)} \quad , \tag{7}$$

where $Df(z)$ denotes the complex derivative or tangent map of f at the point $z \in \mathbf{C}^n$.

In the following, analytic will always mean $\mathbf{C}$-analytic.

4.2 One asks whether f can be linearized in a neighbourhood of 0, that is whether the formal conjugacy h defined by Proposition 1 defines a germ of analytic diffeomorphism at 0. When $n = 1$, this problem is also called the Schröder equation ([S 0]).

4.3 It is an easy result due to H. Poincaré ([P 5, t. I, p. XXXVI-CXXIX], see also [K 2], [F 3], [P 4] for references before 1912, and for other references [D 1]), that the problem has a positive answer if (7) holds and :

$$\sup_j |\lambda_j| < 1 \quad \text{or} \quad \sup_j |\lambda_j^{-1}| < 1 \,. \tag{8}$$

A matrix A satisfying (8) is said to be in the Poincaré domain. The proof is elementary, using, for instance, majorant series. It does not require that A is diagonizable. Moreover, if (8) holds, (7) is violated but nevertheless f is formally linearizable, then f is analytically linearizable. When (8) is violated, the matrix A is said to be in the Siegel set, which have non empty interior for $n \geq 2$. For $n = 1$, the Siegel set is $\{\lambda, |\lambda| = 1\}$.

4.4 When (7) holds and :

$$|\lambda_j| \neq 1 \,, 1 \leq j \leq n \,, \tag{7'}$$

then, by Sternberg's theorem ([S 8], [S 9]) and its generalization by Chaperon ([C 4]), f is C^∞-linearizable.

When (7') holds (but not necessarily (7)), f is topologically linearizable by Hartman-Grobman's theorem ([C 4], [P 10]). Note that when (7) and (7') hold but (8) is violated, the germ of conjugacy (as a C^∞-diffeomorphism) is not unique.

When $n = 1$ and $Az = \lambda z$, with $|\lambda| = 1$, then f is analytically linearizable if it is topologically linearizable (see §13 and §23).

4.5 When (7) holds but (8) is violated, the question of analytic linearization is much more delicate. The problem was certainly known by H. Poincaré ([P 5, t. IV, p. 36-59 and especially p. 43]), and probably, when $n = 1$, by many other mathematicians at the end of the last century ([P 4]) ; the question was explicitly asked by Kasner ([K 1]) in 1912. See [W 1], and [C 7] to [C 11] for other historical references.

Proposition 3 shows that arithmetical conditions on the λ_j's are certainly necessary when A is in the Siegel set.

4.6 We assume that A satisfies (*). For $m \in \mathbf{N}$, $m \geq 2$, we define :

$$\Omega(m) = \inf_{\substack{2 \leq |k| \leq m \\ 1 \leq j \leq n}} |\lambda^k - \lambda_j| \,.$$

Condition (*) means that $\Omega(m)$ is non zero for $m \geq 2$. If A is in the Siegel set, one has :

$$\lim_{m \to +\infty} \Omega(m) = 0 \,. \tag{9}$$

This relation, which gives birth to the so called "small divisors", is what makes it difficult to prove the convergence of h when A is in the Siegel set.

4.7 When $n = 1$ and (7) is satisfied, the first published example of a germ f with a non convergent h was given by Pfeiffer in 1917 ([P 3]).

The known cases where, given A, one can find a germ f with a non convergent h are summarized in the following proposition. It follows from a simple and elegant result of Il'yachenko ([I 1]). The divergence of h under the hypothesis (10) was first obtained by H. Cremer ([C 11]) when $n = 1$, and then generalized by Brjuno ([B 6]). The divergence under the hypothesis (11) follows from a result of J.C. Yoccoz ([Y 5]) and requires no other hypothesis on λ_1 than $|\lambda_1| = 1$.

PROPOSITION 3. *Suppose that A satisfies (*) and one of the conditions* (10), *or* (11) *:*

$$\limsup_{m\to+\infty}\left(-\frac{1}{m}\operatorname{Log}\Omega(m)\right) = +\infty\ ; \tag{10}$$

$$A \quad \text{has a Jordan block} \quad \begin{pmatrix}\lambda_1 & 1\\ 0 & \lambda_1\end{pmatrix} \quad \text{with} \quad |\lambda_1| = 1\ ; \tag{11}$$

then there exists an analytic germ f with $f(0) = 0$, $Df(0) = A$, such that the formal conjugacy h defined by Proposition 1 has radius of convergence equal to 0.

Using [I 1], it is sufficient, to prove Proposition 3, to check that the linear operator :

$$\eta \to A\eta - \eta \circ A \tag{12}$$

on the space of holomorphic germs η from $(\mathbf{C}^n, 0)$ to itself which satisfy $\eta(0) = D\eta(0) = 0$, is not surjective. Observe that (12) is the linearization of (3) at $f = A$, $h = \text{id}$.

The non surjectivity of (12) is immediate when (10) holds. The matrices A satisfying (10) form a G_δ-dense subset of the Siegel set.

5. ARITHMETIC CONDITIONS.

(For diophantine approximation, the reader can consult [S]).

5.1 Let $A \in GL(n, \mathbf{C})$ satisfying condition (*).

DEFINITION. *The matrix A satisfies the Brjuno condition if :*

$$\sum_{k=0}^{+\infty} 2^{-k}\operatorname{Log}(\Omega^{-1}(2^{k+1})) < +\infty\ . \tag{13}$$

We refer to [B 6] and [R 4] for numerous equivalent formulations of (13).
If there exists $\beta \geq 0$, $\gamma > 0$ such that :

$$\Omega^{-1}(m) \leq \gamma^{-1} m^{\beta}\ ,\ \forall\, m \geq 2 \tag{14}$$

we say that A satisfies a *diophantine condition (of exponent β).*

When $\beta > n$, almost every n-tuple $(\lambda_1, \cdots, \lambda_n) \in \mathbf{C}^n$ satisfies a diophantine condition of exponent β.

5.2 Let $n = 1$; we write $\lambda = \exp(2\pi i\alpha)$, with $\alpha \in \mathbf{R} - \mathbf{Q}$.

Let $\alpha = a_0 + 1/(a_1 + 1/(a_2 + \cdots))$ the continued fraction of α, and $\left(\frac{p_n}{q_n}\right)_{n \geq -2}$ the convergence of α : $p_{-2} = 0$, $q_{-2} = 1$, $p_{-1} = 1$, $q_{-1} = 0$, and $p_n = a_n p_{n-1} + p_{n-2}$, $q_n = a_n q_{n-1} + q_{n-2}$ for $n \geq 0$; one alway has

$$q_n \geq 2^{\frac{n-1}{2}} \, , \, n \geq 1 \, . \tag{15}$$

The continued fraction of α determines its approximation by rationals (cf. [H 3, ch. V]) ; one has :

$$(a_{n+1} + 2)^{-1} q_n^{-2} \leq |\alpha - \frac{p_n}{q_n}| \leq a_{n+1}^{-1} q_n^{-2} \, , \, n \geq 0 \, . \tag{16}$$

The Brjuno condition on λ is equivalent to :

$$\sum_{n=0}^{+\infty} q_n^{-1} \operatorname{Log} q_{n+1} < +\infty \, , \tag{17}$$

and we then say that α is a Brjuno number, writing B for the set of Brjuno numbers. Condition (10) is equivalent to $\limsup_{n \to +\infty} q_n^{-1} \operatorname{Log} q_{n+1} = +\infty$. A diophantine condition of exponent β for λ is equivalent to $q_{n+1} = O(q_n^\beta)$ (so $\beta \geq 1$) ; we write DC for the set of $\alpha \in \mathbf{R}$ such that λ satisfies a diophantine condition.

We say that α is of constant type, and write $\alpha \in CT$, if λ satisfies a diophantine condition of exponent 1.

The set $L = \mathbf{R} - (DC \cup \mathbf{Q})$ is by definition the set of *Liouville numbers* ; it is a G_δ-dense subset of $\mathbf{R}$ and has Haussdorff dimension 0 (hence a fortiori Lebesgue measure 0).

5.3 When $n \geq 2$, nothing as simple as continued fractions exists. The construction of matrices A which satisfy (13) but not (14), with all eigenvalues of modulus 1, is left to the reader.

When $n = 1$, one has $\alpha \in DC$ (resp. $\alpha \in B$) if and only if $-\alpha \in DC$ (resp. $-\alpha \in B$). For $n \geq 2$, however, we only take into account for (13) or (14) the quantities $|\lambda^k - \lambda_j|$, with $k \in \mathbf{N}^n$ ($|k| \geq 2$) ; this is different from considering all quantities $|\lambda^k - \lambda_j|$, with $k \in \mathbf{Z}^n$, and $|k| = \sum |k_j| \geq 2$.

6.

THEOREM 1. *Let f be a germ of analytic diffeomorphisms of $(\mathbf{C}^n,0)$, of the form :*

$$f(z) = Az + O(z^2)\ ,\ A \in GL(n,\mathbf{C}).$$

We assume that A is diagonalizable, and satisfies (*) and Brjuno condition (13). Then the formal conjugacy h of Proposition 1 defines a germ of analytic diffeomorphism in a neighbourhood of 0.

The above theorem was first proved by C.L. Siegel in 1942 ([S 4]), when $n = 1$ and A satisfies a diophantine condition (i.e. (14)). Still under the diophantine condition (14), it was generalized by Siegel ([S 5, Band III, p. 178-187]) to vector fields in $\mathbf{C}^n$ near singular a point, and by S. Sternberg ([S 7]) and Gray ([G 4]) to germs of diffeomorphisms of $(\mathbf{C}^n,0)$. Under Brjuno condition (13), it was first proved by Brjuno ([B 5], [B 6]), and afterwards, when $n = 1$, by H. Rüssmann ([R 2]).

Both Siegel's and Brjuno's proofs use majorant series ; this is the most natural method, and gives the best results : the estimates of the radius of convergence are very reasonable, with the right weights for the contributions of the diophantine approximations. The key point, and probably the whole problem, is of arithmetical nature.

These are many other proofs using rapid iteration methods and the so-called KAM techniques. For Newton's method (i.e. the existence of an inverse up to a quadratic term) see H. Rüssmann ([R 3]) and E. Zehnder ([Z 1]) ; for rapid iteration techniques with an infinite number of change of coordinates, see Brjuno ([B 6]), Rüssmann ([R 2]), Siegel and Moser ([S 6]) and Arnold ([A 1]).

We refer to Brjuno ([B 6]), Rüssmann ([R 5]) and J. Pöschel ([P 9]) for various generalizations of Theorem 1.

Theorem 1 was generalized to non-archimedean complete valued fields of characteristic 0 (instead of $\mathbf{C}$) by Sibuya-Sperber ([S 1]) and Herman-Yoccoz ([H 12]) ; we refer to [S 2], [S 3] for applications.

When A is (conjugate to) a unitary matrix, satisfies condition (*) and a suitable diophantine condition, one can deduce Siegel's theorem from results on normal forms of analytic diffeomorphisms of $\mathbf{T}^n$, i.e. Arnold-Moser's theorem ([A 2], [M 7],and also [H 3, Annexe] and [Z 2]) : see [H 4, VIII] and [H 6] for the precise statements. This requires KAM techniques and is very useful to study the boundaries of Siegel singular disk and the global properties of radii of convergence ; see § 19.

Historical comments. To the best knowledge of the author, the first to use, for solving non linear equations in the complex analytic category, the technique of diminishing domains together with Cauchy estimates (which amounts essentially to the same thing as smoothing operators), was M. Gevrey in 1914 ([G 1]). This method of Gevrey is very standard in complex analysis ; it was for instance used in

1940 by H. Cartan ([C 2]) in combination with Newton's method (without being explicitly stated). A.N. Kolmogorov in 1954 ([K 4]) was the first to state[1] that Gevrey's method in conjunction with Newton's method could be used for questions related to small divisors ; various proofs, due to V.I. Arnold and J. Moser, were given in the sixties ; see [B 4] for a recent survey. The technique of introducing an infinite number of change of coordinates was used, in a different context, by Newlander and Nirenberg in 1957 ([N 3]).

In 1919, G. Julia claimed, in an incorrect paper ([J 1]), to disprove Siegel's theorem ; it was rapidly known that this was in fact an open problem (see H. Cremer's paper) until C.L. Siegel settled it in 1942. Most of the difficult questions of convergence involving small divisors where known to H. Poincaré ; he frequently made the incorrect "conjectures" (for example, about the convergence of perturbation series with fixed frequencies, i.e. the existence of KAM invariant torii), but to the best knowledge of the author, never *claimed to disprove* the convergence. I take the opportunity to quote H. Poincaré about the existence of invariant torii ([P 6, t. II, § 149, p. 104-105]) :

"Ne peut-il pas arriver que les séries (2) convergent, quand on donne aux x_i^0 certaines valeurs convenablement choisies ?

Supposons, pour simplifier, qu'il y ait deux degrés de liberté ; les séries ne pourraient-elles pas, par exemple, converger quand x_1^0 et x_2^0 ont été choisies de sorte que le rapport $\frac{n_1}{n_2}$ soit incommensurable, et que son carré soit au contraire commensurable (ou quand le rapport $\frac{n_1}{n_2}$ est assujetti à une autre condition analogue à celle que je viens d'énoncer un peu au hasard) ?

Les raisonnements de ce chapitre ne me permettent pas d'affirmer que ce fait ne se présente pas. Tout ce qu'il m'est permis de dire, c'est qu'il est fort invraisemblable."

See also [P 5, t. XI, p. 69-78].

[1] A.N. Kolmogorov never published a complete proof of [K 4].

7. SOME CONJECTURES.

CONJECTURE 1. *The Brjuno condition is necessary for Theorem 1 to hold ; in other words, assuming A diagonalizable and satisfying condition (*), the Proposition 3 holds when (10) is replaced by :*

(10′) $\qquad$ A *does not satisfy Brjuno condition.*

Observe that for the linearized equation (12), the condition (10) is the best possible (i.e. surjectivity of (12) is equivalent to the negation of (10)).

For very reasonable support of this conjecture, the reader should look at (11.3). Part of the conjecture was claimed (in a stronger form) by T. Cherry ([C 5]) ; but no proof has appeared, and the claim of [C 5] might well be incorrect.

CONJECTURE 2. *If one replaces* **C** *by a locally compact non-trivial complete valued field of* strictly positive *characteristic, Siegel's theorem is usually false, even for polynomials of one variable.*

What goes wrong in such a field is that there are no Brjuno numbers (i.e. satisfying (13)), cf. [H 12].

For lack of space, we refer the reader to the partial survey of V.I. Arnold ([A 1]) about Il'yachenko's, Pyartli's and his own work on the geometric materializations of resonances when A has no eigenvalues of modulus 1. Other references are [A 3], [I 2], [I 3], [M 1], [M 4], [M 5].

We also will not discuss here the infinite dimensional versions of Theorem 1 ; we only give references to N.V. Nikolenko ([N 4], [N 5], [N 6]) and E. Zehnder ([Z 3]).

Chapter II

THE IDEA OF THE PROOF, BY MAJORANT SERIES, OF SIEGEL'S AND BRJUNO'S THEOREMS ; VARIOUS REMARKS

8. Idea of the proof of Theorem 1.

8.1 We describe here the main ideas of the proof, due to Siegel ([S 4]). We only consider the case $n = 1$; the case $n \geq 2$ is essentially similar : see [B 6], [P 9].

Let $f(z) = \sum_{j \geq j} c_j z^j$, with $c_1 = \exp(2\pi i \alpha) = \lambda$, the germ of holomorphic diffeomorphism we are considering ; replacing if necessary $f(z)$ by $\frac{1}{t} f(tz)$ with small $t \neq 0$, we can assume that :

$$|c_j| \leq 1 \, , \, \forall j \geq 1 \, ; \tag{8.1}$$

in particular, f is holomorphic on $\{|z| < 1\}$.

The (formal) conjugacy $h(z) = \sum_{j \geq 1} h_j z^j, c_1 = 1$, satisfies :

$$h(\lambda z) = f(h(z)) = \lambda z + \sum_{j \geq 2} c_j (h(z))^j \, .$$

For $j \geq 2$, let $s_j = |\lambda^j - \lambda|$, and ω_j a strictly positive number such that $0 < \omega_j \leq s_j$. Consider then the formal germ $g(z) = \sum_{j \geq 1} b_j z^j$, with $b_1 = 1$, which satisfies :

$$\sum_{j \geq 2} \omega_j b_j z^j = \sum_{j \geq 2} (g(z))^j \, . \tag{8.2}$$

This gives for $j \geq 2$:

$$b_j = \omega_j^{-1} \sum b_{k_1} \cdots b_{k_l} \, , \tag{8.2'}$$

the sum being taken for $l \geq 2$, $k_p \geq 1$ and $\sum_p k_p = j$.

By induction, one sees that, for $j \geq 1$:

$$|h_j| \leq b_j \, . \tag{8.3}$$

8.2 By Dirichlet principle, we know that :

$$\liminf_{n\to+\infty} ns_n < +\infty .$$

However, if we take $\omega_j = j^{-\beta}$, for some $\beta > 0$, then g diverges ; this illustrates the difficulty, due to the small divisors s_j (or ω_j), in proving the convergence of h (or g).

By (8.6) and (8.8) a necessary (but not sufficient, see (11.3)) condition for the convergence of g is :

$$\limsup_{j\to+\infty} (\omega_2 \cdots \omega_j)^{-1/j} < +\infty . \tag{8.4}$$

Tambs Lyche ([T 1]) and Hardy and Littlewood ([H 2]) have shown that, for any $\alpha \in \mathbf{T}^1 - (\mathbf{Q}/\mathbf{Z})$:

$$\limsup_{j\to+\infty}(s_2 \cdots s_j)^{-1/j} = \limsup_{j\to+\infty} q_{j+1}^{-1/q_j} . \tag{8.4'}$$

When α is a Brjuno number, both *lim sup* are limits and equal to 1 (but this is not equivalent to the Brjuno condition).

The result (8.4') of Tambs Lyche and Hardy and Littlewood is *not* a consequence of Birkhoff's ergodic theorem for a.e. α. It is related to the following question asked by A.Y. Khintchine :

QUESTION 1. *For which $\varphi \in L^2(\mathbf{T}^n, d\theta)$ does one have, for almost all $\alpha \in \mathbf{T}^n$:*

$$\lim_{j\to+\infty} \frac{1}{j} \sum_{l=1}^{j} \varphi(j\alpha) = \int_0^1 \varphi(\theta) d\theta \; ?$$

By Marstrand ([M 2]), this may be false, even when $n = 1$ and φ is bounded ; it is true, by [K 3], when $n = 1$ and $\varphi(\theta) = \mathrm{Log}\,\|\theta\|$ or $\mathrm{Log}\,|\sin 2\pi\theta|$; see also [H 2].

8.3 We take $\omega_n = s_n$, $\delta_1 = 1$, and define δ_j for $j \geq 2$:

$$\delta_j = \omega_j^{-1} \sup \delta_{k_1} \cdots \delta_{k_l} \ , \tag{8.5}$$

the supremum being taken for $l \geq 2$, $k_p \geq 1$, $\sum_p k_p = j$. We certainly have :

$$\delta_j \geq (\omega_2 \cdots \omega_j)^{-1} \ , \ j \geq 2 \ . \tag{8.6}$$

Define $l(z) = \sum_{j \geq 1} l_j z^j$ by $l_1 = 1$ and

$$l(z) = z + \frac{(l(z))^2}{1 - l(z)}$$

which gives :

$$l_j = \sum l_{k_l} \cdots l_{k_l} \ ,$$

the sum being taken as in (8.2'). Induction shows that :

$$b_n \leq \delta_n l_n \ ; \tag{8.7}$$

$$\delta_n \leq b_n \ . \tag{8.8}$$

We conclude that g is convergent if and only if :

$$\sup_j j^{-1} \operatorname{Log} \delta_j < +\infty \ . \tag{8.9}$$

8.4 To obtain (8.9), Brjuno uses the following property of the sequence (ω_j) : there exists a sequence $(v_j)_{j \geq 1}$ in $]0, 1/2]$, with the v_j all distinct and having 0 as accumulation point, and constants $\theta \in]0, 1/2]$, $c > 0$ such that :

$$v_j \leq c\omega_{j+1} \ , \ j \geq 1 \ ; \tag{8.10}$$

$$\sum_{k=1}^{\infty} -t_k^{-1} \operatorname{Log}(\theta v_{t_k}) = d < +\infty \ . \tag{8.11}$$

$$\text{If} \quad v_j < \theta v_{t_k} \ , \ 1 \leq l < t_{k+1}, \quad \text{then} \quad v_{j-l} \geq \theta v_{t_k} \ , \tag{8.12}$$

where the numbers t_k are defined inductively by $t_1 = 1$ and :

$$t_{k+1} = \inf(l \mid v_l < v_{t_k}) \ , \ k \geq 1 \ .$$

Indeed, from (8.10), (8.11), (8.12), one gets, using a counting lemma and clever (but elementary) manipulations of majorants series, that :

$$\sup_{j \geq 1} j^{-1} \operatorname{Log} \delta_j \leq C_1 d + C_2 \ , \tag{8.13}$$

with positive universal constants C_1, C_2.

Remark : Instead of g, Brjuno works with $\eta(z) = z^{-1} g(z) - 1$ which gives a slightly different inductions (but also (8.13)).

8.5 In our case, we take, for $j \geq 1$, $v_j = \|j\alpha\|$, where $\|x\| = \inf_{p\in\mathbf{Z}} |x+p|$ for $x \in \mathbf{R}$. We then have :

$$s_{j+1} = \omega_{j+1} = |\lambda^{j+1} - \lambda| \geq 4v_j \ , \ j \geq 1 \ . \tag{8.14}$$

On the other hand, one has :

$$\|j\alpha\| + \|(j-l)\alpha\| \geq \|l\alpha\| \ .$$

So (8.10), (8.12) hold with $\theta = 1/2$, $c = 1/4$. One has $t_k = q_{k-1}$ if $\alpha \in (0, 1/2)$, and $t_k = q_k$ if $\alpha \in (1/2, 1)$; hence (8.11) follows from (17), using (16).

Let $R(g)$, $R(h)$ the radii of convergence of g, h ; we finally conclude from (8.3), (8.13) that :

$$\mathrm{Log}[R(h)^{-1}] \leq \mathrm{Log}[R(g)^{-1}] \leq C_3 \left(\sum_{k=0}^{\infty} q_k^{-1} \, \mathrm{Log}(\|q_k\alpha\|^{-1}) \right) + C_4 \ , \tag{8.15}$$

where C_3, C_4 are universal constants, under the assumption (8.1) on f.

9. In the special case :

$$f_\alpha(z) = e^{2\pi i\alpha}(z + z^2) = \lambda_\alpha(z+z^2)$$

a slight modification of Brjuno's proof ([B 6]) gives :

$$\mathrm{Log}[R(h_\alpha)^{-1}] \leq \mathrm{Log}[R(g_\alpha)^{-1}] \leq \mathrm{Log}\,\frac{1}{4} + 2\sum q_k^{-1} \, \mathrm{Log}[(2\sin\frac{\pi}{2}\|q_k\alpha\|)^{-1}] \ , \tag{9.1}$$

where the sum starts with $k = 0$ for 0, α, $1/2$ and $k = 1$ if $1/2 < \alpha < 1$.

Idea of the proof : The term Log 1/4 comes from the special type of induction for h_α ; in order to get better constants, one uses the counting argument of [B 6], but with the fuction $\mathrm{Log}[(2|\sin\pi\theta|)^{-1}]$. ∎

When $\alpha = \frac{\sqrt{5}-1}{2}$, (9.1) gives, up to a factor less than 50, the correct value for $R(h_\alpha)$; this is very reasonable.

To get estimates *from above* for $R = R(h_\alpha)$, one observes that $\tilde{h}(z) = \frac{1}{R}h(Rz)$ is univalent on $\{|z| < 1\}$ (see §13). The critical value $-\lambda_\alpha/4$ of f_α cannot belong to the image of this disk, so we obtain an elementary bound for R by Koebe's 1/4-theorem :

$$R \leq 1 \ . \tag{9.2}$$

A much better estimate is obtained by considering :

$$\tilde{h}(z)\left(1 + \frac{4R}{\lambda_\alpha}\tilde{h}(z)\right)^{-1} = z + b_2z^2 + O(z^3) \ ,$$

which is univalent on $\{|z| \leq 1\}$, and applying Bieberbach's inequality ($|b_2| \leq 2$) to get :

$$R \leq \frac{2|\lambda_\alpha - 1|}{|4 - 3\lambda_\alpha|} \leq \frac{4}{7} \ . \tag{9.3}$$

10. RELATION BETWEEN THE RADIUS OF CONVERGENCE AND THE DIOPHANTINE PROPERTIES OF α.

If there exists $\beta \geq 0$, $\gamma > 0$, the estimate (8.15) shows that :

$$R(g)^{-1} \leq C\gamma^{-1} , \tag{10.1}$$

for some universal constant $C > 0$.

Actually, one can get from (8.15) a much better result, giving up to constants the right contribution of the various rational approximations of α.

Suppose that the continuous fraction $1/(a_1 + 1/(a_2 + \cdots))$ of α is such that all a_i are 1, except one, say a_{n+1}, which is large.

Then we get from (8.15) :

$$\operatorname{Log}[R(g)^{-1}] \leq C_1 + C_2 q_n^{-1} \operatorname{Log} q_{n+1} , \tag{10.2}$$

for some universal constants C_1, C_2.

10.3 The following example shows that this type of estimate is optimal. Consider $f(z) = \lambda z + z^{q+1}$, where $\lambda = \exp(2\pi i\alpha)$, α is above, and $q = q_n$. The conjugacy h has the form :

$$h(z) = z + h_{q+1}z^{q+1} + h_{q+2}z^{q+2} + \cdots$$

with $h_{q+1} = (\lambda^{q+1} - \lambda)^{-1}$.

The image by h of its disk of convergence $D(h)$ is contained in $\{|z| \leq 2^{1/q}\}$, because $\lim_{n\to+\infty} |f^n(z)| = +\infty$ when $|z| > 2^{1/q}$. As h is injective on $D(h)$ (see §13), we obtain :

$$\text{Area } (h(D(h))) = \pi \sum_{j\geq 1} |h_j|^2 j[R(h)]^{2j} \leq \pi 2^{2/q} ,$$

and therefore :

$$\begin{aligned} \operatorname{Log}[R(h)^{-1}] &\geq \frac{1}{2q} \operatorname{Log}(q 2^{-2/q} |\lambda^q - 1|^{-2}) \\ &\geq C_3 + C_4 q_n^{-1} \operatorname{Log} q_{n+1} \end{aligned}$$

for universal constants C_3, C_4.

11. For $\alpha \in \mathbf{T} - (\mathbf{Q}/\mathbf{Z}), \lambda = \exp(2\pi i\alpha)$, let $l(\alpha)$ be the radius of convergence of the function g_α defined by (8.2), with $\omega_n = |\lambda^n - \lambda|$.

It is not difficult to see that :

$$l(\alpha) = 0 \quad \text{on a dense } G_\delta \text{ subset of } \quad \mathbf{T}^1 \; ; \tag{11.1}$$

the measurable function $:\alpha \to [l(\alpha)]^{-1}$is in the weak L^1-space of $(\mathbf{T}^1, d\theta)$. (11.2)

Indeed, with $|E|$ denoting the Lebesgue measure of $E \in \mathbf{T}^1$, we have, for $\beta > 0$, $\gamma > 0$:

$$\left|\left\{\alpha, |\alpha - (p/q)| \leq \frac{\gamma}{q^{2+\beta}} \quad \text{for some} \quad p/q\right\}\right| \leq C(\beta)\gamma \; , \; C(\beta) > 0 \; ,$$

and from this and (10.1), we deduce that :

$$|\{\alpha \in \mathbf{T}^1, l(\alpha) \leq \gamma\}| \leq C\gamma \, .$$

J.C. Yoccoz has shown ([Y 4]) that :

$$\text{if} \quad l(\alpha) \neq 0, \quad \text{then } \alpha \text{ is a Brjuno number} \, . \tag{11.3}$$

This strongly supports Conjecture 1 (at least if one does not believe in wild cancellations due to the fact that $\lambda^n - \lambda$ are complex).

In counterpart, determining *exactly* $l(\alpha)$ seems to be, in view of (11.1), (11.2), an untractable and unreasonable problem. It probably requires the exact knowledge of the continued fraction of α and of all possible cancellations !

We let the reader try to calculate $R(h_\alpha)$ or even give a reasonable lower bound for it, when $\alpha = \pi$ or $\alpha = 2^{k+(1/k)}$, $k \geq 3$, $k \in \mathbf{N}$.

QUESTION 2. *Is it possible to find an algorithm to decide, given a small $\epsilon > 0$, if $R(g_\alpha) \geq \epsilon$ from the base 2 expansion of α (as computers suggest) ?*

For numbers α of constant type, we found [H 8] a very simple general method which applies to almost all small divisors problems ; it gives in particular a very simple proof of Siegel's theorem ([H 4]), and yields, for more difficult problems, very reasonable constants ([H 8, vol. 2, ch. VII]). The constants depend only on the calculation of the logarithms and sines of a couple of numbers, and a pocket calculator is more than enough !

For a remarkably simple minoration of $R(h)$, due to J.C. Yoccoz, in the special case $f(z) = e^{2\pi i\alpha}(z + z^2)$, we refer to 18.4 ; see also [L 1] when α is the golden mean.

Using the simple idea that one can calculate the Taylor coefficients of h and then conjugate f by appropriate truncations of h, various authors ([L 2], [L 3]) have claimed much better estimates for $R(h)$ when α is the golden mean. Unfortunately, this approach requires large computers and a huge amount of numerical work, so the claims are not easily checked. These authors have also used the fact that one can follow Newton's method on computers, a point of view which was first adopted by O. Hald and Braess and Zehnder ([B 7]).

12. GENERALIZATIONS AND REMARKS ON THE PROOF.

The conditions (8.11), (8.12) we have required in 8.4 on the sequence $(v_j)_{j\geq 1}$ may be replaced by the following less restrictive conditions, due to Brjuno ([B 6], see also [P 9]) : there exists $\theta \leq 1/2$ and integers $1 = p_0 < p_1 < p_2 < \cdots$ such that we have, with $\check{\Omega}(m) = \inf_{1\leq k\leq m} v_k$:

$$\sum_{k=0}^{+\infty} -p_k^{-1} \operatorname{Log}[\theta\check{\Omega}(p_{k+1})] = d_1 < +\infty \; ; \tag{8.11'}$$

$$\text{If} \quad v_n < \theta\check{\Omega}(p_k) \quad \text{and} \quad l < p_{k+1}, \quad \text{then} \quad v_{n-l} \geq \theta\check{\Omega}(p_k) \; . \tag{8.12'}$$

These conditions together with (8.10) imply :

$$\operatorname{Log}[R(g)^{-1}] \leq C_3 d_1 + C_4 \; , \tag{8.13'}$$

for universal constants $C_3, C_4 > 0$. When $v_n = \|n\alpha\|$, (8.11') is equivalent to the Brjuno condition, see [B 6]. One uses the sequence (v_j) instead of (ω_j) because (8.12) might not hold for (ω_j).

C. L. Siegel, in [S 4], makes the following assumptions : there exist $\gamma \in]0,1[$ and $\nu > 0$ such that :

$$0 < \omega_n^{-1} < \gamma^{-1}(n-1)^\nu, \quad \text{for} \quad n \geq 2 \; ; \tag{12.1}$$

$$\min(\omega_p^{-1}, \omega_q^{-1}) < \gamma^{-1}(q-p)^\nu \quad \text{for} \quad 1 < p < q \; . \tag{12.2}$$

He then shows that :

$$\sup_{n\geq 1} n^{-1} \operatorname{Log} \delta_n \leq \gamma^{-1} L, \tag{12.3}$$

where δ_n is defined by (8.5) and L is a universal constant.

The conditions (12.1) or (8.11') alone are *not* sufficient to obtain (12.3) ; the conditions (8.12') or (12.2) imply that the ω_n^{-1} are not frequently large, and this is the crucial point.

The conditions (8.11') and (8.12') (or (12.1) and (12.2)) appear in many other problems of small divisors.

12.4 Both Siegel and Brjuno prove that g converges and therefore (by (8.3)), so does h. The other existing proofs of the convergence of h (for instance, using rapid iteration methods, [S 6]) do not show that g converges !! See § 32 and § 34.

Chapter III

THE BOUNDARIES OF SIEGEL SINGULAR DISKS

13. Study of the boundaries of Siegel singular domains.

13.1 For this study to make sense, we will require global assumptions on f.

D. Sullivan asked the following question :

Assume that U is a simply connected domain with compact closure $\bar{U}$, containing 0, and that f is an analytic diffeomorphism of U, extending continuously to $\bar{U}$, such that $f(0) = 0$ and $f'(0) = e^{2\pi i\alpha}$, with $\alpha \in \mathbf{T}^1 - (\mathbf{Q}/\mathbf{Z})$; does this imply that ∂U is a Jordan curve ?

Couterexamples were given by Moeckel ([M 6], see also [P 7]) and independently in [H 5]. In the example of [H 5] (which is adapted from one of M. Handel [H 1]), f extends to a C^∞-diffeomorphism of $\mathbf{C}$ and ∂U can be taken as the "pseudo-circle".

13.2 For the sake of simplicity, we assume that $f = f_\lambda = \lambda g$, with g as in § 3 and :

$$\lambda = e^{2\pi i\alpha} \ , \ \alpha \in \mathbf{T}^1 - (\mathbf{Q}/\mathbf{Z}) \ . \tag{13.3}$$

We assume that f is linearizable at 0 ; this is the case if α is a Brjuno number.

Let U ($\neq \emptyset$) be the maximal connected open set containing 0 on which the family $(f^n)_{n\geq 0}$ is normal. We have $f(U) = U$, and, by a result of P. Fatou (see [F 2]), $U \neq \mathbf{C}$. By the maximum principle, U is simply connected.

Let $h_1 : \mathbf{D} = \{|z| < 1\} \to U$ the conformal representation of U which satisfies $h_1(0) = 0$, $h_1'(0) = t > 0$. By Schwarz's lemma, we have :

$$h_1^{-1} \circ f \circ h_1(z) = \lambda z \ ,$$

so $h(z) = h_1\left(\frac{z}{t}\right)$ is univalent and satisfies the same equation on $\{|z| < t\}$; moreover $h'(0) = 1$ and t is the radius of convergence of h.

We call U a *singular domain (or disk)*. (We add "singular" because "Siegel domains" classically refer to the symmetric spaces $Sp(\mathbf{R}^{2n})/U(n)$.)

Question 3.

a) *If U has compact closure, does there exist a critical point of f on ∂U ?*

b) *If f has no critical points, is U unbounded ?*

Clearly a positive answer to a) implies a positive answer to b).

QUESTION 4. *If U has compact closure, is it true that :*
a) ∂U is a Jordan curve (i.e. a simple closed curve)?
b) ∂U is a quasicircle?

(A quasicircle is the image of the standard circle $\mathbf{S}^1 = \{z \in \mathbf{C}, |z| = 1\}$ by a quasi-conformal homeomorphism of $\mathbf{C}$, hence is a Jordan curve.)

When f is polynomial, U has always compact closure. These questions were first asked by A. Douady (1980) and D. Sullivan (1981) for rational functions ([S 10]).

One of the reasons for asking Question 3.a) is the following result of P. Fatou ([F 1], [H 7]) : let SV be the set of singular values of f, i.e. the critical or asymptotic values of f ; then the ω-limit set of SV by f (i.e. $\overline{\cap_{n\geq 0}\cup_{k\geq n} f^k(SV)}$) contains the boundaries of all singular domains of f. When f is a polynomial, SV is just the set of critical values of f.

14. We assume in this section that :

$$\alpha \in DC \quad \text{(see (5.2))} . \tag{14.1}$$

14.2 Under hypothesis (14.1), E. Ghys ([G 2]) has shown that a positive answer to Question 4.a) implies a positive answer to question 3.a). This was generalized by the following theorem ([H 7]) :

14.3

THEOREM 2. *If U has compact closure, $f_{|\partial U}$ is injective and (14.1) holds, then there is a critical point of f in ∂U.*

Hence, to answer positively to Question 3.a), under the hypothesis (14.1), supposing that there is no critical point of f on ∂U, one has to check that $f_{|\partial U}$ is injective ; curiously enough, it is not easy at all, and we have only been able to do that in special examples : see [H 7] where the following theorem is proved.

THEOREM 3. *If f is a polynomial with only one critical point c, and (14.1) holds, then $c \in \partial U$.*

One an take for instance, $f(z) = \frac{\lambda}{n}((z+1)^n - 1)$, $n \geq 2$; the theorem also applies for periodic elliptic points.

We also obtained in [H 7] many examples with a critical value of f on ∂U.

14.4 Consider the case where $f(z) = \lambda(e^z - 1)$. It was shown in [H 7] that, when (14.1) holds, U is unbounded ; this implies that ∂U is not, in the Riemann sphere, a Jordan curve and is rather complicated : $\{\infty\}$ is contained in the impression of every prime end of U.

This example is the reason why we asked in Question 4 for U to have compact closure.

In the following questions, $f(z) = \lambda(e^z - 1)$.

QUESTION 4.

c) When U is unbounded, does the omitted value $-\lambda$ of f belong to ∂U ?

d) When f is still linearizable at 0, but (14.1) does not hold, is U always unbounded ?

14.5 After [H 7] was obtained, L. Carleson and P. Jones gave a simpler proof of Theorem 3, showing that $c \in \partial U$ for a.e. $\lambda \in \mathbf{S}^1$; they use the ingredients of J.C. Yoccoz's proof of Siegel's theorem for this particular class of polynomials (see § 18).

15. In both Ghys' partial result ([G 2]) and in the proof of Theorem 2, the main ingredient is the following result.

THEOREM 4. *Let f be a **R**-analytical diffeomorphism of the circle, with rotation number $\alpha \in DC$; then f is **R**-analytically conjugated to the rotation $R_\alpha : \theta \to \theta + \alpha$.*

This theorem was first proved by the author in 1975 for $\alpha \in CT$, in 1976 for a.e. α ([H 3]), and was generalized to $\alpha \in DC$ by J.C. Yoccoz in 1982 ([Y 1], see also [Y 2]). By Denjoy's theorem, f is topologically conjugated to R_α. One first shows that the conjugacy h is C^1, then that it is C^∞ ; the proof is more direct and natural than the usual techniques in KAM theory (which, anyway, give only perturbative or local results). Finally, one shows that if h is C^∞ and $\alpha \in DC$ then h is **R**-analytic : for this we used in [H 3] an improvement of a theorem of Arnold and Moser ([H 3, annexe]), but this can be avoided for a.e. α by adapting [H 6].

The author does not know of any other *global* result than Theorem 4 in small divisors theory ; in fact, it could well be the only simple one, cf. [H 9] and [H 3, XIII].

QUESTION 5. *Does Theorem 4 hold when α is a Brjuno number ?*

A positive answer would imply that Theorems 2 and 3 are still valid when $\alpha \in B$. The *local* conjugacy theorem, for $\alpha \in B$, is true and is proved by adapting [R 3] and [H 3, annexe].

16. AN EXAMPLE OF APPLICATION OF THEOREM 3.

Let $f(z) = \lambda(z+z^2)$, with $\lambda = \exp(2\pi i\alpha)$; we assume that $\alpha \in DC$. Let $c = -\frac{1}{2}$ the critical point of f_λ ; a theorem of Fatou ([F 1]) says that ∂U is contained in the closure of the orbit of c under f.

Using Theorem 3 we conclude that :

(16.1) $\qquad (f_\lambda^n(c))_{n\geq 0}$ is dense in ∂U .

This, as the following example shows, can force ∂U to be geometrically complicated.

16.2 *Example :* Let $n \geq 1$, $p \in \mathbf{N}^* \cup \{\infty\}$; suppose that the continued fraction of α satisfies $a_i = 1$ if $i \neq n$ and $a_n = p$ (when $p = \infty$, this means that the continued fraction stops at stage $(n-1)$). We suppose that n and p are very large, and write f_p to indicate the dependance on p of $f(z) = e^{2\pi i\alpha}(z+z^2)$.

When $p = 1$, α is the golden mean. Given $\epsilon > 0$ and $h \in \mathbf{N}^*$, if n is very large (independently of p), the distance between $f_1^l(c)$ and $f_p^l(c)$ will be less than ϵ for $l \leq k$.

On the other hand, α is rational when $p = \infty$, so by a result of G. Julia and P. Fatou ([F 1]) we have $\lim_{l\to+\infty} f_\infty^l(c) = 0$. Given $k_1 \in \mathbf{N}^*$ and $\epsilon > 0$, the orbit $(f_\infty^l(c))_{l\leq k_1}$ is ϵ-close to $(f_p^l(c))_{l\leq k_1}$ is p is large enough (note that here, "large" depends on n). As p is still finite, α is of constant type and (16.1) applies, showing wild oscillations for ∂U.

This example shows that one cannot conclude anything from the numerical computations of $(f^n(c))_{n\geq 0}$ if one does not control the error terms ; on the other hand, to keep track of these terms seems difficult, as c is on the boundary of the basin of ∞.

17. ARITHMETIC CONDITIONS ARE NECESSARY.

17.1 Recall that in the result of E. Ghys (14.2) as well as in Theorems 2 and 3, we assume that $\alpha \in DC$.

The following theorem ([H 10]), which is obtained using Ghys's construction ([G 2]), shows that an arithmetic hypothesis is necessary.

THEOREM 5. *There exists* $\alpha \in \mathbf{T}^1 - (\mathbf{Q}/\mathbf{Z})$ *such that* $f(z) = e^{2\pi i\alpha}(z+z^2)$ *is linearizable at* 0, *and its singular Siegel disk* U *satisfies :*

i) ∂U *is a quasicircle ;*

ii) no point $f^n(c)$, $n \geq 0$ *lies on* ∂U.

17.2 Fatou's theorem says that $\partial U \subset \overline{\{f^n(c), n \geq 0\}}$. By Theorem 3, the number α in Theorem 4 cannot belong to DC. In [H 10], it is also shown that Question 3.b) has a negative answer. Theorem 5 shows that Ghys's result 14.2 is false for some Liouville number α, and that most results of [H 7] are false without arithmetic assumptions.

17.3 Taking into account § 15, at least one of the following statements is true :
a) Question 5 has a negative answer ;
b) The number α in Theorem 5 is not a Brjuno number.
If b) holds, Cherry's claim in [C 5] is incorrect.

QUESTION 6. *Which of these statements is true ?*

QUESTION 7. *In Theorem 5, can one find α such that one can replace i) by one of the following statements :*
a) ∂U is a C^k-submanifold for some $1 \leq k \leq \infty$?
b) ∂U is a Jordan cruve, but not a quasi-circle?

17.4 Ghys's result 14.2 shows that ∂U cannot be a C^1-submanifold when $\alpha \in DC$

For any α such that f is linearizable at 0, and any $z_0 \in \partial U$, the intersection of ∂U with a neighbourhood of z_0 cannot be an analytic arc. Otherwise, using Schwarz's reflection principle, the conjugacy equation, and the minimality of $z \to \lambda z$ on $|z| = \text{constant}$, one would be able to extend the conjugacy to a bigger disk, in contradiction with the maximality of U.

18. YOCCOZ'S PROOF OF SIEGEL'S THEOREM FOR $f(z) = \lambda(z+z^2)$ ([Y 3]).

For $|\lambda| \leq 1$, let $f_\lambda(z) = \lambda(z+z^2)$. Let $h_\lambda(z) = z + O(z^2)$ the formal conjugacy :

$$f_\lambda \circ h_\lambda(z) = h_\lambda(\lambda z) \,. \tag{18.1}$$

Poincaré's result shows that for $|\lambda| < 1$ (including $\lambda = 0$), the radius of convergence $R(\lambda)$ of h_λ is strictly positive ; the image L_λ by h_λ of its circle of convergence is a Jordan curve, analytic except at $c = -1/2$ where it has a right angle. One has, for $|\lambda| < 1$, the following elementary facts :

$$h_\lambda(\{|z| < R(\lambda)\}) \subset \{|z| \leq 2\} \,. \tag{18.2}$$

$$u(\lambda) = \lim_{n \to +\infty} \lambda^{-n} f_\lambda^n(c) \quad \text{exists} \;; \tag{18.3}$$

$$u \quad \text{is analytic in} \quad \{|\lambda| < 1\} \;; \tag{18.4}$$

$$|u(\lambda)| \leq 2 \;; \tag{18.5}$$

$$h_\lambda(u(\lambda)) = c \;; \tag{18.6}$$

$$|u(\lambda)| = R(\lambda) \;; \tag{18.7}$$

$$u(0) = -\frac{1}{4} \,. \tag{18.8}$$

(To show (18.5) one uses the maximum principle and that $(f_\lambda^n(c))_{n \geq 0}$ is bounded (by P. Fatou's and G. Julia's result), which imply $|f_\lambda^n(c)| \leq 2$ when $|\lambda| = 1$.)

18.9

LEMMA 1. *If $(\lambda_i)_{i\geq 0}$ is a sequence such that :*

i) $$|\lambda_i| < 1, \lim_{i\to+\infty} \lambda_i = 1 ;$$

ii) $$R(\lambda_i) \geq \delta > 0, \quad \textit{for some} \quad \delta ;$$

then a subsequence of $(h_{\lambda_i})_{i\geq 0}$ converges to a function $H_\lambda(z) = z + O(z^2)$ analytic on $\{|z| < \delta\}$ which satisfies on this disk $f_\lambda \circ H_\lambda(z) = H_\lambda(\lambda z)$.

Using (18.2) it is a straightforward application of Montel's theorem. Unicity shows that $H_\lambda = h_\lambda$ and $R(\lambda) \geq \delta$.

By P. Fatou's theorem, there is a function $U \in L^\infty(\mathbf{T}^1, d\theta)$ such that the radial limits :

$$\lim_{\substack{t\to 1\\ 0<t<1}} u(te^{2\pi i\alpha}) = U(\alpha), \quad \text{for a.e. } \alpha . \tag{18.10}$$

By a theorem of F. Riesz, the function Log $|U|$ belongs to $L^1(\mathbf{T}^1, d\theta)$, so we have :

$$U(\alpha) \neq 0, \quad \text{for a.e. } \alpha . \tag{18.11}$$

Siegel's theorem, for a.e. α, now follows from (18.7), (18.10), (18.11) and Lemma 1. We actually conclude from Lemma 1 that :

$$R(\lambda) \geq \limsup_{\substack{\lambda_i\to\lambda\\ |\lambda_i|<1}} R(\lambda_i). \tag{18.12}$$

J.C. Yoccoz proves more ([Y 3]) : for every λ with $|\lambda| = 1$ one has that radial limit :

$$\lim_{\substack{t\to 1\\ 0>t>1}} |u(t\lambda)| \quad \text{exists and is equal to} \quad R(\lambda) . \tag{18.13}$$

The function u has curious properties, see [Y 3]. By (18.5), it belongs to $H^\infty(\mathbf{D})$; by (11.2) it is an outer function.

18.14 From (18.8), the remarkably simple following observation follows : *there exists a subset of $\mathbf{S}^1$ of positive Lebesgue measure for which $R(\lambda) \geq 1/4$.*

On the other hand, by Proposition 2 and (8.12), $|U(\alpha)| = R(\lambda)$ vanishes on a dense G_δ subset of $\mathbf{S}^1$.

19. PROPERTIES OF THE RADIUS OF CONVERGENCE.

19.1 Let $\lambda = e^{2\pi i\alpha}$, with $\alpha \in DC$; we denote O_λ the space of entire functions f which satisfy $f(0) = 0$, $f'(0) = \lambda$ equipped with the compact open topology ; it is a complex codimension 1 affine subspace of the space of all entire functions.

For $f \in O_\lambda$, let h_f the analytic function defined near 0 by :

$$f \circ h_f(z) = h_f(\lambda z), \ h_f(0) = 0, \ h'_f(0) = 1 \ ; \tag{19.2}$$

let $R(h_f)$ the radius of convergence of h_f, and $g(f) = \mathrm{Log}[R(h_f)^{-1}]$.

We have $g(f) = -\infty$ if and only if $f(z) = \lambda z$.

THEOREM 6. *The function g on O_λ is continuous and plurisubharmonic (i.e. subharmonic on any complex line in O_λ).*

Idea of the proof :
Lower semi-continuity : this is proved as in Lemma 1, using that the functions h_f, univalent on a disk $\{|z| < r\}$ and satisfying $h_f(0) = 0$, $h'_f(0) = 1$, form a compact set for the compact open topology.
Upper semi-continuity : this is more delicate and requires the appropriate generalization of the theorem of Arnold and Moser on diffeomorphisms of the circle : see [H 4], [H 6].
Plurisubharmonicity : the Taylor coefficients $h_n(f)$ of h_f depend analytically on f and one has :

$$g(f) = \limsup_{n\to+\infty} \frac{1}{n} \mathrm{Log}\,|h_n(f)| \ . \ \blacksquare$$

19.3 Define :

$$U_\epsilon = \{f \in O_\lambda \mid \sup_{|z|\leq 4} \left|f(z) - \lambda(z+z^2)\right| < \epsilon\}.$$

If ϵ is small enough, g is pluriharmonic on U_ϵ ; to see this, one uses Douady-Hubbard's theory of polynomial-like mappings and adapts, using [H 4, VIII], Yoccoz's proof (§ 18).

19.4 An example :

Let $f_b = \lambda(z + bz^2) + z^3$, for $b \in \mathbf{C}$, and denote $g(f_b)$ simply by $g(b)$. One has $g(b) > -\infty$ for any $b \in \mathbf{C}$.

PROPOSITION. *The function g is not harmonic.*

Proof : The conjugacy h_b has the form :

$$h_b = z + h_2(b)z^2 + O(z^3) ,$$

with $h_2(b) = b(\lambda-1)^{-1}$ satisfying $|h_2(b)| \leq 2e^{g(b)}$ by Bieberbach's inequality. This shows that $\operatorname{Log}|b| - g(b)$ is bounded from above ; if g were harmonic, one should have $g(b) = \operatorname{Log}|b| +$ constant, in contradiction with $g(0) > -\infty$. ■

From 19.3, considering $bf_b(b^{-1}z)$, one sees that g is harmonic for large b and that :

$$\lim_{b\to\infty} (g(b) - \operatorname{Log}|b|) = g(f_\lambda) ,$$

with $f_\lambda(z) = \lambda(z + z^2)$.

By F. Riesz's theorem ([T 2]), one can write :

$$g(b) = l(b) + \int_{\mathbf{C}} \operatorname{Log}|b-u|\, d\mu(u) ,$$

where μ is a Radon measure with compact support and l is harmonic ; l is actually constant, because $l(b) \leq c_1 \operatorname{Log}|b|$ where $|b|$ is large (then, an entire function k with modulus e^l is a polynomial and does not vanish). The measure μ has no atoms, and its support K has strictly positive capacity.

CONJECTURE 3. *K is not a countable union of C^1-embedded arcs.*

Possibly K is not even locally connected. The structure of K is related to the set of parameters b where the orbits of the two critical points of f_b interact with each other.

20. A FORMULA FOR THE RADIUS OF CONVERGENCE.

Let $f \in O_\lambda$; we assume that the Siegel singular domain of f is bounded. From (19.1) we have :

$$\frac{1}{n}\sum_{j=0}^{n-1} \operatorname{Log}|f^j(u)| = \frac{1}{n}\sum_{j=0}^{n-1} k(\lambda^j z) ,$$

where $u \in U-\{0\}$, $u = h_f(z)$ and $k(z) = \operatorname{Log}|h(z)|$ is harmonic and bounded from above on $\{0 < |z| < R(h_f)\}$. As $z \to \lambda z$ is uniquely ergodic on $\{|z| = \text{constant}\}$, the right-hand side converges to $\int_0^1 k(e^{2\pi i\theta}|z|)d\theta = \operatorname{Log}|z|$ when $n \to +\infty$. Taking radial limits, and applying the ergodic theorem to $\lim_{t\to 1} k(tz)$, $|z| = R(h_f)$, we conclude :

$$\operatorname*{Lim}_{n\to+\infty} \frac{1}{n}\sum_{j=0}^{n-1} \operatorname{Log}|f^j(u)| \leq \operatorname{Log} R(h_f)$$

for every $u \in U$, and :

$$\operatorname*{Lim}_{n\to+\infty} \frac{1}{n}\sum_{j=o}^{n-1} \operatorname{Log}|f^j(u)| = \operatorname{Log} R(h_f) , \tag{20.1}$$

for almost every $u \in \partial U$, with respect to harmonic measure.

QUESTION 8. *When does (20.1) hold for every $u \in \partial U$?*

If Question 4.a) has a positive answer, the unique ergodicity of $z \to \lambda z$ on $\{|z| = R(h_f)\}$ implies that question 8 has a positive answer. If moreover Question 4.b) has positive answer (with estimates on the quasiconformal constants), one can use (20.1), [H 3, VI 3.2] and the classical estimates on univalent functions ([P 8]) to get numerical values for $R(h_f)$ (see remarks at the end of § 16).

21. Some questions on non linearizable fixed points.

21.1 Consider the special case $f(z) = f_\alpha(z) = e^{2\pi i\alpha}(z+z^2)$, for $\alpha \in \mathbf{T}^1 - (\mathbf{Q}/\mathbf{Z})$: there exists a dense G_δ-subset of $\mathbf{T}^1 - (\mathbf{Q}/\mathbf{Z})$ such that f_α is not linearizable at 0.

A. Douady ([D 3]) has shown that one can find α such that f_α is non linearizable at 0 and the Julia set $J(f_\alpha)$ (which here consists of the points with bounded orbits) is not locally connected. In his example, the critical point $c = -1/2$ is not accessible in $\mathbf{C} - J(f_\alpha)$. He asked :

Question 9. *Can $J(f_\alpha)$ have positive Lebesgue measure when f_α is not linearizable at 0 ?*

One can also ask Question 9 when f_α is linearizable. The following question seems to be important in order to understand the Siegel singular domain of f_α.

Question 10. *Can one find $\alpha \in \mathbf{T}^1 - (\mathbf{Q}/\mathbf{Z})$ such that the orbit of the critical point is dense in the Julia set ?*

Problem. *Calculate, or at least find, reasonable estimates of*
$\sup_\alpha \sup_{n\geq 0} |f_\alpha^n(c)|$.

21.2 We mention that, in the general problem of classification of the germs of the form $f(z) = e^{2\pi i\alpha}z + O(z^2)$, $\alpha \in \mathbf{T}^1 - (\mathbf{Q}/\mathbf{Z})$, Naǐshul' has shown ([N 1]) that $\pm\alpha$ is an invariant of topological conjugacy.

22. A non-archimedean example

Let $(\mathbf{Q}_2, |\ |_2)$ be the 2-adic field with its standard absolute value, defined by $|2|_2 = \frac{1}{2}$ and $|p|_2 = 1$ for every odd prime.

For $\lambda \in \mathbf{Q}_2$, with $|\lambda| = 1$, consider $f_\lambda(z) = \lambda(z + z^2)$; the critical point $c = -1/2$ satisifes :

$$\lim_{n\to+\infty} |f_\lambda^n(c)|_2 = +\infty .$$

By [H 12], f_λ is linearizable at 0 if λ is not a root of unity. This shows that the answers to Questions 3 and 4 certainly depend on the base field.

Chapter IV

SOME EXAMPLES ON $\mathbf{C}^n$, $n \geq 2$.

23. We consider an entire mapping $f : \mathbf{C}^n \to \mathbf{C}^n$; we assume that $f(0) = 0$ and $Df(0) = A$ is unitary.

If f is linearizable at 0 (in particular, if A satisfies Brjuno condition (13)), one can define as in § 13.2 the *Siegel singular domain* U of f at 0. Then one defines for $z \in U$:

$$h(z) = \operatorname*{Lim}_{n \to +\infty} \frac{1}{n} \sum_{j=0}^{n-1} A^{-j} f^j(z) , \tag{23.1}$$

(a standard formula due to Bochner and Martin ([B 3], see also [P 1])), and one has :

$$h(0) = 0 \, , \; Dh(0) = \mathrm{id}$$

$$h(f(z)) = Ah(z) \, , \; z \in U \, .$$

Adapting Fatou's arguments ([F 1], see also [D 0]), one concludes that $f_{|U}$ is a diffeomorphism of U. The closure of $(f^n_{|U})_{n \geq 0}$ for the compact open topology is a compact group, isomorphic to $\mathbf{T}^n$ when A satisfies condition (*).

QUESTION 11. *Is $h : U \to h(U)$ a diffeomorphism ?*

I have only been able to prove this when the jacobian of f is constant and A satisfies (*) ; then the jacobian of h is constant and equal to 1.

23.2 When A is diagonal and satisfies condition (*), the $(A^n)_{n \in \mathbf{Z}}$ form a dense subgroup of the standard action of $\mathbf{T}^n$ on $\mathbf{C}^n$, defined by :

$$(\theta_1, \cdots, \theta_n) \cdot (z_1, \cdots, z_n) = (e^{2\pi i \theta_1} z_1, \cdots, e^{2\pi i \theta_n} z_n);$$

hence $h(U)$ is a Reinhardt domain.

QUESTION 12. *Describe which Reinhardt domains one can obtain in this way, up to biholomorphic diffeomorphisms.*

23.3 A theorem of Cartan and Thullen, solving a conjecture of G. Julia, says that U is polynomially convex (i.e. a Runge domain) : see [V 1, § 24.8, p. 207].

When A is diagonal, and Question 11 has a positive answer, $h(U)$ is the domain of normal convergence of h^{-1} ; then $h(U)$ is a pseudoconvex complete Reinhardt domain ; complete means that $(\lambda_1 z_1, \cdots, \lambda_n z_n) \in h(U)$ if $(z_1, \cdots, z_n) \in h(U)$ and $|\lambda_j| \leq 1$).

24. One can have $U = h(U) = \mathbf{C}^n$ even when f is not linear. Indeed, $\mathbf{C}^n$ has a large group of biholomorphic diffeomorphisms by which one can conjugate A to define f. Examples of such diffeomorphisms are :

$$h(z_1,\cdots,z_n) = (\exp(\varphi_1(z_2,\cdots,z_n))z_1 + \varphi_2(z_2,\cdots,z_n), z_2,\cdots,z_n)$$

where φ_1, φ_2 are entire functions.

25. **Example 1 :** Let λ_1, $\lambda_2 \in \mathbf{C}$, $|\lambda_1| = |\lambda_2| = 1$, and $E \in (-2,2)$. Define a biholomorphic diffeomorphism of $\mathbf{C}^3$ by :

$$f(z_1,z_2,z_3) = \left(\lambda_1 B(z_3)\begin{pmatrix} z_1 \\ z_2 \end{pmatrix}, \lambda_2 z_3\right),$$

where $B(z_3) = \begin{pmatrix} E+z_3 & -1 \\ 1 & 0 \end{pmatrix}$. We have :

$$Df(0) = A = \begin{pmatrix} \lambda_1 E & -\lambda_1 & 0 \\ \lambda_1 & 0 & 0 \\ 0 & 0 & \lambda_2 \end{pmatrix}.$$

For a.e. $(\lambda_1,\lambda_2,E) \in \mathbf{S}^1 \times \mathbf{S}^1 \times (-2,2)$, A satisfies Brjuno condition and we showed in [H 9] that $U = \{(z_1,z_2,z_3) \in \mathbf{C}^3 \mid |z_3| < 1\}$; in this case ∂U is an analytic manifold, whose Levi form vanishes (we recall that this is not possible in one variable, cf. 17.4).

Example 2 : Let $n \geq 2$, $\varphi : \mathbf{C}^{n-1} \to \mathbf{C}$ an entire function satisfying $\varphi(0) = 0$, $D\varphi(0) = 0$, and $\lambda_1,\cdots,\lambda_n$ complex numbers of modulus 1. Define a biholomorphic diffeomorphism f of $\mathbf{C}^n$ by :

$$f(z_1,\cdots,z_n) = (\lambda_1 z_1 + \varphi(z_2,\cdots,z_n), \lambda_2 z_2,\cdots,\lambda_n z_n) .$$

We assume that condition (*) is satisfied by the λ_j's. The unique formal conjugacy of Proposition 1 has the form :

$$h(z_1,z_2,\cdots,z_n) = (z_1 + \eta(z_2,\cdots,z_n), z_2,\cdots,z_n)$$

and η is solution of :

$$\eta(\lambda_2 z_2,\cdots,\lambda_n z_n) - \lambda_1\eta(z_2,\cdots,z_n) = \varphi(z_2,\cdots,z_n) .$$

Using Baire category arguments (cf. [H 12]) one can consruct the (λ_j) and a sequence $(k_j)_{j\geq 0}$ in $\{0\} \times \mathbf{N}^{n-1}$ with $|k_j| \to +\infty$ such that :

$$|\lambda_1 - \lambda^{k_j}| \leq |k_j|^{-|k_j|},\ j \geq 0 .$$

If we then choose :

$$\varphi(z) = \sum_j (\lambda^{k_j} - \lambda_1) z^{k_j}$$

(with $z = (z_2, \cdots, z_n)$) we will have :

$$\eta(z) = \sum_j z^{k_j} .$$

The Siegel singular domain of f at 0 in this case has the form $\mathbf{C} \times V$, where V is a Reinhardt domain different from $\mathbf{C}^{n-1}$.

26. Example 3 : For $\lambda_1, \lambda_2 \in \mathbf{C}$, with $|\lambda_1| = |\lambda_2| = 1$, we consider the biholomorphic diffeomorphism of $\mathbf{C}^2$:

$$f(z_1, z_2) = (\lambda_1 z_1 + z_2, \lambda_2 z_2 + (\lambda_1 z_1 + z_2)^2).$$

We assume that $\lambda_1 \neq \lambda_2$ and that (λ_1, λ_2) satisfy Brjuno condition. Then f has a Siegel singular domain U at 0.

PROPOSITION. *The closure of U in $\mathbf{C}^2$ is compact.*

Idea of the proof : As f has constant jacobian, we know from 23.2, 23.3 that there exists a biholomorphic diffeomorphism g from a pseudoconvex complete Reinhardt domain R onto U which satisfies :
- $f(g(z)) = g(\lambda_1 z_1, \lambda_2 z_2)$ for $z = (z_1, z_2) \in R$;
- The Taylor series of g at 0 is normally convergent. With $g = (g_1, g_2)$, we obtain :

$$g_2(z) + \lambda_1 g_1(z) = g_1(\lambda_1 z_1, \lambda_2 z_2) \; ; \tag{26.1}$$

$$(g_1(\lambda_1 z_1, \lambda_2 z_2))^2 = g_1(\lambda_1^2 z_1, \lambda_2^2 z_2) - (\lambda_1 + \lambda_2) g_1(\lambda_1 z_1, \lambda_2 z_2) + \lambda_1 \lambda_2 g_1(z_1, z_2) \, . \tag{26.2}$$

If $\tilde{R} = \{(z_1, z_2) \mid |z_1| \leq r_1, |z_2| \leq r_2\}$ is included in R, we get from (26.2) that $|g_1| \leq 4$ on $\tilde{R}$ and then from (26.1) that $|g_2| \leq 8$ on $\tilde{R}$; we conclude that $U \subset \{(z_1, z_2) \mid |z_1| \leq 4, |z_2| \leq 8\}$. ∎

Keeping the same notations, R is logarithmically convex ([V 1]) and g_1, g_2 are bounded and analytic on R ; for each complex line L passing through 0, at least one of g_1, g_2 has non constant restriction to $L \cap R$, so this intersection has the form $\{|z| < r\}$, with $r < +\infty$ and z a coordinate on L (vanishing at 0). We conclude that R is bounded and $\partial R \cap \{(z_1, z_2), z_1 z_2 \neq 0\}$ is a topological manifold.

QUESTION 13. *Is ∂U a C^∞-submanifold of $\mathbf{C}^2$?*

QUESTION 14. *Is ∂U a topological submanifold of $\mathbf{C}^2$? When it is, is it a locally flat submanifold ?*

CONJECTURE 4. *For almost every (λ_1, λ_2) (with $|\lambda_1| = |\lambda_2| = 1$), ∂U is not a C^∞-submanifold of $\mathbf{C}^2$.*

Observe that f is a diffeomorphism, so has no critical point. J. H. Hubbard ([H 13]) has a more geometrical approach to examples where the proposition holds.

The fact that ∂U is compact is probably related to the fact that normal forms for diffeomorphisms of $\mathbf{T}^2$ are only local (cf. [H 9]).

27. Remarks on Fatou-Julia theory on $\mathbf{C}^n$, $n \geq 2$.

(See also [N 7] ; this paper claims, p. 367, using [J 1], that Siegel's theorem is incorrect !)

27.1 For an entire mapping $f : \mathbf{C}^n \to \mathbf{C}^n$, one defined the *Julia set* $J(f)$ as follows : $x \in \mathbf{C}^n - J(f)$ if and only if there is a neighbourhood U of x such that $(f^n_{|U})_{n\geq 0}$ is normal with values in $\mathbf{C}^n \cup \{\infty\}$ (i.e. we allow $\{\infty\}$ as a limit function). It follows from the definition that $J(f)$ is closed and that one has :

$$f(J(f)) \subset J(f) ;$$
$$f^{-1}(J(f)) = J(f) .$$

27.2 The reader should consult [H 7] for the properties of $J(f)$ when $n = 1$; for $n \geq 2$, the properties of $J(f)$ are quite different :

- $J(f)$ can be empty even if f is not linear : see § 24.
- Periodic points are not always dense in $J(f)$: see Examples 1 and 2 above.
- One can have $\operatorname{int}(J(f)) \neq \emptyset$ but $J(f) \neq \mathbf{C}^n$, and $f_{|J(f)}$ does not always have a dense orbit : Examples 1, 2.
- $J(f)$ can have isolated points : for instance, take for f a Cremona diffeomorphism of $\mathbf{C}^2$, with a repulsive linearizable fixed point at 0, whose basin B of attraction of 0 by f^{-1} is a Poincaré-Fatou- Bierberbach domain (i.e. $\mathbf{C}^2 - B$ has non-empty interior) ; see [D 1], [P 5, t. IV, p. 537-582] ; observe that for any small neighbourhood V of 0, $\cup_{n\geq 0} f^n(V)$ is not dense in $\mathbf{C}^2$.

27.3 If f is a biholomorphic diffeomorphism of $\mathbf{C}^2$, and x_0 is a fixed point of f not in $J(f)$ or $J(f^{-1})$, then f is linearizable at x_0 and $Df(x_0)$ is conjugate to a unitary matrix : this is implied by Bochner-Martin's formula (23.1) and the fact that $((Df(x_0))^n)_{n\in\mathbf{Z}}$ is bounded in $\mathcal{L}(\mathbf{C}^n, \mathbf{C}^n)$.

27.4 One can construct ([H 11]) a biholomorphic diffeomorphism f of $\mathbf{C}^2$ with the following properties :

- $f(0) = 0$;
- $Df(0)$ *is a diagonal, unitary and satisfies condition (*) :*
- 0 *is the unique periodic point of* f ;
- f *has a dense orbit in* $\mathbf{C}^2$.

Question 15. *Does there exist a polynomial Cremona diffeomorphism of* $\mathbf{C}^2$ *which preserves Lebesgue measure and has an infinite number of periodic points outside its Julia set ?*

Using S. Newhouse's techniques ([N 2]) one can probably construct polynomial Cremona diffeomorphisms f of $\mathbf{C}^2$ (preserving $\mathbf{R}^2$) with an infinite number of periodic sinks.

Chapter V

CENTER MANIFOLDS AND THEIR RELATION TO INVARIANT CIRCLES OF TWIST-MAPS. EXAMPLES

28. CENTER MANIFOLDS.

Let $f(z) = Az + O(z^2)$ a germ of analytic diffeomorphism of $\mathbf{C}^n$. When the matrix A does not satisfy condition (*), usually f is not formally linearizable ; when, for instance, $A = \begin{pmatrix} \lambda & 1 \\ 0 & \lambda \end{pmatrix}$ with $|\lambda| = 1$, λ is not a root of unity, f is formally linearizable but usually not analytically linearizable.

Nevertheless, under suitable arithmetical hypotheses on the eigenvalues of A, one can frequently find, even when condition (*) is not fulfilled, germs of analytic submanifolds through 0 which are invariant under f and on which f acts like a diagonal matrix. We refer to J. Pöschel ([P 9]) for some general results and other references.

We want here to give some examples (slightly different from Pöschel's) and show the strong relation between these invariant submanifolds and the invariant torii of symplectic diffeomorphisms.

29. POINCARÉ-LINDSTEDT PERTURBATION SERIES[1].

We only look at simple examples and refer to H. Poincaré ([P 6]) and J. Moser ([M 9]) for a more general (but elementary) approach.

Let φ a $\mathbf{Z}$-periodic real valued C^∞ function, satisfying $\int_0^1 \varphi(\theta)d\theta = 0$. We consider, fot $a \in \mathbf{R}$, the diffeomorphism f_a of $\mathbf{T}^1 \times \mathbf{R}$ defined by :

$$f_a(\theta, r) = (\theta + r, r + a\varphi(\theta + r)) \ . \tag{29.1}$$

Let $\alpha \in DC$, and C_a a C^∞ simple closed curve, homotopic to $\{r = 0\}$, such that f_a leaves C_a invariant and $f_{a|C_a}$ has rotation number α. One can then (as $f_{a|C_a}$ is smoothly conjugated to a rotation) choose a parametrization of C_a of the form :

$$\theta \to (\theta + \eta_a(\theta), \alpha + l_a(\theta))$$

where η_a, $l_a \in C^\infty(\mathbf{T}^1)$ satisfy :

$$\eta_a(\theta + \alpha) = \eta_a(\theta) + l_a(\theta) \ ; \tag{29.2}$$

[1] H. Poincaré attributes to Lindstedt the discovery of these series ; but most of Poincarés attributions are not quite correct, since he frequently forgets his own contributions.

$$L_\alpha \eta_a(\theta) = a\varphi(\theta + \eta_a(\theta)) \ , \tag{29.3}$$

the linear map L_α being defined by :

$$L_\alpha \eta_a(\theta) = 2\eta_a(\theta) - \eta_a(\theta + \alpha) - \eta_a(\theta - \alpha) \ .$$

One can expand η_a as a formal power series :

$$\eta_a(\theta) = \sum_{n \leq 1} b_n(\theta) a^n \ ,$$

with the b_n in $C^\infty(\mathbf{T}^1)$, $\int_0^1 b_n(\theta)d\theta = 0$: these are the Poincaré -Lindstedt perturbation series (with fixed frequency).

We restrict ourselves to the *standard map*, obtained by taking $\varphi(\theta) = \frac{1}{2\pi}\sin(2\pi\theta)$. One shows inductively that b_n is an odd trigonometric polynomial of degree n. D. Goroff showed me, and many authors have noticed, ([G 3], [G 5], [R 1]) how to calculate inductively the b_n. We define :

$$\exp(2\pi i(\theta + \eta_a(\theta))) = \sum_{n \geq 0} c_n(\theta) a^n \ ; \tag{29.4}$$

then we have :

$$c_0(\theta) = e^{2\pi i\theta} \ ; \tag{29.5}$$

$$c_n = 2\pi i n^{-1} \sum_{k=1}^{n} k b_k c_{n-k}, \quad \text{for} \quad n \geq 1 \ ; \tag{29.6}$$

$$b_n = \frac{1}{4\pi i} L_\alpha^{-1}(c_{n-1} - \bar{c}_{n-1}), \quad \text{for} \quad n \geq 1 \ . \tag{29.7}$$

Relation (29.6) is obtained from (29.4) differentiating with respect to a ; in (29.7), L_α is invertible because α is irrational and $c_{n-1} - \bar{c}_{n-1}$ is a trigonometric polynomial without constant term.

When $\alpha \in DC$ and φ is $\mathbf{R}$-analytic, the series for η_a converge ; in fact, for $\epsilon > 0$ sufficiently small, the function : $(\theta, a) \to \eta_a(\theta)$ extends to an analytic map on $\{|\mathrm{Im}\theta| < \epsilon\} \times \{|a| < \epsilon\}$ (where $\theta \in \mathbf{C}/\mathbf{Z}, a \in \mathbf{C}$). This is shown adapting a proof of E. Zehnder ([Z 2], see also [M 8]).

The most natural proof (for the standard map) of the convergence of the Poincaré-Lindstedt perturbation series (with fixed frequency) would be to deduce it directly from the relations (29.4)-(29.7).

Unfortunately, this is a delicate problem, as explained now.

The relations (29.6), (29.7) allow to express the coefficients of the trigonometric polynomial c_n as polynomials in the coefficients of the c_j with $j < n$. These polynomials have real coefficients. If one replaces these real coefficients by their absolute value (as one does in Siegel's theorem), the majorant series obtained in this way for η_a are divergent ! The reason is the following : for $l(\theta) = \sum_{k\neq 0} l_k e^{2\pi i k\theta}$, one has :

$$L_\alpha^{-1} l(\theta) = \sum l_k v_k^{-1} e^{2\pi i k\theta} ,$$

with $v_k = \sin^2 k\pi\alpha$, and in particular :

$$v_{q_k}^{-1} \geq (4\pi^2)^{-1} q_k^2 ,$$

if q_k is the denominator of a convergent of α. In the expression for the coefficient of c_n, with $q_k < n < q_{k+1}$, $v_{q_k}^{-1}$ appear in some terms with an exponent which is larger than approximately $\frac{1}{2}(n - q_k)$, and this is too much for convergence.

This means that to prove the convergence of the Poincaré-Lindstedt perturbation series (with fixed frequency), one has to take into account the cancellations due to the signs of the different terms in the induction. Eliasson claims ([E 1], [E 2]) to overcome in part this difficulty. His arguments are (and have to be) very delicate ; the author of these lines is far from understanding the whole story, but is convinced that eventually this type of approach will give a remarkably simple, natural and beautiful new proof of the results of KAM-theory in the **R**-analytic case.

30. We indicate how *some* of the coefficients of the b_n (for the standard map) are closely related to the Siegel's theorem and the induction (8.2).

Let u_n be the coefficient of $e^{2\pi i n\theta}$ in $2\pi i b_n$ (this is the higher degree term). Then one has :

$$u_{n+1} = \frac{1}{2} v_{n+1}^{-1} t_n, \quad \text{for} \quad n \geq 0 , \tag{30.1}$$

$$\text{with} \quad t_0 = 1 , t_n = \frac{1}{n} \sum_{k=1}^{n} k u_k t_{n-k}, \; n \geq 1 .$$

This gives :

$$u_n = \frac{1}{(n-1)v_n} \sum_{k=1}^{n-1} k u_k u_{n-k} v_{n-k} . \tag{30.2}$$

This implies :

$$t_n > 0, \; u_n > 0 \quad \text{for} \quad n \geq 1 ; \tag{30.3}$$

$$u_n \geq (2v_n)^{-1} u_{n-1} \quad \text{for} \quad n \geq 2 ; \tag{30.4}$$

$$u_{2n} \geq \frac{1}{2} v_n v_{2n}^{-1} u_n^2 \quad \text{for} \quad n \geq 1 . \tag{30.5}$$

31. THE STANDARD MAP (cf. [G 5]).

We consider the biholomorphic diffeomorphism G of $(\mathbf{C}/\mathbf{Z}) \times \mathbf{C}$ defined by :

$$G(\theta, r) = (\theta + r, r - (2\pi i)^{-1} e^{2\pi i(\theta+r)}) .$$

Let $\alpha \in \mathbf{R} - \mathbf{Q}$; we define the u_n, for $n \geq 1$, by (30.1), (30.2) and put :

$$\begin{aligned} q_\alpha(z) &= (2\pi i)^{-1} \sum_{n\geq 1} u_n z^n , \\ \tilde{\eta}_\alpha(\theta) &= q_\alpha(e^{2\pi i\theta}) . \end{aligned} \tag{31.1}$$

Then, we obtain :

$$L_\alpha \tilde{\eta}_\alpha = (4\pi i)^{-1} e^{2\pi i(\theta+\tilde{\eta}(\theta))} ,$$

which we rewrite, with $\lambda = e^{2\pi i\alpha}$, as :

$$2q_\alpha(z) - q_\alpha(\lambda z) - q_\alpha(\lambda^{-1} z) = \frac{1}{2} z e^{q_\alpha(z)} . \tag{31.2}$$

When α is a Brjuno number (see [B 6], the remark after (8.13), and (12.3) when $\alpha \in DC$), the series (31.1) for q_α define an analytic function. Let $R(\alpha)$ be its radius of convergence, $\delta(\alpha) = -\operatorname{Log} R(\alpha)$ and $D^\star_\alpha = \{\theta \in \mathbf{C}/\mathbf{Z}, \operatorname{Im}(\theta) > \delta(\alpha)\}$: then $\tilde{\eta}_\alpha$ is defined on $D^\star_\alpha$,which is a biholomorphic image of $\mathbf{D}^\star = \{0 < |z| < 1\}$. The mapping S_α defined by :

$$S_\alpha(\theta) = (\theta + \tilde{\eta}_\alpha(\theta), \alpha + \tilde{\eta}_\alpha(\theta + \alpha) - \tilde{\eta}_\alpha(\theta))$$

from $D^\star_\alpha$ to $(\mathbf{C}/\mathbf{Z}) \times \mathbf{C}$, satisfies :

$$G(S_\alpha(\theta)) = S_\alpha(\theta + \alpha) . \tag{31.3}$$

As α is irrational, S_α is injective on $D^\star_\alpha$ and $\tilde{\eta}_\alpha$ cannot be extended beyond $D^\star_\alpha$.

31.4 Relation with center manifolds.

Consider the biholomorphic diffeomorphism F of $\mathbf{C}^2$ defined by :

$$F_1(z_1, z_2) = \left(z_1 e^{2\pi i z_2}, z_2 - \frac{1}{4\pi i} z_1 e^{2\pi i z_2} \right).$$

Then, with $h(\theta, r) = (e^{2\pi i\theta}, r)$, the diagram :

$$\begin{array}{ccc} (\mathbf{C}/\mathbf{Z}) \times \mathbf{C} & \xrightarrow{G} & (\mathbf{C}/\mathbf{Z}) \times \mathbf{C} \\ h\downarrow & & \downarrow h \\ \mathbf{C}^2 & \underset{F}{\longrightarrow} & \mathbf{C}^2 \end{array}$$

is commutative. Each point $(0, z_2)$ is fixed by F ; the image $S_\alpha(D_\alpha^\star)$ is associated to the *center manifolds* passing through the point $(0, \alpha)$, and tangent at this point to $\{z_2 = \alpha\}$.

The analogy between $S_\alpha(D_\alpha^\star)$ and a $\mathbf{R}$-analytic invariant curve R of rotation number α, for a $\mathbf{R}$-analytic twist diffeomorphism f of $\mathbf{T}^1 \times \mathbf{R}$, appears when one complexifies R and f to some domaim $\{(z_1, z_2) \in (\mathbf{C}/\mathbf{Z}) \times \mathbf{C}, \max_j |\mathrm{Im}(z_j)| < \delta\}$; this is possible for small $\delta > 0$, giving f_1, R_1 and a ring R_2 invariant under f_1, satisfying $R \subset R_2 \subset R_1$, on which f_1 is conjugate to the translation by α.Observe that the restriction to R_2 of Df_1 is parabolic : R_2 is a center manifold of f_1 just as $S_\alpha(D_\alpha^\star)$ is for F.

PROBLEM. *To describe globally in the complex domain the (complexified) invariant curves of twist maps (29.1), when φ extends to an entire function.*

There are delicate questions of global analytic continuation.

32. We list here some of the properties of the center manifolds $S_\alpha(D_\alpha^\star)$, which may be considered as a simple model for the invariant circles of twist maps (with power series involved instead of trigonometric series).

32.1 For every Brjuno number α, the series defining q_α converge and the center manifold exists.

32.2 For a dense G_δ-subset of $\alpha \in \mathbf{T}^1 - (\mathbf{Q}/\mathbf{Z})$, the series defining q_α has radius of convergence 0 ; this follows from (8.4') since we have from (30.4) :

$$u_n^{1/n} \geq \frac{1}{2}(v_1 \cdots v_n)^{-1/n}. \tag{32.3}$$

32.4 An a priori estimate.

Let α be such that $R = R(\alpha) > 0$; for $0 < r < R$, let M_r be the maximum modulus of q_α on $\{|z| \le r\}$. By (30.3), we have $M_r = q_\alpha(r)$; from (31.2), we obtain :

$$4\ M_r \geq \frac{1}{2} r e^{M_r} , \tag{32.5}$$

from which we deduce :

$$r \leq 8 M_r e^{-M_r} \leq 8 \max_{x \geq 0} x e^{-x} = 8e^{-1} ,$$

$$R(\alpha) \leq 8e^{-1}$$

$$R(\alpha) \leq 8 \sup_{r<R} M_r e^{-M_r} .$$

Moreover M_r increases with r, so we get from the above relations :

$$M_R = \sup_{r<R} M_r < +\infty ; \tag{32.6}$$

$$R \leq 8 M_R e^{-M_R} . \tag{32.7}$$

We conclude from (30.3), (32.6) and Abel's elementary Tauberian theorem that the series defining q_α converge absolutely uniformly on $\{|z| \le R\}$; using (31.3) we see that $S_\alpha(\{z \in \mathbf{C}/\mathbf{Z}, \mathrm{Im} z = \delta(\alpha)\})$ is an *embedded Jordan curve.*

32.7 For any $\epsilon > 0$, the set $\{\alpha, R(\alpha) > \epsilon\}$ is *nowhere dense*. Indeed, if $R(\alpha) > \epsilon$ we get from (32.7) :

$$M_{R(\alpha)} \leq c(\epsilon),$$

$$u_n \leq c'(\epsilon)\epsilon^{-n}$$

where $c(\epsilon)$, $c'(\epsilon)$ only depend on ϵ. But the last inequality is violated on an open and dense set of α (see (32.3)).

This property is analogous to the fact that generic twist maps have invariant curves only for a nowhere dense set of rotation numbers : see [H 8, chap. I].

32.8 It is not difficult to show that $R(\alpha) \leq C\,|\alpha|^2$, using (30.4) and (30.5). For the similar property property for twisp maps, see [H 8, ch. II].

32.9 One can adapt J.C. Yoccoz's proof ([Y 4]) of (11.3) to show that if $R(\alpha) > 0$ *then α is a Brjuno number.*

From this, we conclude that if the perturbation series for the standard map define an $\mathbf{R}$-analytic function :

$$a \to \eta_a \in L^2(\mathbf{T}^1, d\theta)$$

for small $|a|$, then α is a Brjuno number.

(For twist maps an Aubry-Mather sets of fixed rotation number α, η_a is of bounded variation so is in L^∞).

The reader should *not* conclude that this implies that the standard map admits only Brjuno numbers as rotation numbers of its invariant circles (J.C. Yoccoz actually remarked that this is false for the generic C^∞ twist map) ; the only thing one can say is that these invariant circles whose rotation numbers are not Brjuno numbers cannot be obtained by analytic perturbation techniques.

33. Remark : Most of what we have said for the standard map is still true when in (29.1) the function φ is $\mathbf{R}$-analytic, extends to an entire function and its Fourier series has the form :

$$\varphi(\theta) = \sum_{n\geq 1} a_n \sin(2\pi n\theta), \ a_n \geq 0 \ .$$

34. AN EXAMPLE WITH *A* PARABOLIC (for related examples, see [P 2]).

For $\alpha \in \mathbf{T}^1 - (\mathbf{Q}/\mathbf{Z})$ and $\lambda = \exp(2\pi i\alpha)$, we consider the biholomorphic diffeomorphism f_α of $\mathbf{C}^2$ defined by :

$$f_\alpha(z_1, z_2) = (\lambda(z_1 + z_2), \lambda z_2 - \lambda^2(z_1 + z_2)^2) \ . \tag{34.1}$$

We look for an invariant (complex) curve through 0, tangent to $\{z_2 = 0\}$ at 0, invariant under f and such that the restriction of f to it is linearizable. This means that we look at a germ $S_\alpha = (\eta_\alpha, l_\alpha)$ of analytic map from $(\mathbf{C}^2, 0)$ with $\eta_\alpha(z) = z + O(z^2)$, $l_\alpha(z) = O(z^2)$ and satisfying :

$$\eta_\alpha(\lambda z) = \lambda(\eta_\alpha(z) + l_\alpha(z)) \ ; \tag{34.2}$$

$$(\eta_\alpha(z))^2 = 2\eta_\alpha(z) - \bar\lambda\eta_\alpha(\lambda z) - \lambda\eta_\alpha(\bar\lambda z) \ . \tag{34.2'}$$

With $\eta_\alpha(z) = \sum_{n\geq 1} b_n z^n$, $b_1 = 1$, we obtain the induction relation :

$$b_n = (v_{n-1})^{-1} \sum_{k=1}^{n-1} b_k b_{n-k}, \ n \geq 2 \ , \tag{34.3}$$

where v_n is as in 29.

We see that :

$$b_n > 0 \quad \text{for} \quad n \geq 1 \ . \tag{34.4}$$

34.5 It follows from (8.9) and (11.3) that the radius of convergence of the series defining η_α is strictly positive if and only if α is a Brjuno number. Let $R(\alpha)$ be this radius of convergence, and for $r \in [0, R(\alpha))$, define :

$$M_r = \eta_\alpha(r) = \max_{|z| \le r} |\eta_\alpha(z)| \quad \text{(see (34.4))} .$$

From (34.2'), we have $M_r^2 \le 4M_r$, hence :

$$M_r \le 4 . \tag{34.6}$$

34.7 A consequence of (34.6) is that $R(\alpha)$ is finite and the series defining η_α are absolutely uniformly convergent on the closed disk $\{|z| \le R(\alpha)\}$. Furthermore, from (34.2) and (34.5), S_α is an embedding of this disk into $\mathbf{C}^2$.

34.8 The set of $\alpha \in \mathbf{T}^1 - (\mathbf{Q}/\mathbf{Z})$ for which $R(\alpha) > \epsilon$ is nowhere dense in $\mathbf{T}^1$, for any $\epsilon > 0$. This follows from (34.6), Cauchy's inequality and the estimate :

$$b_n \ge (v_1 \cdots v_n)^{-1} .$$

Caution : If one replaces in (34.1) λ by $t\lambda$, with $0 < t < 1$, we obtain an infinite radius of convergence for the series defining η ; compare with §18.

QUESTION 16. *Is η_α univalent on $\{|z| < R(\alpha)\}$? In other terms, is the image of S_α a graph over $\{z_2 = 0\}$?*

A similar question occurs for the semi-standard map.

BIBLIOGRAPHY

Some surveys :

- *Work before 1912* : [P 4].
- *History of the center problem before 1942* : [C 7], [C 1], [W 1] (see also [K 5]).
- *On Brjuno's work* : [M 3], cf. also [P 9].
- *Iteration of rational functions on the Riemannian sphere* : [B 2], [B 8], [C 6], [D 2], [H 4]. The best reference for the classical work is [F 1].
- *Iteration of entire functions of one complex variable* (classical results of Fatou and I.N. Baker) : [H 7].
- *Iteration of entire mappings of* $\mathbf{C}^n$, $n \geq 2$: [D 1], see also [C 3, vol. I] and [B 3] (there seems to be very few general results).
- *Materializations of resonances* : [A 1, chap. 6], [I 3] ; see also a related topic : [M 1] and [M 3], [M 4].
- *On Sternberg's theorems* : [B 9], [C 4].
- *On* KAM : [B 4].
- *Diophantine approximation* : [S].

[A 1] V.I. ARNOLD, Geometrical methods in the theory of ordinary differential equations, Springer Verlag (1983), *translated from Russian.*

[A 2] V.I. ARNOLD, Small denominators I ; on the mappings of a circumference onto itself (1961), Amer. Math. Soc. Translations, 2nd series, **46**, 213-284.

[A 3] V.I. ARNOLD, Bifurcations of invariant manifolds of differential equations and normal forms in a neighbourhood of elliptic curves, Funct. Analysis and its Appl. **10** (1976), 249-259.

[B 1] I.N. BAKER and P.J. RIPPON, Iteration of exponential functions, Ann. Acad. Sci. Fennicae, ser. AI **9** (1984), 49-77.

[B 2] P. BLANCHARD, Complex analytic dynamics on the Riemann sphere, Bull. Amer. Math. Soc. **11** (1984), 85-141.

[B 3] S. BOCHNER and W.T. MARTIN, Several complex variables, Princeton Univ. Press, Princeton (1948).

[B 4] J.B. BOST, Tores invariants de systèmes dynamiques hamiltoniens, Séminaire Bourbaki, exposé 639, Astérisque **133-134** (1986), 113-157.

[B 5] A.D. BRJUNO, Convergence of transformations of differential equations to normal forms, Dokl. Akad. Nauk. SSSR **165** (1965), 987-989.

[B 6] A.D. BRJUNO, Analytical form of differential equations, Transactions Moscow Math. Soc. **25** (1971), 131-288 ; **26** (1972), 199-239.

[B 7] D. BRAESS and E. ZEHNDER, On the numerical treatment of a small divisor problem, Numer. Math. **39** (1982), 269-292.

[B 8] H. BROLIN, Invariant sets under iteration of rational functions, Ark. Math. **6** (1966), 103-144.

[B 9] F. BRUHAT, Travaux de Sternberg, Séminaire Bourbaki, exposé 217, vol. 1960-61, Benjamin, New York.

[C 1] L. CARLESON and P. JONES, letter 1984.

[C 2] H. CARTAN, Sur les matrices holomorphes de n variables complexes, J. Maths. Pures et Appl. **19** (1940), 1-26, *also in* [C 3], vol. II, p. 539-564.

[C 3] H. CARTAN, Oeuvres, vol. **I** to **III**, Springer Verlag (1979).

[C 4] M. CHAPERON, Géométrie différentielle et singularités des systèmes dynamiques, Astérisque **138-139** (1986).

[C 5] T.M. CHERRY, A singular case of iteration of analytic functions : a contribution to the small divisors problem, Nonlinear problems of engineering, ed. W.F. Ames, Academic Press, New York (1964), 29-50.

[C 6] H. CREMER, Über die Iteration rationaler Funktionen, Jahresber. Deutsch. Math. Verein **33** (1925), 185-210.

[C 7] H. CREMER, Zum Zemtrumproblem, Math. Ann. **98** (1928), 151-163.

[C 8] H. CREMER, Über das Zentrumproblem. (Mit besonderer Berücksichtigung der Lückenreihen), Ber. Math. Phys. Klasse der Sächs. Akad. Wiss. Leipzig **82** (1930), 243-250.

[C 9] H. CREMER, Über die Schrödersche Funktionalgleichung und das Schwarzsche Eckenabbildungsproblem, ibid, **84** (1932), 291-324.

[C 10] H. CREMER, Überkonvergenz und Zentrumproblem, Math. Ann. **110** (1935), 739-744.

[C 11] H. CREMER, Über die Häufigkeit der Nichtzentren, ibid. **115** (1938), 573-580.

[D 0] G. DELLA RICCIA, Iteration of analytic mappings of several complex variables, Bull. Amer. Math. Soc. **75** (1969), 340-342.

[D 1] P.G. DIXON and J. ESTERLE, Michael's problem and the Poincaré-Fatou Bieberbach phenomenon, Bull. Amer. Math. Soc. **15** (1986), 127-187.

[D 2] A. DOUADY, Systèmes dynamiques holomorphes, Séminaire Bourbaki, exposé 599, Astérisque **105-106** (1983), 39-63.

[D 3] A. DOUADY, private communication 1983.

[D 4] A. DOUADY and J.H. HUBBARD, On the dynamics of polynomial like mappings, Ann. Sc. E.N.S., 4ème série, **18** (1985), 287-343.

[E 1] L.H. ELIASSON, Generalization of an estimate of small divisors by Siegel, manuscript 1986.

[E 2] L.H. ELIASSON, Small divisors and Siegel's method, manuscript 1986.

[F 1] P. FATOU, Sur les équations fonctionnelles, Bull. S.M.F. **47** (1919), 161-271 ; (1920), 33-94, **48** (1920), 208-304.

[F 2] P. FATOU, Sur l'itération des fonctions transcendantes entières, Acat Math. **47** (1926), 337-370.

[F 3] P. FATOU, Substitutions analytiques et équations fonctionnelles à deux variables, Ann. Sc. E.N.S. **41** (1924), 67-142.

[G 1] M. GEVREY, Oeuvres, Editions du CNRS, Paris (1970), 243-304 and 353-368.

[G 2] E. GHYS, Transformations holomorphes au voisinage d'une courbe de Jordan, C.R. Acad. Sc. Paris t. **289** (1984), 383-388.

[G 3] D. GOROFF, manuscript 1983.

[G 4] A. GRAY, A fixed point theorem for small divisor problems, J. Diff. Eq. **18** (1975), 346-365.

[G 5] J.M. GREENE and I.C. PERCIVAL, Hamiltonian maps in the complex plane, Physica **3D** (1981), 530-548.

[H 1] M. HANDEL, A pathological area-preserving C^∞ diffeomorphism of the plane, Proc. Amer. Math. Soc. **86** (1982), 163-168.

[H 2] G.H. HARDY and J.E. LITTLEWOOD, Notes on the theory of series XXIV. A curious power series, Proc. Cambridge Phil. Soc. **42** (1946), 85-90.

[H 3] M.R. HERMAN, Sur la conjugaison différentiable des difféomorphismes du cercle à des rotations, Publ. Math. IHES **49** (1979), 5-233.

[H 4] M.R. HERMAN, Exemples de fractions rationnelles ayant une orbite dense sur la sphère de Riemann, Bull. S.M.F. **112** (1984), 93-142.

[H 5] M.R. HERMAN, Construction of some curious diffeomorphisms of the Riemann sphere, *to appear in* J. London Math. Soc.

[H 6] M.R. HERMAN, Simple proofs of local conjugacy theorems for diffeomorphisms of the circle with almost every rotation number, Bol. Soc. Bras. Mat. **16** (1985), 45-83.

[H 7] M.R. HERMAN, Are there critical points on the boundedness of singular domains ? Comm. Math. Phys. **99** (1985), 593-612.

[H 8] M.R. HERMAN, Sur les courbes invariantes pour les difféomorphismes de l'anneau, vol. I, Astérisque **102-103** (1983) ; vol. II, Astérisque **144** (1986).

[H 9] M.R. HERMAN, Une méthode pour minorer les exposants de Lyapounov et quelques exemples montrant le caractère local d'un théorème d'Arnold et de Moser sur le tore de dimension 2, Comm. Math. Helv. **58** (1983), 453-502.

[H 10] M.R. HERMAN, Conjugaison quasi-symétrique des difféomorphismes du cercle et applications aux disques singuliers de Siegel, manuscript 1986.

[H 11] M.R. HERMAN, Construction de difféomorphismes ergodiques, manuscript 1978.

[H 12] M.R. HERMAN and J.C. YOCCOZ, Generalizations of some theorems of small divisors to non archimedean fields, Lect. Notes in Math. **1007**, Springer Verlag (1983), 408-447.

[H 13] J.H. HUBBARD, The Hénon mapping in the complex domain, *in* "Chaotic dynamics and fractals", Editors M.F. Barnsley and S.C. Demenko, Academic Press, New York (1986), 101-111.

[I 1] Y.S. IL'YACHENKO, Divergence of series reducing an analytic differential equation to linear normal form at a singular point, Funct. Analysis and Appl. **13** (1979), 227-229.

[I 2] Y.S. IL'YACHENKO, Embedding of positive type of elliptic curves in complex surfaces, Trans. Moscow Math. Soc. **44** (1984), 35-67.

[I 3] Y.S. IL'YACHENKO and S. PYARTLI, Materialization of Poincaré resonances and divergence of normalizing series, J. Soviet Math. **31** (1985), 3053-3092.

[J 1] G. JULIA, Oeuvres, vol. **I**, Gauthier-Villars, Paris (1969), 231-232.

[K 1] E. KASNER, Conformal geometry, Proc. 5th Int. Congress of Mathematicians, Cambridge (1919), vol. **II**, 81-87.

[K 2] G. KOENIGS, Recherches sur les équations fonctionnelles, Ann. Sc. E.N.S. **1** (1884), supplément, 3-41.

[K 3] J.F. KOKSMA, A diophantine property of some summable functions, J. Indian Math. Soc. (N.S.) **15** (1951), 87-96.

[K 4] A.N. KOLMOGOROV, On the preservation of conditionally periodic motions, Dokl. Akad. Nauk. SSSR **96** (1954), 527-530.

[K 5] M. KUCZMA, Functional equations in a single variable, PWN-Polish Sci. Publishers, Warsaw (1968).

[L 1] R. de la LLAVE, A simple proof of a particular case of C. Siegel's center theorem, J. Math. Phys. **24** (1983), 2118-2121.

[L 2] R. de la LLAVE and D. RANA, Proof of accurate bounds in small denominators problems, preprint, Princeton University 1986.

[L 3] C. LIVERANI and G. TURCHETTI, Improved KAM estimates for Siegel radius, preprint, Univ. Bologna 1986.

[L 4] M. LJUBICK, The investigation of the stability of rational functions, *in* "Theory of functions", Kharkov (1981) **42**, p. 89 (in Russian).

[M 1] B. MALGRANGE, Travaux d'Ecalle et de Martinet-Ramis sur les systèmes dynamiques, Séminaire Bourbaki, exposé 582, Astérisque **92-93** (1982), 59-73.

[M 2] J.M. MARSTRAND. On Khintchine's conjecture about strong uniform distribution, Proc. London Math. Soc. **21** (1970), 540-556.

[M 3] J. MARTINET, Normalisation des champs de vecteurs, d'après A.D. Brjuno, Séminaire Bourbaki, exposé 564, Lect. Notes in Maths. **901**, Springer Verlag (1981), 55-70.

[M 4] J. MARTINET et J.P. RAMIS, Problèmes de modules pour les équations différentielles non linéaires du premier ordre, Publ. Math. IHES **55** (1982), 391-404.

[M 5] J. MARTINET et J.P. RAMIS, Classification analytique des équations différentielles non linéaires résonantes du premier ordre, Ann. Sc. E.N.S., 2ème série, **16** (1983), 571-621.

[M 6] R. MOECKEL, Rotations of the closures of some simply connected domains, Complex variables **4** (1985), 285-294.

[M 7] J. MOSER, A rapidly convergent iteration method, II, Ann. Scuola Norm. Sup. di Pisa, Ser. III, **20** (1966), 499-535.

[M 8] J. MOSER, Convergent series expansions for quasi-periodic motions, Math. Ann. **169** (1967), 136-176.

[M 9] J. MOSER, Stability theory in celestial mechanics, *in* "The stability of the solar system and of small stellar systems", Inter. Astronomical Union Symp. **62**, Editor Y. Kozai, Reidel Dorchecht (1974), 1-10.

[N 1] V.I. NAĬSHUL', Topological invariants of analyic and area-preserving mappings and their applications to analytic differential equations in $\mathbf{C}^2$ and $\mathbf{C}P^2$, Trans. Moscow Math. Soc. **42** (1983), 239-250.

[N 2] S.E. NEWHOUSE, The creation of non-trivial recurrence in the dynamics of diffeomorphisms, Les Houches session XXXVI, Chaotic behaviour of deterministic systems, Editor G. Iooss et al., North Holland (1983), 381-442.

[N 3] A. NEWLANDER and L. NIRENBERG, Complex coordinates in almost complex manifolds, Ann. Math. **65** (1957), 391-404.

[N 4] N.V. NIKOLENKO, Complete integrability of the nonlinear Schrödinger equation, Soviet Math. Dokl. **17** (1976), 398-402.

[N 5] N.V. NIKOLENKO, On complete integrability of the nonlinear Schrödinger equation, Funct. Analysis and its Appl. **10** (1976), 209-220.

[N 6] N.V. NIKOLENKO, Invariant asymptotically stable tori of the perturbed Korteweg-deVries equation, Russian Math. Surveys **35** (1980), 139-207.

[N 7] T. NISHINO and T. YOSHIOKA, Sur l'itération des transformations rationnelles de l'espace à deux variables complexes, Ann. Sc. E.N.S., 3ème série, **82** (1965), 327-376.

[P 1] N. PASTIDÈS, Sur les fonctions à centre, C.R. Acad. Sc. Paris, **237** (1953), 957-959.

[P 2] I.C. PERCIVAL, Chaotic boundary of a hamiltonian map, Physica **6D**, (192), 67-77.

[P 3] G.A. PFEIFFER, On the conformal mapping of curvilinear angles. The functional equation $\varphi[f(x)] = a_1\varphi(x)$, Trans. Amer. Mat. Soc. **18** (1917), 185-198.

[P 4] S. PINCHERLE, Equations et opérations fonctionnelles, Encyclopédie des Sc. Math. Pures et Appl., II_5 **1**, Paris, Leipzig (1912), II.26.

[P 5] H. POINCARÉ, Oeuvres, t. **I** à **XI**, Gauthier-Villars, Paris, (1928-1956).

[P 6] H. POINCARÉ, Les méthodes nouvelles de la Mécanique Céleste, t. **I** à **III**, Gauthier-Villars, Paris (1892-1899).

[P 7] Ch. POMMERENKE and B. RODIN, Intrinsic rotations of simply connected regions, II, Complex variables **4** (1985), 223-232.

[P 8] Ch. POMMERENKE, Univalent functions, Vandenhoeck and Ruprecht, Göttingen, (1975).

[P 9] J. PÖSCHEL, On invariant manifolds of complex analytic mappings near fixed points, Exp. Math. **4** (1986), 97-109 ; Addendum p. 1 to 4.

[P 10] C.C. PUGH, On a theorem of P. Hartman, Amer. J. Math. **91** (1969), 363-367.

[R 1] E. ROSENGAUS and R.L. DEWAR, Renormalized Lie theory, J. Math. Phys. **23** (1982), 2328-2338.

[R 2] H. RÜSSMANN, Über die Iteration analytischer Funktionen, J. Math. Mech. **17** (1967), 523-532.

[R 3] H. RÜSSMANN, Kleine Nenne, II : Bemerkungen zur Newtonschen Methode, Nachr. Akad. Wiss. Göttingen, Math. Phys. Kl. (1970), 1-10.

[R 4] H. RÜSSMANN, On the one dimensional Schrödinger equation with a quasi-periodic potential, Ann. New York Acad. Sci. **357** (1980), 90-107.

[R 5] H. RÜSSMANN, On the convergence of power series transformations of analytic mapping near a fixed point into normal form, preprint, I.H.E.S. 1977.

[S] W.M. SCHMIDT, Diophantine approximation, Lect. Notes in Math. **785**, Springer-Verlag (1980).

[S 0] E. SCHRÖDER, Über iterierte Funktionen, Math. Ann. **3** (1871), 296-322.

[S 1] Y. SIBUYA and S. SPERBER, Convergence of power series solutions of $\wp$-adic nonlinear differential equations, Proc. of the Conf. on Recent Advances in Diff. Eq., Trieste 1978, Academic Press.

[S 2] Y. SIBUYA and S. SPERBER, Some new results on power series solutions of algebraic differential equations, Proc. Conf. on Singular Perturbations and Asymptotics, Academic Press (1980), 379-404.

[S 3] Y. SIBUYA and S. SPERBER, Arithmetic properties of power series solutions of algebraic differential equations, Ann. of Math. **113** (1981), 111-157.

[S 4] C.L. SIEGEL, Iteration of analytic functions, Ann. Math. **43** (1942), 807-812 *also in* [S 5], p. 265-269.

[S 5] C.L. SIEGEL, Gesammelte Abhandlungen, Band **I** to **III**, Springer Verlag, (1966).

[S 6] C.L. SIEGEL and J. MOSER, Lectures on Celestial Mechanics, Springer Verlag (1971), *an expanded and enlarged translation of* "Vorselunger Himmelsmechanik" by C.L. Siegel (1956).

[S 7] S. STERNBERG, Infinite Lie groups and the formal aspects of dynamical systems, J. Math. Mech. **10** (1961), 451-474.

[S 8] S. STERNBERG, The structure of local homomorphisms, II, Amer. J. of Math. **80** (1958), 623-631.

[S 9] S. STERNBERG, Local contractions and a theorem of Poincaré, Amer. J. Math. **79** (1957), 809-824.

[S 10] D. SULLIVAN, Quasi-conformal homeomorphisms, I, Solutions of the Fatou-Julia problem on wandering domains, Annals of Math. **122** (1985), 401-418 ; III, preprint I.H.E.S. 1983.

[T 1] R. TAMBS LYCHE, Sur une formule qui généralise la formule du binôme, Avh. Utg. av. Det. Norske Vid. Akad. i Oslo, Math. Naturv. Kl. **6** (1928).

[T 2] M. TSUJI, Potential theory in modern function theory, Chelsea, New York.

[V 1] V.S. VLADIMIROV, Les fonctions de plusieurs variables complexes, *French translation from Russian*, Dunod (1967).

[W 1] B. WARE, The Siegel Center theorem in Hilbert space, preprint Graz research Center 1975.

[Y 1] J.C. YOCCOZ, Conjugaison différentiable des difféomorphismes du cercle dont le nombre de rotation vérifie une condition diophantienne, Ann. Sc. E.N.S., 4ème série, **17** (1984), 333-359.

[Y 2] J.C. YOCCOZ, C^1-conjugaison des difféomorphismes du cercle, Lect. Notes. in Math. **1007**, Springer Verlag (1983), 814-827.

[Y 3] J.C. YOCCOZ, Siegel's theorem for quadratic polynomials, manuscript 1985.

[Y 4] J.C. YOCCOZ, A remark on Brjuno condition, manuscript 1985.

[Y 5] J.C. YOCCOZ, A remark on Siegel's theorem for non diagonalizable linear part, manuscript 1978.

[Z 1] E. ZEHNDER, A simple proof of a generalization of a theorem by C.L. Siegel, Lect. Notes in Math. **597**, Springer Verlag (1977), 855-866.

[Z 2] E. ZEHNDER, Generalized implicit function theorems with applications to some small divisor problems, I, Comm. Pure Appl. Math. **28** (1975), 91-140 ; II, ibid. **29** (1976), 49-111.

[Z 3] E. ZEHNDER, C.L. Siegel's linearization theorem in infinite dimensions, Manuscripta Math. **23** (1978), 363-371.

A Model for Crystallization: A Variation on the Hubbard Model

Elliott H. Lieb
Departments of Mathematics and Physics
Princeton University
Jadwin Hall, P.O. Box 708
Princeton, NJ 08544

Abstract

A quantum mechanical lattice model of fermionic electrons interacting with infinitely massive nuclei is considered. (It can be viewed as a modified Hubbard model in which the spin-up electrons are not allowed to hop.) The electron-nucleus potential is "on-site" only. Neither this potential alone nor the kinetic energy alone can produce long range order. Thus, if long range order exists in this model it must come from an exchange mechanism. N, the electron plus nucleus number, is taken to be less than or equal to the number of lattice sites. We prove the following: (i) For all dimensions, d, the ground state has long range order; in fact it is a perfect crystal with spacing $\sqrt{2}$ times the lattice spacing. A gap in the ground state energy always exists at the half-filled band point (N = number of lattice sites). (ii) For small, positive temperature, T, the ordering persists when $d \geq 2$. If T is large there is no long range order and there is exponential clustering of all correlation functions.

Introduction

This lecture concerns joint work with Tom Kennedy [4], first announced in [10]. It also contains some new work on the Peierls instability, not previously reported.

Much attention has been paid to the question of proving the existence of long range order in model statistical mechanical systems in which the basic atomic constituents interact with short range forces. An important example is a lattice spin system in which the spin at each site represents the localized spin of an atom located at that site and where the short range, pairwise interaction (Ising or Heisenberg) reputedly comes from an interatomic exchange energy. Another problem - so far unsolved - is the existence of periodic crystals which are supposed to come from short range (e.g. Lennard- Jones) interatomic potentials.

In the real world, however, these interactions are not given a-priori; it is ultimately itinerant electrons and their correlations that give rise to the long range ordering. In other words, a deep unsolved problem is to derive magnetism or crystallization from the Schrödinger equation - or some caricature of it. The construction of a simple model based on itinerant electrons, and the rigorous derivation of ordering from it, is a challenge for mathematical physicists.

A lattice model of itinerant electrons that is believed to display ferro and antiferromagnetism - if it could be solved - is the Gutzwiller-Hubbard-Kanamori model [1-3]. We are also unable to solve it, but we have succeeded in proving that a simplified version of it does display crystallization. It is a toy model but it is, to our knowledge, the first example of this genre. Roughly, it has the same relation to the Hubbard model as the Ising model has to the quantum Heisenberg model. Here we shall give a brief report of our results, the full details of which will appear elsewhere [4].

The Hubbard model, which is the motivation for our model, is defined by the second-quantized Hamiltonian

$$H^H = \sum_{\sigma} \sum_{x,y\in\Lambda} t_{xy} c^\dagger_{x\sigma} c_{y\sigma} + 2U \sum_{x\in\Lambda} n_{x\uparrow} n_{x\downarrow} \tag{1}$$

with the following notation: $\sigma = \pm 1$ denotes the 2 spin states of the electrons; Λ is a finite lattice; $c_{x\sigma}$ is a fermion annihilation operator for a spin σ electron at $x \in \Lambda$; $n_{x\sigma} = c^\dagger_{x\sigma} c_{x\sigma}$ is the number operator for spin σ at x. Electrons interact only at the same site with an energy $2U$, and $t_{xy} = t_{yx}$ is the hopping amplitude from x to y.

The crucial assumption will be made that Λ is the union of two sublattices $A \cup B$ such that $t_{xy} = 0$ unless $x \in A, y \in B$ or $x \in B, y \in A$. The number of sites in Λ, A, and B are denoted by $|\Lambda|, |A|$ and $|B|$. The two sublattices need not be isomorphic. Thus, for example, a face-centered cubic lattice is allowed with A=face centers and B=cube corners. Λ is said to be *connected* if every $x, y \in \Lambda$ can be joined by a "path" through nonzero t's.

In our model we assume that one kind of electron (say $\sigma = -1$) does not hop. One can say that these electrons are infinitely massive. The Hamiltonian is then

$$H = \sum_{x,y\in\Lambda} t_{xy} c^\dagger_x c_y + 2U \sum_{x\in\Lambda} n_x W(x) \tag{2}$$

with $n_x = c^\dagger_x c_x$ (the subscript σ is omitted since the dynamic electrons have $\sigma = +1$) and with $W(x) = +1$ if a fixed electron ($\sigma = -1$) is at x and $W(x) = 0$

otherwise. It will be recognized immediately that (2) is just a fancy way to say that the movable electrons are independent, with a single particle Hamiltonian

$$\tilde{h} = T + V, \tag{3}$$

with T being the $|\Lambda|$-square matrix t_{xy} and $V_{xy} = 2UW(x)\delta_{xy}$. It is convenient to write $\tilde{h} = h + U$ with

$$h = T + US \tag{4}$$

with $S_{xy} = s_x\delta_{xy}$, and $s_x = 1$ (resp. -1) if x is occupied (resp. unoccupied). The $\{s_x\}$ are like Ising spins.

We shall henceforth call the movable particles "electrons" and the fixed particles "nuclei". This terminology is most appropriate if $U < 0$, for then H does represent a lattice system of electrons and nuclei in which all Coulomb interactions except the on-site electron-nucleus attraction and the on-site infinite nuclear repulsion are regarded as "screened out". This conforms with the spirit of the original Hubbard model. The electron number, the nuclear number and the total particle number, all of which commute with H, are, respectively

$$N_e = \sum_{x\in\Lambda} n_x, \quad N_n = \frac{1}{2}\sum_{x\in\Lambda}[s_x + 1] = \sum_{x\in\Lambda} W(x), \quad \mathcal{N} = N_e + N_n. \tag{5}$$

It is to be emphasized that W does *not* represent a disordered potential. We take the "annealed", not the "quenched" system. The ground state for fixed N_e and N_n is defined by taking the ground state of H (with respect to the electrons) for each W and then minimizing the result with respect to the location of the nuclei. The ground state energy will be denoted by $E(N_e, N_n)$. Likewise, for positive temperature, we take $Tre^{-\beta H}$ with respect to *both* the electron variables and the nuclear locations.

Since $t_{xy} = 0$ for x, y on the same sublattice, the spectra of H and H^H are invariant under $t_{xy} \to -t_{xy}$ (all x, y). There is also a hole-particle symmetry. If $c_x^\dagger \to c_x, c_x \to c_x^\dagger, n_x \to 1 - n_x, t_{xy} \to -t_{xy}$, then $H(U) \to H(-U) + 2UN_n$. If $W(x) \to 1 - W(x)$, then $H(U) \to H(-U) + 2UN_e$. A similar symmetry holds for H^H. Thus, the $U > 0$ and $U < 0$ cases are similar — from the mathematical point of view.

Our results are of two kinds. The first concerns the ground state which we prove always has perfect crystalline ordering and an energy gap (defined later). The second concerns the positive temperature ($1/kT = \beta < \infty$) grand canonical state. For large β and dimension $d \geq 2$, the long range order persists. For small β it disappears and there is exponential clustering of the nuclear correlation functions.

The Ground State

Theorem *1*: (a) *Let $U < 0$. Under the condition $\mathcal{N} \equiv N_e + N_n \leq 2|A|$, the ground state (i.e. we minimize $E(N_e, N_n)$ over the set $N_e + N_n \leq 2|A|$) occurs for $N_e = N_n = |A|$ and a minimizing nuclear configuration is $W(x) = W_A(x) \equiv 1(x \in A), 0(x \notin A)$ Under the condition $\mathcal{N} \leq 2|B|$, the ground state is $N_e = N_n = |B|$ and the B sublattice is occupied ($W = W_B$). If Λ is connected, these are the only groundstates, i.e. if $|A| > |B|$ the ground state is unique; if $|A| = |B|$ it is doubly degenerate. No assumption is made about the sign or magnitude or periodicity of the t_{xy} other than $t_{xy} = 0$ for $x, y \in A$ or $x, y \in B$.*

(b) *Let $U > 0$. Under the condition $\mathcal{N} \geq |A| + |B|$, there are two ground states: $N_e = |A|, N_n = |B|, W = W_B$ and $N_e = |B|, N_n = |A|, W = W_A$. If Λ is connected, these are the only ground states.*

The condition $\mathcal{N} = |\Lambda|$ is called the *half-filled* band. If $|A| = |B| = |\Lambda|/2$, the crystal occurs at the half-filled band. If Λ is a cubic lattice, for example, this means that the ground state is a cubic lattice of period $\sqrt{2}$ oriented at 45° with respect to Λ.

Theorem 1 relies heavily on the fact that the electrons are fermions. The ground state would be completely different if they were bosons. For bosons and for Λ a cubic lattice, the nuclei would all be clumped together in the ground state instead of being spread out into a crystal. By using rearrangement inequalities it is possible to describe this clumping quantitatively.

Next, we define the *energy gap*. Actually two different definitions are of interest. First, let

$$E(\mathcal{N}) \equiv \min\{E(N_e, N_n) | N_e + N_n = \mathcal{N}\}. \tag{6}$$

The *chemical potential* is defined by

$$\mu(\mathcal{N}) \equiv E(\mathcal{N} + 1) - E(\mathcal{N}). \tag{7}$$

We say there is a *gap of the first kind* at $\mathcal{N}$ if

$$\mu(\mathcal{N}) - \mu(\mathcal{N} - 1) \geq \varepsilon_1 > 0 \tag{8}$$

with ε_1 being independent of the size of the system. We say there is a *gap of the second kind* at N_e, N_n if

$$E(N_e + 1, N_n) + E(N_e - 1, N_n) - 2E(N_e, N_n) \geq \varepsilon_2 > 0. \tag{9}$$

In other words, the nuclear number is fixed in the second definition.

A gap is one indication that the system is an insulator, for it implies that it costs more energy to put a particle into the system than is gained by removing one. The first kind of gap is relevant if one views our model as an approximation to the Hubbard model; the second is relevant from the "electrons and nuclei" point of view.

Theorem *2: Assume that Λ is not only connected but that every $x, y \in \Lambda$ can be connected by a chain with $|t_{ab}| \geq \delta$ for every a, b on the chain. Also, assume that $\|T\| \leq \tau$. The following energy gaps exist with $\varepsilon_2 \geq \varepsilon_1 > 0$ and depending only on δ, τ and U:*

$U < 0$: First kind at $\mathcal{N} = 2|A|$ and at $\mathcal{N} = 2|B|$. Second kind at $N_e = |A|, N_n = |A|$ and at $N_e = |B|, N_n = |B|$.

$U > 0$: First kind at $\mathcal{N} = |A| + |B|$. Second kind at $N_e = |A|, N_n = |B|$ and at $N_e = |B|, N_n = |A|$.

In order to give the flavor of our methods, the proof of Theorem 1 will be given here. The proof of Theorem 2 is more complicated.

Proof of Theorem 1: Let $\lambda_1 \leq \lambda_2 \leq \ldots$ be the eigenvalues of h in (4). They depend on the nuclei. For N_e electrons the ground state energy, E, of H satisfies

$$E - UN_e = \sum_{j=1}^{N_e} \lambda_j \geq \sum_{\lambda_j < 0} \lambda_j = \frac{1}{2}[Trh - Tr|h|]. \tag{10}$$

But $Trh = U\sum s_x = 2UN_n - U|\Lambda|$ and $|h| = \{T^2 + U^2 + UJ\}^{1/2}$ with $J_{xy} = t_{xy}[s_x + s_y]$. Since the function $0 < x \to x^{1/2}$ is concave, $f(y) = Tr\{T^2 + U^2 + yUJ\}^{1/2}$ is concave in $y \in [-1, 1]$. But $f(-1) = f(1)$ (since spec(h) is invariant under $T \to -T$), so $f(1) \leq f(0)$, with equality if and only if $J \equiv 0$. Thus

$$E \geq U\mathcal{N} - \frac{1}{2}U|\Lambda| - \frac{1}{2}Tr(T^2 + U^2)^{1/2}. \tag{11}$$

If Λ is connected, the only ways to have $J \equiv 0$ are either $W = W_A$ or W_B. Consider $U < 0$ and $\mathcal{N} \leq 2|A|$, whence, from (11),

$$E \geq U(2|A| - \frac{1}{2}|\Lambda|) - \frac{1}{2}Tr(T^2 + U^2)^{1/2}. \tag{12}$$

If $W = W_A$ then, as is easily seen, h has precisely $|A|$ negative and $|B|$ positive eigenvalues. Thus, if $W = W_A$ and $N_e = |A|$, then (12) is an equality. The other cases are similar. □

Grand Canonical Ensemble

First, we define the partition function Ξ. A nuclear configuration is denoted by $S = \{s_x\}, s_x = \pm 1$, and the λ_j are the eigenvalues of h in (4). If μ_n, μ_e are the nuclear and electronic chemical potentials,

$$\Xi = \sum_S \exp[\frac{1}{2}\beta\mu_n(\sum_x s_x + |\Lambda|)] \prod_{j=1}^{|\Lambda|} \{1 + \exp[-\beta(\lambda_j + U - \mu_e)]\}. \qquad (13)$$

The product in (13) is just the well known Fermi-Dirac grand canonical partition function for the electrons. We want to choose μ_e, μ_n so that $\langle N_e \rangle = \frac{1}{2}|\Lambda|$ and $\langle N_n \rangle = \frac{1}{2}|\Lambda|$, or $\langle \sum s_x \rangle = 0$. From the fact that if $T \to -T$, spec$(h) \to$ spec(h), one has that when $S \to -S$, spec$(h) \to -$spec(h). It is then easy to see that the desired chemical potentials are $\mu_e = \mu_n = U$. Since $\sum \lambda_j = U \sum s_x$, (13) becomes in this case (after dropping an irrelevant factor $2^{|\Lambda|} e^{\beta U |\Lambda|/2}$)

$$\Xi = \sum_S \exp[-\beta F(S)] \qquad (14)$$

with

$$\begin{aligned} -\beta F(S) &= \sum_{j=1}^{|\Lambda|} \ell n \cosh(\frac{1}{2}\beta\lambda_j) = Tr \ell n \cosh[\frac{1}{2}\beta h] \\ &= Tr \ell n \cosh[\frac{1}{2}\beta(T^2 + U^2 + UJ)^{1/2}]. \end{aligned} \qquad (15)$$

Thus, (14) is like an Ising model partition function but with a complicated, temperature dependent "spin-spin" interaction, $F(S)$, given by (15) in terms of the eigenvalues of h. With respect to this "spin measure" we can talk about the presence or absence of long range nuclear order in the thermodynamic limit. In order to discuss this limit we henceforth restrict ourselves to a translation invariant nearest neighbor hopping on a cubic lattice in d dimensions.

What we are able to prove is summarized in the schematic figure below and in

Theorem *3: For all U and sufficiently large β there is long range order for $d \geq 2$ (the same kind as in the ground state). For all U and sufficiently small β there is none; indeed there is exponential decay of all nuclear correlation functions.*

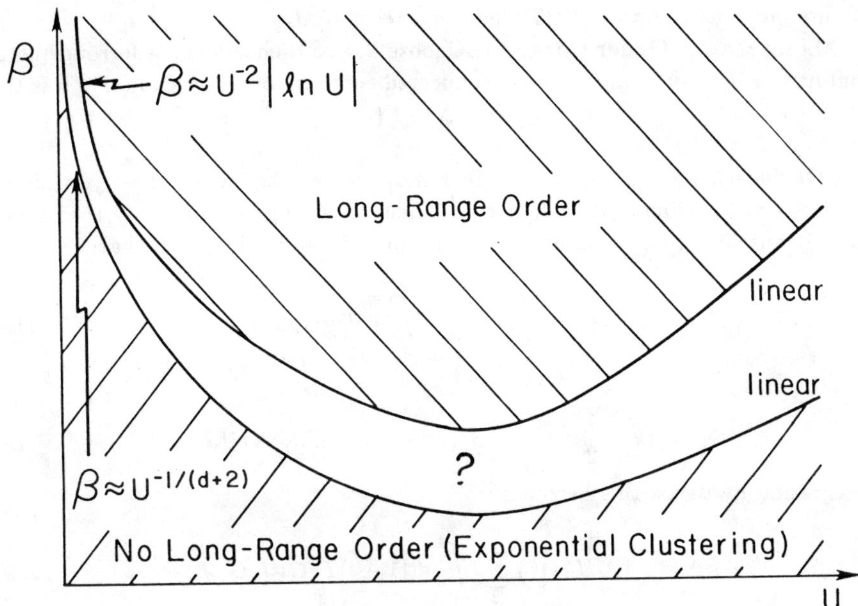

Presumably there is no intermediate phase, but we cannot prove this. For large U, β_c is clearly linear in U. For small U, we have the bound on the lower $\beta_c \sim U^{-1/(2+d)}$ and the bound on the upper $\beta_c \sim |\ell n U| U^{-2}$. Our guess is that the true state of affairs is $\beta_c \sim U^{-2}$.

For large β we use a Peierls argument; for small β we use Dobrushin's uniqueness theorem. A sketch of our proof - omitting many important details - is the following. For simplicity, we here consider only large $U > 0$.

Define, for $x > 0$,

$$P(x) = \ell n \cosh x^{1/2}. \tag{16}$$

[We note in passing that P is concave, and we see from the last expression in (15), using the proof in Theorem 1, that $F(S)$ has its minima at precisely the same values of W (or S) as in Theorem 1.] P is a Pick (or Herglotz) function with the representation

$$P'(x) = \frac{1}{2} x^{-1/2} \tanh x^{1/2} = \sum_{k=0}^{\infty} [(k + \frac{1}{2})^2 \pi^2 + x]^{-1}. \tag{17}$$

We are interested in $x = \frac{1}{4}\beta^2(T^2 + U^2 + UJ)$ with $J_{xy} = t_{xy}(s_x + s_y)$.

Long Range Order (large β): Choose S, and then define antiferromagnetic contours in the usual way. If γ is a connected contour component and $\Gamma + \gamma$ is the whole contour, we want to prove that $\Delta \equiv F(\Gamma + \gamma) - F(\Gamma) \geq C|\gamma|$ for a suitable constant $C = C(U)$.

Obviously, $J_{\Gamma+\gamma} = J_\Gamma + J_\gamma$. To remove γ, we change s_x to $-s_x$ inside γ. For $0 \leq t \leq 1$, define $J(t) = J_\Gamma + tJ_\gamma$. Then, assuming for simplicity that Γ lies entirely outside γ, we have, by differentiating (15) and using (17), that

$$-\beta\Delta = \frac{1}{4}\beta^2 U \sum_{k=0}^{\infty} \int_o^1 dt Tr(G_k J_\gamma), \tag{18}$$

with

$$G_k = [(k + \frac{1}{2})^2\pi^2 + \frac{1}{4}\beta^2(T^2 + U^2 + UJ_\Gamma + tUJ_\gamma)]^{-1}. \tag{19}$$

Integrating by parts, this becomes

$$-\beta\Delta = -(\beta^4 U^2/16)\sum_k \int_o^1 dt(1-t) Tr G_k J_\gamma G_k J_\gamma + \ldots \tag{20}$$

[The ... terms in (20) come from $t = 0$ in the partial integration. They are small and easily bounded for large U, but they have to be treated more judiciously when U is small.] If A, B, D are matrices with $A \geq B \geq 0$, then $TrAD^\dagger AD \geq TrBD^\dagger BD$. Also, $A \geq B > 0 \Rightarrow 0 < A^{-1} \leq B^{-1}$. Moreover, $T^2 \leq (2d)^2$ and $UJ \leq T^2 + U^2$. Using this, we have the matrix inequality

$$G_k \geq \left[(k + \frac{1}{2})^2\pi^2 + \frac{1}{2}\beta^2\left((2d)^2 + U^2\right)\right]^{-1}. \tag{21}$$

Summing on k,

$$\Delta \geq \text{(const.) } \mathrm{U}^2\left((2\mathrm{d})^2 + \mathrm{U}^2\right)^{-3/2}\mathrm{Tr}(\mathrm{J}_\gamma)^2.$$

Clearly $Tr(J_\gamma)^2 =$ (const.)$|\gamma|$. Thus, for U large, $C(U) \geq$ (const.)U^{-1}, and thus long range order exists in $d \geq 2$ if β/U is large enough.

Absence of Long Range Order (small β): Dobrushin's uniqueness theorem [5-7], together with the modification in [8], gives the following criterion for exponential clustering. We have to bound the change in F when we change the spins at x and y (taking the worst case with respect to the other spins). Call this f_{xy}. The requirement is that for some $m > 0$ and all x,

$$\beta \sum_y f_{xy} \exp[m|x-y|] < 1. \tag{22}$$

By an argument similar to the preceding (large U)

$$f_{xy} \cong (\beta^3 U^2/16) \sum_k T_r G_k J_x G_k J_y$$
$$\cong (\beta^3 U^2/16) \sum_k |G_k(x,y)|^2. \tag{23}$$

Here, $J_x = \delta_x T + T\delta_x$ and $G_k(x,y)$ is the x,y matrix element of (19).

To implement (22) we now require an *upper* bound on $G_k(x,y)$ that has exponential decay. For this purpose the Combes-Thomas argument is ideal. Let Q be the matrix with elements $Q_{xy} = \delta_{xy} e^{n \cdot x}$ and with $|n| = 1$. Then

$$QG_k^{-1}Q^{-1} = G_k^{-1} + R_k \equiv L_k. \tag{24}$$

The "remainder" R_k can be bounded for large U: $\|R_k\| \le C\beta^2 U$ for some constant C. Similar to (21), for large U

$$G_k^{-1} \ge [(k+\frac{1}{2})^2\pi^2 + \frac{1}{8}\beta^2 U^2] \equiv \alpha_k. \tag{25}$$

Thus,

$$\|L_k^{-1}\| \le [\alpha_k - C\beta^2 U]^{-1}. \tag{26}$$

Since $|(L_k^{-1})_{xy}| \le \|L_k^{-1}\|$, we have from (24), $|(QG_kQ^{-1})_{xy}| \le \|L_k^{-1}\|$, and thus

$$|(G_k)_{xy}| \le \exp[n\cdot(y-x)]\{\alpha_k - C\beta^2 U\}^{-1}. \tag{27}$$

This holds for all n, so summing on k,

$$f_{xy} \le \text{ (const.) } U^{-1} e^{-2|x-y|}.$$

Hence, (22) holds with $m = 1$ if β/U is small enough.

The Peierls Instability

Another itinerant electron model in which translation invariance is partially broken was introduced by Peierls [11], see also [14]. Consider a linear chain of $2N$ atoms (arranged in a ring) and let $w_i (i = 1, \ldots, 2N)$ denote the distance from atom i to atom $i+1$ (with $2N+1 \equiv 1$). If $d > 0$ is the equilibrium distance, the distortion energy associated with the w_i is taken to be

$$U(\{w\}) = \kappa \sum_{i=1}^{2N} (w_i - d)^2. \tag{28}$$

Here $\kappa > 0$ is a measure of stiffness.

Now suppose there are $2N$ (spin $\frac{1}{2}$) electrons ($\frac{1}{2}$ filled band) which can hop between nearest neighbor atoms, with an amplitude $t_i = t(w_i)$ to hop from atom i to $i+1$ and from $i+1$ to i. The function $t(w)$ will be explained shortly. The ground state energy of these $2N$ electrons is

$$K(\{w\}) = 2\sum_{j=1}^{N} \lambda_j(T) \tag{29}$$

where $\lambda_1 \leq \lambda_2 \leq \ldots \leq \lambda_{2N}$ are the eigenvalues of the $2N \times 2N$ matrix T having matrix elements $T_{i,i+1} = T_{i+1,i} = t_i = t(w_i)$ and $T_{ij} = 0$ otherwise. The total Hamiltonian is

$$H(\{w\}) = K(\{w\}) + U(\{w\}) \tag{30}$$

and the problem is to determine the ground state energy

$$E = \min_{\{w\}} H(\{w\}) \tag{31}$$

and the configuration(s) $\{w\}$ that attains it.

The function $t(w)$ should be something like ae^{-cw}, but here we shall adopt the choice of Su, Schrieffer and Heeger [12]:

$$t(w) = b - c(w - d) \tag{32}$$

with $b, c > 0$.

One guess for a minimizing $\{w\}$ is $w_i = W = \text{const}$. However, Peierls observed that if we take $w_i = W + (-1)^i\delta$ (a *dimerized configuration*) the energy will be lowered. The reason is that when $\{w\}$ has period two a gap in the spectrum of T opens up at $k = \pm\pi/2$; since there are just enough electrons to fill the lower band, the K energy will be lowered by an amount (const.) $\delta^2 ln|\delta|$ for small δ. (This comes from the fact that the perturbation theory is singular at the band edges.) On the other hand, U depends on δ like $\kappa\delta^2$, so it is favorable to have $\delta \neq 0$. This breaking of translation invariance is called the Peierls instability. The interesting question, which was open for a long time, is this: Is the translation invariance merely broken from period one to period two, or does something more complicated happen? Conceivably the minimizing $\{w\}$ could have a longer period - or perhaps no periodicity at all.

Our answer to this question is that period two always holds.

Theorem 4: *For all b, c, d, κ, N there are precisely two minima for E*

$$w_i = W + (-1)^i\delta \text{ or } w_i = W - (-1)^i\delta, \tag{33}$$

with W and δ being constants depending on κ, b, c, N. Moreover, there is a gap, i.e. a jump in the chemical potential:

$$\Delta\mu \equiv [E(2N+1) - E(2N)] - [E(2N) - E(2N-1)] > \varepsilon, \tag{34}$$

with ε depending on κ, b, c but independent of N, d. (Here $E(2N+1)$ is defined as in (30), (31) but with K replaced by $2\sum_{j=1}^{N}\lambda_j(T) + \lambda_{N+1}(T)$ and, for $E(2N-1)$, K is replaced by $2\sum_{j=1}^{N}\lambda_j(T) - \lambda_N(T)$.)

As in the proof of Theorem 1, the proof of Theorem 4 uses the concavity of $Tr\sqrt{X}$ with respect to X, and with $X = T^2$. The crucial observation is that for the dimerized configuration (33), T^2 is translation invariant.

It turns out that if the U in (28) is modified, an "integrable system" is obtained. In terms of t, using (32), we have

$$(w_i - d)^2 = \alpha + \beta t_i + \gamma t_i^2 \tag{35}$$

with α, β, γ constants. The modification is

$$U(\{w\}) = \alpha + \sum_{j=1}^{2N}[-\beta ln t_i + \gamma t_i^2] \tag{36}$$

with α, β, γ arbitrary constants (but (32) is unchanged). This system is presumably not very different from ours (but there is no proof). It was investigated in [13]. However, the solution in [13] covers all electron numbers *except* the half-filled band ($2N$ electrons), which is the most interesting case since it is only here that one expects to see the instability manifest itself in the form proposed by Peierls - namely the partial breaking of translation invariance into a period two invariance.

The support of the U.S. National Science Foundation, grant PHY-8515288, is gratefully acknowledged.

References

1. M. C. Gutzwiller, Phys. Rev. Letters *10*, 159-162 (1963), and Phys. Rev. *134*, A923-941 (1964), and *137*, A1726-1735 (1965).
2. J. Hubbard, Proc. Roy. Soc. (London), Ser. A*276*, 238-257 (1963), and *277*, 237-259 (1964).
3. J. Kanamori, Prog. Theor. Phys. *30*, 275-289 (1963).

4. T. Kennedy and E. H. Lieb, An itinerant electron model with crystalline or magnetic long range order, Physica A (1986), in press.
5. R. L. Dobrushin, Theory Probab. and Its. Appl. *13*, 197-224 (1968).
6. L. Gross, Commun. Math. Phys. *68*, 9-27 (1979).
7. H. Föllmer, J. Funct. Anal. *46*, 387-395 (1982).
8. B. Simon, Commun. Math. Phys. *68*, 183-185 (1979).
9. J. M. Combes and L. Thomas, Commun. Math. Phys. *34*, 251-270 (1973).
10. Proceedings of the Groningen Conference on Statistical Mechanics and Field Theory, Springer Lecture Notes in Physics (to appear).
11. R.E. Peierls, *Quantum Theory of Solids*, Clarendon Press, Oxford (1955), p. 108.
12. W.P. Su, J.R. Schrieffer and A.J. Heeger, Phys. Rev. Lett. *42*, 1698-1701 (1979).
13. S.A. Brazovskii, N.E. Dzyaloshinskii and I.M. Krichever, Sov. Phys. JETP *56*, 212-225 (1982).
14. H. Fröhlich, Proc. Roy. Soc. *A223*, 296-305 (1954).

DETERMINISTIC THEORY OF TURBULENCE IN HYDRODYNAMICS (COHERENT STRUCTURES)

V.P.Maslov

Department Math. Acad. of Sc. of USSR - Moscow

§1. Asymptotical solution of gas dynamics equations for large Reynolds numbers.

For the sake of simplicity we consider isotropic gas:

(1) $$\frac{\partial u}{\partial t} + \langle u, \nabla \rangle u = \frac{1}{Re} \Delta u - \frac{1}{\rho} \nabla P,$$

(2) $$\frac{\partial \rho}{\partial t} + \operatorname{div} \rho u = 0, \qquad \rho = \Phi(P).$$

We set $\varepsilon h^2 = 1/Re$, where Re is the Reynolds number, (h is a small parameter). We shall seek asymptotical solution in the form:

(3) $$u = V^0(x,t) + a^0(\eta,x,t) + hV^1(x,t) + ha^1(\eta,x,t) + h^2V^2(x,t) + ha^2(\eta,x,t) + \ldots$$

(4) $$P = P^0(x,t) + h\pi^0(\eta,x,t) + hP^1(x,t) + h^2\pi^1(\eta,x,t) + \ldots,$$

$$\eta = g(x,t)/h, \; V^0 = (V^{01}, V^{02}, V^{03}), \; a^0 = (a^{01}, a^{02}, a^{03}), \; x = (x^1, x^2, x^3),$$

where $a^0, a^1, \ldots, \pi^0, \pi^1, \ldots$ are periodical functions of η , which do not contain a non-oscillating term, i.e.

$$\int_0^{2\pi} a^0 d\eta = \int_0^{2\pi} a^1 d\eta = \int_0^{2\pi} \pi^0 d\eta = \int_0^{2\pi} \pi^1 d\eta = 0;$$

V^0 and V^1, P^0 and P^1 are the non-oscillating parts respectively. Since $\eta = g(x,t)/h$ the functions u and P oscillate rapidly.

We set:

(5) $$\langle a^0, \nabla g \rangle = 0,$$

(6) $\frac{\partial g}{\partial t} + \langle V^\circ, \nabla g \rangle = 0.$

Then we substitute (3) and (4) into (2). The term at h^{-1} in the continuity equation vanishes. We equate to zero the term at h^0, and obtain:

(7) $\frac{\partial \rho^\circ}{\partial t} + \operatorname{div} \rho^\circ V^\circ = 0, \qquad \rho^\circ = \varphi(P^\circ),$

(8) $\rho^\circ \langle \nabla g, a^1_\eta \rangle + \operatorname{div}(\rho^\circ a^\circ) = 0,$

at h^1 we obtain:

$$\frac{\partial (P^1 \varphi'(P^\circ))}{\partial t} + \operatorname{div} \rho^\circ V^1 + \operatorname{div}(V^\circ P^1 \varphi'(P^\circ)) +$$

$$+ \rho^\circ \langle \nabla g, a^2_\eta \rangle + \operatorname{div} \rho^\circ a^1 + \frac{\partial \pi^\circ}{\partial \eta} \langle a^1, \nabla g \rangle \varphi'(P^\circ) + \varphi'(P^\circ) \frac{\partial \pi^\circ}{\partial t} +$$

(9) $+ \operatorname{div} V^\circ \varphi'(P^\circ) + \varphi'(P^\circ)(\pi^\circ + P^1) \langle a^1_\eta, \nabla g \rangle + \{\operatorname{div}[\varphi'(P^\circ)(\pi^\circ + P^1) a^\circ]\} = 0.$

We denote by

$$\int_0^\eta \mathcal{F} d\eta = \int_0^\eta \mathcal{F} d\eta - \frac{1}{2\pi} \int_0^{2\pi} d\eta \int_0^\eta \mathcal{F} d\eta = \int_0^\eta \mathcal{F} d\eta + \frac{1}{2\pi} \int_0^{2\pi} \eta \mathcal{F} d\eta,$$

so that

$$\int_0^{2\pi} \int_0^\eta \mathcal{F} d\eta = 0.$$

The solution of equations (8) and (9) for a^1 and a^2 due the condition: $\int_0^{2\pi} a^1 d\eta = \int_0^{2\pi} a^2 d\eta = 0$

can be expressed in terms of the integral given above. We integrate (9) from 0 to 2π and obtain:

(10) $\frac{\partial (P^1 (\varphi'(P^\circ))}{\partial t} + \operatorname{div} \rho^\circ V^1 + \operatorname{div}(V^\circ P^1 (\varphi'(P^\circ)) + \frac{1}{2\pi} \int_0^{2\pi} \operatorname{div}(a^\circ \pi^\circ \varphi'(P^\circ)) d\eta.$

Then we substitute (3) and (4) into (1) and denote $(a^\circ + V^\circ)$ by u°, and $(a^1 + V^1)$ by u^1. Since, according to (5) $\langle \frac{\partial u}{\partial \eta}, \nabla g \rangle = 0,$

we have from (6), that the term at h^{-1} vanishes. We equate to zero the term at h^0:

$$(11)\quad \frac{\partial u^0}{\partial t}+\langle u^0,\nabla\rangle u^0+\langle u^1,\nabla g\rangle\frac{\partial u^0}{\partial\eta}=\varepsilon(\nabla g)^2u^0_{\eta\eta}-\frac{1}{\rho^0}P^0-\frac{1}{\rho^0}\nabla g\frac{\partial\pi^0}{\partial\eta}.$$

Since due to (8):

$$\frac{\partial u^0}{\partial\eta}=\frac{\partial a^0}{\partial\eta}\quad\text{and}\quad\langle a^1,\nabla g\rangle=-\frac{1}{\rho^0}\int_0^\eta\operatorname{div}\rho^0a^0d\eta,$$

then for V^0 and a^0 we obtain the equations:

$$\frac{\partial a^0}{\partial t}+\langle V^0,\nabla\rangle a^0+\langle a^0,\nabla\rangle V^0-\varepsilon(\nabla g)^2a^0_{\eta\eta}+\frac{\nabla g}{\rho^0}\pi^0_\eta+$$

$$(12)\quad +\langle V^1,\nabla g\rangle a^0_\eta+\Big[\langle a^0,\nabla\rangle a^0-\frac{1}{\rho^0}\int_0^\eta\operatorname{div}\rho^0a^0d\eta\frac{\partial a^0}{\partial\eta}\Big]_{osc},$$

where

$$(13)\quad [F]_{osc.}=F-\frac{1}{2\pi}\int_0^{2\pi}F\,d\eta,$$

$$(14)\quad \begin{aligned}&\frac{\partial V^0}{\partial t}+\langle V^0,\nabla\rangle V^0+\frac{1}{2\pi}\int_0^{2\pi}\Big[\langle a^0,\nabla\rangle a^0-\frac{1}{\rho^0}\int_0^\eta\operatorname{div}\rho^0a^0d\eta.a^0_\eta\Big]d\eta+\frac{1}{\rho^0}\nabla P^0=\\&=V^0_t+\langle V^0,\nabla\rangle V^0+\frac{1}{2\pi}\int_0^{2\pi}\Big[\langle a^0,\nabla\rangle a^0+\frac{1}{\rho^0}a^0\operatorname{div}\rho^0a^0\Big]d\eta+\frac{1}{\rho^0}\nabla P^0=0.\end{aligned}$$

We equate to zero the term at h^1:

$$(15)\quad\begin{aligned}&\frac{\partial u^1}{\partial t}+\langle u^1,\nabla\rangle u^0+\langle u^0,\nabla\rangle u^1+\langle u^1,\nabla g\rangle u^1_\eta+\langle u^2,\nabla g\rangle u^0_\eta+\\&+\frac{\nabla g}{\rho^0}\frac{\partial\pi^1}{\partial\eta}+\frac{1}{\rho^0}\nabla\pi^0+\frac{1}{\rho^0}\nabla P^1-\Big(\frac{1}{(\rho^0)^2}\nabla g\frac{\partial\pi^0}{\partial\eta}-\varepsilon\Delta u^0+\\&+\frac{1}{(\rho^0)^2}\nabla P^0\Big)\varphi'(P^0)(P^1+\pi^0)=0,\end{aligned}$$

since the condition $\rho=\varphi(P)$ yields $\rho=\rho^0+h\varphi'(P^0)(P^1+\pi^0)+\ldots$

Hence, by expressing $\langle a^1,\nabla g\rangle$ and $\langle a^2,\nabla g\rangle$ from (8) and (9), we obtain:

$$\frac{\partial a^1}{\partial t}+\langle V^0, \nabla a^1\rangle+\langle a^1, \nabla\rangle V^0+\langle a^0, \nabla\rangle V^1+$$

$$+\langle V^1, \nabla a^0\rangle+\langle V^1, \nabla y\rangle a^1_\eta+\langle V^2, \nabla y\rangle a^0_\eta+$$

$$+\frac{\nabla y}{\rho^0}\pi^1_\eta+\frac{1}{\rho^0}\nabla\pi^0-\frac{\nabla\rho^0}{(\rho^0)^2}\Phi'(\rho^0)\pi^0-\varepsilon a^1_{\eta\eta}(\nabla y)^2-$$

$$-\varepsilon\Delta a^0-\frac{\nabla y}{(\rho^0)^2}\Phi'(\rho^0)\pi^0_\eta\pi^0+\Big[\Xi a^0_\eta-$$

$$-\frac{a^0_\eta}{\rho^0}\int_0^\eta \operatorname{div}(\rho^0 a^1)\,d\eta+\langle a^1, \nabla\rangle a^0+\langle a^0, \nabla\rangle a^1-$$

$$-\frac{1}{\rho^0}\int_0^\eta \operatorname{div}(\rho^0 a^0)\,d\eta\, a^1_\eta\Big]_{osc.}=0,$$

$$\frac{\partial V^1}{\partial t}+\langle V^0, \nabla\rangle V^1+\langle V^1, \nabla\rangle V^0+\frac{1}{\rho^0}\nabla P^1-$$

$$-\frac{\Phi'(\rho^0)\rho^1}{(\rho^0)^2}\nabla\rho^0-\varepsilon\Delta V^0+\int_0^{2\pi}\Big\{\langle a^1, \nabla\rangle a^0+$$

$$+\langle a^0, \nabla\rangle a^1+\frac{a^0}{\rho^0}\operatorname{div}(\rho^0 a^1)+\frac{a^0}{\rho^0}\operatorname{div}(\rho^0 a^0)+$$

$$+\frac{a^1}{\rho^0}\operatorname{div}\rho^0 a^0+\Xi a^0_\eta\Big\}\,d\eta=0,$$

$$\Xi = \langle a^2, \nabla g\rangle + \frac{1}{\rho^0}\int_0^\eta \operatorname{div}(\rho^0 a^1)\,d\eta =$$

$$= \Phi'(P^0)\frac{\partial \pi^0}{\partial t} + \operatorname{div}(V^0\pi^0\Phi'(P^0)) -$$

$$-\frac{1}{\rho^0}\int_0^\eta \{\pi^0_\eta\langle a^1, \nabla g\rangle \Phi'(P^0)(\pi^0 + P^1)\langle a^1_\eta, \nabla g\rangle +$$

$$+\operatorname{div}(\pi^0 + P^1)a^0\}\,d\eta = \Big[\frac{(\pi^0 + P^1)}{(\rho^0)^2}\Phi'(P^1)\cdot$$

$$\cdot\int_0^\eta \operatorname{div}\rho^0 a^0\,d\eta\Big]_{osc.} - \frac{1}{\rho^0}\int_0^\eta \operatorname{div}\{\Phi'(P^0)(\pi^0 + P^1)a^0\}\,d\eta +$$

$$+\Phi'(P^0)\frac{\partial \pi^0}{\partial t} + \operatorname{div}(V^0\pi^0\,\Phi'(P^0)) =$$

$$= \Big[\frac{(\pi^0 + P^1)}{(\rho^0)^2}\Phi'(P^0)\int_0^\eta \langle a^0, \nabla \rho^0\rangle\,d\eta\Big]_{osc.} - \frac{1}{\rho^0}\int_0^\eta \langle a^0, \nabla\Phi'(\pi^0 + P^2)\rangle\,d\eta +$$

(18) $$+\Phi'(P^0)\frac{\partial \pi^0}{\partial t} + \operatorname{div}(V^0\pi^0\,\Phi'(P^0)).$$

Here $\pi^0 \triangleq \pi^0(a^0, V^0, g)$ and thus:

(18') $$\frac{\partial \pi^0}{\partial t} = \mathcal{F}(a^0, V^0, V^1, g).$$

Evidently, if equations (5)-(9) and (12)-(18) hold, then after the substitution of (3), (4) into equations(1), (2) only the terms of order $\mathcal{O}(h^2)$ remain.

However, there is an indeterminate function $\langle V^2, \nabla g\rangle$ in equations (12)-(18), and by setting different values of it, we obtain for the same initial data some expressions, which differ by a value of the unity order, and after the substitution of which into the system of gas dynamics equations we obtain a small right-hand part. Thus, the term at $O(h^2)$ in (3), (4) effects the terms at h^1.

§2. Problem of averaging.

We have a chain of relations connected together, which define the values a^o, a^1, a^2,...,V^o, V^1, V^2,...,
Due to this chain there is non-uniqueness, and small correction terms effect zero terms.

Let L be a character length associated with the initial problem. Consider a sphere Ω^{x^0} with the centre at x^o, such that $\Omega^{x^0}/L^3 \sim \delta$, $h \ll \delta \ll 1$. For example, $\delta \sim h^{1/10}$. We average the values u and P over this sphere. Evidently,

$$u_{av.} = \frac{1}{\delta}\int_\Omega u\,dx = V^o(x^o,t) + O(h^{9/10}),$$

if all the functions in (3) and (4) are smooth enough. By analogy, time average is: $\int_t^{t+\delta} u\,dt = V^o(x,t) + O(h^{9/10})$, i.e. time average is equal to space average as $h \to 0$. Thus, the average of u, ρ and P is $V^o(x,t)$, $\Omega^{\,o}_{x}(t)$ and $P^o(x,t)$. It turns out, that for any different rapidly varying solutions of problem (1)-(4) (conditions (3),(4) can be considered only at t=0, i.e. as initial conditions) with the same initial conditions the average values coincide. So that, the experiment, which may be different for the same data, must coincide at the average. We prove this statement.

LEMMA 1. Equation (14) is invariant with respect to the change $a^o(x,t,\eta) \to a^o(x,t,\eta+\mathcal{F}(x,t))$, where $\mathcal{F}(x,t)$ is any smooth function.

PROOF. Denote by $[[R(a^o)]]$ the difference between $R(a^o(\eta+\mathcal{F})) - [R(a^o(\eta))]\big|_{\eta=\eta+\mathcal{F}}$ We have:

$$[[\langle a^o(\eta+\mathcal{F}),\nabla\rangle a^o(\eta+\mathcal{F}) - \frac{1}{\rho^o}\int_0^\eta \operatorname{div}\rho^o a^o(\eta+\mathcal{F})\,d\eta\, a^o_\eta(\eta+\mathcal{F})]] =$$

$$= \sum_i a^o_i \mathcal{F}_{x_i} a^o_\eta - \int_0^\eta \frac{\partial a^o_i}{\partial\eta}\mathcal{F}_{x_i}\,d\eta\, a^o_\eta = 0. \tag{19}$$

Lemma 1 is proved.

LEMMA 2. If $\frac{d\mathcal{F}}{dt} = \frac{\partial \mathcal{F}}{\partial t} + V^0 \Delta \mathcal{F} = -\langle V^1, \nabla g \rangle$,
then the function $a^0(\eta + \mathcal{F}(x,t), x, t)$ satisfies equation (12), where $V^1 = 0$ (i.e. the term $\langle V^1, \nabla g \rangle a^0_\eta$ is absent).

PROOF. By substituting instead of $a^0(\eta)$ the expression $a^0(\eta + \mathcal{F})$ we obtain due to (19) an additional term of the form $\frac{d\mathcal{F}}{dt} a^0_\eta$. Hence, we have the statement of our Lemma. Now we express π^0_η from equations (12) and (5); according to (6) we obtain:

$$(20) \quad \frac{\pi_\eta}{\rho_0} = \frac{1}{(\nabla g)^2} \sum_{i,j=1}^{3} \left\{ a^0_i a^0_j \frac{\partial^2 g}{\partial x_i \partial x_j} - 2 a^0_i \frac{\partial g}{\partial x_j} \frac{\partial V^0_j}{\partial x_i} \right\}.$$

(Note, that if at initial time $|\nabla g| \neq 0$, it doesn't vanish for any t. Really, we consider $P(x_0,t) = \nabla g(X(x_0,t),t)$ where $X(x_0,t)$ is obtained from equation $\dot{X} = V^0$, $X(x^0_i, 0) = x^0$. We obtain, that $\dot{P}_i = \sum_{j=1}^{3} P_j \frac{\partial V^0_j}{\partial x_i}$ which means, if $|P(x_0,\bar{t})| = 0$ for any $t = \bar{t}$, then $|P(x_0,t)| \equiv 0$ for all t. And since $|P(x^0,0)| \neq 0$, then also $|P(x^0,t)| = |\nabla g(X(x^0,t),t)| \neq 0$ and therefore $|\nabla g(x,t)| \neq 0$).

§3. Problem with initial and boundary conditions.

Now we consider equation (1), (2) in a sphere $\Omega \subset \mathbb{R}^3$ which is not necessary bounded, and set boundary conditions. Let $\Gamma = \partial\Omega$ be the boundary of the domain Ω, $u_n|_\Gamma$ be a velocity component normal to the boundary at the boundary points, $u_\ell|_\Gamma$ - a tangent component. We set impenetrability conditions:

$$(21) \qquad u_n|_\Gamma = 0, \qquad u_\ell|_\Gamma = \mathcal{R}(x),$$

where $\mathcal{R}$ is the velocity of the boundary Γ. The initial conditions are the same as above, for example:

$$(22) \begin{cases} u|_{t=0} = u_0(x,h) = \varphi(x, g_0(x)/h), \\ \rho|_{t=0} = \rho_0(x), \end{cases}$$

where $\varphi(x,\eta)$ is a function periodical in η,

$$g^0|_\Gamma = 0. \tag{23}$$

We shall seek the solution in the form (3), (4). We apply a smoothing operator to the series (3), for example:

$$\bar{\bar{u}} = \frac{1}{\sigma^{3/2}} \int e^{-(x-\xi)^2/\sigma} u(\xi, t, h)\, d\xi. \tag{24}$$

If σ is small (for example, $\sigma = h^{1/10}$), then by integrating by parts we obtain for $\bar{\bar{u}}$ the expansion:

$$\bar{\bar{u}} = \frac{1}{\sigma^{3/2}} \int e^{-(x-\xi)^2/\sigma} V^0 + hV^1 + o(h). \tag{25}$$

The function $\bar{\bar{u}}$ may also be considered as an average. The impenetrability condition on the boundary yields the relation: $V_n^0|_\Gamma = 0$. Now since $\frac{\partial g}{\partial t} + V_n \frac{\partial g}{\partial R} = 0$ on Γ, the relation $\frac{\partial g}{\partial t} = 0$ holds, i.e. due to (23) $g(x,t)|_\Gamma = 0$. Then (26) yields $a_n^0|_{\eta=0,\Gamma} = 0$, and (26) yields also, that $(a_e^0 + V_e^0)|_\Gamma = R$.

Further, if we set $V_n^1|_\Gamma = 0$, and hence, $a_n^1|_{\eta=0,\Gamma} = 0$, $(a_e^1 + V_e^1) = 0$, then we obtain, that the average velocity V^0 and pressure P^0 are independent of the value of V^1, and hence, they are the same for the whole set of solutions of our problem. Really, equation (11) contains only the term $\langle V^1, \nabla g \rangle$. On the boundary Γ it is, evidently, proportional to V_n^1, i.e. it is equal to zero. Therefore, the function $\mathcal{F}$ used in Lemmas 1 and 2 is equal to a constant on Γ. From condition (23) for t = 0 we have, that this constant is equal to zero. Thus, the value of a^0 on the boundary Γ is independent of V^1. So, for the initial-boundary problem the average values of velocity and density are independent of a solution of this problem which was chosen from the infinite set of its solutions. For all these solutions the average values are the same. We can write, in order to obtain these average values, an equation for some fictitious function Ψ , which is, generally speaking, not associated with any solution of our problem, but the equation for it is compatible with equations for V^0 and P^0, these equations form a closed system, which defines V^0 and P^0, i.e. the average values of our initial problem. We write these equations.

Notations:

$\Psi = \Psi(x, \eta, t)$ is a fictitious function, $g(x,t)$ is the phase of it, $\Psi = (\Psi_1, \Psi_2, \Psi_3)$, $x = (x_1, x_2, x_3) \in \Omega \subset \mathbb{R}^3$, $\bar{u} = \bar{u}(x,t)$ is the average velocity, $\bar{u} = (\bar{u}_1, \bar{u}_2, \bar{u}_3)$, $\bar{u}_n(\Psi_n)$ is normal to the surface $\Gamma = \partial\Omega$, $\bar{u}_e(\Psi_e)$ is a tangent component of velocity (of fictitious function), $\bar{\rho} = \bar{\rho}(x,t)$ is average density, $\bar{P} = \bar{P}(x,t)$ is average pressure.

Initial and boundary conditions:

$$g|_{t=0} = g^0(x), \quad g^0|_\Gamma = 0,$$

$$\bar{u}(x,0) + \Psi\left(x, \frac{g^0(x)}{h}, 0\right) = u_0(x, h),$$

$$\bar{P}(x,0) = P_0(x);$$

$$\bar{u}(x,0) = \frac{1}{2\pi} \lim_{\theta\to\infty} \int_\theta^{2\pi+\theta} u_0(x, g_0(x)/\eta)\, d\eta,$$

$$\bar{u}_n\Big|_\Gamma = 0, \quad (\bar{u}_e + \Psi_e)\Big|_{\Gamma, \eta=0} = R, \quad \Psi_n\Big|_{\Gamma,\eta=0} = 0. \tag{26}$$

Equations for average values:

$$\frac{\partial\bar{u}}{\partial t} + \langle\bar{u}, \nabla\rangle\bar{u} + \frac{1}{2\pi}\lim_{\theta\to\infty}\int_\theta^{2\pi+\theta}\left[\langle \varkappa, \nabla\rangle\varkappa + \frac{1}{\bar{\rho}}\operatorname{div}(\bar{\rho}\varkappa)\right]d\eta = -\frac{1}{\bar{\rho}}\nabla\bar{P}, \tag{27}$$

$$\frac{\partial\bar{\rho}}{\partial t} + \operatorname{div}\bar{\rho}\bar{u} = 0, \tag{28}$$

(29) $\bar{p} = \Phi(\bar{\rho})$.

Equations for fictitious vector-function $\varkappa$ and for its phase:

(30) $\frac{\partial g}{\partial t} + \langle \bar{u}, \nabla g \rangle = 0$,

(31) $$\frac{d\varkappa}{dt} + \langle \varkappa, \nabla \rangle \varkappa + \langle \varkappa, \nabla \rangle \bar{u} - \frac{1}{\bar{\rho}} \int_0^{\eta} \operatorname{div} \bar{\rho} \varkappa \, d\eta \, \frac{\partial \varkappa}{\partial \eta} +$$

$$+ \frac{\nabla g}{(\nabla g)^2} \sum_{i,j=1}^{3} \varkappa_j \left[\varkappa_i \, g_{x_i x_j} - 2 g_{x_i} \bar{u}^i_{x_j} \right] -$$

$$- \frac{1}{2\pi} \lim_{\theta \to \infty} \int_{\theta}^{2\pi+\theta} \Big[\frac{1}{\bar{\rho}} \varkappa \operatorname{div}(\bar{\rho} \varkappa) - \langle \varkappa, \nabla \rangle \varkappa +$$

$$+ \sum_{i,j=1}^{3} \frac{\nabla g}{|\nabla g|^2} g_{x_i x_j} \varkappa_i \varkappa_j \Big] d\eta = \varepsilon |\nabla g|^2 \frac{\partial^2 \varkappa}{\partial \eta^2},$$

where $\frac{d}{dt} = \frac{\partial}{\partial t} + \langle \bar{u}, \nabla \rangle$.

Thus, we solve this problem with a fictitious function $\varkappa$, which is "similar" to a^0, i.e. to real behaviour of the problem solution, but generally speaking, it is not a solution (it has another phase), and we obtain average values of our problem. As for the fictitious vector-function $\varkappa$, we can say, that there is such a function $\mathcal{F}(x,t)$, that $\varkappa(\eta + \mathcal{F}(x,t), x, t)$ for $\eta = g(x,t)/h$ will be an asymptotics to the true fluctuation correction to the mean velocity V^0. Thus, the average with respect to the period from all the powers of $V^0 + \varkappa$ and all possible products of their components will be asymptotics for the average values of the powers of velocity.

We note, that the same properties are obtained for any waves, which carry along the particles of fluid or gas: in imcompressible fluid, in the theory of convection, in plasma, in dynamics of chemical reactions, in the theory of formation and movement of clouds and so on, i.e. everywhere if there is vortex turbulence.

Now we consider the case of oscillating waves which are weak with respect to amplitude. We shall seek the solution of problem (1), (2) in the form similar to (3), (4).

$$u = V^{0}(x,t) + h a^{0}(\eta,x,t) + h V^{1}(x,t) + h a^{1}(\eta,x,t) + \dots,$$

$$p = P^{0} - h^{2}\pi^{0}(\eta,x,t) + h^{2}P^{1}(x,t) + \dots,$$

$$\eta = (\eta_1, \eta_2, \eta_3),\ V^{0} = (V^{01}, V^{02}, V^{03}),\ a^{0} = (a^{01}, a^{02}, a^{03}),$$

where $\eta_i = \frac{g_i(x,t)}{h}$, a^0, a^1, π^0, π^1 are functions periodical in η with the average equal to zero. Then V^0 will be defined by the Euler equations, and a^0 will be defined by the three-dimensional Navier-Stokes equation in lagrangian coordinates:

$$\begin{aligned} &\frac{da^{0}}{dt} + \langle a^{0}, \nabla\rangle V^{0} + \langle a^{0}, \tilde{\nabla}_{\eta}\rangle a^{0} = \varepsilon(\tilde{\nabla}_{\eta})^{2}a^{0} - \frac{1}{\rho}\tilde{\nabla}_{\eta}\pi^{1}, \\ &\langle \tilde{\nabla}_{\eta}, a^{0}\rangle = 0, \end{aligned} \tag{32}$$

where $\frac{d}{dt} = \frac{\partial}{\partial t} + \langle V^{0}, \nabla\rangle$ and $\tilde{\nabla}_{\eta} = A\nabla_{\eta}$, $A = \left\| \frac{\partial g_i}{\partial x_j} \right\|$. This equation coincides with the equation obtained by the author in [1,2,3].

§4. Theory of turbulent pulsations of compressible gas with heat transport.

In a volume Ω bounded by a surface Γ a system of hydrodynamical equations with heat transport is considered:

$$\rho\left(\frac{\partial u_i}{\partial t}+u_k\frac{\partial u_i}{\partial x_k}\right)=-\frac{\partial P}{\partial x_i}+\vec{g}+\frac{1}{Re}\frac{\partial}{\partial x_k}\left\{\mu\left(\frac{\partial u_i}{\partial x_k}+\right.\right.$$

$$(33)\qquad \left.\left.+\frac{\partial u_k}{\partial x_i}-\frac{2}{3}\delta_{ik}\frac{\partial u_\ell}{\partial x_\ell}\right)\right\}+\frac{1}{Re}\frac{\partial}{\partial x_i}\left(\xi\frac{\partial u_\ell}{\partial x_\ell}\right),$$

$$(34)\qquad \frac{\partial \rho}{\partial t}+\operatorname{div}(\rho u)=0,$$

$$\rho T\left(\frac{\partial S}{\partial t}+\langle u,\nabla S\rangle\right)=\frac{1}{Re}\operatorname{div}(\varkappa\nabla T)+\frac{1}{Re}\frac{\gamma}{2}\left(\frac{\partial u_i}{\partial x_k}+\right.$$

$$(35)\qquad \left.+\frac{\partial u_k}{\partial x_i}-\frac{2}{3}\delta_{ik}\frac{\partial u_\ell}{\partial x_\ell}\right)^2+\frac{\xi}{Re}(\operatorname{div}u)^2,$$

$$(36)\qquad \rho=f(P)\varphi(S),\quad T=\psi(\rho,S,P),\quad u|_\Gamma=0,\quad T|_\Gamma=\mathcal{T}(x,t).$$

Here $u = (u_1, u_2, u_3)$ is the vector of velocity, ρ is density, P is pressure, T is tempeture, S is entropy, $\vec{g}$ is gravitational force, $\mu, \xi, \varkappa$ are respectively the viscosity and heat conduction coefficients normed by the Reynolds number, δ_{ik} is the Kronecher's symbol, over all the repeated indexes here and below summation is assumed to be carried out, $x = (x_1, x_2, x_3)$, French quotes denote scalar product in R^3.

In order the system (33)-(36) be closed, it is necessary to have one more relation connecting the temperature T with functions in ρ, T and S. However, our calculations are independent of this relation, and thus we make no assumptions about it.

We denote $h = 1/\sqrt{Re}$ and consider equations (33)-(36) for large Reynolds numbers, i.e. we assume, that $h \ll 1$ is a small parameter.

Let at the initial time the velocity, density and entropy be rapidly oscillating functions, and the pressure be a slowly varying function:

$$u|_{t=0} = u^0\left(\frac{\Phi^0(x)}{h}, x\right), \quad \rho|_{t=0} = \rho^0\left(\frac{\Phi^0(x)}{h}, x\right),$$

(37)
$$S|_{t=0} = S^0\left(\frac{\Phi^0(x)}{h}, x\right), \quad P|_{t=0} = P^0(x),$$

where $\Phi^0(x)$, $P^0(x)$ are some smooth functions, $u^0(\eta, x)$, $\rho^0(\eta, x)$, $S^0(\eta, x)$ are smooth periodical for $\eta \to \infty$ with respect to η functions with the period τ: $u^0(\eta+\tau, x) - u^0(\eta, x) \to 0$, similarly for ρ^0, S^0 for $\eta \to \infty$ uniformly with respect to x.

Our aim is to construct an asymptotical with respect to the parameter h solution, which describes the dynamics of average with respect to rapid oscillations values of the functions u, ρ, S. In order to construct this asymptotics we use the same method as in [3 - 6].

We shall seek a self-similar asymptotical solution of problem (33-37) in the form similar to (37):

$$u = V\left(\frac{\Phi(x,t)}{h}, x, t\right) + hV_1\left(\frac{\Phi(x,t)}{h}, x, t\right) + O(h^2),$$

$$\rho = R\left(\frac{\Phi(x,t)}{h}, x, t\right) + hR_1\left(\frac{\Phi(x,t)}{h}, x, t\right) + O(h^2),$$

(38)
$$S = S\left(\frac{\Phi(x,t)}{h}, x, t\right) + hS_1\left(\frac{\Phi(x,t)}{h}, x, t\right) + O(h^2),$$

$$P = \pi(x,t) + h\pi_1\left(\frac{\Phi(x,t)}{h}, x, t\right) + O(h^2).$$

Here the functions $\Phi(x,t)$, $\pi(x,t)$, $V(\eta,x,t), \ldots, \pi_1(\eta,x,t)$, which are sought, are smooth with respect to all their arguments, and $V(\eta,x,t), \ldots, \pi_1(\eta,x,t)$ for $\eta \to \infty$ are periodical with respect to the variable η with the period τ. By $O(h^2)$ we denote the functions, the value of which has the order h^2.

We substitute expansion (38) into the Navier-Stokes equation. By denoting $1/Re = h^2$ and carrying out the same constructions as in [3], we obtain the Hamilton-Jacobi equation:

$$(39) \quad \frac{\partial \varphi}{\partial t} + \langle \bar{V}, \nabla \varphi \rangle = 0$$

and the condition:

$$(40) \quad \langle [V], \nabla \varphi \rangle = 0.$$

Here, as before, $\bar{V}$ means the average value of velocity, $[V]$ are oscillating components of velocity:

$$(41) \quad \bar{V} = \frac{1}{\tau} \lim_{\theta \to \infty} \int_{\theta}^{\theta+\tau} V(\eta, x, t)\, d\eta, \quad [V] = V - \bar{V}.$$

Evidently, the initial value of the phase φ is equal to φ°:

$$(42) \quad \varphi\big|_{t=0} = \varphi^{\circ}(x).$$

We substitute expansion (38) into equations (33) - (35). It is easily seen, that after the substitution by omitting the terms of order $O(h)$ we obtain the following equations:

$$(43) \quad R\left\{\frac{dV}{dt} + \langle [V], \nabla \rangle V + \langle V_1, \nabla \varphi \rangle \frac{\partial V}{\partial \eta}\right\} = -\nabla \pi + \vec{g} - \nabla \varphi \frac{\partial \pi_1}{\partial \eta} + |\nabla \varphi|^2 \frac{\partial}{\partial \eta}\left(\mu \frac{\partial V}{\partial \eta}\right),$$

$$(44) \quad \frac{dR}{dt} + \langle [V], \nabla R \rangle + \langle V_1, \nabla \varphi \rangle \frac{\partial R}{\partial \eta} + R \frac{\partial}{\partial \eta} \langle \nabla \varphi, V_1 \rangle + R \operatorname{div} V = 0,$$

$$RT\left\{\frac{dS}{dt}+\langle[V],\nabla S\rangle+\langle V_1,\nabla\varphi\rangle\frac{\partial S}{\partial\eta}\right\}=$$

(45) $$=|\nabla\varphi|^2\frac{\partial}{\partial\eta}\left(\varkappa\frac{\partial T}{\partial\eta}\right)+\mu|\nabla\varphi|^2\cdot\left|\frac{\partial V}{\partial\eta}\right|^2,$$

where

(46) $$\frac{d}{dt}=\frac{\partial}{\partial t}+\langle\bar{V},\nabla\rangle$$

is the total derivative along the characteristics of the equation (39). We seek the function $\langle\nabla\varphi,[V_1]\rangle$. For this purpose we note, that the continuity equation (34) can be rewritten in the form:

(47) $$\left\{\frac{\partial}{\partial t}+\langle u,\nabla\rangle\right\}\ln\rho=-\operatorname{div}u.$$

We take the logarithm of (36) and apply the operator $\frac{\partial}{\partial t}+\langle u,\nabla\rangle$ to the obtained equality. By (42) and (35) we obtain:

$$\left\{\frac{\partial}{\partial t}+\langle u,\nabla\rangle\right\}\ln f(P)=\operatorname{div}u-$$

(48) $$-\frac{h^2}{\rho T}\frac{\partial\ln\Phi(S)}{\partial S}\left\{\operatorname{div}(\varkappa\nabla T)+\frac{\mu}{2}\left(\frac{\partial u_i}{\partial x_k}+\frac{\partial u_k}{\partial x_i}-\frac{2}{3}\delta_{ik}\frac{\partial u_\ell}{\partial x_\ell}\right)^2+\right.$$
$$\left.+\zeta(\operatorname{div}u)^2\right\}.$$

We substitute into (48) the expansion (38). By (39) and (40) we have:

$$\left\{\frac{\partial}{\partial t}+\langle V+hV_1+O(h^2),\nabla\rangle\right\}\ln f(\pi(x,t)+$$
$$+h\pi_1\left(\frac{\varphi}{h},x,t\right)+O(h^2))=\left\{\frac{\partial}{\partial t}+\langle V,\nabla\rangle\right\}\ln f(\pi(x,t))+$$

$$+\frac{f'(\pi)}{f(\pi)}\left\{\frac{\partial \varphi}{\partial t}+\langle V,\nabla\varphi\rangle\right\}\frac{\partial \pi_1}{\partial \eta}+O(h)=$$

(49)

$$=\frac{d\ln f(\pi)}{dt}+\langle [V],\nabla\rangle \ln f(\pi)+O(h),$$

where $f'(\pi)=\frac{\partial f(\pi)}{\partial \pi}$ is the total derivative d/dt defined by the formula (46). By using this equality and omitting the terms of order $O(h)$ in (48) we obtain the relation:

$$\frac{d\ln f(\pi)}{dt}+\langle [V],\nabla\rangle \ln f(\pi)=\frac{\partial}{\partial \eta}\langle \nabla\varphi, V_1\rangle+$$

(50)

$$+\operatorname{div} V-\frac{|\nabla\varphi|^2}{RT}\frac{\partial \ln \varphi(s')}{\partial s}\left\{\frac{\partial}{\partial \eta}\left(\varkappa\frac{\partial T}{\partial \eta}+\mu\left|\frac{\partial V}{\partial \eta}\right|\right\}.$$

We separate in (50) the oscillating summands:

$$\frac{\partial}{\partial \eta}\langle \nabla\varphi, V_1\rangle=\langle [V],\nabla\rangle \ln f(\pi)-\operatorname{div}[V]+$$

$$+|\nabla\varphi|^2\cdot\left[\frac{1}{RT}\frac{\partial \ln \varphi(s)}{\partial s}\left\{\frac{\partial}{\partial \eta}\left(\varkappa\frac{\partial T}{\partial \eta}\right)+\mu\left|\frac{\partial V}{\partial \eta}\right|^2\right\}\right].$$

The latter equality yields the relation:

$$(51)\quad \langle \nabla\varphi,[V_1]\rangle=\int_0^\eta G\,d\eta+a(x,t),\quad a(x,t)\big|_\Gamma=0,$$

where

$$G=\langle [V],\nabla\rangle \ln f(\pi)-\operatorname{div}[V]+$$

$$+|\nabla\varphi|^2\cdot\left[\frac{1}{RT}\frac{\partial \ln \varphi(s')}{\partial s}\left\{\frac{\partial}{\partial \eta}\left(\varkappa\frac{\partial T}{\partial \eta}\right)+\mu\left|\frac{\partial V}{\partial \eta}\right|^2\right\}\right].$$

At last, by averaging (50) with respect to the variable η we obtain for the pressure $\pi(x,t)$ the equation:

$$(52)\quad \frac{d}{dt}\ln f(\pi) = \operatorname{div}\bar{V} - |\nabla\varphi|^2\left\{\frac{1}{RT}\frac{\partial \ln \varphi(S)}{\partial S}\overline{\frac{\partial}{\partial\eta}\left(\varkappa\frac{\partial T}{\partial\eta}\right) + \mu\left|\frac{\partial V}{\partial\eta}\right|^2}\right\}.$$

We substitute the expression for $\langle\nabla\varphi, [V]\rangle$ into equations (43), (45):

$$(53)\quad R\left\{\frac{dV}{dt} + \langle[V],\nabla\rangle V + \frac{\partial V}{\partial\eta}\left(\int_0^\eta G\,d\eta + a(x,t)\right)\right\} = \nabla\pi - \nabla\varphi\frac{\partial\pi_1}{\partial\eta} + \vec{g} + |\nabla\varphi|^2\frac{\partial}{\partial\eta}\left(\mu\frac{\partial V}{\partial\eta}\right) - R\langle\bar{V},\nabla\varphi\rangle\frac{\partial V}{\partial\eta},$$

$$(54)\quad RT\left\{\frac{dS}{dt} + \langle[V],\nabla\rangle S + \frac{\partial S}{\partial\eta}\int_0^\eta G\,d\eta + a(x,t)\frac{\partial S}{\partial\eta}\right\} = |\nabla\varphi|^2\frac{\partial}{\partial\eta}\left(\varkappa\frac{\partial T}{\partial\eta}\right) + \mu|\nabla\varphi|^2\cdot\left|\frac{\partial V}{\partial\eta}\right|^2.$$

Like in [3 - 6] we obtain lack of uniqueness for asymptotical expansions, which depend in our case on an arbitrary function $a(x,t)$, which vanishes on the boundary Γ. However, like in [3 - 6] it turns out, that the pressure π and the average $\bar{u}, \bar{\rho}, \bar{T}, \bar{S}$ for all this set of asymptotical solutions satisfying the same initial and boundary conditions will coincide. We obtain for them and for some fictitious functions (which differ from the true asymptotics by a shift of the phase) a closed system of equations.

We introduce fictitious functions R', V', S', T', π' which differ from the functions $R, V, \ldots$ by a shift of the argument η by the value $\mathcal{F}(x, t)$, in particular, $V'(\eta, x, t) = V(\eta + \mathcal{F}(x,t), x, t)$. It is clear, that in the consequence of such a change the values of functions $R, V, \ldots$ remain unchanged:

$$\bar{V}'=\bar{V},\quad \bar{R}'=\bar{R},\quad \bar{S}'=\bar{S},\quad \bar{\pi}'=\bar{\pi},\quad \bar{T}'=T.$$

We rewrite equations (53), (54) with respect to the new functions $V', R', \ldots$

$$R'\left\{\frac{dV'}{dt}+\langle[V'],\nabla\rangle V'+\frac{\partial V'}{\partial\eta}\int_0^\eta G'd\eta\right\}+\nabla\pi+ \tag{55}$$
$$+\nabla\varphi\frac{\partial\pi_1'}{\partial\eta}-\vec{g}-|\nabla\varphi|^2\frac{\partial}{\partial\eta}\left(\mu\frac{\partial V'}{\partial\eta}\right)=-R'\left\{\frac{d\mathcal{F}}{dt}+a(x,t)\right\}\frac{\partial V'}{\partial\eta},$$

$$R'T'\left\{\frac{dS'}{dt}+\langle[V'],\nabla\rangle S'+\frac{\partial S'}{\partial\eta}\int_0^\eta G\,d\eta\right\}- \tag{56}$$
$$-|\nabla\varphi|^2\frac{\partial}{\partial\eta}\left(\varkappa\frac{\partial T'}{\partial\eta}\right)-\mu|\nabla\varphi|^2\cdot\left|\frac{\partial V'}{\partial\eta}\right|^2=-R'T'\left\{\frac{d\mathcal{F}}{dt}+a(x,t)\right\}\frac{\partial S'}{\partial\eta}.$$

Let the function $\mathcal{F}$ satisfy the equality

$$\frac{d\mathcal{F}}{dt}+a(x,t)=0,$$

and

$$\mathcal{F}\Big|_{t=0}=0$$

Hence, in particular, we have:

$$\mathcal{F}\Big|_{\Gamma}=0. \tag{57}$$

Then the right-hand sides of equations (55), (56) vanish. The fictitious functions according to (58) satisfy the same initial conditions as the true functions.

We exclude from equation (55) the unknown function π_1', then we devide equation (55) by R and consider in it only oscillating terms. After scalar multiplication by $\nabla\varphi$ we have:

$$(58)\ \left[\frac{1}{R'}\frac{\partial \pi_1'}{\partial \eta}\right]|\nabla\varphi|^2=-\langle\nabla\varphi,\frac{d[V']}{dt}+[\langle[V'],\nabla\rangle V']\rangle+$$
$$+\left[\frac{1}{R'}\right]\langle\nabla\varphi',\vec{g}-\nabla\pi\rangle.$$

After easy transformations of the first summand in the right-hand side of (58) we finally obtain $\left[\frac{1}{R'},\frac{\partial \pi_1'}{\partial \eta}\right]=\Xi$:

$$(59)\ \Xi=-\frac{1}{|\nabla\varphi|^2}\frac{\partial\varphi}{\partial x_\alpha}\left[[V_\beta]\frac{\partial}{\partial x_\beta}(V_\alpha'+\bar{V}_\alpha')\right]+\left[\frac{1}{R'}\right]\frac{\langle\nabla\varphi,\vec{g}-\nabla\pi\rangle}{|\nabla\varphi|^2}.$$

We substitute the obtained expression into (55). By denoting $V'=u$, $R'=\rho$, $S'=S$, $\bar{u}=\mathcal{U}$, $\pi=P$, $T'=T$, we finally obtain:

$$\frac{\partial u}{\partial t}+\langle u,\nabla\rangle u+\frac{\partial u}{\partial\eta}\int_0^\eta\theta\,d\eta+\frac{1}{\rho}\nabla P+$$
$$+\frac{\nabla\varphi}{|\nabla\varphi|^2}\left\{\frac{\partial\varphi}{\partial x_\alpha}\left[(u-\mathcal{U})_\beta\frac{\partial}{\partial x_\beta}(u-\mathcal{U})_\alpha\right]+\frac{1}{\rho}\langle\nabla\varphi,g-\nabla P\rangle\right\}=$$
$$=\vec{g}+|\nabla\varphi|^2\frac{\partial}{\partial\eta}\mu\frac{\partial u}{\partial\eta},$$

$$\rho T\left\{\frac{\partial S}{\partial t}+\langle u,\nabla S\rangle+\frac{\partial S}{\partial\eta}\int_0^\eta G\,d\eta\right\}=$$

(60)

$$=|\nabla\varphi|^2\frac{\partial}{\partial\eta}\left(\varkappa\frac{\partial T}{\partial\eta}\right)+\mu|\nabla\varphi|^2\cdot\left|\frac{\partial u}{\partial\eta}\right|^2,$$

$$\frac{f'(P)}{f(P)}\left\{\frac{\partial P}{\partial t}+\langle\mathcal{U},\nabla\rangle P\right\}-\operatorname{div}\mathcal{U}=|\nabla\varphi|^2\left\{\frac{1}{\tau}\lim_{\theta\to\infty}\int_\theta^{\theta+\tau}\left\{\frac{1}{\rho T}\cdot\right.\right.$$
$$\left.\left.\cdot\frac{\partial\ln\varphi(S)}{\partial S}\frac{\partial}{\partial\eta}\varkappa\frac{\partial T}{\partial\eta}+\mu\left|\frac{\partial u}{\partial\eta}\right|^2\right\}d\eta\right\},$$

$$G = \langle u - \mathcal{U}, \nabla \rangle \ln P - \operatorname{div} \langle u - \mathcal{U} \rangle +$$
$$+ |\nabla \varphi|^2 \left[\frac{1}{\rho T} \frac{\partial \ln \varphi(S)}{\partial S'} \left\{ \frac{\partial}{\partial \eta} \varkappa \frac{\partial T}{\partial \eta} + \mu \left| \frac{\partial u}{\partial \eta} \right|^2 \right\} \right],$$

$$\rho = f(P)\varphi(S), \; T = \Psi(\rho, S, P), \; u|_{\Gamma, \eta=0} = 0, \; T|_{\Gamma, \eta=0} = \mathcal{T}(x,t), \; x \in \Gamma,$$

$$\mathcal{U} = \frac{1}{\tau} \lim_{\theta \to \infty} \int_{\theta}^{\theta+\tau} u \, d\eta, \quad u|_{t=0} = u^0(\eta, x), \rho|_{t=0} = \rho^0(\eta, x),$$

$$S'|_{t=0} = S'^0(\eta, x), \quad P|_{t=0} = P^0(x),$$ and $u^0(\eta+\tau, x) - u^0(\eta, x) \to 0$, $\rho^0(\eta+\tau, x) - \rho^0(\eta, x) \to 0$, $S'^0(\eta+\tau, x) - S'^0(\eta, x) \to 0$ as $\tau \to 0$ uniformly in x.

We note, that if u^0, ρ^0, S^0 as $\eta \to \infty$ stabilize, i.e. tend to functions independent of η, what is a particular case of our condition, than system (60) becomes a system of Prandtl equations for boundary layer and of ideal gas equations.

REFERENCES

1. Maslov V.P. Resonance processes in the wave theory and self-focalization, M., 1983, VINITI.
2. Maslov V.P. Deterministical theory of hydrodynamics turbulence, Theor. and Mathem. Phys., 1985, v.65, N3.
3. Maslov V.P. Deterministic and indeterministic theory of coherent structures in turbulence, M., Acad. of Sci., Preprint N1, Depart. Math., 1986.
4. Maslov V.P. Disturbance of causality principle for unstationary equations of two-dimensional and three-dimensional gas dynamics for sufficiently large Reynolds numbers, Theor. and Mathem. Phys., 1986, v.69, N3.
5. Maslov V.P. Coherent structures, resonances and asymptotical non-uniqueness for Navier-Stokes equations for large Reynolds numbers, Uspehi Mathem. Nauk, 1986, v.41, N6.
6. Maslov V.P. Rapidly oscillating solutions of a system of compressible gas dynamical equations, Atmosphere and Ocean Phys., 1986, v.22, N12.

SUPERSTRINGS

A. Neveu,
CERN, Geneva, Switzerland

Abstract : Superstrings are the latter-day superstars for physicists in search of the Theory Of Everything (TOE) unifying all known interactions and particles. They were developed in a very different context, and it is only slowly that the discovery of their remarkable properties, briefly presented here, has given rise to such ambitions.

Superstrings are at present a very active field in particle theory. A relativistic string is a generalization of the concept of a point particle. It is the simplest extended object compatible with the axioms of special relativity. Superstrings are the first and so far the only candidates for a consistent quantum theory unifying gravitation together with strong, weak, and electromagnetic interactions, and furthermore having some chance of describing the real world.

Historically, strings were not introduced for such an ambitious enterprise, but appeared in a rather roundabout way at the end of the sixties, to describe some properties of hadronic interactions.

In the charge-exchange scattering of π^- mesons on protons for example, $\pi^- + p \rightarrow \pi^0 + n$, the low-energy cross-section exhibits peaks (Fig. 1) corresponding to the formation of a highly unstable baryonic resonance (Δ, N^*, etc.), according to the Feynman diagram of Fig. 2a. At high energy, the cross-section becomes smooth, and scattering is well described as being dominated by the exchange of the lightest mesons (ϱ, A_2, etc.) which are the origin of the forces between the pion and the proton (diagram of Fig. 2b). This is not surprising; in quantum electrodynamics, in e^+–e^- scattering, one can also observe at very low energies the effect of positronium bound states, whilst at high energies, the electron and the positron fly past each other so quickly that they do not have time to exchange more than one photon.

What is remarkable for strong interactions is the following experimental fact: as one goes higher in energy, denser and denser baryonic resonances conspire to reproduce precisely the exchange of mesons; similarly, coming down from high energies, and taking into account the exchange of heavier mesons, one reconstructs in the cross-section the peaks due to the formation of baryon resonances. This empirical

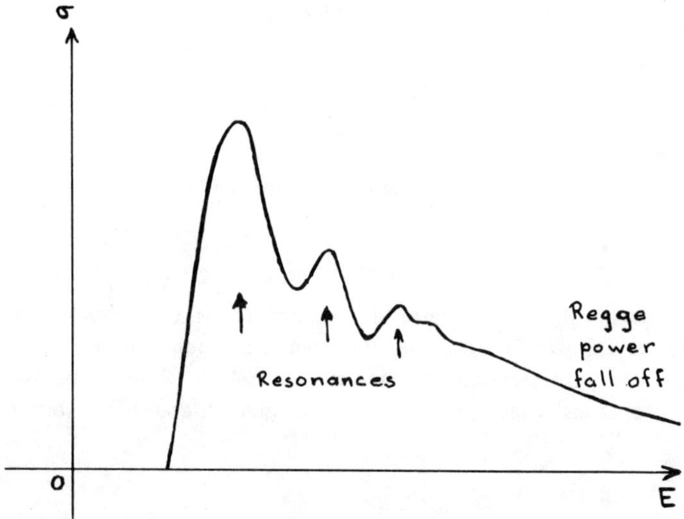

Fig. 1 Behaviour of the cross-section σ for the scattering $\pi^- p \to \pi^0 n$, as a function of the total centre-of-mass energy E

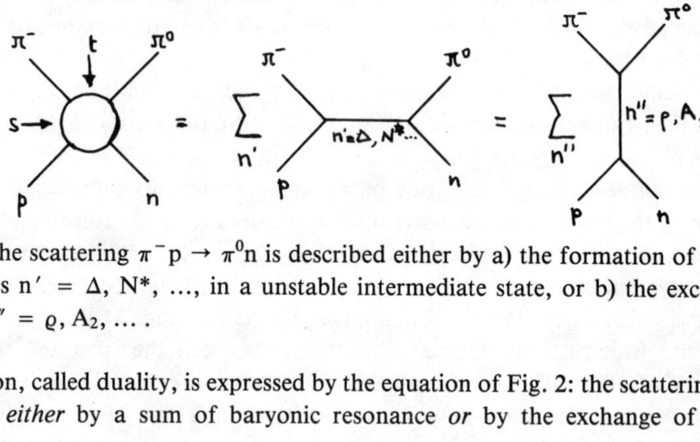

Fig. 2 The scattering $\pi^- p \to \pi^0 n$ is described either by a) the formation of baryonic resonances $n' = \Delta, N^*, \ldots$, in a unstable intermediate state, or b) the exchange of mesons $n'' = \varrho, A_2, \ldots$.

observation, called duality, is expressed by the equation of Fig. 2: the scattering can be described *either* by a sum of baryonic resonance *or* by the exchange of mesonic resonances.

Actually, to obtain the experimentally observed duality equation of Fig. 2, one must have an infinite set of resonances, and an infinite set of exchanged particles. The first example of duality was found in a simplified theory containing only mesons. This is the Veneziano formula [1], which was the historic starting point of string theories. It describes the scattering amplitude of two identical mesons of mass m and four-momenta p_1^μ, p_2^μ, with two other mesons, identical to the first ones, with four-momenta $-p_3^\mu$ and $-p_4^\mu$. By Lorentz invariance, this amplitude depends only on the kinematic invariants $s = -(p_1 + p_2)^2$ and $t = -(p_2 + p_3)^2$ (Fig. 2); s and t are related to the total energy E and the scattering angle θ in the centre-of-mass frame by ($c = 1$)

$$s = E^2, \quad t = -1/2\,(E^2 - 4m^2)(1 - \cos\theta) \quad .$$

The Veneziano amplitude is ($\hbar = 1$)

$$F(s, t) = g^2\,\{[\Gamma(\alpha' m^2 - \alpha' s)\Gamma(\alpha' m^2 - \alpha' t)]/\Gamma(2\alpha' m^2 - \alpha' s - \alpha' t)\} \quad ,$$

where g is a coupling constant, α' is the slope of Regge trajectories ($\alpha' \sim 1\ \mathrm{GeV}^{-2}$ for hadrons, and Γ is the Euler Γ function).

One trivially has $F(s, t) = F(t, s)$. The Γ function has poles for negative values of its argument, and one can expand, for example, on the s-channel poles:

$$F(s, t) = (g^2/\alpha') \sum_{n=0}^{\infty} [(-1)^n/n!] \times [1/m^2 + n - (s/\alpha')](\alpha' m^2 - \alpha' t - 1)(\alpha' m^2 - \alpha' t - 2)\ldots(\alpha' m^2 - \alpha' t - n) \quad .$$

According to Fig. 2a, these poles represent resonances with mass $M_n^2 = m^2 + (n/\alpha')$, $n = 0, 1, 2, 3, \ldots$. The residue of the nth pole is a polynomial in t, hence $\cos\theta$, of degree n. Decomposing this polynomial on the Legendre polynomials in $\cos\theta$, one thus obtains at the nth pole a set of resonances of maximal orbital momentum (spin) n. Hence, the Veneziano formula uses in a crucial way the experimentally observed fact (Fig. 3) that light quark mesons lie on linear Regge trajectories: their spin increases like the square of their mass.

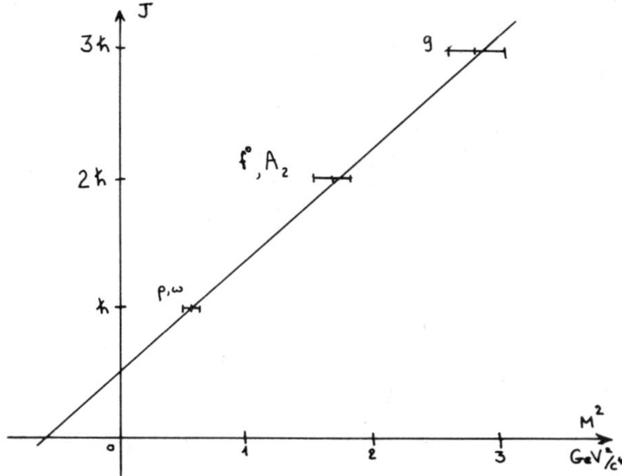

Fig. 3 Example of linear Regge trajectories

The Veneziano formula and the duality condition of Fig. 2 were quickly generalized to many-body processes, in order to describe hadron production. It was then found that the set of particles of the model was nothing but that of the quantized motions of a relativistic string [2], the simplest generalization of a point particle. For a point particle, the action is the length of the space–time path between initial and final positions. Minimizing this action naturally gives straight line propagation with constant velocity. Similarly, for a string, the action is the area of the surface swept by the string as it moves in space–time (Fig. 4). For a string with fixed ends, separated by a distance L, the minimal action over a time interval T is just LT, the area of a rectangle with sides L and T. Hence the energy of a string with fixed ends is proportional to its length, which means that the only dimensional parameter of the theory is the (constant) tension T_0 of the string at rest (if quarks are pictured as quantum numbers at the end of the string, confinement is trivial). In general, minimizing the area (the Minkowskian analogue of the soap-bubble problem), one obtains in an appropriate coordinate choice [2], precisely the same equation,

$$[(\partial^2/\partial\tau^2)X^\mu] - [(\partial^2/\partial\sigma^2)X^\mu] = 0 \quad .$$

One also obtains boundary conditions. For an open free string, these mean that the ends move at the velocity of light c. The simplest motion of an open free string is the solid rotation of a line segment of length 2L around its centre, with constant angular velocity such that the ends move with velocity c; L can vary, only T_0 is constant. The energy of such a motion is found equal to

$$E = \pi T_0 L$$

and its angular momentum

$$J = (\pi T_0/2c)L^2 \quad .$$

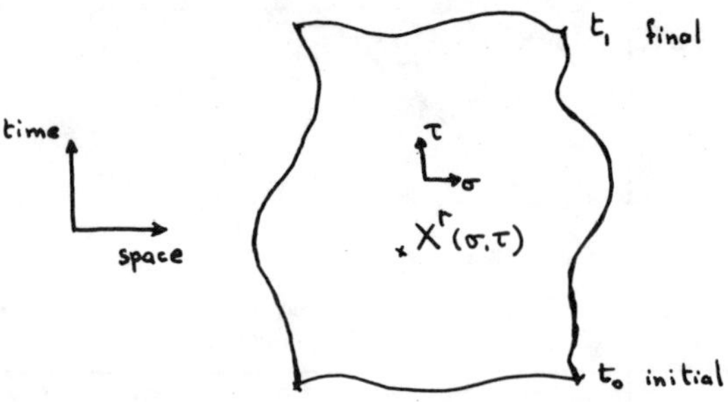

Fig. 4 The space–time worldsheet of the string

Quantizing J in integer units of $\hbar$, one finds a discrete spectrum of particles, with spin $n\hbar$, and mass

$$M_n^2 = E_n^2/c^4 = (2\pi\hbar T_0/c^3)n \quad .$$

Numerically, for strongly interacting particles, a slope (Fig. 3)

$$\alpha' \equiv J/\hbar E^2 = 1/2c\pi\hbar T_0$$

of 1 GeV^{-2} corresponds to a tension of 13 tons.

Interactions between strings are described in a geometrical fashion (Fig. 5): two open strings can join by the ends to form a single one, which can later break in two. This process can repeat itself. Twelve years ago, it was shown that integration over all intermediate string configurations corresponds exactly to the summation over the resonances of Fig. 2a. In this sense, the theory is purely geometrical. This is the source of its beauty and conceptual simplicity. It also makes it very rigid, any modification being generally the source of intractable complications.

The string just described, the simplest possible one, is also called the bosonic string, because all the particles of its spectrum have integer spin. The problem of half-integer spin has been solved by the so-called Neveu–Schwarz–Ramond model [3], or fermionic string. The bosonic string has only orbital degrees of freedom, corresponding to its position in space. Together with these, a fermionic string also has spin degrees of freedom. These can be interpreted as a distribution of half-integer spins along the string. Depending on the boundary conditions, one may have an odd number of these spins, and one obtains a fermion (Ramond sector), or an even number, and one obtains a boson (Neveu–Schwarz sector). To preserve the geometrical simplicity of the theory, and its compatibility with relativity, it is necessary that orbital and spin degrees of freedom be connected through a special symmetry,

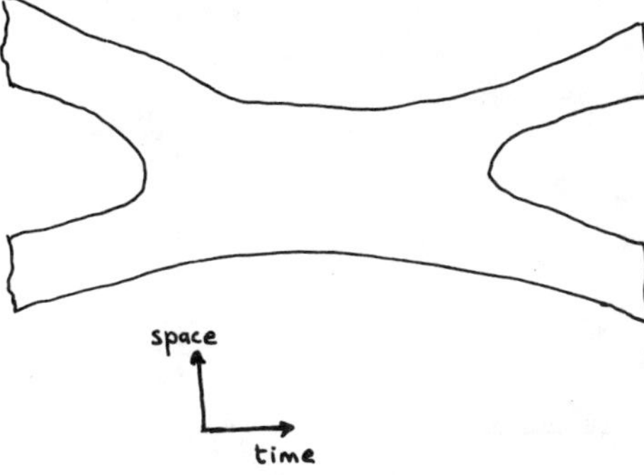

Fig. 5 The surface corresponding to the amplitude of Fig. 2a, considered as the interaction of strings in space–time

called supersymmetry, using anticommuting numbers. Supersymmetry, which appeared for the first time on this occasion, later had applications in other branches of physics and in mathematics. Like bosonic strings, fermionic strings have linear Regge trajectories, and interact in the same fashion (Fig. 5). Superstrings are fermionic strings for which only states of a given chirality have been kept.

Strings have remarkable properties not shared by ordinary point-particle field theories. The most spectacular is probably that their quantization is possible in a natural and consistent way in space–times of fixed and *a priori* rather mysterious dimensions: 26 for the bosonic string and 10 for fermionic strings. Unfortunately, there exists no simple explanation of these numbers which we could propose here. They also all have in their spectrum massless particles of spin one (for open strings) and two (for closed strings) (Fig. 6). Together with other features, this had made it clear about 12 years ago that they would not form the basis of a fundamental theory of strong interactions. Nevertheless, they still provide excellent phenomenological parametrizations. The present candidate for a fundamental theory of strong interactions is quantum chromodynamics; its numerical simulations are judged on their ability to reproduce the linear Regge trajectory spectrum characteristic of strings.

Scherk and Schwarz [4] then turned the above drawbacks of strings as theories of strong interactions, into crucial assets towards a quite different and much more ambitious aim. They suggested to reinterpret them and turn them into a unified theory of all interactions, including gravitation, thanks to the massless spin-two particle identified with the graviton. This is anologous to what happened for non-Abelian gauge theories, initially invented by Yang and Mills for strong interactions, which now form the basis of the unification of weak and electromagnetic interactions (Weinberg–Salam model). The mass scale of the theory is then changed from 1 GeV—characteristic of strong interactions—to 10^{17} GeV, which is the Planck mass, an energy where the gravitational interaction of two electrons becomes comparable

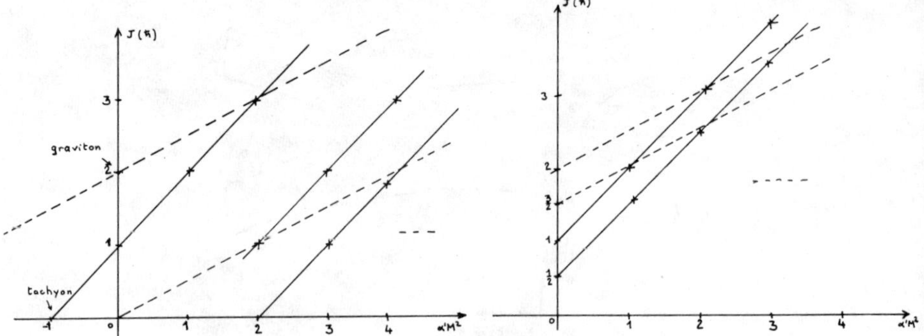

Fig. 6 a) Spectrum of open (solid lines) and closed (dotted lines) bosonic strings. b) Spectrum of open and closed fermionic strings.

with their electromagnetic interaction (according to Einstein, gravitation is coupled to the matter *energy*-momentum tensor, whilst electromagnetism is coupled to the charge, a Lorentz- invariant quantity).

Scherk and Schwarz also proposed that the extra dimensions, from 4 to 10 (or 26), are actually of very small extent (compactified), and hence invisible at presently available energies. The only observable particles are the massless ones. A given string model uniquely predicts a spectrum of massless particles and their interactions, and among them there is always gravitation and the graviton. These suggestions are interesting not only because they unify gravity with the other interactions (which supergravity also does), but also because it seems to tame the non-renormalizable infinities which appear when general relativity, and even supergravity, are quantized. Some string theories have indeed no infinity, order by order in perturbation theory.

Although the above-mentioned nice properties of string models had been known or suspected for some time, they were not the trigger of their present revival. It was the discovery in September 1984 [5] that superstrings in 10 dimensions, which naturally distinguish left from right (like weak interactions; they are called chiral theories), do not necessarily present the inconsistencies generally plaguing chiral theories. These inconsistencies, also called chiral anomalies, which appear in first order of perturbation theory, are absent in superstrings if and only if their internal symmetry group is either SO(32) (the rotation group in 32 Euclidean dimensions) or the direct product $E_8 \times E_8$ of two exceptional Lie groups. It is the first time that the cancellation of chiral anomalies selects in a natural way one (or two) internal symmetry group(s) which, in the case of $E_8 \times E_8$, contains most groups already proposed for unified theories of strong, weak, and electromagnetic interactions.

Another very interesting development concerns the compactification procedure, and is due to two mathematicians, Frenkel and Kac [6]. The initial suggestion of Scherk and Schwarz may seem *ad hoc:* what principle could one invoke to fix the size and shape of the compactified dimensions? Also, compactification would break the group of rotations in the compactified directions to (at best) some much smaller, discrete, crystallographic group. Frenkel and Kac have shown that actually this need not be so, and that one can actually generate a rank d continuous internal symmetry group by cleverly compactifying d spatial directions. This is illustrated in the case d = 2 and the group SU(3) in Fig. 7: space is taken to be the rhombus with sides $\vec{L}_1$ and $\vec{L}_2$ ($|\vec{L}_1| = |\vec{L}_2|$), with periodic boundary conditions. Imagine a bosonic open string moving in the two-dimensional compact space, together with the remaining non-compact 24 dimensions. If the size of the rhombus is chosen correctly in terms of α', one may have, according to the mass formula of Fig. 6a, eight massless scalar states, consisting, as indicated in Fig. 7, of the clockwise and counterclockwise rotations in the plane, with angular momentum one, and of six possible directions for a tachyon moving in the plane. Frenkel and Kac have shown that these eight states actually fall into the adjoint representation of SU(3), and that not only the massless states but also the full string spectrum has this SU(3) symmetry. The Frenkel–Kac

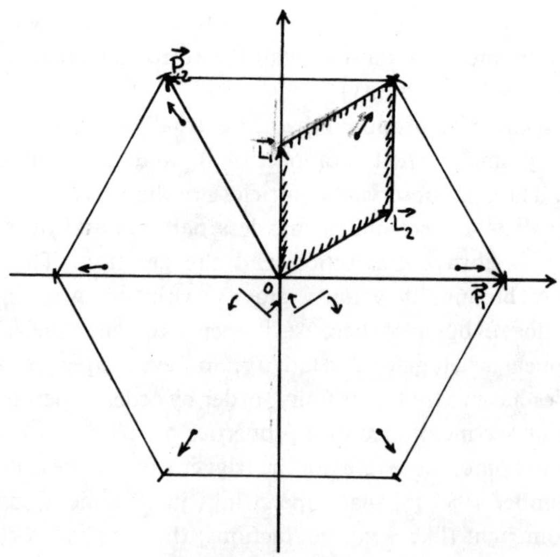

Fig. 7 The Frenkel-Kac construction for the group SU(3)

construction is actually a crucial ingredient for the $E_8 \times E_8$ superstring mentioned above.

At present, investigations are proceeding in different directions. On the formal side, string theories remain poorly understood, and one does not yet have at one's disposal a Lagrangian formulation that is as explicit and powerful as for ordinary quantum field theories. In particular, one just begins to understand their gauge invariances, which allow for the existence of high spins without ghosts; and an eventual fundamental geometric principle, analogous to the principle of equivalence in general relativity, remains to be discovered. Phenomenologically, one must understand how the dimension is effectively reduced from 10 to 4 at low energies. The reduction determines the spectrum of observable particles (leptons, quarks, gluons, intermediate bosons, etc.) and hence the predictive power of the theory. The Frenkel-Kac construction, which requires the existence of the tachyon, does not work for superstrings, and most of the other proposed schemes seem rather artificial, compared with the elegant simplicity and uniqueness of the theory to which they are applied.

REFERENCES

[1] G. Veneziano, Nuovo Cimento **57A,** 190 (1968).

[2] P. Goddard, J. Goldstone, C. Rebbi and C.B. Thorn, Nucl. Phys. **B56,** 109 (1973).

[3] P. Ramond, Phys. Rev. **D3,** 2415 (1971).
A. Neveu and J.H. Schwarz, Nucl. Phys. **B31,** 86 (1971).

[4] J. Scherk and J.H. Schwarz, Nucl. Phys. **B81,** 118 (1979).
T. Yoneya, Prog. Theor. Phys. **51,** 1907 (1974).

[5] M.B. Green and J.H. Schwarz, Phys. Lett. **149B,** 117 (1984).

[6] I.B. Frenkel and V.G. Kac, Inv. Math. **62,** 23 (1980).
P. Goddard and D. Olive, *in* Vertex operators in mathematics and physics (Publication No. 3 of the Mathematical Sciences Research Institute) (Springer Verlag, Berlin, 1984).

TWO DIMENSIONAL SCHRODINGER OPERATOR AND SOLITONS. 3-DIMENSIONAL INTEGRABLE SYSTEMS

S.P. NOVIKOV

Moscow

....

ABSTRACT

We shall consider the inverse spectral problems and soliton-like equations connected with two-dimensional Schrödinger operator and some fixed energy level $\varepsilon = \varepsilon_o$:

$$H\psi = \varepsilon_o\psi \quad , H = -\sum_\alpha(\partial_\alpha - iA_\alpha)^2 + u(x,y) \tag{1}$$

The first ideas of that theory were developped in the works [1], [2] in 1976. Modern development begun in 1984 from the works [3], [4] (see also [5], where the main problem of [1],[2] was solved for the periodic potentials). The analogous theory for rapidly decreasing potential was constructed very recently in the papers [6], [7], [8]. There is now many papers in Moscow concerning the different aspects and generalizations of that theory.

1. SOLITON EQUATIONS AND H :

The well-known integrable soliton systems (P.D.E's like KdV) connected with 1-dim Schrödinger operator $H_1 = -\partial_x^2 + u$ have the Lax form (2) :

$$\partial H_2/\partial t = [A,H_1] \Longleftrightarrow \partial u(x,t)/\partial t = f(u,u_x,\ldots) \tag{2}$$

where (KdV corresponds to m = 1)

$$A_0=\partial_x, f_0=u_x \; ; \; A_1= \partial_x^3 +(3/4)(u\partial_x+\partial_x u),\ldots,A_m=\partial_x^{2m+1}+\ldots \quad (3)$$

In general, $A = \Sigma_{j\geqslant 0} c_j A_j$, $j \geqslant 0$.

Equations (2) have always the Hamiltonian form

$$u_t = \partial_x \, \delta\mathcal{H}/\delta u(x) \quad (4)$$

where the Hamiltonian $\mathcal{H}\{u\}$ depends in fact only on the spectrum of the Schrödinger operator $H_1 = -\partial_x^{\,2} + u$

$$\mathcal{H}\{u(x)\} = \mathcal{H}\{\text{spectrum of } H_1\} \quad (5)$$

The important conclusion is : any functional of the form (5) has the infinite symmetry group (2).

This approach was developped firstly in the work [9] in 1974 ; It finds later some important applications in solid state physics (see [10]). The potentials u(x) satisfying the Euler-Lagrange equation $\delta\mathcal{H} = 0$ have some remarkable ("finite-gap") spectral properties in the case (5). The equation $\delta\mathcal{H}= 0$ is equivalent to

$$[A,H_1] = 0 \quad (6)$$

No direct generalizations of these results to dimensions $n \geqslant 2$ is possible ! The Lax equations (2) for $n \geqslant 2$ are trivial for $H = H_n$, $n \geqslant 2$. The most natural indirect generalization of this theory, which is possible only for n = 2, was found in [1], [2]. Consider the two equations

$$H\Psi=0 \;, (\partial H/\partial t - [A,H]\psi = 0 \quad (7)$$

for some linear P.D.O. A and n = 2 i.e. the coordinates are (x,y,t). The equations (7) are for some P.D.O.'s A,B "usually" equivalent to

$$\partial H/\partial t = [A,H] + BH \;, H\psi = 0 \quad (8)$$

We determine two dimensional analogs of the familiar multisoliton solutions of KdV (2) and finite-gap potentials as the solutions of the following equations

$$[L_\alpha, L_\beta] = D_{\alpha\beta} L_o \ , \ L_o = H \qquad (9)$$

for some P.D.O.'s $(L_\alpha, D_{\alpha\beta})$ $\alpha=\beta=0,1,2$.

Equations (9) are the exact 2-dimensional analog of the commutativity equation (6), describing reflexionless and finite-gap potentials for $n = 1$.

Definition : The Schrödinger operator $H = L_o$ satisfying equations (9) is called the "algebraic operator" corresponding to one energy level.

Suppose $z = x+iy$, $\partial = \partial_x - i\partial_y$. The general Schrödinger operator (1) for $n = 2$ may be reduced by the "gauge" transformation $A_\alpha \to A_\alpha + \partial_\alpha \varphi$ to the canonical form (10)

$$H = H_2 = -\partial\bar{\partial} + A\bar{\partial} + u(z,\bar{z}) \qquad (10)$$

The most interesting ones are the purely potential operators, i.e.

$$H = -\partial\bar{\partial} + u, \quad A = 0 \qquad (11)$$

with a real smooth potential $u(x,y)$ and two types of boundary conditions :

a) rapidly decreasing potentials (the scattering theory) :

$$u \to 0, \ r^2 = x^2+y^2 \to \infty$$

b) periodic potentials (Bloch's theory) :

$$u(x+T_1,y) = u(x,y+T_2) = u(x,y)$$

The general equation (8) found in [2] determines the evolution of the general operator (10) :

$$A_t = F(A,u,\nabla A, \nabla u, \ldots) \ , \ u_t = G(A,u,\nabla A, \nabla u, \ldots)$$

Its most interesting reduction (see [3], [4] determines the

time-evolution of all real purely potential Schrödinger operators (11):

$$\partial H/\partial t = [\mathbf{D}_m+\bar{\mathbf{D}}_m, H] + B_m H \qquad (12)$$

$$\mathbf{D}_m = \partial^{2m+1} + v\partial^{2m-1} + \ldots \quad , \text{order } B_m \leqslant 2m-1$$

For example, we have

$$\mathbf{D}_o = \partial \ , B_o = 0$$

$$\mathbf{D}_1 = \partial^3 + v\partial \ ; \quad B_1 = \ldots$$

The equation (12) for m = 1 has the form (13):

$$\partial u/\partial t=(\partial^3+\bar{\partial}^3)u+\partial(vu)+\bar{\partial}(vu)=(\partial^3+\partial v)u+c.c \qquad (13)$$

$$\partial v = 3\,\bar{\partial} u$$

This equation may be considered as the most natural two dimensional variant of KdV. The solution independent of the variable y reduces to KdV. The well-known physical KP-equation may be obtained as some limit of (13). So it describes all physical situations as KP, but mathematically the equation (13) has better properties than KP. The equations (12), (13) are interesting also as the infinite symmetry group of the spectral theory of 2-dim Schrödinger operator corresponding to one energy level.

2. THE INVERSE SPECTRAL PROBLEM FOR PERIODIC POTENTIALS. ALGEBRA-GEOMETRIC SOLUTIONS. CNOIDAL WAVES.

The spectral theory of periodic operators is familiar in solid state physics. It is based on the class of Bloch's eigenfunctions:

$$H\psi=\varepsilon\psi,\ \psi(x+T_1,y) = \exp(ip_1T_1)\psi(x,y)$$

$$\psi(x,y+T_2) = \exp(ip_2T_2)\psi(x,y) \qquad (14)$$

$$\psi = \psi(p_1,\, p_2,\, \varepsilon,x,y)$$

There is a complex analytic equation

$$\Phi(\varepsilon, p_1, p_2) = 0 \qquad (15)$$

whose solution determines the multivalued function ("dispersion relation")

$$\varepsilon = \varepsilon_n(p_1,p_2)\ , \quad n = 0,1,2, \dots \qquad (16)$$

For the real "quasimomenta" (p_1,p_2) we have a countable sequence of dispersion relations (16). In the generic case (for n = 2) we have $\varepsilon_q(p_1, p_2) \neq \varepsilon_m(p_1,p_2)$, $q \neq m$. Each dispersion relation $\varepsilon_m(p_1,p_2)$, determines a vector U_1–bundle η_m over the torus T^2 whose fiber contains all vectors $(\lambda\psi(\varepsilon_m, p_1, p_2))$ in the Hilbert space $\mathcal{H}(p_1,p_2)$. For the real purely potential operators (13) we have the involution

$$\varepsilon(-p_1,-p_2) = \varepsilon(p_1,p_2)\ , \ \psi^*(\varepsilon,-p_1,-p_2)=\psi(\varepsilon,p_1,p_2) \qquad (17)$$

Therefore the corresponding Chern class $C_1\ (\eta_m)$ is equal to zero. The equation (15) has also an anti-involution

$$\bar{\varepsilon}\ (p) = \varepsilon(\bar{p}) \qquad (18)$$

The analytical properties of the family of Bloch's functions (poles, asymptotics) will be our base for the inverse spectral theory. The Riemann surface $\Gamma=\Gamma(\varepsilon_0)$:

$$\Phi(\varepsilon,p_1,p_2) = \varepsilon_0 \qquad (19)$$

is by definition the "<u>complex Fermi surface</u>".

<u>Definition</u> : The periodic Schrödinger operator is algebraic or finite-zoned corresponding to one energy level ε_0 if the complex Fermi surface Γ has some finite genus $g(\Gamma)<\infty$.

<u>Conjecture</u> : A generic Schrödinger operator has $g(\Gamma) = \infty$ for any level ε_0 but the family of all algebraic operators for one given level ε_0 is probably a dense family in the space of all periodic potentials.

This is rigorously proven for $n = 1$. The algebraic 1-dimensional operators have by definition all Riemann surface (15) with finite genus $g < \infty$. The number g for $n = 1$ is equal to the number of Bloch zones or the number of gaps in the spectrum of H_1 acting on the Hilbert space $L_2(-\infty, +\infty)$ (the states of electron in some lattice). We shall describe now the analytical properties of algebraic operators corresponding to one energy level $\varepsilon = \varepsilon_0$ for $n = 2$. The Riemann surface Γ ("complex Fermi Surface") is generically nonsingular. It has some finite genus $g(\Gamma) < \infty$. The Bloch's function $\psi(x,y,P)$ for $P \in \Gamma$ should be a meromorphic function of the variable P except at two "infinite" points $\infty_1, \infty_2 \in \Gamma$ with some local parameters $w_1 = k_1^{-1}$ and $w_2 = k_2^{-1}$ near ∞_1, ∞_2.

(a) ψ has exactly g simple poles $Q_1, \ldots, Q_g \in \Gamma$ whose positions are independent (x,y)

(b) ψ has asymtotic expansions near ∞_1, ∞_2 :

$$\psi \to c_1(x,y)\exp(\kappa_1 z)\,(1+\Sigma_{j\geq 1}\mu_j(x,y)k_1^{-j}) \tag{20}$$

$$\psi \to c_2(x,y)\exp(\kappa_2 \bar{z})(1+\Sigma_{j\geq 1}\nu_j(x,y)k_2^{-j})$$

The properties (a), (b) determine $\psi(x,y,P)$ uniquely up to the gauge transformations

$$(C_1, C_2) \to (fC_1\,, fC_2),\ \text{where}\ \ f=f(x,y) \tag{21}$$

By using (21) we may choose $C_1 \equiv 1$, $C_2 = C(x,y)$.

There is (only one) operator $L = H$ such that

$$H\psi=0,\ H = \partial\bar{\partial} + A\bar{\partial} + v \tag{22}$$

$$A = \partial \ln c/\partial z\ ,\ v = \partial\mu_1/\partial\bar{z}$$

The Θ-functional formulas for (A,v) and Ψ may be found in the work [1].

The most interesting reduction, $C(x,y) \equiv 1$, leading to the class of purely potential operators was found in [3], [4].

<u>Definition</u> : The collection $\{\Gamma, \infty_1, \infty_2, w_1 = k_1^{-1}, w_2 = k_2^{-1}, Q_1, \ldots, Q_g\}$ are the spectral data for the operator H corresponding to one energy level.

Suppose now the surface Γ has the $\mathbb{Z}_2 \times \mathbb{Z}_2$-group generated by the involution $\sigma^2 = 1$ and anti-involution $\tau^2 = 1$, $\sigma\tau = \tau\sigma$, such that

$$\sigma w_\alpha = -w_\alpha \ , \ \tau w_1 = \bar{w}_2 \ , \alpha = 1,2, \ j = 1,2, \ldots, g \qquad (23)$$

$$\tau Q_j = Q_j \ , \ \sigma D + D \simeq K + \infty_1 + \infty_2$$

Here $D = Q_1 + \ldots + Q_g$ is the "divisor" of poles, K means the divisor of the holomorphic differential 1-forms on Γ and $\simeq$ means the so-called "linear equivalence" of divisors.

In that case we have

$$C(x,y) \equiv 1, \ A \equiv 0 \ , H = \partial\bar{\partial} + u(x,y) \qquad (24)$$

and the potential $u(x,y)$ is real and quasiperiodic. There is a formula for it :

$$u(x,y) = -2\Delta \ln \Theta (Uz + \bar{U}\bar{z} + \xi_o) + \varepsilon_o \qquad (25)$$

The Θ-function here depends on $h = g/2$ variables and (U, ξ_o) are h-vectors. The vector ξ_o is a quantity which only depends on the poles Q_j. The next formula also determines the real potential, corresponding to a change of the local parameters in the spectral data and to a replacement of the vector ξ_o by $i\xi_o$:

$$u(x,y) = -2\Delta \ln \Theta(Uz - \bar{U}\bar{z} + i\xi_o) + u_o \qquad (26)$$

(it was pointed out to the author by Natonson and Taimanov). Probably all real and <u>smooth</u> algebraic potentials can be described by (25), (26). The algebraic curve $\Gamma_o = \Gamma/\sigma$ of genus $h = g/2$ has an anti-involution τ_o generated by $\tau : \Gamma \to \Gamma$. The set of fixed points of τ_o should divide Γ_o in 2 pieces.

The operator $H = -\partial\bar{\partial}+u$ is positive $(H\varphi,\varphi)>0$ if and only if the anti-involution τ has exactly $g+1=2h+1$ fixed point-ovals $(a_1,\ldots, a_h, b_1,\ldots,b_h,c)$ such that $\sigma(a_q)=b_q$, $\sigma(c) = c$, $D = Q_1 + \ldots + Q_g$:

$$Q_j \in a_q \quad \text{for } j = 1, \ldots, h \tag{27}$$

$$Q_j \in b_q \text{ for } j = q+h = h+1, \ldots, g$$

The fixed point-ovals of the involution $(\sigma\tau)$ correspond to the "<u>real Fermi-surface</u>" : $\sigma\tau(P) = P \Longleftrightarrow |\psi(x,y,P)| \leqslant \text{const} < \infty$ for all $(x,y) \in R^2$.

<u>Conjecture</u> :

The approximation of any generic smooth periodic potential by algebraic ones should be such that, asg $\to \infty$; the topology of real Fermi-surface and any finite part of Γ stabilizes . Only near the points (∞_1,∞_2) will appear new small handles with new poles of ψ constructed like in (27) for all big numbers j. (There is a new work of Kriechener studying this process. He constructed also some difference analog of that theory).

Trivial examples of algebraic operators for $n = 2$ are the sum of two finite gap one-dimensional potentials

$$u(x,y) = u_1(x)+u_2(y) \tag{28}$$

They are algebraic corresponding to <u>any</u> level ε_0. We don't know any algebraic operator different from (28) with that property. The genus of Γ for $u(x,y)$ is equal to $2(g_1+g_2)$ for (28), $h=g_1+g_2$ (g_1,g_2 are respectively the number of gaps for $u_1(x)$ and $u_2(y)$).

For the solutions of the soliton equations (12), (13) we have the formula

$$u(x,y,t)=-2\Delta\ln\Theta(Uz\pm\bar{U}\,\bar{z}+W_m t+\sqrt{\pm 1}\xi_o)+\varepsilon_o \qquad (29)$$

In the most interesting case m=1 (the two-dimensional KdV), h=g/2=2, and any generic real positive 2x2-matrix $B_{pq} = B_{qp}$, any vector $U=U_R+iU_I$ and any real "phase vector" ξ_o determine the cnoidal wave". The vector W_1 and constant $u_o = \varepsilon_o$ will be calculated as some functions of (B_{pq},U) (see the work [11] ; the explicit formulas may be obtained from the direct substitution of (29) in the equation (13), but the calculations are non trivial).

Functions of the form (28) have always by (13) trivial time evolution

$$u(x,y,t) = u_o(x,t)+u_1(y) \qquad (30)$$

Here $u_o(x,t)$ should satisfy the ordinary KdV equation.

3. RAPIDLY DECREASING POTENTIAL. SOLITONS.

The scattering matrix $S(\lambda,\lambda')$ for the two-dimensional purely potential Schrödinger operator and n = 2 defined by

$$H\psi=\varepsilon\psi\ ,\ \psi=e^{i(k.x)}+S(\lambda,\lambda')\,e^{i|k|r}/\sqrt{r}+O(1/r) \qquad (31)$$

$$\vec{x}=(x,y),\ r^2=x^2+y^2\ ,\ \lambda=\lambda_R+i\lambda_I, |\lambda|=1$$

$$|\lambda'|=1\ ,\ k^2=\varepsilon>0, \vec{k} = (k\lambda_R, k\lambda_I),\ r\lambda'=z=x+iy$$

We shall use only one energy level $\varepsilon=\varepsilon_o>0$.

Consider now the new eigenfunctions $\psi^{\pm}$:

$$\psi^{+}(\lambda,\varepsilon_0)=e^{i(k.x)}+W^{(+,+)}(\lambda,\lambda',\varepsilon_0)e^{i|k|r}/\sqrt{r}+W^{(+,-)}(\lambda,\lambda',\varepsilon_0)\, e^{-i|k|r}/\sqrt{r}+O(1/r)$$

$$\psi^{-}(\lambda,\varepsilon_0) = e^{i(k.x)}+ W^{(-,+)}\, e^{i|k|r}/\sqrt{r} + W^{(-,-)}\, e^{-i|k|r}/\sqrt{r} + O(1/r) \qquad (32)$$

$$H\psi^{\pm}=\varepsilon_0\psi^{\pm}\ ,\ k^2=\varepsilon_0$$

such that :

$$W^{(+,+)},W^{(+,-)} = 0 \qquad \text{if}\ \ \vec{\lambda}\times\vec{\lambda}' = \lambda_R\lambda'_I - \lambda_I\lambda'_R < 0$$

$$W^{(-,+)},W^{(-,-)} = 0 \qquad \text{if}\ \ \vec{\lambda}\times\vec{\lambda}' = \lambda_R\lambda'_I - \lambda_I\lambda'_R > 0 \qquad (33)$$

We construct a new matrix $R(\lambda,\lambda')$ connecting these eigenfunctions

$$\psi^{+}(\lambda) = \psi^{-}(\lambda) - \oint R(\lambda,\lambda')\psi^{+}(\lambda')d\lambda' \qquad (34)$$

$$|\lambda| = |\lambda'| = |\lambda''| = 1\ \ ,\ \ \psi^{-}=(1+R)\psi^{+}$$

The operators $[S,R,W^{(+,+)}\ ,\ W^{(+,-)}\ ,\ W^{(-,-)}\ W^{(-,+)}]$ act in the space of functions on the circle S^1 and satisfy the following relations

$$(1+W^{(-,-)})\,(1+S) = (1+W^{(+,-)}) \qquad (1+W^{(-,-)})\,(1+R) = (1+W^{(+,-)}) \qquad (35)$$

$$(1+W^{(+,-)})\,(1+S) = (1+W^{(+,+)}) \qquad (1+W^{(-,+)})\,(1+R) = (1+W^{(+,+)})$$

All operators W are "cyclically triangular".

Continuation of the functions ψ^{+} and ψ^{-} to domains $\mathbf{D}^{+}$ ($|\lambda|>1$) for ψ^{+} and to $\mathbf{D}^{-}$ ($|\lambda|<1$) for ψ^{-} needs new "data". There is a continuous function $T(\lambda,\bar{\lambda})$ except for $S^1\subset(|\lambda|=1)$ such that

$$\partial\psi/\partial\bar{\lambda} = T(\lambda,\bar{\lambda})\,\psi(-1/\bar{\lambda}, -1/\lambda\ , z,\bar{z}) \tag{36}$$
$$\psi=\psi^{+} \ \text{ for } |\lambda| > 1$$
$$\psi=\psi^{-} \ \text{ for } |\lambda| < 1$$

This type of continuation by using equations like (36) was first introduced by Ablowitch and Fokas in the theory of KP - equation (but the contours in KP - theory are different). A quantity like R in (34) had also first appeared in the KP-theory in the papers of Manakov. Here we need a combination of their ideas. Let us introduce the quantity

$$\mathbf{U}(\lambda,\bar{\lambda}) = (\mathrm{sgn}(|\lambda|^2-1)/\pi\bar{\lambda}\ )T(\lambda,\bar{\lambda}) \tag{37}$$

and two involutions (σ,τ) like in periodic theory (see § 2,(23)).

$$\sigma\,(\lambda) = -\,\lambda \quad , \tau(\lambda) = -1/\bar{\lambda} \tag{38}$$

For $\lambda \to 0,\infty$, we have asymptotics like in the periodic case (see §2, $0=\infty_1$, $\infty=\infty_2$) :

$$\psi = e^{-i\lambda z\kappa}(1+\mu_1(z)\lambda^{-1}+ \ldots)\ ,\ \lambda \to \infty \tag{39}$$
$$\psi = c(x,y)e^{-i\kappa z/\lambda}\,(1+\nu_1\lambda+\ldots),\ \lambda \to 0$$
$$4\kappa^2=\varepsilon_o$$

In the case of purely potential operators we have some restrictions on the spectral data (R,T) :

$$\tau\mathbf{U} = \mathbf{U}(-1/\bar{\lambda}, -1/\lambda) = \bar{\mathbf{U}}(\lambda,\bar{\lambda}) \tag{40}$$
$$\sigma\tau(\mathbf{U}) = \mathbf{U}(1/\bar{\lambda}\ , 1/\lambda) = \mathbf{U}(\lambda,\bar{\lambda})$$

As in the periodic case we have :

$c(x,y) = 1$, $U(x,y) = \partial\mu_1/\partial\bar{z}$

Consider now the (σ,τ)-reductions for the operator R. The result is :

$$(1+R)\,(1+\overline{\sigma R}) = 0,$$
$$<(1+R)f\ ,\ (1+R)g> = <f,g> \qquad (41)$$
$$<f,g> = \oint_{S^1} f(\lambda)\,g(-\lambda)d\lambda$$

The reductions (41) for $R(\lambda,\lambda')$ are equivalent to the standard properties of the potential scattering amplitude :

(a) $S(\lambda,\lambda',\varepsilon) = S(-\lambda',-\lambda,\varepsilon)$

(b) unitarity of $(1+S)$.

By the soliton equation, the time dependence of the inverse scattering data (R,T) is given by

$$R(\lambda,\lambda',t) = \exp(-8i[\lambda^3+1/\lambda^3-\lambda'^3-1/\lambda'^3]\,t\,\kappa^3)R \qquad (42)$$
$$T(\lambda,\lambda',t) = \exp(-8i[\lambda^3+1/\lambda^3-\lambda'^3-1/\lambda'^3]t\kappa^3)T$$

For $\lambda\to 0,\ \infty$, the quantity T should decrease very rapidly. The reflexionless potentials (for the energy $\varepsilon=\varepsilon_0$) have $R \equiv 0$.

There are many such potentials (in fact, their existence even among spherically symmetric potentials may be deduced noneffectively from the old results of Sabatier. But it is more difficult to find some of them in effective analytical form. Such examples were found by Grinevitch, in the rational functions ($\varepsilon_0=4\kappa^2=k^2$).

Consider 4n points λ_j, $|\lambda_j|\neq 1$, such that :

$$\lambda_{4j+2} = -\lambda_{4j+1},\quad \lambda_{4j+3} = 1/\bar{\lambda}_{4j+1},\quad \lambda_{4j+4} = -1/\bar{\lambda}_{4j+1} \qquad (43)$$

It means just that the "divisor" $\mathbf{D} = \cup_j\lambda_j$ is invariant under the involutions (σ,τ). Let us construct some $4n \times 4n$-matrix $A=(a_{ij})$:

$$a_{ij} = 1/\lambda_i - \lambda_j \ , \quad i \neq j$$

$$a_{jj} = \gamma_j + \kappa_j(z,\bar{z}) = \gamma_j + (ik/2)(z - \bar{z}/\lambda_j^2) \tag{44}$$

$$\gamma_{2j+1} - \gamma_{2j+2} = 1/\lambda_{2j+1} \ , \ \gamma_{4j-1} = \bar{\lambda}^{2}_{4j-3}\, \bar{\gamma}_{4j-3}$$

The formula for the potential is :

$$u(z,\bar{z}) = -\Delta \ln \det A \tag{45}$$

The corresponding "multisoliton" solution has the form

$u(z,\bar{z},t)$, $\kappa_j(z,\bar{z},t) = i(kz/2 - k\bar{z}/2\lambda_j^2 - 3k^3\lambda_j^2 t + 3k^3 t/\lambda_j^4)$.

For n=1 we have a localized soliton.

<u>Remarks</u> :

1. The datas (R,T) are enough to get solutions of the inverse problem for potentials of small norm. Are there any new degrees of freedom needed for rapidly decreasing potentials with large norm ?

2. We described here the solution of the inverse scattering problem only for positive energy level $\varepsilon_o > 0$.

For the levels $\varepsilon_o < 0$, we should use only T (generically no R). The involution τ should be replaced by $(-\tau)$ in the formulas. They are interesting examples corresponding to the ground state $\varepsilon_0 < 0$ in {6} .

<u>Problem</u> : Find the datas corresponding to negative levels ε_o below the ground state. May be they only include smooth $T(\lambda,\lambda')$ only. It is definitely so for the potenitals with a small norm.

3. The theory of general difference operator of 2nd order in [12] is not enough effective. The best reduction found by Kriechever corresponds to the operators :

$$(L\psi)_{mn} = a_{m,n}\psi_{n+1,m} + b_{n-1,m}\psi_{n-1,m} + b_{n,m}\psi_{n,m+1} + b_{n,m-1}\psi_{n,m-1} \tag{46}$$

$$a_{n+2N,m} = b_{n,m} = a_{n,m+2M}, \quad b_{n+2N,m} = b_{n,m} = b_{n,m+2M}$$

Suppose $H^{(+,-)}$ contains all functions $\Psi_{n,m}$ such that

$$\Psi_{nm} = 0 \text{ if } m + n \equiv (1 \pm 1/2 \bmod 2) \tag{47}$$

We have $L : H^{(+,-)} \to H^{(-,+)}$. The spectral problem $L\Psi = \varepsilon\Psi$ is invariant under the transformation $\varepsilon \to -\varepsilon$

The level $\varepsilon = 0$ has some special properties. All complex 2–manifold of the Bloch's eigenfunctions

$$\Psi_{n+2N,m} = W_1 \Psi_{n,m}, \quad \Psi_{n,m+2M} = W_2 \Psi_{n,m}$$

has the "infinity" : $W_1 = 0$ or ∞, $W_2 = 0$ or ∞.

The algebraic curve Γ $(\varepsilon=0)$ is the union of 2 curves $\Gamma = \Gamma_+ \cup \Gamma_-$:

$\varepsilon=0$, $\det L(W_1,W_2) = Q(W_1,W_2) = Q_+ . Q_- = 0$.

Each $\Gamma_\pm$ has exactly four "infinite" points $P^{\pm,\pm}(W_1^{\pm 1}=0, W_2^{\pm 1} = 0)$. Analytical properties of Ψ on Γ_+ for $\varepsilon=0$ were described by Kriechever (for $\varepsilon\neq 0$ by the author and Kriechever). The Bloch's algebraic curve Γ for $\varepsilon\neq 0$ is more complicated and nondegenerate. It is more like in the continuous theory of §2 for $\varepsilon\neq 0$ ($\varepsilon=0$ is essentially simpler). The involutions have the form

$$\sigma(w_\alpha) = w_\alpha^{-1}, \quad \tau(w_\alpha) = \bar{w}_\alpha$$

REFERENCES

1. Dubrovin B.A., Kriechever I.M., Novikov S.P.
DANSSSR 229 (1976) N1 (transl. Sov. Math. Dokl. 17 (1976 NY, p. 947).

2. Manakov S.V.
UMN (Math Uspechi) 31 (1976) N6 p.237 (in Russian)

3. Novikov S.P., Veselov A.P.
DANSSSR, 279 (1984) p. 784 (transl.)

4. Novikov, S.P., Veselov, A.P.
DANSSSR, 279 (1984) p. 20 (transl.)

5. Novikov, S.P., Veselov, A.P.
Physica 18D (1986) p.267

6. Grinevitch P.G., Novikov R.G.
Func. Anal. and appl., 19 (1985) N4. p.32

7. Grinevitch P.G., Manakov S.V.
Func. Anal. and appl., 20 (1985) N4. p.14

8. Novikov R.G.
Theor. and Math. Phys., 66 (1986), N2, p. 234

9. Novikov S.P.
Func. Anal. and appl., (1974) N3

10. Brasoviskii S.A., Dzyaloshinskii, I.E., Krinechever I.M.
JETP, 83 (1982), N1, p389

11. Taimanov I.A.
DANSSSP, 285 (1985) N5 p. 1067

12. Kriechever I.M.
DANSSSR, 285 (1985) N1 p. 31

INFINITE DIMENSIONAL ALGEBRAS IN MODERN THEORETICAL PHYSICS

David Olive
Blackett Laboratory, Imperial College, London SW7 2BZ.
Dept. de Physique Theorique, University of Geneva.

Physics is the study of symmetries of nature and the mathematical structure formed by combining symmetry operations is a group. It follows that the study of groups and their representations provides an appropriate framework for theoretical physics. Recent developments in fundamental physics have often been related to enlargements of the class of group considered. Historically crystallography involved finite groups and atomic spectroscopy the rotation group which was considered in terms of its Lie algebra whose generators are the three angular momenta satisfying

$$[T^a , T^b] = i\varepsilon^{abc} T^c \quad ; \quad a,b,c = 1,2,3 \qquad (1)$$

The representations relevant to quantum mechanics are unitary and act in a positive definite space. They are classified in terms of the total angular momentum T which takes values

$$T = 0 , 1/2, 1,3/2, \ldots \qquad (2)$$

while the values of T^3 , given T, are

$$m = T, T-1, T-2, .. \quad -T. \tag{3}$$

Thus altogether the allowed values of m are

$$m \in \mathbb{Z} \cup (\mathbb{Z} + 1/2) \tag{4}$$

which constitutes the "weight lattice" of this algebra, (1). More general Lie algebras arise in elementary particle physics. These are usually compact and hence possess commutation relations

$$[T^a, T^b] = if^{abc}\, T^c \quad ; \quad a,b,c = 1,2 \,..\dim g \tag{5}$$

where the structure constants f^{abc} are totally antisymmetric and satisfy the Jacobi identity. The possibilities for g a simple Lie algebra were classified by Cartan and others and are conveniently depicted in terms of Dynkin diagrams. The unitary representations are also classified with weight lattices generalising (4) again

playing a role[1]. Do there exist infinite dimensional generalisations of these finite dimensional Lie algebras which are tractable mathematically and relevant to physics? The answer is indeed yes and its elaboration is the subject of this talk. The infinite algebras we shall discuss are known as affine Kac-Moody algebras and have cropped up in various contexts in theoretical physics starting with the "current algebra" theory of elementary particles interactions of the 1960's[2].

The commutation relations (of an untwisted affine Kac-Moody algebra) take the form

$$[T^a_m, T^b_n] = if^{abc}\, T^c_{m+n} \;+\; k\delta^{ab} m\, \delta_{m+n,o} \tag{6a}$$

$$[k, T^a_m] = 0 \tag{6b}$$

Thus the generators of the Lie algebra (5) acquire an

integer suffix m,n which is "conserved" like a momentum, while k is a central term, commuting with everything else.

In the late 60's, quite independently of physicists, the mathematicians Kac, in Russia, and Moody, in Canada, considered the same algebra (6), recognised its root system and classified the possibilities of this type in terms of generalised Dynkin diagrams, thereby furnishing a natural extension[(3)] of the work of Cartan and others concerning the Lie algebras (5). The mathematicians also defined unitary irreducible "highest weight" representations built up on "ground states". These are classified by highest weights lying on a lattice just as in angular momentum theory, see (2) and (3). This theory is essentially built on the results (2) and (3) and while it is gratifying that angular momentum theory can be extended so far in this way, it is clear that it is also rather disappointing and even boring that the results are so similar. Since I want to emphasise the aspects of the theory which are excitingly novel and different from angular momentum theory I shall say nothing about the theory of roots and weights and merely refer to the review by Peter Goddard and myself which has recently appeared treating all aspects of the theory[(4)].

To me as a physicist, the most exciting aspect is that the theory of these infinite dimensional algebras and their representations constitute a wedding between the theory of finite dimensional Lie algebras and quantum field theory. It is a wedding which is quite unexpectedly successful in view of the disparate partners and one in which the vices of quantum field theory, namely its divergences, are turned into virtues, as I shall try to explain.

The algebra (6) can be regarded as a current algebra in momentum space with the integer suffix labelling momentum which is quantised because of periodic boundary conditions. In coordinate space the central term is the coefficient of the derivative of a delta function, the so-called Schwinger term[5]. This is a manifestation of a quantum field theory divergence and its presence is essential to the theory as we shall see below.

Such current algebras arise in massless quark models in two space time dimensions. The energy-momentum tensor is traceless because of scale invariance and hence possesses two independent components. The Heisenberg equations of motion of the currents and of the tensor can be abstracted to the following commutation relations

$$[L_m , T^a{}_n] = - nT^a{}_{m+n} , \tag{7a}$$

$$[L_m , L_n] = (m-n)L_{m+n} + \frac{c}{12} m(m^2-1)\delta_{m+n,o} \tag{7b}$$

Equation (7b) is known as the Virasoro algebra and it originally arose in the string theory of elementary particles[6]. It always provides an extension of the affine Kac-Moody algebra (6) via (7). c commutes with everything and provides a new central term in addition to k occurring in (6b).

A trivial representation of the Virasoro algebra (7b) is provided by

$$L_m = - z^{m+1} \frac{d}{dz} , \quad c=0 \tag{8a}$$

where z is a complex variable. These differential operators generate infinitesimal conformal transformations in two dimensions such as are used to satisfy Laplace's equation $\nabla^2\phi = 0$ when ϕ is regarded as the real part of an analytic function of z. Conformal transformations can also be made analytically in the complex conjugate variable, $\bar{z}$, yielding a second Virasoro algebra commuting with (8a), namely

$$\bar{L}_m = -\bar{z}^{-m+1}\frac{d}{d\bar{z}} \quad , \quad \bar{c} = 0 \; . \tag{8b}$$

The occurrence of two commuting Virasoro algebras L_m , $\bar{L}_n$ constitute a general feature of conformally invariant quantum field theories in two dimensions, sometimes accompanied by two commuting copies of the Kac-Moody algebra (6), but always with c non-zero, unlike (8) as we shall see.

An example of this is furnished by the theory of massless "quarks" giving rise to currents bilinear in the fermion fields describing the quarks and satisfying (6). This theory is "free" and apparently trivial yet much can be learnt from it. In this situation z and $\bar{z}$ relate to the two light cone variables t±x respectively while the two commuting copies of the affine Kac-Moody algebra (6) correspond to the two light cone components of the currents.

In these physical applications, the generators obey the hermiticity conditions

$$T^a_n{}^+ = T^a_{-n} \quad , \quad L_m{}^+ = L_{-m} \tag{9}$$

and act in a quantum mechanical state space which is positive definite. Thus we are interested in "unitary" representations of the algebra. Furthermore, according as z=x±iy or exp(i(t±x)), L_o and $\bar{L}_o$ are identified as operators H±P or D±S whose spectra is bounded below on physical grounds (H is the Hamiltonian, P the momentum, D the dilation operator and S the spin).

For positive values of the suffix n, T^a_n and L_n lower the L_o eigenvalue (by n) according to (7). In order that the spectrum of L_o be bounded below we must eventually reach "ground states" annihilated by all the T^a_n and L_n for n positive. Thus the representation space of the theory resembles that of quantum field theory in that it is built up on "ground states", possibly degenerate. Furthermore the central terms c and k have to be positive in unitary representations. For example, if $\Psi\rangle$ is a ground state then the state $L_{-m}\ \Psi\rangle$, $(m>0)$ must have a positive norm squared given by

$$\langle\Psi\ L_m L_{-m}\ \Psi\rangle = \langle\Psi\ [L_m, L_{-m}]\ \Psi\rangle = 2m\langle\Psi\ L_o\ \Psi\rangle + \frac{m(m^2-1)}{12}\langle\Psi\ c\ \Psi\rangle. \tag{10}$$

Taking m=1 implies $\langle\Psi\ L_o\ \Psi\rangle \geq 0$ while taking m large implies $\langle\Psi\ c\ \Psi\rangle \geq 0$. Sharpening of these arguments implies that $\langle\Psi\ c\ \Psi\rangle = 0$ only if the Virasoro algebra is represented trivially (by zero). Hence c is strictly positive. Yet further refinement of this sort of argument using ideas inspired by the no ghost theorem of string theory(7) leads to surprisingly specific information for the allowed values of k and c :

$$2k/\psi^2 = 0,1,2,3,\ .. \qquad = x, \quad \text{the level}\ . \tag{11}$$

The quantity x is called the level of the irreducible representation. ψ is the highest root of g, (5), and is usually normalised to have length $\sqrt{2}$ so that k is then the level. (11) is proven by applying the results (2) and (3) to judiciously chosen angular momentum sub-algebras of (6). Furthermore

$$\text{either}\quad c \geq 1\ , \tag{12a}$$

$$\text{or}\quad c = 1 - 6/(m+2)(m+3)\ , \quad m=0,1,2,3,\ .. \tag{12b}$$

Thus the spectrum of c has a threshold at 1 below which

lies a discrete sequence of possible values accumulating at 1, as was first clearly realised by Friedan, Qiu and Shenker[8].

If c takes one of the values (12b) less than 1, the possible values of h, the lowest L_0 eigenvalue, will be finite in number and rational:

$$h = \frac{((m+3)p-(m+2)q)^2-1}{4\,(m+2)(m+3)}\ , \quad 1 \leqslant q \leqslant p\ , \quad 1 \leqslant p \leqslant m+1\ . \qquad (13)$$

These numbers are potentially measurable by studying two-dimensional materials constructed in the laboratory. They relate to the critical exponents describing a second order phase transition. Then the system can be described by a scale invariant quantum field theory with a traceless energy-momentum tensor. Such a theory is automatically conformally invariant with its correlation functions obeying a power law in spatial separation whose exponent is the critical exponent given by $h+\bar{h}$ since these are the eigenvalues of D, the dilation operator $L_0 + \bar{L}_0$[9].

Comparison of (13) with the known values of the critical exponents in certain theories lead FQS to identify m=1 (i.e. c=1/2) with the Ising model, m=2 (i.e. c=7/10) with the tricritical Ising model and so on. The remarkable conclusion is that the determination of the critical exponents involves nothing more than the representation theory of the Virasoro algebra, (7b), but many questions which have been opened up by this insight remain. Is it possible to construct explicitly representations with c and h given by (12b) and (13)? How would this construction relate to the physical models? We shall explain some of the steps made so far.

Progress has followed from an understanding of the other physical examples of conformally invariant theories. In the days of current algebras (in the 1960's) it was argued that the algebraic structure was so fundamental that it should determine the complete dynamics of the theory. This meant that it should be possible to express the energy-momentum tensor in terms of the currents and it was realised that this could be done in two dimensions when the theory was scale invariant by considering an expression quadratic in the currents. Translated into the present notation the construction[10], due to Sugawara and Sommerfield, reads

$$\mathcal{L}_m = \frac{1}{2k+Q_\psi} \sum_{n-\infty}^{\infty} \sum_{a=1}^{\dim g} {}^{\times}_{\times} T^a_{m+n} T^a_{-n} {}^{\times}_{\times} \quad . \tag{14}$$

Notice the normal ordering denoted by crosses which moves T^a_m to the right of T^a_n if m is positive and to the left if m is negative. This is needed to evade a potential divergence in (14) arising from the central term in (6a) when we consider representations built up on ground states in the manner already described. The prefactor is not simply $(2k)^{-1}$ but rather is subtly renormalised to $(2k+Q_\psi)^{-1}$ where Q_ψ denotes the quadratic Casimir operator in the adjoint representation of g, (5). The veracity of (14) in yielding (7a) can be checked by methods developed in string theory and also leads to the following value for the central term in (7b)

$$c^g = \frac{2k \dim g}{2k + Q_\psi} = \frac{x \dim g}{x + \tilde{h}(g)} \quad , \tag{15}$$

at least if g is simple or isomorphic to U(1). If not, then (14) is replaced by a sum of similar terms. $\tilde{h}(g) = Q_\psi/\psi^2$ is called the dual Coxeter number of g and is readily tabulated. In general the c-number (15) is a rational number exceeding the rank of g and hence never less than one i.e. in the range (12b).

The fact that the Virasoro algebra is automatically and exactly satisfied by the Sugawara construction (14) means that of the relations (6) and (7), the Kac-Moody algebra (6) is the most fundamental since the remaining Virasoro structure (7) is guaranteed once we have a positive L_o representation of the type discussed. It also follows that theories possessing a current algebra consisting of two commuting copies of (6) and a Sugawara energy-momentum tensor are automatically conformally invariant. This applies to the Wess-Zumino-Witten model for a bosonic field confined to the manifold of a Lie group G[(11)]. The result is exact to all orders (in Planck's constant).

An important special case occurs when G is abelian i.e. U(1) or a product of U(1)'s. Then Q vanishes as the structure constants do and

$$c = \dim g \quad .$$

In this case the currents are just derivatives of dim g real scalar fields with the Sugawara construction reducing to their standard canonical energy-momentum tensor. The corresponding Dirac energy-momentum for fermi fields yields c=1/2 for each real free massless fermi field so that

$$c = \text{number of bose fields} + (1/2)\ \text{number of fermi fields.} \tag{16}$$

in a free massless theory. This construction includes the cases used in the bosonic and fermionic string theories.

The c-number (15) is in general neither an integer nor half-integer and so must then refer to an "interacting theory". The only value common to the sequences (12b) and (16) is c=1/2 indicating that the Ising model relates to a theory with a single real free massless fermion, at least at the critical temperature. This is now known to be the basis for Onsager's celebrated solution of the model[12].

Further relations between the c-numbers (12b) and (15) require a new idea. Suppose G contains a Lie subgroup H so that the Sugawara construction (14) can be applied to both G and H in order to obtain $\mathcal{L}^G$ and $\mathcal{L}^H$ respectively. Then, by (7a), $\mathcal{L}^G_m - \mathcal{L}^H_m$ must commute with $\hat{h}$ and hence, by (14), with $\mathcal{L}^H_m$. Thus $\mathcal{L}^G_m$ breaks into two mutually commuting pieces, ^H_m and

$$K_m = \mathcal{L}^G_m - \mathcal{L}^H_m \quad . \tag{17}$$

It is not difficult to see that K_m also satisfies the Virasoro algebra (7b), now with c-number

$$c_K = c_g - c_h \geq 0 \quad , \tag{18}$$

where c_g and c_h are given by (15) applied to g and h respectively. This is the "coset" construction of Goddard, Kent and Olive[13] and has many consequences, some indicating novel quantum field theoretic representations of Kac-Moody algebras.

As a first application note that the symplectic group Sp(m+1) possesses a subgroup Sp(1)xSp(m). Taking these to be G and H respectively and starting with a level one representation equations (15) and (18) yield

$$c_K = 1 - 6/(m+2)(m+3)$$

Thus we have explicitly constructed all the unitary representations of the Virasoro algebra with c less than one, (12b), from level one representations of sp(m+1). The same construction can be viewed another way in terms of SU(2) by considering $su(2)_{(m)} \times su(2)_{(1)}/su(2)_{(m+1)}$ where the suffix denotes the level[13]. Thus there is a connection between c<1 representations of the Virasoro algebra (7b) and the higher level representations of su(2) and this has been used to realise unitarily all the h values (13)[14]. Inasmuch as the c<1 theory is "interacting" and the level one representations of sp(m+1) obtainable from free fermions, the coset construction (17) seems to furnish a way of constructing interacting theories from free theories.

Cardy[15] has pointed out that when a two-dimensional conformally invariant theory is defined on a torus there is a highly non-trivial requirement of modular invariance restricting the matching of the irreducible representations of the L_m and $\bar{L}_m$ Virasoro algebras. The above mentioned relations between su(2) representations and c<1 is useful in enumerating solutions and the whole study is undergoing a rapid and exciting development.

Another consequence of (18) is that if, by chance c_K vanishes, then K_n also vanishes because K_o is bounded below and we can apply a previous result. Thus the G and H Sugawara constructions are quantum equivalent in a remarkable way. As our prime example, we shall concentrate on the case that H is T, the maximal torus of G, obtained by exponentiating the Cartan subalgebra, t. As we saw c_T = rank g and we further find, using the fact that $\tilde{h}(g) = (n_L + (S/L)^2 n_S)/\text{rank } g$ where n_L and n_S denote the number of long and short roots of g respectively and S/L the ratio of their lengths

$$c_G - c_T = \frac{(x-1)n_L + (x-(S/L)^2)n_S}{x + \tilde{h}(G)} .$$

As $x \geqslant 1$ this is positive and vanishes if and only if x=1 and the long and short roots have the same lengths i.e. g is simply laced and hence of A,D or E type. Then $\mathcal{L}_G$ and and $\mathcal{L}_T$ must be equal, a result suggesting that the E^α_m (α a root of g) can be constructed from the H^i_m, the generators of $\hat{t}$. This is precisely what is achieved by the vertex operator construction of Frenkel and Kac and of Segal[16].

The generators of $\hat{t}$, H^i_m, satisfy the commutation relations of bosonic harmonic oscillators (i=1,2, .. rank g; m=1,2, ..oo) so that their generating function $\sum_{m=o}^{\infty} z^{-m}H^i_m$ resembles the Fubini-Veneziano momentum field of string theory[17]. This can be integrated to give the coordinate field $Q^i(z)$,

$$iz\frac{dQ^i(z)}{dz} = P^i(z) = \sum_{m=o}^{\infty} z^{-m}H^i_m \qquad (19)$$

The constants of integration q^i are taken to be canonically conjugate to $p^i = H^i_o$. The remaining generators of g, E^α_m are then given by

$$\sum_{m=o}^{\infty} z^{-m} E^\alpha_m = z^{\alpha^2/2} :e^{i\alpha.Q(z)}: c_\alpha \qquad (20)$$

provided the roots α of g are assigned length $\surd 2$. The corroboration of the necessary commutation relations depends in an essential way on the vices of quantum field theory, namely the singularities occurring when quantum field operators are multiplied at the same point. Thus the double dots in (20) denote normal ordering of the bosonic oscillators. c_α is the Klein transformation factor needed to correct the signs in the operator product expansion to obtain the appropriate commutation relations.

The right hand side of (20) is recognised as the

vertex for emitting a tachyonic string state of momentum α[16] and it is bizarre that these particles, undesired physically because they move faster than light, should correspond to the roots of a Lie algebra.

Nevertheless this construction shows that a string moving on the maximal torus of a simply laced group G (a flat manifold) acquires G as a gauge symmetry. This is now the most favoured explanation of the origin of gauge symmetry in nature[18].

There are many ways in which the use of vertex operators can be extended[19]. If $\alpha^2 = 1$ rather than two in (20) we obtain fermion fields i.e. affinised Clifford algebras instead of affinised Lie algebras. Physically this is the old construction of Skyrme (and later Mandelstam and others) for the quantum field operator of the Sine-Gordon soliton[20].

This fermionic construction weds with the $\hat{g}$ construction only if $g = D_r = so(2r)$ since only the weight lattice of D_r contains unit vectors generically. $D_4 = so(8)$ yields an important special case with three different types of unit vector, the weights of the vector, spinor and conjugate spinor representations which are related by the triality symmetry of the Dynkin diagram. This exceptional fact underlies the spacetime supersymmetry of the superstring theory of unified particle interactions[21].

If $\alpha^2 = 3$ in (20), the vertex operators represent supercharges of an extended N=2 supersymmetry algebra which, amazingly, can be realised with only boson fields and no fermions[22].

If we allow the vectors α to lie in a space with a Lorentzian metric instead of an Euclidean metric we can utilise the vertices of string theory for emitting particles moving slower than light. Only the physical state vertices arise and the corresponding algebra is called a Lorentzian algebra. It is much larger than an affine Kac-Moody algebra and possibilities of likely physical relevance were discussed in (19).

The vertices discussed so far pertain only to the bosonic string theory and to simply laced algebras, ones whose simple roots all have the same length. It has been found recently that non simply-laced algebras correspond to the fermionic string theory of Ramond, Neveu and Schwarz in that the tachyonic vertices of these theories yield level one representations of $\hat{B}_r$, $\hat{C}_r$ and $\hat{F}_4$.[23].

I have tried to convey the glimmerings of an exciting and rapidly developing new subject in physics and mathematics. More details can be found in the review article of Peter Goddard and myself, which has recently been published[4], and in the current research literature.

REFERENCES

1) J.E. Humpheys: Introduction to Lie Algebras and Representation Theory, Springer, 1972.

2) S. Adler and R. Dashen: Current Algebras and Applications to Particle Physics, Benjamin, New York, 1968.

3) V. Kac: Infinite Dimensional Lie Algebras - An Introduction, Birkhauser, 1983 and Cambridge U.P., 1985.

4) P. Goddard and D. Olive: Int. Journ. of Modern Phys. A1, 303 (1986).

5) J. Schwinger: Phys. Rev. Lett. 3 296 (1959) and ref. 2 p235.

6) M. Jacob (editor): Dual Theory, North Holland, 1974.

7) R. Brower: Phys. Rev. D6 1655 (1972).
P. Goddard and C. Thorn: Phys. Lett. 40B 235 (1972).

8) D. Friedan, Z. Qiu and S. Shenker: Phys. Rev. Lett. 52 1575 (1984).

9) A.A. Belavin, A.M. Polyakov and A.B. Zamolodchikov: Nucl. Phys. B241 333 (1984).

10) H. Sugawara: Phys. Rev. 170 1659 (1968).
C.M. Sommerfield: Phys. Rev. 176 2019 (1968).

11) J. Wess and B. Zumino: Phys. Lett. 37B 95 (1971),
E. Witten: Commun. Math. Phys. 92 455 (1984).

12) L. Onsager: Phys. Rev. 65 117 (1944).

13) P. Goddard, A. Kent and D. Olive: Phys. Lett. 152B 88 (1985).

14) P. Goddard, A. Kent and D. Olive: Commun. Math. Phys. 103 105 (1986).

15) J. Cardy: Nucl. Phys. B270 [FS16] 186 (1986).

16) I.B. Frenkel and V. Kac: Inv. Math. 62 23 (1980),
G. Segal: Comm. Math. Phys. 80 301 (1981).

17) S. Fubini and G. Veneziano: Nuovo Cim. 67A 29 (1970).

18) D. Gross, J.A. Harvey, E. Martinec and R. Rohm: Phys. Rev. Lett. 54 502 (1985), Nucl. Phys B256 253 (1985).

19) P. Goddard and D. Olive: p51 in "Vertex Operators in Mathematics and Physics", MSRI Publication #3, Springer, 1984.

20) T.H.R. Skyrme: Proc. Roy. Soc. A262 237 (1961),
S. Coleman: Phys. Rev. D11 2088 (1975),
S. Mandelstam: Phys. Rev. D11 3026 (1975).

21) F. Gliozzi, D. Olive and J. Scherk: Phys. Lett. 65B 282 (1976), Nucl. Phys. B122 253 (1977),
M. Green and J. Schwarz: Nucl. Phys. B181 502 (1981),
P. Goddard, D. Olive and A. Schwimmer: Phys. Lett. 157B 393 (1985).

22) G. Waterson: Phys. Lett. 171B 77 (1986).

23) P. Goddard, W. Nahm, D. Olive and A. Schwimmer: Commun. Math. Phys. 107 (1986) 179.
D. Bernard and J. Thierry-Mieg: Meudon preprint.

CONSTRUCTIVE RENORMALIZATION

Vincent Rivasseau

Centre de Physique Théorique
Ecole Polytechnique, F91128 Palaiseau Cedex

In recent years constructive field theory has witnessed dramatic progress (perhaps not unlike the blossoming years around 1973...) and several groups have reached some of their major objectives. Among the main recent acquisitions are the proof of ultraviolet stability for three-dimensional nonabelian gauge theories, the construction of asymptotically free models, the control of nontrivial fixed points, and on a more general level, a deepening of our understanding of renormalization theory, both at the perturbative and nonperturbative level. In these notes I will not try to make a general survey of the new results, but will instead present the method of phase-space localization, as it has been developed and applied by our group at Ecole Polytechnique. I also use this occasion to thank J. Feldman, J. Magnen and R. Sénéor, who were at the core of this group and are to be considered coauthors of this paper, for a most friendly collaboration.

I.The phase space expansion: a way to sum up perturbation theory

Phase space localization, developed in constructive field theory by Glimm and Jaffe, has had for a long time the reputation of a difficult and technical subject; I will try to argue that it is in fact a very natural way to organize perturbative and constructive field theory. Let us start with this most well-known example, the renormalized φ^4 perturbation theory. Why does it diverge? There are two main obstacles to the naive summation of the renormalized Feynman amplitudes. One is dimension-independent: the number of Feynman graphs of order n grows at least like $c^n n!$ for large n. We call this the instanton effect. The other one arises only at the critical dimension (here 4): some individual diagrams become large due to renormalization. We call this the renormalon effect. In both cases phase space analysis leads naturally to the solution. Let us examine the first effect.

The first remark is that the instanton effect is somewhat local; if we consider bounded massive propagators $C(x,y)$ which behave at large $x-y$ as $e^{-m|x-y|}$, and if we divide our volume Λ in a lattice of cubes of side approximatively m^{-1}, the piece of perturbation theory which contains only a bounded number of vertices in each cube will not show any instanton effect. (Hint: use the exponential decay of C to find, for each field, the cube containing the field to which it will contract in Wick's theorem. This costs a constant per field, hence only c^n). Hence the instanton divergence is due to the need in perturbation theory to sometimes have too many vertices "closely packed together". In Euclidean space the propagators are really useful only in showing explicit connections which are necessary to perform the thermodynamic limit (and eventually multiparticle analysis). Since information for these features is only relevant at physical scales, hence lengths of order m^{-1}, the solution to the instanton problem is well known. One should keep most of perturbation theory resummed in a functional integral form, and by an appropriate cluster expansion, show explicitly either decoupling or exponentially decaying connections only between regions made of cubes of side $\approx m^{-1}$. The net effect, from the point of view of combinatorics, is then exactly the same as if only a bounded number of vertices per cube of side m^{-1} had been produced. Provided the functional integral in a single cube is bounded (i.e. the theory is stable), this solves completely the instanton problem, at least when ultraviolet effects are absent or minor (Φ^4_1 or Φ^4_2). The cluster expansion followed by a standard Mayer expansion is then similar to a high temperature expansion in statistical mechanics, and controls nicely the theory, provided there is a small constant around (here the coupling constant) to take care of sloppy bounds.

Now for $d>2$ the real propagator of φ^4 behaves as $|x-y|^{-(d-2)}e^{-m|x-y|}$, and hence ultraviolet problems can occur (even after Wick ordering). Again this occurs when vertices come close to each other. In fact it will seriously perturb the solution of the instanton problem that we have just discussed.

Let us remark that when vertices come close together, the corresponding volumes in integration space are small. Phase space localization is a simple way to chop up the theory in a discete manner so that this idea can be systematically exploited. The fact that the chopping is discrete is important in the sense that many "analytic" problems are reduced to "combinatoric" ones; the price to pay is a certain loss of ability to compute exact constants of the theory, which fortunately is not the main objective in constructive field theory.

What is the simplest way to "discretize" the ultra violet problem? When there is an obvious gaussian measure to perturb around, like in the Φ^4 theory, one can cut the Fourier transform of the propagator into "momentum slices". This amounts to a decomposition of the random variables (the fields) into a sum of orthogonal variables, each distributed according to the covariance of a given slice. (Of course in other situations, like non abelian gauge theories, it might be difficult to isolate a global gaussian measure, and to create momentum slices one might use a more general trick like block-spinning [1]; but when available, the orthogonal decomposition induced by slicing up the covariance seems the simplest.) The "best" slicing is geometrical (for the same reason that the "best" function to control an infinite sum is a decreasing exponential). Let M>1 be a fixed number. We write $C(p)=\Sigma_{i\geq 0}C_i(p)$, hence $\varphi = \Sigma_{i\geq 0}\varphi_i$, where:

$$C_i(p) = \frac{1}{p^2+m^2}\left[e^{-M^{-2(i+1)}(p^2+m^2)} - (1-\delta_{io})\, e^{-M^{-2i}(p^2+m^2)}\right] \qquad (1)$$

It is easy to check that in direct space (c being the generic name of a constant, and the dimension being fixed to 4):

$$C_i(x,y) \leq c\, M^{2i} \exp(-cM^i |x-y|) \qquad (2)$$

A momentum assignment is a choice $\mu=\{i_l\}$ of an index for each line l of a graph G. Apart from renormalization and from inessential numerical factors, the Feynman amplitude for G is obtained by integrating the product of the covariances $C_{il}(x_l, y_l)$ of all internal lines, with the positions x_v of all internal vertices v running over $\mathbb{R}^4$. (x_l and y_l are the positions of the vertices at the ends of l). Then one sums over all μ's. A key rôle in optimally organizing the integration over the x_v's is played by a particular "assignment-dependent" class of subgraphs, called "almost local" or AL subgraphs. They are the maximal connected subgraphs such that the assignments of every internal line (e.g. i) is higher than the assignment of every external line (e.g. j) (see Fig.1). Intuitively the vertices of AL subgraphs are likely to be packed in a region of size M^{-i} around their center of mass, which is then bound in a looser way through propagators of momentum j to the rest of the graph. AL subgraphs at a given scale i join together to form larger AL subgraphs at lower scales i'; therefore they have a tree structure for the inclusion relation (when the final graph is connected; otherwise they have forest structure). This tree structure is closely related to the "Gallavotti-Nicolo" trees of [2].

To optimize the integration over the x_v's, one should use as much as possible the decrease (2) of the propagators of the highest available energy, and the tree of AL subgraphs allows one to do so without getting lost in the combinatorics [3]. A general bound on amplitudes follows by induction from the following simple consideration: for an AL subgraph of scale i with n vertices, l internal lines and e external lines, the net effect after integration of all vertices save one (the "center of mass"), hence of n-1 vertices, is a factor $M^{-4(n-1)i}M^{2li} = M^{-i(e-4)}$. (We have used the fact that each vertex joins 4 half lines). When e>4, hence when the corresponding graph is "superficially convergent", one gets an exponential decrease in "assignment-space" or momentum space which is dual to the direct exponential decrease of the propagators (2) [3]. This dual picture goes further: the dual of a line which connects two vertices in direct space is a vertex, which connects 4 half lines in assignment space (see Fig.1). This exponential decrease in assignment space is nothing but power counting seen on a logarithmic scale, but it helps to realize that (as long as only convergent subgraphs are present), the system is still a high temperature system of statistical mechanics, but with one extra dimension (or rather "semi-dimension") added (which is discretized), the assignment space. Hence were it not for the possibility of divergent subgraphs, exponential decrease would occur in every direction of "phase space", which is just a name for this enlarged space. Just as in the cluster expansion, this makes all kinds of summations (Mayer expansions, thermodynamic limit etc..) rather easy, provided the expansion under consideration is suitably generalized to take into account this new dimension.

What happens when divergent subgraphs are present (here, subgraphs with e=2 or e=4)? Note that only AL divergent subgraphs really diverge, because non-AL divergent subgraphs have the scale of at least one of their external legs as a natural ultraviolet cutoff. But precisely as their name indicates, AL subgraphs can have their dominant contribution cancelled in an effective way by truly local counterterms, like their 0 momentum value in the BPHZ renormalization scheme. This operation turns out to just restore the ordinary decrease in assignment space of convergent subgraphs, without any bad side effect [4]. In essence this is the reason why renormalization works. Unfortunately counterterms cannot be introduced only for AL subgraphs; to preserve the locality of field theory, counterterms should be introduced irrespective of the scale of their external legs, hence also for "non-AL subgraphs". This is the source of all the evil and trouble in renormalization theory. Firstly, at the perturbative level the non-AL divergent subgraphs, in contrast with the AL ones, do not have a natural tree or forest structure, but can overlap. Therefore the solution of their renormalization is at best inductive (hence somewhat obscure) or expressed by a sum over forests, a technical tool which has never become really popular. Secondly and perhaps worse, at the constructive level, the pieces of counterterms which renormalize non-AL subgraphs are the sole source of the pernicious "renormalon effect"; when n of them are inserted in an innocent-looking convergent graph, they may create amplitudes as large as $c^n n!$ and the corresponding piece of the perturbation expansion, although made of very few graphs, cannot be summed [4]. However in Φ^4 like theories it is interesting to remark that only the non-AL counterterms (also called "useless" counterterms) for the 4 point subgraphs can create renormalons. Mass and wave function renormalizations do not generate renormalons; neither do they generate "forest" problems because two one-particle-irreducible 2-point subgraphs cannot overlap in a Φ^4 theory [5].

Let us return to the definition of the phase space expansion. According to the principle above, a generalized cluster expansion is performed to create explicit connections between all the cubes in phase space, hence direct connections which are propagators joining cubes of the same scale, and dual ones, which are vertices joining cubes of different scales, one contained into the other. Then one should renormalize all the AL divergent subgraphs. This is the best one can do to keep an expansion of the high temperature type, with exponential decrease in every direction. However at this stage we discover two problems which are inescapable "remnants", respectively of the instanton and of the renormalon problem. In a sense this is reassuring because as we shall see their complete solution is model-dependent, whereas so far our discussion has been completely general.

- As a remnant of the instanton problem, one has to show that the functional integral restricted to an isolated cube is bounded. This was already true before and depends on stability of the theory. But furthermore, as an inescapable consequence of momentum slicing, while at the scale of their highest leg vertices produced by the expansion do behave as if they were "in bounded number per cube", they can violate this condition in larger cubes corresponding to some of their lower momentum legs. Gaussian integration of these low momentum legs would then create bad factorials through Wick's theorem. The solution, called "domination" uses the decrease of the interaction at large fields to improve the Wick bound (an example is sketched in the next section). Hence we find as a first requirement that the model should have an effective potential which is stable and quickly decreasing at infinity.

- The second problem is how to deal with the pieces of

counterterms for non-AL subgraphs. To introduce them would create renormalons, so the only solution is to "hide" them in the definition of "running" parameters. Hence the best expansion to use is neither the bare one, which has no counterterms, nor the renormalized one, which has too many, but an intermediate effective expansion, where only the "useful" renormalizations are performed. The price to pay is that the expansion is no longer a power series in a single parameter, but in an infinity of effective coupling constants, one per slice (by the remarks above, only the flow for the coupling constant is really crucial). It remains to check that these parameters remain uniformly small in terms of the renormalized one, so that one still has "a small parameter around" and this is only true in the class of models called asymptotically safe, which is our second requirement on the model. (Unfortunately it is not satisfied by the usual stable Φ^4_4 model).

Of course this rapid sketch ignores deliberately a number of interesting technical problems, such as the way in which to define precisely the generalized cluster expansions. These are Taylor expansions with integral remainders, performed with respect to interpolating parameters, which roughly control the coupling between pairs of regions in <u>phase space</u>. To define such parameters, keeping in mind the requirements of positivity of the measure, and of compatibility with the operations of renormalization and domination, is still the most painful and least transparent part of the work. Such technicalities are shared by any method at this level. However the principles of the construction do not incorporate any significant idea beyond those discussed above. Hence the whole method can be considered as a natural solution to the problem of summing perturbation theory.

II The construction of models

In principle with these methods, one should be able to add to the standard collection of superrenormalizable models of constructive field theory a new class of models which are either renormalizable and asymptotically free (gaussian fixed point) or perturbatively non-renormalizable with a non-gaussian fixed point but "close" to a gaussian. This class includes various fermionic models and σ-models in

2 dimensions, non-abelian gauge theories in 4 dimensions in a finite volume, models in 2+ε dimensions, models at three dimensions and large number N of components, planar models, etc... In practice the $-g\Phi_4^4$ planar model, the massive Gross-Neveu model in 2 dimensions and the infrared Φ_4^4 model have been built by our group (for other constructions of these models, see in particular [6],[9]-[10]), and programs are under way to build the massive Gross-Neveu model in 3 dimensions and large N, and non-abelian pure gauge theories in 4 dimensions and finite volume. Let us make some remarks about each of them.

The planar $-g\Phi_4^4$ model is obtained by keeping only planar graphs in the Φ^4 theory and is linked to the theory of random surfaces and of many component matrix models. Although it is not really a full interacting field theory it is pedagogically interesting because there are few planar grahs, and hence no "instanton problem". It is therefore the ideal place to watch in detail how the bare or the renormalized series can be transformed into the effective one which has only useful renormalizations, and how the renormalon problem then disappears (the theory being asymptotically free because the coupling constant has the "wrong" sign). One can start from the bare theory (as in [6]), then add and subtract only "useful" counterterms, and hide the subtracted ones in the definition of effective constants. But this requires using the correct ansatz for the bare couplings (up to two loops in the β function) to end up on the desired renormalized couplings. Or one can start from the renormalized series [7] and just hide the "useless" counterterms in the definition of effective constants; this introduces some technicalities like the use of forests, but it is instructive to see explicitly in this simple case how the effective series communicates both with the bare and the renormalized ones.

The massive 2 dimensional Gross Neveu model [8]-[9] has the main advantage of being a full fledged field theory for which the O.S. axioms can be verified. Here again, the instanton problem does not really occur due to the compensations of sign in the fermionic amplitudes; our rule of "not too many vertices per cube" is here enforced by Pauli's principle. But since the graphs are the same as in the Φ^4 theory, there are still $c^n n!$ of them at large n, and it is therefore important that the expansion preserves the determinant structure responsible for the cancellations. This introduces some technical constraints, but there is no domination problem. In fact the theory can truly be constructed as the ultraviolet limit of the sum of a single absolutely convergent power series, i.e. the one in powers of the bare coupling constant with bare wave function, and renormalized mass [8]. The radius of convergence of this series shrinks as the ultraviolet cutoff is removed, but thanks to asymptotic freedom, the bare constant shrinks at the same speed and remains inside the disk of convergence for a fixed small value of the renormalized coupling constant. However to use the renormalized mass, hence a full ("useful" and "useless") mass renormalization, one has to make visible the one-particle structure of the theory (not only the relations of connectedness), hence to push further the cluster and Mayer expansion. This can be done in our formalism, and in fact the generalization of this idea leads to a program for studying multiparticle structure (see the last section).

The "infrared Φ^4_4 model" [10]-[11], although not a field theory but the theory of the critical point of Φ^4_4 with an ultraviolet cutoff, is similar to an ultraviolet model as far as renormalization is concerned. One should remark in particular that the definition of AL subgraphs is not reversed for an infrared problem, as one might naively think. The only difference is that in assignment space there is "a floor" in one case and "a ceiling" in the other; hence this extra "semi-dimension" is oriented differently, but this does not change anything in the general discussion above. In fact infrared Φ^4_4 is richer than Gross-Neveu$_2$ since it has a true "instanton" or domination problem, just as non-abelian gauge theories. I will discuss on a single example how domination improves the factorials of Wick's theorem in this case, without entering into any technicalities. Let us imagine that we have a number M^{4h} of small cubes of scale 0 hence of side 1, which fill exactly a single large cube of infrared scale h, hence of side M^h. (M is assumed to be an integer). Let us suppose that each small cube contains a fixed number k of vertices. Since the conclusion is independent of k, assume k=1. Suppose further that each vertex has only one of its 4 legs at scale 0, and all three others are at the lower scale h. This is in a sense the "worst" case which can appear with our expansion rules. The total number of vertices is $n=M^{4h}$. To evaluate the corresponding contribution by gaussian integration, one remarks first that when integrated in its small cube, each vertex gives a volume factor 1. There is no permutational symmetry and hence no 1/n! factor coming from the expansion of the exponential. Then the gaussian integration of the high momentum legs (of scale 0), as remarked at the beginning, costs only a factor c^n; but the gaussian integration of the 3n low momentum half-lines of scale h costs, by Wick's theorem, $c^n(3n/2)!$. Finally propagators of scale h have a factor M^{-2h} in front (the infrared analog of the factor M^{2i} in (2)). This creates a factor $M^{(-2h)(3n/2)}$, of order $c^n/(3n/4)!$. Hence the total contribution is of order $c^n(3n/4)!$. This is of course not summable. But when the Φ^4 term in the interaction is used instead of the quadratic one for integration of the low momentum fields, the bound of Wick's theorem improves from $c^n(3n/2)!$ to $c^n(3n/4)!$. Hence the final contribution is of order c^n, which is all right.

The massive Gross-Neveu model in three dimensions and large N has a nontrivial fixed point. (Other models of non-renormalizable type have been built [12] but they are not in integer dimensions.) The construction is under way [13], using the Mathews-Salam formalism. After resumming the "chains of bubbles" which are the leading term at N large, the auxiliary field in Mathews-Salam formalism acquires a respectable propagator, which can be sliced. After suitable rescalings, the approach to the fixed point is by a power law, just as in the superrenormalizable case, but there is a small anomalous dimension in 1/N. Since the auxiliary field is bosonic, there is a domination problem, which should be solved by showing that the effective potential has the right decrease.

Finally non-abelian gauge theories in 4 dimensions are by far the richest model, combining all possible kinds of difficulties, including the new ones related to gauge invariance. With a finite volume cutoff the model is still believed to be "without non perturbative effects", i.e. gauge invariant physical quantities should be the Borel sum of their renormalized perturbation expansion (as is true for all preceding examples, with the exception of the Gross-Neveu$_3$ model).

Balaban's program [1] has succeeded in constructing non abelian gauge theories in three dimensions, using a lattice regularization and a presription for block spin averages which in effect creates momentum slices similar to ours. The crucial difficulty lies in the treatment of domination. Indeed the Landau gauge, which gives a well behaved propagator from the point of view of cluster expansions, does not allow direct domination. There are regions in the Landau gauge where the field is large, and nevertheless the action is small. This is known as the Gribov phenomenon. Balaban's solution does not use global gauge fixing but local gauges which are phase-space dependent mixtures of axial-like and Landau-like gauges. His method should in principle apply also to the 4-dimensional case, although the treatment of large field regions is harder there, and the theory is further complicated by the necessity of taking into account the coupling constant flow. However his program as well as Federbush's [14] relies on the use of a lattice regularization. In contrast we hope to provide a complementary approach based on an ordinary gauge-breaking regularization (like Pauli-Villars), hence to show that there is nothing truly unique in the lattice regularization. Although we have not yet a full construction, we have preliminary results [15] which make us confident that at least the small field region can be controlled in this way. We hope to treat the problem of domination in the style of Balaban. (As remarked above, the domination problem being in a sense a local one, it is dual to the propagation problem, and we view this as the main reason for which Balaban's program can work).

III The study of large orders

As noticed already, the instanton and renormalon problem reflect themselves in the large order behavior of perturbation theory. In fact this large order behavior tells us a lot about the possibility of constructing the theory, and its rigorous study is an interesting program, deeply connected with semi-classical analysis and renormalization group analysis. Papers like [16] provided important advances in our qualitative understanding of the structure of perturbation theory, and one should try to put the corresponding ideas on a firm mathematical basis.

The main tool to analyze large orders in the case of Φ^4 is the Borel transform of the perturbation series. Instantons and renormalons appear as singularities in the Borel plane. The situation for Φ^4_3 and Φ^4_4 as expected from [16] is shown in Fig.2, and what has been proved is shown in Fig.3. We will now describe briefly the content of papers [17]-[20].

One can study the instanton problem in Φ^4 with the "Lipatov" method, which is a semi-classical expansion. In its simplest version, this method computes the radius of convergence of the Borel transform of the perturbation series. To illustrate this point, let us imagine a lattice model with finitely many sites. The partition function Z(g) has the formal power series expansion $Z(g) \approx \Sigma\, a_n (-g)^n$, with

$$a_n = \frac{1}{n!} \frac{\int \left[\sum_x \varphi(x)^4\right]^n \exp\left[-1/2 \sum_{y,z} \varphi(y) C(y,z)^{-1} \varphi(z)\right] \prod_x d\varphi(x)}{\int \exp\left[-1/2 \sum_{y,z} \varphi(y) C(y,z)^{-1} \varphi(z)\right] \prod_x d\varphi(x)} \tag{3}$$

To compute the limit as $n\to\infty$ of $(a_n/n!)^{1/n}$ one makes the change of variables $\varphi = n^{1/2}\psi$. This leads to

$$(a_n/n!)^{1/n} = \frac{n^2}{(n!)^{2/n}} \frac{||\exp -S(\psi)||_n}{||\exp -S_F(\psi)||_n} \tag{4}$$

where $S_F(\psi) = (1/2) \sum_{y,z} \psi(y) C(y,z)^{-1} \psi(z)$ and $S(\psi) = S_F(\psi) - \log \sum_x \psi(x)^4$.

By Stirling's formula and the fact that $\lim_{n\to\infty} ||f||_n = ||f||_\infty$, we obtain:

$$\lim_{n\to\infty} (a_n/n!)^{1/n} = \exp \{ -\inf S(\psi) + 2 \} \equiv R_i^{-1} \tag{5}$$

R_i is the radius of convergence in the Borel plane due to "instantons" which appear as cuts on the negative real axis (see Fig.2). For the Φ^4_3 theory, it is shown in [17] that R_i is truly the radius of convergence in the Borel plane of the renormalized series for a relevant quantity like the pressure. In [18], it is shown that the closest singularity to the origin, as expected, is on the real negative axis. Since the model is Borel summable, we know that there is no singularity on the positive real axis. Together with former results in lower dimensions, this is a confirmation that the "Lipatov" analysis, as developed by Brézin, Le Guillou and Zinn-Justin, governs the large order behavior in the superrenormalizable domain, although computing rigorously subdominant behavior should be difficult.

In the 4-dimensional case the situation is more interesting, and phase space analysis becomes indispensible. The closest singularity to the origin for the 2, 4 or 6 point function is expected to be of the renormalon type, on the positive real axis at $R_r=2/\beta_2$, (β_2 being the first non 0 coefficient of the β function) [16]. Hence it prevents Borel summability. A proof of this conjecture is challenging and would shed additional light on the "triviality of Φ_4^4". The ratio R_i/R_r tends to $+\infty$ as N, the number of components of the vector Φ^4 model, tends to $+\infty$, starting at 3/2 for N=1. For N large but finite there is a dominant term responsible for the renormalon singularity, namely the "chain of bubbles", and one can prove the renormalon conjecture [20]. But for N=1 the renormalon dominates over the instanton by only a factor 3/2, and it becomes important to have a sharp upper bound over all the instanton effects. This is accomplished in [19], using phase space analysis and the Sobolev inequality. It is shown in [19] that the piece of the perturbation expansion which does not contain "useless" counterterms (remember that these are responsible for the renormalons) is bounded by $n!R_i^{-n}(1+\varepsilon(n))^n$, where R_i is the exact instanton radius given by (5), and $\varepsilon(n)$ tends to 0 as $n\to\infty$. What is still needed to prove the existence of the first renormalon at N=1 is a lower bound on "renormalons" effects. As usual this is harder than an upper bound since improbable cancellations should be excluded. In [20] an intermediate result is proven, namely that the radius of convergence of the Borel transform of Φ_4^4 is <u>at least</u> the correct conjectured value of $2/\beta_2\equiv R_r$, (a big numerical improvement of the bound [5]). This result is obtained using only upper bounds (both on the renormalon and instanton effects). The key tool is again the definition of an effective expansion, which is then Borel-transformed, and shown to be analytic in the Borel disk $|b|<2/\beta_2$. At $b=2/\beta_2$ plenty of potential singularities arise. However, in contrast with the large N case, we have not found an organizing principle of these singularities into leading and sub-leading ones. Hence miraculous cancellations cannot be yet ruled out. An interesting remark is in order: since the theory is not asymptotically free at $g>0$ and looks unstable at $g<0$, how can the effective expansion be defined? The answer is that the expansion is defined at $g<0$ (otherwise the effective constants would not remain bounded), but a truncated version (up to two loops) of the β-function is used, since the full β function in a given subtraction scheme like BPHZ is ill-defined because of the instanton problem. In our formalism we know exactly which pieces of BPHZ counterterms are responsible for which flow, hence we can define "partial" effective constants which correspond to truncated β functions. After Borel transforming, and before analytically continuing the expansion towards $b=+2/\beta_2$, one should however "undo" the effective constants to check that the Taylor series at the origin in the Borel plane is the correct one.

IV Asymptotic completeness and multiparticle structure

The famous problem of asymptotic completeness in field theory has a long history. It has been realized for a long time that in order to get asymptotic completeness and multi-particle structure analysis, one might exploit the structural equations which describe scattering using r-particle irreducible kernels. For r large it was however not clear which notion of r-irreducibility and which equations to use. Within the formalism of the phase space expansion, there exist natural generalizations of the "0-th" or "1-st" order cluster expansions (which were used in [8],[11] to show connectedness or 1-particle irreducibility). These generalizations yield by inspection <u>both</u> the definition of r-particle irreducible kernels and the set of structural equations, conjectured in [21], which link them [22]. With this tool, it is expected that asymptotic completeness will be shown to hold to arbitrarily high energy, provided the coupling constant is arbitrarily small (the proof for Φ_3^4 is under preparation [23]). For further details, see the contribution of D. Iagolnitzer to these proceedings.

References

[1] T. Balaban, lots of papers in Comm. Math. Phys, hereafter CMP.
[2] G. Gallavotti and F. Nicolo, CMP 100, 545 and 101 (1985)
[3] J. Feldman, J. Magnen, V. Rivasseau, R. Sénéor, CMP 98, 273 (1985)
[4] J. Feldman, J. Magnen, V. Rivasseau, R. Sénéor, CMP 100, 23 (1985)
[5] C. de Calan and V. Rivasseau, CMP 82, 69 (1981)
[6] G. 't Hooft, CMP 88, 1 (1983)
[7] V. Rivasseau, CMP 95, 445 (1984).
[8] J. Feldman, J. Magnen, V. Rivasseau, R. Sénéor, CMP 103, 67 (1986)
[9] K. Gawedzki and A. Kupiainen, CMP 102, 1 (1985)
[10] K. Gawedzki and A. Kupiainen, CMP 99, 197 (1985)
[11] J. Feldman, J. Magnen, V. Rivasseau, R. Sénéor, to appear in CMP
[12] G. Felder,CMP 102,139 (1985); K.Gawedzki,A.Kupiainen,prepint IHES
[13] C. de Calan,P. Faria da Veiga,J. Magnen,R. Sénéor, in preparation
[14] P. Federbush, "By 93 or before" papers.
[15] J. Feldman, J. Magnen, V. Rivasseau and R. Sénéor, in preparation
[16] G. 't Hooft, Lectures at Ettore Majorana School,Erice,Sicily 1977; G. Parisi, Phys. Lett. 76B, 65 (1978), Phys. Rep. 49, 215 (1979).
[17] J. Magnen and V. Rivasseau, CMP 102, 59 (1985).
[18] J. Feldman and V. Rivasseau, to appear in Ann. Inst. H. Poincaré
[19] J. Magnen, F. Nicolo, V. Rivasseau, R. Sénéor, to appear in CMP
[20] F. David, J. Feldman and V. Rivasseau, in preparation
[21] D. Iagolnitzer, Fizika 17, 1985.
[22] D. Iagolnitzer and J. Magnen,preprint Ecole Polytechnique,June 86.
[23] H. Epstein, D. Iagolnitzer, J. Magnen, R. Sénéor, in preparation

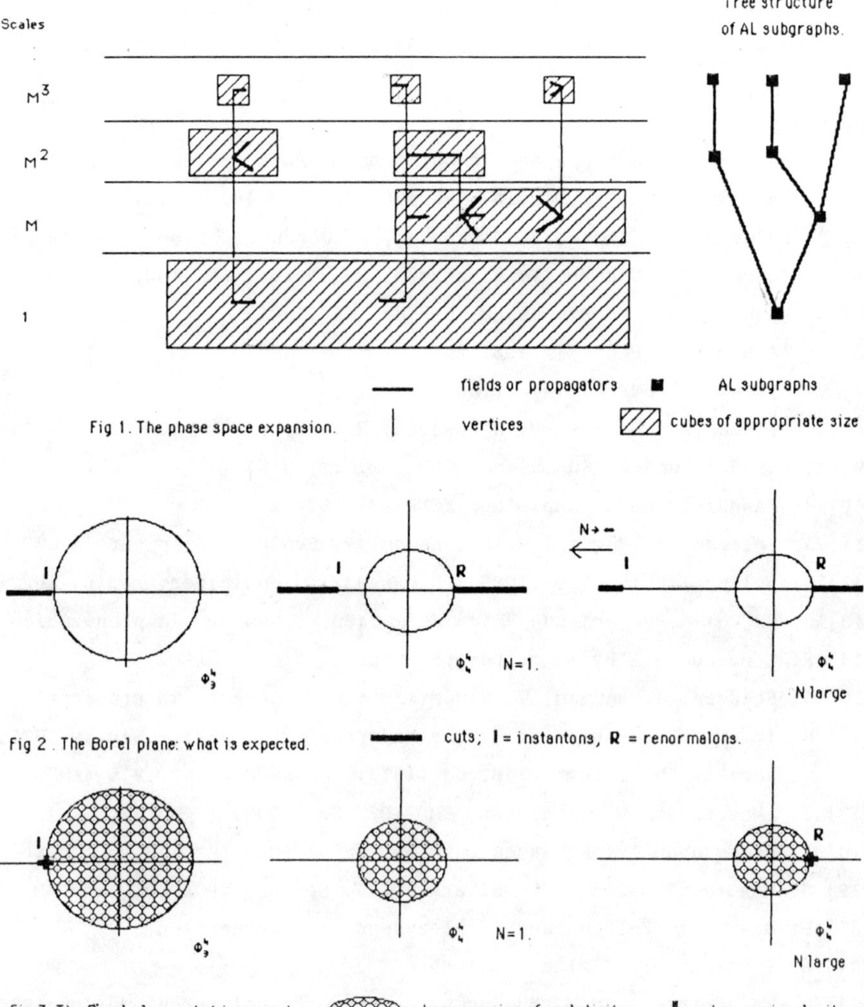

Fig 1. The phase space expansion.

Fig 2. The Borel plane: what is expected.

Fig 3. The Borel plane: what is proved.

THEORY AND EXPERIMENT IN THE ERGODIC STUDY OF CHAOS AND STRANGE ATTRACTORS

David Ruelle

Institut des Hautes Etudes Scientifiques
91440 Bures-sur-Yvette
FRANCE

ABSTRACT

The development of the ergodic theory of differentiable dynamical systems has opened the possibility to investigate moderately excited systems which occur in physics and are already too complicated for a geometric analysis. One can, from the experimental time series of such systems, estimate Liapunov exponents, entropy, information dimension and other ergodic parameters. It is a nontrivial task to extract these parameters from the data, and the talk discusses some problems which arise. The main value of techniques which determine ergodic structure from time series is that they can be applied to interesting naturally occurring systems. There are many potential fields of application, notably astrophysics, neurophysiology and economics. The talk will review these fields and insist on the methodological difficulties. The latter are serious, so that believable results are only just starting to appear.

0. INTRODUCTION

The theory of differentiable dynamical systems has been a remarkably active and successful subject in the last ten or twenty years. This refers both to the purely mathematical aspects, and to the more recent explosion of physical applications to turbulence and chaos. Surprisingly, all this activity has not killed the subject. It is true that the adjective "novel" has been uninhibitedly abused in some of the physics literature, but it is also true that some quite new developments have occurred (and are to be found usually in the less triumphalist literature).

Among the purely mathematical results, I would like to mention a paper by Parry and Pollicott [15] which shows how methods of number theory can be applied to differentiable dynamical systems (the distribution of periods of periodic orbits for Axiom A flows behaves like the distribution of prime numbers). Another remarkable development is due to Yomdin [27] who, in the same paper, solves two important problems (semicontinuity of the entropy, and Shub's entropy conjecture). An account of Yomdin's results has been given by Gromov [12].

Let us now turn to physics, or more generally to applications to natural phenomena. In this area, the main activity for quite a few years has been in a geometric study of strange attractors and bifurcations (in particular the Feigenbaum cascade). This approach has led to the elucidation of various "scenarios" for the onset of chaos, and proved that the quasiperiodic picture of turbulence (Hopf, Landau) had to be replaced by another picture, with strange attractors and sensitive dependence on initial conditions. The geometric approach is however limited to low dimensional attractors which can be "seen" in 2 or 3 dimensional pictures. The access to higher dimension is made possible by the methods of the ergodic theory of differentiable dynamical systems, which we shall discuss below.

At this point it appears that we begin to have the technology needed to analyze interesting moderately excited dynamical systems which occur in nature. This analyzis has barely started, and there is still little to report in terms of results. The following review will thus be mainly about methods and some potential fields of application.

1. ERGODIC THEORY OF DIFFERENTIABLE DYNAMICAL SYSTEMS

In ergodic theory, the central object of study is no longer a geometrical object, like a strange attractor, but a probability measure

ρ invariant under time evolution. This measure describes the average fraction of time spent by the system under consideration in various parts of phase space. One can assume that ρ is ergodic, i.e., it does not have a nontrivial decomposition into different invariant measures.

Various ergodic parameters are associated with the measure ρ, notably the <u>characteristic exponents</u> $\lambda_i = \lambda_i(\rho)$, the <u>entropy</u> $h(\rho)$, and the <u>information dimension</u> $\dim_H \rho$.

The characteristic exponents, or Liapunov exponents, describe the rate at which nearby trajectories diverge. They form a sequence $\lambda_1 \geq \lambda_2 \geq \dots$. The system is said to be chaotic if $\lambda_1 > 0$.

The entropy $h(\rho)$, or Kolmogorov-Sinai invariant is the average rate of creation of information by the system. (Even though the time evolutions we consider are deterministic we may have $h(\rho) > 0$).

The information dimension $\dim_H \rho$ is the smallest Hausdorff dimension of a set of measure 1 with respect to ρ.

A detailed review of the properties of the above ergodic parameters has been given recently (Eckmann and Ruelle [8]) and we shall not repeat this discussion here.

What is important for the present discussion is the possibility of extracting ergodic parameters from experimental time series. The work of Young [28] and especially of Grassberger and Procaccia [10] has shown that this could be done effectively for the information dimension. These authors note that if x is an experimental point in the phase space of the system, and if the number $n_x(r)$ of further experimental points at a distance $\leq r$ of x satisfies $n_x(r) \sim r^\alpha$, then α is the information dimension.

The natural method of determination of characteristic exponents by looking at the rate of divergence of two or a few trajectories yields poor results due to statistical inaccuracy (see [3] and [26]). A recently proposed method (Eckmann and Ruelle [8]) utilizes many points close to a given x to determine the "tangent map" to the time evolution, and then the characteristic exponents. Used with care (see Eckmann, Kamphorst, Ruelle, and Ciliberto [7]), this method seems to hold its promises of much better accuracy.

At this point it should be stressed that there are more ergodic parameters than those mentioned above. In particular, Italian groups [4], [2], and a group centered on Chicago [13], have insisted on the following idea. If an ergodic parameter is defined by a limit time → infinity, this quantity fluctuates and the large fluctuations have no reason to be Gaussian. The function which describes the distribution of large fluctuations is again an ergodic invariant of the system, and deserves study. Kadanoff has conjectured universality properties of this function. A rigorous study of this question, using ideas of statistical mechanics, has been prepared by Collet, Lebowitz and Porzio [6].*)

Other examples of ergodic parameters which can be defined for suitable dynamical systems are resonances (see below) and rotation numbers (see Ruelle [20]).

2. RESONANCES

Let $\rho_{BC}(t)$ be the time correlation function of two observables B and C . The Fourier transform $\hat{\rho}_{BB}(\omega)$ of $\rho_{BB}(t)$ is known as the power spectrum of the observable B . For a quasiperiodic time evolution, the natural periods appear as δ-functions in the power spectrum. On the other hand it is generally admitted (although this is not quite rigorous) that a continuous or broad band spectrum characterizes chaotic systems. The features of a continuous spectrum depend on the observable B considered, unlike the positions of δ-functions in a quasiperiodic spectrum. Therefore it appears that the power spectrum of a chaotic system is not of great interest for obtaining ergodic invariants of the system. Recent results have however modified this conclusion in some cases, as we now explain.

For a class of chaotic differentiable dynamical systems called Axiom A flows, one can show (Pollicott [16], Ruelle [23], [24]) that the power spectrum $\hat{\rho}_{BB}(\omega)$ extends to a meromorphic function of ω in some strip $|\text{Im } \omega| < \alpha$. This means that $\hat{\rho}_{BB}(\omega)$ is analytic in this

*) See also T. Bohr and D. Rand. Preprint.

strip except for pole singularities which we call resonances. If a strip $|\mathrm{Im}\,\omega| < \beta$ is free of poles, we should have exponential decay of correlation functions, which is certainly an important physical property to verify.

It is known that for the geodesic flow on a manifold of constant negative curvature the correlations decay exponentially (Collet, Epstein and Gallavotti [5], Ratner [19]). This probably remains true when the curvature is almost constant (Pollicott, private communications). In general, however, the situation is very unclear, and further study is needed (see Ruelle [21], [22] for a discussion). In some differentiable dynamical systems which are chaotic, but not Axiom A (and in fact with discrete rather than continuous time) there is evidence of non exponential decay of correlations. For an example see Greene, Mac Kay and Stark [11].

The above discussion points to ergodic parameters which describe physically important properties of differentiable dynamical systems : rate of decay of correlations, and position of the resonances. If the correlations of a dynamical system are dominated by a resonance near the real axis, it is obviously desirable to know its real part (frequency) and imaginary part (rate of decay). It must be said however that the determination of these parameters is not very easy, due in particular to the effect of noise (see Ruelle [22]).

3. RECONSTRUCTION OF THE DYNAMICS

Let $(u_i : i = 1,\ldots,N)$ be a time series, corresponding to measurements performed at integer multiples of some unit of time Δt. For simplicity we assume the u_i to be scalar. The vectors $x(i) = (u_i, u_{i+1}, \ldots, u_{i+n-1}) \in \mathbb{R}^n$ give an n-dimensional projection of the dynamics of the original system. (This is the "time-delay method", and n is called "embedding dimension"). Let α_n be the dimension obtained by the method of Grassberger and Procaccia for each embedding dimension n. For a deterministic time evolution on a finite dimensional strange attractor, we shall find that α_n becomes constant for sufficiently large n, the constant value being the information dimension (if all goes well). If random noise is present, α_n will increase without bound with n .

Suppose that we have found that our time series corresponds to a deterministic time evolution (chaotic or not), what is the law governing this time evolution ? This often asked question can be answered only partially, namely only for the evolution restricted to the attractor. If the u_i are all equal, corresponding to a point attractor, nothing can be said of the dynamics outside of this point.

Let us try to be more specific. Suppose that A is a compact set in a Banach space, and that its Kolmogorov dimension $\dim_K A$ is finite. (The Kolmogorov dimension, or capacity, is a dimension concept related to the Hausdorff dimension). A theorem of Mañé asserts that most linear projections of A to $\mathbb{R}^n$ are injective (i.e., one-to-one into) if $n > 2 \dim_K A+1$. Let us assume that the map π_n given by the time-delay method from the attractor A to $\mathbb{R}^n$ is indeed injective for suitable n . (It would be hard to prove such a result from general principles but we can at least, in a given experimental situation and for given n , check that it seems to hold). Thus, every $x(i) = (u_i, u_{i+1}, \ldots, u_{i+n-1}) \in \pi_n A$ corresponds to a unique point on the attractor A , and therefore determines a unique $x(i+1) = $ $= (u_{i+1}, \ldots, u_{i+n})$. Furthermore $x(i+1)$ depends continuously on $x(i)$ on $\pi_n A$ (this is because the injective continuous map π_n is necessarily a homeomorphism of the compact set A to the compact set $\pi_n A$). We have thus a continuous map $\Phi : (u_i, n_{i+1}, \ldots, u_{i+n-1}) \to u_{i+n}$, defined on $\pi_n A$. By the theorem of Stone-Weierstrass we can find polynomials which approximate Φ uniformly, and such polynomials can be obtained from the experimental time series (u_i) . Remember however that Φ is defined only on $\pi_n A$; the polynomial approximants will therefore be largely arbitrary outside of this set.

Finally, note that when $\Delta t \to 0$, the difference equation

$$u_{i+n} = \Phi(u_i, u_{i+1}, \ldots, u_{i+n-1})$$

goes over into a differential equation.

4. POTENTIAL FIELDS OF APPLICATION

The ideas of strange attractors, chaos, etc. have had a considerable impact in hydrodynamics, for the understanding of the onset of turbulence. Other potential domains of application concern various natural phenomena for which one can hope that there is an underlying low dimensional dynamics. Such phenomena occur in Economics, Neurophysiology and Astrophysics.

Before going to specifics let me make some general remarks. First note that it is interesting to know that there is a low dimensional dynamics governing the time evolution of a system, but it is also interesting to know that there is no such low dimensional dynamics. In principle, the distinction between the two cases is easy, using the algorithm of Grassberger and Procaccia, but in practice some expertise is needed, and data of sufficient quality. For a controversial example see for instance Grassberger [9]. The remark about expertise and data of sufficient quality apply even more to the determination of characteristic exponents (joint work with Eckmann and Kamphorst has several times revealed deficiencies of the data which would not appear in a simple determination of information dimension). In any case, at the present time we are still only in position to review problems rather than results.

In Economics (more precisely Macroeconomics) the existence of "business cycles" has been noted a long time ago, and it is a natural question if they correspond to low dimensional dynamics. To my knowledge, the matter is not settled although suggestive results have been obtained by Scheinkman (private communication).

In Neurophysiology, one can study interspike intervals for a single neuron, corresponding to a discrete time dynamical system. The data obtained thus far do not quite have enough precision (P.E. Rapp, private communication). One can also use global electroencephalographic data for which excellent data should soon be available. For a discussion of the present situation see Rapp et al. [17], [18]. Note that the EEG data relevant to the present discussion correspond to a stable regime, and are not of the currently popular variety where one studies the response of the brain to sending a certain stimulus.

Astrophysics provides beautiful dynamical objects, which are the variable stars *). Some of those have non periodic luminosity variations which appear to correspond to low dimensional dynamics (see M. Auvergne and A. Baglin [1]). Progress in this area will come when higher quality data are accessible. There is of course one star on which precise measurements are relatively easy, this is the sun, which exhibits several "chaotic" phenomena. Among those are solar flares which are localized and short duration ($\lesssim$ 1 minute) pulsating structures. These violent manifestations of solar activity produce a radio emission which can be monitored. A study by Kurths and Herzel [14] seems to indicate that the MHD oscillations associated with a flare corresponds to a low dimensional strange attractor.

The sun also has global long period oscillations, corresponding to sunspot cycles. Sunspots statistics have been monitored for only a relatively short time. However, an unexpected and remarkable discovery by G.E. Williams apparently gives us a much longer record of solar activity. Williams [25] has analyzed a precambrian periglacial lake deposit of South Australia. This deposit shows annual layering (varves) modulated by what are apparently solar periods (principally $\sim$ 11 and $\sim$ 22 years). Note that at the time of the deposit ($\sim$ 680 my ago) the earth atmosphere had a rather different composition, with a greatly reduced ozone layer, permitting a greater influence of fluctuations of the solar UV flux on ground level temperatures. Williams has obtained a long record of 1580 cycles, each cycle consisting of 8-16 varves. One may hope that the analysis of these data will eventually lead to a much better understanding of the dynamics of solar cycles.

---

*) I am indebted to J. Cl. Pecker and to Ed. Spiegel for brief introductions to the subject.

REFERENCES

[1] Auvergne, M. and Baglin, A., The dynamics of a variable star. Phase space reconstruction and dimension computation, Preprint.

[2] Badii, R. and Politi, A., Statistical description of chaotic attractors : the dimension function, J. Statist. Phys. 40, 725-750 (1985).

[3] Benettin, G., Galgani, L. and Strelcyn, J.-M., Kolmogorov entropy and numerical experiments, Phys. Rev. A 14, 2338-2345 (1976).

[4] Benzi, R., Paladin, G., Parisi, G. and Vulpiani, A., On the multifractal nature of fully developed turbulence and chaotic systems, J. Phys. A 17, 3521-3531 (1984).
Characterization of intermittency in chaotic systems, J. Phys. A 18, 2157-2165 (1985).

[5] Collet, P., Epstein, H. and Gallavotti, G., Perturbations of geodesic flows on surfaces of constant negative curvature and their mixing properties, Commun. Math. Phys. 95, 61-112 (1984).

[6] Collet, P., Lebowitz, J. and Porzio, A., Dimension spectrum for some dynamical systems, Preprint.

[7] Eckmann, J.-P., Oliffson-Kamphorst, S., Ruelle, D. and Ciliberto, S., Liapunov exponents from time series.

[8] Eckmann, J.-P. and Ruelle, D., Ergodic theory of chaos and strange attractors, Rev. Mod. Phys. 57, 617-656 (1985).

[9] Grassberger, P., Are there really climatic attractors ?

[10] Grassberger, P. and Procaccia, I., Measuring the strangeness of strange attractors, Physica 9 D, 189-208 (1983).

[11] Greene, J.M., MacKay, R.S. and Stark, J., Boundary circles for area-preserving maps.

[12] Gromov, M., Entropy, homology and semialgebraic geometry (after Y. Yomdin), Séminaire Bourbaki June 1986.

[13] Halsey, T.C., Jensen, M.H., Kadanoff, L., Procaccia, I. and Shraiman, B.I., Fractal measures and their singularities : the characterization of strange sets, Preprint.

[14] Kurths, J. and Herzel, H., Characterization of an attractor in a solar time series, Preprint.

[15] Parry, W. and Pollicott, M., An analogue of the prime number theorem for closed orbits of Axiom A flows, Ann. Math. 118, 573-591 (1983).

[16] Pollicott, M., On the rate of mixing of Axiom A flows, Inventiones Math. 81, 413-426 (1985).

[17] Rapp, P.E., Zimmerman, I.D., Albano, A.M., Deguzman, G.C. and Greenbaum, N.N., Dynamics of spontaneous neural activity in the Simian motor cortex : the dimension of chaotic neurons, Physics Lett. 110 A, 335-338 (1985).

[18] Rapp, P.E., Zimmerman, I.D., Albano, A.M., Deguzman, G.C., Greenbann, N.N. and Bashore, T.R., Experimental studies of chaotic neural behavior : cellular activity and electroencephalographic signals, in Nonlinear oscillations in Chemistry and Biology, H.G. Othmer, ed. Springer, New York, To appear.

[19] Ratner, M., The rate of mixing for geodesic and horocycle flows, Preprint.

[20] Ruelle, D., Rotation numbers for diffeomorphisms and flows, Ann. Inst. Henri Poincaré 42, 109-115 (1985).

[21] Ruelle, D., Resonances of chaotic dynamical systems, Phys. Rev. Letters 56, 405-407 (1986).

[22] Ruelle, D., Locating resonances for Axiom A dynamical systems, J. Statist. Phys., To appear.

[23] Ruelle, D., Extension of the concept of Gibbs state in one dimension and an application to resonances for Axiom A diffeomorphisms, J. Diff. Geometry, To appear.

[24] Ruelle, D., Resonances for Axiom A flows, J. Diff. Geometry, To appear.

[25] Williams, G.E., Solar affinity of sedimentary cycles in the late Precambrian Elatina formation, Aust. J. Phys. 38, 1027-1043 (1985).

[26] Wolf, A., Swift, J.B., Swinney, H.L. and Vastano, J.A., Determining Liapunov exponents from a time series, Physica 16 D, 285-317 (1985).

[27] Yomdin, Y., Volume growth and entropy, Addendum : C^k-resolution of semialgebraic mappings, To appear.

[28] Young, L.-S., Dimension, entropy and Lyapounov exponents, Ergod. Th. and Dynam. Syst. 2, 109-124 (1928).

ANDERSON LOCALIZATION FOR ONE-DIMENSIONAL DIFFERENCE SCHROEDINGER OPERATORS WITH QUASI-PERIODIC POTENTIALS

YA.G.SINAI

Landau Inst. for Theor. Phys. Moscow

1. INTRODUCTION :

We consider the operator $H_\omega(\alpha)$ acting in the Hilbert space $l_2(-\infty,\infty)$ by the formula

$$(H_\omega(\alpha)\psi)(t) = -\varepsilon(\psi(t+1)+\psi(t-1)) + V(t\omega+\alpha)\psi(t) \qquad (1)$$

where V is a generic C^2-function on the unit circle. The natural generalization of (1) which covers main applications of the theory of disorded systems is formulated in terms of ergodic theory. Namely, suppose that (M,U,μ) is a measure space with a probability measure μ, V a measurable function on M and T a measure-preserving transformation of the measure space (M,U,μ). We may consider the operator H acting in the same space in the following way :

$$(H\psi)(t) = -\varepsilon(\psi(t+1) + \psi(t-1)) + V(T^t\alpha)\psi(t) \qquad (2)$$

The previous case corresponds to the case of M being unit circle and T being the rotation to the angle ω. The operators (1) and (2) are discrete versions of the one-dimensional Schrödinger operator H acting in the Hilbert space $L^2(-\infty,\infty)$ by the formula

$$H\psi = -\varepsilon\, d^2\psi/dt^2 + V(T^t\alpha)\,\psi(t)\,, \quad -\infty < x\ \infty\,, \qquad (3)$$

where $\{T^t\}$ is a flow in (M,U,μ). We shall deal only with ergodic automorphisms and flows.

The spectral properties of (1)-(3) are of the main interest in this talk. Usually we encounter two different types of eigen functions.

Definition 1. A function $\psi(t)$ is a Bloch function if it can be written in the form

$$\psi(t) = e^{ipt}f(T^t\alpha)$$

where f is a measurable function on M.

The number p which enters in the definition is called a quasi-momentum. The function might take complex values.

Definition 2. A Bloch eigenfunction is a Bloch function ψ such that for some $\lambda \in \mathbb{R}^1$ and almost all α, $\psi(t)$ is an eigenfunction of (2) having the eigenvalue λ.

The set $\Lambda(B) \subset \mathbb{R}^1$ consisting of those λ for which there exist Bloch eigenfunctions is called the Bloch spectrum of H. If $p \neq 0$ then $\overline{\psi(t)} = e^{-ipt}\,\overline{f}(T^t\alpha)$ is also an eigenfunction having the same eigenvalue. Thus usually the Bloch spectrum belongs to the continuous part of the spectrum, has the multiplicity 2 and does not depend on α. A similar definition can be given in the case of flows and (3). Quite a different situation arises in the case of Anderson localization.

Definition 3. The complete Anderson localization holds for (2) if for almost all α one can find a complete sequence of eigenfunctions $\psi_{\alpha,i}(t)$ and eigenvalues $\lambda_{\alpha,i}$ of the operator H.

The essential part of this definition is some dependence of the set $\{\lambda_{\alpha,i}\}$ on α. It is easy to see that the set $\{\lambda_{T^t\alpha,i}\}$ does not depend on t. Thus the spectrum $\{\lambda_{\alpha,i}\}$ is an invariant function with respect to the dynamics and cannot be a measurable function of α. Later we will see in some cases the nature of this non-measurability.

In principle the decay of eigenfunctions might be arbitrary. However the most stable and mostly investigated type of behaviour appears in the case when all eigenfunctions decay exponentially fast at infinity. If this is really the case we shall say that we have the exponential complete Anderson localization.

Apparently in many cases we might only have a non-complete Anderson localization when only a part of spectrum corresponds to localized eigenfunction.

Assume that we have exponential complete Anderson localization. Usually each eigenfunction $\psi_{\alpha,i}(t)$ decays exponentially outside a finite set $Z(\psi_{\alpha,i})$ which we shall call an essential support of $\psi_{\alpha,i}$. The essential support must have two natural properties :

1. $|\psi_{\alpha,i}(t)| \leqslant C(\psi_{\alpha,i})\rho^{\,dist(t,Z(\psi_{\alpha,i}))}$ for some $\rho < 1$;

2. $|Z(\psi_{T^s\alpha,i}(t))| = Z(\psi_{\alpha,i}(t+s)) = Z(\psi_{\alpha,i}(t)) + s.$

Now suppose that we succeeded in defining an essential support for each eigenfunction $\psi_{\alpha,i}$. Then we may introduce two natural functions on the phase space of the initial dynamical system :

1. $\Phi(\alpha)$ is the set of those e.f. $\psi_{\alpha,i}$ for which the essential support contains the point 0 and lies to the right from 0.

2. $\Lambda(\alpha)$ is the set of eigenvalues of eigenfunctions belonging to $\Phi(\alpha)$.

Both $\Phi(\alpha)$, $\Lambda(\alpha)$ are multi-valued functions. It is natural to expect that they are measurable functions on M. In the case of complete Anderson localization the whole spectrum $\{\lambda_{\alpha,i}\} = \bigcup_t \Lambda(T^t{}_\alpha)$. In other words the whole spectrum of H consists of values of the measurable function $\Lambda(\alpha)$ along a single trajectory. This explains the non-measurability of the spectrum. Having $\Phi(\alpha)$, $\Lambda(\alpha)$ one can investigate other properties of H. For example, the integrated density of states is equal to the distribution function of $\Lambda(\alpha)$, in terms of $\Lambda(\alpha)$ one can formulate the problems concerning the limiting distributions for the spacings between the nearest levels and so on.

From what was said above the natural strategy for the investigation of complete Anderson localization follows. Namely, we shall try to construct and investigate the properties of $\Phi(\alpha)$ and $\Lambda(\alpha)$ in various situations. In order to do this we must follow the various processes of resonances and of tunneling which might appear.

2. ANDERSON MODEL :

In Anderson Model T is the Bernoulli shift acting in the space M of infinite sequences $\alpha = \{\alpha_i\}$, μ is the product measure, $V(\alpha) = \alpha_o$. Thus $V(T^n\alpha) = \alpha_n$ is a sequence of independent random variables. The mostly investigated case concerned the situation when the random variables α_n have a distribution with hounded density. Under this condition it was shown that with μ-probability one the complete exponential Anderson localization takes place. The first qualitative explanation is apparently due to Mott and Twose [1]. The idea is quite simple. Let us consider the sequence of recurrent relations :

$$-\varepsilon(\psi(n+1) + \psi(n-1)) + (\alpha_n - \lambda)\,\psi(n)=0, \quad n = 1,2, \dots . \qquad (4)$$

It follows from the famous Fürstenberg theorem that for each λ with μ-probability 1 there exist $\psi^+_\lambda(1)$, $\psi^+_\lambda(0)$ such that if we recursively solve (4) then the corresponding solution $\psi^+_\lambda(n) \to 0$ exponentially fast as $n \to \infty$. Certainly only the ratio $\gamma^+_\lambda = \psi^+_\lambda(1)/\psi^+_\lambda(0)$ is important. In the same manner we can consider the analogous system for $n \leqslant 1$ and get γ^-_λ. The exponentially localized eigenfunctions correspond to those λ for which $\gamma^+_\lambda = \gamma^-_\lambda$. However this simple explanation is not very convincing because the sets of measure zero which appear in this argument depend on λ. Another objection is due to the fact which is not so often emphasized that $\gamma^+_\lambda = \gamma^-_\lambda$ are always everywhere discontinuous functions of λ. Nevertherless the final statement about complete Anderson localization is true. It was first shown in a slightly different situation in the paper by Goldsheid, Molehanov, Pastur [2] and then in the situation under discussion by Kunz, Souillard [3]. The main strategy in the proofs is based on the positivity of

Liapunov exponents. The expository paper by Souillard [4] contains a lot of information about the results obtained in this way.

Quite different approach was suggested several years ago by Fröhlich and Spencer [5] who worked not only in the one-dimensional case and any ε but also in the multi-dimensional case and small ε. Roughly speaking they remarked that if we consider the operator H in two non-overlapping domains with Dirichlet boundary conditions then with a big enough probability the eigenvalues in these domains lie sufficiently far from each other. This gives a possibility to use a version of perturbation theory and to show that in the union of this domains the eigen functions are concentrated only in one of the domains. If we consider a sequence of increasing domains then eventually it leads to the localization. The main estimation for the probability comes from the so-called Wegner lemma.

Our approach to the whole problem was motivated by the Fröhlich-Spencer's paper which explicitly indicates that in order to prove Anderson localization one has to really follow the resonances and the probability with which they appear. The role of resonances and tunneling was also emphasized by Jona-Lasinio, Martinelli, Scoppola [6]. The final results of the whole approach are presented in [7]. The results of [7] can be used for proving the existence of functions $\Phi(\alpha)$, $\Lambda(\alpha)$ but we are still far from complete understanding of their properties in the case of Anderson model. However we shall see that in the case of quasi-periodic potentials $\Phi(\alpha)$, $\Lambda(\alpha)$ is a convenient tool for the analysis.

3. BLOCH SPECTRUM FOR (1) – (3) WITH A QUASI-PERIODIC POTENTIAL.

In [8], [9] the spectral properties of the operator H

$$H\psi = -d^2\psi/dt^2 + \varepsilon(a_1 \cos(\omega_1 t + \alpha_1) + a_2 \cos(\omega_2 t + \alpha_2))\,\psi(t) \quad (4)$$

for small ε were considered. Both works were motivated by the famous results of Novikov when he discovered the deep connections of finite-zone potentials with the Korteweg-de Vries equation. It was shown that for sufficiently small ε the operator (4) has many Bloch eigenfunctions and the corresponding eigenvalues do not depend on (α_1,α_2). The crucial difference with the case of periodic potential is the appearance of infinitely many forbidden zones. These forbidden zones are concentrated near the points λ for which $\pm\sqrt{\lambda}$ are close to the numbers $(n_1\omega_1+n_2\omega_2)$ and their lengths decay very quickly with the increase of $|n_1|+|n_2|$, in fact almost exponentially. Bellissard and al [10] got similar results for the difference Schroedinger operator. The survey paper by J. Avron, B. Simon [11] is dedicated entirely to spectral properties of Schroedinger operators with quasi-periodic potentials. The main method of the papers [8], [9], [10] is a version of KAM-theory. Small divisors which appear in the series of perturbation theory determine the boundaries of forbidden zones. Moser in [12] showed that generically these forbidden zones are persistent. In a sense it means that quasi-periodic potentials give rise to a new kind of spectra which are concentrated on Cantor sets of positive measure. This is one of the "Cantorization" examples of theoretical physics. Apparently the fine structure of Cantor sets can be observed in direct physical experiments. Avron and Simon suggested that the Saturn rings are remnants of some Cantor set which is created by the motion of Saturn and his main satellites. The results which were mentioned are valid for a wider class of potentials.

4. ANDERSON LOCALIZATION FOR SCHROEDINGER DIFFERENCE OPERATORS WITH QUASI-PERIODIC POTENTIALS.

The difference Schroedinger operators with quasi-periodic potentials (1) acquired recently some attention in connection with the discovery of quasi-crystals (see [13], [14] . In these papers the case of potentials taking a finite number of values was treated. One can consult [13], [14] in order to see physical reasons why this class of potentials is interesting. However we shall deal only with C^2-functions V having only two non-degenerate critical points. The first idea that Anderson localization might happen for (1) appeared apparently in the paper by Aubry and Andre [15]. They considered the case of $V(\alpha)=\cos 2\pi\alpha$ and discovered the so-called Aubry duality which implies in particular that for irrational ω and sufficiently small ε the largest Liapunov exponent corresponding to every λ is positive. In other words the behaviour of Liapunov exponents in this model turns out to be the same as in the case of Anderson model. It follows from a theorem by Pastur [16] that the spectrum of H can only be singular.

Now we shall formulate our main result. We assume that

1) the potential V is a C^2-Morse function having one maximum and one minimum : this means in particular that all values except the extremal ones are taken at two points ; in a neighbourhood of one of these points the function V is strictly increasing while it is strictly decreasing in a neighbourhood of another point ;

2) the irrational number ω is typical in the following sense ; let us consider the expansion of ω in continued fraction $\omega = [\kappa_1, \kappa_2, \kappa_3, \ldots, \kappa_n, \ldots]$. We assume that $\kappa_n \leqslant \text{const.}\, n^2$.

Now we can formulate :

Main theorem. Assume that 1), 2) are valid. Then there exists $\varepsilon_0 > 0$ such that for all ε, $|\varepsilon| \leqslant \varepsilon_0$

a_1) the integrated density of states is a Cantor staircase concentrated on a Cantor set C of positive measure

a_2) for almost all α the spectrum of (1) is pure point and $\overline{\{\lambda_{\alpha,i}\}} = C$

Fröhlich and Spencer got a similar result using their method. (Private communication). The fact the integrated density of states is concentrated on a Cantor set is apparently characteristic of quasi-periodic potentials generated by one-incommensurate frequency. Apparently if the number of independent frequencies is larger than one then the integrated density of states is absolutely continuous and the corresponding density is strictly positive.

A more simple situation arises if V is a strictly monotone C^2-function having one point of discontinuity. In this case one can construct uniformly converging series in ε giving the eigenfunctions and eigenvalues (Zitomirskaya, in press).

The method by which we prove the main theorem is based upon the renormalization group ideas in the theory of dynamical system. First we fix s_0 and consider the finite continued fraction approximation $\omega_{s_0} = [\kappa_1, \dots, \kappa_{s_0}]$. In this case an approximate form of the corresponding function $\Lambda^{(s_0)}(\alpha)$ is presented in fig. 1.

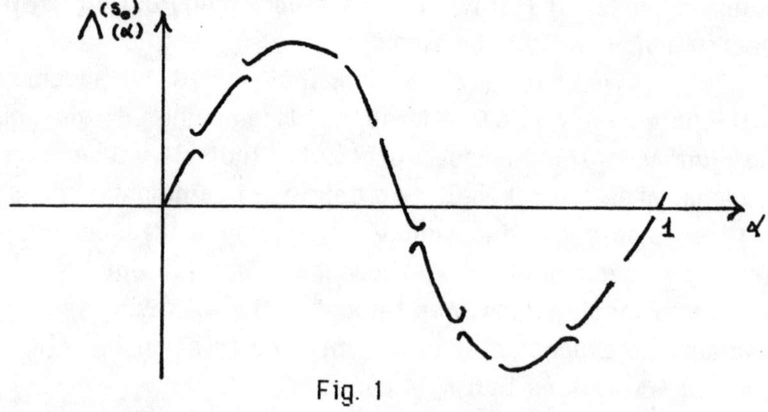

Fig. 1

The main property of this function is the appearence of the intervals where Λ(s) takes two values and of the intervals where Λ(s) has no values, or its value is an empty set. One can also see the appearence of gaps in the spectrum.

In order to construct the limiting function Λ(α) we use some induction on s. Suppose that we already have $\Lambda^{(s)}(\alpha)$ for the s-th approximant $\omega_s = [\kappa_1, \kappa_2, \dots, \kappa_s]$. Then we consider the points α_i where $\Lambda^{(s)}(\alpha_i) = \Lambda^{(s)}(\alpha_i + \omega_{s+1}t)$. For $t \leqslant$ S.const./(ln1/ε) we investigate in detail the change of gaps and the appearence of new gaps under the transition $s \to s+1$.

REFERENCES

1. Mott, N.F. and Twose V.D. ; Adv. Phys., 10, 107(1961), see also Borland R. ; Proc. Roy. Soc., A274, 809(1963).
2. Goldsheid I. Ya., Molchanov S.A., Pastur L.A. ; Funk. Anal. and Appl. , 11, 11 (1977).
3. Kunz M., Souillard B. ; Comm. Math. Phys. 78, 201, (1980).
4. Souillard B. ; Spectral Properties of Discrete and Continuity Random Schroedinger Operators. Preprint. Institut for Mathematics and Applications, 1986. Febr.
5. Fröhlich J. Spencer T. ; Comm. Math. Phys. 88, 151 (1983).
6. Jona-Lasinio G., Martinelli F., Scoppola E. ; Ann. Inst. H. Poincaré, 42, 1(1985)
7. Fröhlich J., Martinelli F., Scoppola E., Spencer T. ; Comm. Math. Physics, 101 (1985).
8. Dinaburg E.I. , Sinai Ya. G. ; Funct. Anal. and Appl. 9, 279(1975).
9. Belokolos E.D. ; Theor. and Math. Phys. , 25, 344 (1975) ; 26,35 (1976).
10. Bellissard J., Lima R., Testard D. ; Comm. Math. Physics, 88, 207, (1983)
11. Avron J., Simon B. ; Bull. Amer. Math. Soc., 6, 81, 1982
12. Moser Ju. ; Comm. Math. Helv. 56, 198, (1981)
13. Kohmoto M., Kadanoff L.P., Chao Tang. ; Phys. Rev. Lett. 50, 1870 (1983).
14. Kalugin P.A., Kutaev A. Yu., Levitov L.S. ; Elektronic Spectrum of One-dimensional Quasi-crystalls. Chernozolovsca, preprint 1986.
15. Aubry S., Andre G. ; Ann. Israel Phys. Soc. , 3, 133 (1980).
16. Pastur L.A. ; Comm. Math. Phys. , 75, 179 (1980).

ASYMPTOTIC COMPLETENESS FOR N-BODY QUANTUM MECHANICS

I.M. Sigal

Department of Mathematics, University of California
Irvine, California 92717, USA

A. Soffer

Division of Mathematics, Physics and Astronomy
California Institute of Technology
Pasadena, California 91125, USA

ABSTRACT

I describe the proof of Asymptotic Completeness for Quantum-Mechanical systems with arbitrary number of particles.

I assume the interactions are of the short range type, namely, decay at infinity faster than the inverse of the distance.

INTRODUCTION

In this talk I describe a proof of Asymptotic Completeness for general N-body Quantum Mechanical systems interacting via short range two body potentials.

Consider an N-body system and let H be its Hamiltonian. One then studies the asymptotic behavior, in time, of the solutions of the corresponding Schrödinger equation, given by $e^{-iHt}\psi$ for ψ orthogonal to the bound states of the system (the eigenfunctions of H). One expects, on physical grounds, that the asymptotic solution approaches a sum of states each of which corresponds to a different breakup of the system to stable, independently moving, clusters of particles.

Rigorous Analysis of Asymptotic Completeness has a long and involved history (see [RS, S, CFKS] for reviews and references).

After more than 35 years of efforts we have a satisfactory understanding of the three-body problem. (Here we mention only the works of [F, GM, E, M]; see [SS] for a comprehensive list.) However, the scattering theory of N-particles with $N \geq 4$ was, at best, only partially understood. Various single channel problems were treated in [H, La, IO]; systems with two-cluster channels were studied in [MS, K]. The case of generic potentials was treated in [S, Hag, HP].

FORMULATION OF THE PROBLEM AND BASIC NOTATION

The dynamics of N-particles (in ν-dimensions) in Q.M. (Quantum Mechanics) is described in terms of the Schrödinger Equation in $L^2(X)$, where X-<u>the configuration space</u> of N-classical particles is

$$X = \{ x \in \mathbb{R}^{\nu N} : \sum_{i=1}^{N} m_i x_i = 0 \} \qquad x \equiv (x_1, \dots x_N) \in \mathbb{R}^{\nu N}$$

with the inner product $\langle x,y\rangle \equiv 2 \sum m_i x_i \cdot y_i$. x_i, m_i are referred to as the position and mass of the i-th particle.

Let the Hamiltonian of the system, H, be given by

$$H = -\Delta + V \quad \text{on } L^2(X) \qquad (1.1)$$

with
$$V \equiv \sum_{i<j} V_{ij}(x_i - x_j) \qquad 1 \leq i, j \leq N \qquad (1.2)$$

Under condition (A) below $D(H) = D(\Delta)$ and H is a self-adjoint operator on $L^2(X)$.

The V_{ij} are assumed to satisfy:

A) V_{ij} are real valued functions of their argument and

$V_{ij}(y)$ is Δ_y-compact (on $L^2(\mathbb{R}^\nu)$)

B) $|y|^{1+\varepsilon}\left|V_{ij}(y)\right|$ is Δ_y-compact.

C) $|y|^{1+\varepsilon}\left|\frac{\partial V_{ij}}{\partial y}\right|$ and $|y|^2\left|\frac{\partial^2 V_{ij}}{\partial y^2}\right|$

are Δ_y-bounded; $\varepsilon > 0$.
When B) is satisfied with some $\varepsilon > 0$, we speak of the short-range case.

We now define some notation needed to describe the different decompositions of N-body systems. Let a,b.. denote partitions of {1,2,..N} to disjoint subsets called clusters. Such a partition a is also called cluster decomposition. $\#(a)$ denotes the number of clusters of a partition a . For a partition b which is a refinement of a we write $b \subset a$. If i,j belong to the same cluster of a partition a we write $(ij) \subset a$, otherwise $(ij) \not\subset a$. Note that $b \subset a$ implies that $\#(b) > \#(a)$.

We define the intercluster interactions of a partition a, I_a , to be

$$I_a = \sum_{\substack{(ij)\not\subset a \\ i<j}} V_{ij}(x_i - x_j) \tag{1.3}$$

and let

$$H_a = H - I_a \tag{1.4}$$

As is easily checked H_a describe the dynamics of the clusters of a, moving independently of each other.

Let H^a be the Hamiltonian of the non-interacting clusters of a , with their center of mass fixed at the origin. Then H^a acts on $L^2(X^a)$ with

$$X^a \equiv X \ominus X_a$$

$$X_a \equiv \{x \in X \mid x_i = x_j \text{ if } (ij) \subset a\}.$$

The eigenvalues of H^a, whenever they exist, are denoted by ε_α where $\alpha \equiv (a,m)$, with m the number of the eigenvalue of H^a, counting multiplicity.

The pairs α label the channels of the system.

(a) channel Hilbert space:

$$\mathcal{H}_\alpha \equiv L^2(X_a)$$

(b) the channel Hamiltonian

$$H_\alpha = \varepsilon_\alpha + T_a \quad \text{acting on } \mathcal{H}_\alpha$$

and $T_a \equiv |P_a|^2$ the kinetic energy of the center of mass motion of the clusters in a.

(c) the channel identification operator

$$J_\alpha : \mathcal{H}_\alpha \to L^2(X) \quad J_\alpha u = \psi^\alpha \otimes u$$

where ψ^α is the eigenvunction of H^a corresponding to ε_α.

We say that a short range many body system is asymptotically complete if for any $\psi \in L^2(X)$ orthogonal to all eigenfunctions of H and for any $\varepsilon > 0$ there exists a finite family A of channels and for each $\alpha \in A$, $\exists u^\pm_{\alpha,\varepsilon} \in \mathcal{H}_\alpha$ such that

$$\lim_{t\to\pm\infty} \| e^{-iHt}\psi - \sum_{\alpha\in A} e^{-iH_\alpha t} u^{\pm}_{\alpha,\varepsilon} \| \leq \varepsilon \quad . (1.5)$$

It can be checked that this definition is equivalent to the standard definition and imply, in particular, that the S-matrix is unitary.

The Strategy of the Proof

The proof is based on the following two steps: At first we decouple the different possible channels from each other. This is achieved by introducing a partition of unity of the phase-space $X \times X'$, X' the dual of X . Each member of the partition projects on essentially one cluster-decomposition. Consequently on its support we can show that H can be asymptotically approximated by some H_a .

The other requirement is that the boundary of each member be localized in regions where one expects, on physical grounds, that the state cannot propagate to.

The second step consists of proving a-priori estimates on the decay of the state in regions which include the "boundary" of the partition of unity constructed in the first step. For this, we use the positivity of the commutator of the Hamiltonian with certain observables supported around such regions.

The notion of Propagation-Set

We want to characterize those regions of phase-space where propagation takes place, depending on the channel and the total energy of system. It is a subset of the phase-space $X \times X'$ called the propagation set. For each channel α

and $E \in \text{spec.}_{\text{cont.}}(H)$ fixed, we define

$$\kappa_\alpha \equiv \sqrt{E - \varepsilon_\alpha} \qquad (2.1)$$

Then the channel α is called open (at energy E) if κ_α is real.

Let

$$\Sigma \equiv \{ \pm\kappa_\alpha \, , \, \alpha\text{-open} \} \qquad (2.2)$$

Then the propagation set at energy E , $PS_E \equiv PS_E^+ \cup PS_E^-$ is

$$PS_E^\pm = \bigcup_{\alpha\text{-open}} \{ (x,k) \in X \times X' \mid x^a = 0 \, ; \, x^b \neq 0 \;\forall\, b \supsetneq a ;$$
$$x_a \parallel \pm k_a \, ; \, |k_a| = \kappa_\alpha \} \qquad (2.3)$$

This definition is motivated by observing that in a fixed channel α , $\frac{x^a}{t} \to 0$ and $\frac{x^b}{t} \to$ const.vector $\neq 0$ for $b \supsetneq a$ as $|t| \to \infty$. Moreover $\frac{x_a}{t} \to \pm k_a$ as one expects the center of mass of the clusters to move freely, asymptotically; by energy conservation we then have

$$k_a^2 = E - \varepsilon_\alpha = \kappa_\alpha^2 .$$

Phase-Space-Operators

We describe the construction of observables (called phase-space operators) to be used in the phase-space analysis of states in the Hilbert space.

Let $\gamma \equiv \frac{1}{2}(\hat{x}\cdot p + p\cdot\hat{x})$ on $L^2(X)$ (3.1)

where $p = -i\nabla$ and $\hat{x} = \frac{x}{\langle x \rangle} \equiv \frac{x}{\sqrt{1+|x|^2}}$.

γ is the generator of the flow of the vector field $\frac{x}{<x>}$ and defines a self adjoint operator in $L^2(X)$.

Definition 1

For $i \mathbb{Z}^+$, let $j_i \; C^\infty(X) \cap L^\infty(X)$, homogenous of degree zero for $|x|>1$.
Let $f_i \in C^\infty(X) \cap L^\infty(X)$ and $\phi_i \in C^\infty(\mathbb{R}) \cap L^\infty(\mathbb{R})$. Assume moreover that $\nabla f_i \in C_0^\infty(X)$ and $\nabla \phi_i \in C_o^\infty(\mathbb{R})$. Let x be the multiplication operator on $L^2(X)$, $p \equiv -i\nabla_x$ and γ as above.

Then

$$J = \sum_{i=1}^{K} j_i(x) f_i(p) \varphi_i(\gamma),$$

$K < \infty$, is called a phase-space operator.

Definition 2

The support of a phase-space operator

$$J = \sum_{i=1}^{K} j_i(x) f_i(p) \varphi_i(\gamma)$$

is the subset of $X \times X'$ given by:

$$\bigcup_i \left\{ \left[(\text{supp.}\, j_i) \times (\text{supp.}\, f_i) \right] \cap \left[\text{supp.}\, \varphi_i \circ \gamma^{\text{class.}} \right] \right\}$$

$$\gamma^{\text{class.}} : X \times X' \to \mathbb{R} \qquad \gamma^{\text{class.}}(x,k) \equiv \hat{x} \cdot k \quad .$$

Definition 3 An operator $O = O(<x>^{-n})$ if $O = B<x>^{-n}$ for some bounded operator B with $[<x>,B]$ is $<x>$-bounded. It is $O_1(<x>^{-n})$ if B is Δ-bounded.

The properties of phase-space operators we use can be summarized as follows:

Localization Lemma

Let J be a phase space operator with empty support.

Then $J = O(\langle x\rangle^{-1})$.

Partitions of Unity

In order to achieve step one of the proof we need to construct a partition of unity which decouples channels.

We have: Channel Decoupling Theorem

Given E, there exists a family of phase space operators J_a s.t.

i) $\sum_a J_a = 1 + O(\langle x\rangle^{-1})$, $2 \leq \#(a) \leq N$. (4.1)

ii) supp. $J_a \subset \{(x,k) \in X \times X' \mid |x|_a > \delta|x|\}$

for some $\delta > 0$ and $|x|_a$ denotes the minimal distance between the clusters in a.

iii) $i[H_a, J_a] = \frac{1}{\sqrt{\langle x\rangle}} j_{a,i} B_a j_{a,i}^* \frac{1}{\sqrt{\langle x\rangle}} + O_1(\langle x\rangle^{-2})$

where each B_a is H-bounded and the $j_{a,i}$ are phase-space operators supported away from the PS_{E^-}.

To construct J_a one first constructs a (finite) partition of unity $\{j_a(x,k)\}$ of $X \times X'$ with $j_a = \sum_i j_i(x) f_i(k)\, \phi_i(\hat{x}\cdot k)$ and s.t. support j_a is in $\{|x|_a > \delta|x|\}$ and with the support of $(\nabla_x j_a)$ is away from PS_E. Moreover, one requires that for $f_i(k)$ which depends on some k_b, supp $j_i(x) \subset \{|x|_b > \delta|x|\}$.

The quantization is done by replacing x by the multiplication operator X, k by $-i\nabla$ and $\hat{x}\cdot k$ by γ, and using the spectral theorem.

We can now complete the proof of step 1 :

Lemma (Existence of Modified Deift-Simon operators).

Assume the $j_{a,i}$ terms, given by the expansion of $i[H_a, j_a]$ are (locally) H-smooth operators. Then the modified wave operators exist for short range two body interactions.

Proof

Let $W_a(t) = e^{iH_a t} j_a e^{-iHt}$.

Then

$$W_a(t) = i\int_0^t e^{iH_a s}(H_a j_a - j_a H)e^{-iHs}\,ds$$

$$i(H_a j_a - j_a H) = i[H_a, j_a] - j_a I_a \ .$$

Since the interactions are of the short range type

$j_a I_a = O(\langle X\rangle^{-1-\varepsilon})$ for some $\varepsilon > 0$ and therefore $j_a I_a$ is locally H-smooth by [PSS]. Since by our assumption $i[H_a, j_a]$ are also locally H-smooth it follows that for all $\psi \in \operatorname{Ran} P_\Delta(H)$, $u \in \operatorname{Ran} P_{\Delta'}(H)$

$$\int_0^t |\langle (H_a j_a - j_a H)e^{-iHs}\psi, e^{-iH_a s} u\rangle|\,ds \le \text{const.}\,\|\psi\|\,\|u\|$$

where $P_\Delta(H)$ is the spectral projection (of H) on any compact interval Δ away from thresholds and eigenvalues of H . This implies the strong convergence of $W_a(t)$ on $\operatorname{Ran} P_\Delta(H)$. It is easily checked that it implies (1.5).

<u>The Second Step of the Proof-Propagation Estimates</u>

As we saw above, the proof of completeness follows if we can show that projections (in phase-space) away from the PS_E are locally H-smooth. In this section we describe the method of <u>local positive commutators</u> that we use to prove such estimates.

The main result can be summarized as follows:

Theorem   Propagation Estimates

Let J be a phase space operator with support away from PS_E. (E is assumed to be away from the thresholds and eigenvalues of H.) Then, there is a small interval Δ around E s.t. for $\psi \in \operatorname{Ran} F_\Delta(H)$

$$\int_{-\infty}^{\infty} \| J \frac{1}{\sqrt{\langle x \rangle}} e^{-iHt} \psi \|^2 dt \leq C \|\psi\|^2 , \quad C < \infty \qquad (5.1)$$

independent of ψ. To prove the theorem we use the idea of local positive commutators: that is, the conclusion of the theorem holds, for a given phase space operator J, if one can construct a bounded operator, F, s.t.

$$F_\Delta(H) \, i[H,F] F_\Delta(H) \geq F_\Delta(H) \frac{1}{\sqrt{\langle x \rangle}} J^2 \frac{1}{\sqrt{\langle x \rangle}} F_\Delta(H) +$$

$$+ \text{ (locally } H\text{-smooth operator)}. \qquad (5.2)$$

Then, one proves the positivity of the commutator by localizing it using $F_\Delta(H)$ and the support of the phase-space operator J.

Consider the special case $J \equiv F_0(\gamma - \gamma_0)$ with $0 \leq F_0(\lambda) \leq 1$, sharply localized around zero and equal to 1 on some neighborhood of zero; the number $\gamma_0 \notin \Sigma$. Then, clearly J is supported away from the PS_E (since $\hat{x} \cdot k \in \Sigma$ on PS_E).

One then uses the formula

$$F_\Delta(H) \, i[H, F(\gamma)] F_\Delta(H) = F_\Delta(H) F'(\gamma) F_{\Delta'}(H) \, i[H,\gamma] F_{\Delta'}(H) F_\Delta(H) +$$

$$+ O(\langle x \rangle^{-2}) ; \quad F'(\gamma) = \frac{dF}{d\lambda}(\gamma) , \quad (\Delta' \supset \Delta)$$

to conclude that for J as above we should choose

$$F(\gamma) = \int_{-\infty}^{\gamma} F_0^2(\lambda-\gamma_0)\,d\lambda \quad .$$

One uses the channel expansion theorem to show that, channel by channel, $i[H,\gamma]$ is positive when localized on the support of $F_0(\gamma)$ (and on the energy-shell $F_\Delta(H)$) .

The treatment of the general case of J, though follows the same strategy, is far more complicated and can not be described here.

Concluding Remarks

The decay estimates one gets by the use of local positive commutators are very weak (at least at certain regions away from the PS_E). On the other hand,they are sufficient to exclude the process in which particles hop between different (open) channels through the (classically forbidden) regions. This hopping process, which for 4-particles or more is a viable possibility has been the major difficulty in extending the known results on the three-body problem to higher particle numbers.

It is a very important mathematical and physical question to find out whether these processes do occur with some small but significant probability. That would introduce new (and mild) singularities in the S-matrix which may be relevant to the study of dispersion-relations at low enough energy (for $N \geq 4$).

Finally, it is hoped that the methods developed here can be used to tackle the spectral and scattering theory of different problems such as spin-wave scattering and field theory.

ACKNOWLEDGEMENTS

One of us (A.S.) would like to thank W. Hunziker for support while visiting ETH-Zürich where this work was completed.

A. Soffer supported in part by USNSF grant DMS 84-07099.

I. M. Sigal supported in part by USNSF grant DMS-85-07040.

REFERENCES

[E] Enss, V. in "Proceeding of the Como Summer School on Schrödinger Equation", Como (1984).

[F] Faddeev, L., "Mathematical Aspects of the Three-body Problem", Israel Program of Sci. translations (1965).

[M1] Mourre, E., Comm. Math. Phys. 78, 391-408 (1981).

[M2] Mourre, E., Comm. Math. Phys. 91, 279-300 (1983); "Operaturs Conjugés II" - Marseille Preprint (1982).

[PSS] Perry, P., Sigal, I.M. Simon, B., Ann. Math. 114, 519-567 (1981).

[RS] Reed, M., Simon, B., "Scattering Theory" III, Academic Press (1979).

[S] Sigal, I.M., "Scattering Theory for Many Body Quantum Mechanical Systems", Springer Lecture Notes in Math. N 1011 (1983).

[SS] Sigal, I.M., Soffer, A., (submitted to Ann. Math.) (1985).

[HP] Hagedorn, G., Perry, P., Comm. Pure App. Math. 36, 213-232 (1983).

[MS] Mourre, E., Sigal, I.M., preprint (1983).

[CFKS] Cycon, H., Froese, R., Kirsch, W., Simon, B., "Lectures on the Schrödinger Equation" (to appear).

[GM] Ginibre, J., Moul, M., in (Combescure), Ann. Inst. Poincaré, 21 (1974) 97-145.

[H] Hepp, K., Helv. Phys. Acta 42 (1969), 425-458.

[K] Krishna, M., Comm. Math. Phys. 101, 129-152 (1985).

[Hag] Hagedorn, G., Trans. AMS 258 (1980), 1-75.

A SURVEY ON THE INTERACTION BETWEEN MATHEMATICAL PHYSICS AND GEOMETRY

Shing-Tung Yau
University of Texas at Austin
Department of Mathematics
Austin, Texas 78712
USA

I was asked to give a survey on a general area between Physics and Geometry. It is a difficult job and the only attempt I will make is to illustrate the interaction by examples. Clearly the examples are chosen based on my prejudice toward some area that is more familiar to me.

In the ancient days, Geometry and Physics were almost indistinguishable. A lot of us are glad to see that, after a long divorce, Geometers and Physicists are working together again. We shall only mention in the following the interactions in the last few centuries.

1. *Geometric optics*: The study of optics influenced strongly the study of the dynamics of geodesics on manifolds. Classical Hamiltonian mechanics initiated the study of symplectic geometry.
2. *Elasticity*: It motivates the study of bending of surfaces in Euclidean space. The problem of rigidity of geometric structure has been an important topic. Isometric embedding is still a difficult problem in geometry.
3. *Electromagnetism*: Maxwell's equation has been rewritten to become harmonic equations for differential forms in Hodge theory. Hodge theory has tremendous influence in both differential geometry and algebraic geometry. The latter development includes the Riemann-Roch formula and the Index formula.
4. *Quantum mechanics*: The introduction of quantum mechanics influences greatly the study of the spectrum of many important classical operators in geometry. Both wave mechanics and the heat equation have been used extensively to study the spectrum. The study of the relationship between

classical mechanics and quantum mechanics has provided a powerful tool for the study of the spectrum. For example, it provides a link between the length of closed geodesics and the spectrum of the Laplacian. The Dirac operator always plays a mysterious role in geometry.

5. *Gauge field theory*: The theory of the Yang-Mills field has been turned into a powerful tool in geometry, especially for questions related to differential topology.
6. *General relativity*: This part of physics has the closest relationship with geometry. The Einstein field equation for vacuum requires the metric to have zero Ricci curvature. It leads to the concept of the Einstein manifold in Riemannian geometry. It has been widely studied in geometry. The singularity theorem in relativity can also be considered as a theorem in geometry. The study of manifolds with positive scalar curvature naturally arose in relativity.

In the following, we shall concentrate on the last three topics.

From the point of view of differential equations, Hodge theory is linear, Yang-Mills is quasi-linear (by fixing the gauge suitably) and the Einstein equation is fully nonlinear. Therefore in many ways there is much more progress in Hodge theory and Yang-Mills theory. However, it is clear that linear theory will be fundamental in understanding the more nonlinear theory.

In Hodge theory, the Laplacian acting on the space of differential forms is a second order elliptic operator with discrete spectrum. The kernel consists of harmonic forms which are dual to the space of cycles which are fundamental topological invariants. Hence the Hodge theory provides a tool to link analytic data to topological data. For example, the vanishing theorems of Bochner-Kodaira demonstrated non-existence of certain cycles by curvature information. (For many cases, it is important to study forms with twisted coefficients.) Both the kernel of the Laplacian and the spectrum of the Laplacian acting on forms are important invariants. However, the spectrum is very poorly understood. A good estimate of the first eigenvalue will be very important in understanding nonlinear elliptic systems in geometry. This is especially true for the Lichnerowicz operator.

Yang-Mills theory can be considered as a nonlinear Hodge theory. Given a complex vector bundle, one needs the concept of connection to define differentiations for sections of V. Every connection gives rise to curvature which measures the twisting of the bundle, or the non-commutativity of the differentiation. It is a Lie algebra valued two form. We shall assume the Lie algebra to be $SU(n)$.

The Yang-Mills connection is a connection which is a critical point of the L^2 norm of the curvature. The Euler-Lagrange equation is a second order equation. If the base manifold is four dimensonal, the L^2-norm of the curvature is invariant under conformal change of the manifold. Hence gauge fields on R^4 can be considered as fields on $S^4\backslash\{\infty\}$. Assuming finite action, K. Uhlenbeck proved that Yang-Mills field can be extended smoothly to be a field over S^4 and hence we are reduced to studying fields over a compact manifold.

It is difficult to study general Yang-Mills fields (even over S^4). By comparing the action $\int \|F\|^2$ with the topological invariant $\int F \wedge F$, one comes up with the concept of self-dual fields (or anti-self-dual fields). If M is four dimensional, a connection is called self-dual or anti-self-dual according to $F = *F$ or $F = - * F$. This class of equations forms a subclass of Yang-Mills fields. It minimizes the action among connections with the same topological invariant $\int F \wedge F$. The equations form a first order system and are much better studied by Twistor's theory when the base manifold has special curvature properties. It transforms the problem to the Cauchy-Riemann problem and in particular to problems in algebraic geometry.

The space of self-dual connections over S^4 is understood by the works of Atiyah, Singer, Ward, Hitchin, Drinfeld, Manin, Donaldson, etc. However, it is still a problem to achieve a better understanding of the topology of the Moduli space. It remains to demonstrate that every $SU(2)$ Yang-Mills field over S^4 is either self-dual or anti-self-dual.

In 1981, C. Taubes, using important estimates by K. Uhlenbeck, demonstrated the existence of $SU(2)$ anti-self-dual connections over any compact four dimensional manifold M whose intersection matrix on the second cohomology is positive definite. Soon afterwards, S. Donaldson found that the Moduli space of the Taubes solution (with Instanton number equal to one) has a deep topological consequence. He demonstrated that no smooth four dimensional manifold exists whose intersection matrix on the second cohomology is $E_8 \oplus E_8$. This is the first significant result on the topology of smooth four dimensional manifolds.

Making use of a new theorem of Taubes, Donaldson has created a new invariant for smooth four dimensional manifold. He made use of some arguments from algebraic geometry and studied the deformation of the Moduli space of self-dual connections when the metric changes. In particular, he gave a counterexample to the h-cobordism theorem for four dimensional manifolds (which holds for every other dimension). More recently, he asserted that if an algebraic surface can be decomposed as the sum of two smooth manifolds, then one of them is homotopic to $\mathbb{C}P^2$ or S^4. [Algebraic surface is a four dimensional manifold in CP^N defined by a set of homogeneous polynomials.] Hence with the powerful arguments from Yang-Mills theory, Donaldson has changed the picture of the theory of smooth four dimensional manifolds completely.

Donaldson's theory motivated the following fundamental question that every four dimensional smooth simply connected manifold can be decomposed as the sum of algebraic surfaces. Would Gauge field theory help to settle this question?

When M is a complex surface with a Kähler metric, anti-self-dual $SU(2)$ connection gives rise to "stable" holomorphic vector bundle over M. Hence Taubes' theorem gives a nontrivial existence theorem for such bundles. (Recently D. Giesecker improved Taubes' theorem by an algebraic geometric argument.) It will be important to understand the generalization of such an existence theorem to higher dimensional algebraic manifolds. Given a complex vector bundle V over a compact algebraic manifold M, when does V admit holomorphic structure? It is necessary that the ith Chern class $C_i(V)$ can be represented by a differential form of type (i, i). If this is the only condition, it would imply the Hodge conjecture that closed forms of type (i, i) can be represented by cycles defined by algebraic equations. Perhaps one has to weaken the conclusion by asking the bundle to be equivalent to the difference of two holomorphic bundles in the sense of topological K-theory. There should be no doubt that an affirmative answer for the Hodge conjecture will be important for Physics.

From the point of view of differential geometry, it is important to ask the following converse of Taubes' theorem. Given a "stable" holomorphic vector bundle with zero first Chern class over a compact Kähler manifold, can we find a Hermitian anti-self-dual connection on this bundle? It should be noted that the holomorphicity of the bundle enables one to define self-duality for manifolds with dimension greater than four. One simply has to require the connection to be Hermitian (i.e., the parallel transport should preserve the holomorphic structure) and the trace of the curvature form should be zero. This definition was first observed by C.N. Yang to be equivalent to self-duality for bundles over $\mathbb{C}^2 = R^4$. Donaldson was also able to prove the existence of Hermitian self-dual connection for stable bundles over an algebraic surface. Uhlenbeck and the author were able to demonstrate the existence for arbitrary stable bundles over a compact Kähler manifold. Last year, Witten was able to apply the existence of such a connection over a three dimensional algebraic manifold to string theory. This was somewhat unexpected as the original motivation for the study of such connections came from consideration of problems in algebraic geometry. (It was used to recapture and generalize some works of Bogomolov.)

From the point of view of partial differential equations, the construction of Yang-Mills connection is a semi-linearization of the construction of Einstein metric. A great deal of activity in geometry was created after Einstein found the relevance of geometry in his theory of gravity. A lot of work was done on finding an exact solution of the Einstein equation with no matter. Geometers studied positive definite metrics which satisfy the corresponding equations. By imposing symmetries on the metric, one can find many exact solutions. However, most of these metrics have little symmetry and it is difficult to answer questions related to them. Both the existence and uniqueness theorems for the Einstein metrics are not well understood. Given a manifold, can one find a complete Einstein metric over such a manifold? If such metric exists, how many are there? When the dimension of the manifold is less than four, Einstein metrics are metrics with constant sectional curvature and they are better understood. When the manifold M is compact and is four dimensional, there is a known topological obstruction (due to Gray and Hitchin) for the existence of Einstein metrics. One has $\chi(M) \geq \frac{3}{2}|\sigma(M)|$ where $\chi(M)$ is the Euler number and $\sigma(M)$ is the signature of M. On the other hand, this is the only known obstruction. It is clearly very desirable to find a general method to understand the corresponding global equation. Will intuitions from Physics help?

The twistor program of Penrose is also able to transform the construction of the self-dual metrics to some problems in complex geometry. It was successful in many cases. However, in general, the transformed problem can be very difficult. When the manifold is Kähler, Calabe observed that the problem of Kähler Einstein metric can be transformed to a problem in solving a nonlinear partial differential equation. This was solved by the author ten years ago.

While the motivation came from the curiosity of solving the Einstein equation in the category of positive definite metrics (Relativists call them Gravitational Instantons), the author found them useful for solving some old problems in algebraic geometry. It was a greater surprise when the compact Ricci flat Kähler manifolds came up in the recent study of string theory. It is hoped that theoretical physics will motivate more fresh ideas in mathematics and vice versa. I have omitted mentioning many exciting developments in mathematics where the ideas come from theoretical physics. The theory of supersymmetry leads Witten to give a new proof for Morse theory which in turn leads to many new developments in geometry and analysis. The interaction of the theory of infinite dimensional Lie algebra and the theory of Moduli space of Riemann surfaces with string theory is another fruitful area in modern mathematics. We hope to see more joint activities by Physicists and Mathematicians.

CONVERGENCE OF MULTISCALE MAYER SERIES

D.C. Brydges [1]
Department of Mathematics
University of Virginia
Charlottesville, VA 22903 USA

T. Kennedy [2]
Department of Physics
Princeton University
Princeton, NJ 08544 USA

Abstract

In these notes we will present a simple self-contained treatment of multiscale Mayer series which follows [1]. Our treatment includes a short proof of convergence of multiscale Mayer series and a new formula for the Ursell coefficients (connected parts).

The first proofs of convergence of the Mayer series were "single scale" in the sense that there was no decomposition of the interaction into pieces that live on different length scales. Göpfert and Mack [2] developed a multiscale Mayer series by iterating the ordinary Mayer series a finite number of times. The Gallavotti-Nicoló tree formalism (see [3] for a review) applied to the case of a field theory with a $\cos\phi$ interaction becomes an infinitely iterated Mayer series. Benfatto [4] used this expansion to give the first direct (non-field theoretic) proof of convergence of the Mayer series for the two-dimensional Yukawa gas for $\beta < 4\pi$. This infinitely iterated Mayer series was generalized and streamlined in [5] by using combinatorial identities of Battle and Federbush [6] and Göpfert and Mack [2]. The treatment presented here is a multiscale Mayer series which avoids having to do iterated

[1] Sloan Fellow. Research partly supported by NSF grant DMS 8500516.

[2] NSF Postdoctoral Fellow. Research partly supported by the Institute des Hautes Etudes Scientifiques and by NSF grant PHY8515288.

Mayer series. The reader should not be misled by the simplicity of our approach; our treatment is just as powerful as iterated Mayer series.

We are given a finite measure space $(\Lambda, d\rho(x))$ (the possible states of a single particle) and a jointly measurable function $v(x,y)$ (the two body potential). The partition function is

$$Z = \sum_{n=0}^{\infty} \frac{z^n}{n!} \int d^n\rho \exp\left[- \sum_{1\le i<j\le n} v_{ij} \right]$$

where $d^n\rho = d\rho(x_1)\dots d\rho(x_n)$ and $v_{ij} = v(x_i, x_j)$. The Mayer series is the expansion of log Z in powers of z

$$\frac{1}{|\Lambda|} \log Z = \sum_{n=1}^{\infty} b_n z^n$$

where $|\Lambda| = \int d\rho(x)$. A standard formula for the b_n is

$$b_n = \frac{1}{|\Lambda|} \frac{1}{n!} \int d^n\rho \; u(x_1, \dots, x_n)$$

where the Ursell functions u are given by

$$u(x_1, \dots, x_n) = \sum_G \prod_{ij\in G} (e^{-v_{ij}} - 1) \tag{1}$$

with G summed over all connected graphs on $\{1, 2, \dots, n\}$.

We assume that the interaction v can be written as

$$v(x,y) = \int_0^T \dot{v}(x,y;s)ds \tag{2}$$

where the interaction $\dot{v}(x,y;s)$ is *stable* in the sense that there exists a function $B(s)$ with

$$\sum_{1\le i<j\le n} \dot{v}(x_i, x_j; s) \ge -B(s)n \tag{3}$$

for all n and all $x_1, \dots, x_n \in \Lambda$. The norm of $\dot{v}(s)$ is defined to be

$$\|\dot{v}(s)\| = \sup_x \int d\rho(y) |\dot{v}(x,y;s)|. \tag{4}$$

We define $v(t)$ by

$$v(t) = \int_0^t \dot{v}(s)ds$$

so that $v(t)$ is a one parameter family of potentials which interpolates between the zero potential and the potential v. We have taken the limits of integration in (2) to be 0 and T, but in some cases it is more convenient to use other limits, e.g., $-\infty$ and 0.

There are many ways to realize eq. (2), the most trivial being to take $\dot{v}(t) = v$ and $T = 1$. This leads to a single scale Mayer series. To obtain multiscale Mayer series we should follow the renormalization group philosophy of turning on the parts of the interaction that live on different length scales one at a time, beginning with the shortest length scales. Thus the most powerful results are obtained when t corresponds to a length scale and $\dot{v}(t)$ lives on that length scale.

By using $v(t)$ in eq. (1) we obtain a one parameter family of Ursell functions $u(x_1, \dots, x_n; t)$. The key to the simplicity of our approach is to find a system of differential equations (with respect to t) that these functions satisfy. By rewriting these equations as integral equations we will obtain bounds that imply convergence of the Mayer series. To verify our formula for the Ursell functions all we have to do is verify that they satisfy the differential equations.

Actually, we will take a slightly different but shorter approach. We *forget about eq. (1) altogether.* Instead, we will write down the differential equations and define the Ursell functions to be the solution to these equations. We then prove an identity, eq. (8), which shows that we have made the correct definition.

To keep the notation simple we shall use the following shorthand. Given $x_1, x_2, \ldots \in \Lambda$ and a finite subset I of $\mathbb{N}$, $u(I;t)$ will denote $u(\{x_i\}_{i\in I};t)$, i.e., the Ursell function of order I evaluated on the points x_i with $i \in I$.

Lemma 1: The system of initial conditions and differential equations

$$u(I;0) = \begin{cases} 1 & \text{if } |I| = 1 \\ 0 & \text{if } |I| > 1 \end{cases} \tag{5}$$

$$\begin{aligned} \frac{du}{dt}(I;t) = &- \sum_{i,j\in I: i<j} \dot{v}_{ij}(t)u(I;t) \\ &- \frac{1}{2}\sum_{J\subset I}\ \sum_{i\in J, j\in I\setminus J} \dot{v}_{ij}(t)u(J;t)u(I\setminus J;t) \end{aligned} \tag{6}$$

(where J is summed over *proper* subsets of I) is equivalent to the system of integral equations

$$\begin{aligned} u(I;t) = -\frac{1}{2}\int_0^t ds \sum_{J\subset I}\ &\sum_{i\in J, j\in I\setminus J} \dot{v}_{ij}(s)u(J;s) \\ &u(I\setminus J;s)\exp\left[-\sum_{k,\ell\in I:k<\ell}\int_s^t \dot{v}_{k\ell}(s')ds'\right] \end{aligned} \tag{7}$$

for $|I| > 1$ and $u(I;t) = 1\ \forall t$ for $|I| = 1$. Furthermore these systems have a unique solution, and it satisfies

$$\exp\left[-\sum_{i,j\in I:i<j} v_{ij}(t)\right] = \sum_{\{I_1,\ldots,I_k\}} \prod_{i=1}^{k} u(I_i;t) \tag{8}$$

where $\{I_1,\ldots,I_k\}$ is summed over all partitions of I.

Proof: The equivalence of the two systems is straightforward and left to the reader. The existence and uniqueness of the solution follows from the system of integral equations and a trivial induction in $|I|$.

To verify eq. (8) let $f(t)$ be the right hand side of (8). The initial conditions (5) imply $f(0) = 1$. Using eq. (6) we find

$$\frac{df}{dt} = \sum_{\{I_1,\ldots,I_k\}} \sum_{j=1}^{k} \prod_{i:i\neq j} u(I_i;t)\left[-\sum_{\ell,m\in I_j:\ell<m} \dot{v}_{\ell m}(t)u(I_j;t)\right.$$
$$\left.-\frac{1}{2}\sum_{J\subset I_j}\sum_{\ell\in J,m\in I_j\setminus J} \dot{v}_{\ell m}(t)u(J;t)u(I_j\setminus J;t)\right]. \tag{9}$$

A little reflection reveals that this just equals

$$-\sum_{\{I_1,\ldots,I_k\}} \prod_{i=1}^{k} u(I_i;t) \sum_{\ell,m\in I,\ell<m} \dot{v}_{\ell m}(t).$$

The terms where ℓ and m belong to the same I_j come from the first term in eq. (9), while the terms where ℓ and m belong to different I_i's come from the second term in eq. (9). Thus

$$\frac{df}{dt} = -\sum_{\ell,m\in I,\ell<m} \dot{v}_{\ell m}(t)f(t)$$

and so eq. (8) follows. □

We now *define the Ursell functions* $u(I;t)$ to be the functions given by the above lemma. It is not hard to show that this solution is given by eq. (1) (the proof appears in [1]), but we shall never need this fact.

Theorem 2: If

$$z\int_0^T \|\dot{v}(t)\| \exp\left[2\int_t^T B(s)ds\right] dt < \frac{1}{e} \tag{10}$$

then the series

$$\sum_{n=1}^{\infty} \frac{z^n}{n!} \int d^n\rho\, u(x_1, \ldots, x_n) \tag{11}$$

converges absolutely and equals $\log Z$.

Proof: Combining the integral equation (7) with the stability bound (3) we have the inequality

$$|u(x_1, \ldots, x_n; t)| \leq \frac{1}{2} \sum_J \sum_{i \in J, j \in J^c} \int_0^t \Big| \dot{v}_{ij}(s) \\ u(J; s) u(J^c; s) \Big| \exp(n \int_s^t B) ds \tag{12}$$

where J is summed over proper subsets of $\{1, \ldots, n\}$ and $J^c = \{1, \ldots, n\} \backslash J$. Define for $t \in [0, T]$

$$c_n(t) = \sup_{x_1} \int d\rho(x_2) \ldots \int d\rho(x_n) |u(x_1, \ldots, x_n; t)|$$

and note that $c_1(t) = 1\ \forall t$. Assume by induction that for $k < n$

$$c_k(t) \leq \left[\int_0^t \|\dot{v}(s)\| \exp\,(2 \int_s^t B) ds \right]^{k-1} k^{k-2} \tag{13}$$

we will use inequality (12) to bound $c_n(t)$.

We integrate (12) over $x_2, \ldots, x_n$ and sup over x_1 to obtain

$$c_n(t) \leq \frac{1}{2} \sum_{k=1}^{n-1} \binom{n}{k} k(n-k) \int_0^t \|\dot{v}(s)\| c_k(s) c_{n-k}(s) \exp\,(n \int_s^t B) ds.$$

Combining this result with the inductive assumption (13) and the combinatorial identity

$$\frac{1}{2} \sum_{k=1}^{n-1} \binom{n}{k} k^{k-1} (n-k)^{n-k-1} = (n-1) n^{n-2}$$

yields

$$c_n(t) \leq (n-1) n^{n-2} \int_0^t \|\dot{v}(s)\| \left[\int_0^s \|\dot{v}(s')\| \exp\left(2 \int_{s'}^s B\right) ds' \right]^{n-2} \\ \exp\left(n \int_s^t B\right) ds.$$

Next we use the trivial bound

$$\exp\left(n\int_s^t B\right) \leq \exp\left(2(n-2)\int_s^t B + 2\int_s^t B\right).$$

The resulting bound on $c_n(t)$ can then be written as

$$\begin{aligned} n^{n-2}\int_0^t \frac{d}{ds}\left[\int_0^s \|\dot{v}(s')\| \exp\left(2\int_{s'}^t B\right)ds'\right]^{n-1} ds \\ = n^{n-2}\left[\int_0^t \|\dot{v}(s)\| \exp\left(2\int_{s'}^t B\right)ds'\right]^{n-1} \end{aligned}$$

which completes the induction. The bound (13) for $t = T$ and Stirlings formula imply that the series (11) converges absolutely when hypothesis (10) holds.

It remains to be shown that this series equals $\log Z$. This follows from eq. (8) in a standard way. Integrate (8) with respect to $d^n\rho$, multiply by $\frac{z^n}{n!}$ and sum over n. After some simple combinatorics this shows that Z is the exponential of the series (11). □

Theorem 3: The Ursell functions are given by

$$u(x_1, \ldots, x_n; t) = (-1)^{n-1} \sum_T \prod_{b \in T} \int_0^t ds_b\, \dot{v}_b(s_b) \exp\left[-\sum_{1 \leq i < j \leq n} \int_{t(i,j)}^t ds\, \dot{v}_{ij}(s)\right] \tag{14}$$

where T is summed over all tree graphs (minimally connected graphs) on $\{1, 2, \ldots, n\}$. b runs over all bonds in T. $\dot{v}_b$ means $\dot{v}_{ij}$ where i and j are the endpoints of b. The function $t(i,j)$ depends on T and the s_b's:

$$t(i,j) = \max\{s_b : b \in \text{unique path in } T \text{ between } i \text{ and } j\}.$$

Proof: We give only a sketch of the proof, refering the reader to section 2 of [1] for details. It suffices to show that the right hand side of eq. (14) satisfies the differential eq. (6) and the initial conditions (5).

When $\frac{d}{dt}$ acts on this expression it can act either on the exponential factor or on one of the ds_b integrals. The first case yields the linear term in the differential equation. In the second case one of the s_b, call it $s_{b'}$, is set equal to t. If i and j are vertices such that the unique path in T between i and j contains b', then $t(i,j)$ will be t and so $\int_{t(i,j)}^t \dot{v}_{ij} = 0$. Thus the integral over the remaining s_b will factor into two factors. After playing around with tree graphs one finds that the result is the nonlinear term in eq. (6) □

Remarks

1. The criterion for convergence given in Theorem 2 is essentially the same as that of [5]. Theorem 2 can also be proved using the explicit formula for $u(x_1, \dots, x_n)$ of Theorem 3. Such a proof appears in [1].

2. The multiscale nature of our treatment may be seen in both Theorems 2 and 3. The important feature of the criterion for convergence (10) is that the exponential contains $\int_t^T B(s)ds$ and not $\int_0^T B(s)ds$. Thus a poor stability bound (large $B(s)$) for the short distance parts of the interaction can be overcome by the weakness of the interaction (small $\|\dot{v}(s)\|$) at these distances. A striking example of this is the two-dimensional Yukawa gas. (See [1] or [5] for details.) In Theorem 3 the important feature is again in the exponential. In expression (14) particles i and j "interact" only on length scales larger than $t(i,j)$.

References

1. Brydges, D.C. and Kennedy, T.: Mayer expansions and the Hamiltonian-Jacobi equation, preprint.
2. Göpfert, M. and Mack, G.: Iterated Mayer expansions for classical gases at low temperatures, Commun. Math. Phys. 81, 97 (1981).
3. Gallavotti, G.: Renormalization theory and ultraviolet stability for scalar fields via renormalization group techniques, Rev. Mod. Phys. 57, 471 (1985).
4. Benfatto, G.: An iterated Mayer expansion for the Yukawa gas, J. Stat. Phys. 41, 671 (1985).
5. Brydges, D.C.: Convergence of Mayer expansions, J. Stat. Phys. 42, 425 (1984).
6. Battle, G. and Federbush, P.: A note on cluster expansions, tree graph identities, extra $1/N!$ factors, Lett. Math. Phys. 8, 55 (1984).

MANY SCALES PICTURE OF FIRST ORDER PHASE TRANSITIONS

K. GAWĘDZKI

C.N.R.S., I.H.E.S., 91440 Bures-sur-Yvette, FRANCE

ABSTRACT. A rigorous theory of first order phase transitions alternative to the Pirogov-Sinai approach is sketched. The theory is based on a renormalization group argument. It works also for complex Hamiltonians and can be used to extend the Gibbs phase rule to this case.

The renormalization group, see [1] and [2], has been invented as an approach to critical systems with infinite (or very large) length scale : the correlation length. Nevertheless, it was noticed quite early that its language of fixed points, scaling fields, scaling dimensions is also useful in heuristic description of first order phase transitions, in which the correlation length stays finite throughout the transition region, see [3] and [4] . On the other hand, it was observed that near first order transitions obstructions may appear to rigorous implementation of simple real space renormalization group schemes [5],[6]. This cast a doubt on the well-founding of the heuristic analysis. There exists however a deep physical reason why a multi-scale approach à la renormalization group should be useful in the description of first order phase transitions : the presence of the second length scale, different from the correlation length, which, unlike the latter, diverges at the transition point. This length scale, the size of a critical droplet in the false vacuum, may be visualized in the equilibrium setup as the minimal size of the box in which the wrong-phase boundary conditions cease to be effectively distinguished from the right ones inside the volume (at the transition point boundary conditions make a difference even in the thermodynamical limit).

The rigorous multi-scale analysis of first order phase transitions which I sketch here works directly with the lowest lying exitations of the system : the Bloch walls (or Peierls contours). A rather natural renormalization group transformation which avoids the obstructions of

[5],[6] may be defined in terms of these excitations. Its iterations are easily controlled leading to a scaling behavior at the transition points predicted in [3], [4]. Besides, the topology of the phase diagram near the transition can be easily established, as in the more traditional approach of Pirogov-Sinai [7], [8] and a convergent low temperature expansion for pure phases follows. As opposed to the original Pirogov-Sinai theory (see however [9], [10] and [11]) , we may also treat the case of complex Hamiltonians and extend the Gibbs phase rule to this case. This is important for the study of metastability in the equilibrium setup .The details of our analysis are contained in [12] and [13].

Suppose that we deal with a system whose low temperature behavior is dominated by a finite number of translationally invariant states q . The statistical sum for such a system may be often reformulated as a sum over collections of contours separating regions in the space where the configuration is equal to one of the states q . We shall consider contours γ which are connected unions of cubes of a unit lattice. The connected components of the boundary $\partial\gamma$ of each contour will be labelled by states q according to the model's dominant configuration on which they border. Each (labelled) contour will contribute its activity $z(\gamma)$ as a factor and each unit cube Δ of the islands of the state q outside the contours the exponential of the minus energy density e_q^Δ of this state. We shall additionally specify the boundary conditions by fixing the configurations of the system to be equal to q_o outside the volume Λ . This results in the following form of the statistical sum

$$Z^{q_o} = \sum_{\substack{\partial \\ \text{compatible}}} \exp[-\sum_q \sum_{\Delta\subset\Lambda^q(\partial)} e_q^\Delta] \prod_{\gamma\in\partial} z(\gamma) \qquad (1)$$

where the collections ∂ of disjoint contours γ in Λ over which we sum are compatible in the sense that the labels of the components of $\partial\gamma$ together with the boundary conditions determine uniquely the configuration of the system outside $\bigcup_{\gamma\in\partial}\gamma$ and consequently the partition of $\Lambda\setminus\bigcup_{\gamma\in\partial}\gamma$ into the sets $\Lambda^q(\partial)$. It is convenient to admit a slight dependence of the energy densities e_q^Δ on the cube Δ to account for the finite volume effects. This dependence will disappear in the thermodynamical limit leading to cube-independent densities e_q .

Relation (1) will be our starting point. It serves also as the starting point of the Pirogov-Sinai approach [7], [8]. The statistical sums of many of the spin systems and euclidean field theory models may be cast into this form. We shall not assume $z(\gamma)$ nor e_q^Δ to be real but only that the contours are strongly suppressed or more exactly that

$$|z(\gamma)\exp[\sum_{\Delta\subset\gamma} e_{q^\gamma}^\Delta]| \leq e^{-\beta_o|\gamma|} \qquad (2)$$

where q^γ is the label on the exterior component of $\partial\gamma$. Inequality (2) is the "Peierls condition" of [7], [8] . It is reasonable to expect it to hold with β_o large in the low temperature regime near the transition region. In our approach we shall treat e_q as parameters. In fact, they can be changed in many systems by change of the values of external fields.

Now suppose that for $q \neq q_{min}$, the state with minimal $\mathrm{Re}\, e_q$,

$$\mathrm{Re}(e_q - e_{q_{min}}) > D \tag{3}$$

where $1 \ll D \ll \beta_o$. We shall say in this case that the energy gap is open above state q_{min}. The inequality (3) implies that q_{min} is the dominant state. This is the situation in which the second length scale discussed above is short (with respect to the correlation length) and the system may be treated by a single scale expansion. Indeed, consider (1) for $Z^{q_{min}}$. Given ∂, call the connected components of $\Lambda \setminus \Lambda^{q_{min}}(\partial) = (\bigcup_{\gamma \in \partial} \gamma) \cup (\bigcup_{q \neq q_{min}} \Lambda^q(\partial))$ polymers X_α. Representation (1) may be easily transformed by fixing the set of polymers X_α and resumming over compatible contour families which lead to a given polymer configuration. The result is :

$$Z_\Lambda^{q_{min}} = e^{-\sum_{\Delta \subset \Lambda} e^\Delta_{q_{min}}} \sum_{\substack{\{X_\alpha\} \\ \text{disjoint}}} \prod_\alpha \rho(X_\alpha) \quad . \tag{4}$$

The sum in (4) is the standard polymer-gas statistical sum. Moreover the polymer activities $\rho(X_\alpha)$ are strongly suppressed, due to (2) and (3), say,

$$|\rho(X_\alpha)| \leq e^{-\frac{1}{2}D|X_\alpha|} \quad . \tag{5}$$

Consequently, a standard high temperature expansion for (4), see e.g. [14] or [15], converges uniformly in the volume. It may be used to establish the existence of the q_{min}-phase in the thermodynamical limit. For real e_q and positive $z(\gamma)$, it is also easy to show that any q-state boundary conditions lead to the same phase.

What to do if (3) does not hold? Let us order the states as $q_{min}, q_1, \dots, q_{Q-1}, q_{max}$ according to the increasing value of $\mathrm{Re}\, e_q$. Let us fix a very big length scale L and limit ourselves to very low temperature by assuming that

$$\beta_o \geq (3CL)^Q D \, , \; \beta_o \geq CL \, \mathrm{Re}(e_{q_{max}} - e_{q_{min}}) \tag{6}$$

for a fixed constant C. We shall say that an energy gap is open above state q_i $(i \geq 1)$ if

$$\mathrm{Re}(e_{q_{i+1}} - e_{q_i}) \geq 2CL \, \mathrm{Re}(e_{q_i} - e_{q_{min}}) \tag{7}$$

and

$$\mathrm{Re}(e_{q_{i+1}} - e_{q_i}) \geq 2CLD \quad . \tag{8}$$

Suppose now that there is a q_i with $i \geq 1$ such that the energy gap above q_i is open but all lower gaps are closed. We may easily eliminate the states q_k with $k > i$ from our description of the system as they are dominated by states q_j, $j \leq i$, and cannot lead to stable phases. To this end, given a term in the representation (1) for $Z_\Lambda^{q_j}$, we consider the connected components of $(\bigcup_{\gamma\in\partial}\gamma)\cup(\bigcup_{k>i}\Lambda^{q_k}(\partial))$ which form (labelled) contours of the system with states q_j , $j \leq i$. This leads to the new representation (1) for $Z_\Lambda^{q_j}$ involving only the states below the gap. Moreover, the new contour activities still satisfy (2) but with $\beta_o \mapsto \frac{1}{2}\mathrm{Re}(e_{q_{i+1}}-e_{q_i})$, say. Due to (7) and (8) , new β_o fulfils

$$\beta_o \geq CLD \quad , \quad \beta_o \geq CL\,\mathrm{Re}(e_{q_{max}}-e_{q_{min}}) \tag{9}$$

where new q_{max} is of course old q_i . We shall also change (diminish) scale L in order that the inequalities (6) continue to hold after elimination of the subdominant states.

The next step in our procedure will consist of a <u>coarse-graining</u> renormalization group transformation. Let us divide the set of (new) contours into the <u>big</u> ones which surround volumes $\geq L^d$ (new L!) and the <u>small</u> ones. Let us limit ourselves in (1) to small contours only (denote the modified statistical sum by $\tilde{Z}_\Lambda^{q_o}$) . $\tilde{Z}_\Lambda^{q_o}$ can be again treated by a simple single scale expansion. The reason is that due to the second inequality in (6) the suppression (2) of the contours is sufficient to control also contributions from their insides. As a result one obtains for any volume $\Lambda_1 \subset \Lambda$

$$\tilde{Z}_{\Lambda_1}^q = \exp[-\sum_{\Delta\subset\Lambda}(e_q^\Delta+s_q^\Delta)] \sum_{\substack{\{C_\alpha^q\} \\ C_\alpha^q\cap\Lambda_1\neq\emptyset \\ C_\alpha^q\not\subset\Lambda_1}} \prod_\alpha u(C^q) \tag{10}$$

where s_q^Δ is a tiny free energy type contribution from the unit cube Δ, essentially constant inside Λ except near the boundary.

$$|s_q^\Delta| \leq e^{-\frac{1}{2}\beta_o} . \tag{11}$$

C_α^q are connected clusters of the expansion lying in Λ which "decorate" the boundary of Λ_1 (surface terms).

$$|u(C_\alpha^q)| \leq e^{-\frac{1}{2}\beta_o |C_\alpha^q|} . \tag{12}$$

Going back to the complete statistical sum as given by (1), let us regroup it by resumming first the collections of small contours for each fixed compatible collection $\bar{\partial}$ of big contours. This leads to

$$Z_\Lambda^{q_o} = \sum_{\substack{\bar{\partial} \\ \text{compatible}}} \prod_{\gamma\in\partial} z(\gamma) \prod_q \tilde{Z}^q_{\Lambda^q(\bar{\partial})}$$

$$= \sum_{\substack{\bar{\partial} \\ \text{comp.}}} \exp[-\sum_q \sum_{\Delta\subset\Lambda^q(\bar{\partial})} (e_q^\Delta + s_q^\Delta)] \prod_{\gamma\in\bar{\partial}} z(\gamma) \prod_q \sum_{\substack{\{C_\alpha^q\} \\ C_\alpha^q\cap\Lambda^q(\bar{\partial})\neq\emptyset \\ C_\alpha^q\not\subset\Lambda^q(\bar{\partial})}} \prod_\alpha u(C_\alpha^q) . \tag{13}$$

Notice that C_α^q decorate in fact the big contours of $\bar{\partial}$.

Next, we build the coarse-grained contours by considering the connected components of the union of the L-lattice blocks which intersect the countours of $\bar{\partial}$ and their decorations $C_{\alpha_q}^q$. L is chosen so that $L^{d-1} \leq L$. A (labelled) coarse-grained contour may be represented as $L\gamma'$ with γ' being the next-scale contour in $\Lambda' = L^{-1}\Lambda$. Resumming in (13) $\bar{\partial}$ and $\{C^q_{\alpha_q}\}$ leading to the same coarse-grained configuration and denoting

$$e'^{\Delta}_q = \sum_{\Delta_1 \subset L\Delta} (e^{\Delta_1}_q + s^{\Delta_1}_q) \quad , \tag{14}$$

we obtain

$$Z^{q_o}_{\Lambda} = \sum_{\substack{\partial' \\ \text{comp.}}} \exp[-\sum_q \sum_{\Delta \subset \Lambda^q(\partial')} e'^{\Delta}_q] \prod_{\gamma' \in \partial'} z'(\gamma') \quad , \tag{15}$$

i.e. representation (1) on the next scale.

What have we gained by the coarse graining ? Firstly, the new contour activities satisfy

$$|z'(\gamma')\exp[\sum_{\Delta \subset \gamma'} e'^{\Delta}_{q_{\gamma'}}]| \leq e^{-\beta'_o|\gamma'|} \tag{16}$$

with $\beta'_o = L\beta_o/O(1)$ (L is due to the distance scaling). The effective temperature goes down : we are driven towards the <u>zero temperature fixed point</u>. Secondly, in the thermodynamical limit, (14) and (11) imply that

$$e'_q = L^d(e_q + s_q) = L^d(e_q + O(e^{-\frac{1}{2}\beta_o})) \tag{17}$$

so that the differences between the q-states energy densities <u>expand</u> : in the renormalization group language they form <u>relevant variables</u>.

Let us now relabel the states as q'_{min}, $q'_1,\dots,q'_{Q-1},q'_{max}$ according to $\mathrm{Re}\, e'_q$. Three possible outcomes of the coarse graining have to be analysed.

1) The energy gap has been open above q'_{min} (equal to q_{min} in this case), see (3). Then the single scale expansion discussed above establishes the existence of the q_{min} phase (and its uniqueness for the case of real energies).

2) The energy gap has been open above q'_i, see (7) and (8). We proceed as before eliminating the states q'_k for $k > i$ from our description as they are dominated by the states q'_j with $j \leq i$. There is one difference in comparison to the previously discussed situation : after the coarse graining the second inequality of (6) may be violated. But since (6) held on the previous scale, still

$$\beta'_o \geq \frac{CL}{O(1)L^{d-1}} \mathrm{Re}(e'_{q'_{max}} - e'_{q'_{min}}) \geq \mathrm{Re}(e'_{q'_{max}} - e'_{q'_{min}}) \tag{18}$$

and (18) is sufficient to show that (2) with $\beta'_o \mapsto \frac{1}{2}\mathrm{Re}(e'_{q'_{i+1}} - e'_{q'_i})$ holds after the elimination of q'_k, $k > i$. Next, L is diminished as before to restore the first inequality of (6) (since the number of states in finite L cannot drop to much if its initial value was sufficiently big) and we proceed to the next coarse graining.

3) No energy gap has been open as a result of the coarse graining. In this case

$$\mathrm{Re}(e'_{q'_{max}} - e'_{q'_{min}}) \leq (1+2CL)^{Q-1}D \leq (3CL)^{Q-1}D \tag{19}$$

and the first inequality of (6) for β_0' implies the second one. We may thus proceed directly to the next coarse-graining step without change (or with an increase) of scale L.

The outcome of our iterative renormalization group analysis is the following description of the phase coexistence. The q_j-phases coexist in the situation when we never open the energy gaps between q_j-states but do open the gap above them at some step. The energy densities of those states in fact approach each other under repeated coarse graining. In the single phase regime on the other hand, the energy gap above the lowest state opens up after some coarse-graining step and the analysis is completed by the single scale expansion. The expansive properties of the renormalization group transformation (17) of the energy densities lead easily to the Gibbs phase rule geometry of the phase coexistence manifolds. I refer the interested reader to [12] and [13] for the details of the arguments, the treatment of the correlation functions and the comparison of the present approach to the Pirogov-Sinai theory.

ACKNOWLEDGEMENTS. The present article is based on results of the joint work with R. Kotecký and A. Kupiainen. I would like to thank both of them for the joy of collaboration.

REFERENCES

[1] K.G. Wilson and J. Kogut, Phys. Rep. 12C (1974), p. 75.

[2] K.G. Wilson, Rev. Mod. Physics 55 (1983), p. 583.

[3] B. Nienhuis and M. Nauenberg, Phys. Rev. Lett. 35 (1975), p. 477.

[4] M.E. Fisher and A. Nihat Berker, Phys. Rev. B26 (1982), p. 2507.

[5] R.B. Griffiths and P.A. Pearce, Phys. Rev. Lett; 41 (1978), p. 917.

[6] R.B. Griffiths and P.A. Pearce, J. Stat. Phys. 20 (1979), p. 499.

[7] S. Pirogov and Ya. G. Sinai, Teor. Mat. Fiz. 25 (1975), p. 1185 and 26 (1976) p. 36.

[8] Ya. G. Sinai, Theory of phase transitions : rigorous results, Pergamon Press, Oxford 1982.

[9] S. Pirogov, Teor. Mat. Fiz. 66 (1986), p. 331.

[10] M. Zahradnik, Commun. Math. Phys. 93 (1984), p. 559.

[11] M. Zahradnik, Contribution to the preceedings of the 8th International Congress on Mathematical Physics, Marseille 1986.

[12] K. Gawędzki, R. Kotecký and A. Kupiainen, Coarse graining approach to first order phase transitions, IHES preprint.

[13] K. Gawędzki, R. Kotecký and A. Kuipiainen, in preparation.

[14] E. Seiler, Gauge theories as a problem of constructive quantum field theory and statistical mechanics, Lecture Notes in Physics, Vol. 159, Springer, Berlin-Heidelberg-New York 1982.

[15] R. Kotecký and D. Preiss, Commun. Math. Phys. 103 (1986), p. 491.

A ROUGHENING TRANSITION INDICATED BY THE BEHAVIOUR OF GROUND STATES

R. Kotecký* and S. Miracle-Sole**
* Dept. of Mathematical Physics, Charles University, V. Holesovičkách 2 - 18000
Praha 8 - Czechoslovakia
** Centre de Physique Théorique, CNRS
Luminy - Case 907, 13288 - Marseille - F

ABSTRACT

Our aim in this contribution is to present some illustrations for the claim that already by looking at the ground states of classical lattice models, one may meet some interesting and non trivial structures.

First of all we shall clarify what we mean by ground states. We consider a finite spin model on a lattice L with configuration space $\Omega = S^L$ and finite range interaction $\{\varphi_\Lambda, \Lambda \subset L\}$. The energy in a finite volume $V \subset L$ under a boundary condition (b.c.) $\bar{\sigma} \in \Omega$ is in an obvious notation

$$H_V(\sigma_V|\bar{\sigma}) = \sum_{\Lambda : \Lambda \cap V \neq \emptyset} \varphi_\Lambda(\sigma_V \cup \bar{\sigma}_{V^c})$$

The collection of the finite volume Gibbs states (a specification)

$$\mu_V^{\beta H}(\sigma_V|\bar{\sigma}) = Z_V(\bar{\sigma})^{-1} \exp(-\beta H_V(\sigma_V|\bar{\sigma}))$$

where

$$Z_V(\bar{\sigma}) = \sum_{\sigma_V} \exp(-\beta H_V(\sigma_V|\bar{\sigma}))$$

determine (by the DLR equations) the set $G(\beta H)$ of Gibbs states on L corresponding to a hamiltonian H and an inverse temperature β. If a Gibbs state $\mu \in G(\beta H)$ happens to equal the limit

$$\mu = \lim_{V \nearrow \infty} \mu_V^{\beta H}(\,.\,|\bar{\sigma})$$

under a fixed b.c.$\bar{\sigma}$ we shall call it the Gibbs state corresponding to a b.c. $\bar{\sigma}$.

Following Dobrushin and Shlosman [1] we introduce the ground states simply as the Gibbs states at $\beta = \infty$; i.e. the Gibbs states determined by the specification

$$\mu_V^{\infty H}(\sigma_V|\bar{\sigma}) = \lim_{\beta \to \infty} \mu_V^{\beta H}(\sigma_V|\bar{\sigma})$$

Clearly

$$\mu_V^{\infty H}(\sigma_V|\bar{\sigma}) = \begin{cases} 1/|M_V(\bar{\sigma})| & \text{if} \quad \sigma_V \in M_V(\bar{\sigma}) \\ 0 & \text{if} \quad \sigma_V \notin M_V(\bar{\sigma}) \end{cases}$$

where

$$M_V(\bar{\sigma}) = \{\tilde{\sigma}_V \,|H_V(\tilde{\sigma}_V|\bar{\sigma}) = \inf_{\sigma_V} H_V(\sigma_V|\bar{\sigma})\}$$

is the set of the ground configurations in V under the b.c. $\bar{\sigma}$. A ground state will be called rigid if it is supported by a single configuration $\sigma \in \Omega$, i.e. if it is a Dirac measure δ_σ on Ω. However, the ground states may often be measures supported by a large set of configurations. Examples of such random ground states are met e.g. for the Ising antiferromagnet on triangular or FCC lattices. We notice that even the problem of describing all the extremal periodic ground states is often non trivial. Let us mention in this connection the case of the three state Potts antiferromagnet on a square or a cubic lattice which is still open. An attempt to clarify it by mapping the ground states onto the Gibbs states of equivalent ferromagnetic models at a particular temperature was made in [2].

Of course, an important problem is to distinguish which ground states are stable in the sense that there exist Gibbs states at low temperatures that are "near" to the ground states in question. Some particular cases of periodic rigid ground states are covered by the Pirogov-Sinaï theory [3] and a class of non translation invariant rigid ground states was tackled in [4]. A general criterium for the stability of rigid ground states has been conjectured by Dobrushin and Shlosman [1]. However, no theory of stability exists for random ground states, although some statements about the thermodynamics involving such ground states were proven in a work by Aizenman and

Lieb [5] about the third thermodynamical principle.

When probing Gibbs states at low temperatures a useful notion may be that of weak ground states [1] describing the effect of a small perturbation added to the hamiltonian. Namely, considering an additional finite range interaction $\{\tilde{\varphi}_\Lambda\}$ and the corresponding hamiltonian $\tilde{H}$, a weak ground state (corresponding to the "direction" $\tilde{H}$) is a Gibbs state with the specification

$$\mu_V^{\infty H,\tilde{H}}(\sigma_V|\bar{\sigma}) = \lim_{\beta\to\infty} \mu_V^{\beta(H+\tilde{H}/\beta)}(\sigma_V|\bar{\sigma}) =$$

$$\begin{cases} = \dfrac{\exp - \tilde{H}_V(\sigma_V|\bar{\sigma})}{\sum_{\sigma_V\in M_V(\bar{\sigma})} \exp - H_V(\sigma_V|\bar{\sigma})} & \text{if } \sigma_V \in M_V(\bar{\sigma}) \\ = \quad 0 & \text{if } \sigma_V \notin M_V(\bar{\sigma}) \end{cases}$$

Let us now inspect two particular examples with an interesting structure of weak ground states. The case of periodic ground states will be discussed for an Ising antiferromagnet in an external magnetic field [1,6], while our main example, the ground states describing an interface and its roughening, will be discussed for an Ising model on a BCC lattice [7].

Considering the Ising antiferromagnet on a square lattice with a nearest neighbour (n.n.) coupling J and an external field h, one easily shows that, for $|h| < 4|J|$ there are the two customary antiferromagnetic rigid ground states (stable according to, say, the Pirogov-Sinaï theory). Inspecting now the border points, say, $h = 4|J|$, we get random ground states living on the set of all configurations for which no nearest neighbours are occupied by a pair of minus spins. Following [1,6] we may look at the lattice sites with minus spins as if they were occupied by a particle and thus we may equivalently think of a hard core lattice gas. Considering now the limit $\beta \to \infty$ along the lines $h = 4|J| + \mu/\beta$ as shown in Fig. 1, i.e. a weak ground state in the directions $\widetilde{H}$ given by the external field μ, we get the hard core lattice gas with a chemical potential -2μ, which is expected to have a critical value $\mu_c \sim -(1/2) \ln 3.8$ (for a rigorous estimate see [6]). This suggests the phase diagram of Fig.1

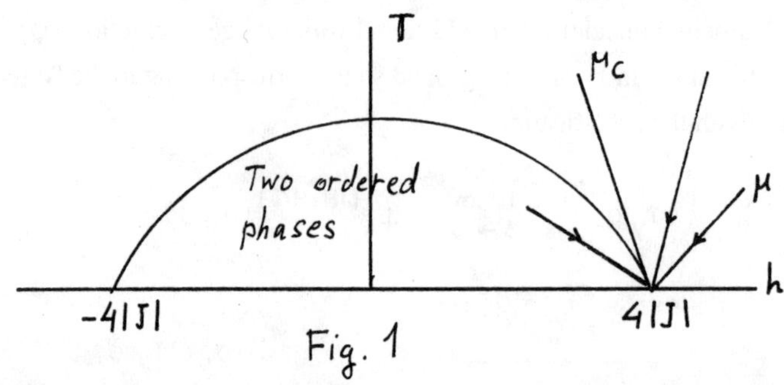

Fig. 1

Finally, let us consider the case of an Ising model on a BCC lattice with a n.n. ferromagnetic coupling $J_o > 0$ and a n.n.n. coupling J. Let us stress right now that we have in mind an isotropic model. Whenever $J > -(2/3) J_o$ there are two stable, rigid, translation invariant ground states of constant magnetization. We shall enforce the (100) interface between these two phases by taking a b.c. $\bar{\sigma}$ with $\bar{\sigma}_i = +1$ if $i = (i_1, i_2, i_3)$ is a lattice site with $i_1 \geqslant 0$ and $\bar{\sigma}_i = -1$ if $i_1 < 0$. For $J > 0$ it is easy to show that this b.c. leads to a rigid ground state supported by $\bar{\sigma}$ itself. Using the method of Dobrushin [8] as generalized in [4] or the method of van Beijeren [9], one may prove that this ground state is stable ; actually one gets the existence of the corresponding Gibbs state with a rigid interface in all the region shaded in Fig.2 (with $\alpha_o = (1/2) \ln(1+\sqrt{2})$ denoting the critical value of the coupling constant for the Ising model on a square lattice). A more interesting situation is obtained for $J = 0$. Let us consider right away the weak ground states corresponding to the directions $\widetilde{H}$ yielded by a n.n.n. interaction of the form $J = \alpha/\beta$, i.e. the weak ground states obtained along the lines shown in Fig.2. One easily observes that the configurations in $M_V(\bar{\sigma})$ contain interfaces with no overhangs. Actually these configurations are exactly those considered by van Beijeren in his body centered solid-on-solid (BCSOS) model [10]. Without going into the details [7] we may refer to his results to get a description of our weak ground state in terms of a six vertex model with the weights

$\omega_1=\omega_2=\omega_3=\omega_4= e^{-\alpha}$, $\omega_5=\omega_6=1$. If $\alpha > \alpha_R = (1/2)\ln 2$, the six-vertex model is in the ferroelectric phase and the interface is rigid ; if $\alpha < \alpha_R$, the results about the six-vertex model are usually interpreted as discribing a rough interface which actually should mean that the corresponding infinite volume Gibbs state of the BCSOS model does not exist and our weak ground state is translation invariant. Even though the above equivalence is exact only in the limit $\beta \to \infty$, one may expect that there is a curve $T_R(J)$ of roughening transitions as shown in Fig.2.

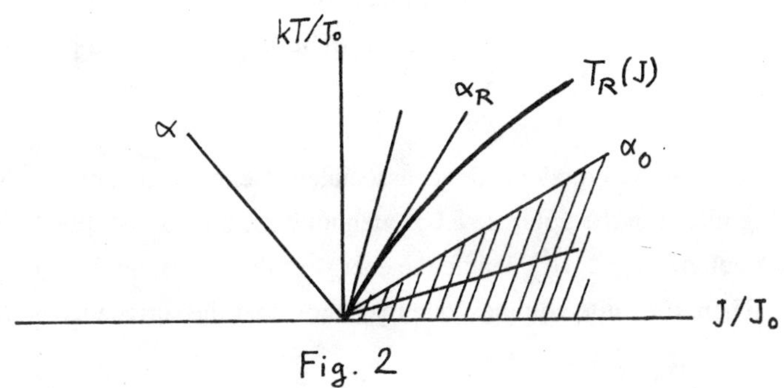

Fig. 2

Let us mention that one may investigate also the ground states with other b.c. $\bar{\sigma}(\vec{K})$ corresponding to general inclined interfaces $(K_1K_2K_3)$ with normal $\vec{K}$. It turns out that for both positive and negative J the ground state corresponding to an interface (110) is rigid and stable while the ground state corresponding to an interface (111) is translation invariant and the interface is rough. To see this last fact one may (for any J in an interval around zero) use a similar equivalence as above to the triangular Ising solid on solid (TISOS) model and then refer to the analysis of this model by Nienhuis, Hilhorst and Blöte [11] who solved it exactly. When one of the conditions $-K_1 \leqslant K_2 + K_3 \leqslant K_1$, $-K_2 \leqslant K_3 + K_1 \leqslant K_2$ or $-K_3 \leqslant K_1 + K_2 \leqslant K_3$ is fulfilled the interface corresponding to the b.c. $\bar{\sigma}(\vec{K})$ is again described in terms of a BCSOS model with the appropriate b.c. which is, in its turn, equivalent to a six vertex model with fixed polarizatons. For normals $\vec{K}$ in the complementary region the corresponding interface may be described in terms of the TISOS model.

To conclude let us comment about the connection of the roughening transition in the above sense and the facet formation in the equilibrium shape of a crystal (a droplet in the Ising model). One expects (see for instance [12]) that the roughening transition corresponds to the rounding of facets while a rigid interface associated to the b.c. $\bar{\sigma}(\vec{K})$ would imply the presence of a cusp in the corresponding direction in the graph of the surface tension as a function of $\vec{K}$, and by the Wulf construction give rise to a plane facet. In fact following Bricmont, El Mellouki and Fröhlich [12] one may use correlation inequalities to prove the existence of a cusp for the (100) facet in the shaded region of Fig.2 and for the (110) facet if T is small enough and $J \geqslant 0$. In what concerns the rounding of the facets some insight may be obtained by introducing the free energy

$$f_V(\vec{K}) = S_V^{-1} \ln Z^{sos}(S_V, \vec{K})$$

of the appropriate SOS model and b.c. associated as explained above to the b.c. $\bar{\sigma}(\vec{K})$ of the Ising model in the volume V (S_V denoting the area of the interface $(K_1K_2K_3)$ inside this volume) let $e_V(\vec{K})$ denote the energy of the corresponding ground state per unit area. Then, for finite volumes, one may show that the surface tension of the Ising model

$$\tau_V(\vec{K}) = S_V^{-1} \ln (Z(V, \bar{\sigma}(\vec{K})) / Z^+(v))$$

behaves asymptoticaly with $T \to 0$ as

$$\tau_V(\vec{K}) = e_V(\vec{K}) + kT f_V(\vec{K})$$

Supposing that this is correct also in the thermodynamic limit, the funciton $\tau_V(K)$ can in principle be computed using the equivalences of the SOS models with b.c. with exact solvable models with fixed polarizations and from it get quantitative information on the equilibrium crystal shape as a function of the coupling constants.

REFERENCES

[1] R.L. Dobrushin and S.B. Shlosman, Soviet Scientific Reviews **C5**, 54, (1985)

[2] R. Kotecký, Phys. Rev. **B31**, 3088 (1985)

[3] S.A. Pirogov and Ya. G. Sinaï, Teor. Math. Phys. **25**, 1185 (1975) ; **26**, 39

(1976). For a review see Ya. G. Sinaï, "Theory of Phase Transitions ; Rigorous Results". (Pergamon, New York, 1982).

[4] P. Holický, R. Kotecký and M. Zahradník, Rigid interfaces for lattice models at low temperatures, Preprint, Marseille, 1986.

[5] M. Aizenman and E. Lieb, J. Stat. Phys. **24**, 279 (1981)

[6] R.L. Dobrushin, J. Kolafa and S.B. Shlosman, Commun. Math. Phys. **102**, 89 (1985).

[7] R. Kotecký' and S. Miracle-Sole, Phys. Rev. **B34**, 2049 (1986)

[8] R.L. Dobrushin, Theory Prob. Appl. **17**, 582 (1972).

[9] H. van Beijeren, Commun. Math. Phys. **40**, 1 (1975)

[10] H. van Beijeren, Phys. Rev. Lett. **38**, 993 (1977)

[11] E. Nienhuis, H.S. Hilhorst and H.W.J. Blöte, J. Phys. **A17**, 3559 (1984)

[12] J. Bricmont, A. El Mellouki and J. Fröhlich, J. Stat. Phys. **42**, 743 (1986).

The wetting transition. New rigorous results for the semi-infinite Ising model.

Charles-Edouard Pfister
Départements de mathématiques et de physique
E.P.F.-L, CH-1015 Lausanne, Switzerland.

We consider a binary system in the two-phase region, with phases + and - . We suppose that the system is in the - phase. If we insert a wall, which adsorbs preferentially the + phase, there is a film of the + phase between the wall and the bulk - phase. If the thickness of this film is microscopic the interface between the adsorbed phase and the bulk phase is localized in the vicinity of the wall : there is partial wetting. On the other hand, if the thickness of the film is macroscopic, the interface is delocalized : there is complete wetting. The wetting transition is the transition from partial wetting to complete wetting. The theoretical work on this subject is reviewed in [1], [2], [3], [4]. We report upon new rigorous results obtained for the Ising model. In the two-dimensional case we already have a lot of information from exact computation of the relevant physical quantities [5] , [6]. Our approach is different. We use correlation inequalities and we have results for dimensions $d \geq 2$.

Let us consider the Ising model on $\mathbb{Z}^d$,

$$H = -\sum_{\langle ij\rangle} K\sigma(i)\sigma(j) - \sum_i \lambda\sigma(i) \tag{1}$$

where $<ij>$ means that we sum over all pairs of points i and j such that $|i-j| = 1$. We insert a wall by setting $\sigma(i) = 1$ for all $i = (i^1,\dots,i^d)$ with $i^d \leqslant 0$. We get a semi-infinite Ising model on the sublattice $\mathbb{L} = \mathbb{Z}^{d-1} \times \mathbb{Z}^+$, with coupling constant K, bulk field λ, and boundary field K. We generalize this model by taking an arbitrary boundary field h and by taking a coupling constant J for the interaction of two spins inside the first layer $\Sigma = \{ i \in \mathbb{L} ; i^d = 1 \}$. Our final hamiltonian is

$$H(J,K,h,\lambda) = -\sum_{<ij>\subset\mathbb{L}} K(i,j)\sigma(i)\sigma(j) - \sum_{i\in\mathbb{L}} \lambda\sigma(i) - \sum_{i\in\Sigma} h\sigma(i) \qquad (2)$$

with $K(i,j) = J$ if $<ij> \subset \Sigma$ and otherwise $K(i,j) = K$. We always choose $K > 0$, $J > 0$ and $h > 0$. The last condition means that the wall adsorbs preferentially the + phase.

We get a first understanding of the model by looking at the structure of the ground states (g.s.), see figure 1. (We consider only g.s. invariant under the lattice translations $a = (a^1,\dots,a^{d-1},0)$.) We have three regions with a unique g.s. On the vertical line $\lambda = 0$, there is coexistence of the g.s. $\sigma(i) \equiv 1$ and $\sigma(i) \equiv -1$ for $0 \leqslant h < K$. At $h = K$ there are infinitely many g.s. : $\sigma(i) \equiv 1$, $\sigma(i) \equiv -1$, and $\sigma(i) = 1$ for the first K layers ($i^d \leqslant k$), $\sigma(i) = -1$ otherwise. For $h > K$ there are still infinitely many g.s, but the g.s. $\sigma(i) \equiv -1$ has disappeared. On the line $h = K - \lambda$, $\lambda < 0$, there is coexistence of the g.s. $\sigma(i) \equiv -1$ and $\sigma(i) = 1$ for $i \in \Sigma$, $\sigma(i) = -1$ otherwise.

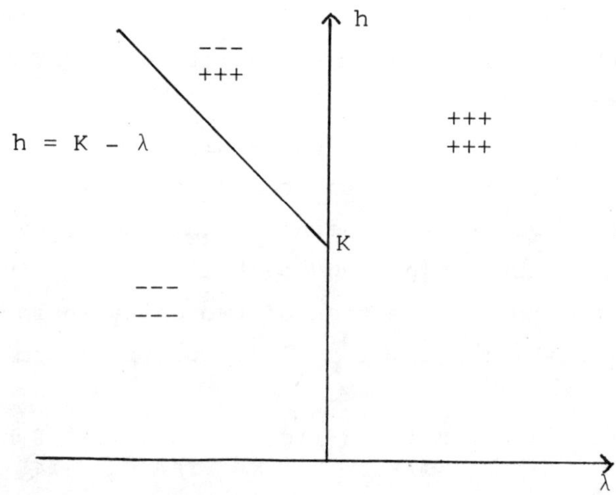

figure 1.
Ground states of the model

Let $\lambda = 0$ and $K > K_c(d)$, the critical coupling constant of the Ising model (1) on $\mathbb{Z}^d$. We consider the two Gibbs states of our model defined by (2), and by the + boundary condition (+ b.c.), respectively the - b.c. They are denoted $\langle \cdot \rangle^+$ and $\langle \cdot \rangle^-$. In the state $\langle \cdot \rangle^+$ the bulk phase and the adsorbed phase are the same, but in the state $\langle \cdot \rangle^-$ they are different. Let $\tau_w^{\pm}(J,K,h)$ be the free energy due to the presence of the interface in the state $\langle \cdot \rangle^-$,

$$\tau_w^{\pm}(J,K,h) = F^-(J,K,h) - F^+(J,K,h) \tag{3}$$

In (3), F^- resp. F^+ are the (surface) free energies of the model defined with - b.c. resp. + b.c.. If there are several bulk phases, the free energy of the model depends on the type of the bulk phase. In our case, if we choose the + b.c. there is no interface, and if choose the − b.c. there is an interface. Therefore we have (3). Let $\tau^{\pm}(K)$ be the free energy of the interface between the + and − phases in the Ising model (1) on $\mathbb{Z}^d$ (with $\lambda = 0$). ($\tau^{\pm}(K)$ is often called surface tension.) Our main result is a proof of the inequality

$$0 \leqslant \tau^{\pm}(J,K,h) \leqslant \tau^{\pm}(K), \quad h \geqslant 0 \tag{4}$$

The basic properties of $\tau_w^{\pm}(J,K,h)$ are

a) It is a monotone increasing function of J, K and h. $\tau_w^{\pm}(J,K,o) = 0$. It is a concave function of h, $h \geqslant 0$

b) $\tau_w^{\pm}(J,K,h) = \tau^{\pm}(K)$ if $J \geqslant K$, $h \geqslant K$

c) $\tau_w^{\pm}(h) = \int_0^h dh' (\langle\sigma(i)\rangle^+(h') - \langle\sigma(i)\rangle^-(h'))$

where i is any point of the first layer Σ.

Let us define $h_w(J,K)$ by

$$h_w(J,K) = \inf \{h : \tau_w^{\pm}(J,K,h) = \tau^{\pm}(K)\} .$$

Since $\tau_w^{\pm}(h)$ is increasing and concave in h, $\langle\sigma(i)\rangle^+(h') - \langle\sigma(i)\rangle^-(h')$ is positive and decreasing in h' for all $i \in \Sigma$. Therefore, if $h > h_w(J,K)$, $\langle\sigma(i)\rangle^+(h) = \langle\sigma(i)\rangle^-(h)$ for all $i \in \Sigma$. We can prove that this implies the uniqueness of the Gibbs state. On the other hand, if $h < h_w(J,K)$, $\langle\sigma(i)\rangle^-(h) \neq \langle\sigma(i)\rangle^+(h)$, and we have at least two Gibbs states. In the former case the thickness of the adsorbed + phase in the state $\langle \cdot \rangle^-$ is macroscopic and we have complete wetting. In the latter case this thickness is microscopic and we have partial wetting. We have obtained a rigorous derivation of the Cahn criterion for the wetting transition [7]. When $h = h_w$, there is a unique Gibbs state if the transition is of 2nd order, and several Gibbs states if it is of 1st order.

It is easy to give a lower bound for h_w. From C we have

$$\tau^{\pm}(K) = \int_0^{h_w} (<\sigma(i)>^+(h') - <\sigma(i)>^-(h'))dh' \leqslant 2h_w$$

For $J \geqslant K$ we also have an upper bound, $h_w(J,K) \leqslant K$ (see b). In figure 2 we have drawn a typical phase diagram in the (h,T) - plane (T, the temperature) when $\lambda = 0$, $d \geqslant 3$ and J/K large enough, so that there is at $h = \lambda = 0$ a surface phase transition at T_S before the bulk transition at $T_c(d)$. If $h = \lambda = 0$ and $T_c(d) < T < T_S$ the layer Σ is ordered, $<\sigma(i)>^+ > 0$, $i \in \Sigma$, but the bulk is disordered. For $T < T_c(d)$, the layer Σ and the bulk are ordered.

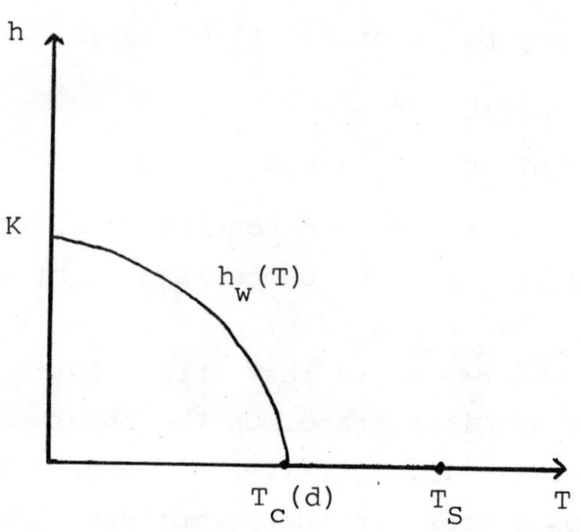

figure 2.

Phase diagram for $\lambda = 0$ in the (h,T)-plane. There are several Gibbs states below $h_w(T)$ and on the segment $[T_c(d), T_S)$.

Let us consider the particular case $J = K = h$. (see the definition of the model). For <u>any</u> $d \geq 2$ there is always a unique Gibbs state at finite temperature. This has been conjectured and proved in the SOS limit in [8]. On the other hand if $J < K$ and $h = K$, we can prove that there is for any $d \geq 2$ a temperature $T^*(J)$, such that below $T^*(J)$ there are several Gibbs states.

Let us now consider $\lambda \neq 0$. In the bulk there is a unique phase and we expect that the free energy of the model is independent of the b.c. (We can prove only partial results). We <u>suppose</u> that $F^+(J,K,h,\lambda) = F^-(J,K,h,\lambda)$. Under this hypothesis there is a unique Gibbs state for $h \geq 0$ and $\lambda > 0$. The situation is different for $\lambda < 0$ and $h \geq 0$ (see figure 1). Let K and J be fixed, $K < 2J$, <u>and let $d \geq 3$</u>. If the temperature is small enough, there is a line $h_{p.w.}(\lambda)$, $\lambda < 0$, of first-order phase transitions. For large values of $|\lambda|$, $h_{p.w.}(\lambda)$ is asymptotically given by $h_{p.w.} = K-\lambda$. When λ tends to zero, $h_{p.w.}$ tends to $h^* \leq K$. (see figure 3). More precisely, if λ and T are fixed, there is a constant $c > 0$, such that

$$<\sigma(i)>^-(h) \leq -C , \quad h \leq h_{p.w.} , \quad i \in \Sigma$$

and

$$<\sigma(i)>^-(h) \geq C , \quad h > h_{p.w.} , \quad i \in \Sigma$$

Moreover, if λ tends to zero, $C(\lambda)$ tends to $C^* > 0$. This means that h^* is a point of first-order phase transition since $\lim_{\lambda \uparrow 0} < \cdot >^-(\lambda) = < \cdot >^-(\lambda = 0)$. We cannot unfortunately locate precisely the point h^*. For $J \geq K$ it is possible that $h^* = h_w$, and this would mean that the wetting transition is of 1st order. On the other hand, if $K/2 < J < K$ we know that $h_w(J,K) > K$ at low temperature therefore we have in this case a prewetting transition at

$h^* \leqslant K$, which is of 1st order, and then the wetting transition, which could be of second order.

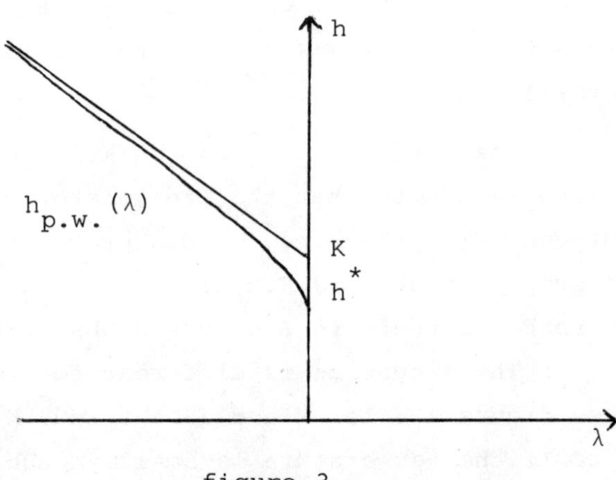

figure 3.

The prewetting transition line. T is fixed.

Proofs, as well as several other results on the phase diagram of the model may be found in [9]. We give below a proof of (4).

Proof of (4)

Let $\Lambda(L,M) = \{i \in \mathbb{L} \; ; \; |i^k| \leqslant L/2, \; 1 \leqslant k \leqslant d-1, \; 1 \leqslant i^d \leqslant M\}$. In the box $\Lambda(L,M)$ we consider the model with hamiltonian (2) and + b.c.. The corresponding partition function is $Z^+(L,M)$. Let $\Omega(L,M) = \{i \in \mathbb{Z}^d \; ; \; |i^k| \leqslant L/2, \; 1 \leqslant k \leqslant d-1, \; -M < i^d \leqslant M\}$. In the box $\Omega(L,M)$ we consider the model with hamiltonian (1) and +b.c.. The corresponding partition function is $Q^+(L,M)$. We define

$$F^+(J,K,h,\lambda) = \lim_{L\to\infty} \frac{-1}{2|\Sigma(L)|} \ell n \frac{(Z^+(L,M))^2}{Q^+(L,M)} \tag{5}$$

where we choose in (5) $M = L^\alpha$, and $|\Sigma(L)|$ is the cardinality of the set $\Sigma \cap \Lambda(L,M)$. We define similarly F^- and $\tau_w^\pm$ by (3). When $\lambda = 0$, we can write

$$\tau_w^\pm(J,K,h) = \lim_{L\to\infty} \frac{-1}{|\Sigma(L)|} \ell n \frac{Z^-(L,M)}{Z^+(L,M)} \tag{6}$$

since in this case $Q^+(L,M) = Q^-(L,M)$ by symmetry. For fixed L we modify the -b.c. by taking $\sigma(i) = 1$ for all $i \notin \Lambda(L,M)$ with $i^d \leqslant M/2$. We get a new partition function $Z^\pm(L,M)$. For fixed L, the difference between (6) and

$$\frac{-1}{|\Sigma(L)|} \ell n \frac{Z^\pm(L,M)}{Z^+(L,M)} \tag{7}$$

is of order $|\Sigma(L)|^{-1} O(M\cdot L^{d-2}) = O(L^{\alpha-1})$, since $M = L^\alpha$. We choose $\alpha < 1$ so that in the limit $L \to \infty$ we get the same result. Taking the derivative of (7) with respect to h, we see that (7) is increasing in h by correlation inequalities. We let $h \uparrow \infty$ in (7). In this limit we get

$$\frac{-1}{|\Sigma(L)|} \ell n \frac{Z^\pm(L,M-1)}{Z^+(L,M-1)} \tag{8}$$

where in (8) the hamiltonian is such that $J = K = h$. In other words (8) is precisely the finite volume expression of the surface tension of the Ising model. Thus in the limit $L \to \infty$, we get (4). It remains to see that we can choose $M = L^\alpha$ with $\alpha < 1$. This follows, e.g. for (8), from the results of [10]. Property b also follows from [10]; a and c are direct consequences of duplicate variables inequalities [11]. (See e.g. [10])

References

1. Binder, K. : Phase transitions and critical phenomena vol 8 Ed. C. Domb and J.L. Lebowitz. Academic Press London 1983, 1-144.

2. Fisher, M.E. : J. Stat. Phys. 34, 667-729 (1984).

3. de Gennes, P.G. : Rev. Mod. Phys. 57, 827-863 (1985).

4. Abraham, D.B. : in Phase transitions and critical phenomena vol 10 Ed. C. Domb and J.L. Lebowitz. To appear in 1986.

5. Abraham, D.B. : Phys. Rev. Lett. 44, 1165-1168 (1980).

6. McCoy, B.M., Wu, T.T. : The two-dimensional Ising model. Harvard University Press, Cambridge, Massachusetts (1973).

7. Cahn, J.W. : J. Chem. Phsy. 66, 3667-3672 (1977).

8. Bricmont, J., El Mellouki, A., Fröhlich, J. : J. Stat. Phys. 42, 743-798 (1986).

9. Fröhlich, J., Pfister, C.E. : in preparation.

10. Bricmont, J., Lebowitz, J.L., Pfister, C.E. : Annals of the New York Academy of Sciences 337, 214-223 (1980).

11. Lebowitz, J.L. : Commun. Math. Phys. 35, 87-92 (1974).

STABLE AND UNSTABLE PHASES IN THE PIROGOV SINAI THEORY

Miloš Zahradník

Faculty of Mathematics and Physics, Charles University

Prague, Sokolovská 83, 186 00 Czechoslovakia

This is a short exposition of some ideas of the P.S. theory. We restrict ourselves to the case of discrete spin models and follow the approach developed in [1]. (This paper should be consulted for more details). The case of continuous spin models is dealt with in [2] for the special case of "all phases being stable". One has therefore to combine [1] with [2] to obtain a more general theory. For other (sometimes parallel) developments of the theory see [3] and the references given there (notably the Imbrie's work was the first to handle the continuous spin case ; Basuev obtained the analyticity of the phase diagram).
A standard reference of the P.S. theory is [4] .

I Introduction

We consider a configuration space

$$X = S^{\mathbb{Z}^{\nu}}$$

with S finite, $\nu \geq 2$. The hamiltonian is given as

$$H(x) = \sum \Phi_A(x_A) \qquad (1)$$

with translation invariant interactions Φ_A, diam $A \leq N$. Usually, the low temperature situation is considered. We incorporate the inverse temperature into the hamiltonian.

Given a finite family $\{x^q,\ q\in Q\}$ of constant configurations ("ground states of the unperturbed hamiltonian") say that a point $t\in \mathbb{Z}^\nu$ is a q-correct point of x if $x_s = x_s^q$ whenever $|s-t|\le r$. The set of all noncorrect points is splitted into connected components which are called (if finite) supports of contours. A contour Γ is then a restriction of x to supp Γ.

II Reformulation of (1)

The hamiltonian (1) can be written, for any volume Λ and any configuration x such that supp $\Gamma_i \subset \Lambda$ for any contour Γ_i of x, in the following way:

$$H(x_\Lambda) = \sum_i \Phi(\Gamma_i) + \sum_q e_q |\Lambda_q| \qquad (1')$$

where $\Phi(\Gamma)$ are suitable "contour hamiltonians", Λ_q denotes the set of all q-correct points of x_Λ and e_q is the density of energy of x^q, $e_q = \sum_{A\ni t} |A|^{-1} \Phi_A(x_A^q)$.

Note. We will write H in the form (1′) everywhere. In fact, it is reasonable to formulate the P.S. machinery starting from the very formulation (1′) of the problem, with "configuration" being defined as a "compatible collection" of "contours". Such an abstract approach is useful e.g. in the study of interfaces. The behaviour of an interface can be described by (1′), not by (1) (Holický-Kotecký-Zahradník [3]).

III How the P.S. theory works with models (1′)

A configuration x_Λ on Λ is called q-diluted if dist(supp $\Gamma \cup$ int Γ, Λ^c) ≥ 2 for any contour Γ of x_Λ and if x_Λ equals to q on the boundary of Λ.

Hence the notion of a diluted partition function

$$Z^{q}(\Lambda) = \sum_{\text{all } q\text{-diluted } x_\Lambda} \exp\left(-H(x_\Lambda)\right)$$

Another "traditional notion" is that of "crystallic partition function" in the volume $V(\Gamma) = \operatorname{supp}\Gamma \cup \operatorname{int}\Gamma$:

$$Z(\Gamma) = \sum \exp\left(-H(x_\Gamma)\right)$$

where the sum is over all x_Γ on $V(\Gamma)$ such that Γ is a contour of x_Γ .

We will compare $Z(\Gamma)$ with the "reference partition function"

$$Z_{\text{ref}}(\Gamma) = \sum \exp\left(-H(x_\Gamma)\right)$$

the sum being taken over all x_Γ on $V(\Gamma)$ which are equal to q on supp Γ and q-diluted on int Γ (assuming that $\Gamma = \Gamma^{q}$ i.e. q is the "exterior colour" of Γ).

The basic notion of a <u>contour functional</u> (the "work needed to install the contour") can be defined from the relation

$$Z(\Gamma) = \exp(-F(\Gamma))\, Z_{\text{ref}}(\Gamma) \quad . \tag{2}$$

The <u>contour model</u> (more precisely, q-contour model) is then defined as a <u>polymer model</u> with partition functions

$$Z^{q}_{\Lambda} = \sum \prod_i k_{\Gamma_i^{q}}$$

the sum being taken over all famillies of q-contours $\{\Gamma_i^{q}\}$ such that $\operatorname{dist}(\operatorname{supp}\Gamma_i, \Lambda^c) \geq 2$ and $\operatorname{dist}(\operatorname{supp}\Gamma_i, \operatorname{supp}\Gamma_{i'}) \geq 2$ .The "activity" k_Γ is

$$k_\Gamma = \exp(-F(\Gamma)) .$$

It can be shown that (2) implies the <u>equivalence</u> of

both the "real" and the contour ensemble: namely

$$Z^{q}(\Lambda) = \exp(-\ell_{q} |\Lambda|)\, Z^{q}_{\Lambda}$$

for any Λ with simple connected components. This is one of the main ideas: to describe the behaviour of the <u>system of external contours</u> of the model in terms of the contour models. One motivation is that quite a lot is known about the behaviour of polymer models (see e.g. [7] for recent references) :

IV Polymer models

<u>Lemma 1</u>. (We take connected sets T instead of contours Γ here). If for some small $\varepsilon > 0$ and for all T,

$$|k_{T}| \leq \varepsilon^{|T|} \tag{3}$$

then

$$|\log Z_{\Lambda} - s|\Lambda|| \leq C\,|\partial\Lambda^{c}|$$

with $s = \lim |\Lambda|^{-1} \log Z_{\Lambda}$, C small $(C = o(\varepsilon),\ \varepsilon \to 0)$.

V Stable and unstable phases

The condition (3) imposes the limitations on the use of contour models. Namely (3) requires that

$$F(\Gamma) \geqslant \tau\, |\mathrm{supp}\, \Gamma| \quad , \tau \text{ large} \quad . \tag{3'}$$

For Φ instead of F, this is the so called <u>Peierls condition</u> (also named Gertzik-Pirogov-Sinai condition), which is valid in most (not all !) low temperature situations having a finite number of "ground states". This condition is assumed to hold in the following.

However, even in this case the condition (3′) may be violated for some $F(\Gamma^{q})$, indicating the "nonstability" of the corresponding q-th phase. This suggests that the notion of contour model must be supplemented by further devices. Pirogov and Sinai used the contour models with a parameter. We follow the alternative approach [1] which gives, as a byproduct, also the completeness of the phase diagram -see [1]- and the analyticity of its strata (Theorem 2).

Definition. A contour Γ^q is called stable if

$$F(\Gamma^q) > \frac{\tau}{3} |\text{supp}\, \Gamma^q| \quad .$$

A stable contour Γ^q satisfying the condition that even any $\tilde{\Gamma}^q$, $\text{supp}\,\tilde{\Gamma}^q \subset \text{int}\,\Gamma^q$ is stable is called small contour. (There is some "freedom" is this and also the forthcoming definition, not affecting the physically meaningful notion of a stable phase below. A suitable modification of these notions is used in the proof of Theorem 2). Denote by $-s_q$ the free energy of the "metastable" contour model defined as a polymer model with $k_{\Gamma^q} = \exp(-F(\Gamma^q))$, Γ^q small. Write $h_q = e_q - s_q$, $h = \min\{h_q\}$, $a_q = h_q - h$ (a_q can be compared to P.S. parameters; they are not exactly equal to them). We say that q is stable if $a_q = 0$.

Theorem 1. (See [1])

i) Nonstable contours satisfy the inequality

$$a_q \,|\text{int}\,\Gamma^q| \geqslant \frac{\tau}{3} |\text{supp}\,\Gamma^q| \quad .$$

In particular, there are no unstable Γ^q for q stable.

ii) $\quad Z^q(\Lambda) \geqslant \exp(-h_q|\Lambda| - C|\partial\Lambda^c|)$

iii) $\quad Z^q(\Lambda) \leqslant \exp(-h|\Lambda| + C|\partial\Lambda^c|)$

where $C = C(\tau)$ is some constant, $\lim_{\tau\to\infty} C = 0$

The <u>interpretation</u> of this result is as follows : "stability of q" means that there is an infinite volume translation invariant Gibbs state which is a small perturbation of the q-th "ground state" ([4]). Nonstability means that there is no such Gibbs state. There are no other translation invariant Gibbs states ([1]).

VI The Phase Diagram

Let $H = H(\vec{\omega})$ depend on some vector parameter $\vec{\omega}$ which we will write in the form ($n \geq 2$, $m \geq 1$)

$$\vec{\omega} = (\vec{\lambda}, \vec{\mu}) = (\lambda^1, \ldots, \lambda^{n-1}, \mu^1, \ldots, \mu^m).$$

Denoting $Q(\vec{\omega}) = \{ q \in Q$: stable for $H(\vec{\omega}) \}$ we will be interested in the behaviour of the set $\{ \vec{\omega} : Q(\vec{\omega}) = \bar{Q} \}$ where $\bar{Q} = \{ q_1, \ldots, q_n \} \subset Q$ is fixed.

<u>Theorem 2</u>. ([5]). Let $\vec{\omega}_0$ be such that $\bar{Q} = Q(\vec{\omega}_0)$ resp. $\bar{q} \in \bar{Q} \Rightarrow e_{\bar{q}}^{\vec{\omega}_0} = \min \{ e_q^{\vec{\omega}_0} \}$. Let

$$\left(\frac{\partial (e_{q_i} - e_{q_1})}{\partial \lambda^j} \right) \quad \begin{array}{l} i = 2, \ldots, n \\ j = 1, \ldots, n-1 \end{array} \tag{4}$$

be invertible (degeneracy removing condition). Assume the real analyticity and translation invariancy of all the mappings $\{ \vec{\omega} \leadsto \Phi_A^{\vec{\omega}}(x_A) \}$ in some neighborhood $\mathcal{V}$ of $\vec{\omega}_0$. Assume that $H(\vec{\lambda}_0)$ satisfies the Peierls condition $\Phi^{\vec{\omega}_0}(\Gamma) \geq \tau\, |\mathrm{supp}\, \Gamma|$ for each Γ, with a sufficiently large τ. Then there is some $\tilde{\mathcal{V}} \ni \vec{\omega}_0$ resp. some $\emptyset \neq \tilde{\mathcal{V}} \subset \subset \mathcal{V}$ and a real analytic function $\{ \vec{\mu} \leadsto \vec{\lambda} \}$ such that $(\lambda^1(\vec{\mu}), \ldots, \lambda^{n-1}(\vec{\mu}), \mu^1, \ldots, \mu^m) = \{ \vec{\omega} \in \tilde{\mathcal{V}} : Q(\vec{\omega}) = \bar{Q} \}$.

(See [4] for a "global" description of the phase diagram)

References

1 Zahradník M., Comm.Math.Phys.93 (1984)

2 Dobrushin R.L., Zahradník M., Phase diagrams for continuous spin models, in Math.Problems of Stat.Phys. and Dyn. ed.by R.L.Dobrushin, D.Reidel 1986

3 Imbrie J., Comm.Math.Phys.82 (1981), 83 (1982)
Bricmont J., Slawny J., contributions to the Proc.of Conf. Stat.Phys.and Field Theory, Groningen, Aug.1985, ed.by N. Hugenholtz and M.Winnink, Lecture Notes in Phys. 1986
Zahradník M., the same volume
Dinaburg E.I., Sinai Ya.G., Comm.Math.Phys.98 (1984)
Basuev A.G.Theor.Math.Phys.58 (1984), 64 (1985)
Bricmont J., El Mellouki A., Fröhlich J., J.St.Phys.42, 5/6(86)
Slawny J., in Phase Trans. and Crit.Phen., C.Domb and J.Lebowitz, vol.10, Academic Press 1985
Bricmont J., Kuroda K., Lebowitz J.L., J.St.Phys.33 (1983)
Holický P., Kotecký R., Zahradník M., Preprint 1986, see also Proc.of Třeboň Conf.Stat.Phys.of Phase Transit.Sept. 1986 to appear in J.Stat.Phys. 1987
Isakov S., this volume

4 Sinai Ya.G. Theory of Phase Transitions : Rigorous Results Pergamon Press, Oxford 1982

5 Zahradník M., to appear in Proc.of Třeboň Conf.(see [3])

6 Gawedzki K., Kotecký R., Kupiainen A., Coarse grain approach to 1.order phase transitions, Preprint 1986

7 Kotecký R., Preiss D., Comm.Math.Phys.103, 491 (1986)

RECENT PROGRESS ON ENTROPY AND RELATIVE ENTROPY

Huzihiro ARAKI
Research Institute for Mathematical Sciences
Kyoto University, Kyoto 606
JAPAN

ABSTRACT

Recent progress on entropy and relative entropy is reviewed, including integral representations by Pusz and Woronowicz and by Kōsaki, various properties of relative entropy, variational definition of entropy by Narnhofer and Thirring, entropy of an automorphism (non-commutative version of the Kolmogorov-Sinai invariant) of Connes, variational formulation of relative Hamiltonian by Donald, and the weak sufficiency condition of Petz.

1. INTRODUCTION

This is a review of recent developments in (mathematical) theory of entropy, following a suggestion of the session organizer.

Entropy was introduced in (phenomenological) thermodynamics by Clausius in 1865 in connection with the second law of thermodynamics. The quantum entropy

$$S(\rho) = - \mathrm{Tr}\rho \log \rho \tag{1.1}$$

for a state with a density matrix ρ was introduced by von Neumann in 1927. In statistical mechanics, the mean entropy

$$s(\phi) = \lim V^{-1} S_V(\phi) \tag{1.2}$$

plays an important role where $S_V(\phi)$ is the entropy, which is given for a quantum system by (1.1), ρ being the density matrix of the restriction of the state ϕ to (the observable algebra for) a finite box V, and the limit of V becoming infinite (in a reasonably uniform manner) is taken. For detailed review and relevant references of these earlier developments up to proof of strong subadditivity of entropy by Lieb and Ruskai, we refer to the article of Wehrl[35].

For a dynamical system consisting of a measure space Ω with a probability measure μ and a measure preserving transformation T, an entropy was introduced as an isomorphism invariant by Kolmogorov and Sinai in 1958 and 1959. For classical statistical mechanics (in one dimension), a state is a probability measure on a certain measure space and the Kolmogorov-Sinai entropy for a spatial translation T by a unit length coincides with the average entropy (1.2). Recently, an entropy $H(\theta,\phi)$ for a W*-dynamical system consisting of a state ϕ of a von Neumann algebra M and an automorphism θ of M leaving ϕ invariant has been introduced by Connes[12]. It coincides with the Kolmogorov-Sinai entropy when M is abelian. For a state ϕ of one-dimensional quantum spin lattice system, it is shown by Connes to be bounded above by the average entropy (1.2) and the two coincide for equilibrium states of quasifree dynamics of finite range according to Connes, Narnhofer and Thirring[13]. The Connes entropy provides a possibility of applying quantum statistical concepts and methods for a general W*-dynamical system. This is the most interesting topics in recent development in theory of quantum entropy.

The quantum relative entropy (also called conditional entropy)

$$S(\phi,\psi) = \mathrm{tr}\, \rho_\psi(\log \rho_\psi - \log \rho_\phi) \tag{1.3}$$

of two (normal) states ϕ and ψ of a type I von Neumann algebra with density matrices ρ_ϕ and ρ_ψ was introduced by Umegaki[34] and extended by Araki[5) 6] to normal states of a general von Neumann algebra M in terms of the relative modular operator. (It is defined $+\infty$ unless $\mathrm{supp}\,\phi \geq \mathrm{supp}\,\psi$.)

For states ϕ and ψ of a C*-algebra $\mathfrak{A}$ (and hence for non-normal states of a von Neumann algebra), we may consider their normal extensions $\tilde{\phi}$ and $\tilde{\psi}$ to the double dual $\mathfrak{A}^{**}$ (the envelopping von Neumann algebra) and define $S(\phi,\psi) = S(\tilde{\phi},\tilde{\psi})$.

Lindblad[26) 25)] is probably the first to realize an important connection of properties of relative entropy and those of entropy. The properties of relative entropy plays an important role for the Connes entropy of an automorphism. We start our review with discussions of properties of relative entropy, which follow easily from variational expressions by Pusz and Woronowicz[32)] and by Kosaki[22) 23)]. We will then describe the entropy $H(\theta,\phi)$ of Connes. Applications of relative entropy including an interesting suggestion of Donald[16)] about variational definition of relative hamiltonian, and application to sufficiency problem by Petz[31)] will be briefly reviewed at the end.

Relative (or conditional) entropies which are different from ours (even for non-commutative finite dimensional algebras) have been discussed by Naudts[28)], Benoist-Gudder-Marchand-Wyss[10) 18)], and Belavkin-Staszewski[8) 9)]. These are not included due to restrictions of allotted time (and space).

2. RELATIVE ENTROPY

The following variational expressions, the first by Pusz and Woronowicz[32)] and the other two by Kosaki[22) 23)], for the relative entropy $S(\phi,\psi)$ of two positive linear functionals ϕ and ψ on a von Neumann algebra M is a convenient starting point of our discussion.

$$\begin{aligned} S(\phi,\psi) &= -\inf\int_0^1 [\psi(\tfrac{1}{t}\, y(t)^*y(t)-x(t)^*x(t)) + \phi(x(t)x(t)^*)]dt \\ &= \sup_{n\in\mathbb{N}^+} \sup\{\psi(1)\log n - \int_{1/n}^{\infty} [\psi(y(t)^*y(t)) + \tfrac{1}{t}\,\phi(x(t)x(t)^*)]\tfrac{dt}{t}\} \\ &= \sup\int_0^{\infty}[\tfrac{\psi(1)}{1+t} - \psi(y(t)^*y(t)) - \tfrac{1}{t}\,\phi(x(t)x(t)^*)]\tfrac{dt}{t}\,. \qquad (2.1)\end{aligned}$$

Here $x(t) \in M$ is varied over all step functions of t with a finite number of steps and $y(t) = 1 - x(t)$.

The following properties can be read off from (2.1).

(1) Lower Semicontinuity of $S(\phi,\psi)$ in (ϕ,ψ) relative to $\sigma(M_*,M)$-

topology. In particular $S(\phi_\nu,\psi)\uparrow S(\phi,\psi)$ if $\phi_\nu\downarrow\phi$ (see (3-1)).

(2) Joint Convexity: For $\lambda_i \geq 0$, $\sum_{i=1}^{n} \lambda_i = 1$,

$$S\left(\sum_{i=1}^{n} \lambda_i\phi_i,\ \sum_{i=1}^{n} \lambda_i\psi_i\right) \leq \sum_{i=1}^{n} \lambda_i\ S(\phi_i,\psi_i)$$

(3) Monotone Properties:

(3-1) If $\phi_1 \leq \phi_2$, then $S(\phi_1,\psi) \geq S(\phi_2,\psi)$. In particular, $S(\phi_1,\phi_2) \geq 0$ and $S(\phi_2,\phi_1) \leq 0$ due to $S(\phi,\phi) = 0$.

(3-2) For a von Neumann subalgebra M_1 of M, $S(\phi|M_1,\ \psi|M_1) \leq S(\phi,\psi)$.

(3-3) If a map $\gamma: M \to N$ between von Neumann algebras M and N is normal, linear, unit preserving $(\gamma(1) = 1)$, and strongly positive $(\gamma(x^*x) \geq \gamma(x)^*\gamma(x)$ for all $x \in M)$, then

$$S(\phi\circ\gamma,\ \psi\circ\gamma) \leq S(\phi,\psi) \quad (\phi,\psi \in N_*^+).$$

(3-2) is a special case of (3-3). If $N = \mathbb{C}1$, (3-2) gives the so-called Peierls-Bogoliubov inequality. If ϕ and ψ are states in addition, it gives the following property because $\phi|N = \psi|N$ and $S(\omega,\omega) = 0$.

(4) Positivity: $S(\phi,\psi) \geq 0$ for states ϕ and ψ.

(5) Martingale Convergence: If $M_\nu\uparrow M$, then $S(\phi|M_\nu,\ \psi|M_\nu)\uparrow S(\phi,\psi)$, where M_ν is an increasing net of von Neumann subalgebras of M with their union dense in M.

(6) Scaling Property: $S(\lambda_1\phi,\ \lambda_2\psi) = \lambda_2 S(\phi,\psi) + \lambda_2\psi(1)\ \log(\lambda_2/\lambda_1)$. In particular it is homogeneous $(S(\lambda\phi,\lambda\psi) = \lambda S(\phi,\psi))$. (Scaling of ϕ is equivalent to scaling of t and hence that of n in the second variational expression.)

The following result of Hiai, Ohya and Tsukada[19)] is not apparent from variational expressions.

(7) $S(\phi,\psi) \geq \|\phi-\psi\|^2/2$ for states ϕ and ψ. This implies that $S(\phi,\psi) = 0$ for states ϕ and ψ if and only if $\phi = \psi$.

The following results on subadditivity of various kinds follow from the above properties and some identities (which are immediate from (1.3) for type I case and which follows from the definition in terms of relative modular operators in general). Details of proof will be discussed elsewhere.

(8) <u>Subadditivity</u>: All sums below are finite sums.

(8-1) $S(\Sigma\phi_i, \Sigma\psi_i) \leqq \Sigma S(\phi_i, \psi_i)$. This is equivalent to (2) via (6).

(8-2) We identify M and $M_\alpha = M \otimes \prod_{i\in\alpha}^{\otimes} N_i$ for $\alpha = 123, 12, 13, 1$, all as von Neumann subalgebras of M_{123}, where $N_i \subsetneq B(\mathbb{C}^{n_i})$, $n_i < \infty$. Let $\phi_{123} \in (M_{123})_*^+$ and $\phi_\alpha = \phi_{123}|M_\alpha$ $(\phi = \phi_{123}|M)$. Let ω_i be any faithful positive linear functional of N_i and $\omega_\alpha = \prod_{i\in\alpha}^{\otimes} \omega_i$. Then

(i) $S(\phi\otimes\omega_{123}, \phi_{123}) + S(\phi\otimes\omega_1, \phi_1) \geqq S(\phi\otimes\omega_{12}, \phi_{12}) + S(\phi\otimes\omega_{13}, \phi_{13})$.

(ii) $S(\phi_1\otimes\omega_{23}, \phi_{123}) \geqq S(\phi_1\otimes\omega_2, \phi_{12}) + S(\phi_1\otimes\omega_3, \phi_{13})$.

The (ii) follows from (i) by specialization $n_1 = 1$ and substitution of M_1 into M, with a new n_1. When M is of finite dimension,

$$-S(\phi\otimes Tr_1, \phi_1) = S(\phi_1) - S(\phi)$$

where Tr_1 is the usual trace on N_1 and $S(\phi_1)$ and $S(\phi)$ are entropies (1.1). It is the difference of the entropy of ϕ for $M\otimes N_1$ and for M and is interpreted as entropy of the state ϕ of the open system N_1 in an environment M. For finite systems, their properties have been investigated by Lieb[23]. The above two inequalities, as well as the following are generalizations and all properties found true by Lieb[23] follow from these.

(iii) $S(\phi_{13}\otimes\omega_2, \phi_{123}) \geqq \max\{S(\phi_1\otimes\omega_2, \phi_{12}), S(\phi_3\otimes\omega_2, \phi_{32})\}$.

(iv) $S(\phi_1\otimes\omega_2, \phi_{12}) \geqq S(\phi\otimes\omega_2, \phi_2)$.

These follow from (3-2).

(v) $S(\phi \otimes Tr_2, \phi_2) + S(\phi|N_1 \otimes Tr_2, \phi|N_1 \otimes N_2) \leqq 0.$

(vi) $|S(\phi \otimes Tr_1, \phi_1)| \leqq S(\phi_1|N_1) \leqq (\log n_1)$ for states ϕ_1.

(8-3) If $\phi = \sum_{i,j,k} \phi_{ijk}$, $\phi_i^{(1)} = \sum_{j,k} \phi_{ijk}$, $\phi_{ij}^{(12)} = \sum_k \phi_{ijk}$ and $\phi_{ik}^{(13)} = \sum_j \phi_{ijk}$, then

$$\sum_{i,j,k} S(\phi, \phi_{ijk}) + \sum_i S(\phi, \phi_i^{(1)}) \geqq \sum_{i,j} S(\phi, \phi_{ij}^{(12)}) + \sum_{i,k} S(\phi, \phi_{ik}^{(13)}).$$

This follows from (i) by specializing $\omega_{123} = Tr_{123}$, $\phi_{123} = \sum_{i,j,k} \phi_{ijk} \otimes \omega_{e_{1i} \otimes e_{2j} \otimes e_{3k}}$ for orthonormal e's in $\mathbb{C}^{n_\ell}$ $(\ell = 1, 2, 3)$.

(8-4) If $\phi = \sum_{j,k} \phi_{jk}$, $\phi_j^{(1)} = \sum_k \phi_{jk}$ and $\phi_k^{(2)} = \sum_k \phi_{jk}$, then

$$\sum_{j,k} S(\phi, \phi_{jk}) \geqq \sum_j S(\phi, \phi_j^{(1)}) + \sum_k S(\phi, \phi_k^{(2)}).$$

This is a special case of (8-3) where the index i in (8-3) takes only one value. It is proved and used by Connes[12)]. It can be directly derived also by the following identity and (8-1).

$$S(\phi,\psi) + \sum_i S(\psi,\psi_i) = \sum_i S(\phi,\psi_i) \quad \text{if} \quad \psi = \sum_i \psi_i. \qquad (2.2)$$

(8-4) has an immediate generalization to a multi-index $i_1 \cdots i_n$ by recursion.

3. ENTROPY OF AN AUTOMORPHISM

In the definition of Kolmogorov-Sinai entropy, a finite partition of the given measure space plays an essential role. This is generalized to the non-commutative case by considering a finite dimensional subalgebra N of a given von Neumann algebra M. For a probability measure space (Ω,μ), $M = L^{\infty}(\Omega)$, N corresponds to the finite dimensional subalgebra of M generated by the characteristic functions for the subsets defining the partition and we are interested in the state $\phi(f) = \int f d\mu$.

In non-commutative case, there is a difficulty of defining the joint partition for a finite number of finite dimensional subalgebras $N_1,\ldots,N_p$, because they may generate an infinite-dimensional subalgebra of M. Connes[12] directly defines the joint entropy function for them by

$$H_{\phi}(N_1,\ldots,N_p) = \mathrm{Sup}\{ \sum_{(i)} \eta(\phi_{(i)}(1)) + \sum_{k} \sum_{i_k} S(\phi|N_k, \phi^{(k)}_{i_k}|N_k)\}$$

where ϕ is a normal state of M under investigation, the index (i) is an abbreviation for $(i) = (i_1 \ldots i_p)$, $\phi = \sum_{(i)} \phi_{(i)}$ is a multi-index partition of ϕ into $\phi_{(i)} \in M_*^+$,

$$\phi^{(k)}_j = \sum_{(i):i_k=j} \phi_{(i)},$$

the supremum is taken over all such partition, and $\eta(x) = -x \log x$. It is a non-negative real quantity bounded by $\sum_{k=1}^{p} S(\phi|N_k)$.

Incidentally, for a trivial partition $p=1$, $N_1=M$, it gives the entropy defined and shown to coincide with (1.1) for M of type I by Narnhofer and Thirring[26].

$$S(\phi) = H_{\phi}(M) = \mathrm{Sup} \sum_i \lambda_i S(\phi,\phi_i)$$

where $\phi = \sum \lambda_i \phi_i$, ϕ's are states, $\lambda_i > 0$, $\sum \lambda_i = 1$ and the supremum is over all such decomposition of the state ϕ.

The Connes entropy of an automorphism Θ of M for a Θ-invariant state ϕ is now defined by

$$H(\Theta,\phi) = \mathrm{Sup}\ h_{\phi,\Theta}(N),$$

$$h_{\phi,\theta}(N) = \lim_{n\to\infty} H_\phi(N,\theta(N),\ldots, \theta^n(N))/(n+1),$$

where the supremum is taken over all finite dimensional subalgebra N of M and the existence of the limit is guaranteed by the upper bound $h_{\phi,\theta}(N) \leq S(\phi|N)$ and the convexity $a_n + a_m \geq a_{n+m}$ for $a_n = H_\phi(N,\theta(N), \ldots,\theta^{n-1}(N))$.

If there is an increasing sequence of finite dimensional subalgebras N_k of M with dense union, then Connes shows that the above superemum can be replaced by

$$H(\theta,\phi) = \lim_{k\to\infty} h_{\phi,\theta}(N_k).$$

In such a case (i.e. for an approximately finite-dimensional M), $H(\theta^q,\phi) = |q|H(\theta,\phi)$. For the lattice translation automorphism θ of a spin lattice system in one dimension and a translationally invariant state ϕ, $H(\theta,\phi)$ is bounded above by the mean entropy (1.2). The question of whether they coincide is investigated by Connes, Narnhofer and Thirring[13].

4. APPLICATIONS

4.1 Relative Hamiltonian

In quantum statistical mechanics of spin lattice systems, relative entropy has been used in the following instances. (Some of them was reviewed earlier[4]. Also see [11].)

(1) Proof of the equivalence of the KMS condition and the variational principle[2].

(2) Proof of uniqueness of equilibrium states for one-dimension[3]. The method used there is applied to proof of non-existence of broken continuous symmetry for 2-dimension[17].

(3) Entropy of open systems is used for proof of local thermodynamical stability of KMS states[7].

In the above applications, the notion of relative hamiltonian[1] of two states plays a crucial role. For $h = h^* \in M$, and for a faithful normal positive linear functional ϕ of M, a perturbed functional ϕ^h is defined and h is called the relative hamiltonian of ϕ^h relative to ϕ. Then $\phi^{h+x\mathbb{1}} = e^x\phi^h$ for $x \in \mathbb{R}$,

$$S(\phi^h,\phi) = \phi(h), \quad S(\phi,\phi^h) = -\phi^h(h), \tag{4.1}$$

$$S(\phi^h,\psi) = S(\phi,\psi) + \psi(h). \tag{4.2}$$

For a state ψ, the minimum of $S(\phi^h,\psi)$ for a state $\hat{\phi}^h = \hat{\phi}^h(1)^{-1}\phi^h$ is 0 and reached by the unique $\psi = \hat{\phi}^h$. Thus $\hat{\phi}^h$ is characterized by

$$S(\hat{\phi}^h,\phi) + \hat{\phi}^h(h) = \inf_{\psi} (S(\phi,\psi) + \psi(h)) \; (= \log \phi^h(1)). \tag{4.3}$$

Donald[16) 15) 14)] has an extension of ϕ^h to a selfadjoint h affiliated with M and lower bounded. It can be formulated as follows.

We extend $S(\phi,\psi)$ for all states ϕ, ψ of M by going over to normal extension of ϕ, ψ to M^{**}. In the variational expression, $x(t)$ is used to produce dense sets in a cyclic representations and can be restricted to be M-valued instead of M^{**}-valued. Thus the variational expressions can be used for general states ϕ and ψ and the properties (1) $\sim$ (6) follow. (We now have $\sigma(M^*,M)$ topology.) The main point of considering all states on M is the compacteness. Suppose h is such that there exists a state ψ of M satisfying $S(\phi,\psi) + \psi(h)<\infty$ for a given normal state ϕ of M. Because $S(\phi,\psi) + \psi(h)$ is bounded below (h is assumed to be bounded below) on the set M_1^* of all states ψ of M, by compactness and lower semicontinuity there exists a state ψ_{min} giving the finite minimum value

$$x = \inf_{\psi} (S(\phi,\psi) + \psi(h)). \tag{4.4}$$

In order that $S(\phi,\psi) < \infty$, the central support of $\tilde{\psi}$ in M^{**} has to be included in that of $\tilde{\phi}$ and hence ψ_{min} has to be normal. Finally, if there are two solutions ψ_1 and ψ_2, the identity (2.2) and (7) imply $S(\phi,\psi) + \psi(h) < x$ for $\psi = (\psi_1 + \psi_2)/2$ unless $\psi = \psi_1 = \psi_2$. Thus there is a unique ψ_{min} and we define

$$\psi^h \equiv e^x \psi_{min}. \tag{4.5}$$

This seems to be a promising tool.

Actually Donald[14)] defines the relative entropy for non-normal states in a different way. As far as the above discussion is concerned, the variational definition seems to work perfectly well.

4.2 Sufficiency

Entropy and relative entropy have statistical significance. (For example, see pp.426-428 of Bratteli and Robinson[11]. Also see Donald[15].) In commutative case, the sufficiency and the relevance of relative entropy arises in the following way. A statistical system is described by a measurable space Ω or equivalently by a commutative von Neumann algebra $M = L_\infty(\Omega)$ and the outcome of measurement is described by states of M, or equivalently by probability measures on Ω. Then a coarser measurement will be represented by a subalgebra R of M and there is a natural problem of whether N can discriminate a set of possible statistical hypotheses about the state of the system, given by a collection S of states on M. N is defined to be sufficient for S if for every $x \in M$, there is $E(x) \in N$ such that E is linear and $\mu(yx) = \mu(yE(x))$ for all $y \in N$ and all $\mu \in S$ (because the measurement of $x \in M$ can be effectively replaced by that of $E(x) \in N$ as far as S is concerned). If S consists of two states ϕ, ψ,

$$S(\phi,\psi) = S(\phi|N, \psi|N) \tag{4.6}$$

is necessary and sufficient condition for N to be sufficient for $\{\mu,\nu\}$, in commutative case.

In non-commutative case, Petz[30] defines a von Neumann subalgebra N of M to be sufficient for a pair of faithful normal states ϕ and ψ of M if (4.6) holds and derives various equivalent conditions, including the condition $(D\phi:D\psi)_t \in N$ and the condition $E_\phi = E_\psi$ for the conditional expectation of Accardi and Cecchini. In this discussion, quasi-entropies[28),29] are skillfully used. Quasi-entropies have also variational expressions by Kosaki[21] which are similar to (2.1) and can be used as in section 2 to derive properties of quasi-entropies. While quasi-entropies of Petz refers to two states and are generalizations of the relative entropy, Wehrl[35] discusses similar quantity for a (single) state as generalizations of the entropy in connection with the order structure of states.

REFERENCES

1. Araki, H., Publ. Kyoto Univ. 9, 165 (1973).
2. Araki, H., Commun. Math. Phys. 38, 1 (1974).
3. Araki, H., Commun. Math. Phys. 44, 1 (1975).
4. Araki, H., "Relative Entropy and Its Applications", p.61 in "Les Méthodes Mathématiques de la Théorie Quantique des Champs", CNRS, Paris, 1976.
5. Araki, H., Publ. RIMS Kyoto Univ. 11, 809 (1976).
6. Araki, H., Publ. RIMS Kyoto Univ. 13, 173 (1977).
7. Araki, H. and Sewell, G.L., Commun. Math. Phys. 52, 103 (1977).
8. Belavkin, V.P. and Staszewski, P., Rep. Math. Phys. 20, 373 (1984).
9. Belavkin, V.P. and Staszewski, P., Ann. Inst. H. Poincaré 37A, 51 (1982).
10. Benoist, R.W., Marchand, J.-P. and Wyss, W., Lett. Math. Phys. 3, 169 (1979).
11. Bratteli, O. and Robinson, D.W., "Operator Algebras and Quantum Statistical Mechanics" vol. II, New York-Heidelberg-Berlin, Springer, 1981.
12. Connes, A., C.R. Acad. Sc. Paris 301, 1 (1985).
13. Connes, A., Narnhofer, H. and Thirring, W., Private communication.
14. Donald, M.J., Commun. Math. Phys. 105, 13 (1986).
15. Donald, M.J., "Further results on the relative entropy" (preprint).
16. Donald, M.J., "The uniqueness of states of minimum free energy" (preprint).
17. Fröhlich, J. and Pfister, C., Commun. Math. Phys. 81, 277 (1981).
18. Gudder, S., Marchand, J.-P. and Wyss, W., J. Math. Phys. 20, 1963 (1979).
19. Hiai, F., Ohya, M. and Tsukada, M., Pac. J. Math. 96, 99 (1981).
20. Hiai, F., Ohya, M. and Tsukada, M., Pac. J. Math. 107, 117 (1983).
21. Kosaki, H., Commun. Math. Phys. 87, 315 (1982).
22. Kosaki, H., "Variational expressions of relative entropy of states on W*-algebras", p.669-671 in Proceedings of the 27th Midwest Symposium on Circuits and Systems, West Virginia University, 1984.

23. Kosaki, H., "Relative Entropy of States, A Variational Expression", MSRI preprint 15511-85 (1986). To appear in J. Operator Alg.
24. Lieb, E., Bull. Am. Math. Soc. 81, 1 (1975).
25. Lindblad, G., Commun. Math. Phys. 33, 305 (1973).
26. Lindblad, G., Commun. Math. Phys. 39, 111 (1974).
27. Narnhofer, H. and Thirring, W., "From relative entropy to entropy" (preprint).
28. Naudts, J., Commun. Math. Phys. 37, 175 (1974).
29. Petz, D., Lecture Notes in Math. 1136, 428 (1985).
30. Petz, D., Publ. RIMS Kyoto Univ., 21, 787 (1985).
31. Petz, D., Commun. Math. Phys. 105, 123 (1986).
32. Pusz, W. and Woronowicz, S.L., Lett. Math. Phys. 2, 505 (1978).
33. Uhlmann, A., Commun. Math. Phys. 54, 21 (1977).
34. Umegaki, H., Kōdai Math. Sem. Rep. 14, 59 (1962).
35. Wehrl, A., Rev. Mod. Phys. 50, 221 (1978).
36. Wehrl, A., "Entropy in quantum mechanics and the order structure of states", p.271 in "Recent Advances in Statistical Mechanics, Proceedings", Central Institute of Physics, Bucharest, 1979.

NON COMMUTATIVE DIFFEOMORPHISMS

Laurent BAULIEU
Physics Department, Rockfeller University
and Laboratoire de Physique Théorique et Hautes Energies*, Paris**, France

ABSTRACT

The notion of local diffeomorphisms is consistently generalized, in order to accomodate situations where the operators playing the roles of partial derivatives form a non commutative Lie algebra. We construct an associated nilpotent BRS operator, and introduce tensor metrics whose transformation laws involve the non-abelian structure. This determines a new type of gauge symmetry. For example, one may carry out this construction for the Virasoro algebra. The corresponding gauge symmetry could be of interest in string field theory.

1) INTRODUCTION

All known internal gauge symmetries acting on local exterior forms can be enlarged into bigger symmetries which include invariance under local diffeomorphisms. In physics, this permits consistent couplings of all types of gauge theories to gravity [1]. The general idea is to express firstly the local gauge fields which vary under the internal symmetries and the associated ghosts as forms $B_p = B_{\mu_1.....\mu_p}(x^\mu)\, dx^{\mu_1}.....dx^{\mu_p}$ where $B_{[\mu_1......\mu_p]}$ is a local fonctional of the space-time coordinates x^μ. Then, one has the possibility of introducing local vector fields $\xi^\mu(x)$, the commutation properties of which are determined by a grading $g(\xi)$ (called ghost number in this paper), and the corresponding contraction operators i_ξ. i_ξ acts as a differential operator on exterior forms and products of forms, and its grading is defined as $g(\xi)-1$. i_ξ is therefore fully determined from its action on 0-forms $f(x)$ and on 1-forms dx^μ. One defines $i_\xi\, f = 0$ and $i_\xi\, dx^\mu = \xi^\mu$. By introducing the graded commutator of i_ξ with the exterior derivative $d \equiv dx^\mu\, \partial/\partial x^\mu$, one then defines the Lie derivative L_ξ :

$$L_\xi \equiv [\, i_\xi \, , d\,] \equiv i_\xi\, d + (-)^{g(\xi)}\, d\, i_\xi \tag{1}$$

By using the graded commutation relations satisfied by the dx^μ and ∂_μ :

$$[\, dx^\mu , dx^\nu] \equiv dx^\mu dv^\nu + dx^\nu dx^\mu = 0$$

$$[\, \partial_\mu \, , \partial_\nu\,] \equiv \partial_\mu \partial_\nu - \partial_\nu \partial_\mu = 0$$

(2)

one can prove that the L_ξ 's build up the Lie algebra of local diffeomorphisms :

$$[L_\xi , L_{\xi'}] \equiv L_\xi L_{\xi'} - (-)^{g(\xi)g(\xi')} L_{\xi'} L_\xi = L_{\{\xi, \xi'\}} \quad (3)$$

where { , } stands for the graded Poisson bracket :

$$\{\xi, \xi'\}^\mu \equiv \xi^\alpha \partial_\alpha \xi'^\mu - (-)^{g(\xi)g(\xi')} \xi'^\alpha \partial_\alpha \xi^\mu \quad (4)$$

It is in fact sufficient to prove eq. (3) on 0-forms and 1-forms. Moreover, since $[i_\xi , i_{\xi'}] = 0$ one can reduce the proof of this equation to that of :

$$[i_\xi , L_{\xi'}] = [i_{\xi'} , L_\xi] = i_{\{\xi, \xi'\}} \quad (5)$$

Observe that we use the unified notations [,] for commutators and anticommutators and { , } for Poisson brackets and anti-brackets. The grading which is relevant is equal to the sum of the usual Lorentz-degree of forms and of the ghost number. One has for example that $\partial\mu$ has grading 0, d and dx^μ have grading 1 and L_ξ has grading $g(\xi)$ while $i_{\{\xi, \xi'\}}$ has grading $g(\xi) + g(\xi')-1$.

The gauge symmetry corresponding to local diffeomorphism is straightforward to construct in term of a nilpotent differential operator s , called BRS operator.[1] s has grading 1 and its action on any p-form $(p \geqslant 0)$ $B_p \equiv B_{\mu_1, ..., \mu_n}(x)\, dx^{\mu_1} \dots dx^{\mu_p}$ is :

$$s\, B_p \equiv [\mathcal{L}_\xi , B_p] \tag{6}$$

where the vector field $\xi^\mu(x)$, with ghost number 1, is called the vector field ghost of diffeomorphisms. One defines the graded commutation rule :

$$[s, d] = 0 \qquad [s, dx^\mu] = 0 \tag{7}$$

so that $[s, \partial_\mu] = 0$. Then $s\,\xi$ is defined by :

$$[s, i_\xi] \equiv i_{s\xi} \tag{8}$$

By applying eq. (8) on the 1-forms dx^μ one finds easily that :

$$(s\,\xi)^\mu = \xi^\alpha \partial_\alpha \xi^\mu = \frac{1}{2}\{\xi, \xi\}^\mu \tag{9}$$

Besides, using the Jacobi identity for graded commutators one gets :

$$[s, \mathcal{L}_\xi] = \mathcal{L}_{s\xi} \tag{10}$$

Using these equations it is then obvious to prove the nilpotency of s, $s^2 = 0$ on all the fields. On B_p, one has :

$$\begin{aligned} s^2 B_p &= s(\mathcal{L}_\xi B_p) = \mathcal{L}_{s\xi} B - \frac{1}{2}[\mathcal{L}_\xi, \mathcal{L}_\xi] B_p \\ &= \mathcal{L}_{s\xi - \frac{1}{2}\{\xi,\xi\}} B_p = 0 \end{aligned} \tag{11a}$$

and :

$$s^2 \xi = \{s\xi, \xi\} = \frac{1}{2}\{\{\xi,\xi\}, \xi\} = 0 \tag{11b}$$

The last equation is also equivalent to $[[L_\xi, L_\xi], L_\xi] = 0$, obviously true from the Jacobi identity of the graded bracket.

The property $s^2 = 0$ expresses the fact that the infinitesimal gauge transformations acting on B_p that one can define by substituting within sB_p the ghost $\xi^\mu(x)$ by a commuting infinitesimal vector field $\epsilon^\mu(x)$ build up a closed algebra with a Jacobi identity. The latter is of course isomorphic to the Lie algebra generated by local infinitesimal diffeomorphisms.

One can proceed, and introduce the tensor metrics $g_{\mu\nu}(x)$, with

$$s g_{\mu\nu} = L_\xi g_{\mu\nu} = g_{\mu\rho}\partial_\nu \xi^\rho + g_{\nu\rho}\partial_\mu \xi^\rho + \xi \partial_\rho g^{\mu\nu} \quad (12)$$

and verify that $s^2 g_{\mu\nu} = 0$.[1]

The above construction relies on the existence of a Lie derivative L_ξ, eq. (1), which satisfies the involution property, eq. (3). The consistency of this equation with the definition of $s\,\xi^\mu$, eq. (9), (which identifies the structure functions of the local diffeomorphism algebra as $\delta^\rho_\alpha \partial_\rho \delta^\gamma_\beta$), relies on the hypothesis that the ∂x^μ's commute. In what follows we shall relax this hypothesis of commutativity and generalize the ∂_μ's into operators obeying a non abelian structure, and prove that in this case the notion of local diffeomorphisms can still be consistently defined. Thereby we shall introduce a new type of gauge symmetry.

2) NON-ABELIAN GENERALIZATION OF LOCAL DIFFEOMORPHISMS

We introduce a set of fermionic operators β^n and their conjugate β^+_n. The indices n run over a set of integers which is possibly infinite. The β^n's are the generalization of the dx^μ's and they span the space of what we call the $\binom{0}{1}$ -forms while the $\beta^+ n$ span the basis of $\binom{1}{0}$ -forms. One assumes that the β^n's and β^+'s can be multiplied together, so that one can generate the space of $\binom{p}{q}$ forms, and one has by definition :

$$[\beta^n, \beta^+_m] \equiv \beta^n \beta^+_m + \beta^+_m \beta^n = \delta^n_m \tag{13}$$

A $\binom{p}{q}$ -form is defined as:

$$B^p_q = B^{n_1 \dots n_q}_{m_1 \dots m_p} \beta^{m_1} \dots \beta^{m_p} \beta^+_{n_1} \dots \beta^+_{n_q} \tag{14}$$

and $B^{m_1 \dots m_q}_{n_1 \dots\dots n_p}$ is called a double antisymmetric tensor of rank $\binom{p}{q}$.

One then assumes the existence of a set of operators ∂_m, acting on $\binom{p}{q}$ - forms as differential operators. Tensors vary under the action of the ∂_m's but not the β^n's and $\beta^+ m$'s :

$$[\partial_m, B^p_q] \neq 0 \quad [\partial_m, \beta^n] = 0 \quad [\partial_m, \beta^+_n] = 0 \tag{15}$$

Besides, the ∂_m are supposed to build up a (possibly infinite dimensional) Lie algebra with structure coefficients $f^p_{m,n}$:

$$[\partial_m , \partial_n] = \frac{1}{2} f^{\rho}_{mn} \partial_\rho \tag{16}$$

The ∂_n's generalize in a non Abelian way the partial derivative operators $\partial\mu$ of ordinary differential geometry. The β^m 's play the role of the dx^μ 's.

Physicists have been recently led to use the above ingredients in order to build a covariant formalism in string field theory [2]. There, the indices m,n are Virasoro indices. The β^n and $\beta^+{}_n$ are ghost coordinate creation and annihilation operators. The operators $\partial_n = L_n - (-)^n L_{-n}$ are isomorphic to the subgroup of reparametrizations generated by $R^n = L^n - L^{-n}$ which constitute the largest subgroup with no central charge of the Virasoro algebra, with $[\partial_m, \partial_n] = (n-m)\, \partial_{m+n} - (-)^m (n+m)\, \partial_{n-m}$. The tensors $B^{m_1 \ldots m_q}_{n_1 \ldots\ldots n_p}$ are the various components of the classical string field and of its ghosts which are relevant for a covariant quantization, and the dependences of these tensors on the string position $X^\mu(\sigma)$ imply that they vary under the action of the ∂_m's while the ∂_m's leave invariant the β^m 's and $\beta^+{}_m$'s as in eq. (15). The BRS structure which reflects the internal gauge symmetry of the string has been found to be remarkably close to that of a system of local p-form gauge fields [2], but the concept of local diffeomorphisms has not yet been introduced for the string.

By going back to the general structure introduced above, it is natural to ask whether one can introduce objects which would generalize the notion of a local vector field in ordinary differential geometry. In order to consistently define a Lie derivative one needs an exterior derivative which would generalize $dx^\mu\, \partial/\partial x^\mu$. The most trivial candidate $\beta^m \partial_m$ is not satisfying, because it is not nilpotent, in view of eqs. (13,16). However it is well known that the operator $\beta^m \partial_m$ can be completed into the following one :

$$d \equiv \beta^m \partial_m + \frac{1}{2} f_{pq}^{m} \beta^p \beta^q \beta^+_m \tag{17}$$

which is nilpotent

$$d^2 = \frac{1}{2} [d, d] = 0 \tag{18}$$

as can be easily verified from eq. (13,16) by using the Jacobi relation satisfied by the f_{mn}^{p}'s.

One must introduce the notion of a vector field ξ^m, with ghost number g(ξ). Such an object is expected to be non invariant under the action of the ∂_m's, $[\partial_m, \xi^n] \neq 0$, and should be defined by the action on forms of the corresponding contraction operators i_ξ :

$$\begin{aligned} [i_\xi, \beta^m] &\equiv \xi^m \\ [i_\xi, \beta^+_m] &\equiv 0 \\ [i_\xi, B^{m_1 \ldots m_q}_{m_1 \ldots m_p}] &\equiv 0 \end{aligned} \tag{19}$$

One imposes that i_ξ acts on forms as a derivative graded by $g(\xi)-1$, so that eqs.(19) define the action of i_ξ on all forms. One has in particular:

$$[i_\xi, i_{\xi'}] = 0 \tag{20}$$

and thus $[i_\xi, \xi'^m] = 0$, $[\xi^m, \xi'^m] = 0$ by application of the Jacobi identity to the equation $[[i_\xi, i_{\xi'}], \beta^m] = 0$.

Having introduced the exterior derivative d, eq. (17), and the operator i_ξ, eq. 19, we can now define the Lie derivative along ξ:

$$L_\xi \equiv [i_\xi, d] = i_\xi d + (-)^{g(\xi)} d\, i_\xi \tag{21}$$

Notice that the property $d^2 = 0$ and the graded Jacobi relation implies the useful relation $[d, L_\xi] = 0$. One must only verify for consistency that the graded commutator of two Lie derivatives is still a Lie derivative:

$$[L_\xi, L_{\xi'}] = L_{\{\xi, \xi'\}} \tag{22}$$

where $\{\xi, \xi'\}$ is some vector field bilinear in ξ, ξ'. The relation $d^2 = 0$ and the graded Jacobi identity, together with the definition $L_\xi = [i_\xi, d]$, allows one to write the following equation, equivalent to eq. (22):

$$[i_\xi, L_{\xi'}] = [i_{\xi'}, L_\xi] = i_{\frac{1}{2}\{\xi, \xi'\}} \tag{23}$$

It is sufficient to prove eq. (23) for $\binom{0}{0}$-forms, $\binom{1}{0}$ and $\binom{0}{1}$-forms, since all forms are build from these building blocks. On a $\binom{0}{0}$-form f, one has :

$$[[i_\xi, L_{\xi'}], f] = [i_\xi, [L_{\xi'}, f]] = [i_\xi, [i_{\xi'}, [d, f]]]$$
$$= [i_\xi, [i_{\xi'}, \beta^m \partial_m f]]$$
$$= [i_\xi, \xi'^m \partial_m f] = 0 \qquad (24)$$

which is consistent with eq. (23). It is trivial to prove eq. (23) for $\binom{1}{0}$ -forms $\beta^+{}_m$, $[[i_\xi, L_{\xi'}], \beta^+{}_m] = 0$. The only non trivial computation is for proving equation (23) on $\binom{0}{1}$-forms β^m. One starts by using the graded Jacobi identity :

$$[[i_\xi, L_{\xi'}], \beta^m] = [i_\xi, [L_{\xi'}, \beta^m]] + [L_{\xi'}, [i_\xi, \beta^m]] \qquad (25)$$

On the other hand, one has

$$[L_{\xi'}, \beta^m] = [i_{\xi'}, [d, \beta^m]] + [d, [i_{\xi'}, \beta^m]]$$
$$= [i_{\xi'}, \tfrac{1}{2} F^m_{pq} \beta^p \beta^q] + [d, \xi'^m]$$
$$= \beta^p (\partial_p \xi'^m + F^m_{pq} \xi'^q) \qquad (26)$$

and

$$[L_{\xi'}, \xi^m] = [i_{\xi'}, [d, \xi^m]] + [d, [i_\xi, \xi^m]]$$
$$= \xi'^p \partial_p \xi^m \qquad (27)$$

so that one has :

$$[[i_\xi, L_{\xi'}], \beta^m] = [[i_{\xi'}, L_\xi], \beta^m] = i_{\{\xi, \xi'\}} \qquad (28)$$

with

$$\{ \xi, \xi' \}^m \equiv \xi^p \partial_p \xi'^m - (-)^{g(\xi)g(\xi')} \xi'^p \partial_p \xi^m + f^m_{pq} \xi^p \xi^q \tag{29}$$

We have thus proven the validity of eq. (22), where the graded generalized Poisson bracket $\{ \xi , \xi' \}$ has now been computed in eq. (29). Comparing eqs. (9) and (29) we see clearly that we have extended the diffeomorphism algebra in a non Abelian way. This structure is presumably related to that of Lie-Cartan pairs [3].

At this point a BRS structure can be easily constructed. The ghost vector field of non Abelian diffeomorphisms is defined as ξ^m with ghost numer $g(\xi) = 1$. The BRS operator s has grading one, and its graded commutation properties with the operators i_ξ, d, ∂_m are defined as :

$$[s, d] \equiv 0$$
$$[s, \partial_m] \equiv 0$$
$$[s, i_\xi] \equiv i_{s\xi} \tag{30}$$

This implies that the β^m and $\beta^+{}_m$ are s-invariant $[s, \beta^m] = [s, \beta^+{}_m] = 0$.

One has also :

$$[s, L_\xi] = [s , [i_\xi , d]] = [[s, i_\xi], d] = L_{s\xi} \tag{31}$$

We now define the action of s on forms B_p and ghost field ξ^m as follows :

$$s\, B_p = [\, L_\xi ,\, B_p\,]$$
$$s\, \xi^m = \tfrac{1}{2}\{\, \xi, \xi\}^m = \xi^p \partial_p \xi^m + \tfrac{1}{2} f^m_{pq} \xi^p \xi^q \tag{32}$$

It is obvious that s is nilpotent. One has indeed :

$$\begin{aligned} s^2 B_p &= s([L_\xi, B_p]) = [L_{s\xi}, B_p] - [L_\xi, [L_\xi, B_p]] \\ &= [L_{s\xi}, B] - \tfrac{1}{2}[[L_\xi, L_\xi], B_p] \\ &= [L_{s\xi - \frac{1}{2}\{\xi,\xi\}}, B_p] = 0 \end{aligned} \tag{33}$$

and the Jacobi identity implies :

$$\begin{aligned} 0 = [L_\xi, [L_\xi, L_\xi]] &= \tfrac{1}{2}[\, L_\xi , L_{\{\xi,\xi\}}] \\ &= \tfrac{1}{2} L_{\{\xi,\{\xi,\xi\}\}} \end{aligned} \tag{34}$$

so that

$$s^2 \xi = 0 \tag{35}$$

A direct computation, starting from eq. 32, can be also done to verify eq. (35).

The BRS operator s, eqs. (30,32), obviously generalizes in a non Abelian way the BRS operator of ordinary local diffeomorphisms eqs. (6,9). As compared to the ordinary case, eq. (9), the new BRS operator is characterized by the presence of the c-numbers f^p_{mn} (which arise from the non commutativity of the ∂_m) in $s\xi^m$.

There is no difficulty to introduce the notion of a metrics g_{mn} in the formalism. Imagine the existence of a 1-form $e^a \equiv e^a{}_m \beta^m$, where a is some index such that the matrix $e^a{}_m$ is invertible. Then one can define, a la Cartan :

$$g_{mn} = \sum_a e^a_m e^a_n \tag{36}$$

Since

$$\begin{aligned} s e^a = [L_\xi, e^a] &= [L_\xi, e^a_m]\beta^m + e^a_m [L_\xi, \beta^m] \\ &= \beta^p [\xi^q \partial_q e^a_p + e^a_q \partial_p \xi^q + f^m_{pq} e^a_p \xi^q] \\ &= (s e^a_p)\beta^p \end{aligned} \tag{37}$$

one has :

$$s e^a_m = \xi^q \partial_q e^a_m + e^a_p \partial_m \xi^p + f^m_{pq} \xi^p e^a_p \tag{38}$$

and :

$$\begin{aligned} s g_{mn} &= \sum_a \left((s e^a_m) e^a_n + e^a_m (s e^a_m) \right) \\ &= g_{mp} \partial_n \xi^p + g_{np} \partial_m \xi^p + \xi^p \partial_p g^{mn} \\ &\quad + f^p_{mq} \xi^q g_{np} + f^p_{nq} \xi^q g_{mp} \end{aligned} \tag{39}$$

Since $s^2 e^a = 0$ and $s^2 \beta^m = 0$ it is obvious that $s^2 e^a{}_m = 0$ and $s^2 g_{mn} = 0$. Eq. (39) is the usual expression of the variation of the metrics tensor under local diffeomorphisms, up to terms depending on the f_{mn}^{p}'s. By replacing the ghost ξ^m by an infinitesimal vector field ϵ^m in the expressions of sB_p, se^a_m, sg_{mn} in eqs. (32, 38, 39) one obtains infinitesimal gauge transformations. The nilpotency of the BRS operator that we have constructed ensures that these corresponding infinitesimal gauge transformations build up a closed algebra with a Jacobi identity. The gauge symmetry expressed in eq. (39) is certainly related to a local SO(N) invariance, where N stands for the number of independent indices n. There is no difficulty to introduce a spin connection field ω_m^{ab} in view of gauging this SO(N), and one can construct field strenghts which are functions of e and ω, generalize the usual torsion and Lorentz curvature of general relativity and transform tensorially under s. This, and the construction of invariants under the action of s will be presented elsewhere.

3) CONCLUSION

We have seen that the notion of local diffeomorphisms can be consistently extended in a non Abelian way. This extension is relevant when one considers a system where the operators playing the role of the partial derivatives in ordinary differential geometry are not commutative but satisfy non abelian Lie algebra relations. Since such a situation occurs in string field theory where only the internal gauge symmetry is known, it may happen that the type of symmetry presented in this work could be used to introduce the notion of local change of coordinates in the case of an extended object such as the string.

Acknowledgements

I wish to thank Bernard Grossman for drawing my attention on the relevance of the invariance of the string action under ∂_n-like operators. I am grateful to the Rockfeller Physics department, and especially to Professor N. Khuri, for their generous and pleasant hospitality.

References

1. L. Baulieu, M. Bellon, Nucl. Phys. B266 (1986), 75.

2. T. Banks, M. Peskins, Nucl. Phys. B264 (1986), 513 ;
A. Neveu, P. West, Phys. Lett. 167B (1986), 307 ;
W. Siegel, B. Zwiebach, Nucl. Phys. B263 (1986), 105 ;
K. Itoh, T. Kugo, H. Kunimoto, H. Ooguri, Prog. Theo. Phys. 75 (1986) 162 ;
E. Witten, Nucl. Phys. B268 (1986), 253.
L. Baulieu, S. Ouvry, Phys. Lett. 171B (1986), 57 ;
L. Baulieu, LPTHE preprint 86/35 submitted to Phys. Lett. B.

3. D. Kastler, R. Stora, preprint CPT 84/1714, LAPP TH 135.

ON PARTICLES, INFRAPARTICLES, AND THE PROBLEM OF ASYMPTOTIC COMPLETENESS

Detlev Buchholz

II. Institut für Theoretische Physik
Universität Hamburg

With regard to the particle aspects of local quantum field theory there exist two fundamental problems which - in spite of many efforts - have not yet been brought to a satisfactory solution.

Firstly, there is the well-known infraparticle problem: particles carrying an electric charge cannot be described by eigenstates of the massoperator since they are inevitably accompanied by clouds of soft photons. (For a simple argument relating this phenomenon to Gauss' law cf. [1].) So in the presence of such "infraparticles" the particle content of a theory cannot simply be read off from the poles of the respective Green' s functions, and standard collision theory does not work. There does not yet exist a general procedure by which the particle spectrum, the collision cross sections etc. can be determined from a theory in these more involved cases. (For an account of some recent work on this problem cf. [2].)

Secondly, there is the problem of asymptotic completeness, i.e. the question of whether a field theory has a complete particle interpretation. Note that in the presence of infraparticles it is not even clear how to formulate this question in appropriate mathematical terms. From a conceptual point of view it would be desirable to find conditions ("axioms") in terms of the local fields, having a simple physical interpretation, which imply that asymptotic completeness holds. One might hope that such conditions could also be useful for the analysis of concrete models. (For a recent investigation of the problem of asymptotic completeness in $\mathcal{P}(\varphi)_2$-theories and other models cf. the contribution of D. Iagolnitzer [3].)

It is the aim of these notes to revive an approach to collision theory, proposed almost twenty years ago by Araki and Haag [4], which in the light of recent results [5] seems to be suitable for a model-independent investigation of both above-mentioned problems. Araki and Haag based their considerations on an algebra $\mathcal{C}$ of operators imitating the gross properties of "detectors". Given the algebra $\mathfrak{A}$ of all local (bounded) observables of a field theory one obtains the algebra $\mathcal{C}$ as follows: let $A \in \mathfrak{A}$ be any operator which is localized in some bounded space-time region of $\mathbb{R}^4$ and let f be any testfunction on $\mathbb{R}^4$ whose Fourier-transform $\tilde{f}$ has compact support in the complement of the closed forward lightcone $\bar{V}$. It follows from these support properties of $\tilde{f}$ and the relativistic spectrum condition that the operator

$$B = \int d^4x \; f(x) \; A(x), \tag{1}$$

where $A(x) = U(x) A U(x)^{-1}$ and $U(x)$ are the unitary space-time translations, annihilates the vacuum vector Ω. The operator $C = B^* B$ may thus be interpreted as some kind of a detector: it is an almost local observable which is sensitive to excitations of the vacuum [4]. Taking all finite linear combinations and products of such operators C one obtains the $*$-algebra $\mathcal{C}$ of detectors.

We mention as an aside that the weak closures of the algebras $\mathcal{C}$ and $\mathfrak{A}$ coincide on all charged superselection sectors [5], so proceeding from $\mathfrak{A}$ to $\mathcal{C}$ one does not lose essential information. But in contrast to $\mathfrak{A}$, the algebra $\mathcal{C}$ admits non-normalizable "improper states". (Note that $\mathcal{C}$ does not contain a unit operator.) This fact will subsequently be of importance.

Focussing attention to asymptotically complete theories of massive particles, Araki and Haag made the interesting observation that the operators $C \in \mathcal{C}$ become "particle counters" at asymptotic times. They showed that the particular spatial averages of C at time t

$$C(t;h) = \int d^3x \; h(\underline{x}/t) \; C(t,\underline{x}), \tag{2}$$

where h is any testfunction on $\mathbb{R}^3$, converge weakly on some dense set of vectors in the limit of large t. Moreover, they established for the matrix elements of $C(t;h)$ the formula (given here for simplicity only for a theory of a single spinless particle)[1)]

$$\lim_{t \to t_{as}} (\Phi, C(t;h)\Phi) = \int d^3p \; \langle \underline{p}|C|\underline{p}\rangle \cdot h(\underline{p}/p_0) \, (\Phi, \rho^{as}(\underline{p})\Phi). \tag{3}$$

Here $\langle \underline{p}|C|\underline{p}\rangle$ are the matrix elements of C between improper single-particle states of momentum $\underline{p}$, p_0 is the energy of these states, and $\rho^{as}(\underline{p})$ is the operator corresponding to the asymptotic momentum space density of the particles. Similar formulae hold for theories with an arbitrary number of species of particles.

---

1) The suffix "as" stands for "incoming" or "outgoing", and t_{as} for $-\infty$ or ∞, respectively.

Since we want to take these results as a guide line, let us briefly digest the information provided by eq. (3) and its generalizations: first of all we see that one can determine from the states Φ in the vacuum sector, say, all single particle states (including the charged ones) without any a priori information about charged fields or the particle spectrum. But in contrast to standard collision theory, where the particles appear as intermediate states in the Green's functions of the (charged) fields, they now manifest themselves as mixtures of improper states $\langle \underline{p} | \, . \, | \underline{p} \rangle$ on the algebra $\mathcal{C}$ of observables. Secondly, as argued in [4], one can use these results directly for the calculation of collision cross sections, although the procedure is tedious compared to standard collision theory. Finally we note that in the present setting the property of asymptotic completeness finds its expression in the fact that the von Neumann algebra $\mathcal{C}^{as}$, which is generated by the operators $\lim_{t \to t^{as}} C(t;h)$ and the unit operator 1, contains a complete set of commuting observables. In the simple cases covered by eq. (3), $\mathcal{C}^{as}$ is the abelian algebra generated by the smoothed out densities $\int d^3p \, f(\underline{p}) \, \rho^{as}(\underline{p})$ and 1, which is maximal abelian since it has cyclic vectors. Note, however, that $\mathcal{C}^{as}$ need not be abelian in general.

Having seen how the particle properties manifest themselves in this setting in the case of massive, asymptotically complete theories, we want to provide now some rigorous model-independent arguments indicating that this approach works equally well in the presence of infraparticles. (Similar ideas have been discussed by Steinmann [6] in the framework of perturbative quantum electrodynamics and, on a more intuitive basis, by Weeks [7].) The first and essential step is the demonstration that the sequences $C(t,h)$ always have limits at asymptotic times t. This is a consequence of the following key lemma which is based on the spacelike commutativity of observables and the relativistic spectrum condition [5].

Lemma: Let $E(\Delta)$ be the spectral projections of the Hamiltonian corresponding to some bounded subset Δ of the spectrum, let $C \in \mathcal{C}$, and let $h \in \mathcal{S}(\mathbb{R}^3)$. Then

$$\| \int d^3x \; h(\underline{x}) \, C(\underline{x}) \, E(\Delta) \| \leq c \cdot \sup_{\underline{x}} |h(\underline{x})|$$

where c is a constant independent of h.

Remark: This result holds also for $h \in L^\infty(\mathbb{R}^3)$.

Making use of this lemma and the fact that the time translations commute with $E(\Delta)$ we obtain for fixed $C \in \mathcal{C}$ and Δ the estimate

$$\| C(t;h) \, E(\Delta) \| \leq c \cdot \sup_{\underline{x}} |h(\underline{x})| \tag{4}$$

which is independent of t. It then follows from standard compactness arguments that for suitable sequences of times t' tending to infinity, the operators $C(t',h)$ converge weakly on the dense domain $\mathcal{D} = \bigcup_{\Delta} E(\Delta)\mathcal{H}$, the limits being closable (unbounded) operators. So the sequences $C(t,h)$ never diverge, but at present we cannot exclude the possibility that they have several limit points. We therefore postpone the question of convergence for the time being and focus attention on the properties of the limit points.

This discussion simplifies if one restricts the operators $C(t,h)$ to the vacuum sector, where one can rely on Lorentz-covariance. As was pointed out, this restriction still provides sufficient information for the investigation of the particle properties of a theory.

Proposition 1: Let $C \in \mathcal{C}$ be selfadjoint, let $h \in \mathcal{S}(\mathbb{R}^3)$ be real, and let $C^{as}(h)$ be any limit point of the sequence $C(t,h)$, $t \to t^{as}$ on the states of finite energy in the vacuum sector. Then

i) $C^{as}(h)$ is essentially selfadjoint.
ii) $C^{as}(h)$ commutes with the space-time translations $U(x)$.
iii) $C^{as}(h) = 0$ if $\operatorname{supp} h \cap \{\underline{x} : |\underline{x}| \leq 1\} = \emptyset$.

Remark: Similar results hold on any superselection sector if one proceeds to suitable time-averages of the operators $C(t,h)$.

These results show that the operators $C^{as}(h)$ are bona fide observables which, at asymptotic times, give rise to a resolution of states of definite energy and momentum. In fact, these observables can be interpreted as (infra) particle counters, namely there holds an "abstract" version of relation (3). In order to explain this statement let us reinterpret the expectation values

$$\sigma_t(C) = (\Phi, C(t,h)\Phi), \quad C \in \mathcal{C} \tag{5}$$

where $\Phi \in E(\Delta)\mathcal{H}$ and $h(\underline{x}) \geq 0$ are kept fixed, as positive linear functionals σ_t on the algebra $\mathcal{C}$. Equipping $\mathcal{C}$ (regarded as a linear space) with the semi-norm

$$|||C||| = \sup_t \|C(t,h)E(\Delta)\|, \quad C \in \mathcal{C} \tag{6}$$

which is finite according to relation (4) we have

$$|\sigma_t(C)| \leq \|\Phi\|^2 \cdot |||C|||. \tag{7}$$

So applying once more standard compactness arguments (Alaoglu's theorem) we conclude that the sequence σ_t of functionals has at asymptotic times t weak limit points σ_{as} which are still positive linear functionals on $\mathcal{C}$ [2].

---

[2] Note that these functionals may be trivial, i.e. $\sigma_{as} = 0$, cf. the discussion below.

We have sufficient control on these limit points in order to establish the following

Proposition 2: Let σ_{as} be any weak limit point of the sequence σ_t. Then

i) $\sigma_{as}(C(x)) = \sigma_{as}(C)$ for $C \in \mathcal{C}$, $x \in \mathbb{R}^4$.
Moreover, the functions $x \rightarrow \sigma_{as}(C_1 C_2(x))$ are continuous for any $C_1, C_2 \in \mathcal{C}$ and their Fourier transforms vanish in the complement of a fixed forward light cone V_p with apex p.

ii) $\int d^3x_1 \cdots d^3x_n \, |\sigma_{as}(C C_1(\underline{x}_1) \cdots C_n(\underline{x}_n)| < \infty$
for any set of operators $C, C_1, \ldots C_n \in \mathcal{C}$ and any $n \in \mathbb{N}$.

The first half of this proposition shows that the functionals σ_{as} on $\mathcal{C}$ have all the properties expected from an incoherent superposition of (improper) energy-momentum eigenstates. In particular one can construct in the GNS-representation associated with any (non-trivial) σ_{as} a continuous, unitary representation of the translations which satisfies the relativistic spectrum condition. The second half of the proposition says that coincidence arrangements of detectors which are placed in widely separated regions are insensitive to the "states" σ_{as}. Thus σ_{as} describes systems consisting of a single localization center at all times (recall that σ_{as} is invariant under translations). This result fits exactly with the intuitive space-time picture of a single particle, cf. the discussion in [8]. So we find that the functionals σ_{as} have the properties of (improper) single particle states and that the expectation values of the operators $C(t;h)$ can be interpreted in terms of such functionals at asymptotic times.

These arguments can be extended to arbitrary products of the operators $C(t;h)$. To give an example, let us consider the expectation value

$$\sigma_t^{(2)}(C_1 \times C_2) = (\Phi, C_1(t;h_1) C_2(t;h_2) \Phi) \qquad C_1, C_2 \in \mathcal{C} \tag{8}$$

where $\Phi \in E(\Delta)\mathcal{H}$ and the functions h_1, h_2 are non-negative and have disjoint supports. As indicated by the notation, we regard these expectation values as functionals on the tensor algebra $\mathcal{C} \times \mathcal{C}$. Again one can show that the sequences $\sigma_t^{(2)}$ have weak limit points $\sigma_{as}^{(2)}$ which are positive linear functionals on $\mathcal{C} \times \mathcal{C}$. These limits have the properties of a tensor product of two single particle states, i.e. they exhibit with respect to each component of the product $\mathcal{C} \times \mathcal{C}$ the features described in the previous proposition. Analogous results hold for the expectation values of products of any number of operators $C(t;h)$, so in this sense the limits $C^{as}(h)$ of these operators can be interpreted as particle counters.

Note that these results have been derived without any ad hoc assumptions about particles. In particular they hold also for theories of infraparticles, as can also be seen by direct computations in simple models, such as the Schroer model [9].

Both, particles and infraparticles, appear in the present setting as improper momentum eigenstates $\langle \underline{p} | \, . \, | \underline{p} \rangle$ on the algebra $\mathcal{C}$. The difference between particles and infraparticles only show up if one tries to proceed from the improper states $|\underline{p}\rangle$ to normalizable states by forming "wave packets". In contrast to the case of ordinary particles this is impossible for infraparticles, where states $|\underline{p}\rangle$ of different momenta cannot coherently be superimposed because they have different asymptotic Coulomb fields (cf. [10] and references quoted there). But at the level of improper states particles and infraparticles can be treated in the same manner. — The present results are not yet sufficient to establish a general algorithm for the calculation of collision cross sections. What is missing is a procedure assigning to prescribed configurations of (infra)particles, characterized by functionals $\sigma_{as}, \sigma_{as}^{(2)}, \ldots$ as above, suitable vectors Φ describing these configurations at asymptotic times. It should be possible to solve this problem by a simulation of monitoring experiments with the help of the operators $C(t;h)$, cf. the discussions in [4], [6], and [7]. For a rigorous implementation of these ideas one can, however, no longer avoid the problem of convergence of the operators $C(t;h)$ in the limit of asymptotic times.

Let us finally comment on the problem of asymptotic completeness in the present setting. As we have seen, all states Φ have, with regard to the observables $C(t;h)$ at asymptotic times, an interpretation in terms of certain configurations of (infra) particles. It may, however, happen that the set of these asymptotic observables does not provide sufficient information about the states Φ. In the extreme case of a theory without any (infra) particle content, such as the generalized free field with continuous mass spectrum, one finds for example that the limits of all operator sequences $C(t;h)$ vanish, and consequently the functionals $\sigma_{as}, \ldots$ are zero. In contrast, in theories with a reasonable particle interpretation there exists for every vector $\Phi \neq c \cdot \Omega$ some $C \in \mathcal{C}$ and some function h (one may take in fact $h(\underline{x}) = 1$) such that

$$\liminf_{t} \left| (\Phi, C(t;h)\,\Phi) \right| > 0 \,, \tag{9}$$

compare eq. (3). So in this case the functionals $\sigma_{as}, \ldots$ are different from zero. Relation (9) thus provides a simple criterion for theories with a non-trivial (infra) particle structure.

For a better understanding of the way how the particle aspects of a theory emerge from its local properties it would be desirable to derive relation (9) from local conditions. Up to now three simple conditions have been isolated which seem to be relevant in this context.

i) Primitive causality [11] : the algebra which is generated by the local observables in a fixed time-slice $\{(t,\underline{x}) \in \mathbb{R}^4 : a < t < b\}$ must be (weakly) dense in the algebra $\mathfrak{A}$ of all local observables.

It is obvious that this condition is satisfied in the presence of a dynamical law (equations of motion).

ii) <u>Split property</u> (cf. the contribution of S.Doplicher, J.E.Roberts [12] and ref. quoted there.) : let $\mathcal{O}_1, \mathcal{O}_2$ be bounded regions such that the closure of $\mathcal{O}_1$ is contained in the interior of $\mathcal{O}_2$. Then there must be a type I factor $\mathcal{M}$ (i.e. a von Neumann algebra which is isomorphic to the algebra of all bounded operators on some Hilbert space) such that

$$\mathfrak{A}(\mathcal{O}_1) \subseteq \mathcal{M} \subseteq \mathfrak{A}(\mathcal{O}_2). \qquad (10)$$

The split property (10) has been established in free field theory, and thereby also in interacting theories which are locally Fock, such as the $\mathcal{P}(\varphi)_2$ models. More recently, it has been shown that the split property holds, whenever the number of local degrees of freedom of a theory (which can be measured with the aid of a "nuclearity index") is not too large [13] .

The third condition, saying that the vacuum must be locally almost a pure state, is actually a consequence of the split property.

iii) <u>Almost purity:</u> For fixed $\mathcal{O}_1$ and sufficiently large $\mathcal{O}_2$ one can choose the type I factor $\mathcal{M}$ in (10) in such a way that

$$\inf \|\Omega - \Psi\| < \varepsilon \qquad (11)$$

where the infimum is to be taken with respect to all vectors Ψ inducing pure states on $\mathcal{M}$. For a quantitative version of this result cf. [14] .

It is not known whether these conditions and results already imply relation (9), but they seem to bring us close to this goal. In order to substantiate this conjecture let us proceed to the following tighter conditions which are met by many non-relativistic field theories and (in a discretized version) also by quantum spin systems, but which are not quite compatible with the postulates of local quantum field theory.

Let $\mathfrak{A}(\underline{\mathcal{O}})$ be the "kinematical algebras" associated with the open regions $\underline{\mathcal{O}}$ of space at a given time.

We assume in analogy to the previous three conditions:

i') For any family of regions $\underline{\mathcal{O}}_i$, $i \in \mathbb{N}$ generating a dense paving $\cup \mathcal{O}_i$ of space, the associated algebra $\bigvee_i \mathfrak{A}(\underline{\mathcal{O}}_i)$ is (weakly) dense in the algebra $\mathfrak{A}$ of all observables.

ii') The algebras $\mathfrak{A}(\underline{\mathcal{O}})$ are type I factors.

iii') The vacuum Ω induces a pure state on each algebra $\mathfrak{A}(\underline{\mathcal{O}})$.

It is easy to show that relation (9) holds under these circumstances: given any region $\underline{\mathcal{O}}$ (we take here for simplicity a cube) there exists according to the latter two assumption a projection $C \in \mathfrak{A}(\underline{\mathcal{O}})$ which annihilates the vacuum and which has, relative to $\overline{\mathfrak{A}(\mathcal{O})}$, co-dimension one. So C has the properties of a "perfect" local detector which indicates any excitation of the vacuum in the region $\underline{\mathcal{O}}$. Now let $\underline{x}_i, i \in \mathbb{N}$ be translations such that the regions $\underline{\mathcal{O}}_i = \underline{\mathcal{O}} + \underline{x}_i$, $i \in \mathbb{N}$ are disjoint and $\bigcup_i \underline{\mathcal{O}}_i$ is dense in $\mathbb{R}^3$. Then it follows from the first assumption that the smallest projection containing all the (commuting) projections $C(\underline{x}_i), i \in \mathbb{N}$ is $1 - P_\Omega$, i.e. the projection onto the orthogonal complement of the vacuum. Hence we obtain for any vector Φ the estimate

$$\int d^3x \, (\Phi, C(\underline{x})\Phi) \geq |\underline{\mathcal{O}}| \cdot \left(\|\Phi\|^2 - |(\Phi, \Omega)|^2 \right), \quad (12)$$

where $|\underline{\mathcal{O}}|$ denotes the volume of $\underline{\mathcal{O}}$. If we replace on the left hand side of this inequality the vector Φ by $U(-t)\Phi$ we arrive at the same lower bound since the vacuum Ω is invariant under time-translations. So with C as above and $h(\underline{x}) = 1$ the criterion (9) is satisfied in these cases.

In view of the fact that the assumptions used in this argument are quite similar to the original conditions, one may hope that relation (9) can also be established in local quantum field theory. This would imply that the asymptotic algebras $\mathcal{C}^{as}$ are non-trivial. It would thus be an important step towards a general answer to the question under which circumstances these algebras contain a complete set of commuting observables (the problem of "asymptotic completeness" in the present setting).

References

1. D. Buchholz: Gauss'law and the infraparticle problem. Phys. Lett. B 174, 331 (1986)

2. G. Morchio, F. Strocchi: Infrared problem, Higgs phenomenon and long range interactions. In: Fundamental problems of gauge field theory. Erice Lecture notes 1985

3. D. Iagolnitzer: Asymptotic completeness and multiparticle analysis in field theories. These proceedings.

4. H. Araki, D. Haag: Collision cross sections in terms of local observables. Commun. Math. Phys. 4, 77 (1967)

5. D. Buchholz: In preparation

6. O. Steinmann: Scattering of infraparticles. Fortschr. Physik 22, 367 (1974)

7. B. Weeks: Cross sections from observable fields. Phys. Rev. D 27, 1340 (1983)

8. V. Enß: Characterization of particles by means of local observables. Commun. Math. Phys. 45, 35 (1975)

   R. Haag: Structural questions in quantum field theory. In: Proceedings of Symposia in Pure Mathematics 38/2, Providence: Am. Math. Society 1982

9. B. Schroer: Infrateilchen in der Quantenfeldtheorie. Fortschr. Physik 11, 1 (1963)

10. D. Buchholz: The physical state space of quantum electrodynamics. Commun. Math. Phys. 85, 49 (1982)

11. R. Haag, B. Schroer: Postulates of quantum field theory. J. Math. Phys. 3, 248 (1962)

12. S.Doplicher, J.E.Roberts: C^*-algebras and duality for compact groups. These proceedings.

13. D. Buchholz, E. H. Wichmann: Causal independence and the energy-level density of states in local quantum field theory. Commun. Math. Phys. 106, 321 (1986)

    D. Buchholz, C. D'Antoni, K. Fredenhagen: in preparation.

14. C. D'Antoni, S. Doplicher, K. Fredenhagen, R. Longo: in preparation.

KAC-MOODY AND VIRASORO ALGEBRAS IN MATHEMATICAL PHYSICS

Peter Goddard

Department of Applied Mathematics and Theoretical Physics
University of Cambridge,
Silver Street, Cambridge CB3 9EW United Kingdom
and
Institute for Theoretical Physics
University of California
Santa Barbara, CA 93106

ABSTRACT

An account is given of the classification and construction of the unitary representations of the Kac-Moody and Virasoro algebras.

We shall discuss aspects of the representation theory of certain infinite dimensional Lie algebras, the Virasoro algebra, $\hat{v}$, and the untwisted affine Kac-Moody algebras, $\hat{g}$. (Much of this discussion extends easily to the case of twisted affine Kac-Moody algebras.)

The Virasoro algebra, $\hat{v}$, has a basis labelled by the integers L_n, $n \in \mathbf{Z}$, together with a central element, c, that is an element which commutes with all the other generators. The algebra is specified by the commutation relations,

$$[L_m, L_n] = (m-n)L_{m+n} + \frac{c}{12}m(m^2-1)\delta_{m,-n} \tag{1}$$

where $m, n \in \mathbf{Z}$, together with $[L_n, c] = 0$. We shall be concerned with representations of this algebra which are unitary in the sense that they satisfy the hermiticity condition,

$$L_n^\dagger = L_{-n}. \tag{2}$$

In an irreducible unitary representation, c may be ascribed a fixed numerical value.

To define an untwisted affine Kac-Moody algebra, $\hat{g}$, we start from a compact finite dimensional Lie algebra g, with basis t^a, $1 \le a \le \dim g$, specified by commutation relations

$$[t^a, t^b] = i f_{abc} t^c \tag{3}$$

where the structure constants f_{abc} are totally antisymmetric. A basis for the associated untwisted affine Kac-Moody algebra $\hat{g}$ consists of generators T^a_m: $1 \le a \le \dim g$, $m \in \mathbf{Z}$, again together with a central element k, and the commutation relations are

$$[T^a_m, T^b_n] = i f_{abc} T^c_{m+n} + k m \delta^{ab} \delta_{m,-n}, \tag{4}$$

where $m, n \in \mathbf{Z}$, together with $[T^a_m, k] = 0$. In this case, we shall be interested in representations which are unitary in the sense that

$$T^{a\dagger}_n = T^a_{-n} \tag{5}$$

and in an irreducible representation of this sort k will take a single numerical value. Note that $\hat{g}$ has a subalgebra, with basis T^a_0, $1 \le a \le \dim g$, which is isomorphic to g and which we will identify with it.

(In most of what follows, to avoid complication, we shall assume that g is simple unless we specify to the contrary. If it is not we can clearly introduce an independent central term k for each simple or $U(1)$ factor. At certain places in our discussion, we shall need to refer to particular non-simple algebras and we shall then explain how the results we establish are generalized straightforwardly to the non-simple case.)

These algebras have applications in a wide variety of different areas of mathematics and physics, including sporadic finite simple groups, the theory of modular forms, "completely integrable" dynamical systems, string theories of fundamental particles and their interactions, conformally invariant two-dimensional field theories, and critical phenomena in two-dimensional statistical systems. The occurrence of these algebras and, associated with their representation theory, modular forms, can be seen as a unifying feature underlying connections between such disparate, or tenuously connected, branches of mathematics and physics.

For a longer and more detailed review of the material covered in these lectures.[1] For other reviews of Kac-Moody algebras and their physical applications consult refs. [2] and [3]. For a synopsis of the mathematical development of the subject see the article of Lepowsky[4], or other contributions in the same volume, the review by MacDonald[5] and the book by Kac.[6]

We can construct infinite-dimensional groups of which $\hat{g}$ and $\hat{v}$ are the Lie algebras, provided that we omit their centers, k and c. Denote the corresponding algebras by $\hat{g}_0$ and $\hat{v}_0$, respectively. If g is the algebra of a compact finite-dimensional Lie group, G, the infinite-dimensional Lie group $\mathcal{G}$ with Lie algebra $\hat{g}_0$ is just the loop group of G, that is the group of (suitably smooth) maps from the circle S^1 to G, with multiplication defined pointwise. To be more explicit, realize S^1 as the unit circle in the complex plane

$$S^1 = \{z \in \mathrm{C} : |z| = 1\}. \tag{6}$$

Then the group product of two elements, $\gamma_1, \gamma_2 \in \mathcal{G}$ is defined by

$$\gamma_1\gamma_2(z) = \gamma_1(z)\gamma_2(z). \tag{7}$$

To construct an infinite-dimensional group with algebra $\hat{v}_0$ consider the group $\mathcal{V}$ of smooth one-to-one maps $S^1 \to S^1$ with the group law in this case defined by composition; that is we define the product of two such maps $\xi_1, \xi_2 : S^1 \to S^1$ by

$$\xi_1\xi_2(z) = \xi_1(\xi_2(z)). \tag{8}$$

There is a natural interconnection between the groups $\mathcal{V}$ and $\mathcal{G}$, since $\mathcal{V}$ can be made to act on $\mathcal{G}$ by way of reparameterization. Thus they go together to form a semi-direct-product, and with the consequent interaction of the algebras:

$$[L_m, T_n^a] = -mT_{m+n}^a. \tag{9}$$

We shall often find it profitable to think of the algebra defined by (1), (4) and (9), that is

$$[T_m^a, T_n^b] = if_{abc}T_{m+n}^c + km\delta^{ab}\delta_{m,-n}, \tag{1.10a}$$

$$[L_m, T_n^a] = -nT_{m+n}^a, \tag{1.10b}$$

$$[L_m, L_n] = (m-n)L_{m+n} + \frac{c}{12}m(m^2-1)\delta_{m,-n}. \tag{1.10c}$$

In physical applications L_0 normally has an interpretation which requires it to have a spectrum which is bounded below (e.g. it corresponds to something like an energy or dilatation). Then it follows that the states for which the eigenvalue of L_0 is least must be annihilated by the generators T_n^a, L_n with $n > 0$, and that an invariant and irreducible subspace of the representation is built up by applying the T_n^a, $n \leq 0$, or L_n, $n < 0$, or both sets [depending on whether we are considering $\hat{g}$ of $\hat{v}$ or the whole algebra (10)] to a "vacuum state" Ψ satisfying

$$T_n^a \Psi = L_n \Psi = 0, \qquad n > 0. \tag{11}$$

A representation built up in this way from vacuum states satisfying (11) is called a highest weight representation.

Let us consider how to label the highest weight representations of the Virasoro and Kac-Moody algebras. For $\hat{v}$ alone, if h is the lowest eigenvalue of L_0, the states of an irreducible unitary highest weight representation can be built up from a single vacuum state $|h\rangle$ satisfying

$$L_0|h\rangle = |h\rangle, \tag{12}$$

$$L_n|h\rangle = 0, \qquad n > 0, \tag{13}$$

by applying the algebra. Thus we can label such representations by the pair of real numbers (c, h). We shall return to the question of for which values of (c, h) such unitary representations exist.

For an untwisted affine Kac-Moody algebra $\hat{g}$, we in general have a space of vacuum states Ψ satisfying

$$T_n^a \Psi = 0, \qquad n > 0, \tag{14}$$

from which $\hat{g}$, or in fact just T_{-m}^a, $m > 0$, generates the whole of a unitary highest weight representation. The vacuum space (14) must form an irreducible representation of $g \cong \{T_0^a\}$; for an irreducible representation of $\hat{g}$ this vacuum representation of g must be itself irreducible as it is clear that $\hat{g}$ would generate orthogonal invariant subspaces by its action on different irreducible vacuum representations. The irreducible unitary highest weight representations of $\hat{g}$ are thus labelled by a vacuum representation, or equivalently its highest weight λ, and the value of k.

It is much easier to establish necessary and sufficient conditions for the existence of unitary highest representations in the Kac-Moody case than in the Virasoro case. Given that g is simple, the conditions for $\hat{g}$ take the form that

$$x = \frac{2k}{\psi^2} \tag{15a}$$

be a non-negative integer, where ψ is a long root of g, and that the highest weight of the vacuum representation satisfy

$$|\alpha \cdot \lambda| \leq k \tag{15b}$$

for each root α of g. The integer x is called the level of the representation. (If $x = 0$ the representation is necessarily trivial, i.e., $T_n^a = 0$.)

To prove the conditions (15) are necessary we only have to consider appropriate $su(2)$ subalgebras of $\hat{g}$. If we write g in a basis consisting of a Cartan subalgebra H^i, $1 \leq i \leq rank\ g$, and step operators, E^α, $\alpha \in \Phi$, where Φ denotes the set of roots of g, we see that $\hat{g}$ has an $su(2)$ subalgebra consisting of $I_+ = E_1^{-\alpha}, I_- = E_{-1}^{\alpha}$, $2I_3 = \frac{2}{\alpha^2}(k - \alpha \cdot H_0)$; that is

$$[I_+, I_-] = 2I_3, \quad [I_3, I_\pm] = \pm I_\pm. \tag{16}$$

Since any unitary representation of $\hat{g}$ must provide unitary representations of this $su(2)$, it follows that the spectrum of $2I_3$ must be integral. Consider any simultaneous eigenvector $|\mu\rangle$ of H_0^i,

$$H_0^i|\mu\rangle = \mu^i|\mu\rangle, \tag{17}$$

so that μ is a weight of g; on $|\mu\rangle$, I_3 takes the value

$$\frac{2k}{\alpha^2} - \frac{2\alpha \cdot \mu}{\alpha^2} \in \mathbf{Z}. \tag{18}$$

But, by the usual theory of roots and weights (for a summary see ref. 1), $2\alpha \cdot \mu / \alpha^2 \in \mathbf{Z}$, implying that

$$\frac{2k}{\alpha^2} \in \mathbf{Z} \quad \text{if} \quad \alpha \in \Phi. \tag{19}$$

Since g is assumed simple, it has at most two distinct squared root lengths and if there are two they are in the ratio 2:1 or 3:1. In any case we see that if (19) holds for a long root it holds for all roots, amounting to the condition that (15) be integral. To obtain the remaining part of (15) take $|\mu\rangle$ to be a vacuum state, so that

$$E_1^{-\alpha}|\mu\rangle = 0 \tag{2.15}$$

and consider the norm

$$\begin{aligned} ||E_{-1}^{\alpha}|\mu\rangle||^2 &= \langle\mu|E_1^{-\alpha}E_{-1}^{\alpha}|\mu\rangle \\ &= \langle\mu|\left[E_1^{-\alpha}, E_{-1}^{\alpha}\right]|\mu\rangle \\ &= \frac{2}{\alpha^2}(k - \alpha \cdot \mu)||\ |\mu\rangle\ ||^2. \end{aligned} \tag{20}$$

Thus

$$k \geq \alpha \cdot \mu \tag{21}$$

for all roots α and all vacuum weights μ. In particular, by changing the sign of α if necessary, we see that $k \geq 0$ and also (15b) follows. This shows that the conditions (15) are necessary. It is not difficult to show that they are sufficient by actually constructing them.

The discussion of unitary height weight representations of the Virasoro algebra is more difficult. It is easy to see, by considering

$$\begin{aligned} ||L_{-n}|h\rangle||^2 &= \langle h|L_n L_{-n}|h\rangle \\ &= \langle h|[L_n, L_{-n}]|h\rangle \\ &= \left\{2nh + n(n^2-1)\frac{c}{12}\right\} ||\,|h\rangle||^2, \end{aligned} \tag{22}$$

and taking $n = 1$ and then n large, that unitarity requires that

$$h \geq 0 \quad \text{and} \quad c \geq 0. \tag{23}$$

To progress further we need to consider the matrix $M_N(c, h)$ of inner products of states of the form

$$L_{-1}^{n_1} L_{-2}^{n_2} \ldots L_{-m}^{n_m}|h\rangle \tag{24}$$

with a given eigenvalue, N, of L_0, i.e., for a given value of $N = \sum_{j=1}^m j n_j$. The representation will be unitary, in the sense of the hermicity condition $L_n^\dagger = L_{-n}$ holding with respect to some positive definite scalar product, provided that the matrix $M_N(c, h)$ is positive semi-definite for each value of N. (If $M_N(c, h)$ has some zero eigenvalues, but its remaining eigenvalues are positive, we can consistently set to zero the linear combinations of vectors (24) corresponding to the zero eigenvalues.)

This problem can be explored case by case for low values of N, but since $\dim M_N(c, h)$ grows like the number of partitions of N, it quickly becomes impractible. However a remarkable formula for det $M_N(c, h)$ was given by Kac[7] and proved by Feigin and Fuchs.[8] From this formula it follows quite easily that the unitarity condition is satisfied if $h \geq 0$ and $c \geq 0$. Analyzing the formula in detail, Friedan, Qiu and Shenker[9] were able to show that unitarity requires, if $0 \leq c < 1$, that c be a number in the series

$$c = 1 - \frac{6}{m(m+1)}, \quad m = 2, 3 \ldots, \tag{25}$$

i.e., c is one of the number $0, \frac{1}{2}, \frac{7}{10}, \frac{4}{5}, \frac{6}{7}, \frac{25}{28}, \ldots$. For each value of c in this series (the "discrete series" of unitary representations of $\hat{v}$) there are $\frac{1}{2}m(m-1)$ possible values of h given by

$$h = h_{p,q}^{(c)} = \frac{[(m+1)p - mq]^2 - 1}{4m(m+1)} \tag{26}$$

where $1 \leq p \leq m-1$, $1 \leq q \leq p$. The first few possible values of (c, h) are thus

$$c = 0; \quad h = 0; \tag{27a}$$

$$c = \frac{1}{2}; \quad h = 0, \ \frac{1}{16} \text{ or } \frac{1}{2}; \tag{27b}$$

$$c = \frac{7}{10}; \quad h = 0, \ \frac{3}{80}, \frac{1}{10}, \frac{7}{16}, \frac{3}{5}, \text{ or } \frac{3}{2}; \tag{27c}$$

$$c = \frac{6}{7}; \quad h = 0, \ \frac{1}{40}, \frac{1}{15}, \frac{2}{5}, \frac{21}{40}, \frac{2}{3}, \frac{7}{5}, \frac{13}{8}, \text{ or } 3. \tag{27d}$$

The first of these possibilities $c = h = 0$, corresponds to the trivial representation $L_n = 0$. The result, that if $c = 0$, for a highest weight unitary representation of $\hat{v}$, then $L_n = 0$, is a very powerful one. It follows from the general analysis of Friedan, Qiu and Shenker,[9] but a simple direct proof has been provided by Gomes.[10]

The representations with $c = \frac{1}{2}$ have been known in the context of string theories since 1971 where they occurred implicitly in the fermionic theory of Neveu and Schwarz[11] and Ramond.[12] Consider a d-component real (i.e. Majorana) fermi field $H^\alpha(z)$, $1 \leq \alpha \leq d$, defined on the unit circle, $|z| = 1$. Since physical quantites involve even products of fermi fields, $H^\alpha(z)$ is either periodic (Ramond case) or periodic (Neveu-Schwarz case). Thus

$$H^\alpha(z) = \sum_r b_r^\alpha z^{-r} \tag{28}$$

where the sum is *either* over $r \in \mathbf{Z}$ (Ramond case) *or* over $r \in \mathbf{Z} + \frac{1}{2}$ (Neveu-Schwarz case). The oscillators b_r^α satisfy the hermicity condition

$$b_r^{\alpha\dagger} = b_{-r}^\alpha, \tag{29}$$

and the canonical anti-commutation relations

$$\{b_r^\alpha, b_s^\beta\} = \delta^{\alpha\beta}\delta_{r,-s}. \tag{30}$$

The field $H_r^\alpha(z)$ acts in a space generated by the b_r^α, $r < 0$, from a vacuum $|0\rangle$ satisfying

$$b_r^\alpha |0\rangle = 0, \; r > 0. \tag{31}$$

In the NS case we may assume that we have a non-degenerate vacuum but in the R case it has to be of dimension $2^{d/2}$, d even, or $2^{(d-1)/2}$, d odd, in order to provide a representation of the Dirac algebra of the b_0^α.

In this space we can write down a representation of the Virasoro algebra using what is essentially the energy-momentum tensor for the fermion field $H^\alpha(z)$,

$$L(z) \equiv \sum_n L_n z^{-n} = \frac{1}{2} z : \frac{dH}{dz} H : + \epsilon d. \tag{32}$$

Here $\epsilon = 0$ in the NS case and $\frac{1}{16}$ in the R case, and the colon denotes normal ordering with respect to the fermion oscillators b_r^α, to be precise

$$\begin{aligned} : b_r^\alpha b_s^\beta : &= -b_s^\beta b_r^\alpha \quad \text{if} \quad r > 0, \\ &= \frac{1}{2}[b_r^\alpha, b_s^\beta] \quad \text{if} \quad r = 0, \\ &= b_r^\alpha b_s^\beta \quad \text{if} \quad r < 0. \end{aligned} \tag{33}$$

From the definition (32) it is straightforward to show that

$$[L_m, L_n] = (m-n) L_{m+n} + \frac{d}{24} m(m^2 - 1) \delta_{m,-n}, \tag{34}$$

i.e., L_n satisfies the algebra (1.1) with $c = \frac{d}{2}$. This construction provides just one value of $c < 1$, namely $c = \frac{1}{2}$ from $d = 1$. We obtain all the corresponding values of h given by (27b): $h = 0$ and $h = \frac{1}{2}$ are provided by the states $|0\rangle$ and $b_{-\frac{1}{2}}|0\rangle$ in the NS case, and $h = \frac{1}{16}$ is provided by the vacuum $|0\rangle$ in the R case. (These are all the highest weight states for $d = 1$.)

In order to construct representations of the Virasoro algebra with values of c in the discrete series (25) other than 0 and $\frac{1}{2}$ we need to consider the interrelation of the Virasoro and Kac-Moody algebras in the semi-direct product (10). Suppose we start with a given unitary highest weight representation of an untwisted affine Kac-Moody algebra, $\hat{v}$. This can always be extended in a certain "minimal" way to provide a representation of the whole semi-direct product by introducing the Virasoro generators $\mathcal{L}_n$ as quadratic functions of the T_n^a,

$$\mathcal{L}_n^g = \frac{1}{2\beta^g} \sum_{m,a} {}^\circ_\circ T_m^a T_{n-m}^a {}^\circ_\circ, \tag{35}$$

where the open does denote normal ordering with respect to the T_n^a,

$$\overset{\circ}{\circ} T_m^a T_n^a \overset{\circ}{\circ} = T_n^a T_m^a, \qquad m \geq 0, \tag{36a}$$

$$= T_m^a T_n^a, \qquad m < 0. \tag{36b}$$

With this definition

$$[\mathcal{L}_m^g, \mathcal{L}_n^g] = (m-n)\mathcal{L}_{m+n}^g + \frac{c^g}{12} m(m^2-1)\delta_{m,-n} \tag{37}$$

and

$$[\mathcal{L}_m^g, T_n^a] = -T_{m+n}^a, \qquad 1 \leq a \leq \dim\, g \tag{38}$$

provided that

$$\beta = k + \frac{1}{2} Q_\psi^g, \tag{39}$$

where Q_ψ^g is the value of the quadratic Casimir operator for g in the adjoint representation, i.e.,

$$f_{abc} f_{abd} = Q_\psi^g \delta_{cd}. \tag{40}$$

The value of the central element c^g in (4.3) is given by the formula

$$c^g = \frac{2k \dim g}{Q_\psi^g + 2k}. \tag{41}$$

We can rewrite this in terms of the level x of the representation of $\hat{g}$ and the quantity

$$\tilde{h} = Q_\psi^g / \psi^2 \tag{42}$$

which is an integer called the dual Coxeter number. In these terms,

$$c^g = \frac{x \dim g}{x + \tilde{h}}. \tag{43}$$

Thus c^g is a rational number. It satisfies

$$\text{rank}\ g \leq c^g \leq \dim\ g \tag{44}$$

so that this construction can not directly provide us with any of the discrete series with $c < 1$.

To get the series values of $c < 1$ we need to consider[13] a subalgebra $h \subset g$ and the corresponding $\hat{h} \subset \hat{g}$. Let us suppose that T_m^a, $1 \leq a \leq \dim\ h$, $m \in \mathbf{Z}$, and k

provide a basis for $\hat{h}$, whilst if $1 \leq a \leq \dim\ g$ we have a basis for $\hat{g}$. Then we can apply the construction (35) to $\hat{h}$,

$$\mathcal{L}_n^h = \frac{1}{2\beta^h} \sum_{m,a=1}^{\dim h} : T_m^a T_{n-m}^a : \tag{45}$$

with

$$\beta^h = k + \frac{1}{2} Q_\psi^h \tag{46}$$

and with the corresponding value of c,

$$c^h = \frac{2k \dim\ h}{Q_\psi^h + 2k}. \tag{47}$$

Then

$$[\mathcal{L}_m^h, T_n^a] = -nT_{m+n}^a, \qquad 1 \leq a \leq \dim\ h, \tag{48}$$

so that if we introduce the difference

$$K_m = \mathcal{L}_m^g - \mathcal{L}_m^h \tag{49}$$

it follows that

$$[K_m, T_n^a] = 0 \qquad 1 \leq a \leq \dim\ h. \tag{50}$$

Since K_m commutes with the whole of $\hat{h}$,

$$[K_m, \mathcal{L}_n^h] = 0 \tag{51}$$

and so

$$[\mathcal{L}_m^g, \mathcal{L}_n^g] = [\mathcal{L}_m^h, \mathcal{L}_n^h] + [K_m, K_n]. \tag{52}$$

Hence K_m satisfies the Virasoro algebra

$$[K_m, K_n] = (m-n)K_{m+n} + \frac{c_K}{12} m(m^2-1)\delta_{m,-n} \tag{53}$$

with

$$c_K = c^g - c^h. \tag{54}$$

This construction yields the entire discrete series of representation of $\hat{v}$ but to describe this we must explain how the construction (35) extends to the case where g is not simple. If

$$g = g_1 \oplus g_2 \oplus \cdots \oplus g_M, \tag{55}$$

with g_j simple, then we replace (35) by

$$\mathcal{L}_n^g = \mathcal{L}_n^{g_1} + \mathcal{L}_n^{g_2} + \cdots + \mathcal{L}_n^{g_M}, \tag{56}$$

with each $\mathcal{L}_n^{g_j}$ given by (4.1), and the corresponding c is

$$c^g = c^{g_1} + c^{g_2} + \cdots + c^{g_M} \tag{57}$$

with each c^{g_j} given by (43).

One way to obtain the discrete series is to take $g = su(2) \oplus su(2)$ and h to be the diagonal $su(2)$ subalgebra.[13] In general for $su(2)$ if we take the root length $\psi = 1$, as usual as in (16), the quadratic Casimir operator takes the value $\ell(\ell+1)$ in the spin ℓ representation and $Q_\psi = 1$. So for a level x representation of $su(2)$,

$$c^{su(2)} = \frac{3x}{x+2}. \tag{58}$$

If we choose a representation of $\hat{g} = \hat{su}(2) \oplus \hat{su}(2)$ which has levels $n-1$ and 1 for the two factors, respectively, it induces a level n representation of the diagonal subalgebra h. Then

$$\begin{aligned} c_K = c^g - c^h &= \frac{3(n-1)}{n+1} + 1 - \frac{3n}{n+2} \\ &= 1 - \frac{6}{(n+1)(n+2)} \end{aligned} \tag{59}$$

providing all of the sequence (25).

It can also be shown that this construction yields all the corresponding values of h given by (26).

ACKNOWLEDGEMENT

I am very grateful to Adrian Kent, Werner Nahm and David Olive for discussions on which many aspects of the presentation here are based. This research was supported in part by the National Science Foundation under Grant No. PHY82-17853, supplemented by funds from the National Aeronautics and Space Administration, at the University of California at Santa Barbara.

REFERENCES

1. Goddard, P. and Olive, D., *Kac-Moody and Virasoro Algebras in Relation to Quantum Physics,* International Journal of Modern Physics A1, 303 (1986).

2. Dolan, L., Phys. Rep. 109, 1 (1984).

3. Julia, B., A.M.S. Lectures in Applied Mathematics, 21, 355 (1985), and in *Vertex Operators in Mathematics and Physics* (ed. J. Lepowsky *et al.*, Springer, 1984) 393.

4. Lepowsky, J., in *Vertex Operators in Mathematics and Physics* (ed. J. Lepowsky *et al.*,Springer, 1984) 1.

5. MacDonald, I., Lecture Notes in Mathematics 901, 258 (1981).

6. Kac, V.G., Infinite Dimensional Lie Algebras (2nd edition, Cambridge University Press, 1985).

7. Kac, V.G., Proceedings of the International Congress of Mathematicians, Helsinki, 299 (1978).

8. Feigin, B.L., and Fuchs, D.B., Funct. Anal. App. 16, 114 (1982).

9. Friedan, D., Qiu, Z. and Shenker, S., Phys. Rev. Lett. 52, 1575 (1984); in *Vertex Operators in Mathematics and Physics* (ed. J. Lepowsky *et al.*, Springer, 1984) 491.

10. Gomes, J.F., Phys. Lett. 171B, 75 (1986).

11. Neveu, A. and Schwarz, J.H., Nucl. Phys. B31, 86 (1971); Phys. Rev. D4, 1109 (1971).

12. Ramond, P., Phys. Rev. D3, 2415 (1971).

13. Goddard, P., Kent A. and Olive, D., Phys. Lett. 152B, 88 (1985); Comm. Math. Phys. 103, 105 (1986).

BOUNDARY CONDITIONS FOR INTEGRABLE SYSTEMS

E.K.SKLYANIN

Steklov Mathematical Institute, Leningrad Branch
Fontanka 27, Leningrad 191011, USSR

The Quantum Inverse Scattering Method [1] has demonstrated the advantage of applying quantum ideas to classical problems. Here is one more example of such influence.

Consider an integrable one-dimensional evolution equation possessing an LM-pair, i.e. equivalent to the relation [2]

$$\dot{L} = M' + [M, L] \quad \forall u \tag{1}$$

for two matrix-valued functions $L(u;x,t)$ and $M(u;x,t)$ The equality (1) is the <u>local</u> integrability condition. To discuss the <u>global</u> integrability one needs to set a range of x and boundary conditions (BC). The case of the infinite interval $x \in (-\infty, +\infty)$ is well studied. The most popular boundary conditions are here various fixed asymptotics and quasiperiodicity condition. In case of the finite interval $x \in [x_-, x_+]$ only the periodicity condition and its variants seem to be investigated.

The problem is to find a more extensive class of BCs on the finite interval compatible with the complete integrability.

To be more specific, consider the nonlinear Schrödinger equation

$$i\dot{\psi} = -\psi'' + 2\bar{\psi}\psi^2 \tag{2}$$

with the density of hamiltonian

$$h(x) = \bar{\psi}'\psi' + \bar{\psi}^2\psi^2 \tag{3}$$

and Poisson brackets

$$\{\psi(x), \bar{\psi}(y)\} = i\delta(x-y) \tag{4}$$

The corresponding LM-pair is

$$L=i\begin{pmatrix}-\frac{u}{2} & \bar{\psi}\\ -\psi & \frac{u}{2}\end{pmatrix},\quad M=i\begin{pmatrix}\frac{u^2}{2}+\bar{\psi}\psi, & -i\bar{\psi}'-u\bar{\psi}\\ -i\psi'+u\psi, & -\frac{u^2}{2}-\bar{\psi}\psi\end{pmatrix} \qquad (5)$$

The generalized periodic boundary conditions

$$\psi(x_+)=e^{i\varkappa}\psi(x_-),\quad \bar{\varkappa}=\varkappa \qquad (6)$$

are well known to be integrable [2]. By analogy with the linear Sturm-Liouville problem it is quite natural to consider also another class of BCs, namely, the mixed BC on each end

$$\psi_\pm' \pm \theta_\pm \psi_\pm = 0,\quad \bar{\theta}_\pm=\theta_\pm,\quad \psi_\pm \equiv \psi(x_\pm) \qquad (7)$$

corresponding to the hamiltonian

$$H=\int_{x_-}^{x_+} h(x)dx + \sum_{\alpha=\pm}\theta_\alpha \bar{\psi}_\alpha \psi_\alpha \qquad (8)$$

Note that the cases $\theta=0$ (Neumann condition) and $\theta=\infty$ (Dirichlet condition) can be reduced to the periodic case $(\varkappa=0)$ with the parity (resp.imparity) restriction.

<u>Question</u>: Is the hamiltonian (8) completely integrable?

<u>Answer</u>: Yes,it is.

To verify the proposition it is necessary to present a generating function $\tau(u)$ of commuting integrals of motion. By analogy with the periodic BCs case (6) where $\tau(u)$ is given by

$$\tau(u)=tr\,K\,T(u),\quad K=e^{i\frac{\varkappa}{2}\sigma_3} \qquad (9)$$

$T(u)\equiv T_{x_-}^{x_+}(u)$ being the transition matrix for the L-operator (5) defined from

$$\frac{\partial}{\partial x_+} T_{x_-}^{x_+}(u) = L(u, x_+) T_{x_-}^{x_+}(u) \quad , \quad T_x^x(u) = 1 \tag{10}$$

it is natural to suppose that $\tau(u)$ for BCs (7) is also some functional of $T(u)$.

<u>Theorem</u> 1. Let $\tau(u)$ be

$$\tau(u) = \operatorname{tr} K_+(u) \mathcal{T}(u) \tag{11}$$

where
$$\mathcal{T}(u) = T(u) K_-(u) T^{-1}(-u) \tag{12}$$

and

$$K_\pm(u) = \begin{pmatrix} u + i\theta_\pm , & 0 \\ 0 & , -u + i\theta_\pm \end{pmatrix} \tag{13}$$

Then

a)
$$\dot{\tau}(u) \equiv \{H, \tau(u)\} = 0 \tag{14}$$

where the hamiltonian H is given by (8)

b)
$$\{\tau(u_1), \tau(u_2)\} = 0. \tag{15}$$

The statement a) follows immediately from the obvious equality

$$\dot{T}(u) = M(u, x_+) T(u) - T(u) M(u, x_-)$$

and the identity

$$K_\pm(u) M(\pm u, x_\pm) = M(\mp u, x_\pm) K_\pm(u) \tag{16}$$

which is straightforward to verify.

The proof of the statement b) uses the classical r-matrix formalism [1, 3] which is based on the relation

$$\{\overset{1}{L}(u_1, x), \overset{2}{L}(u_2, y)\} = [r(u_1 - u_2), \overset{1}{L}(u_1, x) + \overset{2}{L}(u_2, y)] \delta(x-y) \tag{17}$$

where

$$\overset{1}{L} \equiv L \otimes 1 , \quad \overset{2}{L} \equiv 1 \otimes L$$

and for the nonlinear Schrödinger equation

$$r(u) = -\frac{1}{u}\begin{pmatrix} 1&0&0&0 \\ 0&0&1&0 \\ 0&1&0&0 \\ 0&0&0&1 \end{pmatrix}.$$

It follows from (10) and (17) that

$$\{\overset{1}{T}(u_1), \overset{2}{T}(u_2)\} = [r(u_1-u_2), \overset{1}{T}(u_1)\overset{2}{T}(u_2)] \tag{18}$$

It is easy to verify that the matrices $K_\pm(u)$ (13) satisfy the identity

$$\begin{aligned} 0 = {} & [r(u_1-u_2)\overset{1}{K}(u_1)\overset{2}{K}(u_2)] \\ & + \overset{1}{K}(u_1)r(u_1+u_2)\overset{2}{K}(u_2) - \overset{2}{K}(u_2)r(u_1+u_2)\overset{1}{K}(u_1). \end{aligned} \tag{19}$$

Using (18) and (19) one then obtains

$$\begin{aligned} \{\overset{1}{\mathcal{T}}(u_1), \overset{2}{\mathcal{T}}(u_2)\} = {} & [r(u_1-u_2), \overset{1}{\mathcal{T}}(u_1)\overset{2}{\mathcal{T}}(u_2)] \\ & + \overset{1}{\mathcal{T}}(u_1)r(u_1+u_2)\overset{2}{\mathcal{T}}(u_2) - \overset{2}{\mathcal{T}}(u_2)r(u_1+u_2)\overset{1}{\mathcal{T}}(u_1) \end{aligned} \tag{20}$$

and, finally, the equality (15).

Note that due to the obvious symmetry $r(-u)=r(u)$ the generating function (11) contains only deformations of <u>pair</u> integrals of motion for the equ. (2). Indeed, it is easy to see that all the impair integrals like the total momentum do not conserve under BCs (7).

Let us stress that the r-matrix approach described above provides a quite general method to construct integrable BCs for any equation possessing L-operator and r-matrix satisfying the fundamental relation (20). To perform this it is sufficient to find matrices $K(u)$ satisfying (19) for given $r(u)$ and to determine then the BCs from (16). Here are several examples.

<u>Sine-Gordon equation</u>

$$\ddot{\varphi} = \varphi'' - \sin\varphi\,, \quad \varphi'_\pm \pm 2\theta_\pm \sin\frac{\varphi_\pm}{2} = 0\,, \quad \bar{\theta}_\pm = \theta_\pm$$

$$H = \frac{1}{2}\int_{x_-}^{x_+} (\dot{\varphi}^2 + \varphi'^2 - \cos\varphi)\,dx - \sum_{\alpha=\pm} 4\theta_\alpha \cos\frac{\varphi_\alpha}{2}$$

Landau-Lifshitz equation [6].

$$\dot{\underline{S}} = \underline{S} \times \underline{S}'' + \underline{S} \times \hat{J}\underline{S}\,, \quad \underline{S}'_\pm \times (\theta_\pm \underline{e}_3 \pm \underline{S}_\pm) = 0\,, \quad (\underline{S}, \underline{S}) = 1$$

$$H = \frac{1}{2}\int_{x_-}^{x_+} \left((\underline{S}', \underline{S}') + (\underline{S}, \hat{J}\underline{S}) \right) dx + \sum_{\alpha=\pm} \theta_\alpha S^3_\alpha\,, \quad \underline{S} = (S^1, S^2, S^3)$$

Toda chain

$$H = \sum_{n=1}^{N} \frac{1}{2} p_n^2 + \sum_{n=1}^{N-1} e^{q_{n+1} - q_n}$$

$$+ \left(\alpha_1 e^{q_1} + \beta_1 e^{2q_1}\right) + \left(\alpha_N e^{-q_N} + \beta_N e^{-2q_N}\right). \tag{21}$$

The Hamiltonian (21) does not seem to fit in available classifications of integrable Toda chains. The correspon-ding action-angle variables can be constructed in the same manner as in the periodic case [7].

Let us turn now to the quantum case. Though a lot of quantum systems with BCs can be solved by means of the coordinate Bethe ansatz [8, 9, 10], the algebraic approach, like in the periodic case [1], seems to be more systematic. The quantum analogue of the Poisson brackets algebra (20) looks as follows

$$R(u_1 - u_2)\overset{1}{\mathcal{T}}(u_1) R(u_1 + u_2 - \eta) \overset{2}{\mathcal{T}}(u_2) = \overset{2}{\mathcal{T}}(u_2) R(u_1 + u_2 - \eta) \overset{1}{\mathcal{T}}(u_1) R(u_1 - u_2) \tag{22}$$

Here $R(u)$ is a solution to the quantum Yang-Baxter equation [1] and η is a constant depending on $R(u)$. The associative algebra generated by the quadratic relations (22) plays the same role in our case as the algebra

$$R(u_1 - u_2) \overset{1}{T}(u_1) \overset{2}{T}(u_2) = \overset{2}{T}(u_2) \overset{1}{T}(u_1) R(u_1 - u_2) \tag{23}$$

for the systems with periodic BCs [1]. In both cases the BCs are characterized by the simplest representations $\mathcal{T}(u) = K(u)$, or $T(u) = K$ of (22) and resp. (23) which do not contain quantum operators. Some examples of such matrices $K(u)$ were found by Cherednik [4] who considered the problem of factorized scattering on the semi-axis.

Theorem 2. Let $T(u)$ satisfy (23) and $\mathcal{T}(u) = K_-(u)$ satisfy (22) for the same $R(u)$, $K_-(u)$ being purely numerical matrix. Then $\mathcal{T}(u)$ given by

$$\mathcal{T}(u) = T(u)K_-(u)T^{-1}(-u+\eta) \tag{24}$$

also satisfies (22).

Consider for instant the spin-1/2 XXZ ferromagnet [5]. Let

$$T(u) = L_N(u) \dots L_1(u)$$

$$L_n(u) = \begin{pmatrix} \sinh u \cosh\frac{\eta}{2} + \cosh u \sinh\frac{\eta}{2}\sigma_n^3, & \sinh\frac{\eta}{2}\cosh\frac{\eta}{2}(\sigma_n^1 - i\sigma_n^2) \\ \sinh\frac{\eta}{2}\cosh\frac{\eta}{2}(\sigma_n^1 + i\sigma_n^2), & \sinh u \cosh\frac{\eta}{2} - \cosh u \sinh\frac{\eta}{2}\cdot\sigma_n^3 \end{pmatrix} \tag{25}$$

$$R(u) = \begin{pmatrix} a & 0 & 0 & 0 \\ 0 & b & c & 0 \\ 0 & c & b & 0 \\ 0 & 0 & 0 & a \end{pmatrix} \quad \begin{array}{l} a(u) = \sinh(u+\eta) \\ b(u) = \sinh u \\ c(u) = \sinh\eta \end{array}$$

$$K_\pm(u) = \coth\xi_\pm + \sigma_3 \coth\left(u \pm \frac{\eta}{2}\right). \tag{26}$$

Taking $T(u)$ and $K_\pm(u)$ from (25) and (26) construct the matrix $\mathcal{T}(u)$ (24) satisfying (22) and the operator

$$\tau(u) = \operatorname{tr} K_+(u)\mathcal{T}(u) \tag{27}$$

One verifies easily that $\tau(u)$ is a commuting family $[\tau(u_1), \tau(u_2)] = 0$ and can be considered thus as a generating functions of commuting integrals. In particular, $\tau(u)$ contains the Hamiltonian

$$H = \sum_{n=1}^{N-1}\left(\sigma_n^1\sigma_{n+1}^1 + \sigma_n^2\sigma_{n+1}^2 + \cosh\eta\,\sigma_n^3\sigma_{n+1}^3\right) + \sinh\eta\left(\coth\xi_+\sigma_N^3 + \coth\xi_-\sigma_1^3\right)$$

The spectrum of $\tau(u)$ can be found by means of the usual algebraic Bethe ansatz technique [1].

In conclusion, let us note that a lot of problems connected with the algebra (22) remains still unsolved. For instance, it is unknown if $\tau(u)$ given by (27) commutes in general case and what are invariant (Casimir operators) of the algebra (22).

References

1. L.D.Faddeev, in Les Houches Summer School Proc. (1982), Eds. J.B.Zuber, R.Stora, Elsevier Sci.Publ. (1984) 561.
2. R.K.Bullough, P.J.Caudrey (Eds.). Solitions, Topics in Current Physics, 17, Springer, Berlin (1980).
3. E.K.Sklyanin, J.Soviet Math., 19 (1982) 1546.
4. I.V.Cherednik, Teor.Mat.Fiz., 61 (1984) 35.
5. P.P.Kulish, E.K.Sklyanin, Phys.Lett. 70A (1979) 461.
6. E.K.Sklyanin, LOMI preprint, E-3-79 (1979).
7. E.K.Sklyanin, in Non-Linear equations in classical and quantum field theory, ed. by N.Sanchez (Lect.Notes in Phys.) 226 (1985) 196.
8. M.Gaudin, Phys.Rev. A4 (1971) 386.
9. F.Woynarovich, Phys.Lett. A108 (1985) 401.
10. H.Schultz, J.Phys. C18 (1985) 581.

Some personal remarks on nonstandard analysis in probability theory and mathematical physics

by

Sergio Albeverio
Mathematisches Institut
Ruhr-Universität Bochum
D-4630 Bochum 1, (FRG) and
BiBoS-Research Centre, Volkswagenstiftung

1. Introduction

Professor G. Jona-Lasinio asked me to give some personal remarks on the uses of nonstandard analysis in probability theory and mathematical physics. I shall try to do so providing at the same time a very short exposition of the basic ideas of nonstandard analysis.

As well known, nonstandard analysis is in a sense a justification of the classical "method of infinitesimals" (...Archimedes ... Leibniz ... Euler ...) and was founded at the end of the fifties by A. Robinson.[1)] It has at least three basic aspects: on one hand it is a new mathematical technique, to be added to other ones, on the other hand is a new way of looking at the continuum and constructing (mathematical, physical ...) models, in some sense thus a new way of thinking, "plus conforme à l'art d'inventer" (G. Leibniz), thirdly it is a proper mathematical way to justify the kind of heuristic computations that physicists call "formal" and mathematicians call "informal".

Here is a typical example. Problem: give meaning to $-\Delta + \lambda\delta(x)$, λ a constant, as an operator in $L^2(\mathbb{R}^d, dx)$, and study its spectrum, eigenfunctions, scattering theory etc. A typical mathematician's reaction would be: $\lambda\delta(x)$ is not defined as an operator acting in $L^2(\mathbb{R}^d, dx)$, so $-\Delta + \lambda\delta(x)$ can not be given a meaning as sum of operators. He might try then the method of quadratic forms, but soon he will realize that it only works for $d = 1$ (cfr. e.g. [3]) . Eventually, he might define $-\Delta + \lambda\delta(x)$ indirectly as suitable limit of regularized objects (cfr. e.g. [4]). The typical physicist reaction is completely different, more active, "algebraic" and "computational": he will start computing "heuristically" objects related to $-\Delta+\lambda\delta(x)$. E.g. he might recall that if $\lambda\delta(x)$ were a nice potential V he could use Neumann series to compute the resolvent

$$(-\Delta+V-k^2)^{-1} = G_k + \sum_{j=1}^{\infty} (-1)^j (G_k V)^j G_k, \text{ with } G_k \equiv (-\Delta-k^2)^{-1}$$

(say for $\mathrm{Im}k^2$ sufficiently large). Doing the same for $V(x) = \lambda\delta(x)$ he will find, after (heuristic) algebraic manipulations (not justified by "standard mathematics") that the kernel $(-\Delta+\lambda\delta(x)-k^2)^{-1}(x,y)$ is given by $G_k(x,y) - G_k(x)\,(\frac{1}{\lambda}+G_k(0))^{-1}G_k(y)$. Of course $G_k(0)$ is "infinite positive" for $d \geq 2$. To compensate this infinity he will then try to take λ "infinitesimal negative", independent of k, so that $\frac{1}{\lambda}+G_k(0)$ gets to be finite, real, different from zero. He will at this point realize that λ independent of k is only possible for d = 2,3, so his conclusion will be that $-\Delta+\lambda\delta(x)$ can be given nontrivial meaning in $L^2(\mathbb{R}^d,dx)$ for d = 2,3 by a "proper renormalization" of λ. The same ideas he will then go on to apply to operators of the form $-\Delta+\sum_{y\in Y}\lambda_y\delta_y(x)$ in $L^2(\mathbb{R}^d,dx)$ for Y an arbitrary set ("point interactions" at Y) .[2)] This "don't worry principle" turns out to be the correct attitude: by non standard analysis, namely, the whole procedure above can be given direct rigorous meaning. But before going on, let me give a ten minutes exposition of what I understand by nonstandard analysis.

2. What is nonstandard analysis?

It is analysis based on a field of numbers $^*\mathbb{R}$ larger than the usual field of real numbers $\mathbb{R}$, in fact containing infinitesimal and infinite numbers. Here is a short way to get at $^*\mathbb{R}$. Let m be a 0-1-valued <u>finitely</u> additive probability measure ($m(\mathbb{N}) = 1$) defined on all subsets of $\mathbb{N}$ and giving probability zero to all finite subsets of $\mathbb{N}$.[3)] Then $(\Omega,\mathcal{A},m)$ is a finitely additive probability space, with $\Omega = \mathbb{N}$, $\mathcal{A}$ = power set of Ω. Arbitrary maps $f: \mathbb{N}\to\mathbb{R}$ can be thought as real-valued random variables (on (Ω,m)). We call such random variables f,g, <u>equivalent</u> iff f = g m-a.e. To f we might consider in the usual way the equivalence class <f> with representative f (all random variables equal to f modulo zero measure sets). Our basic field of nonstandard analysis $*\mathbb{R}$ is going to be the set of all equivalence classes <f>, with certain algebraic operations and an order defined on them. First we realize that $\mathbb{R}$ can be thoughtas embedded into $*\mathbb{R}$ by identifying $r \in \mathbb{R}$ with the equivalence class <f> of the "constant" random variable f s.t. $f(i) = r \; \forall i \in \mathbb{N}$. Addition (+) and multiplication (·) are defined in $*\mathbb{R}$ in the natural way as addition and multiplication of (equivalence classes modulo zero measure sets of) random variables. 0 and 1 (as embedded in $*\mathbb{R}$) are then the neutral elements for addition resp. multiplication in $*\mathbb{R}$. $*\mathbb{R}$ is obviously a field, moreover it is an ordered one since for any random variables f,g s.t. $\langle f\rangle \neq \langle g\rangle$ we have either $m(f<g) = 0$, in which case we write $\langle f\rangle < \langle g\rangle$, or $m(f>g) = 0$, in which case we write $\langle g\rangle < \langle f\rangle$ (at this point the fact that we have a 0-1-valued measure is decisive). $*\mathbb{R}$ is called the ordered field of "hyperreals". It is immediately clear that $*\mathbb{R}$ contains infinite numbers, like e.g. $\eta \equiv \langle f\rangle$, with f the random variable $f(i) = i \; \forall \; i \in \mathbb{N}$. In fact obviously $r < \langle f\rangle \; \forall \; r \in \mathbb{R}$, since $m(r < f) = 1 - m(r \geq f) = 1$, $m(r \geq f)$ being m evaluated on the finite set $(\{1,2,\ldots [r]\})$ (where [r] is the largest integer $\leqq r$), hence zero. By the fact that $*\mathbb{R}$ is a field we have that e.g. η^2, $\eta + 10$, 3η, e^η, $-\pi\eta$ are infinite numbers: by definition we call infinite numbers all elements s of $*\mathbb{R}$ which are such that $|s| > r \; \forall \; r \in \mathbb{R}$, (where $|s|$ is of course defined as absolute value in the usual way). Inverses of infinite numbers are by definition infinitesimals (also 0 is often called infinitesimal). Thus e.g. $\eta^{-1} = \langle f\rangle$ with $f(i) = \frac{1}{i}$, $\forall \; i \in \mathbb{N}$, is an infinitesimal. Infinitesimals are in absolute value smaller than any strictly positive real number. Of course for any $r \in \mathbb{R}$, $r + \varepsilon$, with ε infinitesimal is also in $*\mathbb{R}$. It is easy to show that $*\mathbb{R}$ consists precisely of

a) all numbers of the form $r + \varepsilon$ with r varying in $\mathbb{R}$ and ε varying in the infinitesimals: these are the "finite numbers in $*\mathbb{R}$",

b) all numbers > 0 (resp. < 0) which are infinite: "positive (resp. negative) infinite numbers".

One calls numbers $x,y \in {}^*\mathbb{R}$ "infinitely near" iff they differ by some infinitesimal and one writes $x \approx y$. (This concept is a suitable substitute for topology in ${}^*\mathbb{R}$). Given x, all numbers infinitely near x form a "monad around x" (in french: "halo"). Numbers of ${}^*\mathbb{R}$ which are also in $\mathbb{R}$ are called standard numbers, the other ones are called nonstandard (e.g. infinitesimals $\neq 0$ are nonstandard, infinite numbers are non-standards, but also e.g. $\frac{1}{\pi} + \varepsilon$ with $\varepsilon \neq 0$ infinitesimal is nonstandard). Given any finite $x \in {}^*\mathbb{R}$ there is a uniquely defined $r \in \mathbb{R}$ s.t. $r \approx x$. r is usually denoted by st x (or ${}^{o}x$) and is called the standard part of x: we have st $x = \sup \{t \in \mathbb{R} \mid t \leq x\}$.

To develop calculus over ${}^*\mathbb{R}$ (or complex analysis over ${}^*\mathbb{C} \equiv {}^*\mathbb{R} + i{}^*\mathbb{R}$) we also need e.g. to extend functions f over $\mathbb{R}$ to functions *f over ${}^*\mathbb{R}$. A natural definition is $*f(x) = y$, with $y = f(x)$ m-a.e. Similarly we associate to subsets $A \subset \mathbb{R}$ naturally subsets ${}^*A \subset {}^*\mathbb{R}$ by ${}^*A = \{x \in {}^*\mathbb{R} \mid m\{i \mid x(i) \in A\} = 1\}$. *f, *A are called *-standard quantities. They "come" directly from quantities defined in $\mathbb{R}$. Are there functions on ${}^*\mathbb{R}$ or subsets of ${}^*\mathbb{R}$ which are in a sense natural and are not *-standard? Intuitively it will be so, e.g. the set I of all infinitesimals is not * of any set $A \subset \mathbb{R}$ (what could A possibly be?), hence its characteristic function will not be *-standard. What about a set of the type $B \equiv \{x \in {}^*\mathbb{R} \mid a \leq x \leq b\}$, with $a,b \in {}^*\mathbb{R} - \mathbb{R}$? Again intuitively it is not of the form *A for any $A \subset \mathbb{R}$.

For elementary analysis, these are all types of sets and functions one encounters. Subsets of ${}^*\mathbb{R}$ like B which are of the form $B = \{\langle f\rangle \in {}^*\mathbb{R} \mid f_i \in B_i\}$, with B_i subsets of $\mathbb{R}$ for m.a.e. i (in above example: $B_i = \{y \in \mathbb{R} \mid a(i) \leq y \leq b(i)\}$ are called "internal". *-standard sets can be looked upon as special internal sets $B = (B_i)$ with all components B_i equal. Sets like I which cannot be written in the above form are called external. Internal sets and functions are exactly the ones for which "all statements"[4)] valid for standard sets and functions "can be transferred" without loosing validity, e.g. every internal subset of ${}^*\mathbb{R}$ which has on upper bound has a lowest upper bound (this is not true for I, but I is precisely not internal). In a sense internal quantities are basic quantities of nonstandard analysis, like open sets and continuous functions are for topology or measurable functions are for measure theory. For extensions of higher analysis (functional analysis, measure theory, probability theory ...) one also needs objects like spaces of functions, spaces of measures, dual spaces etc. Is there a way to put stars on them getting * -standard entities , and perhaps again introduce the distinction between internal and external entities? The answer is yes. To make it precise it would take however more time than I have at disposal (this is in a sense the part of the formalism of nonstandard analysis where some care and some practice are needed, to get an intuitive grasp on the objects involved although precise definitions can be given quite quickly, see e.g. [5], [6]). The definitions are such that internal entities are elements of *-standard entities, [5], [6] . An important property of internal entities is given by the following saturation theorem, which was the starting point for a real breakthrough in the application of nonstandard analysis in probability theory (and measure theory):

<u>Theorem.</u>

If A_n, $n \in \mathbb{N}$, are internal entities, $A_{n+1} \subset A_n$ and $A_n \neq \emptyset$ $\forall\, n \in \mathbb{N}$, then $\bigcap_{n\in\mathbb{N}} A_n \neq \emptyset$.

An immediate corollary is that internal subsets of $\mathbb{R}$ are either finite or uncountable. Above theorem indicates that internal entities are likely to play a role somewhat similar to compact sets in standard analysis. As an elementary illustration for $\bigcap_{n\in\mathbb{N}} A_n \neq \emptyset$, let us take $A_n = {}^*(0, \frac{1}{n})$ (which are *-standard hence, in particular, internal subsets of ${}^*\mathbb{R}$). Without * we would have a void intersection but with the star of course we have $\bigcap_n A_n$ = positive infinitesimals.

<u>3. Some applications (to probability theory / mathematical physics)</u>

The consequences of the above saturation principle are amazing, as first realized by P. Loeb in 1975: From any internal (non standard) finitely additive probability space $(X, \mathcal{A}, \nu)$ [6] one can obtain a σ-additive probability space $(X, \sigma(\mathcal{A}), L(\nu))$, with $\sigma(\mathcal{A})$ the σ-algebra of subsets of X generated by $\mathcal{A}$ and L(ν) defined as the σ-additive probability measure on $(X,\sigma(\mathcal{A}))$ defined as extension of the finitely additive [0,1]-valued measure $^{o}\nu$ given by $^{o}\nu(A) \equiv \text{st}\ \nu(A)\ \forall\ A \in \mathcal{A}$. L(ν) is called Loeb measure to ν. For any $B \in \sigma(\mathcal{A})$, L(ν)(B) is approximated by the ν measure of elements in $\mathcal{A}$ in the sense that $\exists\ A_B \in \mathcal{A}$ s.t. $L(\nu)(B) \approx \nu(A_B)$. The proof is an immediate application of Caratheodory's extension theorem, the saturation property providing trivially the required continuity. What does this say? $(X,\mathcal{A},\nu)$ behaves like a "finite probability space" ("discrete probability") (good combinatorial properties), whereas $(X,\sigma(\mathcal{A}), L(\nu))$ instead can be looked upon as an abstract probability space in the usual sense, with a good integration and measure theory. Hence one has a reduction of abstract measure theory to combinatorial finite probability. A typical example is the following: let $X = T \equiv \{0, \Delta t, 2\Delta t, \ldots, \eta!\Delta t\}$, with η an infinite positive number, $\Delta t \equiv \frac{1}{\eta!}$, a positive infinitesimal [7]. T is a so called hyperfinite set, an "hyperdiscrete" (infinitesimal spacing!) model of the standard internal set [0,1].

Let ν be counting measure on T i.e. $\nu(A) = |A|/|X| = |A|\ (\eta!+1)^{-1}\ \forall A \subset T$, A internal. Then the Loeb measure L(ν), as given by above theorem, is nothing but a realization of Lebesgue measure on [0,1], in the sense that $|C| = L(\nu)\ (\text{st}^{-1}(C))$ for any Borel subset C of [0,1], with |C| its Lebesgue measure. $\text{st}^{-1}(C)$ is the subset of T mapped by the standard part map into [0,1].
Similarly the Lebesgue integral of a function f can be expresses as integral with respect to Loeb-measure of an associated nonstandard counterpart of f (and the latter integral, by definition of Loeb measure, is in fact infinitely near a sum over T: thus we get a representation of integrals as "sums with infinitesimal steps"). A probability space like (T,L(ν)) is said to be hyperfinite.

A second important example is given by "Brownian motion as infinitesimal random walk". Let us consider a random walk B going in infinitesimal time Δt left or right with equal probability on the hyperreal line $^{*}\mathbb{R}$ by steps of length $\sqrt{\Delta t}$ i.e.

$$B(\omega,\tilde{t}) \equiv \sum_{\substack{s=0 \\ s\in T}}^{\tilde{t}-\Delta t} \omega(s)\ \sqrt{\Delta t}$$

with $\tilde{t}\in T, \omega(s) = \pm 1$ with equal probability, $\omega\in\Omega = \{-1,+1\}^{|T|}$. The underlying internal probability space is here $X = \Omega$, $\nu = \bigotimes_{\tilde{t}\in T} \nu_{\tilde{t}}$, with $\nu_{\tilde{t}}$ the symmetric Bernouilli distribution $B_2(\frac{1}{2})$. B is called the Anderson's hyperfinite random walk: Anderson proved (in '76) that "its standard part is Brownian motion", more precisely st $B(\omega,\tilde{t}) = b(\omega,t)$, $t = \text{st}\ \tilde{t}$, with b a realization of Brownian motion. [8] $L(\nu)$ is a realization of Wiener measure. This result leads to many applications. E.g.

1) Local times: (Perkins '81, '83 [7]; Stoll '84 [8]).

To $t\in[0,1]$ there corresponds $\tilde{t}\in T$ and to $x\in\mathbb{R}^d$ we let correspond $\tilde{x}\in\sqrt{\Delta t}\ {}^*\mathbb{Z}^d$ s.t. st $\tilde{t} = t$, st $\tilde{x} = x$. Let

$$L(\tilde{t},\tilde{x}) \equiv \sum_{\tilde{s}<\tilde{t},\ \tilde{s}\in T} \delta_{\tilde{x},B(\tilde{s})} (\Delta t)^{(1-\frac{d}{2})},$$

where B is now the d-dimensional hyperfinite random walk whose standard part is d-dimensional Brownian motion b. Then st $L(\tilde{t},\tilde{x}) = \ell(t,x)$ is the "measure local time of Brownian motion". For d = 1 it has a density with respect to Lebesgue measure which is continuous in t, x and one has $\ell(t,x) = \int_0^t \delta(x-b(s))ds$. As an example of applications, Levy's formula (d=1) has been proven for the first time by non standard methods [7]: $\lim_{\delta\downarrow 0} \sup_{\substack{t\leq 1 \\ x\in\mathbb{R}}} |m(t,x,\delta)\ \delta^{-1/2} - 2\sqrt{\frac{2}{\pi}}\ \ell(t,x)| = 0$, a.s.

with $m(t,x,\delta)$ the Lebesgue measure of the set of points within $\delta/2$ of the level set $\{s\leq t\ |\ b(s) = x\}$.

2) Polymer measures ([5], [8])

A construction of measures on $(C[0,t])^2$ of the type

$$d\nu_2(b_1,b_2) \equiv Z^{-1}\exp(-\lambda\int_0^{t_1}\int_0^{t_2} \delta(b_1(s)-b_2(s'))ds\,ds')dP(b_1)dP(b_2),$$

with $(b_i,P(b_i))$, i=1,2 independent Brownian motions, Z a normalization, has been given for $d\leq 5$. For d = 4,5, λ infinitesimal negative is required (for non triviality of the construction). The method uses a non standard ("hyperfinite") Feynman-Kac formula, together with a construction of stochastic Hamiltonians of the form discussed at the end of Sect. 1 (realized e.g. as $-\Delta + \lambda L(\tilde{t},\tilde{x})$ in a hyperfinite setting, with Δ the Laplacian on $\sqrt{\Delta t}\ {}^*\mathbb{Z}^d$). From limits of densities in measures of the type ν_2 one can get polymer measures of the type $d\nu_1(b) = Z^{-1}\exp(-\lambda \int_0^{t_1}\int_0^{t_2}\delta(b(s)-b(s')dsds')dP(b)$, discussed in Yor's lecture at this conference. (For d=2 a non standard construction has been given by A. Stoll [8], who has also a partial non standard version of well known Westwater's result for d=3). Measures of the types ν_i, i=1,2 also occurr in Symanzik's approach to the φ_d^4-model and one gets partial results on non triviality for $\lambda\varphi_4^4$ with λ infinitesimal negative [5].

3) On a more general line, non standard methods give the possibility to provide (infinitely) <u>discrete versions</u> of the theory of continuous Markov processes, the associated potential theory and the theory of Dirichlet spaces, replacing abstract "compactifications" by an hyperfinite setting [5]. Moreover certain global problems in the theory of random fields can be handled most conveniently in such a setting [5], [18].

4) Also non standard methods give a natural tool for <u>stochastic analysis.</u> E.g. a stochastic integral $\int_0^t f(\omega,s)db(\omega,s)$ is realized pathwise as

$\sum_{s<t, s\in T} F(\omega,s)\ \Delta B(\omega,s)$ with $\Delta B(\omega,s) \equiv \omega(s)\ \sqrt{\Delta t}$. By this, a new type of stochastic differential equation was solved strongly on a Loeb space, (e.g. $X(t) = \int_0^t g(s,X(s))db(s)$, with g bounded progressively measurable, a.s. for each t). Also one has "<u>universality properties</u>" : α) weak solutions of stochastic equations are automatically strong solutions on a hyperfinite Loeb probability space; β) for any (polish-valued) stochastic process X there exists a random walk X with same state space, on a hyperfinite probability space, with the same distribution as X. Moreover any two random walks on a hyperfinite probability space having the same finite dimensional distributions are related by an automorphism and this in an "adapted way", see [5], [9], [10].

5) Another advantage of the non standard approach is that there is no difference" between finite dimensional and infinite dimensional stochastic calculus, e.g. replacing, in the formula for B, $\omega\sqrt{\Delta t}$ by $\omega_j e_j \sqrt{\Delta t}$ and summing over j = 1,...d, with e_j the unit vector in the j-th direction, yields an hyperfinite realization of Brownian motion in d-dimensions, whether d is finite or not. For d positive infinite, B is an internal object which realizes the standard Brownian motion associated with some standard separable Hilbert space. Also Gaussian measures on Hilbert spaces can be realized as $(2\pi)^{-d/2}(\text{Det } A)^{1/2} \exp(-\frac{1}{2}(x,Ax))^* dx$, for d a positive infinite integer and A a matrix in ${}^*\mathbb{R}^d$ (cfr. [5]). Unfortunately very little has been done as of yet to exploit these facts and justify heuristic computations of indices (some motivations can be gained from the lecture of Bismut at this Meeting). On the positive side, let us mention an application [5] of this approach to the realization of (continuum) quantum fields as Loeb measures of Euclidean lattice models with infinitesimal lattice spacing δ. The lattice field $\Phi_\delta(x)$, $x \in \delta\,{}^*\mathbb{Z}^d$ is defined pointwise and realizes the continuum field $\Phi(x)$, $x \in {}^*\mathbb{R}^d$, which itself is not defined pointwise. (The non standard construction gives an existence result for (standard) quantum fields, the question about triviality / non triviality remains open for d = 4; here better general techniques on the "coming back" from a non standard quantity to standard ones would be desirable). Also we mention Kessler's work on global Markov fields [18].

4. Concluding remarks

Non standard analysis gives, in our opinion, two types of better models of the continuum, (cfr. [19]): a) a richer one: ${}^*\mathbb{R}^d$, which permits e.g. a1) to find regular models of singular interactions (like $-\Delta + \lambda\delta(x)$ realized as $-\Delta + \lambda_\varepsilon \delta_\varepsilon(x)$ with $\delta_\varepsilon \in C^\infty$ in the non standard sense, ε infinitesimal, λ_ε negative infinitesimal; or in the study of singular Sturm-Liouville problems, see [5]); a2) to find new classes of solutions (e.g. for nonlinear partial differential equations, like in the recent exciting work of Arkeryd on Boltzmann equation and of Hurd on the BBGKY hierarchy, see [5], [12]); b) a "fine discrete" one: $\varepsilon\,{}^*\mathbb{Z}^d$ with ε infinitesimal, like in the examples of infinitesimal random walks we discussed above.

Are these models closer to the needs of physics and mathematics?

In my opinion the answer is yes. Here are some reasons:

a) Explicitely solvable but ill defined models can be given a direct meaning (like for the point interactions or else computations "heuristically correct but involving infinite quantities", arising naturally in physics).

b) Also it permits to think naturally of dynamical systems with infinitesimal (or infinite) forces and parameters, like e.g. $\varepsilon\ddot{x} + (x^2-1)\dot{x} + a - 1 = 0$, with ε and a-1 infinitesimals (this is "modified von der Pool equation" and gives rise to the "canards", see e.g. [5], [13]: this is just an instance of a new concept of "natural problems" coming into play).

c) Replaces the operation of taking limits by the study of ideal elements (pretty much like the theory of generalized functions: one advantage of the ideal elements of non standard analysis, as compared with distribution is that they have good algebraic properties).

e) It gives a natural tool for probability theory, regaining the combinatorial core of it (a point stressed particularly by Ed Nelson [3], [14]). It leads to such results as the above universality, substituting "universal" hyperfinite models to complicated often ad hoc topological completions. As areas in which more applications could be expected let us mention e.g. the study of fractal phenomena (see [15]), stochastic analysis in infinite dimensions (see [5]) differential geometry, nonlinear p.d.e. Let me close with a word about difficulties encountered when starting to work with non standard methods. There are educational/historical/psychological problems (infinitesimals used to be excluded from rigorous mathematics). But more substantially the ways of "coming back" from result about non standard objects to results on standard ones should be improved resp. better ways should be found to interpret implications of non standard deductions. Certainly Loeb's construction points in the good direction (can an analogue of regularity results for p.d.e. be found for Loeb solutions of equations?). Personally I do think that many hard core applications have yet to come. Not so many probabilists, analysts and physicists have as of yet worked seriously with these methods, for a long time the main users were mathematical logicians (because of the historical origins of the techniques). Although foundational questions are still very interesting (cfr. [14], [16]), the theory in itself has already both "constructive" [5] and "axiomatic" [3], [17]) versions complete in themselves and ready for applications. It is not difficult to learn and apply, and a strong control on rigour is " inbuilt". Applications can often become rather technical, despite the fact that the initial steps and the basic ideas are extremely simple, intuitive and fascinating. Fascination plays a big role in the subject, in a sense one should both be fascinated and resist easy temptation (to avoid superficial excitement and go deeper into the subject). Problems involving many scales, like those arising in several problems of mathematical physics are in a sense ideal for a non standard treatment, the theory providing a sort of "universal receptacle" for all kinds of limits .

Acknowledgements. It is a pleasure to thank Prof. G. Jona-Lasinio and the Organizers for the very kind invitation. My talk is based on a long standing collaboration with Profs. J.E. Fenstad, R. Høegh-Krohn and T. Lindstrøm, and I am very grateful to them for all they taught me.

Footnotes

1) See e.g. the historical remarks in [1], and references in [2]

2) For rigorous discussion of such operators and many applications see [4]

3) The existence of at least one such m (but in fact of a continuum of such) is ensured by Zorn's Lemma (actually less than this is needed). The measures are in 1-1 correspondence with ultrafilters extending the filter of cofinite subsets of $\mathbb{N}$, see e.g. [5], [6]

4) Formalizable by 1. order logic (i.e. with quantifiers only on numerical variables).

5) E.g. the above B is an element of the *standard quantity $^*P(\mathbb{R})$, with $P(\mathbb{R})$ the family of all subsets of $\mathbb{R}$.

6) A natural object to consider in non standard analysis, as apposite to σ-additive probability spaces.

7) Of course the choice $\Delta t = 1/\eta\,!$ is just to make sure that T has an " hyperinteger " number of points, namely $\eta! \equiv 1.2....\eta$, $\eta \in {}^*\mathbb{N} - \mathbb{N}$ (η "behaves" like an integer but is infinite, like e.g. $\eta = \langle f \rangle$, $f(i) = i \;\forall\, i \in \mathbb{N}$).

8) A Brownian motion with continuous paths can also be constructed, using simple non standard tools, cfr. e.g. [5], [6].

9) Known models can be recovered (for d = 2,3 see [5]; for d = 4 see [11]).

References

[1] D. Laugwitz, Zahlen und Kontinuum, Bibl. Inst., Zürich (1986)

[2] S. Albeverio, An introduction to nonstandard analysis and applications to quantum theory, Proc. Udine CISM Conference on "Stochastic Control and Quantum Mechanics", Ed. S. Diner et al. (1986)

[3] E. Nelson, Internal set theory, Bull. AMS 83, 1165-1198 (1977)

[4] S. Albeverio, F. Gesztesy, R. Høegh-Krohn, H. Holden, Solvable models in quantum mechanics, Springer Verlag (1987)

[5] S. Albeverio, J.E. Fenstad, R. Høegh-Krohn, T. Lindstrøm, Nonstandard methods in stochastic analysis and mathematical physics, Academic Press, New York (1986)

[6] K.D. Stroyan, J.M. Bayod, Foundations of infinitesimal stochastic analysis, North-Holland, Amsterdam (1986)

[7] E. Perkins, Stochastic processes and nonstandard analysis, in A.E. Hurd, Ed., "Nonstandard Analysis - Recent Developmets", Lect. Notes Maths. 983, Springer, Berlin (1983)

[8] A. Stoll, Invariance principle for Brownian intersection local time and polymer measures, IMA Prepr., Minnesota (1986)

[9] H.J. Keisler, An infinitesimal approach to stochastic analysis, Mem. AMS 297 (1984)

[10] N. Cutland, T. Lindstrøm, Edts., Proceedings Hull Conference on Nonstandard Analysis, Cambridge Univ. Press (1987)

[11] S. Albeverio, R. Høegh-Krohn, Euclidean Markov fields and relativistic quantum fields from stochastic partial differential equations in four dimensions, Phys. Letts. B 177, 175-179 (1986)

[12] L. Arkeryd, On the Boltzmann equation in unbounded space far from equilibrium, at the limit of zero mean free path, Comm. Math. Phys. 105, 205-219 (1986)

[13] A.K. Zvonkin, M.A. Shubin, Nonstandard analysis and singular perturbations or ordinary differential equations, Russ. Math. Surv. 39, 69-131 (1984)

[14] E. Nelson, Radically elementary probability theory, Princeton (1984)

[15] T. Lindstrøm, Nonstandard energy forms and diffusions on manifolds and fractals, Proc. Ascona Conf. "Stochastic Processes in Classical and Quantum Systems", Edts. S. Albeverio, G. Casati, D. Merlini, Lect. Notes Phys., Springer (1986)

[16] E. Nelson, Predicative Arithmetic, Princeton Preprint (1985)

[17] N.J. Cutland, Nonstandard measure theory and its applications, Bull. London Math. Soc. 15, 529-589 (1983)

[18] C. Kessler, Example of extremal lattice fields without the global Markov property, Publ. RIMS 27, 877-888 (1985)

[19] J.E. Fenstad, The discrete and the continuous in mathematics and the natural sciences, Proc. Roma (1986)

POINT PROCESSES AND QUANTUM PHYSICS : SOME RECENT DEVELOPMENTS AND RESULTS

Ph. Combe*, M. Sirugue, M. Sirugue-Collin**
Centre de Physique Théorique
C.N.R.S. - Marseille
and
* Université d'Aix-Marseille II, Marseille
** Université de Provence, Marseille

ABSTRACT

In the recent years, there has been a renewal of interest for the use of Point Processes in Quantum Physics. We review some works concerning various domains such as quantum time evolution of spinless particles, models of emission and absorption of quanta, quantum time evolution and stochastic mechanics of spin 1/2 particles.

1. INTRODUCTION

Stochastic aspects of quantum physics have been of major importance during the past decades. The very basic Feynman's idea (1) which describes quantum evolutions via path integrals can find a proper mathematical status within the theory of stochastic processes. Already long time ago diffusion processes were recognized as a very useful tool for imaginary time. They are also important for Nelson stochastic mechanics.

Besides the fact that they appeared from time to time in various domains (see eg. ref. 2,3) it was only recently recognized that point processes are extremely relevant in many problems such as, for example, real time description of spinless non relativistic particles quantum evolutions or spin systems. They give an alternative and in some sense complementary way for a better understanding of quantum physics. The perturbed evolution of a free evolution by gentle, possibly velocity dependent, potential can be described in terms of the expectation with respect to a Poisson process on a group. This applies, via Weyl quantization, to usual quantum mechanics but also to Fermi and spin systems (see eg. ref. 4,5). The method extends an original remark by Maslov and Chebotarev (6,7).

As an example we can consider the evolution of Wigner functions

$W_t(q,p)$ (8) which are defined on phase space ($q,p\in R^{2N}$) in the framework of Weyl quantization by

$$< \psi_t\, Q(f)\psi_t> = \int dpdq\, f(q,p)\ W_t\,(q,p)$$

where the left hand side is the mean value in the state $|\psi_t>$ of the quantum observable Q associated with the classical one $f(q,p)$.

For a hamiltonian $H(q,p) = (1/2m)\,p^2 + 1/2\, m\omega^2\, q^2 + V(q)$

$W_t(q,p)$ is such that

$$W_t(q,p)=\exp\,[\,2t\mu(R^{2N})/\hbar]\ \mathbb{E}\,\{e^{iS(t)/\hbar}\, W_o(X_{q,p,o}(t)Y_{q,p,o}(t))\}$$

where the processes X and Y are defined by the 2N stochastic integral equations

$$X^i_{q,p,o}(t)=q^i-1/m_i\int_0^t Y^i_{q,p,o}(\tau)d\tau + \int_0^t\int_{R^{2N+1}} q'^{\,i}\nu(d\tau,dq',dp',du')$$

$$Y^i_{q,\,p,o}(t) = p_i+m_i\omega_i^{\,2}\int_0^t X^i_{q,p,o}\ (\tau)d\tau + \int_0^t\int_{R^{2N+1}} p'^{\,i}\nu(d\tau,\,dq',\,dp',du')$$

and ν is a Poisson measure in 2N+1 dimensions tied to the potential. S is a functional of the processes reminiscent of an action.

With the same methods it is possible to deal with evolutions of quantum observables (9) and get deep insights into field theories (10,11,12).

It is important to emphasize that, using point processes, new fields of interest arise either in mathematics or in theoretical physics as it appears when considering the links between classical and quantum physics (13, 14, 15, 16). Some of these results have been developpped in previous conferences so I shall elaborate now on some more recent works.

II. DIRAC PARTICLES :

Many attempts have been made to give a probabilistic treatment of spin 1/2 particles (see eg 17, 18, 19, 20, 21, 22). Already in Feynman and Hibbs's book there is a remark that there exists a natural path integral representation of the time evolution of the wave function of a Dirac particle in 1+1 space-time dimensions. Typical paths are those of a particle moving on the light cone and reversing its velocity at Poisson distributed random times.

Further progress was made these past years (21, 23, 24, 25, 26,27,28,29). An essential tool is to put on the same footing space and spin variables along the line which was already used in the probabilistic treatment of generalized Pauli equation (30,31).

In order to be more precise, let me describe a probabilistic treatment of Dirac equation given in reference 28.

Dirac equation in 1+1 dimension rewrites :

$$(i\hbar\, \partial_t - qA_0)\, \Phi(x,t) - \sigma_3\, c(\hbar/i\, \partial_x - qA_1)\Phi(x,t) \qquad (2.1)$$
$$-mc^2\sigma_1\, \Phi(x,t) = 0$$

where σ_1, σ_3 are the usual Pauli matrices, m is the mass of the particle, q its charge, c the light velocity and $\hbar$ the Planck constant divided by 2π. $A_i(x,t)$, $i = 0,1$, are the two components of the electromagnetic potential.

We are concerned with solutions of equation (2.1) satisfying the initial condition

$$\lim_{t\to 0_+} \Phi(x,\sigma,t) = \varphi_0(x,\sigma) \qquad (2.2)$$

Introducing a new variable $y\in R$ and a change in time (chosing T > 0) we introduce the new function

$$\Psi(x,y,\sigma,t) = \exp\left[-\frac{mc^2}{\hbar}(T-t)\right] e^{-iy}\, \Phi(x,\sigma,T-t)$$

which satisfies the equivalent equation, $t \leqslant T$:

$$0=\frac{\partial}{\partial t}\Psi(x,y,\sigma,t) + \frac{mc^2}{\hbar}\left[\psi(x, y+\pi/2, -\sigma,t) - \psi(x,y,\sigma,t)\right] \qquad (2.3)$$
$$-c\sigma\frac{\partial}{\partial x}\psi(x,y,\sigma,t) + \frac{q}{\hbar}\left[A_0(x,T-t) - c\sigma A_1(x,T-t)\right]\frac{\partial}{\partial y}\psi(x,y,\sigma,t)$$

and the final condition

$$\lim_{t\nearrow T}\Psi(x,y,\sigma,t) = \exp(-iy)\,\varphi_0(x,\sigma)$$

In fact (2.3) has the canonical form of a backward Kolmagorov equation whose solution has an obvious probabilistic representation. Indeed if we introduce the following processes :

$$\text{(a)}\quad X_{x,\sigma,t}(s) = x - c\sigma\int_t^s (-1)^{N_\tau - N_t}\, d\tau$$

$$\text{(b)}\quad Y_{y,\sigma,t}(s) = y + \frac{\pi}{2}(N_s - N_t) + \frac{q}{\hbar}\int_t^s \left[A_0(X_{x,\sigma,t}(\tau), T-\tau)\right. \qquad (2.4)$$
$$\left. - c\sigma(-1)^{N_\tau - N_t} A_1(X_{x,\sigma,t}(\tau), T-\tau)\right] d\tau$$

$$\text{(c)}\quad \Sigma_{\sigma,t}(s) = \sigma(-1)^{N_s - N_t} \qquad t \leqslant s \leqslant T$$

where N_t is the Poisson process with intensity $\frac{mc^2}{\hbar}$ starting at 0, the solution of Dirac equation has the Feynman-Kàc type representation

$$\Phi(x,\sigma,t) = \exp\left(\frac{m\,c^2}{\hbar}\,t\right) \mathbb{E}\,\{\, \varphi_o(X_{x,\sigma,o}(t)\,,\, \Sigma_{\sigma,o}(t))$$

$$\exp\left(-i\frac{\pi}{2}N_t - \frac{iq}{\hbar}\int_o^t d\tau\,[A_o(X_{x,\sigma,o}(\tau),t-\tau) - c\sigma(-1)^{N_\tau}A_1(X_{x,\sigma,o}(\tau),t-\tau)\,]\,)\}. \tag{2.5}$$

The typical paths of X_t are of the form

$$X_{x,\sigma,o}(t)(\omega)=x-c\sigma(t-t_n)+c\sigma(t_n-t_{n-1})+...+(-1)^n c\sigma(t_2-t_1)+(-1)^{n+1}c\sigma t_1 \tag{2.6}$$

where the t_i's are the times of jump of ω up to time t, which is precisely what can be expected.

Let us stress that the previous method works as well for imaginary time (and imaginary c).

If we consider the Dirac equation in more than one space dimension the algebra of Dirac matrices prevents us to treat Dirac equation using cartesian coordinates. Nevertheless if there is only one scalar potential which is spherically symmetric one can use the symmetry to reduce the problem (29).

The Dirac equation rewrites as :

$$i\hbar\partial_t\psi = \{-i\hbar c\alpha_r(\partial_r - (n-1)/2r) - i\,\hbar c/r\;\alpha_r\beta K \;+ mc^2\beta + V(r)\}\psi \tag{2.7}$$

where $\quad \psi^i \equiv \psi^i(t,r,\theta_1, ...,\theta_{n-1}),\ \alpha_r=(1/r)\,\Sigma_i\, x_i\alpha_i\ ,\quad i = 1, ...\ N,$

K is the generalized total angular momentum.

It can then be shown that the component ω of ψ in any eigenspace of K depends on one dichotomic variable σ and has the following representation :

$$\omega(t,r,\sigma) = \exp\left(\frac{m\,c^2}{\hbar}\,t\right) \mathbb{E}\,\{\omega_o(R_{r,\sigma,o}(t),\, \Sigma_{\sigma,o}(t)) \tag{2.8}$$

$$\exp\left[-\frac{i}{\hbar}\int_o^t d\tau W(R_{r,\sigma,o}(\tau)\right](-i)^{N_t}\exp\left[-\int_o^t L_n(G(R_{r,\sigma,o}(\tau)))dN_\tau\right]$$

where $W(r)$ and $G(r)$ are known functions of r, $R_{r,\sigma,o}(t)$ is the stochastic process previously described (2.4,a) , killed once it reaches 0. N_t is the Poisson process whose jumps are those of R. We have to point out that (2.8) is not really a Feynman-Kàc formula because it involves a stochastic integral.

III. STOCHASTIC MECHANICS OF SPIN 1/2 PARTICLES :

Following the Nelson's ideas on stochastic quantization it is also very interesting to deal with the case of spin 1/2 particles.

In the Nelson quantization procedure the classical equations of motion lead to stochastic differential equations which have to be solved in a class of stochastic processes, the choice of this class being equivalent to the quantum hypothesis added to the classical model.

The stochastic treatment of Dirac equation given by De Angelis et al (32-33) follows this line : In 1+1 space-time dimensions the probability density $\rho(t,x,\omega)$ associated with the ω-component of the normalized Dirac spinors ($\omega \in So$; $So = \{\omega \in R ; |\omega| = 1\} = \{+1, -1\}$) satisfies the quantum mechanical continuity equation :

$$\partial_t \rho(t,x,\omega) = -\omega c \partial_x \rho(t,x,\omega) - \frac{4Mc^2}{\hbar} \operatorname{Im} \{\psi(t,x,\omega)\overline{\psi(t,x,-\omega)}\} \tag{3.1}$$

This can be rewritten in the form of a forward Kolmogorov equation

$$\partial_t \rho(t,x,\omega) = -\omega c \partial_x \rho(t,x,\omega) + q(t,x,-\omega)\, \rho(t,x,-\omega) - q(t,x,\omega)\, \rho(t,x,\omega)$$

with $q(t,x,\omega) = \frac{Mc^2}{\hbar}(|\lambda| - \operatorname{Im}\lambda)$, $\lambda = \psi(t,x,-\omega)\,[\psi(t,x,\omega)]^{-1}$,
for a random process

$$t \to \xi_t = \xi_0 + c\omega_0 \int_0^t (-1)^{X_s} ds \tag{3.2}$$

where X_s is a point counting process which "counts" the jump of the velocity $\dot{\xi}_t = c\omega_0(-1)^{X_s}$. $q(t,x,\omega)$ is the probability per unit time of a jump in the velocity from $c\omega$ to $-c\omega$ at the space-time point (x,t).

Given the initial probability distribution $\rho(0,x,\omega)$ it is possible to construct ξ_t such as :

$$\text{Prob.}(\xi_t \in A \; ; \; \dot{\xi}_t = c\omega) = \int_A |\psi(t,x,\omega)|^2\, dx$$

for any mesurable region A of the real line.

This can be extended to d dimensions in the following way : take S_{d-1} the unit sphere on R_d, $\omega \in S_{d-1}$, B any subset of S_{d-1}, the generalized Kolmogorov equation is

$$\int_B (\partial_t + k\Sigma\omega_r\partial_r)\rho(t,x,\omega)\Omega(d\omega) = \int_{S^{d-1}} p(t,x,\omega\; ; B)\, \rho(t,x,\omega)\Omega(d\omega) \tag{3.3}$$

where $\rho(t,x,\omega)$ is a probability density on $R_d \times S_{d-1}$ corresponding to a stochastic process $t \to (\xi_t, k^{-1}\dot{\xi}_t)$ which represents a random motion with speed of fixed magnitude k, and

$$p(t,x,\omega,B) = \lim_{\varepsilon \to 0} \frac{1}{\varepsilon}\{\text{Prob}(k^{-1}\dot{\xi}_{t+\varepsilon} \in B \mid k^{-1}\dot{\xi}_t = \omega, \xi_t = x) - \delta_\omega(B)\}$$

$$\delta_\omega(B) = \begin{cases} 1 & \omega \in B \\ 0 & \text{otherwise} \end{cases}$$

The problem to solve can be formulated as follows : given $\psi(t,x)$ a normalized solution of Dirac equation in d space dimensions (with electromagnetic field), construct a process

$\{\xi_t, k^{-1}\dot{\xi}_t\}$ such that

$$\mathrm{Prob}(\xi_t \in A) = \int_A \psi(t,x)^* \, \psi(t,x)dx \qquad (3.4)$$

at any time and for every measurable subset A of R_d.

Choosing $\rho(t,x,\omega) = \psi(t,x)^*(\mathbb{1} + \omega.\alpha)\psi(t,x)$

$$= J^{\circ}(t,x) + \omega.J(t,x)$$

where $J^{\nu}(t,x)$ is the Dirac probability current, ρ satisfies a continuity equation which can be identified with Kolmogorov eq.(3.3) only if

(i) $k = cd \quad (k > c \text{ if } d > 1)$

or (ii) ψ is stationary

Then it is possible to find an explicit form of $p(t,x,\omega,B)$ such as the continuity equation is identical to the Kolmogorov equation and the properties of p linked to probability theory are satisfied.

It has then been shown that Dirac equation admits a "classical" probabilistic (gauge invariant) description in any number of space dimensions in terms of random motion with speed greater than c if $d > 1$. Nevertheless it is possible to describe stationary states by (stationary) stochastic processes of speed c. The same procedure as previously can then be used with $k = c$.

IV. STOCHASTIC THEORY OF ABSORPTION AND EMISSION OF QUANTAS.

For many years there was no explicit stochastic description of emission and absorption of quantas. Cini and Serva have proposed a new stochastic model involving point processes (34). This model is based on the following idea : for any normal mode of a given classical field the corresponding action variable becomes a stochastic variable assuming only integer values in units of h.

The method can be described using the fixed source model in one dimension (in a box l). In this simple case the field decouples into a sum of independent harmonic oscillators which are treated separately.

For the normal mode ω $\qquad H = 1/2(p^2+\omega^2 q^2)+vq.$

If v is time independent any solution $|\psi(t)\rangle$ of the Schrödinger equation can be developed on the basis of eigenstates $|n\rangle$ of the harmonic oscillator and

$$i\frac{d}{dt}\langle n|\psi(t)\rangle = n\omega\langle n|\psi(t)\rangle + \lambda\omega(\sqrt{n+1}\langle n+1|\psi(t)\rangle + \sqrt{n}\langle n|\psi(t)\rangle) \qquad (4.1)$$

$$\lambda = \frac{v}{\omega\sqrt{2\omega}}$$

The ground state of H is just the coherent state

$$|-\lambda)_{coh} = |0)_H = \exp(-\lambda^2/2)\Sigma\, [(-\lambda)^n/\sqrt{n!}\,]\, |n\rangle$$

The density $\rho(n,t) = |\langle n|\psi(t)\rangle|^2$ satisfies a continuity equation :

$$\frac{d}{dt}\rho(n,t) = \mathrm{Im}\,\Delta_-(n+1,t)\,\rho(n+1,t) + \mathrm{Im}\,\Delta_+(n-1,t)\,\rho(n-1,t) \qquad (4.2)$$
$$-\mathrm{Im}(\Delta_+(n,t) + \Delta_-(n,t))\,\rho(n,t)$$

with

$$\Delta_+(n,t) = \lambda\omega\sqrt{n+1}\,\frac{\langle\psi(t)|n+1\rangle}{\langle\psi(t)|n\rangle}\ ,\quad \Delta_-(n,t) = \lambda\omega\sqrt{n}\,\frac{\langle\psi(t)|n-1\rangle}{\langle\psi(t)|n\rangle} \qquad (4.3)$$

As previously this is the starting point of the stochastic interpretation, the continuity equation being interpreted as the forward Kolmogorov equation for a discontinuous Markov process in the state space of positive integers

$$\frac{d}{dt}\rho(n,t) = p_-(n+1,t)\,\rho(n+1,t) + p_+(n-1,t)\,\rho(n-1,t)$$
$$-\,[p_+(n,t) + p_-(n,t)\,]\,\rho(n,t) \qquad (4.4)$$

where $p_\pm(n,t) = |\Delta_\pm(n,t)| + \mathrm{Im}\Delta_\pm(n,t)$ are the jump probabilities per unit time from n to $n \pm 1$.

Moreover if we define the phase $\langle n\,|\,\psi\,(t)\rangle = \rho^{1/2}(n,t)\exp(iS(n,t))$

$$-\frac{d}{dt}\,S\,(n,t) = n\omega + \mathrm{Re}\,\{\Delta_+(n,t) + \Delta_-(n,t)\,\} \qquad . \qquad (4.5)$$

$\Delta_\pm$ being functions of ρ and S the equations (4.4) and (4.5) define a stochastic process independently of the underlaying quantum theory.

If we consider the ground state case $\psi(t) = |0)_H$ we have

$$\begin{cases} p_+(n) = \lambda^2\omega \\ p_-(n) = n\omega \end{cases} \quad .$$

By the fact two different processes govern the emission and absorption of quantas :

- one is a Poisson process with constant probability rate
- the other one is a point process with probability rate proportional to n.

Moreover using (4.4) and (4.5) it is possible to find the transition probability $P(n,t\,;\,n_o,t_o)$ in an explicit form

$$P(n,t;n_o,\,o) = \{(e^{\omega t}-1)^{-1}\,d/d\,\lambda^2 \;+\; 1\}^{n_o}\,P(n,t\,;\,o,o)$$

$$P(n,t;o,o) = \exp(\lambda^2(e^{-\omega t}-1))\,\frac{[-\lambda^2\,(\,e^{-\omega t}-1)]^n}{n!}$$

The same methods can be developed if v is time dependent. With $v(t)=e^{i\omega_o t}\,v_o$ and $\psi(0) = |n_o\rangle$ it is possible to recover, by calculating the jump probabilities, the standard results for the Einstein probability rates of photon emission and absorption from an electric dipole source in the presence of radiation.

For any potential $V(t)$, the hypothesis $p_+(n,t)=p_+(t)$

$$p_-(n,t)=np_-(t)$$

corresponds to the case where the initial state is a coherent state and allows to calculate explicitely $p_\pm(t)$ and the transition probability. It is interesting to notice that the transition probability from the vacuum to the state with n quanta corresponds to the expression given by Feynman for the transition amplitude of the same process.

I want to point out that the formulation of quantum models in terms of stochastic processes taking values on the discrete eigenvalues of a complete set of orthonormal functions in the Hilbert space of the system has been used also for the solution of Schrödinger equation of a boson field interacting with a quantum particle (11) and also to link classical mechanics and stochastic mechanics via variational principle (35).

As a matter of fact, in this short review, we have selected some special new applications of jump processes which seemed to us particularly representative among the very interesting developments which are going on.

REFERENCES

[1] R.P. Feynman. Space-time approach to non-relativistic quantum mechanics. Rev. Mod. Phys. **20**, 367-387 (1948).

[2] J. Ginibre. Reduced density matrices of the anisotropic Heisenberg model. Comm. Math. Phys. **10**, 140-154 (1968).

[3] M. Fukushima. On the spectral distribution of a disordered system and the range of a random walk. Osaka Journal of Mathematics **11**, 73-85 (1974).

[4] Ph. Combe, R. Hoegh-Krohn, R. Rodriguez, M. Sirugue,M. Sirugue- Collin. Poisson processes on groups and Feynman path integral. Comm. Math. Phys. **77**, 269-288 (1980).

[5] J. Bertrand, G. Rideau. Stochastic jump processes in the phase space representation of quantum mechanics. Lecture Notes in Physics, vol. **173**, Springer Verlag, (1982) p. 13-18.

[6] V.P. Maslov, A.M. Chebotarev. Jump-type processes and their applications in quantum mechanics. Transl. Journal of Soviet Math. **13**, 315-357 (1980).

[7] A.M. Chebotarev, V.P. Maslov. Processus à sauts et leurs applications en mécanique quantique. Proceedings of the conference on "Feynman Path Integrals" Marseille, 1978. Lecture Notes in Physics, vol. **106**, Springer Verlag (1979), p. 58-72.

[8] Ph. Combe, R. Hoegh-Krohn, R. Rodriguez, M. Sirugue,M. Sirugue- Collin. Quantum dynamical time evolutions as stochastic flows on phase space. Physica A **124 A**, 561-574 (1984).

[9] J. Bertrand, G. Rideau. Stochastic processes and the evolution of quantum observables. Letters in Math. Phys. **7**, 327-337 (1983).

[10] S. Albeverio, Ph. Blanchard, Ph. Combe, R. Hoegh-Krohn, M. Sirugue. Local relativistic flows for quantum fields. Comm. Math. Phys. **90**, 329-351 (1983).

[11] J. Bertrand, B. Gaveau, G. Rideau. Quantum fields and Poisson processes : interaction of a cut-off boson field with a quantum particle. Letters in Math. Phys. **9**, 73-82 (1985).

[12] Ph. Combe, R. Hoegh-Krohn, R. Rodriguez, M.Sirugue, M. Sirugue- Collin. Zero mass, two dimensional, real time Sine-Gordon model without U.V. cut-off. Annales Inst. H. Poincaré **A37**, 115-127 (1982).

[13] R. Azencott, H. Doss. L'équation de Schrödinger quand h tend vers 0 : une approche probabiliste. Lecture Notes in Mathematics, Vol. **1109**, Springer Verlag, (1985), p 1-17.

[14] Ph. Combe, R. Rodriguez, M. Sirugue, M. Sirugue-Collin. Quantum deviation from classical paths in the classical limit. Public. Univ. Nantes, p. 87-90 (1985).

[15] Ph. Blanchard, M. Sirugue. Large deviations from classical paths : Hamiltonian flows as class limit of quantum flows. Comm. Math. Phys. **101**, 173-185 (1985).

[16] Ph. Blanchard, Ph. Combe, M. Sirugue, M. Sirugue-Collin. Large deviations from classical paths and the classical limit of quantum stochastic flows. Proceeding Vilnus Conference on Probability Theory (1985) to appear.

[17] G.V. Riazonov. The Feynman path integral for the Dirac equation. Soviet Physics JETP **6**, 1107-1113 (1958).

[18] G.F. De Angelis, D. De Falco, F. Guerra. Probabilistic ideas in the theory of Fermi fields : I Stochastic quantization of the Fermi oscillator. Phys. Rev. D **23**, 1747-1751, (1981).

[19] G. Rosen. Formulation of classical and quantum theories. Mathematics in Science and Engineering, Vol **60**, Academic Press (1969).

[20] T. Jacobson, L.S. Schulman. Quantum stochastics : the passage from a relativistic to a non relativistic path integral, Journal of Physics A, **17**, 375-383, (1984).

[21] T. Jacobson. Spinor chain path integral for the Dirac equation. Thesis University of Austin (1983).

[22] B. Gaveau, J. Vauthier. Integrales oscillantes stochastiques : l'équation de Pauli. Journal of functional Analysis, **44**, 388-400 (1981).

[23] B. Gaveau, T. Jacobson, M. Kàc, L.S. Schulman. Relativistic extension of the analogy between quantum mechanics and brownian motion. Phys. Rev. Letters, **53**, 419-422 (1984).

[24] B. Gaveau. Representation formulas of the Cauchy problem for hyperbolic systems generalizing Dirac equation. Journal of functional Analysis, **58**, 310-319 (1984).

[25] T. Ichinose. Path integral for the Dirac equation in two space-time dimensions. Proceeding Japan Academy of Science, **58** A, 290-293 (1982).

[26] T. Ichinose. Path integral formulation of the propagator for a two dimensional Dirac particle. Proceeding of the International Conference of the IAMP in Boulder, August 1983, Physica **124** A, 419-426, (1984).

[27] T. Ichinose, H. Tamura. Propagation of a Dirac particle : a path integral approach. Journal of Math. Phys. **25**, 1810-1819 (1984).

[28] Ph. Blanchard, Ph. Combe, M. Sirugue, M. Sirugue-Collin. Probabilistic solution of the Dirac equation. Preprint Bibos Mai 85/44.

[29] Ph. Blanchard, Ph. Combe, M. Sirugue, M. Sirugue-Collin. Stochastic jump processes associated with Dirac equation. Proceeding Ascona Como Conference, June 85, to appear.

[30] G.F. De Angelis, G. Jona Lasinio. A stochastic description of a spin 1/2 particle in a magnetic field. Journal of Physics A, **15**, 2053-2080 (1982).

[31] G.F. De Angelis, G. Jona Lasinio, M. Sirugue. Probabilistic solutions of the Pauli type equation. Journal of Physics A, **16**, 2433-2444 (1983).

[32] G.F. De Angelis. Are Dirac electrons faster than light ? . Proceeding Nato Workshop, Villa Olmo 1985 (to appear).

[33] G.F. De Angelis, G. Jona Lasinio, M. Serva, N. Zanghi. Stochastic mechanics of a Dirac particle in two space-time dimensions. Journal of Physics A, **19**, 865-871 (1986).

[34] M. Cini, M. Serva. Stochastic theory of emission and absorption of quanta. Journal of Physics A, **19**, 1163-1177 (1986).

[35] F. Guerra, R. Marra. Discrete stochastic variational principles and quantum mechanics. Phys. Rev. D, **29**, 1647-1655 (1984).

F. Guerra, R. Marra. Stochastic mechanics of spin 1/2 particles. Phys. Rev. D, **30**, 2579-2584 (1984).

The Degree Theory of the Nicolai Map

Ezra Getzler
Society of Fellows and Department of Mathematics
Harvard University
Cambridge, MA 02138, USA

In his thesis, Nicolai showed that one distinguishing feature of supersymmetric field theories is that their functional integral measures may be obtained from a gaussian measure by a nonlinear change of variables - at least, in the sense of perturbation theory [1].

Later, it was observed by Cecotti and Girardello [2] that in the simplest case, of supersymmetric quantum mechanics, the Nicolai map is not one-to-one - in fact, it has a degree, equal to index of the corresponding Hamiltonian. Here, we will describe the application of the degree theory, based on the Malliavin calculus, developed in [3] to the Nicolai map considered by Cecotti and Girardello. For more details, refer to the papers [3,4].

1 - The Model

The Hamiltonian of $N=2$ supersymmetric quantum mechanics is familiar from the work of Witten on Morse theory [5]. The Hilbert space of this model is the space of L^2 differentiel forms on $\mathbb{R}^N$, and the Hamiltonian H equals QQ^*+Q^*Q, where

$$Q = e^{-V} \circ d \circ e^{V} \tag{1}$$

(here, V is a real function on $\mathbb{R}^N$, the "superpotential", and d is the exterior derivative operator).

This Hamiltonian corresponds to the $N=2$ superspace action

$$S(\Phi) = \int_T \{\tfrac{1}{2}\,\overline{D}\Phi.D\Phi + V(\Phi)\}\, d\overline{\theta}d\theta dt, \tag{2}$$

where $D = \frac{\partial}{\partial\theta} + \overline{\theta}\,\frac{\partial}{\partial t}$, and T is the domain of integration for the variable t. Written out more prosaically, this action splits into a bosonic piece, corresponding to an anharmonic oscillator, and a fermionic piece, which is quadratic in the complex fermionic field ψ. Indeed, writing $\Phi = \varphi + \bar\theta\psi + \theta\bar\psi + \bar\theta\theta F$, and integrating out the auxiliary variable F, we obtain

(3) $$S(\varphi,\bar\psi,\psi) = S_B(\varphi) + \int_T \bar\psi\left(\frac{d}{dt} + \nabla^2 V(\varphi)\right)\psi \, dt,$$

where $$S_B(\varphi) = \frac{1}{2}\int_T \left\{\left|\frac{d\varphi}{dt}\right|^2 + |(\nabla V)(\varphi)|^2\right\} dt.$$

In fact, in performing these transformations, we have ignored any boundary conditions, and from here, we shall assume that T is a circle.

Following Parisi and Sourlas [6] and Cecotti and Girardello [2], consider the nonlinear mapping from the space of loops in $\mathbb{R}^N$ to itself defined by

(4) $$(A(\varphi))(s) = \frac{d\varphi}{dt}(s) + (\nabla V)(\varphi(s)).$$

This map is linked to the action $S(\varphi,\bar\psi,\psi)$ by two properties :

a) integration by parts shows that $S_B(\varphi) = \frac{1}{2}\int_T |A(\varphi)|^2 \, dt$;

b) the operator $\frac{d}{dt} + \nabla^2 V(\varphi)$ entering into the fermionic part of the action is the differential $\nabla_\varphi A$.

Thus, formally at least, the pullback of the volume form on the loop space of $\mathbb{R}^N$ given by the white noise measure $d\kappa = Z^{-1} \exp\{-\frac{1}{2}\int_T |\varphi|^2 \, dt\} \, D\varphi$, by the map A equals

(5) $$A^*(d\kappa) = Z^{-1} \det\left(\frac{d}{dt} + \nabla^2 V(\varphi)\right) \exp\{-S_B(\varphi)\} \, D\varphi,$$

which is the functional integral for the action $S(\varphi,\bar\psi,\psi)$ after the fermionic fields have been integrated out.

It is important to note here that usually, measures are <u>pushed forward</u>; it is differential forms which are pulled back. Thus, to understand the Nicolai map A, we must understand volume forms in infinite dimensions, and how to pull them back. Once that is done, it is easy to define the degree of a map : take any volume form ω, then

$$(6) \qquad \int A^*\omega = \deg(A). \int \omega.$$

It follows from de Rham's theorem that $\deg(A)$ is well defined : if ω_1 and ω_2 are such that $\int \omega_1 = \int \omega_2$, then there exists α such that $\omega_1 - \omega_2 = d\,\alpha$, and hence

$$(7) \qquad \int A^*(\omega_1 - \omega_2) = \int d(A^*\alpha) = 0.$$

On letting the volume form ω converge to the delta function at a regular value y of the map A, we deduce in the standard way

$$(8) \qquad \sum_{A(x)=y} \operatorname{sgn} \det(\nabla_x A) = \deg(A).$$

The final ingredient of degree theory is Sard's theorem, which says that the set of critical values (those that are not regular) of A has measure zero.

In fact, it is possible to carry out the extension of the above précis of degree theory to an infinite dimensional setting, which includes as a special case the Nicolai map A. This theory rests on a change of variables formula which is valid in infinite dimensions, and which we shall describe in the next section.

2 - Ramer's Change of Variables Formula

In [7], Ramer gave a change of variables formula for Gaussian measures which is rather general. To state it, we must recall a few notions of calculus on spaces with a gaussian measure. The reader should bear in mind the following philosophy : if the statement of a theorem makes no mention of the dimension of the space, and the theorem is true for finite dimensal spaces with a gaussian measure, then it is true for an arbitrary Banach space with gaussian measure [Maybe this philosophy should be reexamined in the light of S. Albeverio's talk on nonstandard analysis].

Thus, let B be a Wiener space, that is, a Banach space with gaussian measure $d\mu$. Then, there is a Hilbert space H embedded in B

such that if α is a linear form on B, then

(9) $$\int_B e^{\alpha(x)}\, d\mu(x) = e^{-|\alpha|_H^2/2} \;;$$

here, we identify α with an element of H. In fact, H is the object of interest, and B is a "thickering" of H the details of which are of little interest, so long as it carries a measure $d\mu$ satisfying (6).

It is now possible to define a closed operator $\nabla : L^p(B) \longrightarrow L^p(B;H)$, the gradient operator, and its adjoint $\nabla^* : L^p(B;H) \longrightarrow L^p(B)$. We also have Sobolev spaces $L_k^p(B)$:

$$L_k^p(B) = \{f \in L^p(B) \mid (\nabla^*\nabla)^{k/2} f \in L^p(B)\}.$$

It is a nontrivial (and important) result that for $1<p<\infty$, and with constants independent of the dimension of B, the operators ∇ and ∇^* are bounded with loss of one derivative.

The maps from B to itself for which there is a change of variables formula are those which are of the form

(10) $$A(x) = x + F(x),$$

where F is a C^1-map from B to H. Thus, in one sense, A is a gentle perturbation of the identity : on the other hand, if B is finite dimensional, then A is an arbitrary C^1-map.

If B is finite dimensional, it is easy to give a formula for $A^*(d\mu)$:

(11) $$A^*(d\mu) = \det(I+\nabla_x F)\, e^{-(x,F(x))_H - |F(x)|_H^2/2}\, d\mu.$$

But this formula does not make sense in infinite dimension; not only is $(x,F(x))_H$ meaningless, since x is not in H, but in general $\nabla_x F$ is a Hilbert-Schmidt operator on H, so that $\det(I+\nabla_x F)$ is not defined.

This is remedied by using the regularized determinant $\det_2(I+\nabla_x F) = \det(I+\nabla_x F)\exp(-\mathrm{Tr}\,\nabla_x F)$, which is well defined for $\nabla_x F$ Hilbert-Schmidt. Using the fact that in finite dimensions,

$$(12) \qquad \nabla^* F = -\mathrm{Tr}\,\nabla F + (x,F)_H,$$

we arrive finally at Ramer's formula

$$(13) \qquad A^*(d\mu) = \det_2(I+\nabla_x F)\, e^{-\nabla^* F - |F|_H^2/2}\, d\mu.$$

Using this formula, it is possible to develop a degree theory for Wievier maps (satisfying a few technical conditions). The main result is as follows :

Theorem 1. For $d\mu$ - a.e point $y \in B$, the set $A^{-1}(y)$ is finite, and there is an integer $\deg(A)$ such that

$$\sum_{A(x)=y} \operatorname{sgn} \det_2(I+\nabla_x A) = \deg(A). \qquad \square$$

To apply this result to the Nicolai map (4), we have to rewrite it a little so that it maps from one space to itself. In fact, we consider the operator

$$(14) \qquad \Lambda^{-1} \circ A : C^\alpha(T,\mathbb{R}^N) \longrightarrow C^\alpha(T,\mathbb{R}^N),$$

acting on α-Holder loops in $\mathbb{R}^N$. Here, Λ is the operator $d/dt+1$, which is a linear map, so has degree equal to 1; thus, the degree of A, if it exists, will be equal to the degree of $\Lambda^{-1} \circ A$.

The loop-space $C^\alpha(T,\mathbb{R}^N)$ is a Wiener space if $\alpha < \frac{1}{2}$, with the gaussian measure (which is known as the Ornstein-Uhlenbeck measure)

$$(15) \qquad d\mu = \Lambda^*(d\kappa) = Z^{-1} \exp\left(-\tfrac{1}{2}\int\{|\dot\varphi|^2 + |\varphi|^2\}\right) D\varphi.$$

It is a fairly standard exercise to show that the map $\Lambda^{-1} \circ A$ satisfies the hypotheses of Theorem 1. Thus, we see that the Nicolai map A has a well defined degree.

3 - The Index of H

We must now link the measure $A^*(d\kappa)$, which was proved to exist in the last section, to the supersymmetric Hamiltonian H (1). In fact, it is possible to prove the following result.

Theorem 2. If V is such that $\lim_{x\to\infty} |\nabla V(x)| = \infty$ and $|\nabla^m V| \leq C_n(|\nabla V|+1)$ for all $m \geq 2$ (so that H is Fredholm), then

$$\int_{C^\alpha(T,\mathbb{R}^N)} f_1(\varphi(t_1))\ldots f_n(\varphi(t_m))A^*(d\kappa)$$
$$= \mathrm{Str}\,\{e^{-t_1H/2}\, f_1\, e^{-(t_2-t_1)H/2} \ldots f_n\, e^{-(1-t_n)H/2}\}$$

where $f_i \in C_0^\infty(\mathbb{R}^N)$ and $0 < t_1 < t_2 < \ldots < t_m < 1$. □

The proof of this result is obtained by comparing the formula for left-hand side obtained through Ramer's formula (13) with the formula for the right-hand side obtained from the Freynman-Kac formula for the operator H. The technical details are a little tricky, but well known to practioners of Yukawa models in two space-time dimensions.

It is a special case of Theorem 2 that

$$(16) \qquad \int_{C^\alpha(T,\mathbb{R}^N)} 1\, A^*(d\kappa) = \mathrm{Str}\,\{e^{-H/2}\}.$$

But by what we have proved in Section 2, it is seen that this can be translated to read :

Theorem 3. The degree of A equals the index of the corresponding Hamiltonian H. □

All that remains is to calculate the index of H . In fact, this can be done very rapidly using the Atiyah-Singer index theorem, but there are more elementary methods available to us here, using path integrals. Let H_ε be obtained from H by replacing V by $\varepsilon^{-1}V$, and let $d\nu_\varepsilon$ be the measure on $C^\alpha(T,\mathbb{R}^N)$ defined by

$$\int_{C^{\alpha}(T,\mathbb{R}^N)} f_1(\varphi(t_1)) \,\ldots\, f_n(\varphi(t_n))\, d\nu_\varepsilon \tag{15}$$
$$= \operatorname{Str}\{e^{-\varepsilon t_1 H_\varepsilon/2} f_1 \ldots f_n\, e^{-\varepsilon(1-t_n)H_\varepsilon/2}\}\ .$$

Then it is rather easy to prove the following result (by using the Feynman-Kac formula) : if V is a Morse function, then

$$\lim_{\varepsilon\to 0} d\nu_\varepsilon = \sum_{\nabla_x V=0} \operatorname{sgn}\det(\nabla_x^2 V)\,.\delta(\varphi(t)-x). \tag{16}$$

To calculate the index of H, we now perform the following sequence of steps :

a) deform V into a Morse function - this does not change the index of H;

b) by the McKean-Singer formula, the index of H equals $\operatorname{Str}\{e^{-H}\}$;

c) replacing H by $\varepsilon H_\varepsilon$ leaves this supertrace constant ; thus

$$\begin{aligned}\text{index }(H) &= \lim_{\varepsilon\to 0} \operatorname{Str}\{e^{-\varepsilon H_\varepsilon/2}\}\\ &= \lim_{\varepsilon\to 0}\int_{C^{\alpha}(T,\,\mathbb{R}^N)} 1\, d\nu_\varepsilon\\ &= \sum_{\nabla_x V=0} \operatorname{sgn}\det(\nabla_x^2)\int_{C^{\alpha}(T,\,\mathbb{R}^N)} \delta(\varphi(t)-x)\\ &= \text{index }(\nabla V : \mathbb{R}^N \longrightarrow \mathbb{R}^N).\end{aligned}$$

Thus, by this semiclassical calculation, we see that the degree of A may be calculated by restricting to the space of constant loops $\mathbb{R}^N \longrightarrow C^{\alpha}(T, \mathbb{R}^N)$, on which A equals the map $x \longmapsto \nabla V(x)$.

Example : If $N = 1$, then

$$\deg(\varphi \longmapsto d\varphi/dt + \varphi^n) = \begin{cases} 1 & \text{if } n \text{ is odd}\\ 0 & \text{if } n \text{ is even.}\end{cases}$$

BIBLIOGRAPHY

[1] H. Nicolai.- On a new characterization of scalar supersymmetric theories. Phys. Lett. B 89 (1980), 341-346.

[2] S. Cecotti and L. Girardello.- Functional measure, topology and dynamical supersymmetry breaking. Phys. Lett. B 110 (1982), 39-43.

[3] E. Getzler.- Degree theory for Wiener maps. To appear, J. Functional Analysis.

[4] E. Getzler.- The degree of the Nicolai map in supersymmetric quantum mechanics. To appear, J. Functional Analysis.

[5] E. Witten.- Supersymmetry and Morse theory. J. Differential Geom. 17 (1982), 661-692.

[6] G. Parisi and N. Sourlas.- Supersymmetric field theories and stochastic differential equations. Nuclear Phys. B 205 (1982), 337-344.

[7] R. Ramer.- On nonlinear transformations of Gaussian measures. J. Functional Analysis 15 (1974), 166-187.

SOME RECENT STUDIES OF BROWNIAN PATHS INTERSECTIONS

Marc YOR

Laboratoire de Probabilités, Université P. et M. Curie, Paris

1. INTRODUCTION.

Let $B = (B_t, t \geq 0)$ denote a Brownian motion with values in $\mathbb{R}^d$, $d \geq 2$, and define, for $k \geq 2$:

$$\Delta_k = \{(s_1,\ldots,s_k) \in \mathbb{R}_+^k : B_{s_1} = B_{s_2} = \ldots = B_{s_k}, \text{ and } s_i < s_{i+1}\}.$$

Dvoretzky-Erdös-Kakutani and Taylor ([4], [5], [6]) proved that Δ_k is almost surely non-empty if and only if :

(1.a) $\quad d = 2$, and k is arbitrary, or $d = 3$, and $k = 2$.

Since, for any $h > 0$, $(B_{t+h} - B_h \,;\, t \geq 0)$ is a Brownian motion independent of $(B_u, u \leq h)$, the previous result is equivalent to the analogous result involving k independent d-dimensional Brownian motions $B^{(1)},\ldots,B^{(k)}$, with :

$$\tilde{\Delta}_k = \{(s_1,\ldots,s_k) \in \mathbb{R}_+^k : B^{(1)}_{s_1} = B^{(2)}_{s_2} = \ldots = B^{(k)}_{s_k}, \text{ and } s_i < s_{i+1}\}.$$

In the sequel, we shall refer to the first case, resp : the independent case, as Case 1, resp : Case 2. A tilda shall be added to the notation of objects which arise in the study of Case 2, while a star is used for objects which arise in the study of either case 1 or 2.

2. EXISTENCE OF LOCAL TIMES OF INTERSECTION.

For the sake of clarity, we restrict ourselves to $k = 2$, and dimensions $d = 2$ or 3. Let $T(t) = \{(s_1,s_2) \in \mathbb{R}_+^2 : 0 \leq s_1 \leq s_2 < t\}$, and B and B' be two independent d-dimensional Brownian motions.

Dynkin [7] and Geman-Horowitz-Rosen [10] have shown that the random measure :

$$\nu^*(\omega \,;\, dz) : A(\in \mathcal{B}(\mathbb{R}^d)) \longrightarrow \int_{T(t)} ds_1\, ds_2\, 1_A(B_{s_2} - B^{(')}_{s_1})$$

is absolutely continuous with respect to Lebesgue measure on $\mathbb{R}^d$.

More precisely, there exist random kernels $\alpha^*(z \,;\, ds_1\, ds_2)$ on $\mathbb{R}^d \times \mathcal{B}(T(\infty))$ such that :

(2.a) $$\int_\Gamma ds_1\, ds_2\, f(B_{s_2} - B^{(')}_{s_1}) = \int dz\, f(z)\, \alpha^*(z \,;\, \Gamma)$$

(2.b) for every $z \in \mathbb{R}^d$, $\alpha^*(z \,;\, ds_1\, ds_2)$ is carried by $\{(s_1,s_2) : B_{s_2} - B^{(')}_{s_1} = z\}$.

(2.c) *(2.c.1)* $\quad z \to \alpha(z \,;\, ds_1\, ds_2)$ is vaguely continuous on $\mathbb{R}^d \setminus \{0\}$;

moreover, for every Borel set $\Gamma \subset T(t)$, with $d(\Gamma, T(t)) > 0$, $z \to \alpha(z\ ;\ \Gamma)$ is continuous on $\mathbb{R}^d$, but $\alpha(0\ ;\ T(t)) = \infty$.

(2.c.2) $z \to \tilde{\alpha}(z\ ;\ ds_1\ ds_2)$ is vaguely continuous on $\mathbb{R}^d$.

The explosion : $\alpha(0\ ;\ T(t)) = \infty$ may be explained by the great number of intersections which occur when s_1 and s_2 are very close.

3. *TANAKA-ROSEN FORMULA FOR LOCAL TIMES OF INTERSECTION.*

We use the notation $\alpha^*(z\ ;\ t)$ for $\alpha^*(z\ ;\ T(t))$.

d = 2 Rosen [21] (see also [29]) showed, for $y \neq 0$, that,

$$\text{(3.a)} \qquad \int_0^t ds\ \{\log|B_t - B_s - y| - \log|y|\} = \int_0^t dB_u \cdot \int_0^u ds \frac{B_u - B_s - y}{|B_u - B_s - y|^2} + \pi\alpha(y\ ;\ t).$$

In case 2, the following variant holds for any y :

$$\int_0^t ds\ \{\log\ |B_t - B'_s - y| - \log|B_s - B'_s - y|\}$$

$$= \int_0^t dB_u \cdot \int_0^u ds \frac{B_u - B'_s - y}{|B_u - B'_s - y|^2} + \pi\tilde{\alpha}(y\ ;\ t)$$

d = 3 The same authors obtained, for $y \neq 0$:

$$\text{(3.b)} \qquad \int_0^t ds\ \{\frac{1}{|B_t - B_s - y|} - \frac{1}{|y|}\} = -\int_0^t dB_u \cdot \int_0^u ds \frac{B_u - B_s - y}{|B_u - B_s - y|^3} - 2\pi\alpha(y\ ;\ t)$$

and, in case 2, there is a variant of this formula which holds for every y.

4. *RENORMALIZATIONS OF VARADHAN (d = 2) AND WESTWATER (d = 3). CENTRAL LIMIT THEOREMS (d ≥ 3).*

In the sequel, we shall often use the notation $\{X\} = X - E(X)$, when X is an integrable random variable.

The following theorem is easily deduced from formula *(3.a)*.

Theorem 1 : *(Varadhan's renormalization)* **d = 2**

1) (Dynkin [8] ; Rosen [21] ; Le Gall [13] ; [27]).

Both L^2-limits exist :

$$\gamma(t) \equiv \lim_{y \to 0} (\alpha(y\ ;\ t) - \frac{t}{\pi} \log \frac{1}{|y|}) \ ; \qquad \overline{\gamma}(t) \equiv \lim_{y \to 0} \{\alpha(y\ ;\ t)\}.$$

2) (Varadhan [24]).

Let $f_n(\cdot) = n^d f(n\cdot)$, when $f : \mathbb{R}^2 \to \mathbb{R}_+$ is bounded, continuous, has compact support, and verifies $\int dx\ f(x) = 1$. Then :

$$\int_{T(t)} ds_1\, ds_2 \{f_n(B_{s_2}-B_{s_1})\} \xrightarrow[L^2]{(n\to\infty)} \bar{\gamma}(t).$$

Both statements in theorem 1 can be amplified :

1') (Rosen [22] ; [28])

The process $\gamma(y\ ;\ t) = \pi\alpha(y\ ;\ t) - t \log \frac{1}{|y|}$, defined on $(\mathbb{R}^2 \setminus \{0\}) \times \mathbb{R}_+$, has a continuous extension to the entire space $\mathbb{R}^2 \times \mathbb{R}_+$, which, moreover, satisfies :

$$\overline{\lim_{\delta\to 0}} \frac{1}{\delta(\log\frac{1}{\delta})^{7/2}} \sup_{\substack{s\le t\\ |x-y|\le\delta}} |\gamma(x,s) - \gamma(y,s)| < \infty\ ;$$

2') (Dynkin [9] ; Rosen [22])

For any finite order of multiplicity $k \in \mathbb{N}$,

$$\int_{\{0\le s_1\le s_2\ \dots s_k\le t\}} ds_1\, ds_2 \dots ds_k \{f_n(B_{s_2}-B_{s_1})\}\dots\{f_n(B_{s_k}-B_{s_{k-1}})\}$$ converges in L^2,

as $n \to \infty$.

The following theorem can be obtained from formula (3.b).

Theorem 2 ([26]) : $d = 3$

1) $$\left(B_t\ ;\ \frac{1}{(\log\frac{1}{|y|})^{1/2}}\left[2\pi\alpha(y\ ;\ t) - \frac{t}{|y|}\right]\ ;\ t \ge 0\right) \xrightarrow{(d)} (B_t\ ;\ 2\beta_t\ ;\ t \ge 0)$$

where β *denotes a one-dimensional Brownian independent of* B.

2) *Consequently, if* $f : \mathbb{R}^3 \to \mathbb{R}$ *satisfies the same hypothesis as in theorem 1, the above convergence in distribution holds jointly with that of :*

$$\left(\frac{1}{(\log n)^{1/2}}\left(\int_{T(t)} ds_1\, ds_2\, f_n(B_{s_2}-B_{s_1}) - \frac{tn}{2\pi}\int \frac{dy}{|y|} f(y)\right)\ ;\ t \ge 0\right)$$

towards $(\frac{1}{\pi}\beta_t\ ;\ t \ge 0)$.

Also in dimension 3, Westwater [25] has obtained a different very interesting renormalization result :

let P be the distribution on $C([0,1]\ ;\ \mathbb{R}^3)$ of Brownian motion starting at 0, and define, for any $g \ge 0$, and $y \ne 0$, the new probability :

$$Q_y^g = \frac{\exp - g\,\alpha(y\ ;\ 1)}{E[\exp - g\,\alpha(y\ ;\ 1)]} \cdot P.$$

Then, Westwater showed that, as $y \to 0$, Q_y^g converges weakly towards a probability, which we denote by Q^g. Moreover, for $g_1 \ne g_2$, Q^{g_1} and Q^{g_2} are mutually singular, and Kusuoka [12] has proved that, for every g, a process with distribution Q^g is the sum of a Brownian motion and of a process with zero quadratic variation. Although there are no double points in dimension $d \geq 4$, the following theorem may be of some interest

<u>*Theorem 3*</u> ([3]) : $\boxed{d \geq 4}$

Let c_d *be the volume of the unit ball in* $\mathbb{R}^d$

If $f : \mathbb{R}^d \to \mathbb{R}$ *satisfies the same hypothesis as in theorem 1, then :*

$$(B_t \ ; \ \frac{1}{n^{d-3}} \int_{T(t)} ds_1 \, ds_2 \, f_n(B_{s_1} - B_{s_2}) - tn \, c_d \int \frac{dy}{|y|^{d-2}} f(y) \ ; \ t \geq 0)$$

converges in distribution towards a gaussian process, the components of which are multiples of linked Brownian motions.

5. SOME ASYMPTOTICS FOR THE WIENER SAUSAGE.

The so-called Wiener sausage is defined, for $\varepsilon > 0$, and $t \geq 0$, as the ε-neighborhood of the Brownian trajectory $(B_s \ ; \ s \leq t)$, precisely :

$$S_\varepsilon(t) = \bigcup_{s \leq t} (B_s + y \ ; \ |y| < \varepsilon).$$

The next two theorems are devoted to first and second order asymptotics for the volume $m(S_\varepsilon(t))$ of the Wiener sausage as $\varepsilon \to 0$.

<u>*Theorem 4*</u> : (*Le Gall [14] ; Kesten-Spitzer-Whitman (see [11], p. 252)*)

$\boxed{d = 2}$ $$\lim_{\varepsilon \to 0} (\log \frac{1}{\varepsilon}) \, m(S_\varepsilon(t)) = \pi t$$

$\boxed{d \geq 3}$ $$\lim_{\varepsilon \to 0} \frac{1}{\varepsilon^{d-2}} m(S_\varepsilon(t)) = (\frac{d}{2} - 1) c_d \, t$$

where c_d *is the volume of the unit ball in* $\mathbb{R}^d$*, and the limits hold both a.s. and in* L^p.

<u>*Theorem 5*</u> : (*Le Gall [17]*)

$\boxed{d = 2}$ $$(\log 1/_\varepsilon)\,((\log 1/_\varepsilon) m(S_\varepsilon(t)) - \pi t) \xrightarrow[(\varepsilon \to 0)]{L^2} \frac{\pi}{2}(1 + C - \log 2) - \pi^2 \,\overline{\gamma}(t)$$

where C *is Euler's constant, and* $\overline{\gamma}$ *is defined in theorem 1.*

$\boxed{d = 3}$ $$(\frac{1}{\varepsilon^2 (\log \frac{1}{\varepsilon})^{1/2}} (m(S_\varepsilon(t)) - 2\pi\varepsilon t) \ ; \ t \geq 0) \xrightarrow{(d)} (4\pi \, \beta_t \ ; \ t \geq 0)$$

where β *is a one-dimensional Brownian motion.*

$\boxed{d \geq 4}$ $$(\frac{1}{\varepsilon^{d-1}} (m(S_\varepsilon(t)) - (\frac{d}{2} - 1) c_d \, \varepsilon^{d-2} t \ ; \ t \geq 0)$$

converges in distribution towards a multiple of Brownian motion.

The results in theorem 5 are closely linked with, respectively, those featured in theorems 1, 2, and 3.

Next, we present two limit theorems for the volume of the intersection of two independent Wiener sausages.

Theorem 6 (*Le Gall* [15]) :

$$\boxed{d = 2} \qquad (\log \frac{1}{\varepsilon})^2 \, m(S_\varepsilon(t) \cap S'_\varepsilon(t)) \xrightarrow[(\varepsilon \to 0)]{L^2} \frac{\pi^2}{2} \tilde{\alpha}(0 \,;\, t)$$

$$\boxed{d = 3} \qquad \frac{1}{\varepsilon^2} \, m(S_\varepsilon(t) \cap S'_\varepsilon(t)) \xrightarrow[(\varepsilon \to 0)]{L^2} 4\pi^2 \tilde{\alpha}(0 \,;\, t).$$

Theorem 7 (*Le Gall* [16]) :

1) $(\frac{1}{c} m(S_1(e^{ct}) \cap S'_1(e^{ct})) \,;\, t \geq 0) \xrightarrow[c \to \infty]{(f.d)} (\pi^2 \Gamma_t \,;\, t \geq 0)$

2) *If* $f : \mathbb{R}^4 \to \mathbb{R}$ *satisfies the same hypothesis as in theorem 1, then* :

$$\frac{1}{c} \int_{T(e^{ct})} ds_1 \, ds_2 \, f(B_{s_2} - B'_{s_1}) \xrightarrow[c \to \infty]{(f.d)} (\frac{1}{8\pi^2} \int dx \, f(x) \, \Gamma_t \,;\, t \geq 0).$$

In both statements, $(f.d)$ *indicates the convergence of finite dimensional marginals and* $(\Gamma_t \,;\, t \geq 0)$ *is an increasing process with independent increments such that, for each* t, $\frac{1}{t} \Gamma_t$ *is distributed as the square of a centered gaussian variable, with variance* 1.

Theorem 6 was first used by Le Gall [14] to prove a conjecture of S. Taylor [23] concerning the Hausdorff measure functions for Brownian multiple points.

Using different techniques, Le Gall [18] has now obtained the exact Hausdorff measure functions for all orders of multiplicity k in dimensions 2 and 3.

Le Gall [19] presents a survey of the topics discussed in this paragraph, as well as their relationship with asymptotics for random walks.

6. APOLOGY.

Lack of space prevent me from dealing with further studies of Brownian paths intersections, in particular the work of Aizenman [2].

Albeverio [1] offers a more complete picture of the physicists interests in Brownian paths intersections.

REFERENCES :

1 S. ALBEVERIO : Some points of interaction between stochastic analysis and quantum theory. Proceedings Bonn Meeting. ed. Helmes (1985).

[2] M. AIZENMAN : The intersection of Brownian paths as a case study of a renormalization group method for quantum field theory. Comm. Math. Phys. 97, 91-110 (1985).

[3] J.Y. CALAIS, M. YOR : Renormalisation et converveŗgence en loi pour certaines intégrales multiples associées au mouvement Borwnien dans $\mathbb{R}^d$. To appear in Sém. Probas. XXI. Lect. Notes in Maths. Springer (1987).

[4] A. DVORETZKY, P. ERDŐS, S. KAKUTANI : Double points of paths of Brownian motion in the plane. Acta Sci. Math. (Szeged) 12, 74-81 (1950).

[5] A. DVORETZKY, P. ERDŐS, S. KAKUTANI : Multiple points of Brownian motion in the plane. Bull. Res. Council Israêl Sect. F 3, 364-371, 1954.

[6] A. DVORETZKY, P. ERDŐS, S. KAKUTANI : Triple points of Brownian motion in 3-space. Proc. Cambridge Philos. Soc. 53, 856-862 (1957).

[7] E.B. DYNKIN : Additive functionals of several time-reversible Markov processes. J. Funct. Anal. 42, 64-101, 1981.

[8] E.B. DYNKIN : Random fields associated with multiple points of the Brownian motion. J. Funct. Anal. 62, 397-434, 1985.

[9] E.B. DYNKIN : Regularized self-intersection local times of the planar Brownian motion. Sém. Probas. XX. Lect. Notes in Maths. 1204. Springer (1986).

[10] D. GEMAN, J. HOROWITZ, J. ROSEN : A local time analysis of intersections of Brownian paths in the plane. Ann. Proba. 12, 86-107 (1984).

[11] K. ITÔ, H.P. Mc KEAN : Diffusion processes and their sample paths. Springer (1965).

[12] S. KUSUOKA : On the path property of Edwards model for long polymer chains in three dimensions. Proc. Bielefeld Conference in Infinite dimensional analysis and stochastic processes. S. Albeverio, ed. Res. Notes Math. 124. Pitman (1985).

[13] J.F. LE GALL : Sur le temps local d'intersection et la méthode de renormalisation de Varadhan. Sém. Probas XIX. Lect. Notes in Maths. 1123. Springer (1985).

[14] J.F. LE GALL : Sur la saucisse de Wiener et les points multiples du mouvement Brownien. To appear in Annals of Proba. (1986).

[15] J.F. LE GALL : Propriétés d'intersection des marches aléatoires, I. Convergence vers le temps local d'intersection. Comm. Math. Phys. 104, 471-507 (1986).

[16] J.F. LE GALL : Propriétés d'intersection des marches aléatoires, II. Etude des cas critiques. Comm. Math. Phys. 104, 509-528 (1986).

[17] J.F. LE GALL : Fluctuation results for the Wiener sausage. Preprint (1986).

[18] J.F. LE GALL : The exact Hausdorff measure of Brownian multiple points. To appear in the Seminar on Stochastic Processes. Birkhauser (1986).

[19] J.F. LE GALL : Some intersection properties of Brownian paths and random walks. To appear in the Proceedings of the Tashkent Congress of the Bernoulli Society (1986).

[20] J. ROSEN : A local time approach to the self-intersections of Brownian paths in space. Comm. Math. Phys. 88, 327-338, 1983.

[21] J. ROSEN Tanaka's formula and renormalization for intersections of planar Brownian motion. To appear in Ann. Proba.

[22] J. ROSEN : A renormalized local time for multiple intersections of planar Brownian motion. Sém. Probas XX. Lect. Notes in Maths. 1204. Springer (1986).

[23] S.J. TAYLOR : Sample paths properties of processes with independent increments. In : Stochastic Analysis ; D. Kendall, G. Harding, eds. Wiley (1973).

[24] S.R.S. VARADHAN : Appendix to : Euclidean Quantum Field Theory, by K. Symanzik. In : Local Quantum Theory ; R. Jost, ed. Academic Press (1969).

[25] M.J. WESTWATER : On Edwards model for long polymer chains. Comm. Math. Physics 72, 131-173, 1980.

[26] M. YOR : Renormalisation et convergence en loi pour les temps locaux d'intersection du mouvement brownien dans $\mathbb{R}^3$. Sém. Probas XIX. Lect. Notes in Math. 1123. Springer (1985).

[27] M. YOR : Sur la représentation comme intégrale stochastique des temps d'occupation du mouvement brownien dans $\mathbb{R}^d$. Sém. Probas XX. Lect. Notes in Math. 1204, Springer (1986).

[28] M. YOR : Précisions sur l'existence et la continuité des temps locaux d'intersections du mouvement brownien dans $\mathbb{R}^2$. Sém. Probas XX. Lect. Notes in Math. 1204. Springer (1986).

[29] M. YOR : Compléments aux formules de Tanaka-Rosen. Sém. Probas XIX. Lect. Notes in Maths. 1123. Springer (1985).

HAMILTONIAN FORMALISM FOR FINITE-ZONE SOLUTIONS OF NON-LINEAR INTEGRABLE EQUATIONS.

SOLOMON J. AL'BER

and

MARK S. AL'BER

The authors have found infinite sequences of series of integrable non-linear equations and by the method of recurrence relations the authors have studied the corresponding Hamiltonian systems. The equations of the Korteweg-de Vries (KdV) type, the equations of C. Neumann problem, the equations of the inverse KdV type, the non-linear Schrödinger equations and the Sine-Gordon equations, the equations of the Jacobi problem are only the initial members of the found sequences.

These sequences have the following generating equations

$$\partial U/\partial t = -\mathcal{B}'''_D/2 + 2\mathcal{B}'_D U + \mathcal{B}_D U' \equiv \mathfrak{M}[\mathcal{B}_D] \tag{1}$$

which are equivalent to the Lax system

$$L\Psi = 0\ , \ ((\partial L/\partial t) + LA)\Psi = 0\ , \ A = \mathcal{B}_D d/dx - \mathcal{B}'_D/2 \tag{2}$$

and are connected to the opérator

(3) $L=-d^2/dx^2+U(x,t,\lambda)$

The stationnary generating equations can be written in the form

(4) $\mathfrak{m}_0[\mathcal{B}] = \mathcal{B}''\mathcal{B} - \mathcal{B}'^2/2 - 2\mathcal{B}^2 U = 2C(\lambda)$

The equations (4) are the main equations as they not only give the system of differential equations of the stationary problem but also the complete system of first integrals. Moreover the space of solutions of these equations gives the domains of definition for the corresponding dynamical systems with into generating equations (1) which can be transformed

(5) $\partial\mathcal{B}/\partial t = \mathcal{B}_D\mathcal{B}' - \mathcal{B}'_D$

Assume that the potential U has the form

$$U(x,t,\lambda)=Z(\lambda)\left(\frac{M(\lambda)}{Z(\lambda)}\right)'+(M(\lambda))^2+\left(\frac{1}{2}\Sigma\frac{m_K\lambda'_K}{\lambda-\lambda_K}\right)'$$

(6)

$$+\left(\frac{1}{2}\Sigma\frac{m_K\lambda'_K}{\lambda-\lambda_K}\right)^2-R(\lambda)\,Z(\lambda)$$

where

$$Z(\lambda)=\prod_{K=1}^{N}(\lambda-\lambda_K)^{m_K},$$

(7) $$M(\lambda)=\sum_{K=1}^{N}\sum_{\ell=1}^{m_k}\mu_{K,\ell-1}\,\varphi_{k,\ell}(\lambda),$$

$\varphi_{K,\ell}(\lambda)$ are the Lagrange-Hermite polynomials.

<u>Definition 1</u>. We choose $\mathcal{B}=G_n(\lambda)/Z(\lambda)$, $\mathcal{B}_D=(\Sigma A_j\,G_{n_j}(\lambda))/Z(\lambda)$

where $G_m(\lambda)=\sum_{s=0}^{m}g_s\lambda^{m-s}$ are polynomials in λ.

Then the solutions $\mathcal{B}$, $\mathcal{B}_D$ are called finite-zone or Abelian solutions.

<u>Example 1</u>: If $U=u(x,t)-\lambda$, $\mathcal{B}_D=\sum_{K=0}^{n}B_K\lambda^{n-K}$, $N=0$,

then (1) is a generating equation for the equations of KdV type.

Example 2 : If $U = u(x,t) - \lambda$, $\mathcal{B} = \sum_{j=1}^{n+1} (\prod_{s\neq j} \lambda - \ell_s))\, q_j^2$, N=0, then (4) is a generating equation of C. Neumann systems for the motion of particles on a sphere.

Example 3 : If $U = u(x,t)/\lambda$, $\mathcal{B}_D = \sum_{K=0}^{n} B_K \lambda^{n-K}$, N = 0 , then (1) is a generating equation for the equations of the inverse KdV type.

Example 4 : If $U = u(x,t)/\lambda$, $\mathcal{B} = \sum_{j=1}^{n+1} (\Pi_{s\neq j}(\lambda - a_s))q_j^2$, N = 0 , then (4) is a generating equation for Jacobi systems for geodesics on quadrics.

Example 5 : If $\mu_1 = i(\bar{\psi} - \psi)/2$, $\lambda_1 = -(\bar{\psi} + \psi)/2$, N = 1 , then (1) is a generating equation for equations of the non-linear Schrödinger type.

Example 6 : If $\mu_1 = -i\,(u' + \dot{u})/4$, $\lambda_1 = \exp(-iu)$, N = 1 , then (1) is a generating equation for the equations of the Sine Gordon type.

The authors show that the systems with generating equations (4) and (5) and with potential (6) are Hamiltonian systems.

Introduce the momenta canonically conjugate to M as the coefficients of the formal series.

$$(8) \qquad W_G = 1/Z(\lambda)\,[-(G'/2G) + \sum_{K=1}^{N} \sum_{\ell=1}^{m_K} ((\partial^{\ell-1}/\partial\lambda^{\ell-1})(G'/2G))_{\lambda=\lambda_K} \varphi_{K,\ell}(\lambda)]\,.$$

<u>Theorem 1</u>. The equation (4) with potential (6) is equivalent to a canonical system with Hamiltonian

$$(9) \qquad \mathcal{H} = \Delta_{-1}[GW_G^2\, Z(\lambda) + G(\lambda)\, Z(\lambda)/G - 2GW_G M(\lambda) + GR(\lambda)]\,.$$

The system has a complete system of first integrals H_K.
The dynamical equation (5) can be also represented as a Hamiltonian system. Now go over to the root variables $G = \Pi(\lambda-\gamma_j)$ and momenta W_j^{γ} canonically conjugated to γ_j. In root variables the Hamiltonians become

$$(10)\qquad \mathcal{H} = \sum_{j=1}^{n} \frac{((W_j^{\gamma})^2 + C(\gamma_j))Z(\gamma_j)}{\prod_{\ell \neq j} (\gamma_j - \gamma_\ell)}$$

$$(11)\qquad \mathcal{H}_D = \sum_{j=1}^{n} \frac{((W_j^{\gamma})^2 + C(\gamma_j)}{\prod_{\ell \neq j} (\gamma_j - \gamma_\ell)} \quad (\Sigma A_K G_{n_K}(\gamma_j))$$

Example 4:: (Jacobi problem). The geodesics Hamiltonian is

$$\mathcal{H} = -2\,\Sigma \frac{(W_j^{\gamma})^2}{\gamma_J} \quad \frac{\prod_\ell (\gamma_j - a_\ell)}{\prod_{s \neq j} (\gamma_j - \gamma_s)}$$

Example 5 : (Non-linear Schrôdinger equation). The Hamiltonians are

$$\mathcal{H} = \sum_{j=1}^{n} \frac{((W_j^{\gamma})^2 + C(\gamma_j))(\gamma - \lambda_1)}{\prod_{\ell \neq j} (\gamma_j - \gamma_\ell)},$$

$$\mathcal{H}_D = \sum_{j=1}^{n} \frac{((W_j^{\gamma})^2 + C(\gamma_j))(1 - G_2(\gamma_j))}{\prod_{\ell \neq j} (\gamma_j - \gamma_\ell)}$$

Example 6 : (sine Gordon equation). The Hamiltonians are

The following system of first integrals has been found

(12) $$(W_j^{\gamma})^2 + C(\gamma_j) = 0$$

Definition 2, A Riemann surface defined by an algebraic equation $W^2 = -C(\gamma)$ where γ is a complex variable, is called a spectrum of the problem. The variables γ_j are complex variables on this surface.

Finally let us go to the "action-angle" variables H_K, W_K^H. Introduce the action function

(13) $$S(\gamma_1,\ldots,\gamma_n\,;H_1,\ldots,H_n) = \sum\int_{\gamma_j^{\,o}}^{\gamma_j} W_j^{\gamma}\, d\gamma_j$$

and momenta W_K^H canonically conjugate to H_K

(14) $$W_K^H = -\,\partial S/\partial H_K$$

Denote

(15) $$J_{n,k} = -\frac{i}{2}\sum_j \int_{\gamma_j^0}^{\gamma_j} \frac{\gamma_j^{\,n-1-K}}{\sqrt{C(\gamma_j)}}\, d\gamma_j$$

Example 4 . (Jacobi problem). The Hamiltonian system for geodesics on quadrics is reduced to a Jacobi integral system of inversion

(16) $$J_{n,K} = -\,\delta_0^{\,K}(x-x_0)+2W_K^{\,H}(x_0)\,,\ K = 0, 1,\ldots, n-1\,.$$

This system for $K=0$ contains the Abelian integrals of second type. But by a change of variables this system can be transformed into the problem of inversion with only first type the Abelian integrals.

<u>Example 5</u>. (Non-linear Schrödinger equation). The problem of finite-zone solutions is reduced to two integral systems of inversion

(17) $J_{n,K} = -\delta_1^K(x-x_0)+W_K^H(x_0)$, $K=1,\ldots,n-1$

(18) $J_{n,K} = -\delta_2^K(t-t_0)+W_K^H(t_0)$, $K=1,\ldots,n-1$.

Notice that the number n of variables γ_j is more then the number $(n-1)$ of equations.

<u>Example 6</u>. (Sine-Gordon equation). The problem of finite-zone solutions is also reduced to two integral systems of inversion.

(19) $J_{n,K} = -(\delta_1^K + \delta_n^K)(x-x_0)+W_K^H(x_0)$, $K= 1,\ldots,n$

(20) $J_{n,K} = -(\delta_1^n - \delta_n^K)(t-t_0)+W_n^H(t_0)$, $K= 1,\ldots,n$

Discrete system have been investigated by the same method of recurrence relations.

Integrable Hamiltonian systems are deeply connected with corresponding quantum operators of Laplace-Beltrami type. The series of quasi-classical (WKB) eigenvalues and eigenfunctions of operators of Laplace-Beltrami type have been found by M.S. Al'ber.

(21) $\nabla^K \nabla_K u + \omega^2 u(V-E) = 0$. where $u = u(\gamma_1, \ldots, \gamma_n)$.

Example 1. (Equations of KdV type). The corresponding tensor has the form

$$g^{jm} = \frac{\delta_j^m}{\prod_{\ell \neq j} (\gamma_j - \gamma_\ell)} \quad \text{and} \quad V = \sum_{j=1}^{n} \frac{G(\gamma_j)}{\prod_{\ell \neq j} (\gamma_j - \gamma_\ell)}$$

Example 2. (C. Neumann problem). The corresponding tensor is

$$g^{jm} = \frac{4 \prod_{s=1}^{n+1} (\gamma_j - \ell_K)}{\prod_{K \neq j} (\gamma_j - \gamma_\ell)} \delta_j^m \quad \text{and}$$

$$V = -1/2 \sum_{K=1}^{n} \gamma_K + 1/2 \sum_{s=1}^{n+1} \ell_s$$

Example 3. (Equations of inverse KdV type)

$$g^{jm} = \frac{\delta_j^m}{\prod_{\ell \neq j} (\gamma_j - \gamma_\ell)} \qquad \text{and} \qquad V = \sum_{j=1}^{n} \frac{G(\gamma_j)/\gamma_j}{\prod_{\ell \neq j} (\gamma_j - \gamma_\ell)}$$

Example 4. (Jacobi problem for geodesics on quadrics). The corresponding Riemann metrical tensor has the form

$$g^{jm} = -\frac{4 \prod_{s=1}^{n+1} (\gamma_j - a_s)}{\gamma_j \prod_{\ell \pm j} (\gamma_j - \gamma_\ell)} \delta_j^m \quad \text{and} \quad V = 0 \;.\; E = 1$$

Example 5. (Jacobi problem for geodesics in domains bounded by quadrics). The corresponding Riemann metrical tensor is

$$g^{jm} = -\frac{4 \prod_{s=1}^{n} (\gamma_j - a_s)}{\prod_{\ell \pm j} (\gamma_j - \gamma_\ell)} \delta_j^m \quad \text{and} \quad V = 0 \;.\; E = 1$$

With the following boundary conditions $u|_{Q^{n-1}} = 0$ or

$\partial u / \partial \gamma_j |_{Q^{n-1}} = 0$.

Using in (21) the standard substitution $u = A\exp(i\omega S)$ one obtains the quasi-classical system

(22)
$$\nabla^K(A^2\nabla_K S) = 0 \quad \text{amplitude equation}$$
$$\nabla^K S.\nabla_K S + V = E \quad \text{eikonal equation}$$

In this system the action function $S(y_1,\dots,y_n)$ and the amplitude $A(y_1,\dots,y_n)$, where y_j are the root variables, are considered on the corresponding Jacobi variety.

Here we describe the results for the Jacobi problem for geodesics on quadrics. In this case the action function has the form

$$S = \sum \int_{y_j^0}^{y_j} W_j^{\gamma} dy_j = i/2 \sum \int_{y_j^0}^{y_j} \sqrt{\frac{y_j\, M(y_j)}{\prod_{\ell=1}^{n+1}(y_j - a_\ell)}}\, dy_j \ ,$$

where $M(y) = M_0(y^{n-1} + \dots M_{n-1}) = M_0 \prod_{K=1}^{n} (m_K - y)$

Here y_j varies along real closed cycles ℓ_j on the two sheeted Riemann surface with ends :

(24) m_j, a_j for $j < j_0$; a_{j_0+1}, a_{j_0} for $j = j_0$; a_j, m_j for $j > j_0$

We denote $(\gg\tau_1, \ldots, \gg\tau_j, \ldots, \gg\tau_n), j \neq j_0$ the sheet of the covering manifold of the Jacobi variety. Notice that is a number of points of tangency of caustic

The component S_j of S depending on γ_j on this sheet is

(25) $S_j = (-1)^{\tau_j} \gg L_j + 2\tau_j L_j$

where $\tau_j = [\gg\tau_j + 1)/2]$, $\gg [\tau_j = 0, \ldots, \gg K_j$

and $L_j = \oint_{\ell_j} W_j^{\gamma} dy$, $\gg L_j = \int_{\gamma_j}^{\gamma_j} N_j^{\gamma} d\gamma_j$.

Here the initial points of integration are chosen as $\gamma^o{}_{j_0} = a_{j_0+1}$, $\gamma_j^o = m_j$ for $j \neq j_0$. The quasi-classical eigenfunctions have the form

$$(26) \qquad U = c \sum_{\substack{(\tau_1,\dots,\tau_n) \\ 0 \leqslant \tau_j \leqslant K_j}} \frac{L}{\prod_{k=1}^{n-1} \prod_{j=1}^{n} (m_k - \gamma_j)^{1/4}}$$

$$\exp\left[\frac{-i\pi}{2} \sum_{\substack{j=1 \\ j \neq j_0}}^{n} \varkappa_j \tau + i\omega \sum_j ((-1)^{j\varkappa_j \tau} \varkappa_j + 2\tau_j L_j) \right]$$

where $\varkappa K_j$ and the eigenvalues ω_ℓ are defined by the quantum conditions

$$(27) \qquad 4\omega/\pi = (\varkappa K_j + 4N_j)/K_j$$

$$\omega L_{j_0}/\pi = N_{j_0}/K_{j_0} \quad \text{where} \quad j = 1,\dots,n,(j \neq j_0); \; K_j = [\varkappa K_j + 1]/2.$$

The results are extended to the complex case. Analogous formulas are obtained for the problem on geogedesics in domains bounded by quadrics. Results in cases of dimensions 2 and 3 coïncide with the classical results of J.B. Keller and Rubinow.

The WKB-solutions for operators corresponding to equations of KdV and inverse KdV types and C. Neumann problem have also been found.

Remark. The well-known problem of equality of roots of the spectrum was solved by a change of variables and taking a limit.

We would like to express our deep gratitude to Professeur Henri Cartan, Professor Joël Lebowitz and Professor Norman Zabusky.

BIBLIOGRAPHY

1. J. Al'ber, Investigation of equations of KdV type by the method of recurrence relations, Preprint, I. Chem. Phys., USSR, Chernogolovka, (1976).

2. S.J. Al'ber, Investigation of equations of Korteweg-de Vries type by the method of recurrence relations, J. London Math. Soc.(2)19 (1979), 467-486.

3. S.J. Al'ber, On stationary problems for equations of Korteweg-de Vries type, Comm. Pure Appl. Math. 34(1981), 259-272.

4. S.J. Al'ber and M.S. Al'ber, Stationary problems for equations of KdV type and geodesics on quadrics, preprint, I. Chem. Phys., USSR, Chernogolovka, (1984).

5. S.J. Al'ber and M.S. Al'ber, Hamiltonian formalism for finite-zone solutions of integrable equations, C.R. Acad. Sc. Paris, 301, serie t, (1985), 777-780.

6. S.J. Al'ber and M.S. Al'ber, Hamiltonian formalism for non-linear integrable equations and systems, Deposit manuscript, I. Sci. Tech. Inf. (VINITI), (1986).

7. S.J. Al'ber and M.S. Al'ber, Hamiltonian formalism for non-linear Schrödinger and sine-Gordon equations, Preprint, I. Chem. Phys. USSR, Chernogolovka, (1985).

RELEVANCE OF CLASSICAL CHAOS IN QUANTUM MECHANICS

Giulio Casati
Istituto di Fisica - Università degli Studi di Milano
Via Celoria 16, 20133 MILANO
ITALY

ABSTRACT

We study the excitation and ionization of highly excited hydrogen atoms irradiated by microwaves. Our studies reveal some unexpected new features of quantum dynamics. In particular we show the existence of a large ionization peak at frequencies much below those required for the conventional one-photon photoelectric effect. The frequency-width of this peak is jointly determined by two independent effects: the classical chaotic threshold and the quantum delocalization border.

The recent progress in understanding the behaviour of classical dynamical systems and in particular the discovery of the so-called chaotic motion in deterministic systems has greatly contributed to our understanding of non-equilibrium statistical mechanics. A main feature of these systems is exponential separation of initially close orbits for almost all initial conditions which leads to exponential loss of memory, decay of correlations and therefore provide the possibility of deriving classical statistical mechanics without additional assumptions and/or recourse to the so-called thermodynamic limit. For example, the Fourier law of heat conduction has been recently verified[1)] in a purely dynamical deterministic system.

Turning now to quantum mechanics, we are faced at once with a striking difference. Despite the fact that quantum mechanics is an essentially statistical theory, there is nothing in the type of evolution described by the Schroedinger equation, so complex as the structure of classical orbits. Indeed, the discretness of the energy spectrum of finite particle, conservative quantum systems leads to almost periodicity in time of the motion so that at most, ergodic behaviour is allowed. Still, it is possible that the complexity of the classical motion reappears here in some particular feature of the energy level statistics, or in the eigenfuntions[2).

However perhaps more illuminating is the consideration of systems under external periodic perturbations since they may allow for a continuous spectrum of the motion (the so-called quasi-energy spectrum). A particular example in this class of systems which has been extensively studied[3)], is the so-called δ-kicked rotator - a rotator subject to external time-periodic pulses. The analysis of this model has revealed that quantization places severe limitations to the classical chaotic motion.

If the above limitation is of a general nature, then it should be possible to devise experiments which give evidence of this lack of chaotic behaviour in quantum systems. In order to illustrate the actual possibility of such experiments, we shall now briefly describe a more realistic system, which provides a model for the behaviour of an hydrogen atom in a microwave field.

To this end, let us consider the ionization mechanisms for the hydrogen atom when a linearly polarized monochromatic electric field induces transitions from initial states having principal quantum number $n_0 \gg 1$. For simplicity, we restrict ourselves to the study of very elongated quantum states having parabolic quantum numbers $n_1 = n_0 - 1$, $n_2 = 0$, and magnetic quantum number m=0. Since to a good approximation these wave functions have nonzero values only along the direction of the applied field, we are at liberty here to treat the hydrogen atom as it if were one dimensional, having the Hamiltonian

$$H = p^2/2 - 1/x + \varepsilon x \cos(\omega t), \qquad x > 0 \tag{1}$$

where ε and ω are the microwave electric field strength and frequency, respectively, in atomic units. The validity of this one-dimensional approximation is due to the small value of matrix elements for transitions having $\Delta n_2 \neq 0$. As a consequence, the atom remains one dimensional during the relevant interaction times.

Classical analysis[4)] reveals that for microwave field strengths $\varepsilon_0 = \varepsilon n_0^4$ larger than the critical value

$$\varepsilon_c \approx 1/50\,\omega_0^{1/3} \tag{2}$$

a chaotic excitation of the hydrogenic electron occurs which obeys a diffusion law, where $n_0 \gg 1$ denotes the classical action of the initially excited state and where $\omega_0 = \omega n_0^3$ is required to be greater than unity.

A recent analysis[5] of the quantum behaviour of the microwave-driven hydrogen atom has revealed the existence of a critical field value $\mathcal{E}_q$, called the <u>quantum delocalization border</u>, below which quantum effects suppress diffusive excitation. However, for field values $\mathcal{E}_0$ above this border, the quantum excitation proceeds in much the same way as the classical. The critical field value has been shown to be given[5] by

$$\mathcal{E}_q \approx \omega_0^{7/6} / \sqrt{6 n_0} \qquad (\omega_0 \geq 1) . \tag{3}$$

The model (1) was the object of extensive numerical investigations, which fully confirm the main theoretical predictions[5,6]. Namely, three distinct quantum regimes can be observed according to the particular choice of $\mathcal{E}_0$, ω_0 and of the initial state n_0. If $\mathcal{E}_0$ lies below both the classical stochasticity border $\mathcal{E}_c$ and the delocalization border $\mathcal{E}_q$, then both the classical and the quantum model will exhibit localization in action space.

If $\mathcal{E}_0$ lies in between the thresholds $\mathcal{E}_c$ and $\mathcal{E}_q$, we observe classical chaotic motion and quantum localization; here the classical atom would ionize diffusively.

Finally, if $\mathcal{E}_0$ exceeds both threshold, we observe a qualitative agreement between classical and quantum results. Here a quantum ionization mechanism is at work, that cannot be understood in terms of standard perturbation theory. Indeed, this mechanism is effective at much lower frequencies than those ($\omega_0 > n_0/2$) predicted by the perturbative theory of the photoelectric effect.

Due to the delocalization phenomenon, the dependence of the ionization probability W_I on frequency at <u>fixed field intensity</u> displays a number of unexpected features (Fig. 1). In particular the truly striking result here is not merely that the quantum diffusive ionization occurs at frequencies low compared to the one-photon threshold but that its rate is much higher.

The present high level of experimental technique encourages the hope that this phenomenon can be observed in the laboratory.

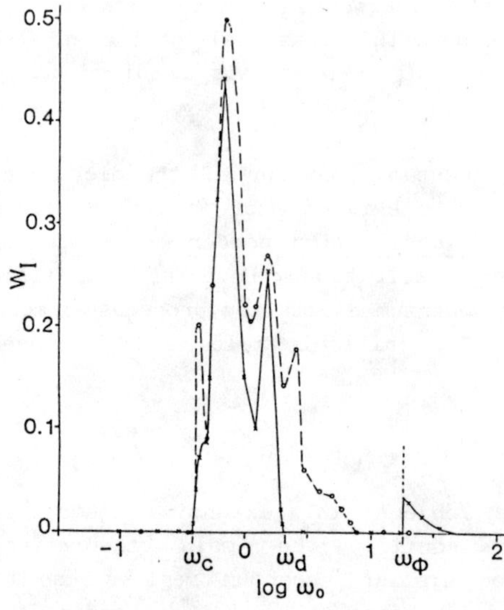

FIG. 1 Ionization probability $W_I = \sum_{n>\bar{n}} |C_n|^2$ vs field frequency ω_o after a time $\tau = 40\ \omega_o$ which corresponds to the same real physical time t for all frequencies. We have set $n_o = 66$, $\varepsilon_o = 0.05$, $\bar{n} = 99$. Crosses, quantum theory; circles, classical theory. Notice that ω_ϕ is here somewhat less than $n_o/2$ because, in our definition of the ionization probability, the contribution of states with $n > \bar{n}$ is also included.

REFERENCES

1. G. Casati, J. Ford, W.M. Visscher, F. Vivaldi: "One-dimensional Classical Many-Body System having a normal Thermal Conductivity" Phys. Rev. Lett. 52, 1861 (1984).

2. Proc. International Conference on "Quantum Chaos", Como 1983 (Plenum, 1985). Edited by G. Casati.

3. B. V. Chirikov, F.M. Izrailev, D.L. Shepelyansky, Soviet Scientific Review, 2C, 209 (1981).

4. N.B. Delone, V.P. Krainov, and D.L. Shepelyansky, Usp. Fiz. Nauk. 140, 355 (1983) (Sov. Phys. Usp. 26, 551 (1983)); R.V. Jensen Phys. Rev. A 130, 386 (1984).

5. G. Casati, B.V. Chirikov, and D.L. Shepelyansky, Phys. Rev. Lett. 53, 2525 (1984).

6. G. Casati, B.V. Chirikov, D.L. Shepelyansky, I. Guarneri, Phys. Rev. Lett. 57, 823 (1986).

PATTERN SELECTION IN HYDRODYNAMICAL EQUATIONS

P.Collet[(1)] and J.-P.Eckmann[(2)]

ABSTRACT

We present some recent results about the existence, selection and stability of propagating solutions (fronts) for nonlinear parabolic equations which appear in various physical problems.

Nonlinear parabolic equations are used to model a rather large number of physical systems mostly related to hydrodynamics and heat conduction. In particular, recent developments have focused on questions of solidification (and metallurgical applications) [6] , flames propagations [7], chemical reactions etc. Related problems are also found in the study of porous media and for diffusion limited aggregation. All these examples correspond to physical situations where the underlying space is unbounded (or at least very large) as opposed to the compact situation which is usually considered. The main problem is to study the dynamics of the transition from one stationary solution (for example the undercooled liquid) to another one (a solid phase for example). Contrary to the case where the evolution is described by a linear parabolic equation, in the nonlinear case the transition has in general a non zero speed, i.e. the "interface" (which is not sharp) propagates at a non zero speed. The first problem is to prove that this speed is non zero and eventually to compute its value. This is however obscured by the fact that in general several propagation speeds are possible and one wants to know which one is selected. The second problem is related to the nature of the new phase left behind the front. There are in general several possibilities among which one is chosen. In other words which pattern is selected? This second question is intimately related to the problem of speed selection mentioned above. Finally a more difficult question is to analyze the structure of the interface which may be very complicated (eventually fractal).

(1) Centre de Physique Théorique, Ecole Polytechnique F91128 Palaiseau Cedex (France)

(2) Département de Physique Théorique, Université de Genève 32 Bd d'Yvoy CH1211 Genève4 (Switzerland)

One should also mention that numerical simulations are rather difficult because there are marginal continuous spectra and therefore very long transients. The above questions are best understood in dimension one although some recent progress has been made in higher dimensions [1][7]. We shall first introduce a precise definition of a front. We shall consider a quasi linear parabolic equation for the state function u given by

$$\dot{u} = \mathrm{P}(x,\partial)u + f(u,\partial u,\cdots,\partial^j u) \tag{1}$$

where $\mathrm{P}(x,\partial)$ is an elliptic partial differential operator, $\partial = \partial/\partial x$ and the index j is smaller than the degree of P. The dot denotes time derivation. Let u_1 and u_2 be two stationary solutions of (1). A front from u_1 to u_2 at speed c is a function v of two variables x_1 and x_2 such that

i) $v(x,-\infty) = u_2$, $v(x,+\infty) = u_1$,

ii) $v(x, x-ct)$ is a solution of (1) .

In other words the phase u_2 is propagating in the phase u_1. If u_1 and u_2 are constant one may look for fronts which are independent of x_1. A major result is due to Aronson and Weinberger [1]. For simplicity we shall state their theorem for equations of the form

$$\dot{u} = u'' + g(u) \tag{2}$$

where u is positive, $g(0) = g(1) = 0$, g is C^2 and g is strictly positive on $]0,1[$. Using the maximum principle they prove that for some $c_0 > 0$, and any $c \in [c_0, +\infty]$, (2) has a front at speed c from 0 to 1. Moreover any reasonable initial condition will converge asymptotically (in time) to a translate of one of these fronts (see also [2]). For a given initial fonction u_0, the front to which it converges is selected by the exponential decay at infinity. In particular all initial functions with compact support will propagate asymptotically at the same speed. In [3] we have investigated similar questions for the following equation (see [5] for previous results and motivations)

$$\dot{u} = \epsilon u - (1+\partial^2)^2 u - \epsilon u^3 \tag{3}$$

where ϵ is positive and small. u_2 is equal to zero and u_1 is a periodic function. It is not possible anymore to use the maximum principle. One can prove the existence of fronts using singular perturbation theory in ϵ (see [8] for the nonlinear WKB method). Using this technique in the moving frame one can control the solution from $-\infty$ to a position of order $O((\log \epsilon)^5)$. From there to $+\infty$ one uses a matching technique. The free constants of integration at $-\infty$ are chosen so that the solution belongs to the stable manifold (of infinite codimension) of the trivial solution zero. Using similar ideas one can analyze the stability of these front solutions (see[4]).

REFERENCES.

[1] D.G.Aronson, H.Weinberger. Multidimensional nonlinear diffusions arising in population genetics. Adv. Math. 30, 30 (1978).

[2] M.Bramson. Convergence of solutions of the Kolmogoroff equations to travelling waves. Memoirs AMS 285 (1983).

[3] P.Collet, J.-P.Eckmann. The existence of dendritic fronts. Preprint University of Geneva (1986).

[4] P.Collet, J.-P.Eckmann. The stability of dendritic fronts (in preparation).

[5] G.Dee, J.S.Langer. Propagating pattern selection. Phys. Rev. Letter 50, 383 (1983).

[6] J.S.Langer. Instabilities and pattern formation in crystal growth. Rev. of Mod. Phys. 52, 383 (1980).

[7] J.S.Langer. Existence of needle crystals in local models of solidification. Phys. Rev. A33, 435 (1986).

[8] V.Maslov, G.Omel'yanov. Asymptotic soliton-form solutions of equations with small dispersions. Russ. Math. Survey 36, 73 (1981).

[9] J.Wesfried, S.Zaleski. Cellular structures in instabilities. Lecture Notes in Physics 210. Springer Verlag, Berlin Heidelberg New-York (1984).

SOME GENERAL PROPERTIES OF EVOLUTION PROCESSES

QUASI-DETERMINISTIC APPROXIMATION

Mark Freidlin
Moscow

ABSTRACT

We consider general properties of evolution processes, i.e. the growing with time of the number of carriers of a character changes due to small random perturbations. The number of the carriers of the character x changes at a rate which depends on x and on the number of carriers already available. This rate may have either sign. We are interested in tunneling transitions and wave front propagation.

This paper considers general properties of evolution processes. By evolution we do not necessarily mean biological evolution. For example, this may be evolution of some physical objects or evolution of languages. The only essential feature is that one studies the growing with time of the number of carriers of a character x. This character changes due to small random perturbations. The number of the carriers of the character x changes with a rate which depends on x and on the number of the carriers already available. This rate may have either sign. We take interest in tunnel transitions and wave front propagation. To be more descriptive, we will resort to biological terminology. In particular, x will be called genotype. The number of carriers of a character x will be termed the number of specimens with the genotype x.

So, consider a set of objects and let the state of each of them definiteness we put $X=R^r$.The phase state of every object changes with time independently of other objects. We will assume that if at time $t=0$ an object had a state (genotype)x, then at time $t>0$ it will have the genotype $X_t^\varepsilon = x + \varepsilon W_t$, where W_t is the standard Wiener process in R^r and $\varepsilon>0$ is a small real parameter. Of course the choice of the phase space and the law of genotype change (mutation process) is somewhat arbitrary. However we will be interested in the properties which are to some extent invariant with respect to this choice. To be more exact, the numerical values to be found are certainly connected with the choice of the phase space and the type of the process X_t^ε, but a number of qualitative implications remain valid under quite broad assumptions on the structure of the phase space and the process X_t^ε.

Let $u_t(x) = u^\varepsilon(t,x)$ be the density of the number of specimens with a genotype x at time t. The value

$$u_t^{-1}(x)\ \frac{\partial u_t(x)}{\partial t} = c(x,\ u_t(x),\ldots),$$

which characterizes the multiplication rate of a specimen with the genotype x, is called the fitness coefficient. Generally speaking, this value may depend on the number of the specimens with other genotype, i.e. on $u_t(y)$ for $y = x$. For the moment we assume (for the sake of simplicity) that c does not depend on the number of specimens with other genotypes: $c = c(x,u)$. With x being fixed, consider the fitness coefficient as a function of its second argument u. There are three possible types of the function.

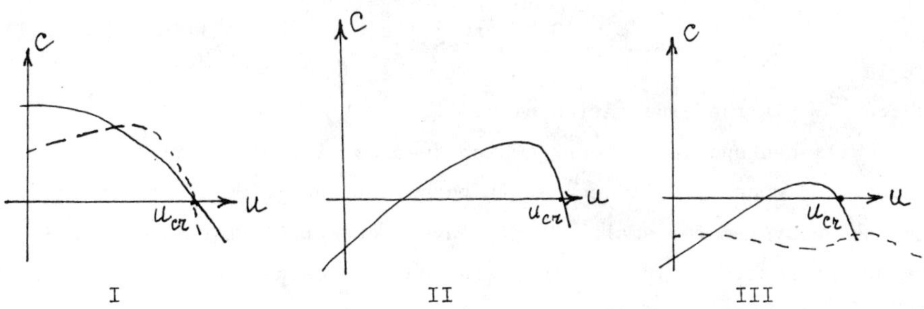

Fig. I

We will write $c = c(x,.) \in$ I (II or III) in order to indicate that c belongs to one of the classes shown in Fig. I. The basic feature of class I is that $c(0) > 0$. In class II, the function $c(u)$ is negative for small u, but afterwards it becomes positive and $\int_0^u uc(u)du < 0$. As is seen in Fig. I, in all cases one assumes that a $u_{cr} > 0$ exists

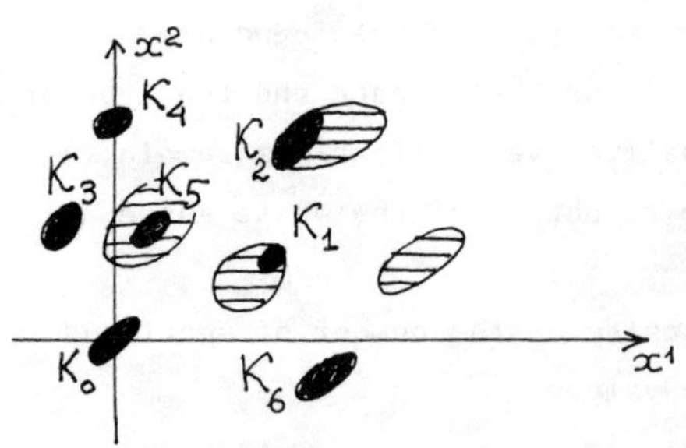

Fig. 2

such that $c(u) < 0$ for $u > u_{cr}$.

We suppose that at most points $X \in R^r$ (for the majority of genotypes) the fitness coefficient pertains to class III. In Fig. 2 the blacked regions stand for those parts of the phase space where $c(x,.) \in I$, the shaded regions - for $c(x,.) \in II$, and at all other points $c(x,.) \in III$. For simplicity we first suppose that there are no shaded regions, that is $R^r = I \cup III$ and that the regions $K_i \in I$ have small diameters $\delta_i < \delta$. Later on we will dwell on those effects which appear in the regions of class II.

A set of specimens whose genotype belong to one and the same K_i will tentatively be called species.

Suppose that at the initial time $t=0$ there were only specimens with genotypes $x \in K_o$: $u(0,x) = 0$ for $x \in K_o$ and $u(0,x) > 0$ at the interior points of K_o. The evolution with time of the density $u^\varepsilon(t/\varepsilon,x) = \hat{u}^\varepsilon(t,x)$ is described by the equation.

$$\hat{u}^\varepsilon(t,x) = E_x u^\varepsilon(0,X_t^\varepsilon)\exp\{\frac{1}{\varepsilon}\int_0^t c(X_s^\varepsilon,\hat{u}^\varepsilon(t-s,X_s^\varepsilon))ds\} \qquad (I)$$

where E_x is the expectation sign.

Will new species appear? Which species will be the first to appear and when will it occurs? It turns out that time evolution of the above system for small ε displays a number of essentially non-random properties. In particular in the general case one can find an increasing sequence of non-random times $T_1^*, T_2^*, \ldots$ and a sequence of numbers $i_o = 0, i_1, i_2, \ldots, i_j, \ldots$ such that with probability approaching 1 for small ε, the K_{i_j} species will be the j-th to appear and the appearance of this species will occur at time T_j^*. Therefore the order of appearing new species and the times when these species appear are not random to a great extent. The tree of species formation is not random either. (By a tree of species formation we mean a graph consisting of arrows. At every of newly formed species points an arrow which begins at the generating species). Before time $T^*_{i_{n+1}}$ the density of the number of specimens outside of the regions $K_o, K_{i_1}, \ldots, K_{i_n}$ is close to zero. So the process of appearing new species can in a way be regarded as tunnelling.

As will be seen below, inside of the regions of type II (and to some extent of type I) the domain of large density will propagate according to the Huygens principle for the velocity field which is defined by equation (I). Of course the times T^*_i, the sequence $K_{i_1}, K_{i_2}, \ldots$ and the wave front velocities depend essentially on the structure of the phase spece and on the nature of the mutation process X^ε_t. But the existence of a strictly defined sequence of appearances of species takes place under quite broad qualitative hypotheses on the phase space and the process X^ε_t.

We put $c(x) = c(x,0)$ for $x \notin K_o$ and $c(x) = 0$ for $x \in K_o$. Consider the functional $R_{OT}(\phi)$, $T > 0$, $\phi \in C_{OT}(R^r)$:

$$R_{OT}(\phi) = \int_0^T (c(\phi_s) - \frac{1}{2}|\dot{\phi}_s|^2)ds, \quad \dot{\phi}_s = \frac{d\phi s}{ds},$$

for absolutely continuous $\phi \in C_{OT}(R^r)$; $R_{OT}(\phi) = -\infty$ for all other $\phi \in C_{OT}(R^r)$. The functional $R_{OT}(\phi)$ is upper semi-continuous in the uniform topology.

We put $T_o = 0$, $i_o = 0$. Define T^*_1 and i_1 from the relations

$$\gamma_{ij}(T) = \sup \{R_{OT}(\phi) : \phi \in C_{OT}(R^r),\ \phi_o \in K_j,\ \phi_T \in K_i\},$$

$$T_{ij} = \min \{T : \gamma_{ij}(T) = 0\}, \tag{2}$$

$$T^*_1 = \min_j T_{oj} = T_{oi_1}.$$

In the general case i_1 is defined in a unique way. We put

$$T_2 = \min_{j : j \neq i_1} T_{oj} = T_{o,i_2},\ i_2 \neq i_1.$$

<u>Theorem I</u>. Suppose that the general situation is considered so that the minimum in (2) is attained only for $j = i_1$. If $\lim \varepsilon \cdot t(\varepsilon) < T^*_1$, then [$(K_o)$ is the interior and $[K_o]$ is the closure of K_o]:

$$\lim_{\varepsilon\downarrow 0} u^{\varepsilon}(t(\varepsilon),x) = \begin{cases} u_{cr}(x), \text{for} & x \in (K_o) \\ 0, \quad \text{for} & x \notin [K_o] \end{cases} \tag{3}$$

If $T_1^* < \underline{\lim}_{\varepsilon\, 0}\, t(\varepsilon)\cdot\varepsilon \leq \overline{\lim}_{\varepsilon\, 0}\, \varepsilon\cdot t(\varepsilon) < T_2$, then there is an open set ξ_{i_1} K_{i_1} such that

$$\lim_{\varepsilon\, 0} u^{\varepsilon}(t(\varepsilon),x) = \begin{cases} u_{cr}(x) & \text{for} \quad x \in K_o \\ u_{cr}(x) & \text{for} \quad x \in \xi_{i_1} \\ 0 & \text{for} \quad x \notin K_o \cup K_{i_1} \end{cases} \tag{4}$$

Theorem I means that at time T_1^* a new species K_{i_1} will appear (at first only a part of a genotypes of this species). The new species arises by way of tunnelling from K_o. The species K_o should be looked upon as a generating one for K_{i_1}.

We only outline the proof of Theorem I. Using (I) and the properties of the action functional we deduce that for $\overline{\lim}_{\varepsilon\to 0}\, t(\varepsilon)\cdot\varepsilon \leq T < T_1^*$ and for small ε

$$\varepsilon \ln u^{\varepsilon}(t(\varepsilon),x) \leq \sup\{R_{OT}(\phi) : \phi_o \in K_o,\ \phi_T = x\}.$$

For $T < T_1^*$ the right-hand side is negative everywhere but K_o. Therefore (3) and the second part of (4) are valid. The proof of the first part of (4) is similar to that of Theorem I in (1), see also (2), ch. 6.

We emphasize that the specimens of the species K_{i_1} to appear at time $T_1^*\varepsilon^{-1}$ will have genotypes x with certain properties. Namely these genotypes x will be close to that point of K_{i_1} from which starts the extremal giving the supremum to the expression defining γ_{oi_1}. Afterwards the specimens of all other genotypes $x \in K_{i_1}$ will appear. The domain of large numbers of $u^{\varepsilon}(t\varepsilon^{-1},x)$ may propagate over the set K_{i_1} gradually by way of continuous movement of the wave front which divides the regions where $u^{\varepsilon} \cong u_{cr}$ and $u^{\varepsilon} \cong 0$ (see below). The domain of large numbers may expand inside of K_{i_1} by leaps (see examples in (1)). After a new species K_{i_1} appeared at time $T_1^*\varepsilon^{-1}$ the next species may arise directly from K_o and also from K_{i_1}. One should bear in mind that as soon as $u^{\varepsilon}(t\varepsilon^{-1},x)$ is equal to $u_{cr}(x)$, we have $c(x,u^{\varepsilon}(t\varepsilon^{-1},x)) = 0$. Hence, the functional $R_{OT}(\phi)$ characterizes the principal logarithmic term of $u^{\varepsilon}(t\varepsilon^{-1},x)$ only provided the supremum in the definition of γ_{oi2} is attained for a function $\hat{\phi}_s$, $s \in (0,T_2)$ such that $u(s\varepsilon^{-1},\hat{\phi}_s)$ is close to 0. In particular if $\hat{\phi}_s \notin K_{i_1}$ for $s \in (T_1,T_2)$ (Condition A), then $T_2^* = T_2$. "The next" species K_{i_2} arises at time $T_2\varepsilon^{-1}$ and will be generated by the species K_o. If $\hat{\phi}_s$ touches K_{i_1} somewhere for $s \in (T_1,T_2)$, then the situation is more complicated. In this case the new species may be generated by the species K_{i_1}. We will put

$$c(x) = \begin{cases} c(x,0) & \text{for } u \notin K_o \cup K_{i_1} \\ 0, & \text{for } u \in K_o \cup K_{i_1} \end{cases}$$

Denote by $\hat{R}_{OT(\phi)}$ the functional $R_{OT}(\phi)$ with $c(x)$ replaced by $c(x)$. We will define $\hat{\gamma}_{ij}$ (T) and $\hat{T}_{ij}$ with the help of the same formulas as γ_{ij} (T) and T_{ij}, R being substituted for $\hat{R}$. Let $\hat{T}_{i2} = \min\limits_{j:j\neq o,j\neq i_1} \hat{T}_{ij} = \hat{T}_{1i_2}$. Then by analogy to Theorem I one can prove that there is T'_{i2} such that $|T'_{i2} - \tilde{T}_{i2}| < A\cdot\delta$, where A is a constant expressed through $\max c(x,0)$, and by time $\varepsilon^{-1}(T_1^* + T_{i2}) = T_2\varepsilon^{-1}$ a new species K_{i_2} will appear (i.e. $u^{\varepsilon}(t\varepsilon^{-1},x) \cong u_{cr}(x)$ for certain $x \in K_{i_2}$, $t > T_2$). In this case $T_2^* = T_2$.

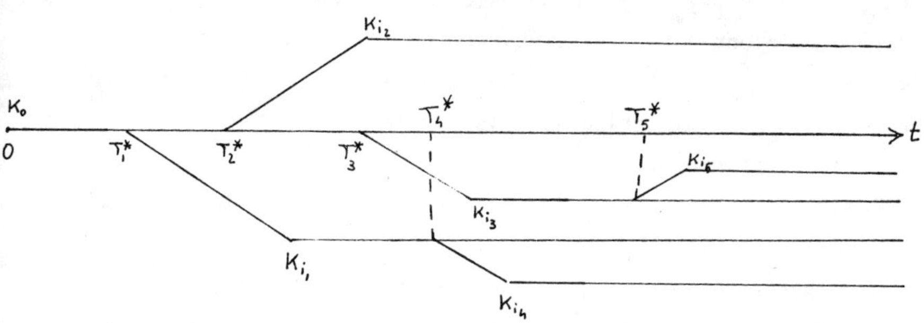

Fig.3

If Condition (A) is fulfilled, then at time $T_2^* = T_2$ the species K_{i_2} arises. Moreover in this case one can claim that K_{i_2} is generated by K_o. Unless Condition (A) is valid, a new species K_{i_2} is generated by K_{i_1} at time $T_2^*\varepsilon^{-1} = (T_1^* + T_{i_2})\varepsilon^{-1}$.

Henceforth new species may be generated by the species K_o, K_{i_1}, K_{i_2} and so on. Therefore the following general rules coming from the model are worth noting.

I. New species are generated in a time of order ε^{-1}, where $\varepsilon = \sqrt{E(X_1^\varepsilon - X_o^\varepsilon)^2}$.

2. New species are generated by way of tunnelling. The number of specimens with intermediate genotypes is exponentially small ($\exp\{-\text{const.}\ \varepsilon^{-1}\}$).

3. With probability approaching I as $\varepsilon \downarrow 0$ the evolutionary tree can be calculated in a unique way: the order of appearing the new species is defined and for every species one can indicate the generating species (Fig. 3).

4. With probability approaching I as $\varepsilon \downarrow 0$ one can indicate the times $T_1^*\varepsilon^{-1}, T_2^*\varepsilon^{-1}, \ldots, T_n^*\varepsilon^{-1}$ of appearing new species.

Remark I. The above model does not take into account competition. If we want to, it is necessary to keep in mind that each time after appearing a new species the fitness coefficient changes. At time $T_1^*\varepsilon^{-1}$ a species K_{i_1} arises and after the time $T_1^*\varepsilon^{-1}$ a new fitness coefficient $c_{T_1^*}(x,u)$ should be used. The time $T_2^*\varepsilon^{-1}$ of appearing the third species K_{i_2} should be calculated with a glance to this new fitness coefficient. For this, obvious changes should be made in the definition of T_2^*. At time T_2^* the fitness coefficient will again change and so on. In particular this may lead to the disappearance of species arisen previously.

Remark 2. Suppose that in the phase space there is a region of type II bordering with a region of type I (for example, in Fig.I it is the region bordering with K_1). Then after the species K_1 arises, specimens with genotypes from the shaded region will begin appearing. The domain of large values of $u^\varepsilon(t,x)$ will propagate over the shaded region according to the Huygens principle with the velocity of order ε. The velocity field is defined by the function $c(x,u)$ (1). We emphasize that the appearance of the specimens with genotypes from a shaded region can lead to changing the fitness coefficient, in particular, to killing the specimens with genotype from the blacked region without which the species of the corresponding shaded region could not arise. If a shaded region in R^r has no bordering region of type I, then the species which corresponds to it will never appear.

Remark 3. In the process of evolution, determinate stabilizing factors may arise such that in the vicinity of some sets $K_i \in I$, the process X_t^ε defining mutations has the form: $\dot{X}_t^\varepsilon = b(X_t^\varepsilon) + \varepsilon \dot{W}_t$, where $b(x)$ is the vector field pointing in the direction of K. The appearance of such a field $b(x)$ stabilizing the species K_i causes the change of the rate of the appearance of new species. Now it will take time of order ε^{-2} rather than that of order ε^{-1} to form a new species generated by such a species K_i.

Remark 4. It is not difficult to modify the model so that it can take into account the non-isotropy and non-homogeneity (in space and in time) of the mutation process, the assumption on the Markov nature being preserved. Then the main claims will remain true. If periodic components are allowed in the mutation process, then new effects can appear.

Now we turn to a somewhat modified model. For every genotype $x \in R^r$ there is a critical value $u_{cr}(x) \geq 0$ such that for $u > u_{cr}(x)$ the fitness coefficient $c(x,u)$ is negative. If $u_{cr}(x)$ is large enough, then in time t the density of specimens with the genotype x can be of order $\exp\{\text{const.}\ \varepsilon^{-1}\}$. Presumably one can suppose that the fitness coefficient $c(x,u)$ gradually changes as the density grows, i.e. it depends on ε in $u^\varepsilon(t,x)$: $c = c(x, \varepsilon \text{ in } u^\varepsilon)$, where $c(x,v)$ is a continuous bounded function.

Let at the initial time $u^\varepsilon(0,x) = \exp\{\alpha(x)\cdot\varepsilon^{-1}\}$. If the numbers change due to the above mechanism, then for the function $u^\varepsilon(t,x) = u^\varepsilon(t\varepsilon^{-1},x)$ we have the Cauchy problem

$$\frac{\partial u^\varepsilon(t,x)}{\partial t} = \frac{\varepsilon}{2}\Delta u^\varepsilon + \frac{1}{\varepsilon} c(x,\varepsilon \ln u^\varepsilon)u^\varepsilon,$$
$$u^\varepsilon(o,x) = \exp\{\alpha(x)\cdot\varepsilon^{-1}\}. \tag{5}$$

We put $v^\varepsilon(t,x) = \varepsilon \ln u^\varepsilon(t,x)$. It follows from (5) that $v^\varepsilon(t,x)$ is the solution of the problem

$$\frac{\partial \delta^{\varepsilon}}{\partial t} = \frac{\varepsilon}{2} \Delta\delta^{\varepsilon} + \frac{1}{2}|\nabla\delta^{\varepsilon}|^2 + c(x,\delta^{\varepsilon}),$$
$$\delta^{\varepsilon}(o,x) = \alpha(x). \tag{6}$$

To evaluate $\lim_{\varepsilon\downarrow 0} v^{\varepsilon}(t,x)$ consider the following equation with respect to the unknown function $v(t,x)$:

$$\delta(t,x) = \sup\{\alpha(x) - \frac{1}{2}\int_0^T|\dot{\phi}_s|^2 ds + \int_0^T c(\phi_s,\delta(s,\phi_s))ds:$$
$$\delta \in C_{ot}(R^r), \phi_t = x\}. \tag{7}$$

To the variational problem (7) one can associate the Hamilton-Jacobi equation

$$\frac{\partial\delta}{\partial t} = \frac{1}{2}|\nabla\delta|^2 + c(x,\delta),\ \delta(o,x) = \alpha(x). \tag{8}$$

Theorem 2. Suppose that the function $c(x,u)$ is Lipschitz continuous and let problem (7) have a continuous solution $v(t,x)$, $t > 0, x \in R^r$. Then for the solutions of problems (5) and (6) the relations

$$\lim_{\varepsilon\downarrow 0} \varepsilon \ln u^{\varepsilon}(t,x) = \lim_{\varepsilon\downarrow 0} \delta^{\varepsilon}(t,x) = \delta(t,x)$$

hold uniformly in $x \in R$, $t \in (0,T)$.

We sketch the proof of the theorem. Let Z_t^{ε} be the Markov corresponding to the operator $\frac{\varepsilon}{2}\Delta$ in R^r. Then

$$\delta^{\varepsilon}(t,x)-\delta(t,x) = \varepsilon \ln E_x \exp\{\varepsilon^{-1}(\alpha(Z_t^{\varepsilon})+\int_o^t c(Z_s^{\varepsilon},\delta^{\varepsilon}(t-s,Z_s^{\varepsilon}))ds)\}-$$
$$-\varepsilon \ln E_x \exp\{\varepsilon^{-1}(\alpha(Z_t^{\varepsilon}) + \int_o^t c(Z_s^{\varepsilon},\delta(t-s,Z_s^{\varepsilon}))ds)\}+ \tag{9}$$
$$+\varepsilon \ln E_x \exp\{\varepsilon^{-1}(\alpha(Z_t^{\varepsilon}) + \int_o^t c(Z_s^{\varepsilon},\delta(t-s,Z_s^{\varepsilon}))ds\}-\delta(t,x)=$$
$$= \varepsilon\ln \frac{E_x \exp\{\varepsilon^{-1}(\alpha(Z_t^{\varepsilon}) + \int_o^t c(Z_s^{\varepsilon},\delta^{\varepsilon}(t-s,Z_s^{\varepsilon}))ds\}}{E_x \exp\{\varepsilon^{-1}(\alpha(Z_t^{\varepsilon}) + \int_o^t c(Z_s^{\varepsilon},\delta(t-s,Z_s^{\varepsilon}))ds\}} + 0_{\varepsilon}(1),\ \varepsilon\downarrow 0.$$

Here we used the fact that

$$\lim_{\varepsilon \downarrow 0} \varepsilon \ln E_x \exp\{\varepsilon^{-1}(\alpha(Z_t^\varepsilon) + \int_o^t c(Z_s^\varepsilon, \delta(t-s, Z_s^\varepsilon))ds)\} =$$
$$= \sup\{\alpha(\phi_t) + \int_o^t c(\phi_s, \delta(t-s, \phi_s))ds - \frac{1}{2}\int_o^t |\dot{\phi}_s|^2 ds:$$
$$\phi \in C_{ot}(R^r), \phi_o = x\} = \delta(t,x).$$

which follows from (7) and from the properties of the action functionals (see (3), Ch.3).

Consider the first term on the right-hand side of (9). Denote by ξ the random variable under the expectation sign in the denominator. Then this term can be written as

$$\varepsilon \ln \frac{Ex\ \xi \exp\{\varepsilon^{-1}\int_o^t (c(Z_s^\varepsilon, \delta^\varepsilon(t-s, Z_s^\varepsilon, \delta(t-s, Z_s^\varepsilon))ds\}}{E_x \xi}$$

Applying the Holder inequality we obtain from (9):

$$|\delta^\varepsilon(t,x) - \delta(t,x)| \leq 0_\varepsilon(1) + \varepsilon \ln \frac{(E_x \xi^p)^{1/p}}{E_x \xi} +$$
$$+ \frac{\varepsilon}{q} \ln E_x \exp\{\frac{q}{\varepsilon} \int_o^t (c(Z_s^\varepsilon, \delta^\varepsilon) - c(Z_s^\varepsilon, \delta))ds\}, \tag{10}$$

where $p,q > 0$, $p+q = I$. From this for $\delta_\varepsilon(t) = \sup_{x \in R^r, o \leq s \leq t} |v^\varepsilon(s,x) - v(s,x)|$ $(s,x)|$ we derive the inequality

$$\delta_\varepsilon(t) \leq 0_\varepsilon(1) + A(p,\varepsilon) + K \int_o^t \delta_\varepsilon(s)ds, \tag{11}$$

where K is the Lipschitz constant for $c(x,u)$ and $A(p,\varepsilon)$ is the second term on the right-hand side of (10). From (11) we deduce

$$\delta_\varepsilon(t) \leq (0_\varepsilon(1) + A(p,\varepsilon))e^{kt}.$$

From this noting that $\lim_{p \to 1} A(p,\varepsilon) = 0$ for every ε, we arrive at the claim of Theorem 2.

One can prove that if problem (8) has a solution, then this solution satisfies (7). Theorem 2 yields that the solution of problem (7) is unique. Certainly the behavior of the solution of problem (7) is defined by the form of the function $c(x,v)$. Of interest are the functions which for a fixed x have the following form:

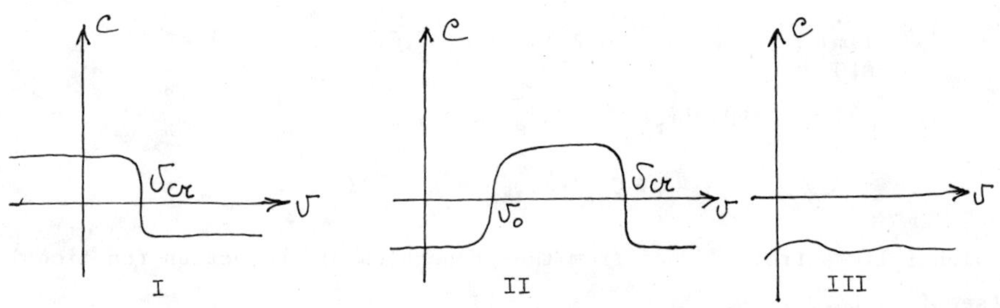

Fig. 4

If one assumes that in Fig. 2 the blacked regions correspond to the coefficients $c(x,v)$ of type I, the shaded ones to type II and all other to type III, then the qualitative behavior of the solution $v(t,x)$ will be as follows. Suppose as before that for $t = 0$ there is a high level density only on K_o, that is $v(0,x)\big|_{(K_o)} > 0$ and $v(0,x)\big|_{x\notin K_o} < 0$. Then the domain where $v(t,x)$ is positive will expand with time over a neighborhood $D_\Gamma \supset K_o$ in accordance with the Huygens principle (Fig. 5). The dashed line in Fig. 5 shows the positions of the boundaries of the domains of positiveness at different times. At a time T_Γ the domain of positiveness will occupy the whole region D bounded by the curve Γ. At a time T_1^* which can be both large and smaller than T_Γ, a new domain of positiveness K_{i_1} will take fire. This domain of positiveness will expand as far as a neighborhood $D(K_{i_1})$. Which of the blacked regions K_i will be the first to take fire from K, over what neighborhoods (and with what velocity) will the excitation expand - this is certain to depend on the form of the function $c(x,v)$.

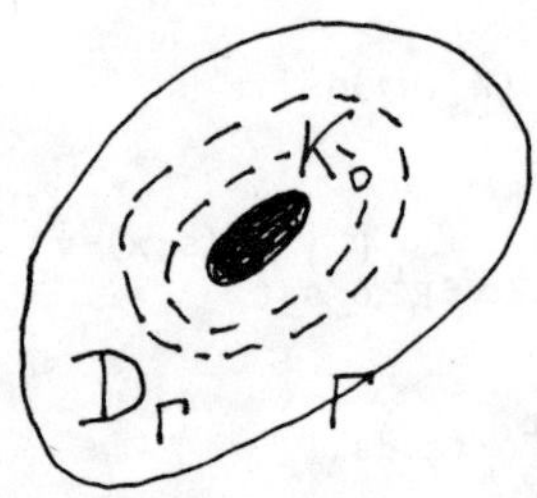

Fig. 5

Suppose that inside of the neighborhood D_Γ (or the corresponding neighborhood of another region K) there is a region G of type II and at a point of this region, $v(t,x)$ exceeds $v_o(x)$ (see Fig. 4). Then the domain of positiveness will expand over the whole G and its neighborhood. Therefore in the model considered in Theorem 2, a region of type II may take fire even if it does not border with a region of type I.

References

1. M. Freidlin, Limit theorems for large deviations and reaction-diffusion equations, The Ann. of Prob., v.13, N3 (1985), pp 639-676.
2. M. Freidlin, Functional Integration and Partial Differential Equations, Princeton Univ. Press, 1985.
3. M.I. Freidlin and A.D. Wetzell, Random Perturbations of Dynamical Systems, Springer-Verlag, 1983.

DIFFUSIVE AND SUBDIFFUSIVE BEHAVIOR IN MECHANICAL AND STOCHASTIC MODELS

Sheldon Goldstein
Department of Mathematics
Rutgers University
New Brunswick, N.J. 08903
USA

ABSTRACT

Let $X(t)$ be the position at time t of a test particle (tp) in a "random environment", which, for an infinite mechanical system, may represent the velocity of the tp and the positions and velocities of all the other particles, while for a stochastic model it may represent an inhomogeneous medium in which the tp is diffusing. We observe the motion of the tp on macroscopic length and time scales, i.e., we consider the process $X_A(t) \equiv X(At)/A^{\gamma}$, in the limit $A \to \infty$. When for $\gamma = 1/2$ the limit is a (nontrivial) Brownian motion, we have diffusive behavior; if the limit is nontrivial for $\gamma < 1/2$, we have subdiffusive behavior. Diffusive behavior has been established for stochastic models with a "symmetric" random environment, as well as for the one-dimensional system of Newtonian hard points, which may interact with a bounded external potential, when the tp has the same mass as the other particles. Subdiffusive behavior arises when the tp undergoes a random walk on a fractal such as the "Sierpinski lattice" or when the tp belongs to a one-dimensional system of colliding hard points whose "free motion" is sublinear.

1. DIFFUSIVE AND SUBDIFFUSIVE BEHAVIOR

Roughly speaking, the motion $X(t) \in \mathbb{R}^d$ is diffusive if $|X(t)| \sim t^{1/2}$ and is subdiffusive if $|X(t)| \sim t^{\gamma}$, $\gamma < 1/2$. Note, however, that under the more precise formulation given in the abstract, diffusive behavior requires that the motion be asymptotically (nontrivial) Brownian. Detailed asymptotic information is not involved in our notion of subdiffusive behavior. Indeed, among the models covered in Section 4 are some with $\gamma = \ln 2 / \ln 5$, which also occurs in Section 5, but with an asymptotic motion of a very different character: in Section 4 the limit is a nonmarkovian Gaussian process while in Section 5 it is a Markov process with single time distribution "orthogonal" to Gaussian.

2. SYMMETRIC RANDOM MOTIONS IN RANDOM ENVIRONMENTS [1],[2]

We have established diffusive behavior for antisymmetric functions of a reversible (i.e., time-symmetric) Markov process, a result which covers a wide variety of random motions in random environments, including random walks on the infinite cluster of the two-dimensional bond percolation model above the percolation threshold. Our formulation leads naturally to a formula for the diffusion matrix which easily yields useful inequalities.

3. NEWTONIAN HARD POINTS OF EQUAL MASS [3],[4]

We have generalized the results of Harris[5] and Spitzer[6] on the motion of a tp in a uniform one-dimensional system of point particles undergoing elastic collisions to the case where there is also an external potential $U(x)$ (which acts on all of the particles). When $U(x)$ is periodic or random (bounded and statistically translation invariant) we have diffusive behavior, while if $U(x)$ is itself changing on a macroscopic scale, i.e., $U(x) \equiv U_A(x) \equiv \tilde{U}(x/\sqrt{A})$, then the limit process (obtained using $\gamma = 1/2$) is a spatially dependent diffusion. In all cases the limit is described by the (Stratonovich) stochastic differential equation $dx_t = (\hat{\rho}(x_t))^{-1} \circ dz_t$, where $\hat{\rho}(x)$ is the macroscopic density profile and z_t is Brownian motion with variance $\rho_{min} \langle|v|\rangle t$. A key element of our analysis for this model, and for that of the next section, is approximation by the "crossing process" $n(t)$, the signed number of crossings of the origin by particles before time t.

4. GENERALIZED COLLISION PROCESS [7]

The preceding model, with $U = 0$, may be generalized to allow the particles to undergo any "equivalent" independent (stochastic) processes, <u>free motions</u>, in the absence of "collisions", whose sole effect is to maintain the order of the particles. We find quite generally that $\gamma = \gamma_o/2$, where γ_o describes the free motion, with limit process the fractional Brownian motion with parameter $\alpha \equiv \gamma_o$, the mean zero Gaussian process with covariance proportional to $t^\alpha + s^\alpha - |t-s|^\alpha$.

5. RANDOM WALKS ON FRACTALS [8]

One expects subdiffusive behavior to arise naturally for random walks on fractals. We have shown that for the symmetric nearest neighbor random walk on the "Sierpinski lattice"

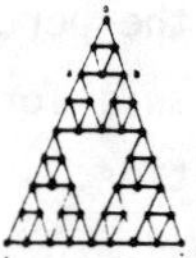

in dimension d, we have (essentially) subdiffusive behavior with $\gamma = \ln 2 / \ln(d+3)$. (The limit, naturally, exists only along subsequences $A = \text{const.}\ (d+3)^\ell$, $\ell=1,2,3,\ldots$, with corresponding spatial rescalings proportional to 2^ℓ). The limit process is a "diffusion" on the Sierpinski gasket

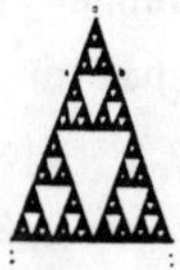

a Cantor set of Lebesgue measure zero. The analysis is based on a simple "renormalization group"-type argument involving self-similarity and "decimation invariance".

6. NEWTONIAN HARD POINTS WITH TP OF A DIFFERENT MASS

It has been shown[9],[10] that the mean square displacement of a tp in a system of Newtonian hard points ($U = 0$) grows linearly in t (corresponding to $\gamma = 1/2$) even if the tp has a mass different from that of the other particles. However, it seems that we don't have diffusive behavior in this case: simulations by the Sinai group suggest that the motion of the tp is not asymptotically Brownian.

REFERENCES

1] De Masi, A., Ferrari, P.A., Goldstein, S. and Wick, D.W., "Invariance Principle for Reversible Markov Processes with Application to Diffusion in the Percolation Regime", Particle Systems, Random Environments, and Large Deviations, A.M.S., 1985.

2] De Masi, A., Ferrari, P.A., Goldstein, S. and Wick, D.W., "An Invariance Principle for Reversible Markov Processes, with Applications to Random Motions in Random Environments", submitted.

3] Dürr, D., Goldstein, S. and Lebowitz, J.L., "Self-diffusion in a Nonuniform Model System", submitted.

4] Dürr, D., Goldstein, S. and Lebowitz, J.L., "Self-diffusion in a Nonuniform One-dimensional System of Point Particles with Collisions", submitted.

5] Harris, T.E., J. Appl. Prob. 2, 323 (1965).

6] Spitzer, F., J. Math. Mech. 18, 973 (1968).

7] Dürr, D., Goldstein, S. and Lebowitz, J.L., Comm. Pure Appl. Math. 38, 573 (1985).

8] Goldstein, S., "Random Walks and Diffusions on Fractals", Pecolation Theory and Ergodic Theory of Infinite Particle Systems, IMA Volumes in Mathematics and its Applications, Springer.

9] Sinai, Ya. G. and Soloveichik, M.R., Commun. Math. Phys. 104, 423 (1986).

10] Szász, D. and Toth, B., Commun. Math. Phys. 104, 445 (1986).

COLLECTIVE PHENOMENA IN STOCHASTIC INTERACTING PARTICLE SYSTEMS

Errico PRESUTTI
Dipartimanto Matematico, Università di Roma
P.le A. Moro, 00185. Roma. Italia.

ABSTRACT

Microscopic and macroscopic descriptions of stochastic particle systems are compared in the long time asymptotic regime.

Two classes of models are considered. The first one contains some one dimensional asymmetric simple exclusion processes with n.n jumps ($p > 1/2$ being the probability of right jumps). For $\varepsilon<o$ let μ^{ε} be the product measure on the phase space $\{0,1\}^{\mathbb{Z}}$, with density $\rho(\varepsilon x)$ at $x \in \mathbb{Z}$, where $\rho : \mathbb{R} \to [0,1]$ is any piecewise smooth function. Let $\rho(r,t)$ be the entropic solution to the unviscid Burgers equation

$$\partial_t \rho + 2(p-1/2) \quad \partial_r (\rho(1-\rho)) = 0 \qquad (1.)$$

with initial condition ρ. Denote by $\mu_t^{\varepsilon,a}$ the measure at time t evoluted from μ^{ε} and shifted to the left by a, then at all (r,t) which are not singularity points for $\rho(r,t)$

$$\lim_{\varepsilon \to o} \mu^{\varepsilon,[\varepsilon^{-1} r]}_{\varepsilon^{-1} t} = \nu_{\rho(r,t)} \qquad (2.)$$

ν_{ρ} being the product measure on $\{0,1\}^{Z}$ with density ρ and [a] the integerpart of a. The travelling wave solutions to (1) are step functions from any ρ_- to any $\rho_+>\rho_-$, their velocity being $v = 1 - \rho_- - \rho_+$. They are stable solutions to (1) except for one neutral direction corresponding to rigid shifts of the wave profile. The microscopic counterparts of such statement are i) the appearence of a non trivial superposition of Bernoulli measures at the step discontinuity reflecting the occurrence of random shifts of the whole profile, ii) the existence of spatially inhomogeneous invariant measures as seen from a suitable [randomly] moving observer (the different asymptotic densities to its right and left being the macroscopic densities after and before the front wave).

A viscosity term $\sim \frac{\partial^2}{\partial \mu^2} \rho$ appears in the r.h.s. of (1) at large times, $t \sim \varepsilon^{-1}$, <u>only</u> for very small asymmetries $p - 1/2 \sim \varepsilon$. In such case the travelling wave solution can also be obtained from the invariant measure mentioned in ii) above, by considering the density profile of such measure when $p - 1/2 \sim \varepsilon$ in the limit $\varepsilon \to o$.

The second class of models concerns Stirring+Glauber one dimensional spin processes (the Stirring being suitably speeded in the continuum limit). The corresponding macroscopic equation is the following Reaction-Diffusion equation

$$\partial_t m = \frac{1}{2} \partial_r^2 m - V'(m) \qquad (3)$$

where $V = -\frac{1}{2} \alpha m^2 + \frac{1}{4} \beta m^4$ and α and β are positive coefficients depending on the parameters of the Glauber interaction. Like before propagation of chaos holds and in the continuum limit the measure becomes Bernoulli with mean magnetisation satisfying (3).

We consider the case when $m(r) = 0$, which is a stationary unstable solution to (3). At the microscopic level the following happens. There exists an initial time layer $T(\varepsilon) = \frac{1}{2\alpha} \log \varepsilon^{-1}$ and the microscopic measure at time $T(\varepsilon) + t$ converges, when $\varepsilon \to o$, to a superposition of Bernoulli measures with an absolutely continuous (w.r.t. Lebesgue) weight $\lambda_t(dm)$, which is a statistical solution to (3).

REFERENCES

A commented list of references on the hydrodynamical limit can be found in H. Spohn, "Hydrodynamic limit for systems with many particles", to appear in J. Stat. Phys.

For the asymmetric simple exclusion process cf. A. Benassi, J.P. Fouque to appear in Annals of Probability ;

D. Wick, J. Stat. Phys. 38, 1015 (1985); P. Ferrari, to appear in Annals of Probability;

A. De Masi, E. Presutti, E. Scacciatelli, preprint 1986.
For the second class of models see A. De Masi, P. Ferrari, J.L. Lebowitz, P.R.Letters 19,1947 (1985) and an extended version to appear in J. Stat. Phys. For the escape from the unstable equilibrium cf. A. De Masi, E. Presutti, and M.E. Vares ; to appear in J. Stat. Phys.

C*-ALGEBRAS AND DUALITY FOR COMPACT GROUPS: WHY THERE IS A COMPACT GROUP OF INTERNAL GAUGE SYMMETRIES IN PARTICLE PHYSICS.

Sergio Doplicher and John E. Roberts

Università di Roma "La Sapienza" and Universität Osnabrück

1. DUALS OF COMPACT GROUPS

Given a category whose objects are finite-dimensional Hilbert spaces and whose arrows are linear mappings between those spaces fulfilling suitable axioms, the classical theorems of Krein and Tannaka (cf. e.g. [1]) enable one to construct a compact group whose representation theory is realized on these spaces in such a way that the arrows are just the intertwining operators for the representations.

In Quantum Field Theory, the analysis of superselection sectors and statistics based on the local observables alone [2] yields a more abstract structure expected to be isomorphic to the finite-dimensional representation theory of a compact group. This group would be the exact internal gauge symmetries of the theory defined by the local observables.

The classical Krein-Tannaka theorems are too weak to be applied to this situation since the intertwiners are not explicitly given as linear operators on the correct representation spaces. Instead only the abstract structure of the category is known.

The solution of this problem has led us to formulate a duality theory for compact groups locating the abstract version of a representation theory within a category defined as follows.

Let $\mathcal{A}$ be a C*-algebra with centre $\mathbb{C}I$ and let End $\mathcal{A}$ denote the category whose objects are unit-preserving endomorphisms ρ of $\mathcal{A}$ and whose arrows between a pair ρ and ρ' of objects are the intertwiners in $\mathcal{A}$:

(1) $$(\rho,\rho') = \{T \in \mathcal{A} : T\rho(A) = \rho'(A)T , \quad A \in \mathcal{A} \}.$$

Besides the composition of arrows, End $\mathcal{A}$ has a strict monoidal structure given on objects by the composition of endomorphisms which form a semigroup with unit ι, the identity automorphism of $\mathcal{A}$. The monoidal product of arrows is given by

(2) $T \in (\rho_1, \rho_2)$, $T' \in (\rho_1', \rho_2') \to T \times T' \in (\rho_1\rho_1', \rho_2\rho_2')$ where
$T \times T' = T\rho_1(T') = \rho_2(T')T$.

The monoidal operations $\rho, \rho' \to \rho\rho'$ and $T, T' \to T \times T'$ provide an abstract model for the tensor product of group representations and intertwiners.

However such tensor products have symmetry properties under permutation of factors. Explicit consideration of these properties would motivate for the reader the following

1. Definition. Let Δ be a semigroup of endomorphisms of the C*-algebra $\mathcal{A}$. A permutation symmetry for Δ is a map ε from $\Delta \times \Delta$ to unitary operators in $\mathcal{A}$ satisfying

(3) $\varepsilon(\rho,\rho') \in (\rho\rho', \rho'\rho)$

If $S \in (\sigma,\sigma')$, $R \in (\rho,\rho')$ then

(4) $\varepsilon(\rho',\sigma')\; R \times S = S \times R\; \varepsilon(\rho,\sigma)$

(5) $\varepsilon(\rho,\rho')\; \varepsilon(\rho',\rho) = I_{\rho'\rho}$

(6) $\varepsilon(\rho\rho',\rho'') = \varepsilon(\rho,\rho'') \times I_{\rho'}\; I_{\rho} \times \varepsilon(\rho',\rho'')$

(7) $\varepsilon(\rho,\iota) = \varepsilon(\iota,\rho) = I_{\rho}$

We then say that (Δ,ε) is a permutation symmetric semigroup of endomorphisms. The corresponding subcategory of End $\mathcal{A}$ is a strict symmetric monoidal category.

A finite-dimensional representation $\mathcal{U}$ of a compact group has a conjugate representation $\bar{\mathcal{U}}$ and $\mathcal{V} = \mathcal{U} \oplus \bar{\mathcal{U}}$ has determinant one.

The last property means that the totally antisymmetric subspace of $\mathcal{V}^{\otimes d}$, d the dimension of $\mathcal{V}$, carries the trivial representation. It can thus be expressed in terms of intertwiners and motivates the following

2. Definition. Let (Δ,ε) be a permutation symmetric semigroup of endomorphisms of $\mathcal{A}$; $\rho \in \Delta$ is special if there is an integer d and an isometry R with

(8) $R \in (\iota, \rho^d)$

(9) $R^*R = I$, $\quad RR^* = \sum_{p \in \mathbb{P}_d} \operatorname{sign}(p)\; \varepsilon_\rho(p)$

(10) $R^* \rho(R) = (-1)^{d-1} \frac{1}{d} I$.

Here $p \to \varepsilon_\rho(p)$ is a unitary representation of the group $\mathbb{P}_\infty$ of finite permutations of the integers derived canonically from ε and permuting the powers of ρ.

Existence of conjugates is bound to play a central role in a duality theory of compact groups. It turns out to be a technically convenient substitute to require an abundance of representations with determinant one. We are hence led to the

3. Definition. Let (Δ,ε) be a permutation symmetric semigroup of endomorphisms of $\mathcal{A}$; (Δ,ε) is specially directed if, given any finite subset $\{\rho_1,\ldots,\rho_n\} \subset \Delta$, there is a special $\rho \in \Delta$ dominating each ρ_i, $i = 1,2,\ldots,n$. Namely, for each i, there is a finite family of partial isometries in $\cup\{(\rho_i,\rho^m)\ ;\ m = 0,1,2,\ldots\}$ with orthogonal initial domains summing up to I.

Now given a permutation symmetric specially directed semigroup Δ of endomorphisms of a C*-algebra $\mathcal{A}$ with centre $\mathbb{C}I$, the full subcategory of End $\mathcal{A}$ with objects Δ and arrows all their intertwiners is precisely the dual object of a compact group G, i.e. the abstract version of a representation category for G.

These statements are made explicit by the theorem below where the group G and a spatial representation category are constructed from (Δ,ε). The group G appears as a group of automorphisms of a C*-algebra $\mathcal{B}$ and the representation theory as a collection of G-stable finite-dimensional Hilbert spaces in $\mathcal{B}$.

Recall that a Hilbert space ([3,4]) in a C*-algebra $\mathcal{B}$ is a closed subspace $H \subset \mathcal{B}$ with trivial left annihilator such that $\psi,\psi' \in H$ implies $\psi^*\psi' \in \mathbb{C}I$. The scalar product (,) in H is then given by

$$(\psi,\psi')I = \psi^*\psi', \qquad \psi,\psi' \in H. \tag{11}$$

The C*-algebra generated by H is a universal object, the Cuntz algebra $\mathcal{O}_d$, d = dim H [5].

If $\psi_1,\psi_2,\ldots,\psi_d$ is a (finite) orthonormal basis in H then setting

$$\rho(B) = \sum_i \psi_i B \psi_i^*, \qquad B \in \mathcal{B}, \tag{12}$$

defines an endomorphism ρ of $\mathcal{B}$ such that

$$\psi B = \rho(B)\psi, \quad \psi \in H, \quad B \in \mathcal{B}. \tag{13}$$

If $H \subset B$ is globally stable under a group of automorphisms of $\mathcal{B}$ then H also induces an endomorphism of the fixed-point subalgebra. Moreover $g \to g|_H$ is a unitary representation. All representation theories of a compact group G can be realized within a C*-algebra $\mathcal{B}$ as we discuss briefly below.

If H, H' are finite-dimensional Hilbert spaces in $\mathcal{B}$, HH' and

$H'H^*$ span isomorphic images of $H \otimes H'$ and $\mathcal{L}(H,H')$ respectively. There is a canonical symmetry θ in $\mathcal{B}$ interchanging the order in a tensor product:

$$\theta(H,H') = \sum_{i,j} \psi'_j \psi_i \psi'^*_j \psi^*_i ,$$

where the sum is taken over orthonormal bases $\{\psi_i\}$, $\{\psi'_j\}$ of H and H' respectively.

After these preliminaries, we can state the main

4. <u>Theorem</u>. Let $\mathcal{A}$ be a C*-algebra with centre $\mathbb{C}I$ and (Δ,ε) a permutation symmetric, specially directed semigroup of endomorphisms of $\mathcal{A}$. There exists a pair $(\mathcal{B},G)$, where $\mathcal{B}$ is a C*-algebra containing $\mathcal{A}$ and G is a compact group of automorphisms of $\mathcal{B}$ such that

(a) $\mathcal{A} = \mathcal{B}^G$; $\mathcal{A}' \cap \mathcal{B} = \mathbb{C}I$

(b) there is a map $\rho \in \Delta \to H_\rho \subset \mathcal{B}$, where H_ρ is a Hilbert space in $\mathcal{B}$ with

(b 1) $\psi A = \rho(A)\psi$, $\psi \in H_\rho$, $A \in \mathcal{A}$

(b 2) $\varepsilon(\rho,\rho') = \theta(H_\rho, H_{\rho'})$, $\rho, \rho' \in \Delta$

(b 3) $\mathcal{B}$ is generated as a C*-algebra by $\mathcal{A}$ and $\{H_\rho : \rho \in \Delta\}$.

Furthermore G is then compact in the strong topology of $\mathcal{L}(\mathcal{B})$ and is precisely the set of those automorphisms of $\mathcal{B}$ leaving $\mathcal{A}$ pointwise fixed.

Moreover if $(\mathcal{B}_1, G_1)$ is another pair satisfying (a) and (b), there is an $\mathcal{A}$-module isomorphism ϕ of $\mathcal{B}$ onto $\mathcal{B}_1$; consequently G and G_1 are isomorphic and homeomorphic via

$$g \in G \to \phi \circ g \circ \phi^{-1} \in G_1$$

and $\phi(H_\rho) = H_\rho^{(1)}$, $\rho \in \Delta$.

<u>Remark</u>. G is naturally expressed as a projective limit of the compact Lie groups $G|_{H_\rho} \subset SU(H_\rho)$, ρ special, $\rho \in \Delta$.

Theorem 4 extends results announced in [6] . Detailed proofs of some of these results may be found in [7,8].

Our current research concerns generalizations to cases where Centre $(\mathcal{A}) \neq \mathbb{C}I$ and Δ is possibly external to $\mathcal{A}$, consisting for example of morphisms from $\mathcal{A}$ to $\mathcal{A} \otimes \mathcal{K}$.

The theorem above provides a natural abstract duality theory for compact groups in the following sense: first, assume that the group G is given together with a representation theory $\mathcal{R}$ containing no 1-dimensional representation other than the identity (*) and specially directed (in a sense analogous to that above). Then we can embed $\mathcal{R}$

(*) This is related to the condition that $\mathcal{A}$ has centre $\mathbb{C}I$.

into a C*-algebra $\mathcal{B}$ by making the representation spaces into Hilbert spaces in $\mathcal{B}$ and the corresponding representations into restrictions to H of an action of G by automorphisms of $\mathcal{B}$. If $\mathcal{A} = \mathcal{B}^G$ is the fixed point subalgebra, the semigroup Δ generated by the endomorphisms $\rho_H | \mathcal{A}$ would fulfill the hypotheses of Theorem 4 and $(\mathcal{B}, G)$ would coincide with the system constructed there(*).

The category $\{\Delta$, intertwiners$\}$ then describes correctly the action on $\mathcal{A}$ of a dual of a compact group.

Moreover, if we are now given an abstract C*-category [9] $\mathcal{T}$ which is strict symmetric monoidal and fulfills natural axioms expressing conjugation symmetry, it can be proved that $\mathcal{T}$ is equivalent to a category of the above type, i.e. the full subcategory of End $\mathcal{A}$ determined by a permutation symmetric, specially directed semigroup of endomorphisms of a suitable C*-algebra $\mathcal{A}$ with centre $\mathbb{C}I$. By the theorem above such a category $\mathcal{T}$ is then equivalent to a spatial representation theory of a compact group.

Theorem 4 is formulated algebraically but it has spatial versions and the following W*-version.

5. Theorem. Let $\mathcal{A}$ be a factor and (Δ, ε) a permutation symmetric, specially directed semigroup of (normal) endomorphisms of $\mathcal{A}$. There is a factor $\mathcal{B} \supset \mathcal{A}$ and a group $G \subset \mathrm{Aut}\, \mathcal{B}$, compact for the strong topology on $\mathcal{L}(\mathcal{B}_*)$, such that conditions (a), (b 1), (b 2) of Theorem 4 are fulfilled together with

(b 3') $\mathcal{B}$ is generated as a W*-algebra by $\mathcal{A}$ and $\{H_\rho : \rho \in \Delta\}$.

Then every automorphism of $\mathcal{B}$ leaving $\mathcal{A}$ pointwise fixed is in G and the system $(\mathcal{B}, G)$ is unique as in Theorem 4.

2. THE GLOBAL GAUGE GROUP, CURRENTS AND FIELDS

The spatial versions of Theorem 4 lead to the solution of the problem in Q.F.T. which motivated the present research. In that context, $\mathcal{A}$ is the C*-algebra (of quasilocal observables) generated by the W*-algebras $\mathcal{A}(\mathcal{O})$ of the observables which can be measured within the bounded space-time regions $\mathcal{O}$ (doubles cones). $\mathcal{A}$ is given as an irreducible representation π_o describing the vacuum superselection sector. The other sectors are obtained composing π_o with the irreducible elements in a semigroup $\Delta \subset$ End $\mathcal{A}$. Here Δ is a semigroup of localized morphisms which turns out to be naturally endowed with a permutation symmetry ε describing the particle statistics determined by the local observables [2,3]. (For a survey, see [10,11]).

(*) In this setting, the classical Tannaka theorem says that, given $\mathcal{B}$, every automorphism of $\mathcal{B}$ leaving $\mathcal{A}$ pointwise fixed is in G [4; 26, Appendix; 27].

If $\rho \in \Delta$ is irreducible, the classes of the representations ε_ρ of $\mathbb{P}_n$ are specified by a sign (distinguishing the Bose or Fermi character of the superselection sector) and by an integer (the order of the parastatistics).(*) Moreover Δ is closed under direct sums and conjugates, where the conjugation of irreducible elements is characterized up to inner automorphisms in Δ by

$$\rho\,\bar{\rho} \cong \iota \oplus \rho'.$$

If we change ε to a new permutation symmetry $\hat{\varepsilon}$ [11, § 5] by "Bosonizing" (which amounts to correct the sign for irreducible $\rho \in \Delta$), we can show that each $\rho \in \Delta$ equivalent to $\rho' \oplus \bar{\rho}'$, for some $\rho' \in \Delta$, is special for $(\Delta, \hat{\varepsilon})$ and $(\Delta, \hat{\varepsilon})$ is specially directed. An appropriate spatial version of Theorem 4 combined with results in [3] gives the following

6. Theorem. There exists a representation π of $\mathcal{A}$ acting on a Hilbert space $\mathcal{H}$ and containing π_o as a subrepresentation on $\mathcal{H}_o \subset \mathcal{H}$, a unitary faithful continuous representation $\mathcal{U}$ of a compact group G (the global gauge group) acting on $\mathcal{H}$ and leaving $\mathcal{H}_o$ pointwise fixed and an inclusion preserving map $\mathcal{O} \to \mathcal{F}(\mathcal{O})$ from double cones in space-time to von Neumann algebras acting on $\mathcal{H}$ (the local algebras generated by field operators) such that

(i) Each irreducible element $\rho \in \Delta$ is equivalent to a subrepresentation of π (π contains all superselection sectors);

(ii) $\mathcal{U}_g$, $g \in G$ induce automorphisms of $\mathcal{F}(\mathcal{O})$ with $\pi(\mathcal{A}(\mathcal{O}))$ as fixed-point algebra, for each double cone $\mathcal{O}$;

(iii) The union $\cup_{\mathcal{O}} \mathcal{F}(\mathcal{O})$ is irreducible;

(iv) If $F_i \in \mathcal{F}(\mathcal{O}_i)$, $i = 1,2$ where $\mathcal{O}_1$, $\mathcal{O}_2$ are spacelike separated double cones and F_i carry definite charges ξ_i, i.e: $F_i \mathcal{H}_o$ belong to irreducible subspaces for π, then

$$F_1 F_2 = \pm F_2 F_1, \tag{14}$$

where (-) occurs if both ξ_i obey para-Fermi statistics, (+) otherwise (normal commutation relations).

The system $(\pi, \mathcal{U}_G, \mathcal{F})$ is unique up to unitary equivalence: if $(\pi_1, \mathcal{V}_{G_1}, \mathcal{F}_1)$ is another such system acting on the Hilbert space $\mathcal{H}_1$, there is a unitary $W : \mathcal{H} \to \mathcal{H}_1$ such that

$$\begin{aligned} W\pi &= \pi_1 W \\ W\mathcal{U}_G &= \mathcal{V}_{G_1} W \\ W\mathcal{F}(\mathcal{O}) &= \mathcal{F}_1(\mathcal{O})\, W, \quad \mathcal{O} \text{ any double cone.} \end{aligned} \tag{15}$$

This general theorem extends results previously obtained in the absence of parastatistics [12]. It is formulated above for the case

(*) Space-time is assumed to have a dimension greater than two.

of "localizable charges", reached from the vacuum by localized morphisms. One is led to consider such sectors by looking at states whose global deviation from the vacuum in the spacelike complement of large but bounded space-time regions becomes negligible [2]. This probably excludes interactions with massless particles.

Although this class would describe e.g. the baryon number in pure massive QCD, topological charges which are not localizable might appear in massive non-abelian gauge theories (cf. K. Fredenhagen, these Proceedings). Buchholz and Fredenhagen [13] proved, without any assumption on spacelike asymptotic properties of states, that a superselection sector where a massive particle appears isolated from the continuum mass spectrum is always generated by states which are strictly localized with respect to the vacuum in arbitrary spacelike cones (i.e. in regularized strings joining a point to spacelike infinity).

Although these representations cannot be described by morphisms of $\mathcal{A}$ into itself, the analysis of [2] was extended in [13] to such more general cases (for a space-time of dimensions greater than three).

The construction leading to Theorem 6 above also covers this more general case and again yields a representation π of $\mathcal{A}$ containing all the superselection sectors (possibly localizable only in spacelike cones), a compact gauge group G unitarily represented on $\mathcal{H}$, $g \to \mathcal{U}_g$ and an inclusion preserving map

(16) $\quad \mathcal{C} \to \mathcal{F}(\mathcal{C})$

from spacelike cones to von Neumann algebras acting on $\mathcal{H}$, so that all conditions stated there are fulfilled with the following modifications:

(ii') $\mathcal{U}_g$, $g \in G$, induce automorphisms of $\mathcal{F}(\mathcal{C})$ with $\pi(\mathcal{A}(\mathcal{C}))^-$ as fixed-point algebra (where $^-$ denotes the weak operator closure);

(iii') For each spacelike cone $\mathcal{C}$, $\bigcup_{x \in \mathbb{R}^4} \mathcal{F}(\mathcal{C}+x)$ is irreducible.

The other statements in the existence and uniqueness assertions can be modified simply by replacing double cones by spacelike cones.

The above theorems establish from first principles for any massive particle theory defined only in terms of local observables the existence of a compact group of global gauge transformations whose representation theory describes the superselection structure.

The statement that one can always choose the field operators to have normal commutation relations despite the occurrence of (an intrinsically defined) parastatistics is put on a similarly firm basis.

The field operators given above are not localizable in general but are associated with regularized infinite strings. Even in such more general theories, we can consider the subalgebra generated by localized field operators, i.e. by all intersections

(17) $\quad F(C_1) \cap F(C_2)$,

where C_1, C_2 are spacelike cones with $C_1 \cap C_2$ bounded. Let $H \subset G$ be the subgroup of those elements which leave all such subspaces (17) pointwise fxed. Then

$$G/H = G_\ell$$

is the gauge group for localizable charges, i.e. Theorem 6 applied only to localizable sectors would yield G_ℓ. The topological charges are necessarily associated with non-trivial representations of H.

In theories where topological charges do not appear we can further ask for local aspects of the superselection rules, e.g. are they generated by local conserved currents as in Lagrangian field theory?

In theories where a further, physically motivated requirement is fulfilled, namely the Split Property, one can indeed establish a weak form of a Quantum Noether Theorem.

As a consequence, the global superselection charge operators derived (as generators or Casimir operators in the case of a Lie group) from the representation $\mathcal{U}$ in Theorem 4, can be proved to be the large volume limits of local charge operators measuring the charges in finite regions by observations made in slightly larger regions.

If G is a Lie group (i.e. if the superselection structure is finitely generated under conjugation, products and reduction of products), the exact current algebra relations hold for local generators which are the analogues of (a suitable perturbation of) the finite volume integrals of the regularized time components of the Noether currents. For space-time symmetries, or possibly supersymmetries, there are local generators fulfilling the exact Lie Algebra relations. It is worth mentioning that these local generators yield representations quasiequivalent to the global ones.

In particular, the local Hamiltonians obtained in this way are positive definite. If one such correct local Hamiltonian in a bounded region were known together with the observables in that region it would be possible to recover just from them the superselection structure (cf [14-17]; for a short review, see [18]).

The Split Property for free scalar fields has been established by Buchholz [19] and for general theories, fulfilling conditions akin to the existence of thermodynamical equilibrium states, by Buchholz and Wichmann ([20]; see also [21,22] and D. Buchholz, these Proceedings). For further results and references the original papers can be consulted and for mathematical developments on split inclusions of von Neumann algebras we refer to [23-25].

The problem of constructing Wightman fields representing conserved currents as the limit of the local currents associated with finite regions as the regions shrink to a point is still open and is an obstacle in the way of establishing a full Quantum Noether Theorem.

3. REFERENCES

[1] Hewitt, E., Ross, K.A.: "Abstract Harmonic Analysis", Springer Verlag, 1970.

[2] Doplicher, S., Haag, R., Roberts, J.E.: "Local Observables and Particle Statistics", I, Comm. Math. Phys. 23, 199 (1971); II, Comm. Math. Phys. 35, 49 (1974)

[3] Doplicher, S., Roberts, J.E.: "Fields, Statistics, and non Abelian Gauge Groups", Comm. Math. Phys. 28, 331 (1972)

[4] Roberts, J.E.: "Cross Products of von Neumann Algebras by Group Duals" Symp. Math. XX, 335 (1976)

[5] Cuntz, J.: "Simple C*-algebras generated by isometries" Comm. Math. Phys. 37, 173 (1977)

[6] Doplicher, S., Roberts, J.E.: "Compact Lie Groups associated with endomorphisms of C*-algebras", Bull. Amer. Math. Soc. 11, 333 (1984)

[7] Doplicher, S., Roberts, J.E.: "A Remark on Compact Automorphism Groups of C*-algebras", Journ. Funct. Analysis 66, 67 (1986)

[8] Doplicher, S., Roberts, J.E.: "Duals of Compact Lie Groups realized in the Cuntz Algebras and their actions on C*-algebras", Journ. Funct. Analysis, to appear

[9] Ghez, P., Lima, R., Roberts, J.E.: "W*-categories", Pacific Journ. of Math. 120 (1985)

[10] Roberts, J.E.: "Statistics and the Intertwiner Calculus", Proceedings of the 1973 Varenna Summer School, D. Kastler ed., Acad. Press (1976)

[11] Doplicher, S.: "The Statistics of Particles in Local Quantum Theories", Proceedings of the Kyoto International Conference on Mathematical Physics, H. Araki ed., Springer lecture notes in Physics 39 (1975)

[12] Doplicher, S., Haag, R., Roberts, J.E.: "Fields, Observables and Gauge Transformations" II, Comm. Math. Phys. 15, 173 (1969)

[13] Buchholz, D., Fredenhagen, K.: "Locality and the Structure of Particle States", Comm. Math. Phys. 84, 1 (1982)

[14] Doplicher, S.: "Local Aspects of Superselections Rules" Comm. Math. Phys. 85, 73 (1982)

[15] Doplicher, S., Longo, R.: "Local Aspects of Superselection Rules" II, Comm. Math. Phys. 88, 399 (1983)

[16] Buchholz, D., Doplicher, S., Longo, R.: "On Noether's Theorem in Quantum Field Theory", Ann. Phys., 170, 1 (1986).

[17] D'Antoni, C., Doplicher, S., Fredenhagen, K., Longo, R.: "Convergence of Local Charges and continuity of W*-inclusions", to appear

[18] Doplicher, S.: "Current Algebra and the Local Nature of Symmetries in Local Quantum Theory", in Trends and Developments in the 80's, S. Albeverio and Ph. Blanchard eds., World Scientific (1986)

[19] Buchholz, D.: "Product States for Local Algebras", Comm. Math. Phys. 36, 287 (1974)

[20] Buchholz, D., Wichmann, E.: "Causal Independence and Energy-level Densities of Localized States in Quantum Field Theory" Comm. Math. Phys. 106, 321 (1986)

[21] Buchholz, D., D'Antoni, C., Fredenhagen, K.: in preparation

[22] Buchholz, D., Junglas, P.: "Local Properties of Equilibrium States and the Particle Spectrum in Quantum Field Theory", preprint

[23] Doplicher, S., Longo, R.:"Standard and Split inclusions of von Neumann Algebras", Invent. Math. 73, 493 (1984)

[24] Longo, R.: "Solution of the Factorial Stone-Weierstrass conjecture. An application of the theory of Standard Split W*-inclusions", Invent. Math. 76, 145 (1984)

[25] Longo, R.: "Simple Injective Subfactors", Adv. Math., to appear

[26] Araki, H., Haag, R., Kastler, D., Takesaki, M.: "Extension of KMS states and Chemical Potential" Comm. Math. Phys. 53, 97 (1977)

[27] Roberts, J.E.: Net Cohomology and its Application to Field Theory", in Quantum Fields, Algebras, Processes, L. Streit ed., Springer Verlag, 1980.

Criteria for Quark Confinement

Klaus Fredenhagen*)

II. Institut für Theoretische Physik, D-2000 Hamburg 50, FRG

Particles with a gauge charge in a massive theory are subject to the following restrictions:

(i) The charge can be measured at arbitrarily large distances due to the existence of a nontrivial string connecting the particle with an antiparticle at spacelike infinity.

(ii) The asymptotic direction of the string is not visible, so it is not possible to observe something like a nontrivial electric flux into a solid angle smaller than 4π .

These restrictions have been derived from the basic principles of quantum field theory by D. Buchholz and myself /1/ in the framework of algebraic field theory /2/. (Compare also our contributions to the proceedings of the IAMP conference in Lausanne (1979) and Berlin (1981).) Property (ii) prevents the existence of nontrivial gauge charges of the additive type (charges which are additively composed out of electric fluxes in different directions); this fact is known as Swieca's theorem /3/. On the other hand, gauge charges of the multiplicative type (like the triality charge of quarks in QCD) may exist, provided the electric fluxes in different directions are strongly enough correlated.

Such strong correlations are actually present in the weak coupling region of gauge theories with a finite gauge group; M. Marcu and myself constructed charged states in the so-called free charge phase of the $\mathbb{Z}_2$ Higgs model /4/, and recently Barata showed the existence of particle like excitations in the charged sector of this model /5/. These particles are the first known examples of unconfined particles with a gauge charge in a massive theory.

*) Heisenberg fellow

Besides the free charge phase the $\mathbb{Z}_2$ Higgs model also has a so-called screening/confinement phase where no charged particles are expected to exist. Actually the method which led in the free charge phase to charged states leads in the screening/confinement phase to uncharged states with a large overlap with the vacuum.

This observation suggests to consider the following parameter as a criterion for confinement /6/. Let Φ_r be a state where charges are separated by a distance 2r and which has been made gauge invariant such that the expectation value of the energy is bounded uniformly in r. An example is

$$\Phi_r = \sigma_3(0)\,\sigma_3(x_r)\,T^r\,\tau_3(L_r)\,\Omega \tag{1}$$

where σ_3 is the time zero Higgs field of the $\mathbb{Z}_2$ Higgs model, T is the transfer matrix, τ_3 the time zero gauge field, L_r a path from 0 to x_r, x_r a point with distance 2r from the origin and Ω the vacuum.

One then defines

$$\rho_1(r) = (\Omega, \Phi_r)\,\|\Phi_r\|^{-1} \tag{2}$$

If the limit state is charged, $\rho_1(\infty) = \lim_{r\to\infty} \rho_1(r)$ must vanish. $\rho_1(\infty)$ may be called the vacuum overlap order parameter. For our choice of Φ_r $\rho_1(r)$ is a quotient of the euclidean expectation values

$$\rho_1(r) = \frac{\langle \sqcap_{2r}\ r \rangle}{\langle \square_{2r \times 2r} \rangle^{1/2}} \tag{3}$$

where the symbolic notation means a product of the gauge fields along the path (in the unitary gauge).

Instead of the vacuum overlap one may consider the expectation value of the charge operator. The $\mathbb{Z}_2$ electric charge in a volume is

$$Q_\Lambda = \prod_{b \in \partial^*\Lambda} \tau_1(b) \tag{4}$$

where $\partial^*\Lambda$ is the set of bonds with one endpoint in Λ and τ_1 denotes the $\mathbb{Z}_2$-electric field at time zero (with the commutation relation $\tau_1(b)\,\tau_3(b) = -\,\tau_3(b)\,\tau_1(b)$). The charge measurement order parameter is

$$\rho_2 = \lim_{\substack{t\to\infty \\ \Lambda\to\mathbb{Z}^{D-1}}} \frac{(\Phi_t,\, Q_\Lambda\, \Phi_t)}{\|\Phi_t\|^2\,(\Omega,\, Q_\Lambda \Omega)} = \lim_{\substack{t\to\infty \\ \Lambda\to\mathbb{Z}^{D-1}}} \frac{\langle \square_{2t} \cdots \partial^*\Lambda \rangle}{\langle \square_{2t} \rangle \langle \cdots_{\partial^*\Lambda} \rangle} \tag{5}$$

where the dotted lines indicate a dual ((D-2)-dimensional) loop in the $\mathbb{Z}_2$ Higgs model. $\rho_2 = -\,1$ (+ 1) then indicates the (non-) existence of charges.

As a third test for the existence of charges one may look at the correlations of electric fluxes which are necessary for the existence of charged particles according to the general analysis. Let Λ be a cube in $\mathbb{Z}^{D-1}$ with side length R, let $\partial^*\Lambda = (\partial^*\Lambda)_e \cup (\partial^*\Lambda)_r$ be the decomposition of $\partial^*\Lambda$ in a left and right hemisphere and let $E_{e,r} = \prod_{b\in(\partial^*\Lambda)_{e,r}} \tau_1(b)$ denote the corresponding electric fluxes. The flux correlation order parameter is

$$\rho_3(R) = \frac{(\Omega, E_e\,\Omega)\,(\Omega, E_r\,\Omega)}{(\Omega,\, E_e E_r\, \Omega)} = \frac{\langle \cdots_R \rangle \langle \cdots_R \rangle}{\langle \cdots_R \rangle} \tag{6}$$

The correlations are considered to be strong (so charges can exist) if $\ln \rho_3$ is proportional to the surface area of $\partial^*\Lambda$

$$\rho_3 \sim e^{-\,\mathrm{const}\, R^{D-2}} \tag{7}$$

and weak (so charges cannot exist) if $\ln \rho_3$ is proportional to the perimeter of the boundary of $(\partial^*\Lambda)_e$

$$\rho_3 \sim e^{-\,\mathrm{const}\, R^{D-3}} \tag{8}$$

In contrast to ρ_1 and ρ_2 ρ_3 can be defined also in the absence of matter fields. Then the denominator in (6) is equal to one, the numerator depends only on the boundary of $(\partial^*\Lambda)_\ell$ and ρ_3 becomes the square of the 't Hooft loop.

In those parts of the phase diagram of the $\mathbb{Z}_2$ Higgs model where convergent cluster expansions are known the proposed parameters show the expected behavior. Up to now only the vacuum overlap order parameter ρ_1 has been investigated in more general situations.

ρ_1 may be considered as a gauge invariant properly normalized 2-point function. Actually ρ_1 is very similar to the Green function in the charged sector whose clustering properties imply the nontriviality of the charge /4/ as well as the existence of charged particles /5/. Thus the nonvanishing of $\rho_1(\infty)$ may be interpreted as long range order for the charged field. There are other gauge invariant 2-point functions which have been proposed as order parameters. Szlachanyi /7/ investigated the 2-point function in the minimal gauge for the $\mathbb{Z}_2$ Higgs model and Kennedy and King /8/ studied the 2-point function in Landau gauge for the noncompact U(1) Higgs model. Both proposals avoid the normalization factor appearing in (2) and (3); this has computational advantages but may be the reason for non reproducing the correct phase diagram in general /7, 9/.

A gauge invariant 2-point function very similar to ρ_1 has been considered by Bricmont and Fröhlich /10/:

$$\rho_{BF}(\tau) = \frac{\langle \overset{\tau}{\longrightarrow} \rangle^2}{\langle \underset{2\tau}{\longrightarrow} \rangle} \tag{9}$$

It has been shown in /4/ that $\rho_{BF}(\infty) = \rho_1(\infty)$ in the screening/confinement phase of the $\mathbb{Z}_2$ Higgs model. In contrast to $\rho_1(\infty)$ $\rho_{BF}(\infty)$ is not expected to vanish in the whole free charge phase. Namely, the nonvanishing of $\rho_{BF}(\infty)$ is equivalent to the existence of a hydrogen like bound state of an external charge with a dynamical charge.

ρ_1 has been determined for several models by analytical and numerical methods. Alonso and Tarancón /11/ showed $\rho_1 \neq 0$ in the 2-dimensional massless Schwinger model. For the compact U(1) lattice Higgs model Kondo /12/ proved $\rho_1 = 0$ in the free charge phase and Barata /13/ $\rho_1 \neq 0$ in the screening/confinement phase. Alessandrini et al. /14/ investigated ρ_1 in slightly more general models by mean field methods. For the noncompact U(1) lattice Higgs model the techniques of Kennedy and King can be used to derive similar results for ϱ_1 as for their parameter.

ϱ_1 has also been tested in Monte Carlo simulations. In the $\mathbb{Z}_2$ Higgs model ϱ_1 shows the expected behavior in a large part of the phase diagram /15/. There is, however, a part of the confinement region where $\varrho_1(r)$ does not reach its asymptotic value for $r \leq 10$; actually it decreases to very small values. This behavior is understood in terms of the fragmentation distance d_f at which the screening of the potential sets in and the string state Φ_r starts to hadronize; d_f is rather large in that part of the confinement region. On the transition between the screening region and the free charge phase the asymptotic value of ϱ_1 can be measured with great precision; it obeys there the same scaling laws as the square of the magnetization in the Ising model. This gives evidence for this transition to be second order /16/.

The Aachen collaboration /17/ measured ϱ_1 for the SU(2) and U(1) Higgs models with similar results: $\varrho_1(\infty) \neq 0$ in the screening region, $\varrho_1(\infty) = 0$ in the free charge phase (which exists only in the U(1) case), no convergence in the confinement region. The SU(2) results are particularly interesting since no analytical result for ϱ_1 is available for nonabelian theories. Azcoiti and Tarancón /18/ measured ϱ_1 at the free charge/Higgs transition in a U(1) Higgs model with Higgs charge two where again the expected behavior was found.

References:

/1/ D. Buchholz, K. Fredenhagen: Locality and the structure of particle states. Commun. Math. Phys. 84, 1-54 (1982)

/2/ R. Haag, D. Kastler: An algebraic approach to quantum field theory. Jour. Math. Phys. 5, 848 (1964)

/3/ J.A. Swieca: Charge screening and mass spectrum, Phys. Rev. D13, 312 (1976);

D. Buchholz, K. Fredenhagen: Charge screening and mass spectrum in abelian gauge theories: Nucl. Phys. B154, 226 (1979)

/4/ K. Fredenhagen, M. Marcu: Charged states in $\mathbb{Z}_2$ gauge theories. Commun. Math. Phys. 92, 81-119 (1983)

/5/ J. Barata, K. Fredenhagen: Charged particles in $\mathbb{Z}_2$ gauge theories. to be published

/6/ K. Fredenhagen, M. Marcu: Confinement criterion for QCD with dynamical quarks. Phys. Rev. Lett. 56, 223 (1986)

/7/ K. Szlachanyi: Non local charged fields and the confinement problem. Phys. Lett. 147B, 335 (1984)

/8/ T. Kennedy, C. King: Symmetry breaking in the lattice abelian Higgs model. Phys. Rev. Lett. 55, 776 (1985)

T. Kennedy, C. King: Spontaneous symmetry breakdown in the abelian Higgs model. Commun. Math. Phys. 104, 327-347 (1986)

/9/ C. Borgs, F. Nill: Gribov copies and absence of spontaneous symmetry breaking in compact U(1) lattice Higgs models. Nucl. Phys. B270 FS 16 , 92-108 (1986)

/10/ J. Bricmont, J. Fröhlich: An order parameter distinguishing between different phases of lattice gauge theories with matter fields. Phys. Lett. 122B, 73-77 (1983)

/11/ J.L. Alonso, A. Tarancon: Study of a recent confinement criterion for QCD with dynamical quarks. Phys. Lett. 165B, 167 (1985)

/12/ K.-I. Kondo: Order parameter for charge confinement and phase structures in the lattice U(1) gauge Higgs model. Prog. Theor. Phys. 74, 152 (1985)

/13/ J. Barata, private communication

/14/ V. Alessandrini, J.L. Alonso, A. Cruz, A. Tarancon: Study of a recent confinement criterion in the lattice U(1) gauge Higgs model. Zaragoza DFTUZ-86.9 (preprint)

/15/ T. Filk, K. Fredenhagen, M. Marcu, to be published

/16/ T. Filk, K. Fredenhagen, M. Marcu: Line of second order phase transitions in 4-dimensional Z_2 gauge theory with matter fields. Phys. Lett. 169B, 405 (1986)

/17/ H.G. Evertz, V. Grösch, J. Jersak, H.A. Kastrup, D.P. Landau, T. Neuhaus, J.-L. Xu: Monte Carlo analysis of gauge invariant 2-point functions in a SU(2) Higgs model. Aachen PITHA 85/24 (preprint)

/18/ V. Azcoiti, A. Tarancon: Monte Carlo study of confinement in the U(1) gauge Higgs model with Higgs charge two. Phys. Lett. 176B, 153 (1986)

Models with Asymptotic Scale Invariance*

O. Piguet
Département de Physique Théorique
Université de Genève
CH-1211 Geneva 4, Switzerland

K. Sibold
Max-Planck-Institut für Physik und Astrophysik
- Werner-Heisenberg-Institut für Physik -
D-8000 München 40 (Fed.Rep.Germany)

1. Preparation

Relativistic quantum field theories in four dimensions are in general plagued by infinities. Although perturbative renormalization theory deals with some of them in a mathematically proper way, many people feel nevertheless uneasy about them and therefore the construction of a finite theory has always been a dream of field theorists. Since their early days supersymmetric models have been considered as chief candidates [1]. The search was successful surprisingly late and not without dispute about the reliability of the result [2]. The main problems are the use of a regularization which preserves at the same time supersymmetry and gauge invariance and the extension of proofs to all orders. In this note we report on some progress which has been made [3] in direction of rigour, scope of validity and simplicity. First of all we do not use any <u>specific</u> regularization or renormalization scheme but by employing the algebraic method of BRS [4] rely only on general principles of field theory, namely locality and power counting. Next we can treat theories which have truely complex representations. And, last, our criterion operates at the one loop level yielding then information on all orders.

---

*) Talk given by K. Sibold.

To begin with we recall some known facts about supersymmetric Yang-Mills theories. It has already been shown with the above mentioned BRS-technique [5] that supersymmetry, BRS-invariance and conformal R-symmetry can be asymptotically established to all orders provided that the representations of the matter fields are suitably restricted. In other words, the classical action

$$\Gamma_{cl} = \Gamma_{YM} + \Gamma_{matter} \,, \tag{1}$$

$$\Gamma_{YM} = -\frac{1}{128 g^2} \operatorname{Tr} \int dS F^\alpha F_\alpha \,, \quad F^\alpha = \bar{D}\bar{D}\left(e^{-\phi} D^\alpha e^{\phi}\right),$$

$$\Gamma_{matter} = \frac{1}{16} \int dV \sum_R \bar{A}_R e^{\phi_i T^i_R} A^R + \int dS\, W(A) + \int d\bar{S}\, \bar{W}(\bar{A}) \,,$$

$$W(A) = \sum_{I=1}^{I_m} \lambda_I W^I(A)$$

which has these symmetries, can be extended to the functional $\Gamma = \Gamma_{cl}$ + loops which generates the vertex functions and satisfies*

$$W_X \Gamma \sim 0 \qquad X = \alpha, \dot{\alpha}, R \tag{2}$$

$$\mathcal{S}(\Gamma) \sim 0 \tag{3}$$

provided $\quad d_{ijk} \operatorname{Tr} T^i T^j T^k = 0 \,. \tag{4}$

(We are working with a simple gauge group and d_{ijk} is a completely symmetric, invariant tensor of it.)

By the same technique it has also been shown [6,7], that there exist BRS-invariant operators $V_{\alpha\dot{\alpha}}$ ("supercurrent", real superfield, dim 3, R-weight 0) and S ("breaking", chiral superfield, dim 3, R-weight -2) which satisfy the "trace equation"

$$\bar{D}^{\dot{\alpha}} V_{\alpha\dot{\alpha}} \sim -2 w_\alpha \Gamma - 2 D_\alpha (S + S_0) \tag{5}$$

---

*) The $\sim$ means "in the asymptotic region", we neglect soft terms.

(here V, S denote the insertions $[V]\cdot\Gamma$, $[S]\cdot\Gamma$; S_0 originates from the gauge fixing procedure). The proof given in [6] uses parity but we have verified that it also holds without assuming parity invariance. The relevance of this equation is the following: not only can the R-current, the supersymmetry current and the energy-momentum tensor be recovered as components of a superfield, namely,

$$V_\mu = R_\mu + i\theta^\alpha(Q_\mu - \tfrac{1}{3}\sigma_\mu\bar\sigma^\nu Q_\nu)_\alpha - i\bar\theta_{\dot\alpha}(\cdots)^{\dot\alpha} + \theta\sigma^\nu\bar\theta(-2(T_{\mu\nu} - \tfrac{1}{3}g_{\mu\nu}T_\lambda{}^\lambda) - \tfrac{1}{2}\varepsilon_{\mu\nu\rho\sigma}\partial^\rho R^\sigma) + \ldots \tag{6}$$

but also all other currents of the superconformal group can be found by the moment construction [7,8], familiar from ordinary conformal symmetry [9]. In particular,

$$D_\mu = x^\nu T_{\mu\nu} \tag{7}$$

$$T_{\mu\nu} = -\tfrac{1}{16}(V_{\mu\nu} + V_{\nu\mu} - 2g_{\mu\nu}V_\lambda{}^\lambda)\big|_{\theta=0}\ , \quad V_{\mu\nu} = [D_\beta, \bar D_{\dot\beta}]\sigma_\mu^{\beta\dot\beta}\sigma_\nu^{\alpha\dot\alpha}V_{\alpha\dot\alpha} \tag{8}$$

It turns out that S yields the respective breaking

$$\partial^\mu V_\mu \sim i w\Gamma + i(DDS - \bar D\bar D\bar S)\ , \tag{9}$$

$$\partial^\mu D_\mu \sim (w_D\Gamma + \tfrac{3}{2}i(DDS + \bar D\bar D\bar S))\big|_{\theta=0}\ , \tag{10}$$

or integrated,

$$W_R\Gamma \sim -\int dx\,(DDS - \bar D\bar D\bar S)\ , \tag{11}$$

$$W_D\Gamma \sim -\tfrac{3}{2}i\int dx\,(DDS + \bar D\bar D\bar S)\ . \tag{12}$$

Since, by dimensional analysis, $i m\partial_m\Gamma = W_D\Gamma$, it is clear that a basis for the r.h.s. of (12) is provided by the supersymmetric differential operators of the theory ("symmetric") meaning: BRS-invariant, commuting with the rigid gauge symmetry and supersymmetry). These are [3]

$$\partial_g,\ \partial_{\lambda_I},\ \partial_{\bar\lambda_I},\ \partial_{a_k} \tag{13}$$

and the counting operators

$$\mathcal{N}_\phi = N_\phi - N_\rho - N_{c_-} - N_{\bar{c}_-} - N_B - N_{\bar{B}} + 2\kappa\partial_\kappa + 2x\partial_x \,,$$
$$\mathcal{N}^R{}_S = N_A{}^R{}_S - N_Y{}^R{}_S + \text{conj.}\,, \quad \mathcal{N}_+ = N_{c_+} - N_\sigma + \text{conj.} \tag{14}$$

So, their coefficients are identified as the usual Callan-Symanzik functions ß and γ . On the other hand we conclude from (9), (10) that the anomalies of an axial current, R_μ, are related to those of the dilatations, D_μ. Now in S participates in particular the SYM Lagrangian $F^\alpha F_\alpha$ which contains $F^{\mu\nu} F_{\mu\nu}$ <u>and</u> $F^{\mu\nu} \tilde{F}_{\mu\nu}$: these observations suggest to prove first non-renormalization theorems for chiral anomalies (like $F^\alpha F_\alpha$) and then to explore their consequences for the dilatations.

2. Non-renormalization theorem

The result of the corresponding investigation is the following theorem: Let the operators c_+^3, c_+^2 be introduced in the theory naively as they are finite operators. Let T be a BRS-invariant chiral insertion of dimension 3 and R-weight -2. Then

$$T \sim \bar{D}\bar{D} J^0 + T^c \,, \qquad B_\Gamma T \sim 0$$
$$J^0 \sim r K^0 + \bar{D}_{\dot\alpha} J^{\dot\alpha} + J_{inv} \tag{15}$$

and r is gauge parameters independent. If furthermore its integral fulfils the Callan-Symanzik equ. up to a BRS-variation and without anomalous dimension, namely

$$C\left(\int dS\, T + \int d\bar{S}\, \bar{T}\right) \sim B_\Gamma \Delta \tag{16}$$

then the coefficient r of K° in (15) is not renormalized i.e. it is precisely of one loop order.

3. Criterion for vanishing ß-functions

The above theorem can be applied to the symmetric insertions L_i given by

$$\nabla_i \Gamma \sim \int dS\, L_i + \int d\bar{S}\, \bar{L}_i \tag{17}$$

(s. (13), (14) for ∇_i).

Similarly it holds for S itself (5). The expansion of S in terms of L_i identifies the functions β, γ and hence we find by comparing the two forms:

$$r = \left(\frac{1}{128 g^3} + r_g \right) \beta_g + \sum_I r_{\lambda_I} \beta_{\lambda_I} - \sum_a \gamma_{oa} r_{oa} \tag{18}$$

(and its conjugate).

The coefficients r are gauge parameters independent and r, r_{oa} have only one-loop contributions:

$$r = \frac{1}{512 (4\pi)^2} \left(-3 C_2(G) + \sum_R T(R) \right), \tag{19}$$

$$r_{oa} = - \frac{1}{256 (4\pi)^2} \sum_R (e_a)^R{}_R\, T(R) \quad , \; T(R)\, \delta^{ij} = Tr (T_R^i T_R^j)$$

γ_{oa} are the anomalous dimensions of those fields whose counting operators satisfy

$$\mathcal{N}_{oa} W^I(A) = 0 \quad \forall I \quad , \; \mathcal{N}_{oa} = \sum_{R,S} (e_a)^S{}_R\, \mathcal{N}^R{}_S \quad . \tag{20}$$

This leads to the following criterion for ß-functions vanishing to all orders:

$$\text{If} \qquad r = r_{oa} = 0 \tag{21}$$

and if the reduction equations [10]

$$\beta_g \frac{d\lambda_I}{dg} = \beta_{\lambda_I} \quad , \quad \beta_g \frac{d\bar{\lambda}_I}{dg} = \beta_{\bar{\lambda}_I} \tag{22}$$

admit power series solutions

$$\lambda_I = \lambda_I(g), \qquad \bar{\lambda}_I = \bar{\lambda}_I(g) \tag{23}$$

non-vanishing for all I, then (19) and its conjugate imply

$$\beta_g(g,\lambda(g)) = \beta_{\lambda_I}(g,\lambda(g)) = \beta_{\bar{\lambda}_I}(g,\lambda(g)) = 0 \quad (24)$$

to all orders of perturbation theory.

4. Asymptotic scale invariance

Suppose (24) is established for a model. The consequence for the Ward-identity of dilatations is that only the $\mathcal{N}$'s and ∂_{a_k} survive as non-naive terms. But those are BRS-variations (cf. App. D in [7]), hence do not contribute between physical states. We may therefore write

$$W_D \Gamma \sim \mathcal{B}_\Gamma \Delta_D \quad (25)$$

i.e. the Green's functions of BRS-invariant operators which have naive dimensions are asymptotically scale invariant. This is the result announced in the title.

To establish the equality sign in (25) would require controlling the limit of vanishing physical masses. This is a non-trivial problem which deserves a separate discussion.

With the help of the analysis performed in [11] we can extend the above also to the special conformal transformations K_μ and the special supersymmetry transformations S_α, $\bar{S}_{\dot\alpha}$. There we found

$$\begin{aligned} W^S_\alpha \Gamma &\sim -12 \int dx\, D_\alpha S \,, \\ W^{\bar S}_{\dot\alpha} \Gamma &\sim -12 \int dx\, \bar{D}_{\dot\alpha} \bar{S} \,, \\ W^K_\mu \Gamma &\sim -3i \int dx\, x_\mu (DDS + \bar{D}\bar{D}\bar{S}) \end{aligned} \quad (26)$$

i.e. all integrated anomalies can be expressed by S, they are indeed BRS-variations. This means

$$W_X \Gamma \sim \mathcal{B}_\Gamma \Delta_X \qquad X = K_\mu, S_\alpha, \bar{S}_{\dot\alpha} \quad (27)$$

In a SYM where (24) is verified the Green's functions of gauge invariant operators are asymptotically superconformally symmetric. For this to happen sect. 3 provides a simple criterion.

References

[1] J. Iliopoulos, B. Zumino, Nucl.Phys. B76 (1974) 310;
W. Lang, J. Wess, Nucl.Phys. B81 (1974) 249.

[2] M.T. Grisaru, M. Rocek, W. Siegel, Nucl.Phys. B159 (1979) 429;
S. Mandelstam, Nucl.Phys. B213 (1983) 149;
M. Sohnius, P.C. West, Phys.Lett. 100B (1981) 245.

[3] O. Piguet, K. Sibold, MPI Preprints 30/86, 31/86.

[4] C. Becchi, A. Rouet, R. Stora in Field Theory, Quantization and Statistical Physics, ed. E. Tirapegui;
D. Reidel Publ. Co. 1981.

[5] O. Piguet, K. Sibold, Nucl.Phys. B247 (1984) 484.

[6] O. Piguet, K. Sibold, Nucl.Phys. B196 (1982) 447.

[7] O. Piguet, K. Sibold, "Renormalized Supersymmetry", Birkhäuser Boston 1986.

[8] T.E. Clark, O. Piguet, K. Sibold, Nucl.Phys. B143 (1978) 445.

[9] S. Coleman, R. Jackiw, Ann. of Phys. 67 (1971) 552.

[10] R. Oehme, W. Zimmermann, Comm.Math.Phys. 97 (1985) 569;
W. Zimmermann, Comm.Math.Phys. 97 (1985) 211;
R. Oehme, K. Sibold, W. Zimmermann, Phys.Lett. 147B (1984) 115.

[11] O. Piguet, K. Sibold, Nucl.Phys. B197 (1982) 247.

THE RELAXATION TERMS OF THE LANDAU-LIFSCHITZ EQUATION AND THE PROBLEM OF SOLITON DAMPING

V.G. Baryakhtar

Institute of Metal Physics
Academy of Sciences of the Ukrainian USSR, KIEV

Let us consider a ferromagnet with magnetization $\vec{M}(t,\vec{x})$ and energy $W\{\vec{\mu}\}$. The time derivative of W is

$$\frac{\partial W}{\partial t} = - \int \vec{F} \cdot \frac{\partial \vec{M}}{\partial t} d^3 x \tag{1}$$

where the effective magnetic field is

$$\vec{F} = - \frac{\delta W}{\delta \vec{M}(t,\vec{x})} \tag{2}$$

Using the Onsager arguments we obtain the following equation for $\frac{\partial \vec{M}}{\partial t}$

$$\frac{\partial \vec{M}}{\partial t} = \gamma_1(\vec{M}) \cdot \vec{F} + (\nabla \gamma_2 \nabla)\vec{F} \tag{3}$$

The second term of the equation allows for the space dispersion of the kinetic coefficients. Here the following notations are used :

$$(\nabla \gamma_2 \nabla)_{ik} = \hat{\Gamma}_{ik} = \gamma_{2,\, ik,lm}\, \partial^2/\partial x_l \partial x_m$$

The antisymmetric parts of both tensors γ and $\Gamma = (\nabla \gamma_2 \nabla)$ describe the dynamics of magnetization. The antisymmetric part of γ_1 leads us to the well-known Landau-Lifschitz equation

$$\frac{\partial \vec{M}}{\partial t} = g\,[\vec{M},\vec{F}] \tag{4}$$

The antisymmetric part Γ has the same origin as γ and gives us only small corrections to Eq. (4). So we don't take it into account.

2. The symmetric parts of γ_1 and Γ describe the relaxation processes of the ferromagnet. For the tensor Γ only the part due to exchange interaction makes sense. In other words, account must only be taken on the part of Γ with the same properties of transformations under rotation as those of $\vec{M}$.
The energy of a ferromagnet in the exchange approximation is

$$W_{ex} = \frac{1}{2}\int\{f(M^2)+\alpha_{ik}a^2 \frac{\partial \vec{M}}{\partial x_i}\cdot\frac{\partial \vec{M}}{\partial x_k}\}\, d^3x \qquad (5)$$

where $f(M^2)$ is an arbitrary function of M of order $f(M^2) \approx M^2(T_c/\mu M)$. T_c is the Curie temperature, and μ is the Bohr magneton ; $\alpha \sim (T_c/\mu M)$ and a is the lattice parameter. Using definition (2), formula (5) and putting $\gamma_1 = 0$ equations (3) , (4) can be rewritten as

$$\frac{\partial M_i}{\partial t} + \frac{\partial}{\partial x_k}\left(\Pi_{ik}^{dyn} + \Pi_{ik}^{rel}\right) = 0 \qquad (6)$$

where $\Pi_{ik}^{dyn} = g\,(\vec{M}\times\alpha_{nk}\frac{\partial \vec{M}}{\partial x_n})_i \,;\, \Pi_{ik}^{rel} = \gamma_{nk}^{ex}\, a^2\, (\partial^2 M_i)/(\partial x_n^2) \qquad (7)$

Obviously, Eq. (6) has the integrals of motion

$$\mathcal{M}_i = \int M_i d^3x \qquad (8)$$

If one takes into account the terms of γ and Γ due to relativistic interactions, the integrals of motion $\mathcal{M}_i$ will be destroyed.

3. To present the tensors γ , Γ and α in an explicit form, it is necessary to use the point symmetric group of a crystal in paramagnetic phase. For simplicity, let us consider a crystal with one-symmetry axis C_∞ of infinite order (axis Z). In this case, to obtain the total energy of the ferromagnet it is necessary to add to

(5) the magnetic anisotropy energy

$$W_a = -\beta/2 \int (\vec{M}\vec{n})^2 d^3x \tag{9}$$

where $\vec{n}$ is the unit vector along the C_∞ axis. For γ_1 we have the expression

$$\gamma_{1,ik} = \gamma_{11} (\delta_{ik} - n_i n_k) + \gamma_{12} m_{\perp i} m_{\perp i} \tag{10}$$

where δ_{ik} is the Kroneker unit tensor, $\vec{m}_\perp = \vec{m} - \vec{n}(\vec{m}\vec{n})$; $\vec{m} = \vec{M}/_M$ and γ_{11}, γ_{12} are the relaxation constants. Formula (10) corresponds to $\mathcal{M}_z$ as integral of motion due to the axis of symmetry C_∞.

The equation of motion of $\vec{M}$ has the form

$$(\partial\vec{M}) / \partial t = g\vec{M} x \vec{F} + \gamma_1 \vec{F}_\perp + \gamma_2 \vec{m}_\perp (\vec{m}_\perp \vec{F}) + (\nabla \gamma^{ex} \nabla) a^2 \vec{F} \tag{11}$$

$$\vec{F} = -(df/(dM^2))\,\vec{M} + (\nabla \alpha \nabla) a^2 \vec{M} + \beta \vec{n} + \beta \vec{n}(\vec{n}\vec{M}) \tag{12}$$

For the dynamical case ($\gamma_1 = \gamma_2 = \gamma^{ex} = 0$), as proved by Borovic (1979) and Sklianin (1979), the Landau-Lifschitz equation is a fully integrable one and many scientists (see references in the book [1]) presented numerous solutions of Eq. (11).

4. If damping parameters are considered, we obtain the relaxation of the solution. Let us demonstrate this statement considering the simplest example of a magnetic bion. The distribution of magnetization for bion takes the form

$$M^2 = M_o^2 = \text{const.}$$
$$\theta = \theta\,((x-vt/(x_o(v))$$
$$\varphi = \omega t + \psi((x-vt/(x_o(v))$$

where v and ω are the velocity and the frequency of bion,

$$x_o(v) = x_o^2 = (\alpha/\beta) a^2$$

Instead of ω and V it is convenient to consider the energy E and dimensionless magnetic moment $N = (M_z - M_o)/M_o$ of bion. The E and N are connected with ω and $u = v^2$ by the formulae ;

$$E = 2E_o(1-u-\tilde{\omega}),\ N = (E_o/\omega_o)\ \mathrm{arsh}\,(E/E_o\Omega),$$

and $\tilde{\omega} = \omega/\omega_o$, $\omega_o = gM_o\beta$, $\Omega^2 = 4u^2+\omega^2$, $E_o =$

In case $(E/E_o) << 1$, the time dependence of E and N can be calculated in an explicit form. Using Eqs. (11), (12), we obtain $N \sim E \sim \exp(-t/t_s)$ where

$$t_s = \begin{cases} (2\gamma_2\omega_o)^{-1}, & v(o) << \omega_o x_o \\ (V_o^4/2\gamma e x \omega_o^5 x_o^4)^{-1}, & v(o) >> \omega_o x_o \end{cases}$$

and v(o) is the velocity of bion at t = 0. More detailed picture of soliton and bion relaxation is given in the works [2,3,4].

REFERENCES

1. A. Akhiezer, V. Baryakhtar, S. Peletminskii "Spinwaves", North-Holland Publishing Company, Amsterdam, 1968, 369 p.

2. V. Baryakhtar, Phenomenological description of Relaxation processes in magnetic materials, Sov. Phys. JETP 60 (4), October 1984, 863-867.

3. A. Kosevic, B. Ivanov, A. Kovaljov Nelinejnie volni namagnicennosty. Dynamiceskie i topologiceskie solitoni. Kiev, Naukova Dumka, 1983, 190 str.

4. V. Barjakhtar, B. Ivanov, T. Soboleva, A. Sukstansky. Teorija relaksatsionich dynamiceskich solitonov w ferromagnetikach, IETP, 1986, v. 91, N4 (10), p. 1552-1565.

Accurate Bounds in K. A. M. Theory

Rafael de la Llave
Mathematics Dept.
Princeton Univ.
Princeton, NJ 08544

David Rana
Physics Dept.
Princeton Univ.
Princeton, NJ 08544

The K. A. M. theory suffers from the fact that the strength of perturbation for which it can be applied is very small in comparison to that suggested by numerical experiments.

We have investigated which strategies of proof can yield results arbitrarily close to optimal. It turns out that the best known among them cannot yield a proof (no matter how much fine tuning is done) for perturbation strengths for which the results are, in fact, true.

Nevertheless, there are strategies of proof without this failing, that is, if the result is true, it can be established by making careful choices of the parameters in the proof. We term these strategies *accurate*. Furthermore some of these accurate strategies are so systematic that the optimization can be left to a computer following the ideas of Lanford [La].

We have implemented some of these strategies in the simplest case, the Siegel Center Theorem, for which bounds on the "strength of perturbation" are given by one parameter, the Siegel radius (r_s), for which we prove lower bounds.

Furthermore, we observed that the theory of univalent functions can be used to obtain upper bounds for r_s and estimates for the speed of convergence of the previously mentioned lower bounds.

For the function $f(z) = \alpha_g z + z^2$, with $\alpha_g = exp(2\pi i \cdot \text{golden mean})$, we can show $.308 \leq r_s \leq .350$. We note that this theorem actually applies to all α close to α_g with Diophantine properties close to those of α_g.

We have also studied which strategies are accurate for the twist mapping theorem and are in the course of implementing some of them.

[La] O.E. Lanford III, *Computer Assisted Proofs in Analysis*, Physica 124A, 465-470, (1984).

FIXED POINTS OF COMPOSITION OPERATORS

J.-P. Eckmann
Département de Physique Theorique,
Universite de Geneve, Geneve, Switzerland
H. Epstein
CNRS and IHES, Bures sur Yvette, France

Talk by H. Epstein (extended version)

ABSTRACT

New proofs are given for the existence of the fixed points of the composition operators occurring in the theory of period doubling of interval maps, and in the theory of circle maps with golden rotation number.

1. INTRODUCTION

This talk will be concerned with the functional equation

$$\phi(x) = -\frac{1}{\lambda}\phi\left(\frac{1}{\lambda^{\nu-1}}\phi(\lambda^\nu x)\right), \quad \phi(0) = 1, \tag{1.1}$$

Here ν is fixed in the closed interval $[1,\ 2]$. For a certain $r > 1$, the solutions are required to be in the set $\mathcal{E}_r$ of the real functions ϕ such that:
• ϕ is continuous on an interval $[0,\ L]$, where $L \geq 1$ may depend on ϕ,
• there is a function f, analytic, strictly decreasing, without critical points on $[0,\ L^r]$, with $f(0) = 1$, such that, for all $x \in [0,\ L]$,

$$\phi(x) = f(x^r). \tag{1.2}$$

Since r is always assumed > 1, elements of $\mathcal{E}_r$ are C^r on $[0,\ 1]$. The constant λ in (1.1) may depend on the solution but is required to be in $(0,\ 1)$.

When $\nu = 1$, the solutions of this equation are restrictions to $[0,\ L]$ of even, unimodal solutions of the Cvitanović-Feigenbaum functional equation [F]. (The theory of Coullet and Tresser [CT] uses a very slightly different formalism leading to different functions, but with the same role in interval dynamics.) When $\nu = 2$, Eq.(1.1) is one form of a functional equation central to the theory of circle maps with golden rotation number formulated in a related spirit by Feigenbaum, Kadanoff, and Shenker [FKS], and by Ostlund, Rand, Sethna, and Siggia [ORSS]. In both theories, these functional equations are fixed point conditions for a renormalization operator conjectured (and in several cases proved) to possess a one dimensional unstable (expanding) manifold, and a codimension one stable (strictly contracting) manifold intersecting at its fixed point. Under repeated action of the renormalization operator, the unstable manifold attracts neighboring curves (i.e. one parameter families in function space) transversal to the stable manifold. Hence the elements of this unstable manifold will possess all properties which are stable under the action of the operator, under taking limits, and which are exhibited by the elements of some neighboring curve. One such property is the basis of the method to be presented here: we will require the inverse function U of the function f mentioned above, and consequently the inverse function u of the function ϕ, to be anti-Herglotz functions.

1.1 Herglotz Functions

Let J be an open, possibly empty interval in $\mathbf{R}$. We denote

$$\mathbf{C}(J) = \{z \in \mathbf{C} : \text{Im } z \neq 0 \text{ or } z \in J\} = \mathbf{C}_+ \cup \mathbf{C}_- \cup J, \tag{1.3}$$

where

$$\mathbf{C}_+ = \{z \in \mathbf{C} : \text{ Im } z > 0\} = -\mathbf{C}_- \,. \tag{1.4}$$

$\mathcal{F}(J)$ is the real Fréchet space of the functions h holomorphic in $\mathbf{C}(J)$, and such that $h^*(z^*) = h(z)$.

$\mathbf{P}(J)$ is the subset of $\mathcal{F}(J)$ consisting of the functions h such that $h(\mathbf{C}_+) \subset \bar{\mathbf{C}}_+$. An element h of $\mathbf{P}(J)$ is called a Herglotz function (and $-h$ is an anti-Herglotz function). It has a once subtracted Cauchy representation:

$$h(z) - h(z_0) = a(z - z_0) + \int d\mu(t) \left[\frac{1}{t - z} - \frac{1}{t - z_0} \right], \tag{1.5}$$

valid for all z and z_0 in $\mathbf{C}(J)$. Here the positive measure μ has support in $\mathbf{R} - J$ and is the boundary value (in the sense of measures) of Im $h/\pi$from the upper half-plane. It satisfies

$$\int \frac{d\mu(t)}{1+t^2} < \infty.$$

The constant a is positive and is called the angular derivative of h at infinity.

Note that if h belongs to $\mathbf{P}(J)$ and is not constant and real in $\mathbf{C}_+ \cup \mathbf{C}_-$, then $h(\mathbf{C}_+) \subset \mathbf{C}_+$. If, moreover, J is not empty, then on J, h' is strictly positive. Finally, the following trivial fact is stated as a lemma because it plays a major part in the following:

Lemma 1

Suppose $J \neq \emptyset$ and $J \neq \mathbf{R}$. Then for any $B > 0$,

$$\mathbf{P}(J) \cap \{h \ : |h(J)| \leq B\} \tag{1.6}$$

is compact in $\mathcal{F}(J)$.

1.2 Remarks

On the basis of much numerical and also rigorous work by many authors, we expect that, for each fixed ν, there will be, for each $r > 1$, a locally (and perhaps globally) unique solution of (1.1) satisfying the above requirements. This solution, and in particular λ should depend analytically on r. (Note that if the requirement of the existence, monotonicity, and analyticity of f is abandoned there appear many other solutions: see e.g. [C]. But we shall always impose this requirement.)

In the very abundant literature devoted to Feigenbaum's theory and to the related theory of circle maps, there appear already several rigorous proofs of the existence of solutions for (1.1) by Collet, Eckmann and Lanford [CEL], Lanford [L1,L2], Campanino, Epstein and Ruelle [CER], Eckmann and Wittwer [EW], Falcolini [Fa], Jonker and Rand [JR], Lanford and de la Llave [LL], Mestel [M]. All of them, and the present one as well, leave something to be desired in terms of generality and of explanatory virtue. The anti-Herglotz property of U was proved in the case $\nu = 1$, $r = 2$, by Epstein and Lascoux [EL].

2. NEW STATEMENT OF THE PROBLEM

The function $U = f^{-1}$ is easily seen to be analytic on $(-\lambda^{-1},\ \lambda^{-2})$. It is convenient to use instead $\psi = U/U(0)$. The problem to be solved now takes the form:

Let ν be fixed in $[1,\ 2]$. *Find real constants r, λ, τ, α and two functions V and ψ satisfying:*

$$0 < \lambda < 1, \quad r > 1, \quad \tau = \lambda^r, \quad \alpha > 1, \tag{2.1}$$

$$\psi \in -\mathbf{P}((-\lambda^{-1},\ \lambda^{-2})), \quad \psi(0) = 1, \quad \psi(1) = 0, \tag{2.2}$$

$$V \in -\mathbf{P}((0,\ \alpha\tau^{-2})), \quad V(1) = 1, \quad V'(1) = -\frac{1}{\lambda}, \tag{2.3}$$

and the identities

$$V(\zeta) = \frac{1}{\tau^{\nu}}\psi\left(\left(\frac{\zeta}{\alpha}\right)^{1/r}\right) \quad \text{for all } \zeta \in \mathbf{C}((0,\ \alpha\tau^{-2})), \tag{2.4}$$

$$\psi(z) = V(\psi(-\lambda z)) \quad \text{for all } z \in \mathbf{C}((-\lambda^{-1},\ \lambda^{-2})). \tag{2.5}$$

Given a solution of this problem, a solution of (1.1) satisfying all our requirements is obtained by setting:

$$\phi = u^{-1}, \quad u(z) = \frac{1}{\lambda^{\nu-1}}[\alpha^{-1}\psi(z)]^{1/r}. \tag{2.6}$$

As already mentioned, all the available information indicates that (for fixed ν) this system will have a locally unique solution for any given r with λ a (smooth) function of r. Although we would like to be able to prescribe r, the more successful version of our method works with a fixed λ. This is the version I shall summarize now. The results of the fixed-r method will be stated later.

3. THE FIXED λ METHOD

In this section, both $\nu \in [1,\ 2]$ and $\lambda \in (0,\ 1)$ are fixed. A function ψ giving a solution of the system (2.1-2.5) will be looked for as a fixed point of an operator M_λ defined as follows:

1) Start with an element ψ_0 of the set

$$\mathbf{E}_0(\lambda) = \{\psi_0 \in -\mathbf{P}((-\lambda^{-1},\ \lambda^{-2}))\ : \psi_0(0) = 1, \quad \psi_0(1) = 0\}. \tag{3.1}$$

A function V is first defined by

$$V(\zeta) = \frac{1}{\tau^\nu}\psi_0\left(\left(\frac{\zeta}{\alpha}\right)^{1/r}\right), \tag{3.2}$$

where the constants τ, α, and r must be determined so that

$$V(1) = 1, \qquad V'(1) = -\frac{1}{\lambda}, \qquad \tau = \lambda^r, \tag{3.3}$$

and must satisfy

$$r > 1, \qquad 1 < \alpha < \tau^{-\nu}. \tag{3.4}$$

2) Then ψ is the solution of

$$\psi(z) = V(\psi(-\lambda z)), \quad \psi(0) = 1, \quad \psi(1) = 0. \tag{3.5}$$

If all these constructions are possible, and if $\psi \in \mathbf{E}_0(\lambda)$, then ψ_0 is in the domain of M_λ, and $\psi = M_\lambda\psi_0$. It turns out that, for $\nu = 1$ (case of the interval), M_λ is defined on the whole of $\mathbf{E}_0(\lambda)$ and maps it into a smaller, compact set. But, for $\nu > 1$, this is only true if $\lambda > \lambda_0(\nu)$. In fact it is expected (see [JR]) that fixed points of M_λ exist iff

$$\lambda^{\nu-1} + \lambda^\nu - 1 \geq 0. \tag{3.6}$$

The condition $\lambda > \lambda_0(\nu)$, which allows us to prove the existence of fixed points by applying the Schauder-Tikhonov theorem, is not as good as (3.6). This means that a smaller domain ought to be used for M_λ, and this is one of the several points where the method will have to be improved. In the next section, an outline of the estimates is given. Details are in [E,EE].

4. DETERMINATION OF THE CONSTANTS

Let $\psi_0 \in \mathbf{E}_0(\lambda)$. If a function V can be defined by (3.2) so as to satisfy (3.3), then $z_1 \equiv \alpha^{-1/r}$ must satisfy

$$\psi_0(z_1) = \tau^\nu, \qquad \frac{z_1\psi_0'(z_1)}{\psi_0(z_1)} = -\frac{r}{\lambda}, \tag{4.1}$$

and since $r = (\log\tau)/(\log\lambda)$, $q(z_1) = 0$, where the function q is defined by

$$q(z) = \frac{\psi_0'(z)}{\psi_0(z)} - \frac{A}{\nu z}\log\psi_0(z), \quad A \equiv A(\lambda) = -1/(\lambda\log\lambda) \geq e, \tag{4.2}$$

for all $z \in \mathbf{C}((-\lambda^{-1},\ 1))$.

It is easy to show that q has a unique zero in (0, 1). We define z_1 as this zero, and:

$$\tau^\nu = \psi_0(z_1) \in (0,\ 1), \quad r = (\log\tau)/(\log\lambda) > 0, \quad \alpha = z_1^{-r} > 1. \tag{4.3}$$

All the subsequent estimates are immediate consequences of the integral representations of the type (1.5) for $-\psi_0$ and $-\log\psi_0$. The function $-\log\psi_0$ is also herglotzian, and

$$\text{Im } z > 0 \Rightarrow 0 < -\text{Im } \log\psi_0(z) < \pi. \tag{4.4}$$

Hence the representation (1.5) for this function takes the form

$$-\log\psi_0(z) = \int \sigma(t)dt\left[\frac{1}{t-z} - \frac{1}{t}\right] \quad \forall z \in \mathbf{C}((-\lambda^{-1},\ 1)), \tag{4.5}$$

where $\sigma \in L^\infty$ has support in $\mathbf{R} - (-\lambda^{-1},\ 1)$, $0 \leq \sigma \leq 1$, and $\sigma(t) = 1$ for all $t \in [1,\ \lambda^{-2}]$. Therefore

$$q(z) = \int \frac{\sigma(t)dt}{t-z}\left[\frac{A}{\nu t} - \frac{1}{t-z}\right] \quad \forall z \in \mathbf{C}((-\lambda^{-1},\ 1)). \tag{4.6}$$

Let $z \in (0,\ 1)$. Then the integrand of (4.6) is easily seen to be positive when t is in

$$\Sigma(\lambda) = \mathbf{R} - (-\lambda^{-1},\ \lambda^{-2}).$$

When $1 < t < \lambda^{-2}$ the integrand is strictly positive if $z \leq 1 - \nu/A$. We conclude

$$z_1 > 1 + \nu\lambda\log\lambda > \lambda^\nu. \tag{4.7}$$

Moreover (for $z \in (0,\ 1)$) a lower (resp. upper) bound for q is obtained by replacing $\sigma(t)$ by 0 (resp. 1) on $\Sigma(\lambda)$, i.e.

$$q_2(z) \le q(z) \le q_3(z), \qquad z \in (0,\ 1), \tag{4.8}$$

where q_2 (resp. q_3) is the function q obtained if ψ_0 is chosen to be ψ_2 (resp. ψ_3), with

$$\psi_2(z) = \frac{1-z}{1-\lambda^2 z}, \qquad \psi_3(z) = \frac{1-z}{1+\lambda z}. \tag{4.9}$$

Both ψ_2 and ψ_3 are elements of $\mathbf{E}_0(\lambda)$, and for $z \in (0,\ 1)$,

$$\psi_3(z) \ \le\ \psi_0(z) \ \le\ \psi_2(z). \tag{4.10}$$

From (4.8) it follows that $z_2 \le z_1 \le z_3$, where z_2 and z_3 are the zeros of q_2 and q_3, respectively. Hence

$$\psi_3(z_3) \ \le\ \psi_3(z_1) \ \le\ \psi_0(z_1) \ \le \psi_2(z_1) \ \le\ \psi_2(z_2), \tag{4.11}$$

i.e.

$$\tau_3 \ \le\ \tau \ \le\ \tau_2, \qquad r_2 \ \le\ r \ \le\ r_3, \tag{4.12}$$

where $\tau_j \equiv \tau_j(\lambda)$ and $r_j \equiv r_j(\lambda)$ are the values of τ and r for the case when $\psi_0 = \psi_j$, $j = 2$ or 3. Thus the quantities z_1, r and τ have upper and lower bounds depending only on λ, and these bounds are actually attained as ψ_0 varies over $\mathbf{E}_0(\lambda)$.

Other straightforward bounds include:

$$r \ >\ \frac{\lambda^3}{1-\lambda^4}, \qquad \alpha\ >\ 1 + \frac{\lambda}{1+\lambda^2}. \tag{4.13}$$

The first of these shows that $r \to \infty$ when $\lambda \to 1$. (Details are in [EE].)

In fact the minimum of r as ψ_0 varies in $\mathbf{E}_0(\lambda)$ is $r_2(\lambda)$. It is easy to see that the necessary and sufficient condition for $r_2(\lambda) > 1$ is

$$(1-\lambda^\nu)(\lambda^{\nu+2} + \lambda^{\nu-1} - 1) + (\lambda^\nu - \lambda^2)\lambda^{\nu-1} \ >\ 0. \tag{4.14}$$

This is always true for $\nu = 1$, but for $\nu > 1$, it is only satisfied for $\lambda > \lambda_0(\nu)$, where $\lambda_0(\nu)$ is the unique zero in $(0,\ 1)$ of the function of λ in (4.14). From now on we assume that λ has been so chosen. Then $r \ge r_2(\lambda) > 1$, so that the function

V, which we now define by the formula (3.2), with τ, r and α as in (4.3), satisfies the conditions (2.3), and $V(0) = \tau^{-\nu}$, $V(\alpha) = 0$.

It is clear that $W = V \circ V$ is a Herglotz function holomorphic in a neighborhood of 1. In fact we have:

$$W \in \mathbf{P}((0,\ \alpha)), \quad W(1) = 1, \quad W'(1) = \lambda^{-2},$$

$$W(\alpha) = \tau^{-\nu}, \quad W(0) = V(\tau^{-\nu}) < 0. \tag{4.15}$$

The last statement holds because (4.7) implies $\alpha < \tau^{-\nu} < \alpha\tau^{-2}$.

It is also useful to consider

$$\hat{V}(\zeta) \equiv 1 - V(1-\zeta), \quad \hat{W} = \hat{V} \circ \hat{V}, \quad \hat{W}(\zeta) \equiv 1 - W(1-\zeta), \tag{4.16}$$

which satisfy

$$\hat{V} \in -\mathbf{P}((1-\alpha\tau^{-2},\ 1)), \quad \hat{W} \in \mathbf{P}((1-\alpha,\ 1)),$$

$$\hat{V}(0) = 0 = \hat{W}(0), \quad \hat{V}'(0) = -\lambda^{-1}, \quad \hat{W}'(0) = \lambda^{-2}, \quad \hat{W}(1) > 1. \tag{4.17}$$

Using the bounds already obtained for the various constants, and those readily obtainable from the representation (1.5) for ψ_0, it is easy to obtain, for $\hat{W}(1)$, an upper bound and a lower bound > 1, depending only on λ.

5. CONSTRUCTION OF ψ

To obtain ψ satisfying (3.5) we first construct a function Ψ satisfying

$$\Psi(z) = \hat{V}(\Psi(-\lambda z)) = \hat{W}(\Psi(\lambda^2 z)), \quad \Psi(0) = 0, \quad \Psi'(0) = 1. \tag{5.1}$$

As it is well known, such a function always exists in a small disk Δ around 0, where it is given by the absolutely and uniformly convergent limit

$$\Psi(z) = \lim_{n\to\infty} \Psi_n(z), \quad \Psi_n(z) = \hat{V}^n((-\lambda)^n z). \tag{5.2}$$

In fact the size of Δ and the convergence of (5.2) are (for a fixed λ) uniform as ψ_0 varies in $\mathbf{E}_0(\lambda)$. Since each Ψ_n is a Herglotz function, and since the sequence is uniformly bounded in Δ, it also converges, by Vitali's theorem, uniformly on any compact, in the whole of $\mathbf{C}_+ \cup \mathbf{C}_- \cup \Delta$, to a Herglotz function. On $\mathbf{R}_+$, Ψ can be extended, using the functional equation, to a strictly increasing, analytic function on an interval $[0, L)$ in which it must take the value 1. Indeed, if it did not, it could be extended to the whole of $\mathbf{R}_+$, then to the whole of $\mathbf{R}_-$. It would thus be entire and Herglotz, hence linear, hence it would coincide with z, contradicting the hypothesis. Thus $\Psi(\gamma) = 1$ for some finite $\gamma > 0$. We now define

$$\hat{\psi}(z) = \Psi(\gamma z), \quad \psi(z) = 1 - \hat{\psi}(z). \tag{5.3}$$

Then ψ is anti-Herglotz, and satisfies (3.5). Moreover

$$\psi(-\lambda^{-1}) = V(0) = \tau^{-\nu}, \quad \psi(\lambda^{-2}) = W(0) = V(\tau^{-\nu}). \tag{5.4}$$

As noted in Section 4, these quantities are bounded in modulus by bounds depending only on λ.

It is easy to check that M_λ is continuous in the Fréchet topology (the restriction of Ψ to Δ depends continuously on $\hat{V}$ and hence by Vitali's theorem so does its restriction to any compact in its cut-plane of analyticity). Therefore, by the Schauder-Tikhonov theorem, there exist fixed points of M_λ for every $\lambda > \lambda_0(\nu)$.

Slightly more detailed estimates show that:

Lemma 2

For each $\lambda > \lambda_0(\nu)$ there exists a finite $C_1(\lambda) > 0$, such that M_λ maps $\mathbf{E}_0(\lambda)$ into the compact convex set

$$\mathbf{E}_1(\lambda) = \{\psi \in \mathbf{E}_2(\lambda) \ : \ |\psi(z)| \leq \ C_1(\lambda) \text{ for all } z \in (-\lambda^{-1}, \ \lambda^{-2})\}, \tag{5.5}$$

where

$$\mathbf{E}_2(\lambda) = \{\psi \in \mathbf{E}_0(\lambda) \ : \ (2z \log \lambda)\psi'(z) \ \geq \ a'[1 - \psi(z)]^3 \text{ for all } z \in [0, \ 1]\}, \tag{5.6}$$

$$a' = \frac{a}{1 + 3a}, \quad a = \frac{1}{12}(1 - r_2(\lambda)^{-2}). \tag{5.7}$$

6. THE LIMIT $\lambda \to 1$

Most of the bounds obtained so far are useless when $\lambda \to 1$. This can be remedied by proving:

Lemma 3

There exists a continuous function $\lambda \to C(\lambda)$ on $(\lambda_0(\nu),\ 1]$ such that for all $\lambda \in (\lambda_0(\nu),\ 1)$

$$M_\lambda(\mathbf{E}_2(\lambda)) \subset \mathbf{E}(\lambda), \tag{6.1}$$

where $\mathbf{E}(\lambda)$ is the compact convex subset of $\mathbf{E}_1(\lambda)$ defined by

$$\mathbf{E}(\lambda) = \{\psi \in \mathbf{E}_0(\lambda)\ :\ |\psi(z)| \le\ C(\lambda) \text{ for all } z \in (-\lambda^{-1},\ \lambda^{-2}),$$

$$(2z\log\lambda)\psi'(z)\ \ge\ a'[1-\psi(z)]^3 \text{ for all } z \in [0,\ 1]\}. \tag{6.2}$$

Theorem 4

For each $\lambda > \lambda_0(\nu)$ there exists at least one fixed point of M_λ in $\mathbf{E}_0(\lambda)$. All such fixed points are in $\mathbf{E}(\lambda)$.

If $\psi \in \mathbf{E}_0(\lambda)$ the function H defined by

$$H(w) = \psi(\lambda^{-w}) = \psi(e^{\beta w}), \quad \beta = \log\frac{1}{\lambda}, \tag{6.3}$$

is holomorphic in the cut-strip

$$\{w \in \mathbf{C}\ :\ |\mathrm{Im}\ w| < \pi/\beta \text{ and } w \notin 2+\mathbf{R}_+\}, \tag{6.4}$$

where $\mathrm{Im}\ w > 0\ \Rightarrow\ \mathrm{Im}\ H(w) < 0$. The function H is decreasing on $(-\infty,\ 2)$, concave on $(-\infty,\ 1)$, tends to 1 at $-\infty$, and vanishes at 0. If, moreover, $\psi \in \mathbf{E}_1(\lambda)$, then on $\mathbf{R}_-$

$$-H'(w)\ \ge\ \frac{a'}{2}[1-H(w)]^3, \quad H(w)\ \ge\ 1-(1-a'w)^{-1/2}. \tag{6.5}$$

If $\psi = M_\lambda\psi_0$, and H_0 is similarly defined by $H_0(w) = \psi_0(\lambda^{-w})$ the map M_λ can be rexpressed in the form

$$V(\zeta) = \frac{1}{\tau^\nu}H_0\left(\frac{\log(\zeta/\alpha)}{\log(1/\tau)}\right), \quad W = V \circ V,$$

$$H(w) = W(H(w-2)), \quad H(0) = 0, \quad H(-\infty) = 1. \tag{6.6}$$

Here again, of course, τ, α and r have to be determined so that (3.3) and (3.4) hold. In this form, using the bounds discussed in this section, it can be shown that the map has a well-defined limit as $\lambda \to 1$. We denote, for each $\lambda \in (\lambda_0(\nu),\ 1)$,

ψ_λ a fixed point of M_λ, and V_λ, W_λ, H_λ, τ_λ, α_λ the corresponding quantities. Then there exists a sequence of values of λ tending to 1 such that these objects all have limits as $\lambda \to 1$. The uniform bounds imply that the limits of H_λ, V_λ, are non-constant, finite functions defining a fixed point of the limiting operator mentioned above, while, because of (6.5), ψ_λ tends, uniformly on every compact in $\mathbf{C}((-1,\ 1))$, to the constant 1. This fact, one of the main results of [EW], and an interesting example of Ecalle's general theory, is now reobtained in a different way, and for all ν, in particular for the circle case.

7. THE FIXED r METHOD (INTERVAL CASE)

The main defect of our method is that, being based on the Schauder-Tikhonov theorem, it does not provide the (local) uniqueness of the solutions and has nothing to say about their continuous dependence upon λ. Thus, given a value of $r > 1$, we cannot assert that there is a corresponding solution. We expect that this situation is temporary, and that a refinement of the method will yield the required continuous dependence. In the meantime, we can try a version of the method in which r, rather than λ, is fixed. In this section, ν is fixed and equal to 1, as this is the only case in which the method gives any results at all, and $r > 1$ is also fixed. A new map S_r is defined as follows.

1) Start with an element ψ_0 of the set $\mathbf{E}_0(\lambda_0)$, where, for the moment, $\lambda_0 \in (0,\ 1)$ may depend on ψ_0. Define a function V by the same formula (3.2) as in the fixed-λ method, the constants λ, τ, α being determined so that (3.3) holds. They must also satisfy

$$0 < \lambda < 1, \quad 1 < \alpha < \tau^{-1}, \tag{7.1}$$

and

$$\tau^{-1} < \alpha\lambda_0^{-2r}. \tag{7.2}$$

2) Then ψ is the solution of

$$\psi(z) = V(\psi(-\lambda z)), \quad \psi(0) = 1, \quad \psi(1) = 0. \tag{7.3}$$

If all these constructions are possible, and if $\psi \in \mathbf{E}_0(\lambda)$, then ψ_0 is in the domain of S_r and $S_r\psi_0 = \psi$. It turns out that the difficulty is to satisfy (7.2) while keeping both ψ_0 and ψ in a compact set. It is shown in [E] that

Lemma 5

Let $r > 1$ be fixed. Suppose that there exist two numbers b and z in $(0,\ 1)$ satisfying the three following inequalities:

$$rb^{r-1}(1-b^2)-(1-b^r)(1-b^{r+2})>0, \tag{7.4}$$

$$z(1-b^2)-r(1-z)^{1-1/r}(1+bz)^{1/r}(1-b^2z)>0, \tag{7.5}$$

$$1-z-(1+bz)(b^2z)^r>0. \tag{7.6}$$

Then there exists a constant $K < \infty$ such that S_r maps into itself the compact convex subset of $\mathcal{F}((-b^{-1},\ b^{-2}))$ given by

$$\{\psi \in \mathrm{E}_0(b)\ :\ |\psi(x)| \leq K \text{ for all } x \in (-b^{-1},\ b^{-2})\}, \tag{7.7}$$

and (hence) S_r has a fixed point in this set. Moreover if $\psi_0 \in \mathrm{E}_0(b)$, then $\lambda < b$ and $z_1 < z$.

It takes no computer at all to verify that the conditions (7.4-7.6) are satisfied, for $r = 2$, by $b = 1/2$, $z = 0.9$ (this is done in [E]). With a little more work, the cases $r = 3$ ($b = 0.65$, $z = 0.9$) and $r = 4$ ($b = 0.7$, $z = 0.9$) can be checked. With the help of a computer, the inequalities can be checked for $r \leq 14.4$, but above this, the method fails. That it must fail above a certain r is clear from (7.6): if $\mathrm{E}_0(b)$ contains a fixed point, the last term in (7.6) is bounded below by τ^2/α, which has a uniform lower bound as $r \to \infty$, while $1 - z \to 0$. This reflects the use of a poor lower bound for λ, and could perhaps be mended, but I believe that trying to improve the fixed λ method holds more promise.

8. UNIVALENCE OF SOLUTIONS

Lemma 6

Suppose that, for a certain value of $\lambda \in (0,\ 1)$, ψ is a fixed point of M_λ in $\mathrm{E}_0(\lambda)$ (with $r > 1$). Then the function ψ is injective in $\mathrm{C}((-\lambda^{-1},\ \lambda^{-2}))$.

Note that if ψ_0 is injective (and in the domain of M_λ) then ψ is also injective. This does not yield a proof of Lemma 6, since injective functions are not a convex subset of $\mathrm{E}_0(\lambda)$. However a direct proof is easily supplied.

REFERENCES

[C] M.Cosnard : Etude des solutions de l'équation fonctionnelle de Feigenbaum. *Bifurcations, théorie ergodique et applications;* Astérisque, **98-99**, 143-62 (1982).

[CEL] P. Collet, J.-P. Eckmann, and O.E.Lanford III: Universal properties of maps on the interval. Commun. Math. Phys. **76**, 211-54 (1980).

[CER] M. Campanino, H. Epstein, and D. Ruelle: On Feigenbaum's functional equation. Topology **21**, 125-9 (1982). On the existence of Feigenbaum's fixed point: Commun. Math. Phys. **79**, 261-302 (1981).

[CT] P.Coullet and C.Tresser: Iteration d'endomorphismes et groupe de renormalisation. J. de Physique Colloque C **539**, C5-25 (1978). CRAS Paris **287 A**, (1978).

[E] H. Epstein: New proofs of the existence of the Feigenbaum functions. Commun. Math. Phys., to appear.

[EE] J.-P. Eckmann and H. Epstein: On the existence of fixed points of the composition operator for circle maps. Commun. Math. Phys., to appear.

[EL] H. Epstein and J. Lascoux: Analyticity properties of the Feigenbaum function. Commun. Math. Phys. **81**, 437-53 (1981).

[EW] J.-P. Eckmann and P. Wittwer: *Computer methods and Borel summability applied to Feigenbaum's equation.* Lecture Notes in Physics 227. Berlin, Springer Verlag 1985.

[Fa] C. Falcolini: to appear.

[F] M.J. Feigenbaum: Quantitative universality for a class of non-linear transformations. J.Stat.Phys. **19**, 25-52 (1978). Universal metric properties of non-linear transformations. J.Stat.Phys. **21**, 669-706 (1979).

[FKS] M.J. Feigenbaum, L.P. Kadanoff, and S.J. Shenker: Quasi-periodicity in dissipative systems: a renormalization group analysis. Physica **5D**, 370-386 (1982).

[JR] L. Jonker and D. Rand: Universal properties of maps of the circle with ϵ-singularities. Commun. Math. Phys. **90**, 273-292 (1983).

[L1] O.E. Lanford III: Remarks on the accumulation of period-doubling bifurcations. Mathematical problems in Theoretical Physics, Lecture Notes in Physics vol.116, pp. 340-342. Springer Verlag. Berlin, 1980. A computer-assisted proof of the Feigenbaum conjectures. Bull.Amer.Math.Soc., New Series, **6**, 127 (1984).

[L2] O.E. Lanford III: A shorter proof of the existence of the Feigenbaum fixed point. Commun. Math. Phys. 96, 521-38 (1984).

[LL] O.E. Lanford III and R. de la Llave: in preparation.

[M] B. Mestel: Ph. D. Dissertation, Department of Mathematics, Warwick University (1985).

[ORSS] S. Ostlund, D. Rand, J. Sethna, and E. Siggia: Universal properties of the transition from quasi-periodicity to chaos in dissipative systems. Physica 8D, 303-342 (1983).

NONLINEAR LOCALIZATION AND AN INFINITE DIMENSIONAL KAM THEOREM

Jürg Fröhlich
Theoretical Physics
ETH-Hönggerberg
CH-8093 Zurich
Switzerland

Thomas Spencer
Courant Institute of
Mathematical Sciences
New York University
New York, New York 10012
USA

Eugene Wayne
Department of Mathematics
The Pennsylvania State University
University Park, PA 16802
USA

The properties of random media have been intensively studied since Anderson's work on the propagation of the electron wave function in a disordered crystal in the 1950's. In the present work we begin a study of how the nonlinearity inevitably present in a real system affects special properties like wave trapping and localization that linear models of these systems possess. In particular we would like to know whether or not the transport of quantities like heat or electricity is suppressed in these systems as is the case in the linearized models. We begin with the simpler question of whether or not localized states persist in the presence of nonlinearities.

For a class of models approximating the classical dynamics of a disordered, anharmonic crystal or atoms deposited on a disordered crystal surface we prove that localized states do exist with large probablity. These states lie on infinite dimensional, compact, invariant tori in phase space. Physically they represent excitations with finite total energy. These states are constructed by extending the methods of Kolmogovov, Arnol'd and Moser to this inifnite dimensional setting. Because one has very little information about how these tori "fit together" in the phase space one cannot yet state whether or not an arbitrary localized initial state will remain localized for all time, or if it does not, how fast it will spread.

The work described here appears in the Journal of Statistical Physics, vol. 42, p.247 (1986). Independently, Bellisard and Vittot (preprint, Marseille) arrived at similar results.

RENORMALIZATION GROUP METHODS FOR CRITICAL CIRCLE MAPPINGS WITH GENERAL ROTATION NUMBER

Oscar E. Lanford III
IHES, Bures-sur-Yvette, France

We will use the term *circle mapping* to denote a strictly increasing mapping f of the real line to itself satisfying

$$f(x+1) = f(x) + 1.$$

The *rotation number* of a circle mapping means

$$\rho(f) = \lim_{n\to\infty} \frac{f^n(x) - x}{n};$$

the limit always exists and is independent of x. We are going to discuss the use of renormalization group methods to analyze the dependence of rotation number on f in the space of circle mappings which are very smooth (analytic) but which have critical points, i.e., places where their first derivatives vanish. Without loss of generality, we can put the origin at a place where the derivative vanishes, and thus we will consider circle mappings satisfying $f'(0) = 0$.

The renormalization group formalism we use is that invented by Ostlund, Rand, Sethna, and Siggia[3]. As they discovered, it is convenient to construct the renormalization operator acting not on the space of circle mappings but on a space of commuting pairs of mappings. Specifically, we will work with commuting pairs (ξ, η) of strictly increasing analytic functions on $\mathbf{R}$ satisfying

$$\xi(x) > x \quad \text{for all } x; \qquad \xi(0) = 1; \qquad \eta'(0) = 0.$$

Such a pair can be constructed for each circle mapping f by

$$f \longmapsto (x+1, -f(-x));$$

commutativity is just a transcription of the identity $f(x+1) = f(x) + 1$. (The minus sign is inserted here for convenience later on.) The notion of rotation number extends to these commuting pairs; the rotation number of a pair (ξ, η) can be defined as the unique real number ρ such that

$$p/q > \rho \Longrightarrow \eta^q \xi^p(x) > x \quad \text{and} \quad p/q < \rho \Longrightarrow \eta^q \xi^p(x) < x$$

for all pairs of integers p, q with $q > 0$ and all x in $\mathbf{R}$. (It is not difficult to prove the existence of a ρ satisfying these conditions, using standard ideas from the theory of circle mappings.)

If $r = 1, 2, \ldots$, and if $\zeta = (\xi, \eta)$ is a commuting pair as above, with rotation number ρ, it is easy to see that $\bar{\zeta} = (\bar{\xi}, \bar{\eta})$ defined by:

$$\bar{\xi}(x) = \frac{1}{\lambda}\eta(\lambda x), \quad \bar{\eta}(x) = \frac{1}{\lambda}\eta^r \circ \xi(\lambda x), \quad \text{where } \lambda = \eta(0),$$

is again a commuting pair and has rotation number $1/\rho - r$. We define an operator $\mathcal{T}_r : \zeta \longmapsto \bar{\zeta}$ on the domain $\mathcal{D}_r$ of all pairs ζ with rotation number strictly between $1/(r+1)$ and $1/r$; by what has just been said, the rotation number of $\mathcal{T}_r\zeta$ is the fractional part of the reciprocal of the rotation number of ζ, i.e., the operator acts on rotation numbers by the *Gauss map* from the classical theory of continued fractions. The domains $\mathcal{D}_r$ are mutually disjoint, so the family of operators $\mathcal{T}_r$ can be regarded as defining a single operator $\mathcal{T}$ (i.e., the domain $\mathcal{D}$ of $\mathcal{T}$ is the union of the $\mathcal{D}_r$, and $\mathcal{T} = \mathcal{T}_r$ on $\mathcal{D}_r$ for each r). Alternatively, the domain of $\mathcal{T}$ can be described as the set of all those commuting pairs whose rotation number is strictly between 0 and 1 and is not the reciprocal of an integer.

Schematically, the space of commuting pairs with rotation number between 0 and 1 can be visualized as a cylinder, with rotation number increasing upwards. This cylinder is divided into a sequence of layers: At the top is $\mathcal{D}_1$, below that is the set of pairs with rotation number 1/2, below that $\mathcal{T}_2$, below that the set of pairs with rotation number 1/3, ... Since $\mathcal{T}$ induces the Gauss map on rotation numbers, the image of any individual $\mathcal{D}_r$ under $\mathcal{T}$ is a set on which the rotation number runs all the way from 0 to 1, i.e., a set which runs the full height of the cylinder. We should therefore expect $\mathcal{T}$ to be expansive in, roughly, the vertical direction. On the other hand, it *appears* that $\mathcal{T}$ is contractive in all other directions. For example, it has a well-studied fixed point in $\mathcal{D}_1$ (corresponding to the golden ratio rotation number) at which its linearization has one expanding eigenvalue and the remainder of its spectrum inside the unit circle. Although this has not yet been checked in a completely convincing way, we propose as a *hypothesis* about the action of $\mathcal{T}$ that it is, in an appropriate sense and on a sufficiently large domain, expansive in one direction and contractive in all others. We will refer to this as the *global* hyperbolicity* hypothesis.

We will not try to give here a precise formulation for this hypothesis, but what it means is roughly as follows: The space of commuting pairs can be parametrized by coordinates (u, v), with u one dimensional and v infinite dimensional, in such a way that the (u, u) component of $D\mathcal{T}$ is uniformly larger than 1 in modulus, the (v, v) component is uniformly smaller than 1 in norm, and the off-diagonal components are uniformly sufficiently small.

If the global hyperbolicity hypothesis turns out to be correct, it will have strikingly strong consequences for the parameter dependence of the rotation number in "general" one-parameter families of circle mappings with critical points. To extract these consequences, we need to introduce the notion of *unstable manifold* for such a globally hyperbolic map. As we have already noted, in terms of our earlier picture of the space of commuting pairs as a cylinder, the image of any $\mathcal{T}_r$ is a thin cylinder running the full height of the original one. For any r', the image of the product $\mathcal{T}_r\mathcal{T}_{r'}$ (i.e., the image of $\mathcal{T}_r$ restricted to the intersection of $\mathcal{D}_r$ with the image of $\mathcal{T}_{r'}$) is an even thinner cylinder, contained in the preceding one and still of full height. More generally: Given any sequence $r_1, r_2, \ldots$ of positive integers, the images of the finite products

$$\mathcal{T}_{r_1}\mathcal{T}_{r_2}\cdots\mathcal{T}_{r_n}$$

form a decreasing sequence of thinner and thinner full-height cylinders. It turns out that what they converge down to is a smooth curve, called an unstable manifold for $\mathcal{T}$. We will denote this curve by $W^u(r_1, r_2, \ldots)$; note that there is such a curve for *each* sequence $r_1, r_2, \ldots$, and that they are in principle all different.

Now consider a one-parameter family ζ_μ which traverses the cylinder in a roughly vertical way. For any $r_1, r_2, \ldots, r_n$, the intersection of this curve with

* The term "global" may be a little misleading here. We use it to emphasize that the hypothesis concerns general rotation numbers and not just special ones like the golden ratio. On the other hand, we don't need to assume that the hypothesis really holds on the whole space of commuting pairs; its validity on a sufficiently large domain suffices.

the domain of the product $\mathcal{T}_{r_1}\mathcal{T}_{r_2}\cdots\mathcal{T}_{r_n}$ is, from an analytical point of view, the set of parameter values where the rotation number lies between $[r_1,\ldots,r_n]$ * and $[r_1,\ldots,r_n+1]$. From a geometrical point of view, it is a small piece of arc which is mapped by $\mathcal{T}_{r_1}\mathcal{T}_{r_2}\cdots\mathcal{T}_{r_n}$ to another curve traversing the whole cylinder. If n is large, this latter curve will be very near to $W^u(r_1,r_2,\ldots)$. It turns out, moreover, that there is a *distinguished* parametrization of the unstable manifold $W^u(r_1,r_2,\ldots)$ and that, except for a linear rescaling, the parametrization of the image curve is very close to this distinguished parametrization of the unstable manifold. Thus, the parameter dependence of rotation number in the original family over the small parameter interval where the rotation number is between $[r_1,\ldots,r_n]$ and $[r_1,\ldots,r_n+1]$ can be read off, in good approximation, from the parameter dependence of the rotation number along the unstable manifold $W^u(r_1,r_2,\ldots)$. The latter dependence is of course "universal"; it does not depend on the family from which we started. Furthermore, the goodness of the approximation can be guaranteed simply by taking n large enough, independent of the choice of the $r_1,\ldots,r_n$. Thus, the family of *all* the unstable manifolds gives quantitative information about the parameter dependence of rotation number of a "general" family for all rotation numbers between 0 and 1.

There are a number of ways of organizing the information which can be obtained by the analysis sketched above. One consequence which is easy to formulate is as follows: Consider two parametrized families $\zeta_\mu^{(1)}$ and $\zeta_\mu^{(2)}$ as above, and let $S^{(i)}[r_1,\ldots,r_n]$, $i=1,2$, denote the length of the parameter interval ("phase locking interval") where the rotation number of $\zeta_\mu^{(i)}$ is $[r_1,\ldots,r_n]$. Then the ratio

$$\frac{S^{(1)}[r_1,\ldots,r_n]}{S^{(2)}[r_1,\ldots,r_n]}$$

is bounded, and bounded away from zero, uniformly in n and $r_1,\ldots,r_n$. Intuitively, this says that the lengths of *all* phase locking intervals are universal up to bounded corrections. This explains in particular the universality of the "fractal dimension" of the complement of the phase locking set discovered numerically by Jensen, Bak, and Bohr[1]. It is possible to derive more detailed information about the distribution of lengths of phase locking intervals, and this information can be expressed in an economical way by a thermodynamic formalism.

* The symbol $[r_1,\ldots,r_n]$ denotes a *continued fraction*; it can be defined recursively by $[r]=1/r$; $[r_1,\ldots,r_n]=1/(r_1+[r_2,\ldots,r_n])$.

It should be stressed that the global hyperbolicity hypothesis is at this point a *conjecture*. This conjecture is supported by a certain amount of indirect evidence, such as the fact that, in every case which has been examined, it appears that $\mathcal{T}_r$ has a hyperbolic fixed point with exactly one expanding direction. It is also appealing because of its consequences, such as the simple explanation it provides for the previously mysterious discovery of Jensen, Bak, and Bohr. It is however still far from being *proved*. In principle, if the conjecture is true, it should be possible to prove it using numerical estimates checked rigorously by computer, but it is not possible at this time to estimate whether such a proof is likely to be feasible in practice.

A more detailed exposition of the ideas sketched above, and a more complete set of references, is given in [2].

References.

1. M. H. Jensen, P. Bak, and T. Bohr, Complete devil's staircase, fractal dimension, and universality of mode-locking structure in the circle map, *Phys. Rev. Lett.* **50** (1983) 1637–1639.
2. O. E. Lanford, Renormalization group methods for circle mappings, to appear in the proceedings of the 1985 Groningen conference on "Statistical Mechanics and Field Theory: Mathematical Aspects", to be published in *Springer Lecture Notes in Physics.*
3. S. Ostlund, D. Rand, J. Sethna, and E. Siggia, Universal properties of the transition from quasi-periodicity to chaos in dissipative systems, *Physica* **8D** (1983) 303–342.

UNIVERSALITY FOR THE BREAKDOWN OF DISSIPATIVE GOLDEN INVARIANT TORI.

David Rand.
Mathematics Institute, University of Warwick,
Coventry CV4 7AL, U.K.

1. INTRODUCTION.

My main aim here is to explain the principles behind renormalisation methods for proving the existence and non-existence of invariant circles and to outline the proof given in [9] that the fixed point of the renormalisation transformation for golden critical circle maps describes the universal properties of the breakdown of golden invariant tori in dissipative systems. I will concentrate on the basic ideas and many technical details such as the precise specifications of domains of definition and function spaces are skipped over. The interested reader can find these details in [9].

Motivation 1. Critical circle maps.

I will be particularly interested in circle mappings, invariant circles etc. with rotation number $gm = (\sqrt{5} - 1)/2$ which is the golden mean (mod 1). Such objects are called *golden*. If ω is an irrational number then its rational approximates p_n/q_n are defined inductively by setting $p_0 = 0$, $q_0 = 1$ and requiring that q_n is the smallest positive integer such that $|\omega - p_n/q_n| < |\omega - p_{n-1}/q_{n-1}|$. These are the rational numbers obtained by truncating the continued fraction expansion for ω. If ω = gm then $p_n = q_{n-1}$ and $q_{n+1} = q_n + q_{n-1}$ so the approximates are given by ratios of successive Fibonacci numbers.

A *critical map* of the circle $\mathbb{T} = \mathbb{R}/\mathbb{Z}$ is an analytic homeomorphism with a single critical point which is cubic. Mappings of $\mathbb{T}$ lift to its universal cover $\mathbb{R}$ to give a map $\bar{f}:\mathbb{R}\to\mathbb{R}$ such that $\bar{f}(x+1) = \bar{f}(x) + 1$ and $0 \le \bar{f}(0) < 1$, and, except in instances where it might lead to confusion, I will identify f and $\bar{f}$. If f is a homeomorphism, the rotation number of f is $\rho(f) = \lim_{n\to\infty} n^{-1}(\bar{f}^n(x) - x)$: the limit exists and is independent of x. As a prototype, consider the 2-parameter family

$$f_{\mu,\nu}(x) = x + \nu - ((1 + \mu)/2\pi)\sin 2\pi x.$$

If $\mu = 0$, $f_{\mu,\nu}$ is critical. If $|1 + \mu| < 1$, so that $f_{\mu,\nu}$ is a diffeomorphism, then

(a) if $f = f_{\mu,\nu}$ is golden then $f^{q_n}(0) - p_n \sim a^n$ where $a = -gm$;

(b) $a^{-n}(f^{q_n}(a^n x) - p_n)$ converges, up to a scale change, to $x \to x + gm$; and

(c) if $\nu = \nu_n$ is such that the rotation number $\rho(f_{\nu,\mu}) = p_n/q_n$ and $|\nu_n - \nu_{n-1}|$ is minimal then $\lim_{n\to\infty}(\nu_n - \nu_{n-1})/(\nu_{n+1} - \nu_n) = \delta$ where $\delta = -gm^{-2}$.

On the other hand, if $\mu = 0$ so that $f_\nu = f_{\mu,\nu}$ is critical then:

(a)' if $f = f_\nu$ is golden then $f^{q_n}(0) - p_n \sim a^n$ where $a = -.776\ldots = -gm^{.527\ldots}$;

(b)' $a^{-n}(f^{q_n}(a^n x) - p_n)$ converges to an analytic function ζ of x^3 as $n \to \infty$; and

(c)' if ν_n is as in (c) with $\mu = 0$ then $\lim_{n\to\infty}(\nu_n - \nu_{n-1})/(\nu_{n+1} - \nu_n) = \delta$ where $\delta = -2.834\ldots = -gm^{-2.164\ldots}$.

Moreover, (d)' there is a neighbourhood U of (0,0) such that if $(\mu,\nu) \in U$ then $\rho(f_{\mu,\nu}) = gm$ if and only if $\nu = \nu_*(\mu)$ and $\mu \le 0$ where the function ν_* is C^∞ on $\mu \neq 0$ and C^2 at $\mu = 0$. If $\mu < 0$ then this circle is C^∞.

This scaling structure appears to be largely independent of the choice of family. If $f_{\mu,\nu}$ is any analytic 2-parameter family in $\mathfrak{D}^\omega$ with a parameterisation such that $f_{0,\nu}$ is critical and which satisfies (a)' – (d)' with the same a, δ and ζ as there then I shall say that $f_{\mu,\nu}$ is *in our universality class.*

Motivation 2. Dissipative diffeomorphisms of the annulus.

To describe the structure for invariant circles of diffeomorphisms, consider as prototype the following family of embeddings of the annulus $R^{-1} < r < R$, $R > 1$:

$$F_{\mu,\nu}(r,\varphi) = (1 + \lambda(r-1) - (\mu/2\pi)\sin 2\pi(\varphi + r), \varphi + r + \nu)$$

where $\lambda > 0$ is small. This map contracts areas by a factor of λ. Let $\bar{F}_{\mu,\nu}$ denote the lift to the universal cover $\mathbb{R}^2$ of the annulus. For $\mu = 0$, $F_{\mu,\nu}$ leaves invariant the circle $r = 1$. In fact, this circle is normally hyperbolic and therefore persists for $|\mu|$ small. Let $f_{\mu,\nu}$ denote the restriction of $F_{\mu,\nu}$ to this invariant circle (when it exists).

Let $I_{p/q}$, $p/q \in \mathbb{Q}$, denote the set of parameter values (μ,ν) such that $\bar{F}_{\mu,\nu}{}^q(\varphi,r) = (\varphi + p, r)$ for some (φ,r). Let $\nu = \nu_*(\mu)$ denote the curve on which $\rho(f_{\mu,\nu}) = gm$ and $\nu = \nu_{p/q}(\mu)$ denote the boundary curve of $I_{p/q}$ nearest $\nu = \nu_*(\mu)$ in the sense of Figure 1.

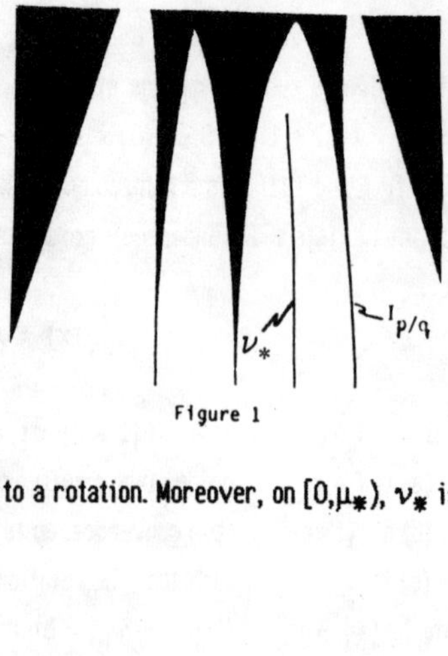

Figure 1

There is strong numerical evidence for the following facts ([2]):

(a) There exists μ_* (= .978837778... when $\lambda = 0.5$) such that if $\mu < \mu_*$ and $\nu = \nu_*(\mu)$ then $F_{\mu,\nu}$ has an analytic invariant circle and $f_{\mu,\nu}$ is analytically conjugate to a rotation. Moreover, on $[0,\mu_*)$, ν_* is analytic.

(b) $F_{\mu_*,\nu_*(\mu_*)}$ has a golden invariant circle, but it is not analytic, The map $f_{\mu_*,\nu_*(\mu_*)}$ can be conjugated to a rotation.

(c) For $\mu > \mu_*$ there is no golden invariant circle.

(d) F has the following scaling structure. (i) If $\nu_n = \nu_{p_n/q_n}(\mu_*)$ then F_{μ_*,ν_n} has a unique (p_n/q_n)–periodic orbit (a saddle node). Let a_n denote the shortest distance between two points on this orbit. Then $\lim_{n\to\infty}(a_n/a_{n+1}) = a$ and $\lim_{n\to\infty}(\nu_n - \nu_{n-1})/(\nu_{n+1} - \nu_n) = \delta$ where the numbers a and δ are exactly as for critical circle maps.

2. CRITICAL CIRCLE MAPS.

It is convenient to construct the renormalisation operator acting not on the space of circle mappings but on a space of commuting pairs of mappings. More specifically, the renormalisation transformation

$$T(\xi,\eta) = a^{-1}\cdot(\eta,\eta\circ\xi)\circ a, \quad a = a(\xi,\eta) = \eta(0) - \xi(\eta(0))$$

is defined on an open subset $\mathcal{O}$ of the bigger space $\mathcal{E}$ consisting of pairs of maps (ξ,η) as in Figure 2 and which satisfy the following conditions:

(a) $0 < \xi(0) = \eta(0) + 1 < 1$;

(b) $\xi''(0) = 0$, $\zeta'(\eta(0)) \neq 0 \neq \eta'(0)$ and $\xi''(\eta(0)) \neq 0 \neq \eta''(0)$; and

(c) $(\xi\circ\eta - \eta\circ\xi)^{(i)} = 0$ for $0 \le i \le 3$.

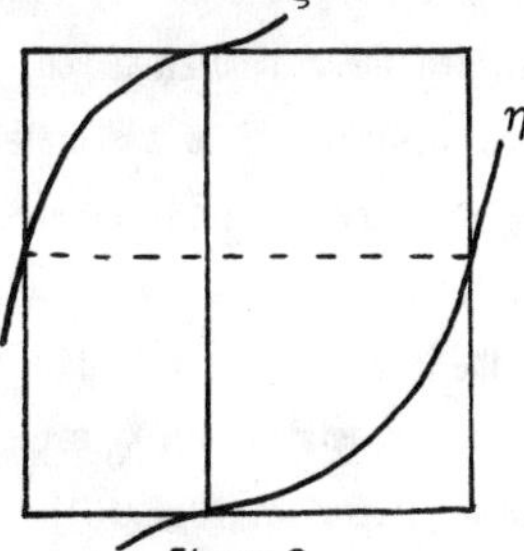

Figure 2.

Actually the condition (c) is only chosen for technical reasons. The natural and important condition is commutation i.e. (c)' $\xi\circ\eta = \eta\circ\xi$ near 0. (ξ,η) is said to be *critical* if $\xi''(0) = 0 = \eta''(0)$.

Note that if $f \in \mathcal{D}^\omega$ than the pair $u_f = (f,R_{-1}\circ f)$ satisfies all these conditions so that $\mathcal{D}^\omega$ is naturally embedded in $\mathcal{E}$. (R_α denotes the rotation $x \to x + \alpha$.) Moreover, if $(\xi,\eta) \in \mathcal{E}$ then the map f $= f_{(\xi,\eta)}$ defined by $f = \xi$ on $x \le 0$ and $f = \eta$ on $x > 0$ defines a continuous map of the circle.

The transformation T has the following properties:

(i) Define $\rho(\xi,\eta) = \rho(f_{(\xi,\eta)})$, if it exists. Then $\rho(T(\xi,\eta)) = \rho((\xi,\eta))^{-1} - 1$. Thus ρ is fixed under T if and only if $\rho = gm = (\sqrt{5} - 1)/2$. and T sends $\rho = p_n/q_n$ to $\rho = p_{n-1}/q_{n-1}$.

(ii) If $f \in \mathcal{D}^\omega$ and $(\xi,\eta) = (f,R_{-1}\circ f)$ then

$$T^n(\xi,\eta) = a_{(n)}{}^{-1}\cdot(f^{q_n} - p_n, f^{q_{n+1}} - p_{n+1})\circ a_{(n)}$$

where $a_{(n)} = a(T^{n-1}(\xi,\eta))\cdot a_{(n-1)}$. This uses the relation $q_{n+1} = q_n + q_{n-1}$.

(iii) If (ξ,η) is critical then so is $T(\xi,\eta)$.

(iv) If (ξ,η) satisfies the commutation condition (c)' then so does $T(\xi,\eta)$.

(v) $\xi = x + gm$, $\eta = x + gm - 1$ is a hyperbolic fixed point of T with a 1-dimensional unstable manifold (with associated eigenvalue $-gm^{-2}$).

The main claim made in [8] and [2] is that a critical hyperbolic fixed point exists. Using a computer and methods similar to Lanford ([6]), Mestel has proved the existence, hyperbolicity etc. of a fixed point in an appropriate space of pairs of analytic functions of x^3. Lanford and de la Llave have also proved the existence. It follows from Mestel's results that (i) for suitable and natural choices of $\mathcal{E}$ and and $\mathcal{O}$ the mapping $T : \mathcal{O} \to \mathcal{E}$ is well-defined and C^∞, (ii) T has a unique fixed point $u_* = (\xi_*,\eta_*)$ in $\mathcal{O}$ with $a = a(\xi_*,\eta_*) \in [-.78, -.77]$. (iii)The spectrum of $dT(u_*)$ has two simple eigenvalues: $\delta \in [-2.9, -2.8]$ and $\gamma = a^{-2}$, with the rest of the spectrum inside the circle $|z| = .53$. The corresponding eigenvectors X_δ and X_γ have the properties described below. (iv) The local unstable manifold of u_* associated with X_δ and X_γ is contained within the space of pairs satisfying the commutation condition (c)' and the local strong stable manifold associated with X_δ consists of pairs which are analytic functions of x^3. (v) $\mathcal{O}_{crit}$, the set of critical pairs in $\mathcal{O}$, is a 1-codimensional submanifold. Both the local stable manifold of u_* and the strong stable manifold associated with X_δ are contained in $\mathcal{O}_{crit}$. (vi) X_δ is positive on $[\eta_*(0),\xi_*(0)]$.

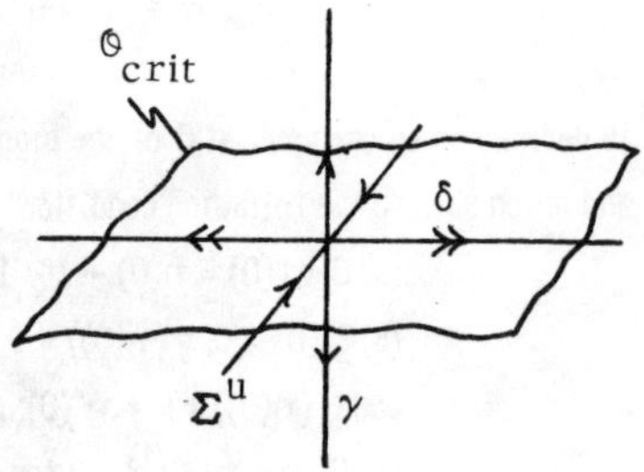

Figure 3.

Note that the fixed point satisfies the equation $\xi_* = a^{-1}\cdot\eta_*\circ a$, $\eta_* = a^{-1}\cdot\eta_*\circ\xi_*\circ a$ where $a = a(\xi_*,\eta_*)$. From this it follows that $\eta_*'(\xi(0)) = a^{-4}$ and $\xi_*'(\eta(0)) = a^{-2}$. Thus f_{u_*} is not analytic. This means that there is a neighbourhood of u_* which contains no pairs of the form u_f, $f \in \mathcal{D}^\omega$. Although this is not a problem in principle, it makes it difficult to deduce that there is any 2-parameter family of maps $f_{\mu,\nu}$ in $\mathcal{D}^\omega$ intersecting the stable manifold of u_*, which is what is required to deduce universality. I shall get over this by showing that there is a sense in which f_{u_*} is analytic. This sort of result will be even more important in the higher dimensional cases and provides a basic connection between commuting pairs of maps and smooth mappings of the circle and annulus.

3. COMMUTING PAIRS AND ANNULUS MAPS.

Clearly if $f \in \mathcal{D}^\omega$ then the pair $(f, R_{-1}\circ f)$ commutes i.e. satisfies condition (c)'. Hence, if p is a diffeomorphism of some appropriate interval into $\mathbb{R}$ then the pair $(p^{-1}\circ f\circ p, p^{-1}\circ R_{-1}\circ f\circ p)$ commutes. I show in this section that all the pairs (ξ,η) of interest are of this sort.

To see this consider $(\xi,\eta) \in \mathcal{E}$, and suppose that g sends some neighbourhood U of $\eta(0)$ diffeomorphically onto a neighbourhood of $\xi(0)$ and is such that $g\circ\eta = \xi$ near 0 and $\eta\circ g = \xi$ near $\eta(0)$. Then (ξ,η) is of the form claimed. There are three special cases to which I shall want to apply this result, namely: (a) when ξ and η commute, η has an analytic inverse near 0 and $g = \xi\circ\eta^{-1}$; (b) when ξ and η commute, $\xi = e(x^3)$ and $\eta = f(x^3)$, f has an analytic inverse near 0 and $g = e\circ f^{-1}$, and (c) to the multidimensional fixed point as explained in Section 7.

The idea behind this result is as follows. Let Δ be a neighbourhood of $[\eta(0),\xi(0)]$ which contains U and gU and such that $f_{(\xi,\eta)}$ is defined on Δ and sends Δ into itself. Now if points in Δ are identified by g then one clearly obtains a topological circle $\Sigma = \Delta/g$. Moreover, Σ aquires the structure of an analytic manifold via the canonical projection $\pi : \Delta \to \Sigma$. Because of (1) it is easy to see that $f_{(\xi,\eta)} : \Delta \to \Delta$ induces a continuous map $\hat{f}: \Sigma \to \Sigma$ i.e. $\hat{f}\circ\pi = \pi\circ f$. But $\hat{f}$ is clearly analytic away from $\pi(0)$ since ξ and η are. Thus using (1) one sees that $\hat{f}(\pi(x)) = \xi(x)$ for x near 0 so that $\hat{f}$ is analytic. Since Σ is a circle its universal cover is $\Pi:\mathbb{R} \to \Sigma$ and $\hat{f}$ lifts to an analytic $f \in \mathcal{D}^{\omega}$. Since Δ is simply connected there is a diffeomorphism $p : \Delta \to \mathbb{R}$ such that $\pi = \Pi\circ p$. Using this one sees that $\xi = p^{-1}\circ f\circ p$ and $\eta = p^{-1}\circ R_{-1}\circ f\circ p$.

I want to emphasize that this argument did not really depend upon the fact that ξ and η where defined on intervals, and works equally well in higher dimensions to give analytic maps of the annulus $A^n = \mathbb{T}\times R^{n-1}$.

4. UNIVERSALITY FOR CIRCLE MAPS.

In this section I indicate the proof of the following theorem.

Theorem 1. There is a 1–codimensional submanifold Σ in $\mathcal{D}^{\omega}$ such that if $f_{\mu,\nu}$ is a 2–parameter family in $\mathcal{D}^{\omega}$ which intersects Σ transversally then $f_{\mu,\nu}$ is in our universality class.

Outline of proof. The first point is that by the result described in Section 3 there exists p_* and $f_* \in \mathcal{D}^{\omega}$ such that $\xi_* = p_*^{-1}\circ f_*\circ p_*$ and $\eta_* = p_*^{-1}\circ R_{-1}\circ f_*\circ p_*$. For $f \in \mathcal{D}^{\omega}$ let z(f) be defined by $f''(z(f)) = 0$ then z(f) is uniquely defined and C^{∞} on some neighbourhood of f_* in $\mathcal{D}^{\omega}$. If f lies in this neighbourhood let $\tilde{f} = f\circ R_{-z(f)}$. Consider the map $\Phi(f) = (p_*^{-1}\circ\tilde{f}\circ p_*,\ p_*^{-1}\circ R_{-1}\circ\tilde{f}\circ p_*)$ defined on some neighbourhood $\mathcal{U}$ of f_*. This sends f_* to u_*. Moreover, if $X_1(x) = \sin 2\pi x$ then $d\Phi(f_*)\cdot X_1$ is clearly transverse to the 1–codimensional subspace of critical pairs $\mathcal{O}_{crit}$. A calculation also shows that the tangent X_2 to the image under Φ of $t \to (p_*^{-1}\circ R_t\circ f\circ p_*,\ p_*^{-1}\circ R_{t-1}\circ f\circ p_*)$ at $t = 0$ is transverse to the stable manifold Σ^s of u_* in $\mathcal{O}$. But this last vector is tangent to $\mathcal{O}_{crit}$. Thus $T_{u_*}\mathcal{O}$ is spanned by $T_{u_*}\Sigma^s$ and the images of X_1 and X_2 i.e. Φ is transverse to Σ^s at u_*. Thus $\Sigma = \Phi^{-1}(\Sigma^s)$ is a submanifold. Now it is clear that any 2–parameter family $f_{\mu,\nu}$ transverse to Σ is sent by Φ to a family $u_{\mu,\nu}$ transverse to Σ^s. The universality for the family $u_{\mu,\nu}$ follows from the usual arguments for such renormalisation saddle points (for example see [8]), and follows for $f_{\mu,\nu}$ because this only differs from $u_{\mu,\nu}$ by the conjugacy p_*. ∎

5. RENORMALISATION STRUCTURE IN HIGHER DIMENSIONS.

The aim is to prove a version of Theorem 1 for dissipative diffeomorphisms of the annulus. Firstly, I construct a renormalisation operator on pairs of commuting maps in $\mathbb{R}^n$. Then I use an argument similar to that of Theorem 1 and results about renormalisation and the existence and non-existence of invariant circles to apply this to analytic annulus maps. The construction of the renormalisation operator uses ideas developed by Collet, Eckmann and Koch [1] for period-doubling. The proof uses the structure of the 1-dimension problem in a fundamental way.

Firstly note an important difference between the 1-dimensional and multi-dimensional cases. In the 1-dimensional case I was able to easily pick out which point to scale about. For critical maps it was the critical point while for diffeomorphisms it did not matter. There is no such obvious choice in the higher dimensional case since I want to treat diffeomorphisms. I could have ignored the choice of origin and allowed transformations of the form $b(x) = \tau_0 + \tau_1 x$ in place of the scale change $x \to ax$. In 1-dimension it was clear that these are the only coordinate changes to worry about. However, in higher dimension one has to allow a slightly larger class of coordinate changes.

In fact I use an operator of the form

$$\mathcal{T}(E,F) = B^{-1} \circ (F, F \circ E) \circ B$$

where $B = B_{E,F}$ belongs to the set $\mathcal{P}$ of polynomial maps of the form

$$B(x,y) = (\tau_0 + \tau_1 x,\ \underline{\tau}_0 + \underline{\tau}_1 x + \underline{\tau}_2 x^2 + \underline{\tau}_3 x^3 + \underline{\mu} \cdot y),$$

where $(x,y) \in \mathbb{R} \times \mathbb{R}^{n-1}$, $\tau_0, \tau_1 \in \mathbb{R}$, $\underline{\tau}_0, \underline{\tau}_1, \underline{\tau}_2, \underline{\tau}_3 \in \mathbb{R}^{n-1}$ and $\underline{\mu} \in \mathrm{Lin}(\mathbb{R}^{n-1}, \mathbb{R}^{n-1})$.

Fix $\alpha \in \mathbb{R}^{n-1}$ such that $|\alpha| = 1$ and define $E_*(x,y) = (\xi_*((x^3 - \alpha \cdot y)^{1/3}), 0)$ and $F_*(x,y) = (\eta_*((x^3 - \alpha \cdot y)^{1/3}), 0)$. (Recall that ξ_* and η_* are analytic functions of x^3.) Then clearly $U_* = (E_*, F_*)$ is a fixed point of the map $\mathcal{T}$ if $B_{E_*,F_*}(x,y) = B_*(x,y) = (ax, a^3 y)$ where $a = a(\xi_*, \eta_*)$. The commutation conditions that are imposed are that $(E \circ F - F \circ E)(0) = 0$ and $(\partial^3/\partial^3 x)(E \circ F - F \circ E)(0) = 0$

Theorem 2. There is a affine map $(E,F) \to B_{E,F}$ into $\mathcal{P}$ such that the transformation $\mathcal{T}$ has the following properties. (i) $B(U_*) = B_*$. (ii) It is well-defined and C^∞ on an appropriate space and

$d\mathcal{T}(U_*)$ is compact. (iii) The eigenvalues of $d\mathcal{T}(U_*)$ of modulus ≥ 1 are δ and $\gamma = a^{-2}$. (iv) The eigenvectors associated with δ and γ are of the form $P_X = (X((x^3 - \alpha \cdot y)^{1/3}), 0)$ where X is the corresponding eigenvector for the 1-dimensional problem.

The proof of this Theorem is rather tedious and I restrict my comments here to trying to give some more insight into the choice of $B_{E,F}$ and further explaining the point made at the beginning of this section. Instead of $\mathcal{T}$ one could consider $S(E.F) = B_* \circ (F, F \circ E) \circ B_*$. Then U_* is a fixed point of S. It turns out that the only eigenvalues of $dS(U_*)$ of modulus ≥ 1 are as follows: (i) δ which is simple and whose eigenvector is of the form P_X; (ii) a^{-2} whose spectral subspace is spanned by an eigenvector of the form P_X and an $(n-1)$-dimensional subspace corresponding to coordinate changes of the form $(0, \underline{\mathcal{I}}_1 x)$; (iii) a^{-3}, a^{-1} and 1 whose spectral subspaces are of dimensions $n-1$, n and n^2-n+1 respectively and which correspond to coordinate changes of the form (4) with $\underline{\mathcal{I}}_1 = 0$.

These extra eigenvalues of modulus ≥ 1 correspond to the extra coordinate changes that have to be factored out. One constructs $(E,F) \to B_{E,F}$ using the spectral projection of $dS(U_*)$ associated with the eigenvalues a^{-3}, a^{-2}, a^{-1} and 1 to pick out the appropriate coordinate change.

6. UNIVERSALITY FOR DISSIPATIVE DIFFEOMORPHISMS OF THE ANNULUS.

I shall say that a 2-parameter family $F_{\mu,\nu}$ of analytic mappings of the annulus $A^n = \mathbb{T} \times \mathbb{R}^{n-1}$ or their lifts to the universal cover $\mathbb{R}^n$ *belongs to our universality class* if for the same numbers a and δ as in Theorem 1 and for some choice of the parameterisation μ, ν and some analytic diffeomorphism P between appropriate domains of $\mathbb{R}^2$, the family has the following properties:

(i) $F = F_{0,0}$ has a golden invariant circle which is the graph of a Lipshitz function.

(ii) There is a sequence B_n of coordinate changes in $\mathcal{P}$ converging exponentially fast to B_* such that if $B_{(n)} = B_1 \circ \ldots \circ B_n \circ P$ then $B_{(n)}{}^{-1} \circ (F^{q_n} - (p_n, 0)) \circ B_{(n)}$ converges to E_* as $n \to \infty$.

(iii) If $\nu = \nu_n$ is such that $F_{0,\nu}$ has a p_n/q_n-periodic point on $x = 0$ and ν_n is minimal then $\lim_{n\to\infty} \delta^n \cdot (\nu_{n+1} - \nu_n)$ exists and is non-zero.

(iv) There is a neighbourhood U of (0,0) such that if $(\mu,\nu) \in U$ then $F_{\mu,\nu}$ has a golden invariant circle if and only if $\nu = \nu_*(\mu)$ and $\mu \leq 0$ where the function ν_* is C^∞ on $\mu \neq 0$ and C^2 at $\mu = 0$. If $a < 0$ then this circle is C^∞.

The rest of this paper is an outline of the ideas behind the proof of the following theorem. However, some of the ideas of these later sections are of more general interest. For example one can use analogues of Proposition 2 to prove existence of invariant circles in other contexts.

Theorem 3. There is a 2-codimensional submanifold Σ in the space of analytic diffeomorphisms of the annulus such that if $F_{\mu,\nu}$ is a 2-parameter family which is transversal to Σ at $(\mu,\nu) = 0$ then $F_{\mu,\nu}$ is in our universality class.

7. COMMUTING PAIRS AND INVARIANT CIRCLES.

Suppose that (E,F) is close to $U_* = (E_*,F_*)$. As in the 1-dimensional case, if there exists a diffeomorphism G from a neighbourhood V of F(0,0) to a neighbourhood of E(0,0) such that $G \circ F = E$ near (0,0) and $F \circ G = E$ near F(0,0) then there exists a diffeomorphism P between appropriate domains and an analytic mapping Φ of the annulus into itself such that $E = P^{-1} \circ \Phi \circ P$ and $F = P^{-1} \circ (\Phi - (1,0)) \circ P$. The particular cases that I shall be interested in are (a) when (E,F) commutes and E and F are invertible on the appropriate domains so that one can take $E \circ F^{-1}$ for G and (b) when $(E,F) = (E_*,F_*)$ and $G(x,y) = ((g((x^3 - \alpha \cdot y)^{1/3})^3 + \alpha \cdot y)^{1/3}, y)$ where $g = e_* \circ f_*^{-1} = \xi_* \circ \eta_*^{-1}$. In the latter case I denote the associated annulus mapping by Φ_*.

Given (E,F) let $K = K_{(E,F)}$ denote the mapping defined as follows: $K = E$ on $x \geq 0$ and $K = F$ on $x < 0$. I say that (E,F) has an invariant circle if $K_{(E,F)}$ does. This is called a C^r, $0 < r \leq \omega$, circle if it is C^r in the annulus obtained by factoring our G.

Proposition 1. $\mathcal{T}(E,F)$ has a C^r golden invariant circle if and only if (E,F) does.

Proof. If σ is a golden invariant circle for (E,F) and $B = B_{(E,F)}$ then $B^{-1}\sigma$ is one for $\mathcal{T}(E,F)$. On the other hand if σ is a golden invariant circle for $\mathcal{T}(E,F)$ and $B = B_{(E,F)}$ then $\sigma_1 \cup \sigma_2$ is one for (E,F) where $\sigma_1 = B\sigma$ and $\sigma_2 = E\sigma$. ∎

8. EXISTENCE OF GOLDEN INVARIANT CIRCLES.

I now want to indicate how to prove the existence of golden invariant circles for pairs (E,F) that are asymptotic to a fixed point U of $\mathcal{T}$. Here the main application of this result is to the case $U = (E_*,F_*)$ but one can also apply it those converging to the trivial fixed point. The same ideas work for arbitrary Diophantine rotation numbers, and in the area-preserving case, an extension of them has been used by Hoidn ([4]) to deduce a version of the Moser Twist Theorem.

Definition. A <u>cyclic orbit segment γ of length</u> N of $K = K_{(E,F)}$ is a sequence of points $z_0, \dots, z_{N-1}$ such that $z_i = K(z_{i-1})$ and such that if $z_i = (x_i, y_i)$ then $x_0 = 0$ and $x_0, \dots, x_{N-1}$ have the same ordering on $\mathbb{R}$ as they would for the rotation R_{gm}. The Lipshitz constant $L(\gamma)$ associated with γ is the least $k > 0$ such that for each i, γ is contained in the cone $C_i(k)$ given by $(y - y_i) \leq k(x - x_i)$.

Lemma. Suppose that (E,F) is invertible. If $\mathcal{T}(E,F)$ has a cyclic orbit segment γ_1 of length q_n then (E,F) has one γ of length q_{n+1}. Moreover, if (E,F) is sufficiently close to (E_*,F_*), $L(\gamma) \le L(\gamma_1)$.

Proof. Let $\gamma' = B\gamma_1$ where $B = B_{(E,F)}$. Then q_{n-1} of the points of γ' lie in $x \le 0$. Let γ'' be the image of these points under E. Then for γ take $\gamma' \cup \gamma''$. To see the last fact note that since γ_1 is contained in the cones $C_i(k)$, $k = \mathrm{Lip}(\gamma_1)$ then γ' is contained in the image $C'_i(k)$ of each of these under B and γ'' is in the images $C''_i(k)$ under E of the $C'_i(k)$ corresponding to points with $x_i \le 0$. Thus a necessary and sufficient condition for a set of cones $C_j(k')$ to suffice is that the one through E(F(0)) (the point at which γ' and γ'' join) contains all the $C'_i(k)$ and the $C''_i(k)$ corresponding to points with $x_i \le 0$. ∎

Proposition 2. Suppose that (E,F) lies in the stable manifold of (E,F). Then (E,F) has a golden invariant circle of the form $y = \psi(x)$ where ψ is a Lipshitz function.

Proof. Since $(E_n,F_n) = \mathcal{T}^n(E,F) \to (E_*,F_*)$ for each $n > 0$ one can construct a cyclic orbit segment $\gamma^{(n)}$ of length q_n for (E,F). Let z_0 be a limit point of the set of initial points of the $\gamma^{(n)}$. Then the orbit $z_0, z_1, \ldots$ of z_0 under $K = K_{(E,F)}$ is ordered as for R_{gm}. Moreover, $L(\gamma^{(n)})$ is bounded since by the Lemma, $L(\gamma^{(n+1)}) \le L(\gamma^{(n)})$ for large n. Thus if $z_i = (x_i,y_i)$ then there is a Lipshitz function ψ such that $y_i = \psi(x_i)$. Using $\mathcal{T}$ this then gives similar orbits $y_i^{(n)} = \psi^{(n)}(x_i^{(n)})$ for (E_n,F_n), $n > 0$.

It remains to check that the x_i are dense in the interval. If this is not the case there exists a complimentary interval J for (E,F) i.e. a subinterval which does not contain an x_i, and hence complementary intervals for each of the (E_n,F_n). Let J_n be the longest such. J cannot contain $x_0 = 0$ and hence not $x_1 < 0$ and $x_2 > 0$. This is the case because the convergence to the fixed point implies that x_0 is a limit point of the x_n, $n > 0$. Moreover, since $|\partial F_n/\partial x| > a_*^{-1}$ for n large (since the same is true of (E_*,F_*)) the longest such interval J_n must be contained in $[x_1^{(n)},x_2^{(n)}]$. Thus (E_{n+1},F_{n+1}) has a complementary interval J which is essentially the image under the B for (E_n,F_n) of J_n. But if (E_n,F_n) is very close to (E_*,F_*), B is close to B_* and $|J|$ is approximately $a_*^{-1}|J_n|$. It follows that $|J_n| > 0.9a_*^{-1}|J_{n-1}|$ for all large n which is clearly a contradiction since $0.9a_*^{-1} > 1$ and $|J_n| < 1$. ∎

Note. With very small changes the above argument can be used to prove that if (E_n,F_n) converges to the simple fixed point $U_{*,0}$ given by $E_{*,0}(x,y) = (x + \alpha \cdot y + gm, 0)$ and $F_{*,0}(x,y) = (x + \alpha \cdot y + gm - 1, 0)$ then (E,F) has an golden invariant circle. A more sophisticated argument shows that in this case the golden invariant circle is C^∞.

9. OUTLINE OF PROOF AND NONEXISTENCE OF GOLDEN INVARIANT CIRCLES.

1. The unstable manifold of U_* defines a 2-parameter family U of retractions onto the

circle $\mathbb{T}$ whose parameterisation by $(\mu,\nu) \in \mathbb{R}^2$ can be chosen so that if $\mu < 0$ then $u_{\mu,\nu} = U_{\mu,\nu}|\mathbb{T}$ is a diffeomorphism of $\mathbb{T}$ (after gluing by g) and $\rho(f_{\mu,\nu})$ = gm if and only if $\nu = 0$. Using Herman's Theorem, ([3]) $u_{\mu,0}$ can be congugated to a rotation whence $\mathcal{T}^n(U_{\mu,0}) \to U_*$ as $n \to \infty$ if $a<0$.

2. Now choose an annular region $A = U_1 \cup U_2 \cup U_3$ in the unstable manifold $U_{\mu,\nu}$ of the form shown in Figure 4 and with the following properties:

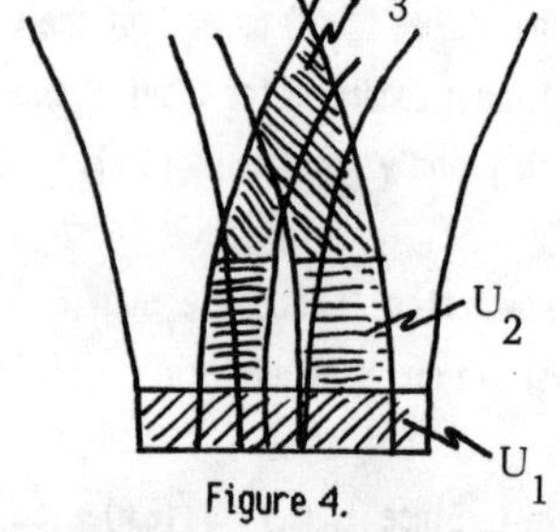

Figure 4.

- If (μ,ν) is close to $(0,0)$ then $\mathcal{T}^n(U_{\mu,\nu}) \in A$ for two consecutive n.
- If $(\mu,\nu) \in U_1$ then $u_{\mu,\nu}$ is a diffeomorphism and $\rho(u_{\mu,\nu})$ = gm if and only if $\nu = 0$.
- If $(\mu,\nu) \in U_2$ then $U_{\mu,\nu}$ has no orbit with rotation number gm.
- If $(\mu,\nu) \in U_3$ then $u = u_{\mu,\nu}$ is such that there is a small subinterval I of $\mathbb{T}$ such that $u^m(I)=\mathbb{T}$ for some $m \geq 0$.

The existence of such regions follows from the theory of circle maps (see [9] for references.).

But then one can extend A to an open subset $\mathcal{A} = \mathcal{U}_1 \cup \mathcal{U}_2 \cup \mathcal{U}_3$ of the full function space such that if U is near the stable manifold Σ^u of U_* then $\mathcal{T}^n(U) \in \mathcal{A}$ for some $n \geq 0$ and such that: .

- There is a 1-codimensional submanifold W of $\mathcal{U}_1$ such that if $U \in \mathcal{U}_1$ than U has a golden invariant circle if and only if $U \in W$. To see that one can extend U_1 to get $\mathcal{U}_1$ note that it follows from the results of [5] and arguments analogous to those of the previous section that there is a neighbourhood $\mathcal{W}$ of $U_{*,0}$ and a 1-codimensional submanifold W' of $\mathcal{W}$ such that U in $\mathcal{W}$ has a golden invariant circle if and only if $U \in W'$. Thus by 1 there is a neighbourhood $\mathcal{U}$ of $\mu < 0$, $\nu=0$ in $\mathcal{O}$ in which the set of U with a golden invariant circle forms a 1-codimensional C^∞ submanifold.
- If $U \in \mathcal{U}_2$ then U has no orbit with rotation number gm.
- If $U \in \mathcal{U}_3$ then U has an overflowing strip i.e. there exists a strip I in the annulus of the form $\theta_0 \leq \theta \leq \theta_0 + c$, $c > 0$ small, such that for some $m > 0$ the projection of $U^m(I)$ into $\mathbb{T}$ covers $\mathbb{T}$.

3. Thus if $U \in \mathcal{A}$ then U has a golden invariant circle if and only if $U \in W$. It follows from Proposition 1 that if U is near Σ^u then U has a golden invariant circle if and only if $\mathcal{T}^n(U) \in W$ for some $n \geq 0$. With a little work it can now be shown that if $U \in \Sigma^u$ then there is a coordinate system (x,y,z) at U with $x,y \in \mathbb{R}$ and z in the unit ball of some Banach space, such that $U_{(x,y,z)}$ has a golden invariant circle if and only if $x = 0$ and $y \leq 0$.

4. The rest of the proof is much the same as that of Theorem 1. As there, one uses conjugation by P_* to construct a map Ξ from a neighbourhood of Φ_* to a neighbourhood of U_* such that

$\Xi(\Phi_*) = U_*$ and such that Ξ is transverse to Σ^u at Φ_*. Let Σ' be the intersection of $\Xi^{-1}(\Sigma^u)$ with this neighbourhood. Then Σ' is a submanifold, at least if one reduces the neighbourhood sufficiently. One then shows that Σ' contains a diffeomorphism Φ. Next one applies 3 to deduce the existence of the coordinate system (x,y,z) at $U = \Xi(\Phi)$. Using the transversality and pulling back by Ξ, gives a similar coordinate system (x',y',z') at Φ with the property that $\Phi_{(x',y',z')}$ has a golden invariant circle if and only if $x' = 0$ and $y' \leq 0$. This completes the outline of the proof. ∎

REFERENCES

[1] Collet, P., Eckmann, J.-P., and Koch, H. "Period-Doubling Bifurcations for Families of Maps on $\mathbb{R}^n$." J. Stat. Phys. 25, (1981) 1.

[2] Feigenbaum, M., Kadanoff, L. and Shenker, S. "Quasi-periodicity in Dissipative Systems: A Renormalisation Group Analysis." Physica 5D, (1982) 370–386.

[3] Herman, M. "Sur la Conjugaison Differentiable des Diffeomorphismes du Cercle a des Rotations." Publ. Math. I.H.E.S. 49, (1979) 5–234.

[4] Hoidn, N. "On Invariant Curves Under Renormalisation." Warwick Preprint, 1985.

[5] Jonker, L. and Rand, D. A. " Universal Properties of Maps of the Circle with ε-Singularities." Comm. Math. Phys. 90, (1983) 273–292.

[6] Lanford, O. "A Computer-Assisted Proof of the Feigenbaum Conjectures." Bull. AMS, 6, (1982) 427–434.

[7] Mestel, B. D. "A Computer Assisted Proof of Universality for Cubic Critical Maps of the Circle with Golden Mean Rotation Number." Warwick University Ph.D. Thesis, 1985.

[8] Ostlund, S., Rand, D. A., Sethna, J. and Siggia, E. "Universal Properties of the Transition from Quasi-Periodicity to Chaos in Dissipative Systems." Physica 8D, (1983) 303. [Also see Phys. Rev. Lett. 49, (1982) 132.]

[9] Rand, D. A. "Universality for Critical Golden Circle Maps and the Breakdown of Dissipative Golden Invariant Tori." LASSP Preprint, Cornell 1984, Submitted to Comm. Math. Phys.

OPERATORIAL QUANTIZATION OF DYNAMICAL SYSTEMS WITH IRREDUCIBLE FIRST AND SECOND CLASS CONSTRAINTS

I.A. Batalin and E.S. Fradkin

Lebedev Physical Institute, Moscow

Abstract

Operatorial version is suggested of the generalised canonical quantisation method of dynamical systems subjected to irreducible first and second class constraints. An operatorial analog of classical Dirac brackets is realised. Generating equations for generalised algebra of first and second class constraints, as well as for the unitarising Hamiltonian are formulated. In the first class constraint sector new generating equations are presented directly in terms of operatorial Dirac brackets.

References

1. E.S. Fradkin and G.A. Vilkovisky, Phys. Lett. 55B (1975) 224.
2. E.S. Fradkin and G.A. Vilkovisky, CERN Report TH-2332 (1977).
3. I.A. Batalin and G.A. Vilkovisky, Phys. Lett. 69B (1977) 309.
4. E.S. Fradkin and T.E. Fradkina, Phys. Lett. 72B((1978) 343.
5. I.A. Batalin and E.S. Fradkin, Phys. Lett. 122B (1983) 157.
6. I.A. Batalin and E.S. Fradkin, Phys. Lett. 128B (1983) 303.
7. I.A. Batalin and E.S. Fradkin, J. Math. Phys. 25 (1984) 2426.
8. I.A. Batalin and E.S. Fradkin, J. Nucl. Phys. (USSR) 39 (1984) 231.
9. I.A. Batalin, J. Nucl. Phys. (USSR) 41 (1985) 278.
10. I.A. Batalin and E.S. Fradkin, Rivista Nuovo Cimento (1986) (in press).
11. E.S. Fradkin, Acta Universitatis Wratislaviensis N 207. Proc. X[th] Winter School of Theoretical Physics Carpacs (1973) 93.
12. I.A. Batalin and E.S. Fradkin, Preprint P.N. Lebedev Inst. (1986) N 132, Phys. Lett. (1986) (in press).

THE INCLUSION OF FERMIONS IN QUANTUM COSMOLOGICAL MODELS

P. D. D'Eath and J. J. Halliwell

Department of Applied Mathematics and Theoretical Physics,
Silver Street,
Cambridge CB3 9EW,
U. K.

1 Introduction

There has been considerable interest in the recent proposal of Hartle and Hawking that one may set boundary conditions on the wave function in quantum cosmology by defining it to be the result of performing a path integral over compact 4-metrics and regular matter fields [1, 2]. One thus writes

$$\Psi = \int_C d[g_{\mu\nu}]\, d[\Phi]\, e^{-I_E} \tag{1.1}$$

where C is the class of compact 4-metrics $g_{\mu\nu}$ and regular matter fields Φ and I_E is the Euclidean action. This proposal has been applied to numerous models, usually of the minisuperspace type [2, 3]. In most of these models, the matter source is taken to be a scalar field. However, most of the theories we know involve other types of matter fields, such as gauge fields, antisymmetric tensor fields and fermion fields, thus it is of interest to include these fields in quantum cosmology. In this paper we briefly describe work done on the inclusion of fermion fields.

Minisuperspace models with a fermion matter source have been considered by Isham and Nelson [4] and more recently, by Christodoulakis and Zanelli [5]. These authors restricted the fermion field to be homogeneous, and the resulting models were really particle theories, rather than field theories, in that they involved a finite, small number of particles. To be realistic, one really needs to consider the full infinite number of modes of the fermion field. This is precisely what we do here.

2 Fermionic Perturbations about Minisuperspace

We consider fermionic perturbations about a homogeneous, isotropic minisuperspace background, namely the massive scalar field model of Hawking [2]. In his model, the 4-metric is restricted to be of the form

$$ds^2 = - dt^2 + e^{2\alpha} d\Omega_3^2 \tag{2.1}$$

where $d\Omega_3^2$ is the metric on the unit 3-sphere. The 4-metric is thus described by a single scale factor e^{α}. The model is driven by a homogeneous massive scalar field ϕ of mass M. The wave function for the model satisfies the Wheeler-DeWitt equation

$$H_0 \Psi = \frac{1}{2} e^{-3\alpha} \left[\frac{\partial^2}{\partial\alpha^2} - \frac{\partial^2}{\partial\phi^2} + e^{6\alpha} M^2 \phi^2 - e^{4\alpha} \right] \Psi = 0 \tag{2.2}$$

One may apply the Hartle-Hawking proposal to (2.2) to set boundary conditions on Ψ and a unique wave function is thus obtained. In the classical limit, it is found to correspond to a superposition of classical solutions, each of which describes an inflationary universe [2].

Bosonic perturations about this model (ie gravitational waves and density fluctuations of the scalar field) have been considered by Halliwell and Hawking [6]. To lowest non-trivial order in the inhomogeneities, the bosonic perturbations decouple from fermionic ones, so we may consider the fermionic perturbations separately.

Proceding as in the bosonic case, one may expand the spatial dependences of the of the fermion field in harmonics on the 3-sphere. We thus introduce a complete set of spinor harmonics $\rho_A^{np}(\underline{x})$, $\bar{\sigma}_A^{np}(\underline{x})$ and their hermitian conjugates $\bar{\rho}_{A'}^{np}(\underline{x})$, $\sigma_{A'}^{np}(\underline{x})$ for the expansion of any field on S^3. These harmonics are eigenfunctions of the Dirac operator on the 3-sphere and each one is labelled by an integer n, corresponding to its eigenvalue, and a degeneracy label p. The most general spin-half field is a massive Dirac field but here, for simplicity of exposition, we will consider a single Weyl spinor χ_A, which can describe massless neutrinos. The classical field χ_A may be expanded in harmonics

$$\chi_A = e^{-3\alpha/2} \sum_{np} \left[s_{np}(t)\, \rho_A^{np}(\underline{x}) + \bar{t}_{np}(t)\, \bar{\sigma}_A^{np}(\underline{x}) \right] \qquad (2.3)$$

where the time-dependent coefficients $s_{np}(t)$, $\bar{t}_{np}(t)$ and their complex conjugates $\bar{s}_{np}(t)$, $t_{np}(t)$ are odd elements of a grassman algebra, ie anticommuting complex numbers.

Inserting (2.3) into the action for the gravity plus homogeneous scalar plus fermion system, using the Dirac procedure, one may derive the Hamiltonian, which vanishes identically. This is the Hamiltonian constraint, and is quantized to yield the Wheeler-DeWitt equation

$$\left[H_0(\alpha,\phi) + \sum_{np} H_{np}(\alpha,\phi,s_{np},\bar{s}_{np},t_{np},\bar{t}_{np}) \right]\Psi = 0 \qquad (2.4)$$

where H_0 is the Wheeler-DeWitt operator of the background minisuperspace model, and H_{np} is the Hamiltonian for the fermionic perturbation mode labelled by n,p. One also finds from the Dirac procedure that the perturbation coefficients satisfy the anticommutation relations

$$\{ s_{np} , \bar{s}_{np} \} = 1 \quad , \quad \{ t_{np} , \bar{t}_{np} \} = 1 \tag{2.5}$$

with all other anticommutators zero. These are satisfied by the representation

$$\bar{s}_{np} \rightarrow \frac{\partial}{\partial s_{np}} \quad , \quad \bar{t}_{np} \rightarrow \frac{\partial}{\partial t_{np}} \tag{2.6}$$

known as the holomorphic representation [7]. The wave function is then a function only of the unbarred grassman variables, s_{np}, t_{np}. The holomorphic representation is the key ingredient for understanding the fermionic perturbations, since it allows them to be treated in a manner almost identical to the bosonic case. Since it plays such a central role, we pause for a brief digression to explain it in a little more detail.

3 The Holomorphic Representation for Fermions

A quantized fermi oscillator of frequency ν is described by the Hamiltonian operator

$$H = \nu \, \hat{a} \, \hat{a}^{\dagger} - \tfrac{1}{2}\nu \tag{3.1}$$

where the operators $\hat{a}$ and $\hat{a}^{\dagger}$ satisfy the anticommutation relations

$$\{ \hat{a} , \hat{a}^{\dagger} \} = 1 \tag{3.2}$$

with all others zero. Each of the Hamiltonians H_{np} is in fact a sum of two terms of the form (3.1), one for s_{np}, $\bar{s}_{np}$ and one for t_{np}.

ϵ_{np}. One may solve this simple quantum mechanical system using the holomorphic representation, in which one introduces the complex grassman number a and its complex conjugate $\bar{a}$, and makes the substitution

$$\hat{a}^{\dagger} \to a \quad , \quad \hat{a} \to \frac{d}{da} \tag{3.3}$$

In this representation, states are represented by analytic functions of a only, not of $\bar{a}$. Since a is a grassman number, there are only two linearly independent such functions, namely $\psi_0(a) = 1$ and $\psi_1(a) = a$. It is easily verfied that $\hat{a}^{\dagger}$ and $\hat{a}$ are creation and annilation operators between ψ_0 and ψ_1, and also that

$$H\,\psi_0 = -\tfrac{1}{2}\,\nu\,\psi_0 \quad , \quad H\,\psi_1 = +\tfrac{1}{2}\,\nu\,\psi_1 \tag{3.4}$$

ψ_0 thus represents the ground state.

The states ψ_0 and ψ_1 are orthonormal with respect to the inner product appropriate to the holomorphic representation, defined for any pair of functions f,g of the variables a:

$$(f,g) = \int \overline{f(\bar{a})}\, g(a)\, e^{-a\bar{a}}\, da\, d\bar{a} \tag{3.5}$$

Integration over a and $\bar{a}$ is performed according to the usual rules of Berezin integration [8].

4 Solution of the Model

The Wheeler-DeWitt equation (2.4) may be solved approximately by the ansatz

$$\Psi = \Psi_0(\alpha,\phi) \prod_{np} \Psi_{np}(\alpha,\phi,s_{np},t_{np}) \tag{4.1}$$

where Ψ_0 is a function only of the background variables α and ϕ, and the Ψ_{np}, for each n,p, are functions of the background variables and also of the perturbation variables s_{np}, t_{np}.

The main point now, is to apply the Hartle-Hawking proposal to set boundary conditions on the perturbation wave functions Ψ_{np}. One thus defines the total wave function to be

$$\Psi = \int_C d[e^a_\mu]\ d[\chi_A]\ d[\bar{\chi}_{A'}]\ e^{-I_E} \qquad (4.2)$$

where C is the class of regular fermion fields χ_A, $\bar{\chi}_{A'}$, and tetrads e^a_μ corresponding to compact 4-metrics. The result of doing this, is that the wave functions Ψ_{np} are found to begin the inflationary phase in the lowest energy eigenstate of the Hamiltonian; that is, they start out in the ground state. This is the main result of the model, and is identical to that found in the bosonic case.

These ground state wave functions may be evolved through the subsequent inflationary phase using the Wheeler-DeWitt equation. We have calculated the resulting particle production during this phase, and the back reaction of the fermionic perturbations on the homogeneous background. Both of these quantities are found to be small. Unlike the bosonic case, the fermionic perturbations have no effect on density perturbations, to the order at whoch we are working; nor do they have any effect on the isotropy of the microwave background. In the limit of an exact de Sitter background, we examined the response of a geodesically moving particle detector to the fermion state picked out by the Hartle-Hawking proposal. We found that the detector experiences a thermal spectrum, at the de Sitter temperature with the expected Fermi-Dirac distribution.

The work described here is described in considerably greater detail in ref. [9].

References

1 S. W. Hawking in: Astrophsical Cosmology, Pontificiae Academiae Scientarium Varia 48, 563 (1982); J. B. Hartle & S. W. Hawking, Phys. Rev. D28, 2960 (1983)

2 S. W. Hawking, Nucl. Phys. B239 257 (1984).

3 P. Amsterdamski, Phys. Rev. D31, 3073 (1985); U. Carow & S. Watamura, Phys. Rev. D32, 1290 (1985); P. F. Gonzalez-Diaz, Phys. Lett. 159B, 19 (1985); J. J. Halliwell, Nucl. Phys. B266, 228 (1986); S. W. Hawking & J. C. Luttrell, Nucl. Phys. B247, 250 (1984); S. W. Hawking & J. C. Luttrell, Phys. Lett. B143, 83 (1984); G. T. Horowitz, Phys. Rev. D31, 1169 (1985); X. M. Hu & Z. C. Wu, Phys. Lett. B149, 87 (1984) I. G. Moss & W. A. Wright, Phys. Rev. D29, 1067 (1984); Z. C. Wu, Phys. Lett. B146, 307 (1984);

4 C. J. Isham & J. E. Nelson, Phys. Rev. D10, 3226 (1974)

5 T. Christodoulakis & J. Zanelli, Phys. Rev D29, 2738 (1984)

6 J. J. Halliwell & S. W. Hawking, Phys. Rev. D31 1777 (1985)

7 L. D. Faddeev & A. A. Slavnov, Gauge Fields: Introduction to Quantum Theory, Benjamin Cummings (1980); L. D. Faddeev, in Methods in Field Theory, Les Houches 1975, Session XXVIII, eds. R. Balian & J. Zinn-Justin (North Holland, 1976)

8 F. A. Berezin, The Method of Second Quantization (Academic Press, NY-London, 1968)

9 P. D. D'Eath & J. J. Halliwell, Fermions in Quantum Cosmology, DAMTP preprint (1986)

STRING CORRECTIONS TO EINSTEIN GRAVITY

S. Deser
Brandeis University
Waltham MA 02254
U.S.A.

ABSTRACT

The quadratic curvature corrections to Einstein gravity resulting from the slope expansion of string models will be obtained and discussed. In particular, we will find that these can be written in terms of the Gauss-Bonnet invariant, but that in accord with the field redefinition theorem the coefficients of the $R^2_{\mu\nu}$ and R^2 contributions are actually undetermined. We will then give the generalization of the Schwarzschild solution in these extended theories and comment on their properties; here the effects of the dilaton will also be included.

A major motivation for studying string theories lies in their inclusion of Einstein gravity as the zero slope limit in an expansion in the parameter α'. While the full theory is very different from our present quantum gravity ideas, it is of some interest to study the first corrections to general relativity which emerge beyond this limit. I will be reporting on very recent work with N. Redlich |1| and D. Boulware |2| which bears firstly on the structure and uniqueness of the quadratic curvature terms that are generated and secondly on the nature of the solutions of these augmented gravitational models, including also the important effects of the dilaton, a new scalar companion of the graviton. There has been some recent discussion regarding the detailed form of the corrections, which our explicit calculations should clarify. Likewise, a general picture of the deviations from the standard Schwarzschild space obtained by solving the extended field equations will provide a "first derivative" picture of string effects even though they can only be trusted to order α' (and therefore include no non-perturbative contributions) and are only valid in the pre-compactification regime.

The spectrum generated by closed strings is known to be ghost-free; on the other hand, models with generic actions

$$I = \int (dx) \sqrt{-g} \left[R + a R_{\mu\nu\alpha\beta}^2 + b R_{\mu\nu}^2 + c R^2 \right] \tag{1}$$

certainly contain ghosts near flat space : In terms of the fields $h_{\mu\nu} \equiv g_{\mu\nu} - \eta_{\mu\nu}$, such actions have quadratic parts of the schematic form $h \Box (1 + \alpha' \Box) h$, with the consequence that they contain a massive ghost besides the usual graviton. The only relevant exception |3| is provided by the Euler-Gauss-Bonnet (EGB) combination

$$\begin{aligned} G &= \int (dx) \sqrt{-g} \left[R_{\mu\nu\alpha\beta}^2 - 4 R_{\mu\nu}^2 + R^2 \right] \\ &\sim \int (dx)\, \epsilon^{\mu\nu\alpha\beta\lambda\cdots} \epsilon^{abcdf\cdots} R_{\mu\nu ab} R_{\alpha\beta cd}\, e_{\lambda f} \cdots \end{aligned} \tag{2}$$

which contains no terms quadratic in $h_{\mu\nu}$ but begins with cubic vertex contributions for dimension $d > 4$, as is most easily seen from the second form of (2). These facts led Zwiebach |4| to conjecture that the string corrections begin precisely as $\alpha' G$, and to verify this for the relevant part of the 3-graviton vertex generated by the string amplitude However, since all external gravitons are on (linear) Einstein mass shell i.e., $R_{\mu\nu}^1 = 0 = R^1$ (where the superscript indicates the power of $h_{\mu\nu}$) there is an ambiguity in that the $R_{\mu\nu}^2$ and R^2 terms in G do not contribute ; no such ambiguity affects the $R_{\mu\nu\alpha\beta}^2$ part, since the curvature (or really Weyl) tensor is not constrained on shell. Alternative approaches to the effective field equations (e.g. from the associated σ-model) yield quadratic terms

$$I_{eff} \sim (2,1,0)\,\alpha' \int (dx) \sqrt{-g}\, R^2_{\mu\nu\alpha\beta} \qquad (3)$$

plus varying amounts of $R^2_{\mu\nu}$ and R^2 |5|, where the coefficients (2,1,0) correspond to the bosonic |6|, heterotic |7| and super-|8| strings respectively. What then does a direct calculation of the relevant string amplitude yield as the "correct" coefficients, and do the $R^2_{\mu\nu}$, R^2 terms matter, when we calculate the 4-graviton part to quartic order in momenta, $O(\alpha' k^4)$ which represent these curvature-squared corrections ?

We will compare the string results first with those due to an action of the form $\int R$ + (2,1,0) α' G and show that they agree ; then we will see explicitly that in fact the $R^2_{\mu\nu}$ and R^2 terms still do not contribute. Thus one way always form the EGB combination and so never see any ghost modes when the external gravitons are on-shell. This result reflects the field-redefinition theorem of 'tHooft and Veltman |9| and in fact it also holds to all orders, i.e., for all dangerous terms of the form $\alpha' \int R (\alpha' \Box)^n R$ in an obvious notation. Thus our calculations |1| check that the string amplitude is reproduced by an effective action whose contributions now include (at this 4-point level) both a contact term (as in the 3-point case) and a Born term with an exchanged off-shell graviton, and that the field redefinition cancellations occur as predicted. We will only give the main formulae here (and omit the dilatons which do not produce any qualitative changes).

In terms of the graviton polarization tensors $e_{\mu\nu}(k_i)$ which are transverse-traceless, with $k_i^2 = 0$, the amplitude is readily obtained for all three models from Schwarz's expression |10|, and has also been studied independently by Cai and Nunez |11| and by Kawai et al. |12|. It is a proportional to an overall product of Γ functions times two kinematic factors K which are bosonic or fermionic (and which contract into the four graviton polarizations). The former contain the famous tachyon pole, expansion of which just yields the (2,1,0) ratio in the bosonic (K_B K_B), heterotic ($K_B K_S$) and supersymmetric ($K_S K_S$) cases. Expansion of this amplitude yields the form

$$A_4(s,t,u) \sim \alpha'(2,1,0)\left[tu(e_1e_2)(e_3e_4) - t^2(e_1e_2e_3e_4) + \text{perm}\right] \quad (4)$$

where perm denotes the usual symmetrization, (s,t,u) are the Mandelstam variables, parentheses mean trace over the polarization indices and we have dropped a series of terms whose role is to maintain gauge invariance of A_4. Next, we calculated the four-point amplitude generated by the Lagrangian $R + \alpha'(2,1,0)\, G$ to this order, both due to the contact h^4 part of G and to a Born exchange of a graviton between h^3 vertices taken one each from R and G. These are the only contributions to order k^4 as is easily seen from the fact that the propagator is just the Einstein $\sim k^{-2}$ one and that (R,G) are of respective orders (k^2, k^4). The calculation is messy but straightforward and indeed reproduces the string result (4). As an interesting check of unitarity, we also mention that (4) cannot be obtained from a purely quadratic curvature action but requires the Einstein part, even though dimensionally a purely quadratic action could suffice. A second remark is that the above methods, when applied to the zero-slope $\mathcal{O}(k^2)$ terms lead to Einstein theory as the corresponding effective action, a result independently obtained by Sannan |13|.

Next, we consider the effect (or lack of it) of the $R_{\mu\nu}^2$ and R^2 terms in the effective action on the amplitude. 'tHooft and Veltman |9| noted, in another connection, that an action of the form $R + aR_{\mu\nu}^2 + bR^2$ would be transformed into a pure Einstein term (plus higher order terms) by a field redefinition $g_{\mu\nu} \to g_{\mu\nu} + aR_{\mu\nu} + (a+2b)(2-d)^{-1} g_{\mu\nu} R$ _if_ the R^2-terms are regarded as "vertex" corrections rather than changes in the propagator, i.e. near Einstein mass shell (essentially $R_{\mu\nu} = 0$ is still a solution of the extended action). This theorem has been used in discussions of effective actions (see e.g. |14|). Here we give a pedagogical discussion of its assumptions and how it emerges in concrete calculations. Consider first the role of assuming that we are near Einstein shell : an action of the form $h_{\mu\nu}\Box(1+\alpha'\Box)h_{\mu\nu}$ transforms, under $\delta h_{\mu\nu} = -\frac{1}{2}\alpha'\Box h_{\mu\nu}$ into $h_{\mu\nu}\Box h_{\mu\nu} + \mathcal{O}(\alpha'^2\Box^2)$. However, near ghost mass shell, where $(1+\alpha'\Box)$ is small, we could transform with $\delta h_{\mu\nu} = \frac{1}{2}(1+\alpha'\Box)h_{\mu\nu}$ to obtain the form $-h_{\mu\nu}(\Box+\alpha'^{-1})h_{\mu\nu}$ which is that of a massive ghost ! So the parameter α' must be small in the first case, and this is just the whole basis of the string's slope expansion. In general, however, as we shall see below, the solution space of the $R+\alpha' R^2$ theories includes solutions which are not analytic in α' , such as de Sitter spaces with radii proportional to α'^{-1}. Returning to the calculation of amplitudes, let us drop the $R^2_{\mu\nu\alpha\beta}$ term for simplicity and consider those generated by the action $\int R + \int(aR^2_{\mu\nu}+bR^2)$. In the spirit of the theorem, we must expand in α' , keeping only the lowest term, and with external legs on Einstein shell. In particular, the propagator is to be expanded ; writing its generic form as $[\Box(1+\gamma\Box)]^{-1} \sim \Box^{-1} - \gamma$ and treating the γ as a two-point insertion. This time we find that all contributions cancel, i.e. that there is <u>no</u> $\alpha' k^4$ contribution at all ! |There is no need to perform any "active" field redefinition as the amplitude is of course invariant.| Thus one cannot tell those actions apart which differ by such terms. This is easily seen to persist for general $R_{\mu\nu}\Box^n R_{\mu\nu}$ terms as well. This means that compatibility with the no-ghost content of the string is guaranteed by the zero slope Einstein shell requirement, at least perturbatively. |Of course, one can also conceive of off-shell actions which are apparently ghost-like, i.e., are of the form $\int h_{\mu\nu}\Box f(\alpha'\Box)h_{\mu\nu}$ with $f \sim 1+\alpha'\Box$, but f has no zeros !|

The additions to the effective action are now known also to R^4 order |11,14,15|, up to (irrelevant) terms involving $R_{\mu\nu}$ and R ; these are outside the scope of this lecture, however.

We now turn to the solutions of the extended actions we have considered, to get some idea of the changes they introduce in the traditional Schwarzschild solution, in particular with respect to horizons, singularities and spaces non-analytic in α'. We shall first neglect dilatons, then see what their effects are ; again we leave the details to |2|. Because the only excitation in $\int R + \alpha' G$ is the graviton, the Birkhoff theorem still holds, so spherically symmetric solutions are time-independent. In Schwarzschild coordinates,

$$ds^2 = g_{00}(r)dt^2 + g_{rr}(r)dr^2 + r^2 d\Omega \qquad (5)$$

we find, as expected, that $-g_{00} = g_{rr}^{-1}$ and the single field equation for this variable is surprisingly easy to solve. One finds

$$-g_{00} = g_{rr}^{-1} = 1 + (r^2/2\tilde{\alpha})\left[1 \mp \sqrt{1 + 8GM\tilde{\alpha}r^{1-d}}\right] \qquad (6)$$

where $\tilde{\alpha} \equiv \alpha'(d-3)(d-4)$ and d is the spacetime dimension. Note that there are two branches, the first of which is asymptotic to the Schwarzschild solution in an expansion in $\tilde{\alpha}$ or 1/r, with

$$-g_{00} \sim 1 - \frac{2GM}{r^{d-3}} + \mathcal{O}(\alpha') \qquad (7)$$

in terms of the integration constant M. For the physically relevant case of positive mass, there can be a naked singularity, but only for the "wrong" sign of α', i.e. that opposite to the positive sign required by the string theory. For positive α', there is always a single event horizon r_H, $g_{00}(r_H) = 0$, which lies closer to the origin than the Schwarzschild one. The singularity at the origin is also weaker than the Einstein case. However, the physics at the horizon is unchanged : the usual Kruskal completion exists, there will be Hawking radiation, etc ...

The other branch of (6) is asymptotic to anti de Sitter space (A d S) for $\alpha' > 0$:

$$-g_{00} \sim \left(1 + \frac{r^2}{\tilde{\alpha}}\right) + \frac{2GM}{r^{d-3}} + \mathcal{O}(\alpha'). \qquad (8)$$

Note the $1/\tilde{\alpha}$ behavior related to absence of asymptotic flatness.

Such solutions are generic to actions involving note than one power of curvature : since the curvature is proportional to the metric, the field equations turn into a single algebraic equation for the cosmological constant which in general can have real solutions. |This result also shows the limitations of the folklore "theorem" that the low frequency limit of extended models is always Einstein gravity|. The above geometry is actually unphysical, however, because it is unstable (it also displays a naked singularity at the origin). This can be seen already from the fact that (8) corresponds (for M > 0) to the Schwarzschild-AdS metric with <u>negative</u> mass ; indeed, small gravitational excitations about AdS are both ghostlike and unstable. So this particular model manages to cure its cosmological problem. The interested reader is referred to the growing literature on various solutions to Einstein-EGB theory |16| for further work.

When the scalar companion (dilaton) to the graviton is included in the string expansion, the Lagrangian is modified and becomes

$$I_{eff} = \int\sqrt{-g}\,R + \int\sqrt{-g}\,\alpha' e^{\phi}(R^2_{\mu\nu\rho\sigma} - 4R^2_{\mu\nu} + R^2) + \alpha'^2\int e^{2\phi}R^3 + \ldots - \frac{1}{2}\int(\partial\phi)^2. \quad (9)$$

This time, de Sitter space is no longer a solution, which is very gratifying. This is easily seen from the fact that the scalar field must be constant on a maximally symmetric space. Thus, the gravitational equations reduce to their previous form (with $\phi = 0$), but there is now an additional constraint from the scalar field equation $\Box\phi = \alpha' e^{\phi} G$ and hence G = 0 ; consequently, R vanishes (by the gravitational equation) and there are no cosmological solutions. In the general case, including the terms $e^{n\phi} R^{n+1}$, there could still be such solutions ; their presence and stability properties might be decidable from the σ-model approach to strings to all orders ; this is a very important question. Explicit solutions to the coupled equations are very hard to obtain even with spherical symmetry. First, because there is now a scalar field, the Birkhoff theorem no longer applies and there could be time-dependence even in the spherically symmetric case ; this may have important consequences for the qualitative properties of the solutions, as has also been speculated by Gross and Witten |14|. We have considered only static solutions, and found that they have a reasonable asymptotic behavior characterized by the mass and a scalar "charge" ; the equations were also consistent with the presence of an event horizon (with regular there), but this is really a global property we could not fully establish.

We emphasize that our solutions are really only valid to order i.e., in a weak curvature regime where higher curvature corrections are neglected. We have also not considered compactification, which presumably sets in at Planck scale. Pre-compactification solutions might be applicable in some transition domain where spacetime is still a meaningfull concept but the breakdown to $d = 4$ has not yet set in. There are also likely to be important non-perturbative effects. So we can only draw some qualitative conclusions such as the fact that the string sign of α' is the desirable one and that inclusion of dilatons seems to improve the purely metric models. Nevertheless, the effective action should be an important tool in understanding the modifications which strings induce in the properties of spacetime, and one may hope that questions such as avoidence of singularities and of horizons may be resolved in this way. Of course deeper challenges, such as the significance of pre-Planck scale physics, of why the signature is hyperbolic, and many others will still remain to be understood.

It is a pleasure to thank my collaborators Norman Redlich and David Boulware. This research was supported in part by NSF grant PHY 82-01094.

REFERENCES

1. Deser, S. and Redlich, A.N., Phys. Lett. B (in press).
2. Boulware, D.G. and Deser, S. Phys.Rev.Lett. 55, 2656 (1985) ; Phys. Lett. B (in press).
3. Zumino, B., Berkeley Report UCB-PTH- 85/13 (to be published) ; Lovelock, D., J.Math.Phys. 12, 498 (1971).
4. Zwiebach, B., Phys.Lett. 156B, 315 (1985) and in Symposium on Anomalies, World Scientific (1985).
5. Fradkin, E.S. and Tseytlin, A.A., Phys.Lett 160B, 69 (1980) ;

Tseytlin, A.A., Lebedev preprints N6, N95 (1986) ; Duff M.J. and Nilsson,B.E.W.,CERN preprint (1986).

6. Friedan, D., Phys.Rev.Lett. 45, 1057 (1980).
7. Callan, C.G., Friedan, D., Martinec, E. and Perry, M.J., Nucl. Phys. B262, 593 (1985).
8. Alvarez-Gaume, L., Freedman D.Z. and Mukhi, S., Nucl.Phys. 134, 85 (1981).
9. 't Hooft, G. and Veltman, M., Ann.Inst.Henri Poincaré 20, 69 (1974); 't Hooft, G., Acta Universitatis Wratislavensis n°368, Proc. of XII Winter School in Theoretical Physics in Karpacz.
10. Schwarz, J.H., Phys. Reports, 89, 223 (1982).
11. Cai, Y. and Nunez, C.A., Texas preprint (to be published).
12. Kawai, H., Lewellen, D.C. and Tye, S.H.H., Nucl.Phys. B269, 1(1986).
13. Sannan, S., Phys.Rev. D (in press).
14. Gross, D.J. and Witten, E., Princeton Preprint (to be published).
15. Grisaru, M.T., Van de Ven,A.E.M. and Zanon, D., Brandeis preprints (in press).
16. Madore, J., Phys.Lett. 110A, 289 ; 111A, 283 (1985) ; Müller-Hoissen, F., ibid 163B, 106 (1985) ; Gibbons, G., DAMTP preprint ; Aragone, C., CERN preprint.

Initial Conditions and Quantum Cosmology

James B. Hartle

Institute for Theoretical Physics
and
Department of Physics
University of California
Santa Barbara, California 93106

ABSTRACT

A theory of initial conditions is necessary for a complete explanation of the presently observed large scale structural features of the universe, and a quantum theory of cosmology is probably needed for its formulation. The kinematics of quantum cosmology are reviewed and some candidates for a law of initial conditions discussed. The proposal that the quantum state of a closed universe is the natural analog of the ground state for closed cosmologies and is specified by a Euclidean sum over histories is sketched. When implemented in simple models this proposal is consistent with the most important large scale observations.

1. INTRODUCTION

The evidence of the observations is that the universe was a simpler place earlier than it is now — more homogeneous, more isotropic, more nearly in thermal equilibrium.[1] In this early simplicity we can find explanations of much of the apparent complexity of the present universe by understanding the dynamical processes that happened over its history. For example we hope to understand stars and galaxies by understanding the evolution of an initial spectrum of perturbations on the prevailing homogeneity and isotropy through the action of gravitational attraction. We hope to understand the abundances of the elementary particles and nuclear species by understanding the evolution of an initial democracy of thermal equilibrium through expansion and fundamental particle interactions. The successes of this program, which I do not need to review here,[1] have naturally raised the question of what is the explanation of the simplicity of the early universe? What is the reason for the homogeneity and isotropy? What is the origin of the primordial spectrum for density fluctuations? In short, can we find a compelling law which specifies the initial conditions of the universe. I want to discuss some proposals for this law here , but first I would like to make a few remarks about this enterprise in general.

(1) A law specifying initial conditions is a different kind of law from those we are used to in physics. In physics we are used to laws which specify evolution. Such laws require boundary conditions and the laws — Einstein's equation and the matter field equations — which govern cosmological evolution are no exception. For most of physics the boundary conditions are determined by observation of the rest of the universe outside the system whose evolution is being considered. If we don't see any incoming radiation, we calculate with no incoming radiation boundary conditions. In cosmology, by definition, "the system" is everything and there is no "rest of the universe" to pass the specification of the boundary conditions off to. As different as it is from our usual occupation, the nature of the cosmological problem forces us to consider the specification of the boundary conditions for cosmological evolution as part of the laws of physics.

(2) Inflation is not enough. We are used to the assertion that inflation explains the homogeneity and isotropy of the universe. Starting from sufficiently regular initial conditions it certainly does that. If the initial conditions are sufficiently irregular, sufficiently anisotropic, sufficiently inhomogeneous, however, it is unlikely for inflation even to act to dominate the expansion of the universe. Such irregular conditions, however, are generic at least as counted today. For, following the argument of Penrose,[2] consider a generic irregular universe today — one very unlike the present universe. This generic universe is likely to become more irregular in the future and likely to have come from a more irregular past. Thus as critical as inflation is in explaining the absence of monopoles, the size of the horizon and the evolution of density perturbations it is unlikely to be an explanation of the homogeneity and isotropy in the context of all initial conditions. A law for initial conditions is still needed.

(3) Quantum Gravity is Important. Classical cosmological spacetimes can be singular, for example the Friedman models, or non-singular, for example, de Sitter space. Depending on the physics of the matter there could either have been a big bang or a small bounce. Singularities are not easy to avoid in general relativity. The singularity theorems of classical general relativity suggest that, with reasonable assumptions on the matter physics, the classical extrapolation of any

universe similar to our own into the past will encounter a big bang singularity.[3] If the classical evolution is singular, then curvatures which vary significantly on the scale of a Planck length will occur and quantum gravitational effects will be significant for the early universe. This is one reason why quantum gravity is the appropriate context in which to search for a theory of initial conditions.

The second reason is that it is difficult to imagine a classical theory of initial conditions. What classical principle for example would single out the *particular* density fluctuations we have today? By contrast, fluctuations occur naturally and fundamentally in a quantum mechanical theory. Of course, one could always imagine a *statistical* classical theory but such a law of initial conditions would contrast with the deterministic character of classical evolutionary laws.

For these two reasons quantum cosmology seems to be the appropriate context to search for a theory of initial conditions. Conversely if one has a theory of quantum gravity the construction of a theory of initial conditions for cosmology would seem to be one of its most fitting applications.

2. KINEMATICS OF QUANTUM COSMOLOGY

2.1 The wave function of the universe.

In quantum mechanics we describe the state of a system by giving its wave function. The wave function enables us to made predictions about observations made on a spacelike surface; it thus captures quantum mechanically the classical notion of the "state of the system at a moment of time." The arguments of the wave function are the variables describing how the system's history intersects the spacelike surface. For example, for the quantum mechanics of a particle, the histories are particle paths $x(t)$. We write for the wave function

$$\psi = \psi(x,t). \tag{2.1}$$

The t labels the hypersurface and the x specifies the intersection of the history with it.

In the quantum mechanics of a closed cosmologies with fixed (for simplicity) spatial topology, say that of a 3-sphere S^3, the histories are the 4-geometries on $S^3 \times \mathbf{R}$. The appropriate notion of a 4-geometry fixed on a spacelike surface is the 3-geometry, ${}^3\mathcal{G}$, induced on that surface. One can think of this as specified by a

3-metric h_{ij} on the fixed spatial topology. Thus for the quantum mechanics of a closed cosmology we write[4]

$$\Psi = \Psi[^3\mathcal{G}] = \Psi[h_{ij}(\mathbf{x})]. \tag{2.2}$$

Note that there is no additional "time" label. This is because a generic 3-geometry will fit in a generic 4-geometry at locally only one place if it fits at all. The 3-geometry itself carries the information about its location in spacetime. This labeling of the wave function correctly counts the degrees of freedom. Of the six components of h_{ij}, three are gauge. If one of the remaining three is time, there are left two degrees of freedom — the correct number for a massless, spin-2 field.

The space of all three geometries is called superspace. Each "point" represents a different geometry on the fixed spatial topology. In the case of pure gravity that we have been describing, the wave function is a complex function on superspace. With the inclusion of matter fields the wave function depends on their configurations on the spacelike surface as well, and we write typically

$$\Psi = \Psi[h_{ij}(\mathbf{x}), \phi(\mathbf{x})]. \tag{2.3}$$

A law for initial conditions in quantum cosmology is a law which prescribes this wave function.

2.2 Interpretation

To make contact with observations we must specifiy the observational consequences of the state of the universe being described by this or another wave function. This is usually called an "interpretation" of Ψ. There is little doubt that what I can say here will not address every issue which can be raised on this fascinating topic and even less doubt that it will not satisfy many who have thought about the subject. I would like, however, to offer some minimal elements of an interpretation which I believe will enable an attribution of Ψ to the universe to be confronted by cosmological observations. These elements are an example of "an Everett interpretation" although the words and emphasis may be different from other interpretations in this broad catagory.[5]

The idea is to take quantum mechanics seriously. One assumes that there is one wave function Ψ defined on a preferred configuration space which contains all the predictable information about observations in the universe. If Ψ is sufficiently peaked about some region in the configuration space we predict that we will observe the correlations between the observables which characterize this region. If Ψ is small in some region we predict the observations of the correlations which characterize this region are precluded. Where Ψ is neither small nor sufficiently peaked we don't predict anything. That's it.

The natural reaction to such an interpretations is to ask "Where is probability?" In response, I can say two things:

First, probabilities for single systems have no direct observational interpretation and the universe by definition is a single system. Second, this interpretation implies the usual probability interpretation of quantum mechanics when applied to ensembles of identically prepared systems: Suppose, for example, the configuration space of a single system is C. The configuration space of an ensemble of N systems is C^N. The wave function of an ensemble of N identically prepared systems is

$$\Psi(q_1, ..., q_N) = \psi(q_1)\psi(q_2)\cdots\psi(q_N), \tag{2.4}$$

where $\psi(q)$ is the wave function of a single system. It was pointed out independently by a number of people[6] in the late 60's that, for large N, such a wave function is sharply peaked in the variable which is the frequency, f_a, that a measurement of q on each member of the ensemble yields the result a. The value of f_a about which Ψ is peaked is $|\psi(a)|^2$.

A more precise statement of this result for ensembles can be given in the familiar Hilbert space formulation of quantum mechanics. Consider a single system described by a wave function ψ. Possible observations correspond to operators in the Hilbert space of states. For the physical interpretation of ψ for a single system assume only the following: *If ψ is an eigenfunction of an observable A then an observation of A will yield (with certainty) the eigenvalue. For those observables of which ψ is not an eigenfunction there is no prediction for the outcome of an observation.* We can then derive the probability interpretation of ψ as follows:

An ensemble of identically prepared systems can be viewed as single system with wave function (2.4). On the Hilbert space on C^N there is an operator $\hat{f}_a$ corresponding to observing q on the first system, q on the second, etc., and then computing the frequency that a given value a occurs. For an infinitely large ensemble of identical systems, each in a state ψ, it is a mathematical fact that the product wave function (2.4) is an eigenfunction of this operator

$$\hat{f}_a \Psi = |\psi(a)|^2 \Psi. \tag{2.5}$$

In this way we deduce the probability interpretation of quantum mechanics from its predictions about individual systems.

An interpretation of this kind means that ones ability to predict in quantum cosmology is limited as it is for any single quantum system. We do not expect, for example, that the wave function will be sharply peaked about the *particular* arrangement of galaxies in the universe. We do hope that it might be peaked about the form of the galaxy-galaxy correlation function at the present epoch.

2.3 The Wheeler-DeWitt Equation

In a theory of quantum spacetime we are not free to propose any wave function as a theory of initial conditions. It must satisfy the constraints which implement the dynamics of the particular theory of quantum gravity we have in mind. For example, in pure Einstein gravity three of the constraints are[4]

$$iD_i \left(\frac{\delta \Psi}{\delta h_{ij}(\mathbf{x})} \right) = 0, \tag{2.6}$$

where D_i is the covariant derivative of the metric h_{ij}. These constraints enforce gauge invariance in the spacelike surface. The additional dynamical constraint is

$$\left[l^2 \nabla^2 + l^{-2} h^{\frac{1}{2}} \left({}^3R - 2\Lambda \right) \right] \Psi = 0, \tag{2.7}$$

where $l = (16\pi G)^{\frac{1}{2}}$ is the Planck length (in units where $\hbar = c = 1$) and ∇^2 is a wave operator in the 3-metric. Explicitly

$$\nabla^2 = G_{ijkl} \frac{\delta^2}{\delta h_{ij}(\mathbf{x}) \delta h_{kl}(\mathbf{x})} \quad + \quad \left(\begin{array}{c} \text{linear terms depending} \\ \text{on factor ordering} \end{array} \right), \tag{2.8a}$$

with

$$G_{ijkl} = \frac{1}{2} h^{-\frac{1}{2}} \left(h_{ik} h_{jl} + h_{il} h_{jk} - h_{ij} h_{kl} \right). \tag{2.8b}$$

This constraint, called the Wheeler-DeWitt equation, reflects the dynamics of general relativity. In a different theory of quantum gravity there would be a different constraint.

One can think of the Wheeler-DeWitt equation as a kind of wave equation in superspace. Finding a law for initial conditions may be viewed as the problem of specifying the boundary conditions which select from its many solutions the one which is the wave function of our universe.

2.4 Semiclassical Approximation

In the present universe we would expect a semiclassical approximation to Ψ to be good, and even where it is not precisely valid this approximation often gives useful qualitative information about the behavior of the wave function. In the semiclassical approximation the wave function is given by

$$\Psi \sim A \cos(S). \tag{2.9}$$

in regions of superspace which are classically allowed, and by

$$\Psi \sim Ae^{I} + Be^{-I}. \tag{2.10}$$

in regions which are classically forbidden. Here S is a real Lorentzian action and I a real Euclidean action.

The utility of this approximation is that it allows direct contact with classical physics. Semiclassically in the classically allowed region the wave function corresponds to an ensemble of classical trajectories determined by the action function S. What we can learn from the gradient of S is the tangent vector to the classical trajectory. What we do not know is the initial position. The various possibilities define the ensemble. Further, in the semiclassical approximation we recover a notion of time — the time of these classical trajectories.[7,11] These properties make the interpretation and comparison with observation a good deal easier in this approximation.

3. PROPOSALS FOR A LAW OF INITIAL CONDITIONS

In recent years a number of different proposals for a law of initial conditions have been put forward. There is Roger Penrose's time asymmetric proposal that the Weyl tensor vanish on initial singularities but not on final singularities.[2] There are the proposals that the universe nucleates spontaneously "from nothing" worked out most explicitly and clearly by Alex Vilenkin.[8] This proposal can be implemented concretely by solving the Wheeler-DeWitt equation with the boundary condition that "at the boundaries of superspace the wave function includes only outgoing modes." There is the proposal of Narlikar and Padmanabhan[9] that one should quantize only the conformal factor and enforce the constraints only semi-classically. There is the proposal of Fischler, Ratra and Susskind[10] that the wave function of the universe is singled out from other solutions of the Wheeler-DeWitt equation by the requirement that the energy in the matter field not diverge at the singularity. There is the proposal of Stephen Hawking and his collaborators[11] that the universe is in the cosmological analog of its ground state, implemented concretely by specifying the wave function as a Euclidean sum over histories. There are no doubt many more.

It would be difficult to review all these proposals, not least, because they are not all sufficiently developed to compare their predictions. To illustrate the ideas I will therefore focus on the ground state wave function proposal because it is the most concrete and the one with which I am most familiar.

The proposal is that the quantum state of the universe is the analog for closed cosmologies of the ground state, or state of minimum excitation. To understand what this means note first that what is meant is not a state of minimum energy. For closed cosmologies there is no natural notion of time, therefore no natural notion of energy, therefore no natural Hamiltonian and therefore no natural notion of a state with the lowest eigenvalue of the Hamiltonian. For systems which possess a Hamiltonian, however, finding the lowest energy eigenstate is not the only way of finding the ground state wave function. One can also calculate it directly as a Euclidean sum over histories. For example, for a particle in a potential $V(x)$ the ground state wave function is

$$\psi_0(x) = \sum_{\text{paths}} \exp\left(-I[x(\tau)]\right), \tag{3.1}$$

where $I = \int dt(m\dot{x}^2/2 + V(x))$ is the Euclidean action and the sum is over all paths which start at the argument of the wave function at $\tau = 0$ and proceeed to a configuration of minimum action in the infinite past. This construction generalizes to the quantum mechanics of closed cosmologies.

For a closed cosmology we shall mean by the ground state wave function:

$$\Psi[{}^3\mathcal{G}, \phi(\mathbf{x})] = \sum_{{}^4\mathcal{G}, \phi(\mathbf{x},t)} \exp\left(-I[{}^4\mathcal{G}, \phi(x)]\right) \tag{3.2}$$

The sum over ${}^4\mathcal{G}$ is over all compact, Euclidean 4-geometries which have a boundary on which the induced three geometry is the argument of the wave function and *no other boundary.* The sum over $\phi(x)$ is over all field configurations which match the argument $\phi(\mathbf{x})$ on the boundary and are otherwise regular. "The boundary conditions for the universe is that it has no boundary." These conditions are the analogs in the particle case of the conditions that the paths start at the argument of the wave function and proceed to a configuration of minimum action in the past. Such a construction, properly implemented, should automatically yield a solution of the Wheeler-DeWitt equation.

Since the remarkable simplicity of the early universe suggests that it is in a state of low excitation it is a natural conjecture that this is the wave function of the universe.

How do we test this proposal? As with any theory we need to compare its predictions with the observations. The cosmological observations which we might hope to explain in a theory of initial conditions might be summarized in a few big cosmological facts.[1]

1) Spacetime is homogeneously four dimensional with Euclidean topology on familiar scales. We usually take this for granted but it is important to remember that all aspects of geometry have an observational basis.

2) The universe is large, old and expanding. This is the observation that the scales of cosmology are not the scales of quantum gravity or of elementary particle physics. The age of the universe, for example, is approximately 10^{60} times larger than the Planck scale. An important test for any theory of initial conditions is whether it explains how the universe can be as big as it is.

3) The universe contains matter and radiation distributed homogeneously and isotropically on the largest scales. A theory of initial conditions should explain the approximate large scale simplicity.

4) The spatial geometry is approximately flat.

5) There is a spectrum of density fluctuations which produced the galaxies. A theory of initial conditions should certainly be expected to produce a spectrum of initial fluctuations which, in amplitude and spectrum, serve as a suitable input for galaxy formation.

6) The entropy of the matter in the universe is low compared to what it might have been and increasing in the direction of expansion. This is another characterization of the fact that the universe is apparently more ordered earlier than it is now.

I would now like to describe how the ground state wave function proposal compares with these observations. To do so I will describe in rough terms the calculations which have been made. Essentially all of them are minisuperspace models or linear perturbations of minisuperspace models. In a minisuperspace model one truncates the infinite number of degrees of freedom to a finite number to obtain a tractable quantum mechanical model. This is usually done by imposing symmetries such as homogeneity and isotropy on the geometries and matter fields which contribute to the sum over histories. One is thus at best obtaining approximate information about the wave function on an infinitesimal region of superspace. The region can be enlarged by considering the linear deviations from exact symmetry. This is a large enlargement from the point of view of number of degrees of freedom, but still small compared to the whole of superspace. The hope is that although this region is small it will contain the region which is compatible with observations. If there is consistency with observations here, the rest of the task will be to show the wave function is small everywhere else. A typical model is the homogeneous, isotropic massive scalar field model of Hawking[11] later extended to include linear perturbations by Halliwell and Hawking.[12] The minisuperspace is obtained by restricting to geometries which are homogeneous and isotropic whose metrics have the form

$$d\hat{s}^2 = -dt^2 + a^2(t)d\Omega_3^2, \tag{3.3a}$$

where $d\Omega_3^2$ is the metric on the unit 3-sphere, and restricting to homogeneous field configurations for which

$$\phi = \phi(t). \tag{3.3b}$$

Euclidean geometries with the same symmetry are obtained by setting $t = -i\tau$ in (3.3a). Geometry and field on a spacelike hypersurface are characterized by just two numbers a and ϕ. The minisuperspace is thus just two dimensional and we write

$$\Psi = \hat{\Psi}(a, \phi). \tag{3.4}$$

The semiclassical approximation to $\hat{\Psi}$ is obtained by making a steepest

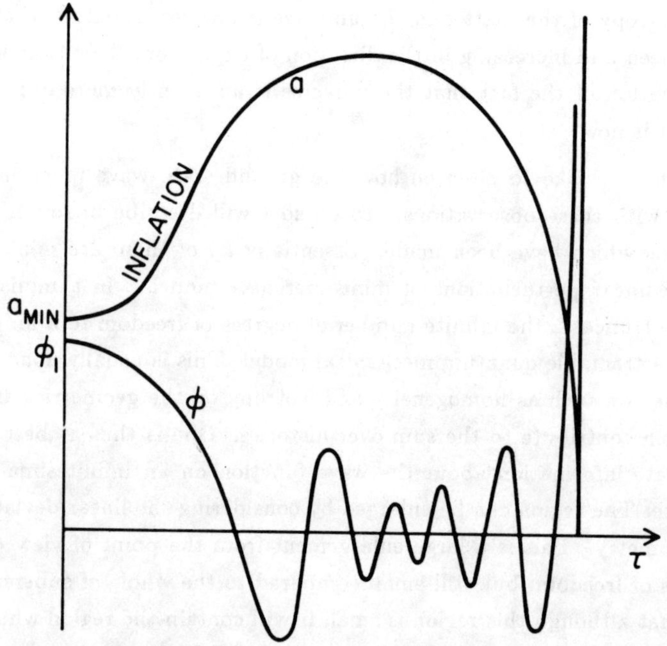

Figure 1: A schematic graph of a typical Lorentzian extremizing configuration. The ground state wave function prescription implies that the solutions start at a minimum radius with $\dot{\phi} \approx 0$. In the domain where ϕ varies slowly the universe follows a de Sitter like inflationary expansion with $H = m\phi_1$. Later the scalar field begins to oscillate and the universe evolves approximately as though matter dominated. Eventually a maximum expansion is reached, the universe recollapses and matter and geometry become singular. A sufficiently large $m\phi_1$ would provide a long enough inflationary period to explain the present large size of the universe and its approximate spatial flatness. The oscillation of the scalar field models the creation of matter.

descents approximation to the defining Euclidean functional integral (3.2). This singles out the appropriate extrema to compute the classical action in either (2.9) or (2.10). In the classically allowed region these extrema solve the classical equations of motion

$$\ddot{\phi} + \frac{3\dot{a}}{a}\dot{\phi} + m^2\phi = 0, \tag{3.5a}$$

$$\left(\frac{\dot{a}}{a}\right)^2 = -\frac{1}{a^2} + \dot{\phi}^2 + m^2\phi^2. \tag{3.5b}$$

They are the possible classical trajectories in the ensemble determined semiclassically by $\hat{\Psi}$. In this way an ensemble of possible classical behaviors for the universe is fixed.

The analysis of which extrema contribute to the semiclassical approximation depends on a detailed analysis of (3.2) which I cannot develop here. Very roughly, however, it proceeds as follows: For small a the steepest descents approximation to (3.2) is given by a compact Euclidean geometry which is a portion of a distorted 4-sphere bounded by a 3-sphere boundary of radius a. The extremizing field configuration is the regular solution of the field equations which assumes the homogeneous value ϕ on the boundary. As a is increased, for sufficiently large ϕ, the Euclidean geometry tries to close and a radius is encountered beyond which it does not exist. Beyond this boundary is the classically allowed region. At this boundary it turns out that $\dot{a} \approx 0$, $\dot{\phi} \approx 0$ so that in the classically allowed region the semiclassical approximation to $\hat{\Psi}$ is determined by solutions to (3.5) with these boundary conditions.

A typical classical trajectory in the ensemble fixed by the ground state wave function is shown in Figure 1. For small t, while $\dot{\phi} \approx 0$ the $m^2\phi^2$ term in (3.5a) acts as an effective cosmological constant. The universe inflates and according to Hawking and Page[13] the ensemble is peaked about inflations of arbitrarily long duration. This is how the universe can get big and still be in its ground state and how it can become arbitrarily close to spatially flat. Eventually the field begins to oscillate and acquires kinetic energy. In the crude terms of the model, this is how the matter of the universe is created.

Halliwell and Hawking[12] have analyzed the linear fluctuations about this minisuperspace model. These are specified by metrics and field configurations of the form

$$ds^2 = d\hat{s}^2 + \epsilon_{\alpha\beta}(\mathbf{x}, t)dx^\alpha dx^\beta, \tag{3.6a}$$

$$\phi = \phi(t) + f(\mathbf{x}, t), \tag{3.6b}$$

where ϕ and $d\hat{s}^2$ are as in (3.3). The linear fluctuations can be expanded in harmonics on the 3-sphere

$$\epsilon_{\alpha\beta}(\mathbf{x}, t) = \sum_n \epsilon^{(n)}(t) Q^{(n)}_{\alpha\beta}(\mathbf{x}), \tag{3.7a}$$

$$f(\mathbf{x},t) = \sum_n f^{(n)}(t) Q^{(n)}(\mathbf{x}). \tag{3.7b}$$

The $h^{(n)}$ and $f^{(n)}$ can be thought of as local coordinates in superspace. We write

$$\Psi = \Psi\left(a, \phi, \epsilon^{(1)}, f^{(1)}, \epsilon^{(2)}, f^{(2)}, \cdots\right). \tag{3.8}$$

The linear modes decouple in linearized perturbation theory. Correspondingly the wave function becomes a product

$$\Psi = \hat{\Psi}(a,\phi) \prod_n \psi^{(n)}\left(\epsilon^{(n)}, f^{(n)}\right). \tag{3.9}$$

In the semiclassical approximation for $\hat{\Psi}$, the Wheeler-DeWitt equation for the $\psi^{(n)}$ becomes an ordinary Schrodinger equation

$$i\frac{\partial\psi^{(n)}}{\partial t} = H^{(n)}\psi^{(n)}. \tag{3.10}$$

The time t being the time of the classical trajectory of the unperturbed problem. The theory of fluctuations in this approximation is thus effectively a field theory in the curved background spacetime which gives the semiclassical approximation to $\hat{\Psi}$. Not surprisingly, the ground state prescription for the wave function requires each mode to start out in its ground state. The wave function is sharply peaked about the observed homogeneity and isotropy and this remains true at later times. There are still quantum fluctuations and these fluctuations, evolved through the inflationary expansion, provide a plausible spectrum of density fluctuations from which to grow galaxies. The physics of this evolution is essentially classical and essentially the same as in familiar inflationary stories. What the ground state wave function proposal provides is the initial condition for the evolution.

The results of this and some other representative* minisuperspace calculations are summarized in Table 1. There one sees that, while the wave function has been calculated on only a limited region of superspace, plausible mechanisms have been provided by which the ground state wave function can explain most of the observed large scale features of the universe. While the enterprise is very different from what we are used to in physics, it is just possible that the remarkable simplicity of the early universe is described by this simplest of all possible quantum states and that

* While I have attempted to be helpful in citing some typical calculations, I have not been able to be exhaustive in what already is a large literature.

all the complexity we see around us arises from its quantum fluctuations and the attractive force of gravity.

ACKNOWLEDGEMENTS

Substantially similar versions of this talk were presented at the Aspen Winter School on the Early Universe, SWOGU/ICOBAN '86, the 26th Liège International Astrophysical Colloquium, and the 13th Texas Symposium on Relativistic Astrophysics and will appear in the proceedings of those conferences. The preparation of this report was supported in part by NSF Grant PHY85-06686 and at the Institute for Theoretical Physics by NSF Grant PHY82-17853, supplemented by funds from NASA.

REFERENCES

1. For reviews of the current status of cosmological observations and models see *Physical Cosmology: Les Houches 1979*, ed. by Balian, R., Audouze, J., and Schramm, D. (North-Holland Amsterdam, 1980); and *Inner Space/Outerspace*, ed. by Kolb, E.W., Turner, M.S., Lindley, D., Olive, K and Seckel, D. (Chicago University Press, Chicago, 1986).

2. Penrose, R. in *General Relativity: An Einstein Centenary Survey,* ed. by Hawking, S.W., and Israel, W., (Cambridge University Press, Cambridge, 1979).

3. See, *e.g.,* Hawking, S.W., and Ellis, G.F.R., *Ap. J.* **152**, 25 (1968) and Geroch, R., and Horowitz, G.T. in *General Relativity: An Einstein Centenary Survey*, ed. by Hawking, S.W., and Israel, W., (Cambridge University Press, Cambridge, 1979).

4. Fur further details on the canonical quantum mechanics of gravity see, *e.g.,* Kuchař, K. in Relativity, Astrophysics, and Cosmology, ed. by Israel, W. (D. Reidel, Dorchecht, 1973).

5. See, *e.g.,* Everett, H. *Rev. Mod. Phys.* **29**, 454 (1957), the many articles reprinted and cited in *The Many Words Interpretation of Quantum Mechanics,* ed. by DeWitt, B., and Graham, N. (Princeton University Press, Princeton, 1973), and the lucid discussion in Geroch R. *Noûs,* **18** 617 (1984).

6. Finkelstein, D. *Trans. N.Y. Acad. Sci.* **25**, 621 (1963); Graham, N. unpublished Ph.D. dissertation, University of North Carolina 1968 and in *The Many Worlds Interpretation of Quantum Mechanics*, ed. by DeWitt, B. and Graham, N. (Princeton University Press, Princeton, 1973); and Hartle, J.B. *Am. J. Phys.* **36**, 704 (1968).

7. Banks, T. *Nucl. Phys.* **B249**, 332 (1985).

8. See, *e.g.,* Vilenkin, A. *Phys. Lett.* **B117**, 25 (1983), *Phys. Rev.* **D27**, 2848 (1983), *Phys. Rev.* **D30**, 509 (1984), *Phys. Rev.* **D32**, 2511 (1985), TUTP preprint 85-7.

9. See, *e.g.,* Narlikar, J.V. and Padmanabhan, T. *Physics Reports* **100**, 151 (1983); Padmanabhan, T. "Quantum Cosmology — The Story So Far," (unpublished lecture notes).

10. Fischler, W., Ratra, B., andd Susskind, L. *Nucl. Phys.* **B259**, 730 (1985).

11. See, *e.g.,* Hawking, S.W. in *Astrophysical Cosmology: Proceedings of the Study Week on Cosmology and Fundamental Physics,* ed. by Brück, H.A., Coyne, G.V. and Longair, M.S. (Pontificial Academiae Scientiarum Scripta Varia, Vatican City, 1982) and Hawking, S.W. *Nucl. Phys.* **B239**, 257 (1984).

12. Hawking, S.W. and Halliwell, J., *Phys. Rev.* **D31**, 1777 (1985).

13. Hawking, S.W. and Page, D.N. *Nucl. Phys.* **B264**, 185 (1986).

14. Wu, Z.C., *Phys. Lett.* **146B**, 307 (1984), *Phys. Rev.* **D31**, 3079 (1985), Hu, X.N. and Wu, Z.C., *Phys. Lett.* **149B**, 87 (1984).

15. Halliwell, J.J., "Quantum Cosmology of the Einstein Maxwell Theory in 6-Dimensions." (preprint)

16. Okada, Y. and Yoshimura, M., *Phys. Rev.* **D323**, 2164 (1986).

17. Hartle, J.B., *Class Quan. Grav.* **2**, 707 (1985).

18. Hawking, S.W. and Luttrell, J.C. *Nucl. Phys.* **B247**, 250 (1984).

19. Horowitz, G., *Phys. Rev.* **D31**, 1169 (1985).

20. González-Diaz, P.F., *Phys. Lett.* **159B**, 19 (1985).

21. Carow, U. and Watamura, S., *Phys. Rev.* **D32**, 1290 (1985).

22. Hawking, S.W. and Luttrell, J.C., *Phys. Lett.* **B146**, 307 (1984).

23. Wright, W. and Moss, I., *Phys. Lett.* **154B**, (1985).

24. Amsterdamski, P., *Phys. Rev.* **D31**, 1169 (1985).

25. Hawking, S.W., *Phys. Rev.* **D32**, 2489 (1985).

26. Page, D.N., *Phys. Rev.* **D32**, 2496 (1985).

STOCHASTIC QUANTIZATION OF GRAVITY AND STRING FIELDS

Helmut Rumpf
Institut für Theoretische Physik, Universität Wien
Boltzmanngasse 5, A-1090 Wien
AUSTRIA

ABSTRACT

The stochastic quantization method of Parisi and Wu is generalized so as to make it applicable to Einstein's theory of gravitation. The generalization is based on the existence of a preferred metric in field configuration space, involves Ito's calculus, and introduces a complex stochastic process adapted to Lorentzian spacetime. It implies formally the path integral measure of DeWitt, a causal Feynman propagator, and a consistent stochastic perturbation theory. The linearized version of the theory is also obtained from the stochastic quantization of the free string field theory of Siegel and Zwiebach.

1. GENERALLY COVARIANT STOCHASTIC QUANTIZATION

The stochastic quantization method of Parisi and Wu[1)] is an extension of the Euclidean path integral formalism just as non-equilibrium thermodynamics is an extension of equilibrium thermodynamics. In order to make the analogy plain consider the prototype of a relaxation process - Brownian motion. The simplest version (due to Einstein and Smoluchowski) of this process is described by the Langevin equation

$$\dot{x} = F(x) + D^{1/2}\, \xi(t) \tag{1.1}$$

$$F(x) = K(x)/m\alpha \tag{1.2}$$

$$D = kT/m\alpha \; . \tag{1.3}$$

Here x is the position of a particle interacting with a medium which is at thermal equilibrium with temperature T, m is the particle mass, α is the friction coefficient, D is Einstein's diffusion constant, and K(x) is an external force field. The stochastic source term $\xi(t)$ is a Gaussian white noise with stochastic averages

$$<\xi(t)> = 0 \tag{1.4}$$

$$<\xi(t)\xi(t')> = 2\delta(t-t') \tag{1.5}$$

and all higher order correlations vanishing. An equivalent description of the same process is provided by the Fokker-Planck equation associated with the Langevin equation (1.1),

$$\frac{\partial P(x,t)}{\partial t} = [D \frac{\partial^2}{\partial x^2} - \frac{\partial}{\partial x} F(x)]P(x,t) \tag{1.6}$$

where P(x,t) is the spatial probability density for finding the particle in x at t. If $K(x) = - \partial V/\partial x$, (1.6) may be used to show that relaxation occurs (under suitable conditions on V(x)) and that the equilibrium distribution is given by

$$P_{eq}(x) \equiv \lim_{t\to\infty} P(x,t) \propto e^{-V(x)/kT} \; . \tag{1.7}$$

Now the functional integral of Euclidean quantum field theory,

$$<f[\Phi]> = \int D[\Phi]f[\Phi]P[\Phi] \tag{1.8}$$

$$P[\Phi] = e^{-S_E[\Phi]/\hbar} / \int D[\Phi] \, e^{-S_E[\Phi]/\hbar} \tag{1.9}$$

with S_E the Euclidean action defines a probability distribution which is a direct generalization of the Boltzmann distribution (1.7). The essence of the Parisi-Wu method is to represent the process $\Phi(x)$ as the equilibrium limit of a relaxation process $\Phi(x,s)$ in a fictitious time parameter s. The analog of the Langevin equation (1.1) is

$$\frac{\partial\Phi(x,s)}{\partial s} = -\frac{\delta S_E[\Phi]}{\delta\Phi(x,s)} + \hbar^{1/2}\,\xi(x,s) \tag{1.10}$$

where the only non-vanishing correlation of ξ is

$$\langle\xi(x,s)\xi(x',s')\rangle = 2\delta(s-s')\delta^{(4)}(x-x') \ . \tag{1.11}$$

The great advantage of this stochastic formulation is that in the case of gauge fields Φ (1.10) is fully gauge-invariant and the equilibrium limit of the expectation value of gauge-invariant observables exists although the equilibrium limit (1.9) of the Fokker-Planck probability distribution does not make sense, as is well-known. In particular no gauge-fixing and associated Faddeev-Popov ghost fields are required, and a new type of perturbation theory is implied by (1.10). Stochastic quantization offers also the prospect of an invariant and non-perturbative regularization[2]. Moreover, Langevin simulation on the lattice is an interesting numerical method which has a wider range of applicability[3] than the Monte Carlo method.

In the equations (1.10) and (1.11) we have suppressed any indices that the field Φ and the source ξ may bear. As long as we are dealing with tensor fields (with an internal symmetry) on the Euclidean background space, the Euclidean metric (and the metric on the internal symmetry group) may always be used to define the additional structure necessary to make (1.10) well-defined. However if we want to quantize the metric tensor field $g_{\alpha\beta}(x)$ itself, the form of the Langevin equation is less obvious. The same problem is also encountered when a non-gravitational field is to be quantized upon a transformation of the field variable, e.g. $\Phi \to e^{\Phi}$ for the scalar field. This corresponds to introducing curvilinear coordinates x in (1.1). Thus a manifestly covariant version of the Langevin equation (1.10) with respect to arbitrary field redefinitions (the analog of general coordinate transformations) is desirable. In the following shall denote the field variable by Φ^A, where the index A comprises also the space-time argument x. The general form of a covariant Langevin equation in the above sense then is

$$\dot{\Phi}^A = - G^{AA'} \frac{\delta S_E[\Phi]}{\delta\Phi^{A'}(s)} + \xi^A(s) \tag{1.12}$$

where $\dot{}$ = d/ds and ħ has been set equal to unity. Covariance requires $G^{AA'}$ to be a metric on field configuration space. The natural generalization of (1.11) is

$$\langle\xi^A(s)\xi^B(s')\rangle = 2 \langle G^{AB}\rangle \, \delta(s-s') \, . \tag{1.13}$$

Note that we have to take the expectation value on the right hand side of (1.13) because G^{AB} will in general depend on Φ. A noise ξ^A satisfying (1.13) can be constructed explicitly as follows. Introduce a Φ-independent reference metric $G^{(o)MN}$ and an associated stochastic vielbein field $E_M{}^A[\Phi(s)]$ obeying

$$G^{AA'} = E_M{}^A[\Phi]E_{M'}{}^{A'}[\Phi]G^{(o)MM'} \, . \tag{1.14}$$

Then

$$\xi^A(s) = E_M{}^A[\Phi(s)]\xi^{(o)M}(s) \tag{1.15}$$

reproduces (1.13), if

$$\langle\xi^{(o)M}(s)\xi^{(o)M'}(s')\rangle = 2G^{(o)MM'} \, \delta(s-s') \tag{1.16}$$

<u>and</u> $E_M{}^A[\Phi(s)]$ is statistically independent of $\xi^{(o)M}(s)$, so that

$$\langle\xi^A(s)\xi^{A'}(s')\rangle = \langle E_M{}^A[\Phi(s)]E_{M'}{}^{A'}[\Phi(s')]\rangle \, \langle\xi^{(o)M}(s)\xi^{(o)M'}(s') \, . \tag{1.17}$$

The latter requirement amounts to the Ito interpretation[4)] of the ambiguous product on the right hand side of (1.15). The adoption of Ito's calculus spoils the covariance of (1.12), however, as $\dot{\Phi}^A$ does not transform as a vector in field configuration space, but

$$\dot{\Phi}'^{A} = \frac{\delta\Phi'^{A}}{\delta\Phi^{B}} \dot{\Phi}^{B} + G^{BC} \frac{\delta^{2}\Phi'^{A}}{\delta\Phi^{B}\delta\Phi^{C}} . \tag{1.18}$$

This is remedied, however, by the following modification of (1.12):

$$\dot{\Phi}^{A} - \Delta_{G}\Phi^{A} = - G^{AA'} \frac{\delta S_{E}[\Phi]}{\delta\Phi^{A'}} + \xi^{A} \tag{1.19}$$

where Δ_G is the Laplace-Beltrami operator with respect to the covariant field metric $G_{AA'}$.

Equations (1.15), (1.16), (1.18) fulfil the requirement of general covariance. It can also be checked that they imply the correct equilibrium limit

$$P_{eq}[\Phi] \propto |G|^{1/2} e^{-S_{E}[\Phi]} \tag{1.20}$$

by using the associated Fokker-Planck equation

$$\frac{\partial Q}{\partial s} = \nabla^{B}\left(\frac{\delta Q}{\delta\Phi^{B}} + \frac{\delta S_{E}}{\delta\Phi^{B}} Q\right) \tag{1.21}$$

where

$$Q[\Phi,s] = |G|^{-1/2} P[\Phi,s] \tag{1.22}$$

$$G = \det(G_{AB}) \tag{1.23}$$

and ∇_A denotes the covariant derivative implied by the Christoffel connection of the metric G_{BC}.

2. EUCLIDEAN GRAVITY

We now specialize to the gravitational field in the standard parametrization of Riemannian metrics, $\Phi^{A} = g_{\alpha\beta}(x)$. It is natural to require that the field metric (i) be local and (ii) that the actions of diffeomorphisms on $g_{\alpha\beta}(x)$ be isometries with respect to the field metric. The properties (i) and (ii) are satisfied by the one-parameter (λ) family[5)]

$$G_{AA'} \equiv G^{\alpha\beta,\alpha'\beta'}(x,x') = \frac{C}{2} g^{1/2}(g^{\alpha\alpha'} g^{\beta\beta'} + g^{\alpha\beta'} g^{\beta\alpha'} + \lambda g^{\alpha\beta} g^{\alpha'\beta'})\delta^{(4)}(x-x') \tag{2.1}$$

$$\lambda \neq -1/2 \tag{2.2}$$

where the constant C is not important and will be chosen as

$$C = (4\kappa)^{-1} \tag{2.3}$$

for convenience. We note that $G_{AA'}$ is pseudo-Riemannian for $\lambda < -1/2$. In this case the Gaussian noise $\xi^{(0)}$ is necessarily complex, and so is the stochastic metric $g_{\alpha\beta}(x,s)$ itself.

The Euclidean Einstein-Hilbert action

$$S_E[g] = -\frac{1}{2\kappa}\int d^4x\, g^{1/2} R \tag{2.4}$$

implies the following form of the covariant Langevin equation (1.19):

$$\dot{g}_{\alpha\beta} + 18\kappa \frac{\lambda+1}{2\lambda+1}\delta^{(4)}(0) g^{-1/2} g_{\alpha\beta} = -2R_{\alpha\beta} + \frac{\lambda+1}{2\lambda+1} g_{\alpha\beta} R + \xi_{\alpha\beta}\ . \tag{2.5}$$

The divergent term on the left hand side vanishes for $\lambda = -1$. This is the parameter value for which the field coordinate $g_{\alpha\beta}(x)$ is harmonic with respect to G_{AB}. Independent reasons for choosing $\lambda = -1$ will be given below when we discuss the linearized theory. But first we note that the formal equilibrium limit implied by (2.5) is characterized, for any λ, by the partition function

$$Z = \int D[g]\, e^{-S_E[g]} \tag{2.6}$$

with the DeWitt path integral measure

$$D[g] = \prod_x \prod_{\alpha\leq\beta} dg_{\alpha\beta}(x) \tag{2.7}$$

since G is independent of $g_{\alpha\beta}$.

Let us now consider the linear approximation for (2.5) defined by splitting the metric and action according to

$$g_{\alpha\beta} = \delta_{\alpha\beta} + 2\kappa^{1/2} h_{\alpha\beta} \tag{2.8}$$

$$S_E[g] = S_E^{(0)}(h) + S^{(int)}[h] \tag{2.9}$$

$$S_E^{(0)}[h] = \frac{1}{2}\int d^4x \, h_{\alpha\beta} V^{\alpha\beta\gamma\delta} h_{\gamma\delta} . \tag{2.10}$$

The selfadjoint differential operator V is most compactly expressed in terms of a complete orthogonal set of spin projection operators[6] P_s^o, $P_s^{o\prime}$, P_s^1, P_s^2 obeying

$$P_s^o + P_s^{o\prime} + P_s^1 + P_s^2 = id_s \tag{2.11}$$

$$P_s^i P_s^j = \delta^{ij} P_s^i \tag{2.12}$$

where id_s denotes the identity on the space of symmetric tensor fields. Then

$$V = -\Box(P_s^2 - 2P_s^{o\prime}) \tag{2.13}$$

and the linearized Langevin equation (2.5) reads

$$\dot{h}_{\alpha\beta} = -G^o_{\alpha\beta,\gamma\delta} V^{\gamma\delta\mu\nu} h_{\mu\nu} + \tilde{\xi}^{(0)}_{\alpha\beta} \tag{2.14}$$

$$G^o_{\alpha\beta,\gamma\delta} = \frac{1}{2}(\delta_{\alpha\gamma}\delta_{\beta\delta} + \delta_{\alpha\delta}\delta_{\beta\gamma} - \frac{\lambda}{2\lambda+1}\delta_{\alpha\beta}\delta_{\gamma\delta}) \tag{2.15}$$

$$\langle\tilde{\xi}^{(0)}_{\alpha\beta}(x,s)\tilde{\xi}^{(0)}_{\gamma\delta}(x',s')\rangle = 2G^o_{\alpha\beta,\gamma\delta}\,\delta^{(4)}(x-x')\delta(s-s') . \tag{2.16}$$

The linear operator acting on h in the drift term of (2.14) is positive for $-2 < \lambda < -1/2$. It is even a projector for $\lambda = -1$, in which case (2.14) becomes[7]

$$\dot{h} = -P_{grav}\, h + \tilde{\xi}^{(0)} \tag{2.17}$$

$$P_{grav} = P^2 - 2P^{o\prime} + \delta(\delta - L) \tag{2.18}$$

$$L_{ab} = \partial_a \partial_b / \Box \, . \tag{2.19}$$

In the indicated range of λ the stochastic perturbation theory[8] for gauge-independent observables is well-defined (it involves as a new feature "stochastic vertices" corresponding to the multiplicative noise), although the Feynman propagator K, obtained in momentum space as

$$\lim_{s\to\infty} \langle h_{\alpha\beta}(k,s) h_{\alpha'\beta'}(k',s)\rangle = (2\pi)^4 \delta^{(4)}(k+k') K_{\alpha\beta,\alpha'\beta'}(k) \tag{2.20}$$

exhibits in its gauge-dependent part a divergence linear in s. Its gauge-independent part

$$K_{\alpha\beta,\alpha'\beta'}(\text{gauge indep.}) = \frac{1}{2k^2}(\delta_{\alpha\alpha'}\delta_{\beta\beta'} + \delta_{\alpha\beta'}\delta_{\beta\alpha'} - \delta_{\alpha\beta}\delta_{\alpha'\beta'}) \tag{2.21}$$

shows that the field metric G with $\lambda = -1$ is indeed distinguished by the linearized Einstein dynamics.

A finite propagator can be obtained using the technique of stochastic gauge-fixing[9]. It is based on the observation that one may modify the Langevin equation (2.14) into

$$\dot{h}_{\alpha\beta} = - G^{o}_{\alpha\beta\gamma\delta} V^{\gamma\delta\mu\nu} h_{\mu\nu} + (\partial_\alpha \Lambda_\beta + \partial_\beta \Lambda_\alpha) + \tilde{\xi}^{(o)}_{\alpha\beta} \tag{2.22}$$

without affecting gauge-invariant expectation values. The most general ansatz for Λ_μ linear in h is

$$\Lambda_\mu = \alpha^{-1}[h_{\mu\nu,\nu} + \frac{\beta}{2} h_{\nu\nu,\mu} + \frac{\gamma}{2} \frac{\partial_\mu}{\Box} h_{\lambda\nu,\lambda\nu}] \, . \tag{2.23}$$

For $\lambda = -1$, $\alpha = 1$, $\beta = -1$, $\gamma = 0$ (2.22) assumes the simple form[10]

$$\dot{h} = - k^2 h + \tilde{\xi}^{(o)} \, . \tag{2.24}$$

For the stochastic process h defined by this equation the limit (2.20) exists and is just the right hand side of (2.21).

We may derive the Fokker-Planck equation corresponding to (2.24) by noting that the probabilistic interpretation of the noise $\tilde{\xi}^{(0)}$ is assured if we set

$$\tilde{\xi}^{(0)} = (id_s - \frac{\delta\delta}{4})\eta + i\,\frac{\delta\delta}{4}\,\eta \tag{2.25}$$

$$\langle\eta(x,s)\eta(x',s')\rangle = 2id_s\,\delta^{(4)}(x-x')\,(s-s') \tag{2.26}$$

which implies (2.16) ($\lambda = -1$). (2.25) implies that $h = h_R + ih_I$ is complex, and the Fokker-Planck equation becomes

$$\dot{P} = \int d^4x\,\{\frac{\delta}{\delta h_R}[k^2h_R + (1-\frac{\delta\delta}{4})\,\frac{\delta}{\delta h_R}] + \frac{\delta}{\delta h_I}[k^2h_I + \frac{\delta\delta}{4}\,\frac{\delta}{\delta h_I}]\}\,P\,. \tag{2.27}$$

This has the stationary solution

$$P_{eq} \propto \delta[\frac{\delta\delta}{4}\,h_R]\delta[(id_s - \frac{\delta\delta}{4})h_I]\cdot e^{-\frac{1}{2}\int h_R k^2(id_s - \frac{\delta\delta}{4})h_R - \frac{1}{2}\int h_I k^2 \frac{\delta\delta}{4}\,h_I} \tag{2.28}$$

which corresponds at the linearized level exactly to the path integration contour of Gibbons, Hawking and Perry[11] (but note that the action appearing in (2.28) is gauge-fixed), as $\delta\delta/4$ is the projector on the conformal modes.

3. LORENTZIAN GRAVITY

The probabilistic interpretation for stochastic Euclidean gravity indicated at the end of the last section is possible only in perturbation theory. For the full nonlinear theory (2.6) does not make sense in general (for special assumptions see[11]), as $S_E[g]$ is not bounded from below. We consider this a serious drawback of the Euclidean quantization program. Another drawback is that even in field theory on a fixed curved background the complexified space-time manifold does in general not possess a Riemannian section. In view of this a generalization of the Parisi-Wu method to the case of Lorentzian metric signature is highly desirable. Such a generalization has indeed been shown to exist[12]. The Langevin equation (1.12) generalizes to

$$\dot{\Phi}^A - \Delta_G \Phi^A = i\, G^{AB} \frac{\delta S[\Phi]}{\delta \Phi^B(s)} + \xi^A(s) \tag{3.1}$$

where S is the action in the pseudo-Riemannian space-time. Because of the complex drift term the process Φ^A is now necessarily complex and will not possess an equilibrium limit in the ordinary sense. However it turns out that in Minkowski space $\lim_{s\to\infty} \langle\Phi(x_1,s) \ldots \Phi(x_n,s)\rangle$ exists in the sense of a tempered distribution of the Cartesian coordinates $x_1, \ldots, x_n$.

Tempered distributions possess a natural analog on pseudo-Riemannian manifolds (being defined via spectral properties of the minimally coupled d'Alembertian[13]). Therefore at least perturbatively (5.1) implies a well-defined equilibrium limit. Remarkably, it even distinguishes two preferred quantum states and thus provides a definite answer to a well-known problem of quantum field theory in curved space-time. For a linear quantum field Φ with

$$\frac{\delta S[\Phi]}{\delta \Phi} = V\, \Phi \tag{3.2}$$

and V self-adjoint the modified stochastic quantization implies a unique Feynman propagator (independent of the special choice within a natural class of initial distributions for $\Phi(x,0)$)

$$K(x,x') = -\, i \lim_{s\to\infty} \langle\Phi(x,s)\Phi^\dagger(x',s')\rangle \tag{3.3}$$

$$K = -\, i \int_0^\infty ds\; e^{iVs} = (V + i0)^{-1} \;. \tag{3.4}$$

The two preferred states $|in\rangle$ and $|out\rangle$ are defined implicitly by K via

$$K(x,x') = -\, i\, \frac{\langle out|T[\Phi(x)\Phi^\dagger(x')]|in\rangle}{\langle out|in\rangle} \tag{3.5}$$

T denoting chronological ordering. In the absence of singularities in the external field these states are Fock space vacua and provide a physically reasonable description of particle creation[14].

The discussion of the two preceding Sections is easily generalized to the case of Lorentzian metric signature by replacing S_E by $-iS$ everywhere. But note that the modified Fokker-Planck equation (1.20) will then no longer possess real solutions. The correct interpretation of $P[\Phi,s]$ in this case is that it belongs to a fictitious real process Φ_R with complex probability[15] yielding the same expectation values as Φ.

4. FREE STRING FIELD THEORY

It is remarkable that the linearized field metric (2.1) with $\lambda = -1$ arises naturally in a gauge-invariant formulation of the free closed bosonic string field theory[7]. As in the gravitational case, the string field action is indefinite. But as we confine ourselves to the free field, we can carry out the stochastic quantization in Euclidean space-time (of dimension D = 26). In the formalism of Siegel and Zwiebach[16] the gauge-invariant Lagrangian of the open string field is given by

$$L_{cl} = \frac{1}{2}\, \phi\, \delta_{T,o} (H - Q_+ T_+^{-1} Q)\, \delta_{T,o}\, \phi \tag{4.1}$$

where ϕ is related to the string field functional $\Phi[X(\sigma), C(\sigma), \tilde{C}(\sigma)]$ via

$$\Phi = \psi + c\,\phi \tag{4.2}$$

c being the zero mode of the ghost coordinate C. $\delta_{T,o}$ is the singlet projector of an SU(1,1) generated by operators $T_\pm$ and T_3 that appear naturally in the BRS first quantization of the string and are bilinear in the ghost and antighost annihilation and creation operators $\tilde{b}_n$, $\tilde{b}_n^\dagger$, b_n, $b_n^\dagger$. The operators H, Q_+ and T_+ are pieces of the first-quantized BRS charge Q,

$$Q = -H \frac{\partial}{\partial c} + T_+ c + Q_+ \tag{4.3}$$

satisfying $Q^2 = 0$. The functional $\delta_{T,o}\phi$ has ghost number zero and contains as components the physical, Stueckelberg and auxiliary fields of gauge-invariant point-field theory. The components are obtained upon expanding $\delta_{T,o}\phi$ into a power series of physical, ghost and antighost creation operators acting on the first quantized string oscillator vacuum and depend on the center-of-mass coordinate x. Thus $\delta_{T,o}\phi$ may be represented as an x-dependent vector in the oscillator Fock space,

$$\delta_{T,o}\,\phi = |\chi(x)> \; . \tag{4.4}$$

In terms of $|\chi(x)>$ the action may be written as

$$S_{cl}[\chi] = \int d^D x \; <\chi(x)|HP|\chi(x)> \tag{4.5}$$

where $P = 1\!\!1 - H^{-1}Q_+T_+^{-1}Q_+$ is a projector because of $Q^2 = 0$. Note that although HP is positive (except in the tachyon sector), the action is not, as the scalar product $<|>$ is indefinite due to the ghost oscillator anticommutation relations

$$\{b_m, \tilde{b}_n^\dagger\} = - \{\tilde{b}_m, b_n^\dagger\} = i\delta_{mn} \; . \tag{4.6}$$

Nonetheless the stochastic quantization of $|\chi(x)>$ is straightforward. The Langevin equation reads

$$\frac{\partial}{\partial s}\,|\chi(x,s)> = -\,HP|\chi(x,s)> + |\xi(x,s)> \; . \tag{4.7}$$

The only natural candidate for the field metric on $|\chi(x)>$ is that implied by the scalar product in Fock space, therefore the two-point correlation of $|\xi>$ is

$$E(|\xi(x,s)> <\xi(x',s')|) = 2\delta^{(D)}(x-x')\delta(s-s')1\!\!1 \; . \tag{4.8}$$

The only consequence of the indefiniteness of the action is that $|\xi(x,s)>$ and hence $|\chi(x,s)>$ have to be complex if their probabilistic interpretation is to be retained.

In order to make contact with Section 2 we now consider the massless sector of the <u>closed</u> bosonic string. The relevant components of the string field are

$$|\chi(x)\rangle = \{\ldots + \frac{1}{2}(h_{\mu\nu}(x) + iA_{\mu\nu}(x))a^{\mu\dagger}_{1,L}\, a^{\nu\dagger}_{1,R} + \\ + \frac{i}{\sqrt{2}}\,\eta\, b^{\dagger}_{1,(R}\, \tilde{b}^{\dagger}_{1,L)} + \ldots\}|0\rangle \, . \tag{4.9}$$

The closed-string analog of the action (4.5) is not diagonal with respect to

$$\chi_o \equiv \begin{pmatrix} A_{\mu\nu} \\ h_{\mu\nu} \\ \eta \end{pmatrix} . \tag{4.10}$$

This is a manifestation of the fact that $\mathrm{Tr}(h_{\mu\nu})$ is an independent physical field, the dilaton. By a linear transformation $\chi_o \to \hat{\chi}_o$ the action can be diagonalized and becomes

$$S = -\int d^Dx\, \hat{\chi}_o^T\, \Box\, \Pi\, \hat{\chi}_o \tag{4.11}$$

with

$$\Pi = \begin{pmatrix} P_a^1 & 0 & 0 \\ 0 & P_s^2 - (D-2)P_s^{o\prime} & 0 \\ 0 & 0 & 1 \end{pmatrix} . \tag{4.12}$$

Thus $\hat{h}_{\mu\nu}$ is the usual linearized gravitational field considered in 2 (cf. (2.13)). P_a^1 is a projection operator on the antisymmetric tensor fields $\hat{A}_{\mu\nu}$. Note that Π is not a projector.

Defining the noises ξ_o and $\hat{\xi}_o$ in terms of $|\xi\rangle$ in an analogous manner as χ_o and $\hat{\chi}_o$ were defined in terms of $|\chi\rangle$ one finds for the diagonalized Langevin equation (4.7) in the massless sector

$$\frac{\partial}{\partial s}\hat{\chi}_o = \Box\, \hat{P}\, \hat{\chi}_o + \hat{\xi}_o \tag{4.13}$$

$$\hat{P} = \begin{pmatrix} P_a^{\perp} & 0 & 0 \\ 0 & P_{grav} & 0 \\ 0 & 0 & 1 \end{pmatrix} \tag{4.14}$$

with P_{grav} being given by (2.18) and

$$E(\hat{\xi}_o \; \hat{\xi}_o^T) = \begin{pmatrix} 1\!\!1_a & & \\ & 1\!\!1_s - \frac{1}{D-2}\,\delta\delta & \\ & & 1 \end{pmatrix} 2\delta^{(D)}(x-x')\delta(s-s') \tag{4.15}$$

i.e. exactly the D-dimensional generalization of (2.17) and (2.16).

REFERENCES

1) Parisi, G. and Wu, Yong-Shi, Sci. Sinica 24, 483 (1981).

2) Breit, J.D., Gupta, S., Zaks, A., Nucl. Phys. B233, 61 (1984); Alfaro, J., Nucl. Phys. B253, 464 (1985).

3) Gausterer, H. and Klauder, J.R., Phys. Lett. 164B, 127 (1985).

4) Arnold, L., Stochastic differential equations (Wiley-Interscience, New York, 1974).

5) DeWitt, B.S., in General Relativity, edited by S.W. Hawking and W. Israel (Cambridge University Press, 1979).

6) van Nieuwenhuizen, P., Nucl. Phys. B60, 478 (1973).

7) Bengtsson, I. and Hüffel, H., preprint CERN-TH.4390-86.

8) Rumpf, H., Phys. Rev. D33, 942 (1986).

9) Hüffel, H. and Rumpf, H., Z. Phys. C 29, 319 (1985).

10) Damgaard, P. and Hüffel, H., Stochastic quantization, to appear in Phys. Rep. C.

11) Gibbons, G.W., Hawking, S.W., Perry, M.J., Nucl. Phys. B138, 141 (1978).

12) Hüffel, H. and Rumpf, H., Phys. Lett. 148B, 104 (1984).

13) Rumpf, H. and Urbantke, H.K., Ann. Phys. (N.Y.) 114, 332 (1978).

14) Rumpf, H., Phys. Rev. D28, 2946 (1983) and references cited therein.

15) Parisi, G., Phys. Lett. 131B, 393 (1983).

16) Siegel, W. and Zwiebach, B., Nucl. Phys. B263, 105 (1986).

TOWARDS A FOUR DIMENSIONAL YANG-MILLS THEORY

Paul Federbush
Department of Mathematics
The University of Michigan
Ann Arbor, Michigan 48109, USA

ABSTRACT

Work is in progress to construct a pure Yang-Mills field in a finite Euclidean volume. One works with a functional integral of continuum fields, but described by a discrete set of group-valued lattice variables living on an infinite sequence of finer and finer lattices. There is a block spin transformation that carries configurations on one lattice to configurations on the next finest lattice. By appropriately modifying the original spin transformation due to Balaban, the "large field" problem has been partially declawed.

The formalism is that of the Phase Cell Cluster Expansion. A convenient set of group-valued variables is defined. One develops a cluster expansion for any expectation value in terms of "connected" contributions of larger and larger finite subsets of these variables. Formally the procedure is remarkably close to the corresponding treatment of the ϕ^4 theory. There are only a finite number of renormalization cancellations to exhibit.

As an introduction to the philosophy of our approach we start with a hasty survey of the corresponding treatment of the ϕ_3^4 theory[5], 2], 16]]. One expands $\phi(x)$ as a discrete sum

$$\phi(x) = \sum_k \alpha_k u_k(x) \tag{1}$$

where the $\{u_k(x)\}$ are an appropriate set of expansion functions. Different choices of the $\{u_k\}$ were made in the study of a Heierachical model by Battle and Federbush[2]] and of the actual theory by Williamson[16]]. At present I would prefer as expansion functions an analog of the choice made by Gawedzki and Kupiainen in their study of lattice $(\nabla\phi)^4$, 12]. In each case the renormalization group is deeply involved in the construction of such $u_k(x)$, functions that live on different scales and provide the "phase space localization" first used by Glimm and Jaffe in their treatment of ϕ_3^4, 13]. This is basically the only use made of the renormalization group in our work.

We then have the α_k given as

$$\alpha_k = \int \phi(x) v_k(x) \tag{2}$$

where the $\{v_k\}$ are a dual basis to the $\{u_k\}$. We describe the continuum field $\phi(x)$ in terms of the discrete set of variables, α_k

$$\phi(x) \xleftrightarrow{1-1} \{\alpha_k\} \tag{3}$$

and the functional integral formally becomes

$$\int d\phi = \prod_k \int d\alpha_k \tag{4}$$

We are describing the theory in terms of variables, the $\{\alpha_k\}$, that are <u>linearly</u> related to the field, by equation (2). We may say that the ϕ^4 theory is a "linear theory". By contrast the Yang-Mills theory will be "an essentially non-linear theory."

To any finite subset of the variables, A, we associate a partial action, S^A, extracted from the formal total action S

$$S = \frac{1}{2} \int (\nabla\phi)^2 + \lambda \int \phi^4 + \delta M^2 \int \phi^2 + \ldots \tag{5}$$

For example, the term $\lambda\int\phi^4$ gives rise to a term

$$\lambda\int \Big(\sum_{k\in A} \alpha_k u_k(x)\Big)^4 \tag{6}$$

in S^A. There are likewise counterterm contributions to S^A. One may then calculate an expectation of an observable p, due to variables in A

$$\langle p \rangle_A = \frac{1}{N_A} \int \Big(\prod_{k\in A} d\alpha_k \Big) e^{-S^A} p \tag{7}$$

One has the prototype convergence theorem

$$<p> = \lim_{A\uparrow\Omega} <p>_A \tag{8}$$

where the limit is taken over any sequence of expanding subsets eventually exhausting all variables. The limit in (8) treats on one footing, the infinite volume limit and the ultraviolet limit. There also is a Glimm-Jaffe-Spencer type cluster expansion[14] for $<p>$

$$<p> = \sum_A K_A(p)\ Z^{A^c}/Z \tag{9}$$

where $K_A(p)$ are the "connected" contributions to the expectation due to variables in A. It is better to write (9) as a polymer expansion, see the review article of Brydges[3] for example, and then the expression for $<p>$ is given <u>with no cutoffs</u>.

There are length scales $\ell_r = 1/2^r$; $r = 0,1,2,\ldots$. Each $u_k(x)$ lives on some length scale $1/2^{r(k)}$. We let $\phi_s(x)$ be as given by the sum in (1), but with the sum restricted to terms, k's, whose length scales $1/2^{r(k)}$ are $\geq 1/2^s$. One has

$$\phi_0 \xleftarrow{BT_1} \phi_1 \xleftarrow{BT_2} \phi_2 \longleftarrow \ldots \phi(x) \tag{10}$$

BT_i , the ith block spin transformation, merely throws away the terms whose length scale is $1/2^i$. We may formally identify $\phi(x)$ with the infinite sequence of functions $\{\phi_i(x)\}$ and mappings BT_i . Each $\phi(x)$ determines such a sequence, each sequence a $\phi(x)$. For reasonable choices of the $\{u_k\}$ one has the precise mathematical statement: <u>There is an injection of the set of continuously differentiable functions</u>, $\{\phi(x)\}$, <u>into the inverse limit</u>[4] (or projective limit) <u>of the set of spaces</u> $\{\phi_i(x)\}$, for each i , and mappings BT_i .

Our phase cell cluster expansion is similar in spirit to the formalism of Magnen and Seneor[15] .

In the Yang-Mills situation we consider an infinite sequence of lattices, L^r , with the edge size in L^r, $1/2^r$; $r = 0,1,2,\ldots$. A Yang-Mills configuration is an assignment to each oriented edge of each lattice of an element of a compact Lie group. The assignments are not totally free, the assignments to L^r determine the assignments to L^{r-1} by an averaging procedure (block spin transformation) due to Balaban[1] . We let F^r be the assignments to L^r . We again have a sequence of mappings.

$$F^0 \xleftarrow{BT_1} F^1 \xleftarrow{BT_2} F^2 \xleftarrow{BT_3} \cdots A_\mu(x) \tag{11}$$

We will say the assignments F^r converge to $A_\mu(x)$ as $r \to \infty$ (the lattice fields converge to the continuum field) if for each edge

$$\left| g(e) - \exp\left(\int_e \vec{A}\cdot d\vec{x}\right) \right| < c\left(\ell(e)\right)^{1+\varepsilon} \tag{12}$$

where $\ell(e)$ is the length of edge e, g(e) the group assignment to edge e , and the bars represent any reasonable norm.

We have proven the following results in the above formalism:

Theorem[6] and 11]. Given a continuously differentiable gauge potential, there is one and only one set of compatible assignments $\{F^r\}_r$ that converges to it.

That is, there is an injection of the space of continuously differentiable gauge potentials into the inverse limit of the system in (11). This provides a duality between the view of the Yang-Mills field as a continuum field, and as a sequence of compatible lattice theories.

Theorem[6] and 11]. (Under mild conditions on the spatial falloff, which we do not here detail) the Wilson actions S^r , of the configuration F^r , converges to the continuum action $1/2 \int (dA)^2$ as $r \to \infty$, for a situation where the $\{F^r\}$ converge to $A_\mu(x)$, a continuously differentiable field.

Theorem[6]. The Small Field Stability Theorem. If we let S^r_Q be the Wilson action computed to quadratic order (in deviation of group element assignments from the identity element) then

$$S^r_Q \geq S^s_Q \ , \quad \text{if} \quad r \geq s \ . \tag{13}$$

Given a continuously differentiable gauge potential, the group elements associated to the edges of the lattices, L^r , are a discrete set of variables, group-valued variables, that (uniquely) describe the field. The variables that are useful to describe the field are related to the field in a very non-linear way, the essential non-linearity of the Yang-Mills field. A plaquette variable $g_{\partial p}$ associated to a plaquette in every L^r may be used to construct a good gauge invariant observable associated to the Yang-Mills field such as $\mathrm{Tr}\left(u(g_{\partial p})\right)$, for example, with u a unitary representation. Expectations of polynomials in such variables may be expected to be well defined, finite, and not generally zero. Actually the variables we use, and decouple in the development of the cluster expansion, are not the group elements associated to the edges, but other group elements associated to these. The construction of the Yang-Mills field in terms of these variables, is to "linear order" as in the ϕ^4 case, but beyond linear order involving a rich world of interesting complexity to take advantage of gauge invariance and the group nature of the variables[9].

We now turn to a discussion of the action of the Yang-Mills theory. From the work of Gross, Politzer,'t Hooft, and Wilczek, we know that the theory is <u>asymptotically free</u>, with a running coupling constant, g_r^2, the effective coupling on L^r, given by

$$\frac{1}{g_r^2} = Ar + B\ \ln(r+1) + C \tag{14}$$

for certain positive constants A and B. The only renormalization cancellations to be exhibited in the cluster expansion are of the running coupling constant. There are <u>only a finite number of "diagrams" to cancel</u>, the expansion is <u>effectively as of a superrenormalizable</u> theory. We take the formal unrenormalized action, S, as the Wilson Action on the "finest lattice, L^∞"

$$\frac{1}{g_\infty^2} S^\infty \tag{15}$$

analagously to (5), and rewrite this via a telescopic sum as

$$\frac{1}{g_0^2} S^0 + \sum_{r=1} \left(\frac{1}{g_r^2} S^r - \frac{1}{g_{r-1}^2} S^{r-1}\right) \tag{16}$$

and then split this into two parts

$$S = S_0 + S_R \tag{17}$$

$$S_0 = \frac{1}{g_0^2} S^0 + \sum_{r=1} \frac{1}{g_r^2} (S^r - S^{r-1}) \tag{18}$$

$$S_R = \sum_{r=1} \left(\frac{1}{g_r^2} - \frac{1}{g_{r-1}^2}\right) S^{r-1} \tag{19}$$

S_0 is the "renormalized action", on "nice" continuum $A_\mu(x)$ it takes a finite value. S_R is the set of "counterterms".

The calculation of expectation values involves integration of the negative exponential of the action over the function space of continuum $A_\mu(x)$. The functional integral becomes an infinite product of integrals over the group manifold, of variables describing the field, with respect to Haar measure. This is much the same as in the work of Balaban. In a general way, the work of Balaban is in a formalism analogous to that of Kupiainen and Gawedzki, and that of our work, in a formalism analogous to that of Magnen and Seneor. In detail, the two treatments differ beyond this in two significant features.

Firstly, the work of Balaban does not involve a choice of variables, to decouple, as in our work, the analog of (1) for the ϕ^4 theory. We are looking at local "excitations" of the field, on all scales at once, in terms of such variables. He instead is seeking the effective modifications of the Wilson action at each level, after smaller scale excitations are integrated out.

Secondly, we have modified the block spin transform of Balaban, at "large field strength", to effectively obtain a stability in the large field region, analogous to (13)[8]. The new block spin transformation is very complicated, but provides a positive contribution to the action S_0 for every large field excitation at every scale. We feel the construction of this block spin transformation is a most significant contribution to the study of the ultraviolet Yang-Mills theory in four dimensions. We feel Balaban will also find this modified block spin transformation of value in his program. (The most aesthetic way to deal with the horrendously complicated modified block spin transformation may be to axiomatize a set of essential properties it satisfies, and work with these properties only --- leaving to a position largely outside the line of computation, the proof, in[8], that some block spin satisfying these properties exists.)

As a final note we will try to give a flavor for the form of stability results that are important in our program. In a pure small field region, the individual excitation, m, has an amplitude that is described by a group element $g(m)$, or Lie Algebra element $A(m)$. We assume that m "lives" at level $r(m)$. We then have proved[10] that one has an estimate

$$S_0 \geq c \sum_m \frac{1}{g^2_{r(m)}} |A(m)|^2 \tag{20}$$

In a general field configuration a similar inequality will be true with contributions to the sum in addition due to "large field" contributions to the action. The modified block spin transformation mentioned above is essential to this result. Work is now in progress proving this stability result for the general field configuration.

REFERENCES

1. Balaban,T., Propagators and Renormalization Transformations for Lattice Gauge Theories, I, Commun. Math. Phys. 95, 17-40 (1984).

2. Battle, G. and Federbush, P., A Phase Cell Cluster Expansion for a Hierarchical ϕ^4_3 Model, Commun. Math. Phys. 88, 263-293 (1983).

   Battle, G. and Federbush, P., A Phase Cell Cluster Expansion for Euclidean Field Theories, Annals of Physics. 142, 95-139, (1982).

3. Brydges, D., A Short Course on Cluster Expansions. In: Les Houches Summer School Notes, 1984. Osterwalder, K. (Ed.)

4. Eilenberg, S. and Steenrod, N., Foundations of Algebraic Topology, Princeton University Press, 1952.

5-11. Federbush, P., A Phase Cell Approach to Yang-Mills Theory, [0]-[VI], Michigan preprints.

12. Gawedzki, K. and Kupiainen, A., A Rigorous Block Spin Approach to Massless Lattice Theories, Commun. Math. Phys. 77, 31-64 (1980).

13. Glimm, J. and Jaffe, A., Positivity of the ϕ_3^4 Hamiltonian, Fort. Phys. 21, 327-376, (1973).

14. Glimm, J. and Jaffe, A., Quantum Physics, Springer-Verlag, New York, (1981).

15. Magnen, J. and Seneor, R., The Infrared Behavior of $(\nabla\phi)_3^4$, Ann. Phys. 152 (1984).

16. Williamson, C., A Phase Cell Cluster Expansion for ϕ_3^4, Missouri preprint.

NON-TRIVIAL RENORMALIZATION GROUP FIXED POINTS

Giovanni Felder

I.H.E.S.
35, Route de Chartres
91440 Bures-sur-Yvette, France

ABSTRACT

After recalling the general wisdom about non-trivial renormalization group fixed points, we review what is rigorously known in this subject. We give some details on the Gallavotti-Nicolò expansion applied to this problem and we report on recent results on the existence of non-trivial fixed points in $2 < d < 4$ dimensions for an approximate renormalization group flow equation.

1. INTRODUCTION.

The most striking results obtained by the renormalization group (RG) theory[1)] are based on the analysis of the RG transformation in the vicinity of its fixed points. In high energy physics one is mostly concerned with trivial (Gaussian) fixed points and their vicinity. In this case, the RG applied to non abelian gauge theories correctly predicts asymptotic freedom, a phenomenon observed experimentally. Non-trivial fixed points become important in statistical mechanics : in fact, as was discovered by Wilson, the behaviour of materials approaching the critical point can be understood in terms of the behaviour of the RG transformation in the vicinity of a (usually non-trivial) fixed point. Moreover, techniques (ε, $\frac{1}{N}$ expansions) have been developed to compute critical exponents.

Non-trivial fixed points in quantum field theory have been less studied, and no application to physical systems has been proven useful. Nevertheless, non-trivial ultraviolet stable fixed points can provide a way to give sense to quantum field models which are perturbatively non-renormalizable[2)] and are thus usually not considered by high energy physics model builders.

But let us recall in the case of a symmetric scalar field theory (or a classical statistical mechanics system with one-component order parameter) what are the expected fixed points of the RG transformation. We plot in fig.1 the dimension of space(-time) d against the (infinite dimensional) space of hamiltonians. For all values of d there is first of all the Gaussian fixed point G. At the thresholds $d_n = 2+2/n$, $n = 1,2,\ldots$, the linearized RG tranformation has non-trivial kernel and a bifurcation occurs. One bifurcation branch at each threshold (pictorially indicated here by a dotted line) is thought to be unphysical being not stable. The fixed point responsible for critical phenomena is denoted by W in fig. 1 and is found by extrapolating

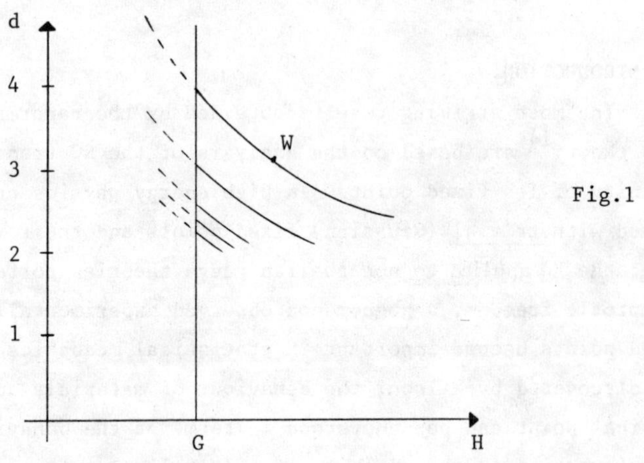

Fig.1

the bifurcation branch form $d = 4$. Its properties can be approximately computed in $\varepsilon = 4-d$ expansion.

From the rigorous point of view, many interesting results have been achieved in the last years concerning the vicinity of trivial fixed points, including the construction of renormalizable asymptotically free models[3]. Non-trivial fixed points are much less understood mathematically. Results have been obtained however over the past decade in toy models. We intend to review some of these results and give some details on the author's contribution in this talk.

2. RIGOROUS RESULTS ON NON-TRIVIAL FIXED POINTS.

The hierarchical model is the first toy model where rigorous results were obtained. It was first introduced by Dyson[4)5)] as a one-dimensional spin system with a long range interaction with hierarchical structure, to show that one-dimensional systems can undergo phase transitions at positive temperature. The same model can be also interpreted as a continuous spin system in d dimensions given by a local perturbation of a Gaussian whose covariance looks at long distances like $(-\Delta)^{-1}_{xy}$ but has a hierarchical structure. This hierarchical property ensures that under a block spin transformation the interaction remains local and can thus be described by a potential $V(\varphi)$, a function of one variable.

In terms of V the RG transformation is

$$R_L = V(\varphi) \mapsto \tilde{V}(\varphi) = -\ln\int \exp[-L^d V(L^{1-d/2}\varphi+z)-z^2/2(L-1)]\frac{dz}{\sqrt{2\pi(L-1)}} \qquad (1)$$

(This is Gallavotti's version of the hierarchical model, somewhat different from the original Dyson's model). The size L of the block is a fixed integer strictly larger than 1, and the (here irrelevant) factors $(L-1)$ are present for later purposes. For the hierarchical model, Bleher and Sinai[6)] proved the existence of a non-trivial fixed point in $d = 4-\varepsilon$ dimensions, for small ε (and also if $d = d_n-\varepsilon$ where $d_n = 2+2/n$ are the thresholds described in the introduction). Their proof was further simplified by Collet and Eckmann[5)], which showed that for the hierarchical model the ε-expansion is asymptotic. A further result on the hierarchical model was obtained by Gawędzki and Kupiainen[7)]; they considered the N-component model ($\varphi \in \mathbb{R}^N$ in (1)) and proved the existence of a non-trivial fixed point in three dimensions provided N is large enough.

Another class of models where results on non-trivial fixed points were obtained, is the class of models without instanton n factorials, i.e. models where perturbation theory in the bare coupling constant (in the presence of cut-offs) has a finite radius of convergence. Examples of such models are planar field theories[8], which are formal $N \to \infty$ limits of models where the field is an $N \times N$ matrix, and whose perturbation series is given by Feynman graphs that can be drawn on a sphere without intersections; and fermionic field theories like the Gross-Neveu model[9]. These models are not trivial to handle, since renormalon $n!$ are present, and renormalized perturbation theory is (presumably) not convergent. See V. Rivasseau's contribution to these proceedings on these issues. Non-trivial fixed points were constructed for the planar scalar field theory in "$4+\varepsilon$ dimensions"[10] and for the Gross-Neveu model in "$2+\varepsilon$ dimensions"[11]. The quotation marks indicate that the models are actually defined in 4 (respectively 2) dimensions, but the covariance is $1/p^{2-\varepsilon/2}$ in momentum space (rather than $1/p^2$) mimicking the power counting of $4+\varepsilon$ $(2+\varepsilon)$ dimensions. As an application, these fixed points provided a way to give a meaning to the perturbatively non-renormalizable theories planar $-\lambda\phi^4_{4+\varepsilon}$ and $GN_{2+\varepsilon}$. The stability problem of the wrong sign of the coupling constant does not arise here since the bare theory is well defined in perturbation theory for positive as well as for negative coupling constant.

We now give a rough idea on how the fixed point for planar ϕ^4 was constructed. For detailed proofs see[10]. See also A. Kupiainen's contribution to these proceedings for a discussion of the GN model. The construction is based on the Gallavotti-Nicolò expansion (See[12] for a review) which we briefly explain in this context. The linearized RG transformation has three eigenfunctions which are relevant (i.e. corresponding to eigenvalues ≥ 1) namely $\int : (\nabla\varphi)^2 : dx$, $\int : \varphi^2 : dx$, $\int : \varphi^4 : dx$. Let P be a projector onto this three dimensional space, and let R_L denote the RG transformation. Define effective potentials $V^{(k)}$ on scale L^k by iterating the RG transformation starting from a bare potential $V^{(N)}$:

$$V^{(k)} = R_L V^{(k+1)} \quad , \quad k \leq N-1 \quad . \tag{2}$$

The Gallavotti-Nicolò expansion is given by the following iterative manipulation

$$\begin{aligned}V^{(k)} &= PV^{(k)} + (1-P)R_L V^{(k+1)} \\ &= PV^{(k)} + (1-P)R_L(PV^{(k+1)}+(1-P)R_L V^{(k+2)}) \\ &= \ldots \\ &= \text{a function of } (PV^{(k)},PV^{(k+1)},PV^{(k+2)},\ldots,PV^{(N)}) \quad .\end{aligned} \tag{3}$$

i.e., the irrelevant parts are recursively expressed in terms of relevant parts on higher scales. Similarly one can express $PV^{(k)}$ in terms of the relevant parts on higher scales :

$$PV^{(k)} = PV^{(k+1)} - \beta_k(PV^{(k+1)},PV^{(k+2)},\ldots,PV^{(N)}) \tag{4}$$

The functional β_k is the GN beta functional. Note that $PV^{(j)}$ is an element of a three-dimensional space and can be parametrized by the coefficients of the relevant eigenfunctions, the <u>running coupling constants</u> (r.c.c.). The point of (3) and (4) is that their perturbative expressions (i.e., the expansion of the r.h.s. in powers of all r.c.c.) are free of renormalons and in the planar case are <u>convergent</u> with convergent radius uniform in N and $|\varepsilon| \leq \varepsilon_o$. The general strategy of the GN method is to first find a solution to (4), then insert this solution in (3) to get all effective potentials. In other words the problem is reduced to a finite dimensional dynamical system (with infinite memory) (4). A naïve way of finding a solution to (4) is to solve it in a power series in the "renormalized coupling constants", the r.c.c. on some fixed low energy scale L^{k_o} . This produces the usual renormalized perturbation theory if $\varepsilon = 0$ (the Taylor operators are here given by (1-P) in (3)), and, if $\varepsilon > 0$, the renormalized perturbation theory (with finite number of counterterms) of a non-renormalizable model, which is well known to diverge as $N \to \infty$ (See[13]) for a discussion of this point).

But there is another method for solving (4) and we are back to the problem of existence of a fixed point : one can make the fixed point ansatz $PV^{(k)} = PV^*$ (all k) and solve

$$\overline{\beta}(PV^*) \equiv \beta_k(PV^*,\ldots,PV^*,\ldots) = 0 \quad . \tag{5}$$

Now $\overline{\beta}$ is a function of three variables which is analytic in a polydisc uniformly in $|\varepsilon| < \varepsilon_o$. The usual bifurcation picture then works and one can show that a non-trivial solution of (5) exists and is in the analycity domain of $\overline{\beta}$ if ε is small enough. Inserting this solution in (3) gives the non-trivial fixed point $V^{(k)} = V^*$ (all k). One can also expand the power series (3), (4) around the non-trivial fixed point. The theory looks then renormalizable and one can proceed to its construction.

We now discuss some more recent developments in hierarchical-type models, concerning the existence of non-trivial RG fixed points when no small parameter (like ε or $1/N$) is present. In a recent study[14), Koch and Wittwer considered the hierarchical model (1) with $L = 2$ in $d = 3$ dimensions and rigorously proved the existence of a non-trivial fixed point with the help of a computer (see H. Koch's notes in these proceedings). Hasenfratz and Hasenfratz[15)] also studied an approximate RG transformation but with continuous scale parameter. They took the local projection of the Wegner-Houghton[16)] RG flow equation (this projection was also considered years ago by Nicoll, Chang and Stanley[17)], who were however only concerned with the perturbative ε-expansion) and got a P.D.E. in two variables, the scale parameter and the field variable. The problem of finding a non-trivial fixed point reduces to finding a non-trivial global solution of an ordinary differential equation. The existence of such a solution is numerically demonstrated in[14)] for $d = 3$.

There exists another renormalization group scheme, besides Wegner-Houghton's, namely Wilson's "exact renormalization group equations"[18)]. The differential equation for the effective hamiltonian on scale e^{-t} expressed as function of the Fourier components of the field variable is

$$\frac{\partial H}{\partial t} = \int_q \left(\frac{d}{2} + q\nabla_q\right)\varphi_q \frac{\delta H}{\delta\varphi_q} + \int_q h(q^2)\left[\varphi_q \frac{\delta H}{\delta\varphi_q} + \frac{\delta^2 H}{\delta\varphi_q \delta\varphi_{-q}} - \frac{\delta H \delta H}{\delta\varphi_q \delta\varphi_{-q}}\right] ,$$

$$h(q^2) = 1 - \frac{1}{2}\eta + 2q^2 \tag{6}$$

(the convention is $\int_q = \int \frac{d^d q}{(2\pi)^d}$, $\int_q \delta_{q-q_o} f(q) = f(q_o)$, $\frac{\delta}{\delta\varphi_q}\varphi_p = \delta_{q-p}$).

We can then write H = Gaussian fixed point +V and, as in Hasenfratz-Hazenfratz, project the r.h.s. onto its local component*). The effective potential V on scale e^{-t} can then be represented by a function of one field variable which obeys the P.D.E. (η is set to zero since wave function renormalization effects disappear in this approximation)

$$\frac{\partial V}{\partial t} = \frac{1}{2}\frac{\partial^2 V}{\partial \varphi^2} - \frac{d-2}{2}\varphi\frac{\partial V}{\partial \varphi} + dV - \frac{1}{2}\left(\frac{\partial V}{\partial \varphi}\right)^2 \tag{7}$$

This equation was also considered by Brydges and Kennedy[19] in the context of the 2-d Coulomb gas. It is interesting to note the (7) can also be obtained as a limit of the hierarchical model (1) as $L \to 1$ (the normalization of the fluctuation covariance $L-1$ was introduced in order for this limit to exist). For this model one can establish some rigorous results, among which <u>the existence of a non-trivial fixed point in</u> $2 < d < 4$ <u>dimensions</u>. We give here only a rough idea of the proof. Details will appear elsewhere[20]. It can be shown that fixed points are <u>global</u> stationary solutions of (7). Two fixed points are obvious, namely the Gaussian massless one, $V = 0$, and the high temperature one, $V = \varphi^2 - \frac{1}{d}$. To search other fixed points we consider the equation for the derivative $\frac{\partial V}{\partial \varphi}$ of the potential which after a trivial rescaling of variables becomes

$$v'' - \frac{x}{\lambda}v' + v - 2vv' = 0 \quad , \quad \lambda = \frac{d+2}{d-2} \tag{8}$$

We look at this equation as a dynamical system in the $(v, w = v')$-phase plane, and look for initial conditions $w(0)$ ($v(0) = 0$ for even potentials) which are such that a global solution exists. To this purpose we divide the phase plane into 6 zones (fig.2) and study the behaviour of the vector field at each $x > 0$ on their boundary. It can be shown that all solutions leave any zone except IV in finite "time" x and that all solutions which get in zone II blow up in finite "time". It is easy to show that there exists a solution w_1 which does blow up and a

*) The local projector P acts on translation invariant monomials as
$P \int_{q_1 \cdots q_n} V(q_1 \cdots q_{n-1})\varphi_{q_1} \cdots \varphi_{q_n}\delta_{q_1+\cdots+q_n} =$
$V(0,\ldots,0) \int_{q_1 \cdots q_n} \varphi_{q_1} \cdots \varphi_{q_n}\delta_{q_1+\cdots+q_n}$.

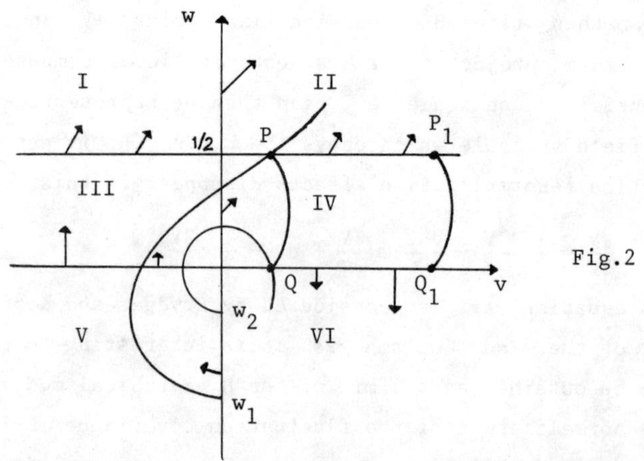

Fig.2

solution w_2 which goes once around the origin to zone VI. This last step is based on linear analysis ($w_2(0)$ is small) and requires $2 < d < 4$. One then constructs a closed interval of initial conditions $[w_1^o, w_2^o]$ which gets mapped after a time x_o into a curve PQ in IV as in fig.2. After a later time x_1 this curve is mapped to a curve P'Q' with $P' \in II$, $Q' \in VI$ but a subset of it is mapped to a curve P_1Q_1 lying in IV. Following this method ("Bleher Sinai argument") one constructs smaller and smaller intervals $[w_1^n, w_2^n]$ of initial conditions which get mapped after times $x_n \to \infty$ to the region IV. Their intersection gives a global solution.

REFERENCES

1. Wilson, K., Kogut J., Phys. Rep. 12, 75 (1974).
   Domb, C., Green, M. (ed.), "Phase transition and critical phenomena", Vol. VI, Academic Press.
2. Parisi, G., Nucl. Phys. B 100, 368 (1975) and in "New developments in quantum field theory and statistical mechanics", Cargèse 1976).
3. Feldman, J., Magnen, J., Rivasseau, V., Sénéor, R., Commun. Math. Phys. 103, 67 (1986).
   Gawędzki, K., Kupiainen, A., Commun. Math. Phys. 102, 1 (1985).
4. Dyson, F.J., Commun. Math. Phys. 12, 91 (1969) and 21, 269 (1971).
5. Collet, P., Eckmann, J.-P., Lecture Notes in Phys. 74 (1978).

6. Bleher, P.M., Sinai, Ya.G., Commun. Math. Phys. 33, 23 (1973) and 45, 374 (1975).

7. Gawędzki, K., Kupiainen, A., Commun. Math. Phys. 89, 191 (1983).

8. 't Hooft, G., in "Progress in gauge field theories" (Cargèse 1983).

9. Gross, D., Neveu, A., Phys. Rev. D 10, 3235 (1974).

10. Felder, G., Commun. Math. Phys. 102, 139 (1985).

11. Gawędzki, K., Kupiainen, A., Nucl. Phys. B 262, 33 (1985).

12. Gallavotti, G., Rev. Mod. Phys. 57, 471 (1985).

13. Felder, G., Gallavotti, G., Commun. Math. Phys. 102, 549 (1986).

14. Koch, H., Wittwer, P., Rutgers University preprint.

15. Hasenfratz, P., Hasenfratz, P., Berne University preprint BUTP-85/26.

16. Wegner, F.J., Houghton, A., Phys. Rev. A 8, 401 (1972).

17. Nicoll, J.F., Chang, T.S., Stanley, H.E., Phys. Rev. Lett. 33, 540; E 33, 1525 (1974).

18. Chapter 11 in Wilson, Kogut (Ref.1); Wegner, F. in Domb, Green (Ref.1).

19. Brydges, D., Kennedy, T., Mayer expansion and Burger's equation, preprint, 1986.

20. Felder, G., in preparation.

RENORMALIZABILITY OF QED_4

J. Feldman, T. Hurd, L. Rosen, J. Wright
Mathematics Department
University of British Columbia
Vancouver, B.C., Canada

ABSTRACT

We outline a simple and complete proof of the perturbative renormalizability of quantum electrodynamics in 4 dimensions, based on the tree expansion of Gallavotti and Nicolò. The renormalization is implemented by guage invariant counterterms in the Lagrangian, but the subtractions must be carried out in a gauge variant way.

In this paper, we outline a simple and complete proof of the (ultraviolet) renormalizability of quantum electrodynamics in 4 dimensions (QED_4) in the context of perturbation theory. Our proof is based on the powerful tree expansion introduced by Gallavotti and Nicolò[1,2] for renormalizing scalar field theories. The question that we asked ourselves was this: Can the GN expansion be successfully applied to a gauge field theory? The difficulty is that there is a basic conflict between the GN procedure and the requirement of gauge invariance; namely, the scale decompositions of the GN renormalization procedure break gauge invariance and seem to necessitate gauge variant counterterms. (For that matter, the decomposition of graphs in any renormalization procedure based on power counting seems to produce the same difficulty.) We now explain how to resolve this apparent conflict (for full details, see Ref. 3).

In terms of the Euclidian fields ψ, $\overline{\psi}$, A_μ where ψ and $\overline{\psi}$ are independent fermi fields and A_μ a vector boson field, the Lagrangian density for QED_4 is given by

$$L = L_p + L_f + L_{int}$$

$$= \left[\frac{1}{4} F^2 + \frac{1}{2} m_0^2 A^2 + \frac{1}{2} (\partial\cdot A)^2\right] + \overline{\psi} (-i\partial\!\!\!/ + m)\psi + e\overline{\psi}A\!\!\!/\psi . \qquad (1)$$

Here (repeated indices are summed over)

$$F^2 = F_{\mu\nu}F_{\mu\nu}, \quad F_{\mu\nu} = \partial_\mu A_\nu - \partial_\nu A_\mu , \quad A^2 = A_\mu A_\mu$$

$$\partial\cdot A = \partial_\mu A_\mu, \quad A\!\!\!/ = A_\mu \gamma^\mu, \quad \partial\!\!\!/ = \gamma^\mu \partial_\mu ,$$

the 4×4 Euclidean Dirac matrices satisfy

$$\gamma^\mu \gamma^\nu + \gamma^\nu \gamma^\mu = -2\delta^{\mu\nu} ,$$

$m > 0$ is the bare electron mass, and e the bare electron charge. Note that we have imposed an infrared cutoff by the inclusion of a photon mass term $\frac{1}{2} m_0^2 A^2$, $m_0 > 0$, and that we are working in the Feynman gauge by the inclusion of the term $\frac{1}{2} (\partial\cdot A)^2$. The resulting free photon measure, $d\mu = \text{const.}\ e^{-\int L_p dx} DA$, has covariance or propagator

$$D_{\mu\nu}(x,y) = \int A_\mu(x)A_\nu(y)d\mu = \frac{\delta_{\mu\nu}}{(2\pi)^4} \int \frac{e^{ik(x-y)}}{k^2 + m_0^2} dk.$$

The free fermi "measure", $d\nu = \text{const.}\ e^{-\int L_f dx} D\psi D\overline{\psi}$, has propagator (we suppress spinor indices)

$$S(x,y) = \int \psi(x)\overline{\psi}(y)d\nu = \frac{1}{(2\pi)^4} \int (k\!\!\!/+m)^{-1} e^{ik(x-y)} dk.$$

The ultraviolet (UV) singularities of the theory occur as singularities of the propagators at coinciding arguments: for $x-y \approx 0$,

$$D(x,y) \sim \frac{1}{|x-y|^2} , \qquad S(x,y) \sim \frac{1}{|x-y|^3} .$$

We resolve these UV singularities by decomposing each field $\Phi_i = \psi,\overline{\psi},A$ into a sum of independent fields $\Phi_i = \sum_{h=0}^{\infty} \Phi_i^{(h)}$ where $\Phi_i^{(h)}$ has length scale M^{-h}, $M > 1$ being a fixed scale parameter. More precisely,

$$D^{(h)}(x,y) = \int A^{(h)}(x)A^{(h)}(y)d\mu \quad \text{and} \quad S^{(h)}(x,y) = \int \psi^{(h)}(x)\overline{\psi}^{(h)}(y)d\nu$$

satisfy

$$|D^{(h)}(x,y)| \leq \text{const.}\ M^{2h} e^{-M^h|x-y|} \quad \text{and}$$
$$|S^{(h)}(x,y)| \leq \text{const.}\ M^{3h} e^{-M^h|x-y|}. \tag{2}$$

The UV-regularized fields are $\Phi_i^{(\leq N)} = \sum_{h=0}^{N} \Phi_i^{(h)}$. We also write

$\Phi_i^{(k,N]} = \sum_{h=k+1}^{N} \Phi_i^{(h)}$, $C_i^{(h)} = D^{(h)}$ or $S^{(h)}$, $C_i^{(<N)} = \sum_{h=0}^{N-1} C_i^{(h)}$, etc.

Formally, the expectation of a function $F(\Phi)$ of the fields is given by

$$\langle F\rangle = \frac{1}{Z} \int F(\Phi)\, e^{V(\Phi)} dP(\Phi) \tag{3}$$

where $dP(\Phi) = d\mu(A)d\nu(\psi,\overline{\psi})$, $V(\Phi) = -\int L_{int} dx = -e \int :\overline{\psi}\, A\!\!\!/\, \psi: dx$, and $Z = \int e^{V(\Phi)} dP(\Phi)$. To make sense of (3) we regularize by replacing Φ by $\Phi^{(\leq N)}$ and we renormalize V with counterterms δV^N of the same form as terms in L: $V \to V_N^N = V + \delta V^N$. The <u>effective potential at scale $k < N$</u> is defined by integrating out the fields $\Phi^{(k,N]}$:

$$V_k^N\left(\Phi^{(\leq k)}\right) \equiv \log \int e^{V_N^N(\Phi^{(\leq N)})} dP\left(\Phi^{(k,N]}\right) + \text{const.} \tag{4}$$

where const. is chosen to cancel the terms independent of $\Phi^{(\leq k)}$ from the first term (formally these terms are infinite but this divergence is easily cured by a volume cutoff). We say that the model is <u>(UV) renormalizable</u> if for each fixed k,

$$V_k\left(\Phi^{(\leq k)}\right) = \lim_{N\to\infty} V_k^N\left(\Phi^{(\leq k)}\right) \tag{5}$$

exists (order by order in perturbation theory).

We now indicate briefly the structure of the Gallavotti-Nicolò tree expansion for V_k^N. Imagine that in (4) we integrate out the fields $\Phi^{(N)}, \Phi^{(N-1)}, \ldots, \Phi^{(k+1)}$ in succession. Integrating out $\Phi^{(N)}$ and Wick ordering the resulting terms gives V_{N-1}^N as a sum over graphs G, $V_{N-1}^N = \sum_G V_{N-1}^N(G)$, where

i) each vertex of a graph G corresponds to a term in V_N^N ;

ii) each line of G corresponds to a propagator which may be "hard" $\left(C^{(N)}\right)$ if it arises from the integration or "soft" $\left(C^{(<N)}\right)$ if it arises from the Wick ordering ;

iii) each external leg of G corresponds to an external field $\Phi_i^{(\leq N-1)}$;

iv) the graphs are connected with respect to the hard lines;

v) the value $V_{N-1}^N(G)$ of the graph G is the product of the propagators corresponding to the lines, the Wick monomial in the external fields $\Phi_i^{(\leq N-1)}$ and an appropriate combinatoric factor.

If we next integrate out $\Phi^{(N-1)}$, $V_{N-2}^N\left(\Phi^{(\leq N-2)}\right) = \log \int e^{\sum_G V_{N-1}^N(G)} dP\left(\Phi^{(N-1)}\right)$ + const., we obtain a similar representation for V_{N-2}^N as a sum of graphs except that now each vertex corresponds to a term in $\sum_G V_{N-1}^N(G)$, each hard connecting line to $C^{(N-1)}$, and external legs to $\Phi^{(\leq N-2)}$. Integrating out all the fields in (4) produces a sum of graphs whose subgraphs satisfy certain nested connectivity conditions.

The key feature of the Gallavotti-Nicolò approach is the organization of this sum into a sum over "trees" τ which specify which subgraphs connect together at a lower scale and a sum over the scales $h_i \in (k,N]$ at which the connections occur. Schematically (see Ref. 1)

$$V_k^N = \sum_\tau \sum_{\vec{h}} \sum_G V(\tau,\vec{h},G) \, , \tag{6}$$

where the inner sum is over graphs G which are connected consistently with τ and $\vec{h}$. For a given order of perturbation theory, the sums over τ and G are finite; hence the existence of the limit (5) reduces to obtaining bounds on the sum over $\vec{h}$ in (6), uniformly in N. The bounds on $V(\tau,\vec{h},G)$ are based on (2) and amount to standard power counting. In particular, the terms in (6) which require renormalization cancellations are those whose external fields have total dimension 4 or less (note the dimensions: $[\psi] = 3/2$, $[A] = 1$, $[\partial] = 1$). In this way we can determine the counterterms δV^N required, tree by tree; we can renormalize the effective potential at each scale, in descending order; and we obtain sums over $\vec{h}$ bounded uniformly in N. The reasoning for QED_4 is no different from that used by Gallavotti and Nicolò for Φ_4^4.

Here's the catch: the local, Euclidean invariant terms of dimension 4 or less that arise as counterterms are

$$F^2,\ \bar{\psi}\psi,\ -i\bar{\psi}\partial\!\!\!/\psi,\ \bar{\psi}A\!\!\!/\psi \tag{7}$$

and

$$(\partial\cdot A)^2,\ A^2,\ A^4. \tag{8}$$

However, the terms in (8) are gauge-variant and are forbidden in the renormalization of QED. Inasmuch as the GN procedure is inherently gauge variant we cannot rule out these forbidden counterterms.

To understand this dilemma let us examine the Ward identities. In terms of the (formal) expectation (3), define

$$W(\eta,\bar{\eta},B) = \log \frac{1}{Z} \int e^{-\int dx(e\bar{\psi}B\!\!\!/\psi + \bar{\psi}\eta + \bar{\eta}\psi)} e^{V(\Phi)} dP(\Phi)$$

where B_μ is a vector source and η and $\bar{\eta}$ spinor sources. Since $S^{-1} = -i\partial\!\!\!/+m$ satisfies

$$e^{-ie\chi}S^{-1}e^{ie\chi} = S^{-1} + e\partial\!\!\!/\chi \tag{9a}$$

we have

$$d\nu(e^{ie\chi}\psi,\ e^{-ie\chi}\bar{\psi}) = e^{-e\int\bar{\psi}\partial\!\!\!/\chi\psi} d\nu(\psi,\bar{\psi}) \tag{9b}$$

and so

$$W(\eta,\bar{\eta},B) = W(e^{-ie\chi}\eta,\ e^{ie\chi}\bar{\eta},\ B + \partial\chi)\ . \tag{10}$$

These are the Ward identities. Note that the Ward identities depend on the relations (9) for $d\nu(\psi,\bar{\psi})$ and not on properties of $d\mu(A)$ (which is not gauge invariant since L_p has a mass and gauge fixing term).

Suppose we knew that the counterterms $\delta V^N(\psi,\overline{\psi},A)$ satisfied the Ward identities (10) (with $\eta,\overline{\eta},B$ replaced by $\psi,\overline{\psi},A$). Then we could conclude that the forbidden counterterms (8) do not occur and that only the counterterms (7) occur (with the last two in the combination $-i\overline{\psi}\partial\!\!\!/\psi + e\overline{\psi}A\!\!\!/\psi$). Unfortunately there seems to be no way of defining the scale propagators $S^{(h)}$ and regularized propagator

$$S^{(\leq N)} = \sum_{h=0}^{N} S^{(h)} \tag{11}$$

so as to satisfy (2) and anything like (9). This is the basic conflict between the GN procedure and the requirements of gauge invariance.

To resolve this conflict we introduce an auxiliary regularization involving S which does not disturb the Ward identities. The regularization we choose is "loop regularization", implemented by means of fictitious spinor fields
ψ_i, $\overline{\psi}_i$, i = 1, 2, 3 (see e.g. Ref. 4). In this regularization, ψ_1 is a fermi field, ψ_2 and ψ_3 are bose fields, and the Lagrangian L of (1) is replaced by

$$L_\Lambda = L + \sum_{i=1}^{3} \overline{\psi}_i(-i\partial\!\!\!/ + M_i + eA\!\!\!/)\psi_i = L_p + L_{f,\Lambda} + L_{int,\Lambda} \tag{12}$$

where $M_1^2 = m^2 + 2\Lambda^2$, $M_2^2 = M_3^2 = m^2 + \Lambda^2$, so that

$$M_1^2 + m^2 = M_2^2 + M_3^2. \tag{13}$$

The effect of these extra terms is to replace each fermi loop in the effective potential by a second difference of loops. When $\Lambda < \infty$ and when the photon propagator is also regularized $\left(D^{(\leq N_p)} = \sum_{h=0}^{N_p} D^{(h)}\right)$ the computation of the (unrenormalized) effective potential then yields finite results.

But the loop regularization cannot be used to define a sufficiently sensitive scale decomposition of S. The difficulty is that one cannot decompose each fermi propagator in a loop independently, as is necessary for the introduction of the correct counterterms. Accordingly, we still decompose S as in (11) (with $N = \infty$) and we are obliged to accept forbidden counterterms at each scale. However, when these forbidden counterterms are summed up over all scales they add up to zero by virtue of the Ward identities (which are not disturbed by the regularizations Λ and N_p). Why do the summed counterterms verify the Ward identities? Let $dP_{N_p,\Lambda}(\Psi,\overline{\Psi},A)$ be the measure dP with the regularizations N_p and Λ in place; here $\Psi = (\psi,\psi_1,\psi_2,\psi_3)$. Let

$$V(\Psi,\Psi,A) = -\int L_{int,\Lambda}(\Psi,\overline{\Psi},A)dx$$

and let $\delta V_n(\Psi,\overline{\Psi},A)$ be the sum of all counterterms given by the GN procedure up to $O(e^n)$. Define

$$W_n(B) = -\log \int e^{(V + \delta V_n)(\Psi,\overline{\Psi},A+B)} dP_{N_p,\Lambda} .$$

Then the sum of the pure boson counterterms of $O(e^{n+1})$ is obtained by taking the local part of the terms in W_n of $O(e^{n+1})$ and dimension ≤ 4. As in (10), $W_n(B)$ satisfies the Ward identities and hence so do the summed counterterms. In this way we can define a renormalized effective potential $V_k^{N_p,\Lambda}$ in which only gauge invariant counterterms are employed but in which cancellations are performed in a gauge variant way. Such a strategy seems to us to be essential in any renormalization scheme for QED based on power counting and BPHZ ideas.

We remark that in renormalizing $V_k^{N_p,\Lambda}$ we use the correct counterterms ($\overline{\psi}\psi$, $\overline{\psi}\not{\partial}\psi$, $\overline{\psi}\not{A}\psi$) for the original spinor fields but not for the fictitious fields ψ_i. The reason is that the loop regularization is a rather delicate subtraction that works only if the masses satisfy (13) and if the terms $\overline{\psi}(-i\not{\partial} + e\not{A})\psi$ and $\overline{\psi}_i(-i\not{\partial} + e\not{A})\psi_i$ in (12) occur with the same coefficient. The correct counterterms for ψ_i would spoil this subtraction.

As a consequence, it is essential that the Λ regularization be removed before the N_p regularization. With $N_p < \infty$ fixed, the graphs in $V_k^{N_p,\Lambda}$ that appear to require fictitious field counterterms $\bar{\psi}_i\psi_i$, $\bar{\psi}_i\not{\partial}\psi_i$ and $\bar{\psi}_i\not{A}\psi_i$ are actually finite, uniformly in Λ. In other words, these counterterms are not needed for renormalization cancellations but only for loop subtractions. Hence, as $\Lambda \to \infty$ and $M_i(\Lambda) \to \infty$ we obtain the convergence $V_k^{N_p,\Lambda} \to V_k^{N_p}$, where $V_k^{N_p}$ contains no reference to fictitious fields.

Using the GN formalism (6) we obtain bounds, uniform in N_p, on each order of the perturbation expansion for $V_k^{N_p}$. The convergence $V_k^{N_p} \to V_k$ as $N_p \to \infty$ is then straightforward. Finally, we remark that the IR cutoff can be removed by an extension of our methods.[3]

Conclusion: The method of Gallavotti and Nicolò can be adapted to prove the renormalizability of QED_4.

References

[1] Gallavotti, G. and Nicolò, F., Renormalization theory for four dimensional scalar fields I and II, Commun. Math. Phys. 100, 545-590 (1985); and 101, 247-282 (1985).

[2] Gallavotti, G., Renormalization Theory and Ultraviolet Stability for Scalar Fields via Renormalization Group Methods, Rev. Mod. Phys.

[3] Feldman, J., Hurd, T., Rosen, L. and Wright, J., QED: A proof of renormalizability, in preparation.

[4] Itzykson, C. and Zuber, J.B., Quantum Field Theory, McGraw-Hill, 1980.

ASYMPTOTIC COMPLETNESS AND MULTIPARTICLE ANALYSIS IN FIELD THEORIES

Daniel Iagolnitzer

Service de Physique Théorique,
CEN-Saclay, 91191 Gif-sur-Yvette Cedex, France

ABSTRACT

Previous proofs of asymptotic completeness and related results on scattering in field theories are restricted to $P(\varphi)_2$ models in the 2 and 3-particle regions. This talk reports recent works, in particular with J. Magnen, which involve new ideas and methods and provide more direct proofs and some generalization of these results: (i) extensions to theories with renormalization such as φ_3^4 and the Gross-Neveu model and (ii) equations closely linked with asymptotic completeness and multiparticle structure in general energy regions.

Once a model has been constructed [0] and the Osterwalder-Schrader (OS) axioms have been established, a next objective in scattering theory is (i) to get information on spectrum (ii) to prove asymptotic completeness (denoted below AC) and (iii) to establish the analytic structure of multiparticle Green functions and collision amplitudes (i.e. of the S matrix) in Minkowski momentum space. (The OS or equivalent Wightman axioms do not provide expected results on (iii): see [1], e.g. Landau singularities are not extricated. Such results have been derived in axiomatic works [2,3] from assumptions on spectrum, AC and further regularity conditions, only so far in the 2 and 3-particle regions).

Starting from 1975, there has been a number of works on the subject in constructive theory [4-11]. These works, as also the recent ones to be reported, apply to weakly coupled theories, with mass gap, and are based on the use of irreducible kernels, in a way which allows one to treat (i) (ii) (iii) in the same time. There might be a more direct approach to AC itself, as in potential theory [12], but this is not known so far. Past results are, however, restricted to low energy regions and to models without renormalization in space-time dimension d=2. In [4], the Bethe-Salpeter kernel B, linked to the connected amputated 4-point Green function F by the BS equation F = B + (F)(B) (where the last term is a Feynman-type integral, or G-convolution in the sense of [2], with 2-point functions on internal lines) is defined in euclidean space-time and shown to satisfy there a property of "4-particle" exponential decay in the 2→2 channel considered, for e.g. an even theory. By Fourier-Laplace transformation B is then analytic, in complex momentum space, in a region that includes euclidean space and goes up to $s=(4m)^2-\varepsilon$ in Minkowski space, where m is the physical mass and s is the squared center-of-mass energy; $\varepsilon\to 0$ as the coupling $\lambda\to 0$. This type of property, satisfied by 2-particle irreducible Feynman graphs, is still called 2-particle irreducibility. The 2-point function is on the other hand analytic up to $s=(3m)^2-\varepsilon$ except for its pole at $s=m^2$. AC follows from these properties in the 2-particle region (up to $(4m)^2-\varepsilon$), as shown in [5] and in more general way in [2a,c] (completed in [13]) in which the unitarity-type discontinuity formula:

$$F_+ - F_- = F_+ * F_-$$

which does characterize AC in the 2-particle region is established if the irreducibility of G, or in theories with renormalization of kernels G_α satisfying regularized BS equations, is assumed (for the recent construction of such kernels in theories at d = 2,3, see later); * denotes on-mass shell convolution and F_+, F_- are the boundary values of F from above and below the cut (shown to start as $s=4m^2$) in the physical sheet. See also a related approach in [6-8].

In the same time, one shows that F can be analytically continued across the cut, and has [9] a pole at s real < (and close to) $(2m)^2$, either in the second sheet, e.g. for φ^4_2, or in the physical sheet e.g. for $(\varphi^6-\varphi^4)_2$: it corresponds in the latter case to a 2-particle bound state. This pole, however small the coupling λ is, is due to specific kinematical factors at d=2,3 (the factor $\sigma^{-1/2}$, $\sigma = 4m^2-s$ at d=2). A new analysis of poles of F at small λ is given in [13] where previous results are reobtained in a more simple and transparent way, with methods that can be extended to theories with renormalization at d = 2,3 (after kernels G_α have been defined).

It has been believed for a long time that extensions of the

results should be obtained successively in more general energy regions, and the derivation of AC in the 3-particle region has indeed been achieved in [10a] (up to some technical limitations) for even $P(\varphi)_2$ theories. It now makes use of a further $3\to3$, 3-particle irreducible kernel and is related to the axiomatic work [2d]. However, the analysis of [14] suggests that the class of irreducible kernels and integral equations introduced in [7,10b,2d] is too limited in more general cases, where new features occur, and structure equations involving more general irreducible kernels have been conjectured there. These equations are generalizations of the Neumann series expansion of F in terms of B. (This series is convergent at small λ in euclidean space and also in Minkowski space, if a neighborhood of $\sigma=0$ is excluded at d=2,3). For each Green function, they are of the form:

$$F = \Sigma_G \ I(G)$$

where I(G) is a Feynman-type G-convolution integral with irreducible kernels at each vertex (or a finite sum of such integrals). The class of graphs G and the degrees of irreducibility of the kernels involved depend on the energy region of interest. The expansion is the most simple one such that, in view of these degrees, the kernels involved can be considered as analytic with respect to that region. In the $3\to3$ case, the 3-particle region and an even theory, terms I(G) are successions of parts  and  . The class of graphs is more complicated in general and kernels with various degrees of irreducibility (and in some cases of reducibility) in various channels are involved. Ref.14 then presents conjectures on discontinuities of individual terms I(G) which, by formal resummations (and some algebraic manipulations in the general case) yield discontinuity formulae for Green functions that characterize AC and other properties in general energy regions, at least for theories without bound states.

Previous methods to define irreducible kernels for $P(\varphi)_2$ models [4,8,11] are interesting but seem indirect and difficult to use for present purposes. In [15], general irreducible kernels that do satisfy structure equations of the type mentioned above have been defined via new cluster expansions (see below) and all results have been extended in [16] to theories with renormalization such as φ_3^4 and the Gross-Neveu model at d=2 (which involves more renormalization). It is hoped that the proof of AC will then be rigorously completed in general energy regions on this basis, at least for theories without bound states and with coupling constants that may have to be smaller and smaller when the energy increases. Work on the subject has started for φ_3^4 with J. Magnen, H. Epstein and R. Sénéor. It should involve the proof of discontinuity formulae and of convergence properties (after some partial resummations) in Minkowski space. An alternative approach would be to first derive from structure equations suitable sets (not known so far) of integral equations.

Details on [15,16] are now given. Ref.15 first introduces a class of new cluster expansions, said "of order p", $p\geq 1$. The choice of p in the application will depend on the region considered: for instance, p=4 in the 2-particle region, for an even theory. These expansions are simple generalizations of previous ones [17] (corresponding to p=1) and are obtained as follows. Green functions are first defined in a box Λ in euclidean space-time as ratios $N_\Lambda(x_1,\ldots,x_n)/Z_\Lambda$. One then introduces a lattice of e.g. unit side in Λ, a set $s = (s_1,s_2,\ldots)$ of auxiliary real variables and functions $C(x,y;s)$ equal to $C(x,y)$ if $s\equiv 1$, where C is the propagator. In fermionic models like the Gross-Neveu model, the most simple choice of variables will be $s = \{s_{\Delta,\Delta'}\}$, with one variable $s_{\Delta,\Delta'}$ for each pair of squares of the lattice. Then $C(x,y;s) = s_{\Delta,\Delta'}\, C(x,y)$ if $x\in\Delta$, $y\in\Delta'$: in particular $C(x,y;s) = 0$ if $s_{\Delta,\Delta'} = 0$. For bosonic models like $P(\varphi)_2$, a more refined procedure is needed: see [17] at p=1 and [15] at p>1. Functions $N_\Lambda(s)$ and $Z_\Lambda(s)$ are then defined by replacing C by C(s) in the definition of N_Λ and Z_Λ. Suitable Taylor expansions of order p (of the form $N_\Lambda \equiv N_\Lambda(1) = N_\Lambda(0)+\ldots$) with respect to the variables s then give the cluster expansions of N_Λ and Z_Λ.

A Mayer procedure is next used to divide N_Λ by Z_Λ. For each p, it gives an expansion of each connected, amputated Green fucntion as a sum of contributions associated with graphs that join squares $\Delta_1,\ldots,\Delta_N$, N arbitrary, of Λ. These graphs include (i) lines to which propagators $C(u_l,v_l)$ are associated (u_l, v_l vary in the respective squares joined by line l), with at most p propagator lines between any pair of squares and (ii) Mayer lines which indicate that the two squares must coincide in Λ. The graphs must be *connected* (with respect to the squares) when all lines are taken into account. The squares $\Delta_1,\ldots,\Delta_N$ vary independently in Λ except that the external points $x_1,\ldots,x_n$ must be contained in some of these squares (and apart from Mayer conditions).

Irreducible kernels are then defined by restricting the sum to graphs with corresponding irreducibility properties with respect to propagator lines (Mayer lines cannot be cut). In contrast to perturbative series, these expansions are convergent at small coupling λ, uniformly with respect to Λ, and the $\Lambda \to \infty$ limit exist. The exponential decay of the propagators $C(u_1, v_1)$ and the properties of the graphs involved yield moreover exponential fall-off properties of these kernels in euclidean space-time that are extensions of the tree-graph decay introduced for connected functions in statistical mechanics in [18] and then in constructive theory in [19]. The latter yield in turn irreducibility properties in the analytic sense in momentum space. (The irreducible kernels thus defined can be shown to be Borel sums of the corresponding perturbative series). Finally, the Bethe-Salpeter and structure equations are checked by direct graphical inspection (using some factorization properties established in [15] at $p>1$).

The extension to theories with renormalization, such as the Gross-Neveu model [0] is not quite trivial because e.g. the usual Bethe-Salpeter kernel B is then infinite. One might first try to start with a theory with cut-off, but the limit cannot be directly controlled. The most convenient method is then to introduce a kernel B_M linked to F via a *regularized* B.S. equation [2a], in which the last term now includes suitable (fixed) analytic cut-off factors on internal lines. Such a kernel and more general ones satisfying regularized structure equations (see [14]) no longer have a simple perturbative content (see [20]), but can be constructed [16] and shown to be irreducible by considering p^{th} order cluster expansions in the slice of lowest momentum that occurs in phase-space analysis and expansions of order 1 in higher slices (whose propagators have much stronger rates of exponential fall-off). The way AC and related results follow is then analogous to above. One might alternatively use (see [20,21]) a "renormalized" kernel B_{ren}, perturbatively the sum of renormalized 2-particle irreducible graphs and satisfying renormalized B.S. equations [22,21,20]. This approach can be made rigorous [23] but is more complicated. (The kernel B_{ren} is of interest [22,21] in connection with Wilson short distance expansion in space-time. For a proof of the latter from phase-space analysis, see [24]).

REFERENCES

[0] Rivasseau, V., in this volume and references there in.

[1] Bros, J., Thesis (1970), Epstein, H., Glaser, V., Stora, R., in *Structural Analysis of Collision Amplitudes,* Les Houches, June 1975, Balian, R., and Iagolnitzer D., eds., North Holland (1976)

[2] a) Bros, J., in *Analytical Methods in Mathematical Physics,* Gordon and Breach, New-York (1970), p.85.
b) Lassalle, M., CMP 36, 185 (1974)
c) Bros, J., Lassalle, M., CMP 54, 33 (1977)
d) Bros, J., Act. Phys. Austr. Suppl.:XXII, 329 (1981), Physica 124A, 145 (1984)

[3] Epstein, H., Glaser, V., Iagolnitzer, D., CMP 80, 90 (1981)

[4] Spencer, T., CMP 44, 153 (1975)

[5] Spencer, T., Zirilli, F., CMP 49, 1 (1976)

[6] Glimm, J., Jaffe, A., Ann. Math. 100, 585 (1974)

[7] Glimm, J., Jaffe, A., in Lecture Notes in Physics 39 (1975)

[8] Glimm, J., Jaffe, A., CMP 67, 267 (1979)

[9] Dimock, J., Eckmann, J.-P., Ann. phys. 103, 289 (1977)

[10] a) Combescure, M., Dunlop, F., Ann. Phys. 122, 102 (1979)
b) Combescure, M., Dunlop, F., CMP 85, 381 (1982)

[11] Cooper, A., Feldman, J., Rosen, L., Ann. Phys. 137, 14, 213 (1981)

[12] Sigal, M., Soffer, A., in this volume and references therein

[13] Bros, J., Iagolnitzer, D., in preparation

[14] Iagolnitzer, D., in *Critical Phenomena*, Brasov 1983, Birkhauser 1985, p.315, and Fizika 17(3), 361 (1985)

[15] Iagolnitzer, D., Magnen, J., preprint Ecole Polytechnique (1986)

[16] Iagolnitzer, D., Magnen, J., preprint Ecole Polytechnique (1986)

[17] Brydges, D., CMP 58, 313 (1978)
Battle, C., Federbush, P., Ann. Phys. 142, 95 (1982)

[18] Duneau, M., Iagolnitzer, D., Souillard, B., CMP 31, 191 (1973), 35, 307 (1974)

[19] Eckmann, J.-P., Magnen, J., Sénéor, R., CMP 39, 259 (1974)

[20] Iagolnitzer, D., CMP 99, 451 (1985), Fizika 17(3), 411 (1985)

[21] Bros, J., Ducomet, D., preprint, Saclay (1986)

[22] Symanzik, K., CMP 34, 7 (1973)

[23] Bros, J., Iagolnitzer, D., Magnen, J., in preparation

[24] Iagolnitzer, D., Magnen, J., Sénéor, R., in preparation

CONSTRUCTION OF RENORMALIZABLE AND NON-RENORMALIZABLE QUANTUM FIELD THEORIES

A. Kupiainen
University of Helsinki
Research Institute for Theoretical Physics
00170 Helsinki, Finland

In the 60's and the 70's a major problem of mathematical physics was the construction of non-trivial quantum field theories without cut-offs - in four dimensions, or more technically, renormalizable ones. Even if the current interest in string theories has raised serious doubts as to whether the fundamental theory of elementary particles is described by local quantum fields at very short distances, it has become clear that (quantum) field theory describes physics on a tantalizing range of scales: starting perhaps from the GUT scale 10^{15} GeV through the 10^2 GeV of the standard model all the way to macroscopic scales of QED and Maxwell theory.

A way in field theory to describe the change of physics as the (momentum-) scale Λ varies is the flow of the Renormalization Group (RG) in the space of effective actions

$$S_\Lambda^{EFF} \underset{R_L}{\rightarrow} S_{\Lambda/L}^{EFF} \tag{1}$$

where S_Λ^{EFF} is a cutoff Λ (eg. lattice of spacing Λ^{-1}) action giving Green functions of momenta $\lesssim \Lambda$ of the theory. R_L results from conditioning on fields with momenta $\lesssim \Lambda/L$, schematically

$$\exp[-S_{\Lambda/L}^{EFF}(\varphi')] = \int \exp[-S_\Lambda^{EFF}(\varphi' + \varphi)]\, D\varphi \tag{2}$$

where $\int D\varphi$ is over fields with momentum $\varepsilon[\frac{\Lambda}{L},\Lambda]$, explicit ways of doing this are mentioned below. We are interested in <u>global flows</u> of R_L i.e. L large. In particular if field theory is to be the fundamental one, we look for <u>continuum limits</u>: we try find a "bare" action S_Λ^{BARE} such that $\Lambda \to \infty$ may be taken, i.e. the limit

$$\lim_{\Lambda\to\infty} R_{\frac{\Lambda}{\tilde{\Lambda}}} S^{BARE}_{\tilde{\Lambda}} = S^{EFF}_{\tilde{\Lambda}} \tag{3}$$

exists for all $\tilde{\Lambda}$, providing the effective actions on all scales.

Thus, from the mathematical point of view, the problem is to find suitable spaces where R acts. In particular the actions in such a space should produce well defined (cutoff) green functions (thermodynamic limit etc.). Furthermore, classification and parametrization of the continuum limits should ensue.

The formal perturbative approach to this problem is well known. S^{BARE}_{Λ} is parametrized by some coupling(s) g_Λ and $S^{EFF}_{\tilde{\Lambda}}$ is computed from (2) as a formal power series in these

$$S^{EFF}_{\tilde{\Lambda}} = \sum_{n=0}^{\infty} g^n_\Lambda S_{n\tilde{\Lambda}}(\Lambda) \tag{4}$$

Typically $S_{n\tilde{\Lambda}}(\Lambda)$ diverge as $\Lambda \to \infty$ but for the <u>renormalizable</u> models one may introduce the corresponding parameter(s) $g_{\tilde{\Lambda}}$ for $S^{EFF}_{\tilde{\Lambda}}$ and after expressing g_Λ as a formal power series in $g_{\tilde{\Lambda}}$

$$g_\Lambda = \sum g_{\tilde{\Lambda}} c_{n\tilde{\Lambda}}(\Lambda) \tag{5}$$

the coefficients $S^R_{n\tilde{\Lambda}}(\Lambda)$ in

$$S^{EFF}_{\tilde{\Lambda}} = \sum_{n=0}^{\infty} g^n_{\tilde{\Lambda}} S^R_{n\tilde{\Lambda}}(\Lambda) \tag{6}$$

have a limit as $\Lambda \to \infty$ (they are typically polynomials in the field φ). For recent proofs of this in the effective action language see [1] and [2].

For <u>non-renormalizable</u> models this procedure will not work: formally an infinite number of bare couplings g_Λ are needed.

The success of perturbative renormalization in the 50's, especially the case of QED, lead to the expectation, that the renormalizable models should be candidates for QFT's without cutoffs. This picture was in the 70's modified with the deepened insight into QFT provided by Wilson's RG theory [3]. Asymptotic freedom, i.e. vanishing of g_Λ as $\Lambda \to \infty$, in the renormalizable case, and more generally the existence of an ultraviolet stable fixed point for R were seen as the conditions for the continuum limit to exist. In particular, some renormalizable models, like ϕ_4^4 and QED, might not exist in the continuum (they would still be perfectly good effective field theories describing physics on a wide range of scales) [4],[5], and some non-renormalizable ones could. In what follows, I will describe some progress made during the past two years in the construction of renormalizables and non-renormalizable QFT's by K. Gawedzki and myself [6]-[10]. Other works on the subject include those of 't Hooft [11], Rivasseu [12], Feldman, Magnen, Rivasseu, Seneor [13], Felder [14] and Balaban and Jaffe [15].

The constructions are based on a detailed control of the flow of R near a fixed point S^*. (here and below we include into R a scale transformation so as to keep the cutoff of S and $R_L S$ the same, thus both belong to the same "space" of fixed cutoff actions). That a fixed point is useful is clear: by letting S_Λ^{BARE} approach the stable manifold of S^*, the effective actions in (3) will end up on the unstable manifold, see figure 1

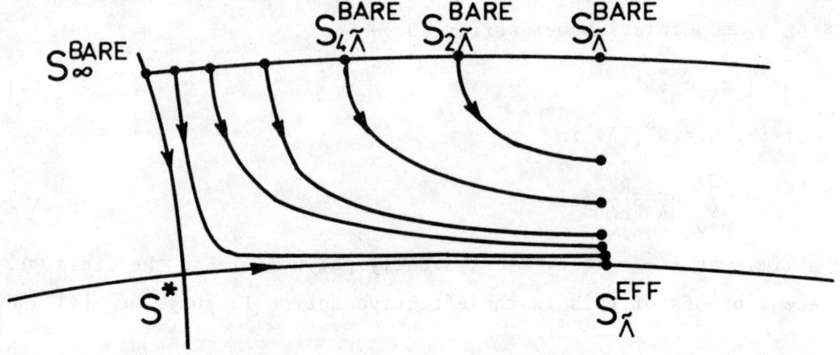

Thus, quantum field theories are flow lines of R on the unstable manifold of a fixed point. The key to the control of R_L is, that for $L = \mathcal{O}(1)$ this is a well behaved operation: the functional integral (2) involves no ultraviolet nor infrared singularities. Indeed, the perturbative expansion (4) with suitable modifications in the large field region will be the core of our construction. The divergences as $L \to \infty$ show up only in the sensitive dependence of the global flow on initial conditions.

The models constructed this way are the Gross-Neveu model in 2 dimensions, the φ^4 model in 4 dimensions with negative coupling and the (hierarchical) non-linear σ-model in 2 dimensions, which are renormalizable, asymptotically free ones, and a non-renormalizable "2+ε dimensional" version of the Gross-Neveu model. I will discuss them separately in what follows.

The Gross-Neveu model

The analysis of R_L turns out to be the simplest in a purely fermionic model. Thus we consider a four fermion interaction in two dimensions with

$$e^{-S_\Lambda^{BARE}} = e^{-S_\Lambda^I} d\mu_\Lambda(Z_\Lambda^{1/2}\bar{\psi}, Z_\Lambda^{1/2}\psi) \tag{7}$$

$$S_\Lambda^I = m_\Lambda Z_\Lambda \int \bar{\psi}\psi\, d^2x + g_\Lambda Z_\Lambda^2 \int (\bar{\psi}\psi)^2 dx \tag{8}$$

and $d\mu_\Lambda$ gaussian "measure" with covariance the fourier transform of

$$G_\Lambda(p) = \not{p}\, p^{-2}\, e^{-p^2/\Lambda^2} \tag{9}$$

Here $\psi(x) \equiv \psi_\alpha^i(x)$, $\alpha = 1,2$ Dirac indices, $i=1,\dots,N$ $N>1$ "flavor" indices, are elements of a Grassmann algebra (see [8] for details). [7]-[9] describe a renormalizable asymptotically free model. RG, i.e. (2), is defined by splitting (9) to high and low momenta

$$e^{-S_{\tilde{\Lambda}}^{EFF}(\bar{\psi},\psi)} = \text{const.}\, d\mu_{\tilde{\Lambda}}(\bar{\psi},\psi) \int d\mu_\Gamma(\bar{\rho},\rho)\, e^{-S_\Lambda^I(\bar{\psi}+\bar{\rho}\;,\psi+\rho)} \tag{10}$$

where

$$\Gamma(p) = G_\Lambda(p) - G_{\tilde{\Lambda}}(p) \tag{11}$$

Remarkably, closed formuli may be given to the bare parameters:

Theorem *Let*

$$Z_\Lambda = 1, \quad m_\Lambda = m\left(\log \frac{\Lambda}{\mu}\right)^{\frac{1-2N}{2N-2}}$$

$$g_\Lambda = g\Big/\left(1+g\,\frac{2(N-1)}{\pi}\log\frac{\Lambda}{\mu} - \frac{g}{\pi}\log\left(1+g\,\frac{2(N-1)}{\pi}\log\frac{\Lambda}{\mu}\right)\right)$$

Then $\lim_{\Lambda\to\infty} S^{EFF}(\tilde{\Lambda})$ *exists for all* $\tilde{\Lambda} \in [0,\infty]$ *and is given by a formula like (7) with*

$$S^I_{\tilde{\Lambda}} = m_{\tilde{\Lambda}} \int \bar{\psi}\psi \, d^2x + g_{\tilde{\Lambda}} \int (\bar{\psi}\psi)^2 \, d^2x + \\ + \sum_{n=1}^{\infty} \int dx_1 \ldots dx_{2n} \, S_{2n\tilde{\Lambda}}(x_1 \ldots x_{2n}) \prod_{i=1}^{n} \bar{\psi}(x_i)\psi(x_{n+i}) \tag{12}$$

where

$$\int dx_2 \ldots dx_{2n} |S_{2n\tilde{\Lambda}}(x_1 \ldots x_{2n})| \equiv \| S_{2n\tilde{\Lambda}} \| \tag{13}$$

are finite (indeed, small). S_{2n0} *are (amputated) Green functions of an interacting Euclidean QFT satisfying the Osterwalder-Schrader axioms with exponential clustering (mass =* $m + \mathcal{O}(g)$*).*

Thus a Wightman field theory exists in Minkowski space. (The physical coupling is $\simeq$ g at the renormalization point μ, an arbitrary scale). The fixed point S^* in this case is gaussian $g_\Lambda = 0$, $g_{\tilde{\Lambda}}$ is a marginally unstable variable:

$$g_{\tilde{\Lambda}/L} = g_{\tilde{\Lambda}} + \frac{2N-2}{\pi} \log L \, g^2_{\tilde{\Lambda}} + \frac{4(N-1)(N-2)}{\pi^2} (\log L)^2 g^3_{\tilde{\Lambda}} + o(g^3_{\tilde{\Lambda}}) \tag{14}$$

and the unstable manifold two dimensional.

The key to the control of the flow is the observation, that R_L for L fixed is an <u>analytic map</u> in the Banach space of effective actions (12),(13). The reason for this is, that the integral (10) may be performed perturbatively in powers of S^I_Λ and the series <u>converges</u>. This happens for two reasons: First, for L finite, $\mathcal{O}(1)$, the integral (in dimensionless units) has fixed $\mathcal{O}(1)$ ultraviolet and infrared cutoffs. Secondly, fermions don't have large field problems: the usual n! divergences of bosonic perturbation expansions don't occur (there are no instantons). Analyticity of R_L allows to make recursions such as (12) rigorous and to establish the full flow pattern near S^*.

<u>The non-renormalizable theory</u>

The preceeding analysis may be extended to a non-renormalizable version of the Gross-Neveu model mimicking the formal "2+ε-dimensional" theory. We replace in (7)-(9) the propagator by a G^ε_Λ which at $\Lambda = \infty$ is

$$p/p^{2-\varepsilon} \tag{15}$$

Then (7)-(9) describe a perturbatively non-renormalizable theory with infinite number of counter-terms needed. Non-perturbatively, however, the situation is very simple: only the couplings m_Λ and g_Λ are needed. Considering e.g. the massless theory where m_Λ is zero (by chiral invariance) and defining a dimensionless coupling $\lambda_\Lambda = \Lambda^{2\varepsilon} g_\Lambda$, (12) gets replaced by

$$\lambda_{\tilde{\Lambda}/L} = L^{-2\varepsilon} \Lambda_{\tilde{\Lambda}} + \beta_2(L)\, \lambda^2_{\tilde{\Lambda}} + o(\lambda^2_{\tilde{\Lambda}})$$

indicating the existance of a fixed point $\lambda^* = \ (\varepsilon)$. Indeed, a simple application of the contraction mapping principle in the Banach space (12)-(13) establishes the existence of a fixed point $\{\lambda^*, S^*_n(x_1 \ldots x_n)\}$ and the flow pattern of figure 2. The gaussian fixed point is stable, and the continuum limit is obtained on the unstable manifold of the non-gaussian one.

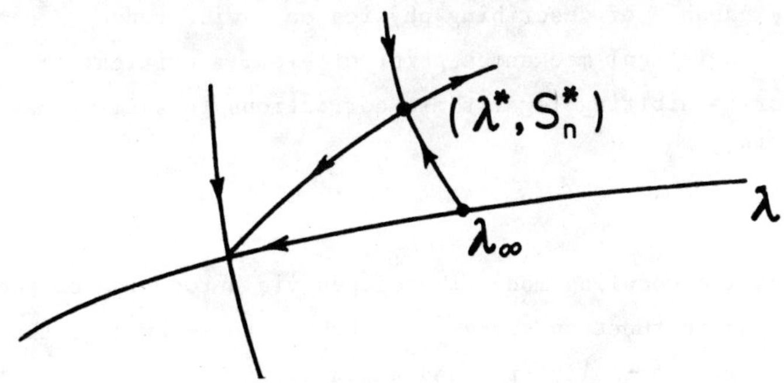

In (7) we choose the bare coupling now as

$$\lambda_\Lambda = \lambda_\infty + (\lambda-\lambda)\left(\frac{\mu}{\Lambda}\right)^{\nu^*}$$

where $L^{\nu^*} > 1$ is the largest eigenvalue of dR_L at S^* (the λ-line in fig. 2 is the line of bare theories (7)). For $\lambda \leq \lambda^*$ (massless) continuum limits of the Euclidean Green functions follow [9]. They are non-trivial and satisfy the OS-axioms except maybe positivity. For $\lambda > \lambda^*$ only massive theories may be constructed, the massless one probably exhibiting dynamical mass generation.

Even though possibly unphysical, this model shows, that non-renormalizable quantum field theories may make perfectly good mathematical sense.

ϕ^4 in four dimensions

Similar results hold for asymptotically free bosonic theories. Consider the φ^4 theory on a four-dimensional lattice of spacing Λ^{-1}:

$$S_\Lambda^{BARE} = \frac{1}{2} Z_\Lambda \int_\Lambda [(\nabla\varphi)^2 + m_\Lambda^2 \varphi^2] + Z_\Lambda^2 g_\lambda \int_\Lambda \varphi^4 \tag{16}$$

Now ([6],[7]) there are two cases

1^0 $g_\Lambda > 0$

The unstable manifold is gaussian in the neighbourhood of gaussian fixed point. Only trivial continuum limits exist there. However, with a fundamental cutoff, φ_4^4 provides still a perfectly good effective field theory capable of describing physics on a wide range of scales.

From a statistical mechanics point of view, a critical (massless) lattice theory exhibiting logarithmic corrections to scaling may be constructed [6].

2^0 $g_\Lambda < 0$

The negative coupling model is defined via a rotation of the integration contour in function space [7]. Like $(\text{Gross-Neveu})_2$, the resulting model is asymptotically free and a non-trivial Euclidean continuum limit can be constructed. The Green functions are not reflection positive; the theory is presumably describing an unstable vacuum (whose lifetime per unit volume by a formal instanton computation [16] is in fact zero!).

In the bosonic case R_L is defined by a block spin transformation and evaluated using a convergent cluster expansion instead of the perturbation expansion which exhibits the usual n! (instanton) divergence of bosonic theories. The main extra difficulty compared to the fermionic theory is, that one has to keep track of the (non-perturbative) large field behaviour of $e^{-S_{\tilde{\Lambda}}^{EFF}}$.

The Non-linear σ-model

The non-linear σ-model is a two dimensional field theory expected to have many of the properties and difficulties (as far as rigorous construction is concerned) of four dimensional non-abelian gauge theories. In particular the space of fields is non-linear (or the space of classical minima in the φ^4 version, see below) making the non-perturbative part of the analysis more difficult.

Here we describe how the ultraviolet problem is solved within a hierarchical approximation to the problem. The action on a $\frac{1}{\Lambda}$-lattice is given by

$$e^{-S_\Lambda^{BARE}(\vec{\varphi})} D\vec{\varphi} = \prod_x \delta(\vec{\varphi}(x)^2 - Z_\Lambda^{-1} f_\Lambda^2) d\mu_{G_\Lambda}(Z_\Lambda^{1/2}\vec{\varphi}) \tag{17}$$

For the standard model $G_\Lambda = (-\Delta)^{-1}$ (on $\frac{1}{\Lambda}$-lattice), here we take G_Λ the hierarchical version of this [10]. The main consequence of this simplification is, that R_L now preserves the locality of the action: S^{EFF} is given by

$$e^{-S_{\tilde{\Lambda}}^{EFF}} D\vec{\varphi} = \exp[-\int_\Lambda \mathcal{L}_{\tilde{\Lambda}}(\vec{\varphi})]\, d\mu_{G_{\tilde{\Lambda}}}(\vec{\varphi}) \tag{18}$$

($Z_\Lambda = 1$ in hierarchical model) and

$$e^{-\mathcal{L}_{\tilde{\Lambda}/L}(\vec{\varphi})} = \text{const}\left[\int e^{-\mathcal{L}_{\tilde{\Lambda}}(\vec{\varphi}+\vec{\rho}) - \frac{1}{2}\vec{\rho}^2} d\vec{\rho}\,\right]^{L^2} \tag{19}$$

The result of iteration of (19) is, that the original δ-function of (17) gets smeared; in fact $\mathcal{L}^{EFF}$ attains exponentially fast a fixed form, only the location of the minimum f_Λ running:

<u>Theorem 2</u> *Let f_Λ be given by*

$$f_\Lambda^2 = f^2 + \beta_2 \log\frac{\Lambda}{\mu} + \frac{\beta_3}{\beta_2}\log(1 + f^{-2}\beta_2\log\frac{\Lambda}{\mu})$$

(where $\beta_2 > 0$ and β_3 are computable constants)

Then $\lim_{\Lambda\to\infty} e^{-\mathcal{L}_{\tilde{\Lambda}}(\vec{\varphi},\Lambda)} = e^{-\mathcal{L}_{\tilde{\Lambda}}(\varphi)}$ *is an entire function on $\mathbb{C}^N$ satisfying*

$$\mathcal{L}_{\tilde{\Lambda}}(\vec{\varphi}) = g^*(|\vec{\varphi}| - \tilde{f}_{\tilde{\Lambda}})^2 + \mathcal{O}(f_{\tilde{\Lambda}}^{-3}(|\vec{\varphi}| - \tilde{f}_{\tilde{\Lambda}})^3) \qquad (g^* = \mathcal{O}(1))$$

for $||\vec{\varphi}| - \tilde{f}_{\tilde{\Lambda}}| < f_{\tilde{\Lambda}}^{\alpha}$, $\vec{\varphi}$ *real and*

$$|e^{-\mathcal{L}_{\tilde{\Lambda}}(\vec{\varphi})}| < e^{-\frac{1}{4}g^*(|Re\vec{\varphi}| - \tilde{f}_{\tilde{\Lambda}})^2 + c(Im\varphi)^2 + 1}$$

elsewhere. Here $\tilde{f}_\Lambda^2 = f_\Lambda^2 + \mathcal{O}(f_\Lambda^{-2})$.

Thus, in the leading order in the logarithmically increasing coupling ("temperature" in the Heisenberg model language) $\tilde{f}_{\tilde{\Lambda}}^{-2}$, the action is just of a φ^4 type concentrated on the sphere

$$\mathcal{L}_{\tilde{\Lambda}} \sim \frac{g}{2\tilde{f}_{\tilde{\Lambda}}^2} (\vec{\varphi}^2 - \tilde{f}_{\tilde{\Lambda}}^2)^2 \tag{20}$$

To get a faster approach to continuum limit, we could take (20) as the bare action: this is an example of the Symanzik improvement program [17].

The main difference of above and the standard σ-model is, that in the latter $Z_\Lambda \neq 0$ (in fact $Z_\Lambda \to \infty$ as $\Lambda \to \infty$) and non-local irrelevant terms appear in the action. There, as in the hierarchical model, the control of $\tilde{\Lambda} \to 0$ i.e. the infrared limit will be more difficult than the ultraviolet one: the coupling $f_{\tilde{\Lambda}}^{-2}$ increases and eventually leaves the essentially perturbative regime where above analysis was done. Indeed, a dynamical mass generation is expected in the σ-model.

It can be said, that a good rigorous understanding of linear asymptotically free models, and some non-renormalizable ones too, now exists. The main open problem for constructive field theory, apart from the σ-model is that of non-abelian gauge theories i.e. QCD. The UV problem there is quite likely doable with the existing techniques, but also quite likely very complicated technically. If simplicity and elegance are to be criteria in mathematical physics as in mathematics, there are still plenty of things for constructivists to learn.

References

1. J. Polchinski, Nucl. Phys. B231, 269 (1984)
2. G. Gallavotti, F. Nicolo, Commun. Math. Phys. 101, 247 (1985)
3. K. Wilson, Rev. Mod. Phys. 55, 583 (1983)
4. J. Fröhlich, Nucl. Phys. B200, 281 (1982)
5. M. Aizenman, Phys. Rev. Lett. 47, (1981)
6. K. Gawedski, A. Kupiainen, Commun. Math. Phys. 99,197 (1985)

7. K. Gawedski, A. Kupiainen, Nucl. Phys. B257, 474 (1985)

8. K. Gawedski, A. Kupiainen, Commun. Math. Phys. 102, 1 (1985)

9. K. Gawedski, A. Kupiainen, Nucl. Phys. B262, 33 (1985)

10. K. Gawedski, A. Kupiainen, Preprint IHES/P/86/21

11. G. 't Hooft, Commun. Math. Phys. 86, 449 (1982)

12. V. Rivasseu, Commun. Math. Phys. 95, 445 (1984)

13. J. Feldman, J. Magnen, V. Rivasseu, R. Seneor, Commun. Math. Phys. 103, 67 (1986)

14. G. Felder, Commun. Math. Phys. 102, 139 (1985)

15. T. Balaban, A. Jaffe, Harvard Preprint HUTMP86/B191

16. A. Kupiainen, Helsinki Preprint HU-TFT-85-58

17. K. Symanzik, in Mathematical Problems - Theoretical Physics, Lecture Notes in Physics, Vol. 153, Eds. R. Schrader et al. Berlin, Springer 1982

SUPERCONFORMAL ALGEBRA AND RELATIVISTIC SUPERSYMMETRIC QUANTUM MECHANICS*

A. Bohm
Center for Particle Theory, Physics Department
The University of Texas at Austin, Austin, Texas 78712

This work is part of a program whose purpose it is to develop a quantum mechanics of relativistic extended objects. Its physical basis are the collective models of molecular and nuclear physics. Its mathematical methods are based on groups and supergroups. These (super) groups do not necessarily describe (super) symmetries[1)] of the S-matrix. They describe a non-trivial mass (or energy) spectrum and are called spectrum generating groups or spectrum supersymmetries. The states of their (infinite) multiplets and supermultiplets are not the elementary particles of our time ((s)leptons, (s)quarks) but different excited states of extended objects. In nuclear physics they are energy eigenstates of several nuclei which differ by one additional nucleon,[2)] in the relativistic model they are hadrons that differ by one additional quark, i.e. baryon and meson resonances.[3)]

Non-relativistic supersymmetric quantum mechanics is usually formulated in terms of[4)]

$$SU(1,1/1) \simeq Osp(2,2) \supset Osp(1,2) \supset SO(2,1)_{\mathcal{H},K,\mathcal{D}} \supset$$
$$\supset SO(2)_{\Gamma_0 = \frac{1}{2}(\mathcal{H}+K)} \qquad (1)$$

and used for the radial excitations, e.g. the radial part of the harmonic oscillator.[5)] Enlarging this by an $SO(3)_{S_{ij}}$ to take the rotational degrees of freedom into account[6)] and making it relativistic leads to the following supersubgroup chain

$$\begin{array}{llll} SU(2,2/1) & \supset Osp(1,4) \supset SO(3,2) & \supset SO(3)_{S_{ij}} & \times SO(2)_{\Gamma_0} \quad (2) \\ \cup & \quad\quad \cup & & \\ SU(2,2) \times U(1) & \quad\quad SO(3,1)_{S_{\mu\nu}} & & \end{array}$$

The representations $D_S(q_0)$ of SU(1,1/1) ("non-typical" which remain irreducible when restricted to Osp(1,2)) reduce into the direct sum of SO(2,1) irreps : $D_S(q_0) \Rightarrow D_+(q_0) \oplus D_+(q_0 + \frac{1}{2})$. Each $D_+(q)$ describes a tower of vibrational levels with μ = eigenvalue $\Gamma_0 = q, q+1, q+2, \cdots$ (the weight diagram of $D_S(q_0)$ is given in Fig. 2 of reference 4a) and $D_S(q_0)$ describes the spectrum of the one-dimensional (radial) supersymmetric oscillator with lowest energy value q_0. In the transition to the relativistic supersymmetric oscillator (SU(1,1/1) → SU(2,2/1)) the $D_S(q_0)$ is replaced by the so-called "mass-less," "positive energy" representation $D_S\ (s_0+1; s_0, 0; s_0+1)$ of SU(2,2/1). These irreps are characterized by one number s_0, they remain irreducible if restricted to Osp(1,4), when they become the irreps $D_S\ (s_0+1, s_0)$.[7] Their weight diagram is given in the Figure for the case $s_0 = 1/2$. j is the angular momentum (hadron spin) and μ is again the vibrational quantum number = eigenvalue Γ_0. Each dot represents a hadron level with quantum numbers (μ, j). Two adjacent columns of this Figure are identical with the weight diagram for $D_S(q_0)$ of SU(1,1/1), so that this irrep of SU(2,2/1) describes indeed a generalization of the states given by SU(1,1/1). Its reduction with respect to the subgroup SU(2,2) is

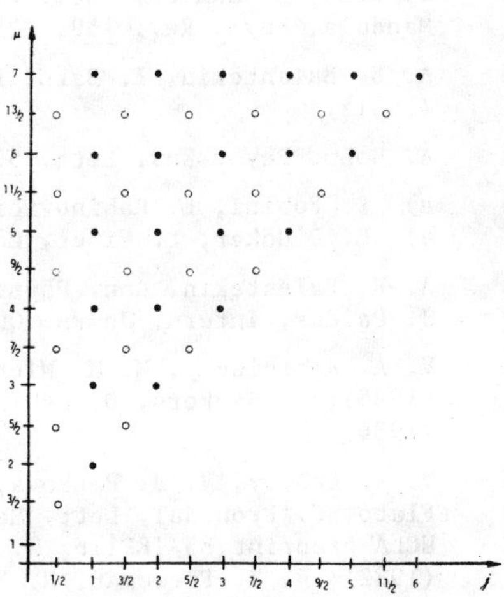

Fig. Weight diagram of the representation D_S (3/2;1/2,0;3/2) of SU(2,2/1).

$$D_S\ (s_0+1; s_0, 0; s_0+1) \Rightarrow D(s_0+1, s_0, 0) \oplus D(s_0+3/2,\ s_0+1/2, 0) \quad (3)$$

where D (s+1, s, 0) describes an infinite hadron multiplet characterized by a number s, where s can be interpreted as the sum of the quark spins.[8] For $s_0 = 1/2$, (3) describes thus a supermultiplet which contains the ρ-meson trajectory (s = 1), the nucleon Regge-trajectory (s = 1/2) and their daughters.

REFERENCES

*) This is a summary of the lecture notes which will be published in the proceedings of the XVth International Conference on Differential Geometric Methods in Physics, H. D. Doebner and J. D. Hennig, editors, World Scientific (1987). University of Texas preprint (1986) CPT.

1) J. Wess, B. Zumino, Nucl. Phys. B70, 39 (1974); S. Coleman, J. Mandula, Phys. Rev. 159, 1251 (1967).

2) A. B. Balantekin, I. Bars, F. Iachello, Phys. Rev. Lett. 47, 19 (1981).

3) A. Bohm, Phys. Rev. Lett. 57, 1203 (1986).

4) a) S. Fubini, E. Rabinovici, Nucl. Phys. B245, 17 (1984);
b) E. D'Hoker, L. Vinet, Commun. Math. Phys. 97, 391 (1985).

5) A. B. Balantekin, Ann. Phys. (N.Y.) 164, 277 (1985); J. Cizek, J. Paldus, Intern. Journ. Quant. Chem. XII, 875 (1977).

6) V. A. Kostelecky, M. M. Nieto, D. R. Truax, Phys. Rev. D32, 2627 (1985); J. Beckers, D. Dehin, V. Hussin, Univ. de Liege preprint (1986).

7) V. K. Dobrev, V. B. Petkova, Phys. Lett. 162B, 127 (1985); M. Flato, C. Fronsdal, Lett. Math. Phys. 8, 159 (1984); B. Binegar, UCLA preprint 85/TEP116; W. Heidenreich, Phys. Lett. 110B, 461 (1982); D. Z. Freedman, H. Nicolai, Nucl. Phys. B237, 342 (1984).

8) A. Bohm, M. Loewe, P. Magnollay, M. Tarlini, R. R. Aldinger, L. C. Biedenharn, H. van Dam, Phys. Rev. 32, 2828 (1985).

HIGHER CONSERVATION LAWS FOR TEN-DIMENSIONAL SUPERSYMMETRIC YANG-MILLS THEORIES

Michael Forger

CERN - Geneva
and
Fakultät für Physik der Universität Freiburg
Hermann-Herder-Str. 3, D-7800 Freiburg

One interesting feature of certain supersymmetric Yang-Mills theories, first noticed by E. Witten for the N=3 and N=4 extended theories in d=4 [1)] and more recently carried over to the N=1 theory in d=10 [2)], is that they admit a natural geometric interpretation in the context of twistor theory. This interpretation emerges from a careful analysis of the constraints which must be imposed on the supercurvature tensor in order to eliminate superfluous degrees of freedom and which, in the cases at hand, are known to put the theory on shell [3)]. More specifically, it turns out that the constraints
a) are equivalent to the equations of motion.
b) are the integrability conditions for an appropriate linear system.
It should perhaps be pointed out that the equivalence in a) has been stated and made plausible, but - contrary to what is often claimed in the literature - not proved, in the original references [1),2)] : the complete proof was only given recently [4)-6)].

Closely related to the twistorial interpretation - for which the reader is referred to the contribution of J. Harnad in these proceedings - is the point of view that supersymmetric Yang-Mills theories, in the cases at hand, may be viewed as higher-dimensional versions of integrable systems. One may therefore wonder whether or to what extent the well-known techniques for handling two-dimensional integrable field theories can be generalized to higher dimensions. In particular, it seems natural to look for higher conservation laws, and for the restrictions that their existence imposes on the dynamics. In fact, there are examples of two-dimensional integrable field theories, such as certain non-linear σ models (e.g. on spheres or on compact simple Lie groups), where these restrictions are sufficiently strong to imply factorization of the S-matrix into two-body amplitudes and to determine completely the two-body S-matrix, so that the following chain of arguments can be established:

$$\text{linear system} \xrightarrow[1]{\text{step}} \text{higher conservation laws (classical)}$$
$$\xrightarrow[2]{\text{step}} \text{higher conservation laws (quantum)}$$
$$\xrightarrow[3]{\text{step}} \text{determination of S-matrix}$$

In higher dimensions, we expect this scheme to break down, for a variety of reasons. One reason is that factorizing S-matrices are trivial in $d > 2$. The other reason is that for an interacting local quantum field theory with a non-trivial S-matrix in $d > 2$, the Coleman-Mandula theorem [7] excludes the existence of higher local conservation laws, and there are partial - though not yet conclusive - results towards a generalized Coleman-Mandula theorem [8] which excludes the existence of higher non-local conservation laws as well. But for supersymmetric Yang-Mills theories, in the cases at hand, the linear system does imply the existence of higher conservation laws at the classical level, which turn out to be non-local [6),9)], and so the scheme above must break down at step 2, i.e., the classical conservation laws, when quantized, are plagued by anomalies.

For the details in the case of N=1 supersymmetric Yang-Mills theories in d=10, the reader is referred to the original paper [6].

References:

1) E. Witten: Phys.Lett. 77 B (1978) 394.

2) E. Witten: Nucl.Phys. B 266 (1986) 245.

3) M. Sohnius: Nucl.Phys. B 136 (1978) 461.

4) J. Harnad, J. Hurtubise, M. Légaré and S. Shnider: Nucl.Phys. B 256 (1985) 609.

5) J. Harnad and S. Shnider: preprint, to appear in Commun.Math.Phys..

6) E. Abdalla, M. Forger and M. Jacques: preprint CERN-TH.4431/86, to appear in Nucl.Phys..

7) S. Coleman and J. Mandula: Phys.Rev. 159 (1967) 1251.

8) D. Buchholz and J.T. Łopuszański: Lett.Math.Phys. 3 (1979) 175; D. Buchholz, J.T. Łopuszański and Sz. Rabsztyn: Nucl.Phys. B 263 (1985) 155.

9) Ch. Devchand: Nucl.Phys. B 238 (1984) 333.

SUPERCONNECTION INTEGRABILITY CONDITIONS AND SUPERSYMMETRIC YANG-MILLS EQUATIONS IN TEN DIMENSIONS[+]

J. Harnad[1*] and S. Shnider[2]

[1] Département de Mathématiques Appliquées, Ecole Polytechnique de Montréal, C.P. 6079, Succ."A",Montréal,Qué.CANADA H3C 3A7.

[2] Department of Mathematics, Ben Gurion University, Beersheva 84105, Israel and McGill University, Montreal, Que.CANADA H3A 2K6

ABSTRACT

An outline is given of a complete proof of the equivalence between the super null-line integrability conditions (constraint equations)and the field equations for $d = 10, N = 1$ supersymmetric Yang-Mills theory.

1. SUPERSYMMETRIC YANG-MILLS THEORY AND SUPERCONNECTION CONSTRAINTS

For supersymmetric Yang Mills theory in dimension $d = 4$, the superconnection constraint equations were originally introduced to provide irreducible supermultiplets(1), and as such were meant to be purely algebraic in content. The fact that for the extended $(N = 3,4)$ type theories field equations were implied(2,3) was regarded as a problem for manifestly supersymmetric quantization in the superpace setting, which requires superfields to be defined on or off shell(4,5). From the "classical" viewpoint (insofar as supersymmetric theories have a classical limit at all), the fact that the field equations turn out to actually be equivalent to the constraints(3,6,7) is of great interest in

[+] Text of invited talk presented by J. Harnad at the topical session on supersymmetry, VIIIth International Congress on Mathematical Physics, Marseille, July 16-25,1986. Research supported in part by NSF Grant DMS-8604189 and the National Science and Engineering Research Council of Canada.

[*] On leave of absence from: Dept. of Pure and Applied Mathematics, Stevens Institute of Technolgy, Hoboken NJ 07030, USA

itself. This permits a supertwistor formulation of the theory[3,8] analogous to the twistor formulation of the self dual Yang Mills equations[9], providing a key step towards reducing the problem to one in (super-) algebraic geometry[10].

It has been known for some time that similar on-shell conditions are implied by the constraints for the $d = 10, N = 1$ case[11]. Witten[12] proposed, and partly proved, that a one-one equivalence again exists between the field equations and the constraints and showed that the latter were also interpretable as super null line integrability conditions, leading to a super twistor correspondence. It is our purpose here to indicate how this result is proved in full, giving an explicit one-one map between the two sets of data: (i) Yang-Mills and spinor fields in 10 dimensional Minkowski space M_{10} satisfying the supersymmetric field equations, and; (ii) superconnections over $(10|16)$ dimensinal super Minkowski space $\tilde{M}_{10}$ with vanishing curvature on super null lines (i.e. satisfying the constraint equations). Complete details of this proof may be found in ref.(13).

2. FIELD EQUATIONS AND CONSTRAINTS FOR d = 10, N = 1.

2.1 Field Equations and Supersymmetry Transformations

The supersymmetric Yang Mills equations in ten dimensions are:

$$\mathcal{D}^{\mu} F_{\mu\nu} = -\frac{1}{2} \Gamma_{\nu AB} \{\psi^A, \psi^B\} \tag{2.1}$$

$$\Gamma^{\mu}_{AB} \mathcal{D}_{\mu} \psi^B = 0. \tag{2.2}$$

Here we are using ten dimensional Cartesian coordinates $\{X^{\mu}\}_{\mu = 0,\ldots 9}$ and expressing covariant differentiation with respect to the Yang-Mills potential (connection form ω on a bundle $E \to M_{10}$) as:

$$\mathcal{D}_{\mu} = \partial_{\mu} + \omega_{\mu} \ ,$$

while the 16-component (half-) spinor field $\{\psi^A\}_{A = 1.,,,16}$ has values in the gauge algebra. The symbol Γ^{μ}_{AB} denotes an off-diagonal block of the usual γ-matrix, under the decomposition of Dirac spinors $(32 = 16 + 16^*)$ and $F_{\mu\nu} = [\mathcal{D}_{\mu}, \mathcal{D}_{\nu}]$ is the field strength (curvature). The supersymmetry transformations leaving (2.1),(2.2) invariant are:

$$\delta\omega_\mu = \varepsilon^A \delta_A \omega_\mu = - \varepsilon^A \Gamma_{\mu AB} \psi^B \tag{2.3a}$$

$$\delta\psi^B = \varepsilon^A \delta_A \psi^B = \frac{1}{2} \varepsilon^A \Sigma^{\mu\nu B}{}_A F_{\mu\nu} \tag{2.3b}$$

where $\{\varepsilon^A\}$ are regarded as anti-commuting "parameters". The superalgebra relations generated by these transformations are:

$$\{\delta_A, \delta_B\} = 2\,\Gamma_{AB}\,\partial_\mu + X_{AB} \tag{2.4}$$

where X consists of terms which vanish on shell, plus an infinitesimal gauge transformation.

2.2 <u>Superspace, Superconnection and Constraints</u>

Extending M_{10} to super Minkowski space $\widetilde{M}_{10}$ with coordinates $\{X^\mu, \theta^A\}$ we can represent the supersymmetry algebra in terms of either left(+) or right (-) translation generators $\{q^\pm_A, \partial_\mu\}$ where

$$q^\pm_A = \frac{\partial}{\partial\theta^A} + \Gamma^\mu_{AB}\theta^B \frac{\partial}{\partial X^\mu} \tag{2.5a}$$

$$\partial_\mu = \frac{\partial}{\partial X^\mu} \tag{2.5b}$$

satisfy:

$$\{q^\pm_A, q^\pm_B\} = \pm 2\,\Gamma^\mu_{AB}\,\partial_\mu \tag{2.6a}$$

$$\{q^+_A, q^-_B\} = 0 \tag{2.6b}$$

We use $\{q^+_A, \partial_\mu\}$ as a basis for our superfields (i.e.a super-frame) and consider $\{q^-_A, \partial_\mu\}$ as the supersymmetry generators. Introducing a superconnection on the super Yang-Mills bundle $\widetilde{E} \to \widetilde{M}_{10}$, with components $\{\widetilde{\omega}_\mu, \widetilde{\omega}_A\}$ relative to some local gauge, the supercovariant derivatives are denoted:

$$\widetilde{\mathcal{D}}_\mu = \partial_\mu + \widetilde{\omega}_\mu \tag{2.7a}$$

$$Q_A = q^+_A + \widetilde{\omega}_A \tag{2.7b}$$

The constraints are:

$$\{Q_A, Q_B\} = 2\,\Gamma^\mu_{AB}\,\widetilde{\mathcal{D}}_\mu \tag{2.8}$$

and appear as integrability conditions for the linear system:

$$\lambda^{\mu}\tilde{\mathcal{D}}_{\mu}\tilde{V}(\lambda) = 0 \tag{2.9a}$$

$$\lambda_{\mu}\Gamma^{\mu AB}Q_{B}\tilde{V}(\lambda) = 0 \tag{2.9b}$$

defining a covariant constant section $\tilde{V}$:

$$\tilde{V}: \tilde{n}(\lambda) \to \tilde{E}_{\tilde{n}(\lambda)}$$

along the super null line with tangent space spanned by $\{\lambda^{\mu}\partial_{\mu}, \lambda^{\mu}\Gamma_{\mu}^{AB}q^{+}_{B}\}$.

3. EQUIVALENCE THEOREM

3.1 Statement of Theorem

Within automorphisms of the respective bundles $E \to M_{10}$ and $\tilde{E} \to \tilde{M}_{10}$ (i.e gauge transformations depending respectively on the $\{X^{\mu}\}$ variables only or on the variables $\{X^{\mu}, \theta^{A}\}$),the following sets of data are equivalent:

(i) The fields $\{\omega_{\mu}, \psi^{A}\}$ in M_{10} satisfying the sypersymmetric Yang-Mills equations (2.1) and (2.2)

(ii) A superconnection $\tilde{E} \to \tilde{M}_{10}$ satisfying the constraints(2.8); i.e. with curvature vanishing on all super null-lines.

3.2 Constraints ⇒ Field Equations

The implication in this sense was proved by Wess(11) and Witten(12). One defines the spinorial superfield $\{\tilde{\psi}^{B}\}$ in terms of the vector-spinor components of the supercurvature:

$$\tilde{\psi}^{A} \equiv \frac{1}{10}\Gamma^{\mu\,AB}[\tilde{\mathcal{D}}_{\mu}, Q_{B}] \tag{3.1}$$

(where $\Gamma^{\mu\,AB}$ is the other off-diagonal block of the γ-matrix) and notes that the Bianchi identities imply the inverse relation:

$$[Q_{A}, \tilde{\mathcal{D}}_{\mu}] = -\Gamma_{\mu\,AB}\tilde{\psi}^{B} \tag{3.2}$$

The constraints (2.8), together with the Bianchi identities further imply:

$$Q_{A}\tilde{\psi}^{B} = \frac{1}{2}\Sigma^{\mu\nu\,B}{}_{A}\tilde{F}_{\mu\nu} \tag{3.3}$$

(where $\Sigma^{\mu\nu\,B}{}_{A}$ is the infinitesimal rotation generator on the 16-spinors).

The set of relations (2.8), (3.2) and (3.3) will be referred to collectively as the "Q-relations", and may be seen as defining the action of a spinorial covariant derivative on each of the three types of superfield components: $\{\tilde{\omega}_A, \tilde{\omega}_\mu, \tilde{\psi}^B\}$ that we have. Expressing (2.8) as:

$$\Gamma^\mu{}_{AB}\tilde{\mathcal{D}}_\mu = \frac{1}{2}[Q_A Q_B + Q_B Q_A],$$

applying this to $\tilde{\psi}^B$, and making use of the Q-relations, gives the super-Dirac equation:

$$\Gamma^\mu{}_{AB}\tilde{\mathcal{D}}_\mu\tilde{\psi}^B = 0. \tag{3.4}$$

Applying the operator $\Gamma_\nu{}^{AC}Q_C$ to this equation and again using the Q-relations gives the remaining superfield Yang-Mills equations:

$$\tilde{\mathcal{D}}^\mu\tilde{F}_{\mu\nu} = -\frac{1}{2}\Gamma_{\nu AB}\{\tilde{\psi}^A, \tilde{\psi}^B\}. \tag{3.5}$$

Taking leading components in a $\{\theta^A\}$ expansion for $\{\tilde{\omega}_\mu, \tilde{\psi}^B\}$ this gives the supersymmetric Yang-Mills equations (2.1) and (2.2).

3.3 The $\mathcal{D}$-Gauge

To complete the proof of equivalence, a method of inverting the passage from the constrained superconnection to the supersymmetric Yang-Mills-Dirac fields must be found. Since the $\{\theta^A\}$-dependence in gauge transformations cannot appear in the latter data on M_{10}, a way must be found to eliminate it. This is done by choosing a suitable family of gauges in $\tilde{M}_{10}$ which uniquely extend any given gauge choice in M_{10}.

Let

$$\mathcal{D} \equiv \theta^A \frac{\partial}{\partial\theta A} = \theta^A q^{\pm}_A \tag{3.6}$$

denote the Euler operator transversal to the embedding

$$M_{10} \hookrightarrow \tilde{M}_{10}$$
$$\{X^\mu\} \hookrightarrow \{X^\mu, 0\} \quad ,$$

i.e. the $\{\theta^A\}$ homogeneity operator. We may always require the following gauge condition to hold:

$$\tilde{\omega}(\mathcal{D}) = \theta^A\tilde{\omega}_A = 0 \tag{3.7}$$

i.e. the components of the connection along $\mathcal{D}$ should vanish. (The geometrical meaning of this condition is that the horizontal lift of $\mathcal{D}$ is tangential to the local section of $\tilde{E} \to \tilde{M}_{10}$ defining our gauge.)

Moreover, it is easily seen that any two choices of gauge

satisfying this condition must be related by a gauge transformation:

$$\tilde{\omega} \to g^{-1}\tilde{\omega}g + g^{-1}dg$$

involving no dependence on the $\{\theta^A\}$ coordinates: $g = g(X^\mu)$. Hence, the extension of gauge choice from M_{10} to $\tilde{M}_{10}$ is unique.

Moreover, contracting the Q-relations (2.8),(3.2),(3.3) with θ^A and using the fact that the gauge condition (3.7) is equivalent to the equality of the covariant and ordinary Euler operator:

$$\mathcal{D} = \theta^A \frac{\partial}{\partial\theta^A} = \theta^A Q_A \ , \tag{3.8}$$

we obtain the following set of relations:

$$(1+\mathcal{D})\tilde{\omega}_A = 2\,\theta^B \Gamma^\mu_{AB}\tilde{\omega}_\mu \tag{3.9a}$$

$$\mathcal{D}\tilde{\omega}_\mu = -\theta^A \Gamma_{\mu AB}\tilde{\psi}^B \tag{3.9b}$$

$$\mathcal{D}\tilde{\psi}^B = \frac{1}{2}\theta^A \Sigma^{\mu\nu\,B}{}_A \tilde{F}_{\mu\nu} \tag{3.9c}$$

These relations have no dynamical content, but rather permit the superfield components $\{\tilde{\omega}_A, \tilde{\omega}_\mu, \tilde{\psi}^B\}$ to be uniquely reconstructed in terms of the leading components$\{\omega_\mu, \psi^B\}$. This follows from the observation that $\mathcal{D}$ is positive semi-definite with integers eigenvalues 0,1,2,..., allowing eqns.(3.4a-c) to be used recursively to determine the superfields. The leading terms are:

$$\tilde{\omega}_B = \theta^A \Gamma^\mu_{AB}\omega_\mu - \frac{2}{3}\theta^A\theta^C \Gamma^\mu_{AB}\Gamma_{\mu CD}\psi^D + \ldots \tag{3.10a}$$

$$\tilde{\omega}_\mu = \omega_\mu - \theta^A \Gamma_{\mu AB}\psi^B + \ldots \tag{3.10b}$$

$$\tilde{\psi}^B = \psi^B + \frac{1}{2}\theta^A \Sigma^{\mu\nu\,B}{}_A F_{\mu\nu} + \ldots \tag{3.10c}$$

3.4 Field Equations ⇒ Constraints

The proof of the above implication proceeds in three steps.

1. First we prove by induction in the $\{\theta^A\}$-homogeneity degree that the ordinary field equations imply the super field equations. This follows by applying the $\mathcal{D}$-operator to both sides of the superfield equations and using the inductive hypothesis.

e.g. $\mathcal{D}[\tilde{\mathcal{D}}^\mu \tilde{F}_{\mu\nu} - \frac{1}{2}\Gamma_{\nu AB}\{\tilde{\psi}^A, \tilde{\psi}^B\}]$

$= \theta^C \{X_{\nu C}\}$ ↑ proportional to the Dirac equation

$= 0$ to the order n+1 by the inductive hypothesis

The computation is formally equivalent to verification of invariance under the supersymmetry transformations.

2. Next we prove, again by induction in the $\{\theta^A\}$-homogeneity, using a similar procedure, that the Q-relations (3.2) and (3.3) are valid. Here the computation is formally equivalent to the verification of the closure of the superalgebra relations.

3. Finally, we prove that the relations (3.2) and (3.3) imply the constraints by applying the positive operator $(2+\mathcal{D})$ to both sides of (2.8), using the relations (3.9a-c) and noting the resulting difference is proportional to the Q-relations (3.2) and (3.3).

The full proof, together with the corresponding dimensional reduction to the $d=4$, $N=4$ or $d=6$, $N=2$ case may be found in ref.(13).

REFERENCES

(1) See,e.g. Gates, S.J.,Jr., Grisaru, M.T.,Rocek,M., Siegel, W.: Superspace. Reading, MA: Benjamin/Cummings (1983).

(2) Sohnius, M., Nucl.Phys. B136, 461 (1978).

(3) Witten, E., Phys. Lett.77B, 394 (1978)

(4) Card, C.T., Davis, R., Restuccia, A. and Taylor, J.G., Phys.Lett. 146B, 199 (1984).

(5) Galperin, A., Ivanov. E., Kalitzin, S. Ogievetsky, V. and Sokatchev. E., Class. Quantum Grav. 2, 155 (1985).

(6) Harnad, J., Hurtubise, J., Legare, M. and Shnider, S., Nucl. Phys. B256, 609 (1985).

(7) Aref'eva, I. Ya. and Volovich, I.V., Class. Quantum Grav. 3, 617 (1986);and: 1984 Proc. 3rd Seminar on Quantum Gravity: (Singapore, World Scientific), p.363.

(8) Manin, Yu., "Flag superspaces and supersymmetric Yang-Mills equations" in: Arithmetic and Geometry, Vol.II, ed.M. Artin and J.Tate, Birkhauser, Boston, p.175 (1983).

(9) Ward, R.S. Phys.Lett. 61A, 81 (1977).

(10) Manin, Yu., Inst.Trudy Steklov 165, 98 (1984).

(11) Wess, J. "Supersymmetric Gauge Theories" in: Unified Theories and Beyond, Proc. 5th Johns Hopkins Workshop, 5 (1981).

(12) Witten,E., Nucl. Phys. B266. 245 (1986).

(13) Harnad, J. and Shnider S., Commun.Math. Phys.106, 183 (1986).

CHERN WEIL FORMS IN SUPERSPACE

G. Girardi, R. Grimm*)

LAPP, BP. 909, 74019 Annecy-le-Vieux Cedex
FRANCE

ABSTRACT

The construction of Yang Mills as well as Lorentz Chern Weil forms in superspace geometry is explained in the context of 16-16 supergravity. Supersymmetry is guaranteed by construction, all the auxiliary fields are present and a superspace action is given.

Chern Weil[1], or Chern Simons[2] forms play a prominent rôle in ten dimensional supergravity[3] and in the Green Schwarz anomaly cancellation mechanism[4]. Whereas the Chern Weil forms constructed out of Yang Mills connections occur in a supersymmetric way, the situation for gravitational Chern Weil forms is not so obvious[4,5]. Another question presently under investigation is the realisation of the algebra of supersymmetry transformations in such theories without assuming the equations of motions[6]. In view of these difficulties we found it instructive to study Chern Weil forms in the context of four dimensional N=1 supergravity. While the Yang Mills type Chern Weil forms can be incorporated in a straightforward way[7,8], the gravitational case turns out to be slightly more complicated[9].

In this note we shall describe in a compact way the geometry[10] of U(1) superspace[8] in the presence of Yang Mills as well as Lorentz type Chern Weil forms, specify to 16-16 supergravity and present an invariant superspace action. A more detailed description of this

*) Talk given by R. Grimm.

superspace geometry as well as a more complete list of references will be contained in a forthcoming article by the present authors.

The basic superfields that we will have to consider are

$$
\begin{aligned}
&E^A = dz^M E_M{}^A && \text{repère mobile} \\
&\phi_B{}^A = dz^M \phi_{MB}{}^A && \text{Lorentz gauge potential} \\
&A = dz^M A_M && \text{chiral U(1) gauge potential} \qquad (1)\\
&\mathcal{A} = dz^M \mathcal{A}_M && \text{Yang Mills gauge potential} \\
&B = \tfrac{1}{2} dz^M dz^N B_{NM} && \text{two form gauge potential.}
\end{aligned}
$$

They transform as usual with respect to local diffeomorphisms and gauge transformations. We define covariant torsion and field strengths

$$
\begin{aligned}
T^A &= dE^A + E^B \phi_B{}^A + w(E^A) E^A A \\
R_B{}^A &= d\phi_B{}^A + \phi_B{}^C \phi_C{}^A \\
F &= dA \\
\mathcal{F} &= d\mathcal{A} + \mathcal{A}\mathcal{A}
\end{aligned}
\qquad (2)
$$

where the chiral weight $w(E^A)$ is defined as follows

$$w(E^a) = 0 \ , \ w(E^\alpha) = +1 \ , \ w(E_{\dot\alpha}) = -1 \qquad (3)$$

The Chern Weil form Q_3 of the Yang Mills potential $\mathcal{A}$ is defined as

$$Q_3(\mathcal{A}) = \mathrm{tr}\left(\mathcal{A} d\mathcal{A} + \tfrac{2}{3}\mathcal{A}\mathcal{A}\mathcal{A}\right) \qquad (4)$$

and has the property

$$dQ_3(\mathcal{A}) = \mathrm{tr}(\mathcal{F}\mathcal{F}) \tag{5}$$

tr stands for some suitably normalized trace. If $\mathcal{A}$ transforms as $\mathcal{A}' = X^{-1}(\mathcal{A}-d)X$, the Chern Weil form changes as:

$$\begin{aligned} Q_3' &= Q_3 - d\Gamma \\ \Gamma &= \mathrm{tr}\left(\mathcal{A}XdX^{-1} + \int_0^1 X_t^{-1} d_t X_t (X_t^{-1} dX_t)^2\right) \end{aligned} \tag{6}$$

X_t interpolates between $X_o = 1$ and $X_1 = X$, $t \in [0,1]$ and d_t is the exterior derivative with respect to t.
Now, the three form field strengths

$$G = dB + Q_3(\mathcal{A}) \tag{7}$$

is invariant if B transforms as

$$B' = B + d\Sigma + \Gamma \tag{8}$$

and the Bianchi identities are

$$dG = \mathrm{tr}(\mathcal{F}\mathcal{F}) \tag{9}$$

Of course similar considerations apply to $\phi_B{}^A$ and A and we shall therefore treat the more general case where

$$dG = \mathrm{tr}(\mathcal{F}\mathcal{F}) + \tau(R_\beta{}^\alpha R_\alpha{}^\beta + R^{\dot\alpha}{}_{\dot\beta} R^{\dot\beta}{}_{\dot\alpha}) + \sigma FF \tag{10}$$

and τ and σ are some constants.

In Ref. 8) it was shown that in U(1) superspace all components of torsion, curvature and chiral U(1) field strengths can be expressed in terms of the superfields R, R^+, V_a and $W_{\gamma\beta\alpha}$, $W_{\dot\gamma\dot\beta\dot\alpha}$ which are in turn subject to further restrictions. In our notations, the non-vanishing torsions are

$$
\begin{aligned}
&T_\gamma{}^{\dot\beta a} = -2i\,(\sigma^a\epsilon)_\gamma{}^{\dot\beta} \\
&T_{\gamma b\dot\alpha} = -i\,\sigma_{b\gamma\dot\alpha} R^+ \quad , \quad T^{\dot\gamma}{}_b{}^\alpha = -i\,\bar\sigma_b{}^{\dot\gamma\alpha} R \\
&T_{\gamma b}{}^\alpha = \tfrac{1}{2}(\sigma_c\bar\sigma_b)_\gamma{}^\alpha V^c \quad , \quad T^{\dot\gamma}{}_{b\dot\alpha} = -\tfrac{1}{2}(\bar\sigma_c\sigma_b)^{\dot\gamma}{}_{\dot\alpha} V^c
\end{aligned}
\tag{11}
$$

and $T_{cb}{}^\alpha$, $T_{cb\dot\alpha}$ which contain the Weyl spinor superfields in their tensor decomposition. Likewise, in this kind of curved superspace the components of the Yang Mills field strengths are completely fixed in terms of the chiral and antichiral spinor superfields λ_α, $\bar\lambda^{\dot\alpha}$.

It remains to specify the non-vanishing components of $G = \frac{1}{3!} E^A E^B E^C G_{CBA}$ compatible with the constraints on torsions and curvatures appearing in eq.(10):

$$
\begin{aligned}
&G_{\gamma\beta a} = 16\tau\,(\sigma_{ac}\epsilon)_{\gamma\beta} V^c R^+ \\
&G^{\dot\gamma\dot\beta}{}_a = -16\tau\,(\bar\sigma_{ac}\epsilon)^{\dot\gamma\dot\beta} V^c R \\
&G_\gamma{}^{\dot\beta}{}_a = -2i\,(\sigma_a\epsilon)_\gamma{}^{\dot\beta}\left(L - 4\tau R R^+ + \tau V^b V_b\right) \\
&\qquad\qquad - 2i\,(\sigma^b\epsilon)_\gamma{}^{\dot\beta}\left(\tfrac{3\nu}{4} - 2\tau\right) V_b V_a
\end{aligned}
\tag{12}
$$

with $\nu = 3\sigma + 4\tau$. Furthermore:

$$
\begin{aligned}
G_{\gamma ba} &= 2(\sigma_{ba})_\gamma{}^\varphi \mathcal{D}_\varphi L - 8\tau R^+ T_{ba\gamma} + 4i\tau V_\gamma{}^{\dot\gamma} T_{ba\dot\gamma} \\
&\quad - \tfrac{3\nu}{4}\left(V_b \mathcal{D}_\gamma V_a - V_a \mathcal{D}_\gamma V_b\right) - \tfrac{i\nu}{2}\left(V_b(\sigma_a\bar\lambda)_\gamma - V_a(\sigma_b\bar\lambda)_\gamma\right)
\end{aligned}
\tag{13}
$$

$$G^{\dot{\gamma}}{}_{ba} = 2(\bar{\sigma}_{ba})^{\dot{\gamma}}{}_{\dot{\varphi}} \mathcal{D}^{\dot{\varphi}} L - 8\tau R T_{ba}{}^{\dot{\gamma}} - 4i\tau V_{\gamma}{}^{\dot{\gamma}} T_{ba}{}^{\gamma}$$
$$- \frac{3v}{4}\left(V_b \mathcal{D}^{\dot{\gamma}} V_a - V_a \mathcal{D}^{\dot{\gamma}} V_b\right) + \frac{iv}{2}\left(V_b (\bar{\sigma}_a \chi)^{\dot{\gamma}} - V_a (\bar{\sigma}_b \chi)^{\dot{\gamma}}\right)$$

The real superfield L is subject to the conditions

$$(\mathcal{D}^{\alpha}\mathcal{D}_{\alpha} - 8R^{+})L = -8\tau W_{\dot{\gamma}\dot{\beta}\dot{\alpha}} W^{\dot{\gamma}\dot{\beta}\dot{\alpha}} + \frac{v}{6}\bar{\chi}_{\dot{\alpha}}\bar{\chi}^{\dot{\alpha}} + \frac{1}{2}\operatorname{tr}(\bar{\lambda}_{\dot{\alpha}}\bar{\lambda}^{\dot{\alpha}})$$
$$(\mathcal{D}_{\dot{\alpha}}\mathcal{D}^{\dot{\alpha}} - 8R)L = -8\tau W^{\gamma\beta\alpha} W_{\gamma\beta\alpha} + \frac{v}{6}\chi^{\alpha}\chi_{\alpha} + \frac{1}{2}\operatorname{tr}(\lambda^{\alpha}\lambda_{\alpha}) \tag{14}$$

with the abbreviations

$$\chi_{\alpha} = \mathcal{D}_{\alpha}R + i\mathcal{D}^{\dot{\alpha}}V_{\alpha\dot{\alpha}}, \quad \bar{\chi}_{\dot{\alpha}} = \mathcal{D}_{\dot{\alpha}}R^{+} + i\mathcal{D}^{\alpha}V_{\alpha\dot{\alpha}} \tag{15}$$

The component G_{cba}, together with further nonlinear contributions, is located in the $\theta\bar{\theta}$ component of the superfield L.

As explained in Ref. 8) this geometrical structure still allows for various cases. In this note we consider only 16-16 supergravity obtained through the following U(1) breaking :

$$L = e^{b\mathfrak{X}} + \Gamma(\tau), \quad \Gamma(0) = 0, \quad \sigma = 0 \tag{16}$$

$$A_{\alpha} = -\mathcal{D}_{\alpha}\mathfrak{X}, \quad A^{\dot{\alpha}} = \mathcal{D}^{\dot{\alpha}}\mathfrak{X}$$
$$A_{\alpha\dot{\alpha}} = \frac{3}{2}V_{\alpha\dot{\alpha}} + \frac{i}{2}[\mathcal{D}_{\alpha}, \mathcal{D}_{\dot{\alpha}}]\mathfrak{X} - iT_{\alpha}T_{\dot{\alpha}} \tag{17}$$
$$T_{\alpha} \equiv \mathcal{D}_{\alpha}\mathfrak{X}, \quad T^{\dot{\alpha}} \equiv \mathcal{D}^{\dot{\alpha}}\mathfrak{X}$$

Here b is a real constant, $\mathfrak{X}$ a real superfield and one may allow for some function $\Gamma(\tau)$ depending on R, R^{+}, V_a which may eventually become useful when the explicit structure of curvature squared terms in the action will be discussed.

The action is given by the volume of superspace,

$$\mathcal{L} = \int_* E \tag{18}$$

and the equations of motion are obtained in the usual way[11], taking into account the particular superspace geometry described above. For $\tau = 0$ one has :

$$\begin{aligned}\delta\mathcal{L} &= \int_* E H_A{}^A(\cdot)^a = \frac{2b}{3b+4}\int_* E\Big[8UR + 8\bar{U}R^\dagger \\ &+ i\,\mathcal{V}^{\varphi\dot{\varphi}}\big(-iV_{\varphi\dot{\varphi}} + bT_\varphi T_{\dot{\varphi}} + \frac{i}{2b}L^{-1}\mathrm{tr}(\lambda_\varphi\lambda_{\dot{\varphi}})\big) \\ &- L^{-1}\mathrm{tr}\Big\{\frac{\Lambda-\bar{\Lambda}}{2b}\big((\mathcal{D}-T)^\varphi\lambda_\varphi - bT^\varphi\lambda_\varphi - bT_{\dot{\varphi}}\bar{\lambda}^{\dot{\varphi}}\big)\Big\} \\ &+ \xi^\varphi\,\Xi_\varphi + \xi_{\dot{\varphi}}\,\Xi^{\dot{\varphi}}\end{aligned} \tag{19}$$

Here U, $\bar{U}$, $\mathcal{V}^{\varphi\dot{\varphi}}$, $\Lambda-\bar{\Lambda}$ and ξ_α, $\xi^{\dot{\alpha}}$ are arbitrary variations. Ξ_φ and $\Xi^{\dot{\varphi}}$ stand for some expression in the basic superfields and their derivatives, which vanish as a consequence of the other superfield equations of motion:

$$R = R^\dagger = 0 \tag{20}$$

$$iV_{\varphi\dot{\varphi}} - bT_\varphi T_{\dot{\varphi}} - \frac{i}{2b}e^{-b\mathcal{X}}\mathrm{tr}(\lambda_\varphi\bar{\lambda}_{\dot{\varphi}}) = 0 \tag{21}$$

$$(\mathcal{D}-T)^\varphi\lambda_\varphi - b(T^\varphi\lambda_\varphi + T_{\dot{\varphi}}\lambda^{\dot{\varphi}}) = 0 \tag{22}$$

For $\tau \neq 0$, e.g. in the presence of Lorentz Chern Weil forms, the situation becomes more complicated. A detailed investigation of this case will be presented elsewhere[12].

REFERENCES

1) Weil A., Géométrie Différentielle des Espaces Fibrés, dans "Oeuvres Complètes" (1949e).

2) Chern S.-S. and Simons J., Proc. Nat. Acad. Sci. USA 68, 791 (1971);
Chern S.-S. and Simons J., Annals of Math. 99, 48 (1974).

3) Chapline G.F. and Manton N.S., Phys. Lett. 120B, 105 (1983).

4) Green M.B. and Schwarz J.H., Phys. Lett. 149B, 117 (1984).

5) Romans L.J. and Warner N.P., Some Supersymmetric Counterparts of the Lorentz Chern-Simons Terms, preprint Caltech CALT-68-1291, August 1985;
Gates Jr. S.J. and Nishino H., Phys. Lett. 173B, 52 (1986).

6) Nilsson B.E.W., Phys. Lett. 175B, 319 (1986).

7) Grimm R., Seminar presented at the NATO Adv. Study Inst., Bonn 1984, preprint LAPP-TH-123, to appear in the proceedings.

8) Müller M., Nucl. Phys. B264, 292 (1986).

9) Cecotti S., Ferrara S., Girardello L. and Porrati M., Phys. Lett. 164B, 46 (1986);
Cecotti S., Ferrara S., Girardello L., Porrati M. and Pasquinucci A., Phys. Rev. D33, 2504 (1986);
Baulieu L. and Bellon M., Phys. Lett. 169B, 59 (1986).

10) Wess J. and Bagger J., Supersymmetry and Supergravity, Princeton Series in Physics, Princeton Univ. Press, Princeton 1983.

11) Wess J. and Zumino B., Phys. Lett. 74B, 51 (1978).

12) Grimm R., Müller M., in preparation.

MINIMAL N = 2 SUPERGRAVITY

Martin Müller

Institut für Theoretische Physik, Universität
Karlsruhe, D-7500 Karlsruhe, Fed. Rep. of Germany

ABSTRACT

Two minimal N = 2 supergravity multiplets with 32+32 degrees of freedom are derived from a geometrical theory in conventional extended superspace. These multiplets are off-shell irreducible although they describe on-shell the coupling of supergravity to a scalar, resp. vector multiplet. One of them is the N = 2 part of a N = 4 supergravity theory.

1. INTRODUCTION

Off-shell formulations of supersymmetric theories without central charges are known only for N = 1 and N = 2. The first off-shell version of N = 2 supergravity, the so-called standard multiplet, was given by Fradkin and Vasiliev[1)] and by de Wit, van Holten, and van Proeyen[2)]. It contains 40 bosonic and 40 fermionic degrees of freedom. Later two other 40+40 supergravity multiplets were found[3)]. Recently we have shown that the standard multiplet is in fact reducible to a minimal 32+32 multiplet[4)]. This will be briefly reviewed in the next section. The remainder of this talk will be concerned with another minimal 32+32 multiplet which has the interesting property to be the N = 2 part of a N = 4 supergravity theory[5)].

2. REDUCTION OF THE STANDARD MULTIPLET

The most natural setting for N = 2 supergravity seems to be conventional extended superspace with the coordinates $z^{\underline{M}} \sim (x^m, \theta^\mu_M, \bar{\theta}^M_{\dot{\mu}})$, M = 1,2. We choose $SL(2,\mathbb{C}) \times U(2)$ as structure group since the constraint take their simplest form in this kind of superspace. (The U(2) will be broken later.) The basic geometric objects are the vielbein $E_{\underline{M}}{}^{\underline{A}}$ and the connection $\Phi_{\underline{MB}}{}^{\underline{A}}$. The covariant tensors are the torsion $T_{\underline{CB}}{}^{\underline{A}}$ and the curvature $R_{\underline{DCB}}{}^{\underline{A}}$.

The huge number of component fields contained in vielbein and connection has to be reduced by covariant conditions. The natural constraints on torsion and curvature are given in ref. 6. The additional constraints that lead to the standard multiplet are the following[7)]. The field strength $F_{\underline{BA}}$ of an abelian gauge potential $A_{\underline{M}}$ is restricted by

$$F_{\beta\alpha}^{BA} = 2\sqrt{2}\,\varepsilon_{\beta\alpha}\,\varepsilon^{BA}\,, \quad F_{\beta A}^{B\dot{\alpha}} = 0\,, \tag{1}$$

and the U(2) symmetry is broken by

$$\Phi_{\gamma}{}^{CB}{}_{A} = \delta_{A}^{B}\,\rho_{\gamma}^{C} - 2\,\delta_{A}^{C}\,\rho_{\gamma}^{B}\,. \tag{2}$$

The various Bianchi identities subject to these constraints reduce the number of independent covariant superfields to one spinor superfield ρ_{α}^{A} and one chiral superfield $X_{(\dot{\alpha}\dot{\beta})}$. The independent component fields form the 40+40 standard multiplet.

The reduction of this multiplet is given by the superfield equations

$$\rho_{\alpha}^{A} = \mathcal{D}_{\alpha}^{A}\,U\,, \quad \bar{\rho}_{A}^{\dot{\alpha}} = \bar{\mathcal{D}}_{A}^{\dot{\alpha}}\,U\,, \tag{3}$$

where U is a real superfield. The component fields reduce as follows:

$$(e_m{}^a,\ \psi_{mA}{}^{\alpha},\ A_m \,|\, \rho_{\alpha}^{A},\ P,\ H_a,\ T_a^{AB},\ X^{AB},\ Y_{ab},\ U_a,\ \Lambda_{\alpha}^{A}) \qquad 40+40 \tag{4}$$

$$\downarrow \qquad\qquad \downarrow \qquad \downarrow \qquad \downarrow \qquad \downarrow$$

$$(e_m{}^a,\ \psi_{mA}{}^{\alpha},\ A_m,\ \varphi_{\alpha}^{A},\ C\,,\ t_{mn}^{AB} \,|\, M^{AB},\ N_{ab},\ b_a) \qquad 32+32 \tag{5}$$

(The bar separates the physical from the auxiliary fields.) The auxiliary fields P, H_a, and Λ_{α}^{A} are eliminated from the multiplet (4) and the new component field $C = -2\,U|$ appears. T_a^{AB} becomes proportional to the dual field strength of three two-form gauge potentials t_{mn}^{AB} (symmetric in AB). The lagrangian and transformation laws of the multiplet (5) are given in ref. 4. On-shell this theory describes the coupling of N = 2 supergravity to a hypermultiplet (φ_α^A, C, t_{mn}^{AB}), where three of the four scalars are replaced by antisymmetric tensors. The remarkable fact is that this coupling is off-shell irreducible.

3. ANOTHER MINIMAL 32+32 MULTIPLET

The constraints that lead to the second minimal supergravity multiplet are the natural constraints on torsion and curvature and the constraints[5)]

$$F_{\beta\alpha}^{BA} = 4\varepsilon_{\beta\alpha}\varepsilon^{BA}\overline{W}\ , \quad F_{\beta A}^{B\dot{\alpha}} = F_{\dot{B}A}^{\dot{\beta}\dot{\alpha}} = 0 \tag{6}$$

on the field strength of a complex abelian one-form potential (equivalent to vector and axial vector). In this case the U(2) symmetry of the structure group cannot be broken completely. Off-shell, a local SO(2) invariance remains.

The consequences of these constraints can be analyzed by solving the Bianchi identities. Like in sect. 2, we find that only two independent superfields remain: U and $W_{(\alpha\beta)}$. The independent component fields form the multiplet

$$(e_m{}^a,\ \psi_{mA}{}^\alpha,\ a_m,\ a'_m,\ \varphi_\alpha^A,\ C,\ t_{mn}\ |\ v_m,\ t'_{mn},\ M,\ b_{aB}^A) \tag{7}$$

with 32+32 degrees of freedom. It contains a vector multiplet (a_m, φ_α^A, C, t_{mn}), where the pseudoscalar is replaced by an antisymmetric tensor. The transformation laws of the multiplet (7) are given in ref. 5. They describe on-shell the truncation of a N = 4 supergravity theory with an antisymmetric tensor potential[8)]. The N = 1 limit is the 16+16 multiplet[9)].

The superfield lagrangian for the multiplet (7) is

$$\mathcal{L} = 6\int d^4\Theta\,\mathcal{E}\,\Phi' R + h.c.\,. \tag{8}$$

$\mathcal{E}$ is a chiral density, Φ' a chiral superfield containing t'_{mn}, and $R \sim R_{\dot{\alpha}\dot{\beta}}^{AA\dot{\alpha}\dot{\beta}}$. The modified lagrangian

$$\mathcal{L} = 6\int d^4\Theta\, \mathcal{E}\Phi'\,(R + gW) + h.c. \tag{9}$$

(g complex) describes N = 2 supergravity with gauged SO(2) × SO(2). Like in the corresponding N = 4 theory with gauged SU(2) × SU(2)[10], the scalar potential is unbounded from below.

Finally we remark that a parameter n like in N = 1 supergravity[9] does not exist in the case N = 2.

REFERENCES

1. E. S. Fradkin and M. A. Vasiliev, Lett. Nuovo Cim. 25, 79 (1979); Phys. Lett. 85B, 47 (1979)
2. B. de Wit and J. W. van Holten, Nucl. Phys. B155, 530 (1979); B. de Wit, J. W. van Holten and A. van Proeyen, Nucl. Phys. B167, 186 (1980)
3. B. de Wit, J. W. van Holten and A. van Proeyen, Nucl. Phys. B184, 77 (1981); B. de Wit, R. Philippe and A. van Proeyen, Nucl. Phys. B219, 143 (1983)
4. M. Müller, Phys. Lett. B172, 353 (1986)
5. M. Müller, Karlsruhe preprint KA-THEP-86-6
6. M. Müller, Karlsruhe preprint KA-THEP-86-4, to appear in Z. Phys. C
7. M. Müller, Z. Phys. C24, 175 (1984)
8. H. Nicolai and P. K. Townsend, Phys. Lett. 98B, 257 (1981)
9. G. Girardi, R. Grimm, M. Müller and J. Wess, Phys. Lett. 147B, 81 (1984); M. Müller, Nucl. Phys. B264, 292 (1986)
10. D. Z. Freedman and J. H. Schwarz, Nucl. Phys. B137, 333 (1978)

CLASSIFICATION OF HOMOGENEOUS KÄHLER MANIFOLDS

H. Römer
Fakultät für Physik der Universität Freiburg
Hermann-Herder-Str. 3, D-7800 Freiburg, FRG

Abstract:
A complete classification of homogeneous Kähler manifolds for semisimple symmetry group G is obtained in terms of the Dynkin diagram of G.

The effective low-energy theory Th' of a quantised field theory Th is given by a field φ which assumes its values in the set V of classical ground states ("vacua") of Th. If a symmetry of Th under a group G is spontaneously broken down to a subgroup $K \subset G$, then, barring "accidental" degeneracy (quasi Goldstone excitations), V is given by the homogeneous space V = G/K. In order to allow for a positive definite kinetic energy, V must have a Riemannian metric. In addition, it is known,[1,2] that if a supersymmetry of Th remains unbroken in Th', V must admit a Kähler structure. More precisely, for space time dimension D = 2,V has a Kähler structure for N = 2 extended supersymmetry and a Hyperkähler structure for N = 4 supersymmetry, whereas D = 4 and minimal N = 1 supersymmetry leads to a Kähler structure of V and D = 4 and N = 2 supersymmetry to a Hyperkähler structure of V. Hyperkähler manifolds are Ricci flat and never homogeneous. We shall not discuss them here except for formulating the following conjecture:

The tangend bundle TM of a homogeneous Kähler manifold with semisimple symmetry group G admits a natural Hyperkähler structure. This was proven[3] locally in a neighbourhood of the zero section of TM for arbitrary M and globally if M is Hermitean symmetric rather than just homogeneous. Here we shall be concerned with homogeneous Kähler manifolds, which in the given setting either arise directly as the set V of ground states or as orbits in V under the symmetry group G. For a complete presentation of the results obtained and a more comprehensive list of references compare ref.[4] by M. Bordemann, M. Forger and

H. Römer.

A Kähler manifold M is by definition given by the following data:

(i) A symplectic structure on M, i.e. a closed nowhere degenerate 2-form ω on M,

(ii) an (integrable) complex structure I: TM $\rightarrow$ TM, I^2=-id on M such that $\omega(IX,IY) = \omega(X,Y)$ and

(iii) the symmetric tensor field g, given by $g(X,Y) = -\omega(IX,Y)$ yields a Riemannian metric on M.

After complexifying TM and T^*M one has in local complex coordinates

$$g = g_{\mu\bar{\nu}}\, dz^\mu d\bar{z}^\nu \quad , \quad \omega = i\, g_{\mu\bar{\nu}}\, dz^\mu \wedge d\bar{z}^\nu \quad .$$

Locally, there exists a Killing potential F such that

$$g_{\mu\bar{\nu}} = \partial^2 F / \partial z^\mu \partial \bar{z}^\nu$$

The Ricci tensor is given by

$$R_{\mu\bar{\nu}} = -\frac{\partial^2}{\partial z^\mu \partial \bar{z}^\nu} \ln \det (g_{\varsigma\bar{\sigma}})$$

M is homogeneous if a group G acts transitively by holomorphic isometries. Homogeneous Kähler manifolds are much more general than Hermitean symmetric spaces. Ref.[4)] gives a complete description of all homogeneous Kähler manifolds with semisimple symmetry group G.

The main results derived there are: Let G be a semisimple Lie group, then

(i) If G/K is a homogeneous Kähler manifold, then K is compact and given by the centraliser of some semisimple element $Z_0 \in \mathfrak{g}$ in the Lie algebra of G: $K = \{ g \in G \mid Ad_g Z_0 = Z_0 \}$

(ii) Conversely, every homogeneous space G/K of this kind admits a pseudo-Kähler metric (Kähler up to signature). It even admits a Kähler metric if G/K is compact or non compact Hermitean symmetric.

(iii) The (pseudo) Kähler metric on G/K linearly depends on a finite number of parameters (squeezing parameters). This number is equal to the dimension of the center of the isotropy group K and equal to unity (uniqueness up to constant multiples) e.g. for G/K Hermitean symmetric.

(iv) The Ricci tensor is independent on the metric. There is always a (up to constant multiples) unique Einsteinian (pseudo) Kähler metric with the defining property $R_{\mu\bar{\nu}} = \text{const}\; g_{\mu\bar{\nu}}$

(v) The Kähler potential can be constructed explicitely. The construction simplifies for the Einsteinian metric.

(vi) A complete list of all homogeneous Kähler manifolds can be constructed by simple graphical rules from the Dynkin diagrams of the symmetry groups G.

Let us briefly sketch the main steps towards these results

1st step: Reduction to algebra. Due to the homogeniety everything can be reduced to the tangent space T_oM at the origin eK. There is a reductive splitting

$$\mathfrak{g} = \mathfrak{k} \oplus \mathfrak{m}$$

where $\mathfrak{k}$ is the Lie algebra of K and $\mathfrak{m}$ can be identified with T_oM. g, ω, I are then given by maps g_o, ω_o, I_o defined on $\mathfrak{m} \times \mathfrak{m}$ and $\mathfrak{m}$ with

(i) $\omega_o([X,Y]_{\mathfrak{m}}, Z) + \omega_o([Y,Z]_{\mathfrak{m}}, X) + \omega_o([Z,X]_{\mathfrak{m}}, Y) = 0$

(ii) $\omega_o(Ad_k X, Ad_k Y) = \omega_o(X,Y)$

(iii) $I_o : \mathfrak{m} \to \mathfrak{m}$, $I_o^2 = -id$, $\omega_o(I_oX, I_oY) = \omega_o(X,Y)$

(iv) $[I_oX, I_oY] - I_o[I_oX,Y] - I_o[X,I_oY] - [X,Y] = 0$

(v) $Ad_k I_o X = I_o Ad_k X$

(vi) $g_o(X,Y) = -\omega_o(I_oX,Y)$

$$(X,Y,Z \in \mathfrak{m},\ k \in K)$$

2nd step: Determination of K and ω_o

The compactness of K follows from (ii), (v), (vi): a positive-definite form is left invariant.

Due to the non-degeneracy of the Killing form (.,.) on $\mathfrak{g}$ (and $\mathfrak{m}$) we can write $\omega_o(X,Y) = (\Phi_o X, Y)$,

where $\Phi_o : \mathfrak{m} \to \mathfrak{m}$ is skew symmetric with respect to (.,.). Extending to $\mathfrak{g}$ by defining $\Phi_o|_{\mathfrak{k}} \equiv 0$ one can show that (i) and (ii) and the non-degeneracy of ω_o imply that there is $Z_o \in \mathfrak{g}$ with $\Phi_o X = [Z_o, X]$. Furthermore, K has to be the centralizer of Z_o. The admissible Z_o span a linear space of dim center K. One can always choose a Cartan subalgebra $\mathfrak{h}^c$ of $\mathfrak{g}^c$ such that $Z_o \in \mathfrak{h}^c$. Taking a root decomposition with respect to this Cartan subalgebra we obtain a splitting $\mathfrak{g}^c = \mathfrak{k}^c \oplus \mathfrak{m}^c$,

$$k^c = \hat{t}^c \oplus \mathfrak{f}'^c \oplus \sum_{\alpha\in\Delta'} g_\alpha \quad , \quad m^c = \sum_{\alpha\in\hat{\Delta}} g_\alpha$$

where Δ' are the roots belonging to k' with the defining property $\alpha(Z_0)=0$, $\hat{\Delta}$ are the roots in m^c with $\alpha(Z_0)\neq 0$, $t_c (\ni Z_0)$ is the subspace of $\mathfrak{f}^c$ annihilated by m^c and $\mathfrak{f}'^c$ the Cartan subalgebra of k.

In a Cartan basis we find

$$\omega_0(E_\alpha, E_\beta) = \alpha(Z_0)\,(E_\alpha, E_{-\alpha})\,\delta_{\alpha,-\beta}$$

3rd step: Determination of I_o.

Form (iii) and (v) we infer that $I_o : m^c \to m^c$ has to obey

$$I_o E_\alpha = i\varepsilon_\alpha E_\alpha \quad \text{with} \quad \varepsilon_\alpha = -\varepsilon_{-\alpha} = \pm 1 , \ \alpha\in\hat{\Delta} \ \text{and}$$

$$\varepsilon_{\alpha+\gamma} = \varepsilon_\alpha \ \text{for} \ \alpha\in\hat{\Delta}, \gamma\in\Delta', \alpha+\gamma\in\hat{\Delta}$$

Condition (iv) leads to $\alpha, \beta, \alpha+\beta \in \hat{\Delta}, \ \varepsilon_\alpha = \varepsilon_\beta \Rightarrow \varepsilon_{\alpha+\beta} = \varepsilon_\alpha$

All complex structures are thus given by Weyl Chambers in $\hat{t}$ and are of the form $\varepsilon_\alpha = \text{sign}\, i\alpha(Z_I)$

One obtains what we call an invariant ordering in $\hat{\Delta}$ by defining a root $\alpha\in\hat{\Delta}$ to be positive if $\varepsilon_\alpha > 0$. Every invariant ordering can be obtained from an ordering in Δ which is compatible with the splitting $\Delta = \Delta' \cup \hat{\Delta}$, and there is one to one correspondence between complex structures and invariant orderings in $\hat{\Delta}$.

The metric is $g_0(E_\alpha, E_\beta) = -i\varepsilon_\alpha \alpha(Z_0)\,(E_\alpha, E_{-\alpha})\,\delta_{\alpha,-\beta}$

For M compact or symmetric definiteness is achieved by choosing $Z_I = \pm Z_0$

4rd step: Classification

The above mentioned constructions carry over to cartesian product groups. Thus, it suffices to take G simple. If G/K is non compact and L$\supset$K a maximal compact subgroup of G, then G/L is non compact Hermitean symmetric and L/K a compact homogeneous Kähler manifold. Hence, for what follows we can restrict ourselves to compact simple symmetry groups. The Dynkin diagram respresents all positive roots with respect to a given ordering of Δ. Take a splitting $\Delta = \Delta' \cup \hat{\Delta}$ and a compatible ordering. Homogeneous Kähler manifolds are obtained from "painted Dynkin diagrams" in the following way: Draw the Dynkin diagram of g and paint some circles in it in black. The black roots give the semi-simple part of k and every white root yields an additional $u(1)$ summand in k.

Example: $E_8 \cong$ o-o-o(o)-o-o-o-o $\rightarrow$ •-•-•(•)-•-o-•-•

$k = \check{u}(1) \oplus \check{u}(1) \oplus s\check{u}(3) \oplus so(8)$

The same k may be obtained in different ways from a painted Dynkin diagram and the resulting homogeneous Kähler manifolds are not always isomorphic.

Isomorphisms are obtained from automorphisms of G. Inner automorphisms lead to conjugate subgroups K and just correspond to a change of the reference point in the orbit. Since Cartan subalgebras are always conjugate we can keep $\mathfrak{f}$ and the invariant ordering fixed. The freedom under inner automorphisms we are left with is given by the Weyl group W, and equivalent homogeneous Kähler sapces are obtained from "admissible" Weyl group elements w, for which also the transformed splitting $w\Delta' \cup w\hat{\Delta}$ is compatible with the ordering.

If w is composed of reflections about black roots, it is of course admissible in the above sense. An admissible Weyl group transformation of a painted subdiagram D_o which is connected to the painted Dynkin diagram D by a white root:

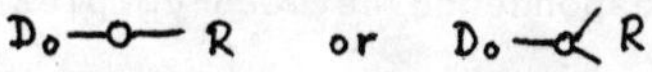

D_o—o— R or D_o—<R

can be extended to an admissible Weyl group transformation on D. (Exterior automorphisms of D_o are not extendable to D). The main application of this lemma is the possibility to shift black $\check{u}(r)$ blocks in $\check{u}(n)$ painted subdiagramms. Other possibilities are

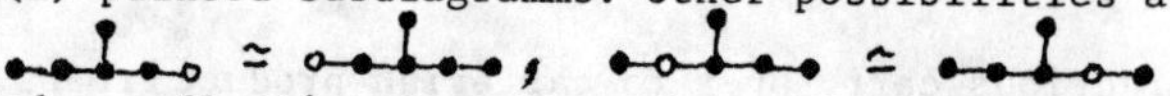

•-•(•)-•-o $\simeq$ o-•-•(•)-•-•, •-o-•(•)-•-• $\simeq$ •-•-•(•)-o-•

and a reflection for the orthogonal algebra D_r with even r. A complete list of all homogeneous Kähler manifolds obtained along these lines is given in ref. [4].

References

1) Zumino, B,. Phys. Lett. 87 B, 203 (79).

2) Alvarez-Gaumé, L. and Freedman, D.Z., Comm.Math.Phys. 80, 443 (81).
Hull, C.M., Gates, S.J. and Roček, M., Nucl.Phys. B248, 157 (84).
Howe, P.S. and Sierra, G., Phys.Lett. 148 B, 451 (84).

3) Burns, D., Univ. of Michigan, Math.Dpt. preprint (1985).

4) Bordemann, M., Forger, M. and Römer, H., Comm.Math.Phys. 102, 605 (1986).

LEVEL COMPARISON THEOREMS AND SUPERSYMMETRIC QUANTUM MECHANICS

B. Baumgartner, H. Grosse

Institut für Theoretische Physik, Universität Wien
Boltzmanngasse 5, A-1090 Wien
AUSTRIA

and

A. Martin

CERN, CH-1211 Geneva 23
SWITZERLAND

ABSTRACT

The sign of the Laplacian of the spherical symmetric potential determines the order of energy levels with the same principal Coulomb quantum number. This recently derived theorem has been generalized, extended and applied to various situations in particle, nuclear and atomic physics. Besides a comparison theorem the essential step was the use of supersymmetric quantum mechanics. Recently worked out applications of supersymmetric quantum mechanics to index problems of Dirac operators are mentioned.

1. HISTORY

Determination of the distribution of energy levels of a Schrödinger Hamiltonian is in general a very difficult problem. Only for a rotational symmetric two-body problem a number of results can be obtained.

Two of us (H.G. and A.M.) started years ago a study of the low lying levels of potential models. Denote by $E_{n,\ell}$ and $u_{n,\ell}$ the energy and reduced wave function of the n-th level with angular momentum ℓ for the potential V fulfilling

$$\left(-\frac{d^2}{dr^2} + \frac{\ell(\ell+1)}{r^2} + V(r) - E_{n,\ell}\right)u_{n,\ell}(r) = 0 \quad \text{on} \quad L^2(\mathbb{R}_+,dr) . \tag{1}$$

At the beginning we found conditions on V(r) in the form of differential inequalities, such that for instance $E_{o,1} \gtrless E_{1,o}$ or $E_{o,2} \lessgtr E_{1,o}$ [1]. Since the nodal structure of the wave functions entered into the proofs a generalization to higher levels was not possible.

A further step was done using a WKB approximation[2]: Although difficult to justify rigorously, it was realized that the Laplacian of the potential controls, within this approximation, the relative order of levels, which are degenerate for the Coulomb case.

Next we obtained a result for all levels by studying perturbations around the Coulomb potential[3]. For the pure Coulomb case there exist simple raising and lowering operators

$$a_\ell^\pm = \pm \frac{d}{dr} - \frac{\ell+1}{r} + \frac{1}{2(\ell+1)} , \tag{2}$$

which are first order differential operators. The Hamiltonian factorizes in terms of these operators

$$h_\ell = - \frac{d^2}{dr^2} + \frac{\ell(\ell+1)}{r^2} - \frac{1}{r} = a_\ell^- a_\ell^+ - \frac{1}{4(\ell+1)^2} . \tag{3}$$

Working out the algebra of a_ℓ^-, a_ℓ^+ and h_ℓ one finds easily the raising and lowering property of $a_\ell^\pm$, while the principal quantum number remains fixed

$$h_{\ell+1}\, a_\ell^+ = a_\ell^+\, h_\ell , \qquad h_\ell\, a_\ell^- = a_\ell^-\, h_{\ell+1} . \tag{4}$$

The reason behind this simple algebraic structure is clearly the underlying O_4 symmetry of the Coulomb problem and a connection of $a_\ell^\pm$ to the Runge-Lenz vector exists. The obtained ladder operators connecting the appropriate wave functions are the first step in the proof of

Theorem 1a: Let $V(r) = - \frac{1}{r} + \lambda v(r)$, then

$$\lim_{\lambda\to 0} (E_{n,\ell} - E_{n-1,\ell+1})/\lambda \gtrless 0 \quad \text{if} \quad \Delta V(r) \gtrless 0 \quad \text{for} \quad r \neq 0 .$$

The second step consists in rewriting the energy difference in terms of an expectation value of the Laplacian of the potential in the $(n-1,\ell+1)$ state. It would be surprising if one could not find an analog for perturbations around the oscillator potential:

Theorem 1b: Let $V(r) = r^2 + \lambda v(r)$, then

$$\lim_{\lambda\to 0} (E_{n,\ell} - E_{n-1,\ell+2})/\lambda \gtrless 0 \quad \text{if} \quad \frac{d}{dr}\frac{1}{r}\frac{d}{dr} V(r) \gtrless 0 .$$

The main questions next were how to generalize these results and how to obtain raising and lowering operators for the general case.

2. MAIN RESULTS

After various attempts the generalization was found[4]. It consists of three steps. The first one is a factorization property (known to Darboux hundred years ago), which is part of supersymmetric quantum mechanics, as we realized afterwards (like Monsieur Jourdain "qui fait de la prose" without knowing it). In a second step we used the Mini-Max principle and finally a technical Lemma was derived, using comparison theorems and the Wronskian techniques.

a) Factorization of Radial Schrödinger Operators

Let $u_\ell = u_{o,\ell}$ be the ground state wave function for fixed ℓ, then

$$H_\ell = -\frac{d^2}{dr^2} + \frac{\ell(\ell+1)}{r^2} + V(r) = A_\ell^- A_\ell^+ - E_\ell \tag{5}$$

where $E_\ell = E_{o,\ell}$ denotes the ground state energy and

$$A_\ell^\pm = \pm\frac{d}{dr} + g_\ell(r) , \qquad g_\ell(r) = -\frac{d}{dr} \ln u_\ell , \tag{6}$$

and the fact that g_ℓ obeys a Ricatti equation is used. Working out the algebra similar as before yields

$$[A_\ell^-,A_\ell^+] = -2g_\ell' , \qquad (H_\ell + 2g_\ell')A_\ell^+ = A_\ell^+ H_\ell , \tag{7}$$

$A_\ell^- A_\ell^+$ and $A_\ell^+ A_\ell^-$ are "essentially isospectral" (as called by Deift), which means that their spectra agree except for zero modes:

$$\sigma(A_\ell^- A_\ell^+) \setminus \{0\} = \sigma(A_\ell^+ A_\ell^-) . \tag{8}$$

Defining a supercharge Q_ℓ and evaluating the anticommutator

$$Q_\ell = \begin{pmatrix} 0 & A_\ell^- \\ 0 & 0 \end{pmatrix} , \quad \{Q_\ell, Q_\ell^\dagger\} = \begin{pmatrix} A_\ell^- A_\ell^+ & 0 \\ 0 & A_\ell^+ A_\ell^- \end{pmatrix} \tag{9}$$

shows the connection to supersymmetric quantum mechanics. Checking in addition the nodal structure of the wave functions involved, shows that (7) implies

$$E_{n,\ell}(V_\ell) = E_{n-1,\ell}(V_\ell + 2g_\ell') , \qquad V_\ell = V + \frac{\ell(\ell+1)}{r^2} , \tag{10}$$

where we indicated explicitly the appropriate potentials. Note that the degeneracy of the Coulomb and the harmonic oscillator problem as well as the solvability of a number of potential problems can be connected to a factorization property. By the inverse method one may construct all isospectral potentials, supersymmetry selects one particular one.

b) Mini-Max Principle

In the Coulomb case we obtained $g_\ell' = (\ell+1)/r^2$, therefore $V_{\ell+1} = V_\ell + 2g_\ell'$ and we jumped by one unit in angular momentum. For a general problem we may only deduce that if

$$V_\ell + 2g_\ell' \gtrless V_{\ell+1} \Rightarrow E_{n,\ell}(V_\ell) \gtrless E_{n-1,\ell+1}(V_{\ell+1}) . \tag{11}$$

c) Comparison Theorem

We continue by using a technical

<u>Lemma</u>: If $\Delta V(r) \gtrless 0 \ \forall r \neq 0$ we get $g_\ell' \gtrless (\ell+1)/r^2$.

The proof uses comparison potentials and the Wronskian techniques.

We note also that the condition $\Delta V < 0$ for $r < R_{max}$ and $dV/dr < 0$ for $r > R_{max}$ are slightly weaker but also sufficient to get the conclusion of the Lemma in the second case.

Alltogether we conclude

Theorem 2a: $\Delta V(r) \gtrless 0$ for $r \neq 0 \Rightarrow E_{n,\ell} \gtrless E_{n-1,\ell+1}$ $\forall n,\ell$.

3. GENERALIZATIONS - APPLICATIONS

Since for the harmonic oscillator the n-th level of angular momentum ℓ is degenerate with the (n-1)th level of angular momentum $\ell+2$, we may ask for a generalization of Theorem 2a. By a variable change we obtained[5)]

Theorem 2b: Assume $V \geq 0$, then

$$D_\alpha V(r) > 0 \quad \text{for} \quad 1 < \alpha < 2 \qquad \Rightarrow E_{n,\ell} > E_{n-1,\ell+\alpha}$$

$$D_\alpha V(r) < 0 \quad \text{for} \quad \alpha > 2 \text{ or } \alpha < 1 \Rightarrow E_{n,\ell} < E_{n-1,\ell+\alpha}$$

where

$$D_\alpha = \frac{d^2}{dr^2} + (5-3\alpha)\frac{1}{r}\frac{d}{dr} + 2(1-\alpha)(2-\alpha)\frac{1}{r^2} .$$

This is not optimal and by approximating the effective potential we conjectured that $E_{n,\ell} < E_{n-1,\ell+\sqrt{\nu+2}}$ for $V = r^\nu$, $\nu > 2$, which has been partly proven recently by A. Martin.

Up to now we have been dealing only with discrete spectra. For Coulomb plus short range potentials V_{sr} we obtained monotonicity conditions in ℓ for the relative scattering phase shift $\delta_\ell(E)$[6)]:

Theorem 3: Suppose $V(r) = -\frac{Z}{r} + V_{sr}$, $\Delta V \gtrless 0$ for $r \neq 0 \Rightarrow \delta_{\ell+1}(E) \gtrless \delta_\ell(E)$. For $Z > 0$, let ε_ℓ denote the ground state energy, then

$$\delta_{\ell+1}(E) \gtrless \delta_\ell(E) + \operatorname{arctg}\frac{Z}{2k(\ell+1)} - \operatorname{arctg}\sqrt{|\varepsilon_\ell|}/k , \qquad k^2 = E .$$

In Theorem 2 the centrifugal barrier term plays a special role; a generalization has been worked out by B. Baumgartner[7)]:

Theorem 4: Denote by $E_n(V)$ the energy of the n-th level for the potential V and let G be such that $2G' = W - V$ and $U = G^2 - G'$. If V fulfills the differential inequality

$$V'' - U'' - \frac{G''}{G'}(V' - U') \gtrless 0 \Rightarrow E_n(V) \gtrless E_{n-1}(W) .$$

The special case $G = -(\ell+1)/r$ yields Theorem 2. Choose $G(x) = \kappa x$ and assume that $V'' \gtrless 2\kappa^2$ yields a bound on the level difference $E_{n+1}(V) - E_n(V) \gtrless 2\kappa$. Generalizations of Theorem 1 have been worked out in Ref. 7) too.

As for applications we mention first quarkonium systems which gave the original motivation for our investigations[1]. The conditions $\Delta V > 0$ and $\frac{d}{dr}\frac{1}{r}\frac{d}{dr} V < 0$ predict the canonical level ordering which is observed in the spectra[8]. These conditions are fulfilled for standard potentials; write the force exerted on a particle as $- Z(r)/r^2$, where $Z(r)$ denotes the effective charge, then $\Delta V > 0$ is equivalent to $Z'(r) > 0$ which agrees with asymptotic freedom for small r. The second condition is guaranteed by lattice QCD which gives an increasing concave potential.

Since the Laplacian of the potential measures the charge density the condition $\Delta V(r) < 0$ can be applied to atomic physics[9]. E.g. in alcaline atoms the outer electron is submitted to a field produced by a nucleus and by an electron cloud. Therefore $\Delta V < 0$ for $r \neq 0$ and $E_{n,\ell} < E_{n-1,\ell+1}$, a fact which is the basis of the Mendeleieff classification.

One of us (B.B.) has worked out the implications of imposing sub- or superharmonicity on V for classical orbits[10]. A "dynamical" Runge-Lenz vector rotates in a definite direction and implies a perihelion shift of a definite sign, depending on the sign of ΔV.

4. FURTHER RESULTS

A. Common has derived a number of bounds on moments of the ground state density for angular momentum ℓ, $\langle r^n \rangle_\ell$, on the position of the maximum of the ground state wave function $r_{M,\ell}$ and on the kinetic energy, assuming $\Delta V > 0$[11]. These results have been generalized recently[12], e.g.

$$V > 0,\ D_\alpha V > 0 \text{ for } 1 < \alpha < 2 \Rightarrow \langle r\rangle_\ell C_{\alpha,\ell} \geq r_{M,\ell} \geq \langle r\rangle_\ell K_{\alpha,\ell}$$

for certain constants $C_{\alpha,\ell}$ and $K_{\alpha,\ell}$ which tend to unity for $\ell \to \infty$. Both-sided bounds on the kinetic energy as well as additional results for pure power potentials are obtained in Ref. 12).

Further work in progress[13)] concerns level ordering for three-body systems. This is a much more difficult problem. On the other hand, recently conjectured three-body mass inequalities by J.M. Richard and P. Taxil have been analyzed by E. Lieb[14)]. Consider the three-body Hamiltonian

$$H^{(3)} = T_1(x_1) + T_2(x_2) + T_3(x_3) + V(x_1 - x_2) + V(x_2 - x_3) + V(x_3 - x_1)$$

where $T_i = p_i^2/2m_i$ denote the kinetic energy operators with masses m_i. Assume that the masses of baryons are given by the ground state energies of $H^{(3)}$ denoted by $E(m_1, m_2, m_3)$. Using the Trotter product formula and assuming that $\exp(-\beta V(x-y))$ is positive semidefinite as a kernel for all $\beta > 0$ it has been shown[14)] that

$$E(m_1, m_2, m_2) + E(m_1, m_3, m_3) \leq 2\, E(m_1, m_2, m_3) ,$$

an inequality, which is related to the Gell-Mann-Okubo mass formula. It is interesting to note that the conjectured inequality $(m_1 = m_2)$ does not hold in general for all potentials.

5. APPLICATIONS TO INDEX PROBLEMS

Recently one of us (H.G.) studied external field problems. For the calculation of an index of a large class of Dirac operators the super-symmetric structure turns out to be essential. The Witten index Δ for a Dirac operator H_D is defined by

$$H_D = \begin{pmatrix} 0 & A^\dagger \\ A & 0 \end{pmatrix} , \qquad \Delta = -\lim_{z\to 0} z\, \mathrm{Tr}\{(A^\dagger A - z)^{-1} - (AA^\dagger - z)^{-1}\} .$$

Δ measures the charge of the fermions interacting with external fields, is topological invariant and has been investigated by us together with the anomaly for one and two dimensions using Krein's spectral shift function[15].

REFERENCES

1) H. Grosse and A. Martin, Phys. Rep. 60, 341 (1980).

2) G. Feldman, T. Fulton and A. Devoto, Nucl. Phys. B154, 441 (1979).

3) H. Grosse and A. Martin, Phys. Lett. 134B, 368 (1984).

4) B. Baumgartner, H. Grosse and A. Martin, Phys. Lett. 146B, 363 (1984).

5) B. Baumgartner, H. Grosse and A. Martin, Nucl. Phys. B254, 528 (1985).

6) B. Baumgartner, H. Grosse and A. Martin, Fysika 17(3), 279 (1985).

7) B. Baumgartner, Ann. Phys. 168, 484 (1986).

8) A. Martin, CERN-TH 4382, Comm. in Nucl. and Part. Phys. (to be published).

9) H. Grosse, A. Pflug and A. Martin, C. R. Acad. Sci. Paris 299 II, 5 (1984).

10) B. Baumgartner, Phys. Lett. 105A, 6 (1984).

11) A. Common, Jour. of Phys. A18, 2219 (1985).

12) A. Common and A. Martin (to be published).

13) A. Martin, J.M. Richard and P. Taxil (work in progress).

14) E.H. Lieb, Phys. Rev. Lett. 54, 1987 (1985).

15) D. Bollé, F. Gesztesy, H. Grosse and B. Simon, preprint KUL-TF-86/5 (to be published).

Wavelets and applications

Ingrid Daubechies
Theoretical Physics
Vrije Universiteit Brussel
Pleinlaan 2
B - 1050 Brussels
Belgium

Thierry Paul
Centre de Physique Theorique
CNRS
Centre de Luminy - Case 907
F - 13288 Marseille Cedex 09
France

The wavelet transform consists in expanding functions in terms of "wavelets", which are constructed from one single function h by means of dilations and translations. More precisely,

$$h^{(a,b)}(x) = |a|^{-1} h(\frac{x-b}{a}) \ , \tag{1}$$

where the function h is the basic wavelet, and where the parameters a , b can be chosen to vary continuously ($a, b \in \mathbb{R}$, $a \neq 0$) or to be restricted to a discrete lattice ($a = a_0^m$, $b = nb_0a_0^m$, with $a_0 > 1$, $b_0 > 0$ fixed, m, n ranging over Z). In both cases, varying a and b corresponds to sweeping phase space. The wavelets (1) can also be used as tools for the construction of operators related to phase space concepts.

The wavelet transform is thus a technique using dilations and translations. Such techniques have in the last decades turned up in different, unrelated fields. In pure mathematics, they were introduced by A. Calderon [1] for the study of singular integral operators. In signal analysis the wavelet transform was first proposed by J. Morlet for the treatment of seismic data [2]. It is now also used for the analysis and synthesis of acoustic signals, by R. Kronland and J.C. Risset [3]. The use of different scales of dilations in the wavelet transform is also related to renormalization group techniques. Contacts with P. Federbush, at this conference, have led to applications of the wavelet transform to constructive quantum field theory [4].

We shall here present several new results concerning wavelets, going from semiclassical expansions in quantum mechanics to applications in signal analysis.

The wavelet transform can be considered as a special case of coherent state analysis. For fixed h, the $h^{(a,b)}$ are called affine coherent states (coherent states based on the affine group), in contrast to the more familiar (canonical) coherent states associated to the Weyl-Heisenberg group. Affine coherent states were first introduced by E. Aslaksen and J.R. Klauder [5]. We have no room here to recall the general framework of coherent states, using square integrable representations [6]. An extensive review of the use of coherent states in physics can be found in [7]. We give here a very short summary of some essential properties.

If the function h is "admissible" [6], i.e. if

$$C_h = \int dy \ |y|^{-1} \ |\hat{h}(y)|^2 < \infty \ ,$$

then the wavelets (1) can be used for a "resolution of the identity". That is, for all $\psi \in L^2(\mathbb{R})$,

$$\psi = (2\pi C_h)^{-1} \int\int da \ db \ a^{-2} h^{(a,b)} \langle h^{(a,b)}, \psi \rangle \ , \qquad (2)$$

where the integral converges strongly. Formula (2) is in fact a "wavelet expansion". An arbitrary function ψ is expressed as a superposition of "wavelets" $h^{(a,b)}$; the coefficients are given by the inner products $\langle h^{(a,b)}, \psi \rangle$. If the function h is reasonably regular and has sufficient decay at ∞, then $h^{(a,b)}$ is well localized in phase space, and (2) corresponds to an expansion of ψ in terms of localized functions in phase space. This is extremely useful for the different applications presented here.

This paper consists of two parts, called, respectively, "Wavelet transform and quantum mechanics" and "Wavelets and phase space localization in signal analysis".

In the first part we describe quantum mechanics on the half line, illustrating it by an exact W.K.B. expression for the hydrogen atom bound states. We also give an algorithm to compute semi-classical expansions for general hamiltonians and to build convergent expansions in powers of $\hbar$ for the Rayleigh-Schrodinger series. We end by a short paragraph describing the construction by Y. Meyer of an orthogonal basis of wavelets (for L^2) which has many miraculous properties.

In the second part we discuss the phase space localization properties of wavelets, mainly in relation with signal analysis. We also give a qualitative comparison of the wavelet approach with the more widely used short-time Fourier transform, which is connected with the Weyl-Heisenberg coherent states.

Both authors would like to thank Alex Grossmann for having initiated them to this subject and its many applications. We are very grateful to him for a very stimulating collaboration, from our scientific childhood to the present day.

A. Wavelet transform and quantum mechanics.

Thierry PAUL

1. INTRODUCTION

We give a few examples of the use of the wavelet transform for quantum mechanics on the half line, with emphasis on the semi-classical aspects. Quantum mechanics on the half line is not only interesting from an abstract and geometrical point of view; it recently has also turned out to be interesting for many physical phenomena. Examples of these are one dimensional models for hydrogen atoms in a microwave electric field, the large N-limit and the labelling of excited levels of Schrodinger operators with radial potentials (see the paper by H. Grosse in the present proceedings).

Wavelets are particularly well suited to semi-classical approximations because they live on and are well localized in phase space. Moreover they provide representation spaces for quantum mechanics consisting of "very nice" functions (e.g. analytic), which often allows to avoid familiar singularities (e.g. turning points) in semi-classical mechanics. This shows that the wavelet transform performs more than just a change of representation (such as the Fourier transform). Finally the wavelet transform gives a Schrodinger equation for the hydrogen atom (with fixed angular momentum l) which is exactly the (first order!) Hamilton-Jacobi equation of the classical associated problem. As a result we can compute quantum perturbation series by means of a classical algorithm and obtain a *convergent* expansion for each coefficient, in powers of $\hbar$.

2. AFFINE COHERENT STATES [8]

Let us consider a quantum particle moving on the half line, or in $\mathbb{R}^n$ with a radial potential after decomposition into partial waves. Let us consider the natural Hilbert space associated to this system, namely $L^2(\mathbb{R}^+)$. Consider a *fixed* number $l > 0$, which will correspond in the radial hamiltonian case to the value of the angular momentum. We call "*affine coherent states*" the family of vectors in $L^2(\mathbb{R}^+)$ obtained, up to normalization, by dilating and translating in Fourier space the fundamental state, with angular momentum l, of the hydrogen atom. Note the analogy with canonical coherent states : the group here consists of shifts and dilations; it acts on the half plane, which is the phase space for the half line, and the "vacuum" is here the fundamental state of the hydrogen atom.

We shall here use a particular set of affine coherent states, given by

$$\phi_z(q) = q^l \exp(-i\bar{z}\, q/\hbar) \qquad \mathrm{Im}z > 0 \ . \tag{A.1}$$

The resolution of the identity (2) becomes in this case

$$\frac{2^{2l}}{2\pi\Gamma(2l)} \int d^2z\, (\mathrm{Im}z)^{2l-1}\, |\phi_z><\phi_z| \;=\; 1\!\!1 \quad \text{in } L^2(\mathbb{R}^+) \ . \tag{A.2}$$

This enables us to build a unitary transform from $L^2(\mathbb{R}^+)$ to the space H_{2l-1} of functions which are analytic on the upper open half plane and square integrable with respect to the measure $(\mathrm{Im}z)^{2l-1}\, d^2z$. More precisely, the function ψ in $L^2(\mathbb{R}^+)$ is mapped to f_l defined by

$$\begin{aligned} f_l(z) &= 2^l\, [2\pi\Gamma(2l)]^{-1/2} <\phi_z, \psi> \\ &= 2^l\, [2\pi\Gamma(2l)]^{-1/2} \int_0^\infty q^l \exp(izq/\hbar)\, \psi(q)\, dq \ . \end{aligned} \tag{A.3}$$

Moreover the vectors ϕ_z are the eigenvectors of a single operator Z^* (which is also used in the lecture by H. Grosse in the present proceedings),

$$Z^*\, \phi_z = \left(-i\frac{d}{dq} - i\, \frac{l\hbar}{q} \right) \phi_z = \bar{z}\, \phi_z \ . \tag{A.4}$$

This implies that the ϕ_z minimize certain "affine Heisenberg inequalities" [8]; it gives the natural classical meaning of z as $z = -p + iLq^{-1}$ with $L = l\hbar$.

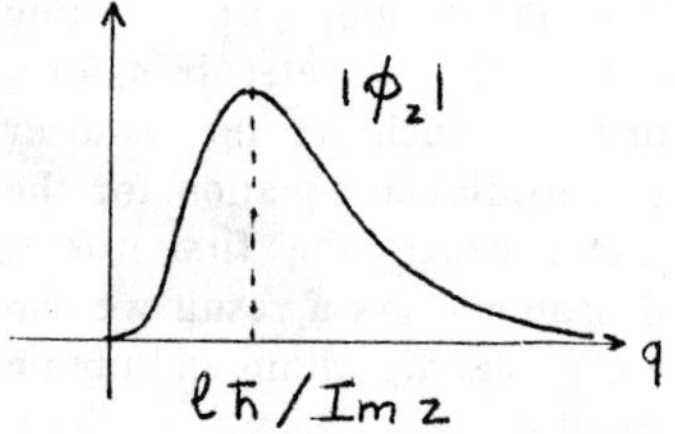

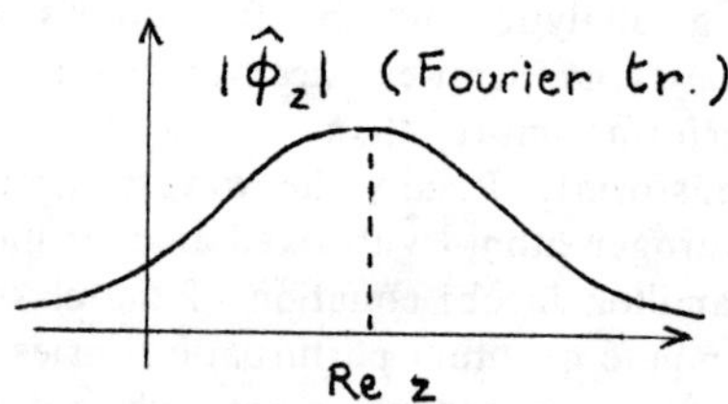

From (A.1) we deduce immediately that

$$Q\,\phi_z = -i\hbar\,\frac{d}{dz}\,\phi_z \qquad \text{where} \qquad (Q\,\psi)(q) = q\,\psi(q) \quad . \tag{A.5}$$

(A.4) and (A.5) imply that the operators Q and Z (adjoint of Z^*) become in H_{2l-1} the operators $-i\hbar\,\dfrac{d}{dz}$ and multiplication by z .

In order to study the hydrogen atom in the space H_{2l-1} (equivalent to Bargmann space) it is important to note that the (time-independent) Schrodinger equation for the hydrogen atom with angular momentum l,

$$\left[\frac{1}{2}\left(-\hbar^2\,\frac{d^2}{dq^2} + \frac{l(l+1)\hbar^2}{q^2}\right) - \frac{\lambda}{q}\right]\psi = E\,\psi \quad , \tag{A.6}$$

can be rewritten in terms of Z and Q as

$$\Big((Z^2 - 2E)\,Q - i(2l+1)\hbar Z - 2\lambda\Big)\,\psi = 0 \quad . \tag{A.7}$$

In H_{2l-1} this becomes immediately

$$-i\hbar\,(z^2 - 2E)\,\frac{d}{dz}\,f_E - [i(2l+1)\hbar Z + 2\lambda]\,f_E = 0 \quad , \tag{A.8}$$

which can be solved by

$$\phi_E = \exp\left[i\int^z \frac{2\lambda + i(2l+1)\hbar z'}{z'^2 - 2E}\,dz'\right] = e^{iS(z)/\hbar}$$

$$= \Big(z - i(-2E)^{1/2}\Big)^{2\lambda(2E)^{-1/2}-l-1}\Big(z + i(-2E)^{1/2}\Big)^{2\lambda(2E)^{-1/2}+l+1} \quad . \tag{A.9}$$

It is amusing to note that, for $E < 0$, the classical trajectory γ_E of energy E in the half plane $z = -p + i\,\dfrac{L}{q}$ is given by a circle γ_E , with center $-iE$ on the imaginary axis.

Computed on this circle, the action $\int_{\gamma_E}^{z} p\,dq = \int_{\gamma_E}^{z} (\mathrm{Im}\,z')^{-1}\,\dfrac{dz'}{L}$ is just $S(z)$ given by (A.9). This means that imposing on f_E to belong to H_{2l-1}. i.e. to be analytic, corresponds to the exact Bohr-Sommerfeld condition

$$\oint_{\gamma_E} p\, dq = n\hbar \quad . \tag{A.10}$$

This gives the exact spectrum $E = -\lambda^2/[\hbar(n+l+1)]^2$.

3. SEMI-CLASSICAL EXPANSIONS [8]

In this paragraph we give an algorithm for semi-classical expansions for bound states or resonances corresponding to small quantum numbers (harmonic approximation and all corrections around a critical point). Although the algorithm gives the same results as a Taylor expansion of the potential around a minimum of the potential ($V'(q_0) = 0$ and $V''(q_0) > 0$), it does *not* use Rayleigh Schrodinger theory and gives an explicit method of computation. Moreover it works perfectly well in the case $V''(q_0) < 0$ and allows to compute the resonances which appear near the top of a barrier (see [8], [9] and [10]).

The starting point is to note that a Schrodinger equation of the form

$$\left[\frac{1}{2}\left(-\hbar^2 \frac{d^2}{dq^2} + \frac{l(l+1)\hbar^2}{q^2}\right) + V(q)\right]\psi = E\,\psi \tag{A.11}$$

becomes in H_{2l-1}

$$\left[-i\hbar\,(z^2-2E)\frac{d}{dz} - i(2l+1)\,\hbar\, z + 2\,(-i\hbar\frac{d}{dz})\,V(-i\hbar\frac{d}{dz})\right] f = 0\,. \tag{A.12}$$

Making the substitution $f(z) = \exp[\frac{i}{\hbar}S(z,\hbar)\ ]$, (A.12) becomes

$$(z^2 - E)\,S'(z) - i(2l+1)z + S'(z)\,V(S'(z))$$

$$+\,\hbar\left[\sum_{k=0}^{\infty} \hbar^k\, R_k\Big(S(z), \ldots, S^{(k)}(z)\Big)\right] = 0 \quad . \tag{A.13}$$

Explicit expressions for the R_k are given in [8]. Imposing that f is analytic leads to

$$\int_{\gamma} S'(z)\, dz = n\hbar \quad , \tag{A.14}$$

for almost every contour γ. Substituting the formal expansion of $S(z,\hbar) = \sum \hbar^m S_m(z)$ into (A.13) and (A.14), one can solve (A.13) recursively in a neighborhood of the classical trajectory of energy E. In the case of a trajectory reduced to one point (bottom or top) the results can be summarized by the following diagrams:

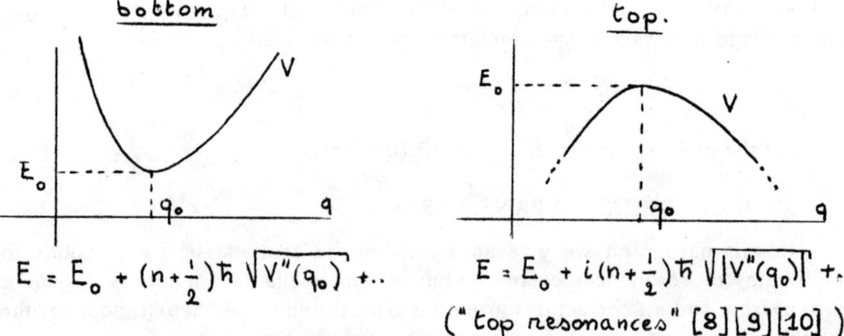

4. CONVERGENT SEMI-CLASSICAL EXPANSION FOR QUANTUM PERTURBATION SERIES.

This paragraph is a report on an article by S. Graffi and the author [11]. We give an algorithm for quantum mechanical perturbation series identical to the canonical perturbation theory in classical mechanics. For the sake of simplicity we restrict ourselves to the case of a perturbation of the one-dimensional hydrogen atom. The more interesting case of perturbation of a multi-dimensional harmonic oscillator is given in [11].

The starting point is the equation (A.11) or (A.12) in which we take

$$V(q) = -\frac{\lambda}{q} + \epsilon\, W(q) \tag{A.15}$$

Then (A.13) is

$$(z^2-E)\, S'(z) - iLz - \lambda + \epsilon \left[S'(z)W(S'(z)) + \hbar \sum_{k=0}^{\infty} \hbar^k\, T_k\Big(S(z),\ldots,S^{(k)}(z)\Big) \right] = 0 \ . \tag{A.16}$$

The series in (A.16) converges for a general class of potentials [11] (for polynomial W it reduces to a finite sum), for each value of z for which $S(z)$ is analytic. We proved [11] that it is possible to apply canonical perturbation theory (see e.g. [12]) to the whole equation (A.16). The terms T_k, involving higher order derivatives, don't present any problem since they are incorporated in the perturbation.

The results can be stated as follows [11] : near an eigenvalue of quantum number n, $E_0(n,\hbar) = -\lambda^2 n^{-2}\hbar^{-2}$, each term $E_k(n,\hbar)$ of the Rayleigh-Schrodinger series

$$E(n,\hbar) = \sum \epsilon^k\, E_k(n,\hbar) \tag{A.17}$$

admits, for $\hbar \to 0$, $n \to \infty$ and $n\hbar \to A$, the following *convergent* expansion

$$E_k(n,\hbar) = H_k(A) + \sum_{r=1}^{\infty} \hbar^r H_k^r(A) \quad , \tag{A.18}$$

where $H_k(A)$ is the corresponding classical term.

5. INCONDITIONAL BASIS FOR BANACH SPACES [13] [14]

A survey of recent progress in wavelets could not be complete without at least a brief description of Y. Meyer's orthonormal basis of $L^2(\mathbb{R})$. This basis is also an inconditional basis for almost all Banach spaces used in functional analysis and should therefore deserve particular attention in mathematical physics.

Consider the function h in $L^2(\mathbb{R})$ defined as follows. The Fourier transform $\hat{h}$ of h is given by

$$\hat{h}(y) = (2\pi)^{-1/2} \exp(iy/2)\, \omega(|y|) \quad , \tag{A.19}$$

where ω is defined by

$$\omega(y) = \begin{cases} 0 & 0 \le y \le \frac{2\pi}{3} \\ \sin\left[\frac{\pi}{2}\, \nu\left(\frac{3y}{2\pi} - 1\right)\right] & \frac{2\pi}{3} \le y \le \frac{4\pi}{3} \\ \cos\left[\frac{\pi}{2}\, \nu\left(\frac{3y}{4\pi} - 1\right)\right] & \frac{4\pi}{3} \le y \le \frac{8\pi}{3} \\ 0 & \frac{8\pi}{3} \le y \end{cases}$$

and where ν is a C^∞-function on $\mathbb{R}$ with the property $\nu(x) + \nu(1-x) = 1$ for all x, and with $\nu(x) = 0$ if $x \le 0$, and $\nu(x) = 1$ if $x \ge 1$.

Then the family of functions h_{jk} defined by

$$h_{jk}(x) = 2^{j/2}\, h(2^j x - k) \quad , \tag{A.20}$$

is an orthonormal basis of $L^2(\mathbb{R})$. Moreover it is an inconditional basis of the spaces $L^p(\mathbb{R})$, $1 < p < \infty$, H_r (Sobolev spaces), ... We have no room here to report on applications of this construction. For this we refer to the literature [13] [14] [15].

B. Wavelets and phase space localization in signal analysis.

Ingrid DAUBECHIES

We present here some results concerning phase space localization, for Weyl-Heisenberg coherent states and for affine coherent states (wavelets). We shall also stress the differences between the two approaches, in several applications.

We start by recalling the respective definitions and basic properties.

Weyl-Heisenberg case	affine case (wavelets)
$g^{(p,q)}(x) = [W(p,q)g](x)$	$h^{(a,b)}(x) = [U(a,b)h](x)$
$= e^{-ipq/2}e^{ipx}g(x-q)$	$= \|a\|^{-1/2}\, h(\frac{x-b}{a})$
$(p,q \in \mathbb{R})$	$(a,b \in \mathbb{R}\ ,\ a \neq 0)$,
	where $C_h = \int dy\, \|y\|^{-1} \|\hat{h}(y)\|^2 < \infty$.

resolution of identity

$\psi = (2\pi \|\|g\|\|^2)^{-1} \times$	$\psi = (2\pi C_h)^{-1} \times$
$\int\int dp\, dq\, g^{(p,q)} <g^{(p,q)},\psi>$	$\int\int da\, db\, a^{-2}\, h^{(a,b)} <h^{(a,b)}, \psi>$.

The main difference between WH coherent states and wavelets is that all the $g^{(p,q)}$ have the same width in x (namely the width of g), while the width of the $h^{(a,b)}$ depends on a. For large a, corresponding to low frequencies (see also below), the $h^{(a,b)}$ are very spread out; for small a, or high frequencies, the $h^{(a,b)}$ are very narrow. As a result of this, wavelets are much better adapted to "zooming in" on singularities or other very localized high frequency components. This will be illustrated by examples below.

In both cases, the varying parameters p, q or a, b permit to "sweep" phase space. This is most effective for choices of g, h which are regular and have fast decay (i.e. are well localized in phase space themselves). This phase space localization, by means of coherent states, will be exploited in the examples below.

1. THE USE OF (CONTINUOUSLY LABELLED) COHERENT STATES TO BUILD OPERATORS.

If an operator T, in quantum mechanics, corresponds to the classical function $t(p,q)$ (e.g. $p^2 + V(q)$), then it can be written, in a first (semi-classical) approximation as $T_{s.c.}$,

$$T_{s.c.}\, \psi = (2\pi)^{-1} \int\int dp\, dq\, g^{(p,q)} <g^{(p,q)}, \psi>\, t(p,q) \qquad \text{(B.1)}$$

This corresponds to the intuition that $<g^{(p,q)}, \psi>$ "measures" the "content" of ψ around the phase space point (p,q) (we assume $||g|| = 1$, $<g,xg> = 0 =$ $<g,-ig'>$). Clearly $t(p,q) = 1$ leads to $T_{s.c.} = \mathbb{1} = T$.

An application of operators of type (B.1) to quantum mechanics can be found in E. Lieb's proof [16; § V.A.1. and 2.] that the (semiclassical!) Thomas-Fermi theory for atoms and molecules is asymptotically exact (for $Z \to \infty$), i.e. that it gives the leading order term. in $Z^{7/3}$, for the true quantum ground state energy.

Among other ingredients, this proof uses operators of type (B.1) to represent $-\Delta$ and $|x|^{-1}$. The remainder terms, $T-T_{s.c.}$, have to be handled by other methods. For the Coulomb potential, this remainder contains the singularity at the origin: operators of type (B.1) cannot reproduce singular potentials.

Similar constructions can be made with affine coherent states. In the treatment of relativistic stability of matter [17] by C. Fefferman and R. de la Llave, a representation of type (B.1) is given, using affine coherent states, for the operator $T\psi = |x|^{-1} * \psi$. In this case the representation is exact: because of the varying support-width of wavelets, integrals of type (B.1), using wavelets, can represent singular potentials (unlike the WH case).

The same idea can be used in signal analysis. Suppose one is especially interested in signal mainly concentrated on the subset S of phase space. In that case it is interesting to construct an operator $\mathbb{P}_S$ which "projects" onto this subset, and to analyze data in terms of the eigenfunctions of $\mathbb{P}_S$. An example of this is given by the (widely used) prolate spheroidal wave functions, which are the eigenfunctions of $P_T Q_\Omega P_T$, where $(P_T f)(t) = \chi_{|t| \le T} f(t)$, $(Q_\Omega f)^\wedge(\omega) = \chi_{|\omega| \le \Omega} \hat{f}(\omega)$; in that case $S = [-T,T] \times [-\Omega,\Omega]$. (See e.g. [18] and the references given there.)

Using coherent states one can construct such $\mathbb{P}_S$ very easily,

$$\mathbb{P}_S f = (2\pi)^{-1} \int\int_{(p,q) \in S} dp\, dq\, g^{(p,q)} \langle g^{(p,q)}, f\rangle \quad .$$

Of course $\mathbb{P}_S$ does not represent a sharp cut-off at the edges of S (which is, in any case, impossible for e.g. compact S); if g is well-localized in phase space, however, then the "tails" of $\mathbb{P}_S$ corresponding to $(p,q) \notin S$ have very fast decay. $\mathbb{P}_S$ is then, in a sense, a "smoothened phase space projection". If S is a disk, $S_R = \{ (p,q) \, ; p^2+q^2 \le R^2 \}$, and if $g(x) = \pi^{-1/4} \exp(-x^2/2)$, then the operator $\mathbb{P}_R \equiv \mathbb{P}_{S_R}$ commutes with the harmonic oscillator Hamiltonian. The eigenfunctions of $\mathbb{P}_R$ are therefore the Hermite functions, $\mathbb{P}_R H_n = \lambda_n(R) H_n$, and the eigenvalues are given by incomplete γ-functions, $\lambda_n(R) = (n!)^{-1} \int_0^{R^2/2} dt\, t^n e^{-t}$ [19]. The eigenfunctions and eigenvalues of $\mathbb{P}_R$ are thus much simpler than the prolate spheroidal wave functions and associated eigenvalues. The behavior, for fixed R, of the $\lambda_n(R)$ in function of n, is very similar to the prolate spheroidal case, and in complete accordance with the concept of a phase space filter [19].

A similar construction can be made for the affine coherent states. The operator which essentially "projects" on S becomes explicit, and easy to characterize, for the choice $\hat{h}(y) = |y|^\beta e^{-|y|}$, and $S = \{ (p,q) \, ; (|p|^{-1}-C)^2 + q^2 \le C^2-1 \}$, for some $C > 1$. This corresponds to a filter which cuts off both very low and very high frequencies, which might be useful in th analysis of optical data [20].

2. THE USE OF DISCRETE LATTICES OF COHERENT STATES IN SIGNAL ANALYSIS.

Two different lattices are used in the two different cases.

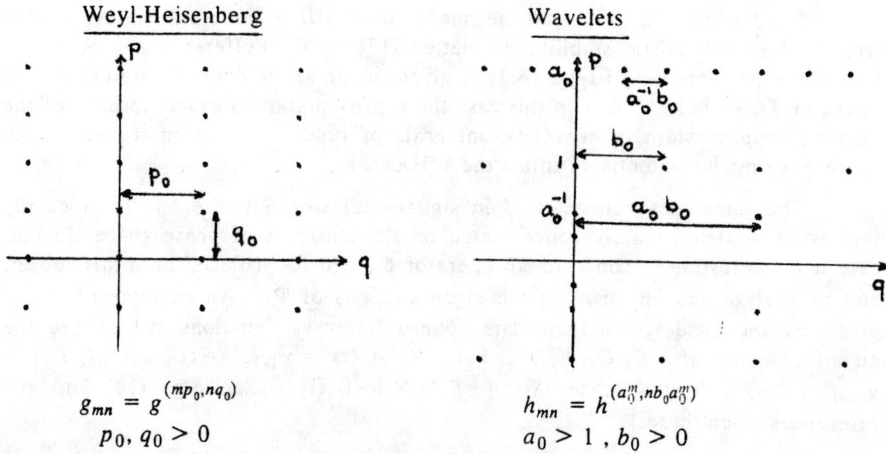

$$g_{mn} = g^{(mp_0, nq_0)} \qquad h_{mn} = h^{(a_0^m, nb_0a_0^m)}$$

$$p_0, q_0 > 0 \qquad a_0 > 1 \, , \, b_0 > 0$$

In both cases one wants to use the scalar products $<\phi_{mn}, f>$ (where ϕ stands for either g or h) to characterize completely a given signal f. The coefficients $<\phi_{mn}, f>$ give an idea of the frequency content of f, *locally in time* (unlike the Fourier transform, where all time-dependence is hidden). The Weyl-Heisenberg case corresponds to the widely used short-time or windowed Fourier transform. The discrete wavelet transform is as yet less widely used. It holds great promise for the treatment of acoustic and visual signals (in this case the signals are 2-dimensional), because it can very efficiently zoom in on short-lived (or very localized) high frequency phenomena (see [21] for an application to edge detection), and because it treats frequencies logarithmically.

In order to have stable numerical reconstruction algorithms for f, from the $<\phi_{mn}, f>$, we impose [22] [23] that there exist $A > 0$, $B < \infty$ such that, for all $f \in L^2(\mathbb{R})$,

$$A \, ||f||^2 \leq \sum_{m,n} |<\phi_{mn}, f>|^2 \leq B \, ||f||^2 \quad . \tag{B.2}$$

If this condition is satisfied, then f is reconstructed as

$$f = \sum_{m,n} \psi_{mn} <\phi_{mn}, f> \, ,$$

where $\psi_{mn} = T^{-1} \phi_{mn}$, $T = \sum_{m,n} \phi_{mn} <\phi_{mn}, \cdot>$. The computation of the ψ_{mn} can be done by a geometric series in $(B/A-1)/(B/A+1)$. If B/A is close enough to 1, then the first approximation $\psi_{mn} = 2\,(A+B)^{-1}\phi_{mn}$ is sufficient. In some cases, for very particular choices of g or h, which are not always suited to the problem at hand, $B/A = 1$ [22].

Let us now specialize to the wavelet case. (Similar results have been obtained for the Weyl-Heisenberg case). For practical purposes it is important to answer the following questions. Given h, for which a_0, b_0 can condition (B.2) be satisfied? Can good estimates for A, B be found (permitting to evaluate whether B/A is close enough to 1)? We have found explicit algorithms to answer these questions [23] [24]. For e.g. $h(x) = (1-x^2)\exp(-x^2/2)$, one finds that $B/A = 1.1$ if $a_0 = 2$, $b_0 = 1$, or $B/A \leq 1.0003$ for $a_0 = 1.25$, $b_0 = .25$. This shows that condition (B.2) is satisfied, and the reconstruction of signals is straightforward, for very realistic values of the parameters (corresponding to what is actually used in computations).

The phase space localization properties of coherent states can be used in the following way, in signal analysis. Suppose that f is mainly concentrated on $[-T,T]$, with frequencies mainly in $[-\Omega,\Omega]$. Intuitively, one would then expect that only those $< h_{mn}, f>$ corresponding to lattice points within the rectangle $[-T,T] \times [-\Omega,\Omega]$ would suffice to approximately reconstruct f. It turns out that this intuition is right. More precisely, for all $\epsilon > 0$, there exist t_ϵ, λ_ϵ such that, if we define $B_\epsilon = \{ (m,n)$; *the corresponding lattice point lies in* $[-(T+t_\epsilon), T+t_\epsilon] \times [-(1+\lambda_\epsilon)\Omega, (1+\lambda_\epsilon)\Omega] \}$, then [23]

$$||f - \sum_{(m,n) \in B_\epsilon} \psi_{mn} < h_{mn}, f > ||$$

$$\leq (B/A)^{1/2} \left[[\int_{|t| \geq T} dt \, |f(t)|^2]^{1/2} + [\int_{|\omega| \geq \Omega} d\omega \, |\hat{f}(\omega)|^2]^{1/2} + \epsilon \right] .$$

(A similar result can be proved for the Weyl-Heisenberg case [23]). As $\epsilon \to 0$, one finds t_ϵ, $\lambda_\epsilon \to \infty$; in order to obtain infinite precision, an infinite number of lattice points is needed. In order to have reasonable values for t_ϵ, and λ_ϵ, h needs to be sufficiently regular, and to have sufficiently fast decay.

REFERENCES

1. See e.g. E. Stein, "Singular integrals and differentiability properties of functions". Princeton Univ. Press (1970).

2. J. Morlet, G. Arens, I. Fourgeau and D. Giard, "Wave propagation and sampling theory", Geophysics **47** (1982) 203-236.

3. R. Kronland and J.C. Risset, in preparation.

4. G. Battle and P. Federbush, "Ondelettes and phase cell cluster expansions - A vindication", submitted for publication.

5. E.W. Aslaksen and J.R. Klauder, J. Math. Phys. **9** (1968) 206, and **10** (1969) 2267.

6. A. Grossmann, J. Morlet and T.Paul, "Integral transforms associated to square integrable representations", J. Math. Phys. **26** (1985) 2473.

7. J.R. Klauder and B.-S. Skagerstam, "Coherent States", World Scientific (Singapore) 1985.

8. T. Paul, "Affine coherent states and the radial Schrodinger equation, I, II and III", submitted to Ann. Inst. Poincare.

9. P. Briet, J.M. Combes, P. Duclos, "On the location of resonances for Schrodinger operators in the semiclassical limit", I, to appear in J. Math. Anal. Appl., and II, to appear in Comm. Part. Diff. Eq.

10. J. Sjostrand, "Semiclassical resonances generated by non degenerate critical points", submitted for publication.

11. S. Graffi, T. Paul, "Schrodinger equation and canonical perturbation theory", to be published in Comm. Math. Phys.

12. G. Gallavotti, "The elements of mechanics", Springer (Berlin, Heidelberg, New York) 1983.

13. Y. Meyer, "Principe d'incertitude, bases hilbertiennes et algebres d'operateurs", Seminaire Bourbaki, 1985-1986, nr. 662.

14. P.G. Lemarie and Y. Meyer, "Ondelettes et bases hilbertiennes", to be published in Rev. Ibero-Americana Mat.

15. P. Tchamitchian, "Calcul symbolique sur les operateurs de Calderon-Zygmund et bases inconditionnelles de $L^2(\mathbb{R})$", C.R. Acad. Sci. Paris **303**, Serie 1 (1986) 215-218.

16. E. Lieb, "Thomas-Fermi theory and related theories of atoms and molecules", Rev. Mod. Phys. **53** (1981) 603-641.

17. C. Fefferman and R. de la Llave, "On the relativistic stability of matter", to be published in Rev. Ibero-Americana Mat. 18. D. Slepian, "On bandwidth", Proc. IEEE **64** (1976) 292-300.

19. I. Daubechies, "Time-frequency localization operators - A geometric phase space approach", submitted for publication.

20. I. Daubechies and T. Paul, in preparation.

21. A. Grossmann, "Wavelet transforms and edge detection", to be published in "Stochastic processes in physics and engineering" (Ph. Blanchard, L. Streit and M. Hazewinkel, Eds.)

22. I. Daubechies, A. Grossmann and Y. Meyer, "Painless nonorthogonal expansions", J. Math. Phys. **27** (1986) 1271-1283.

23. I. Daubechies, "Frames of coherent states - Phase space localization and signal analysis", in preparation.

24. I. Daubechies, "Discrete lattices of coherent states", submitted for publication in the proceedings of the Int. Conf. on Diff. Eq. and Math. Phys., Birmingham (Alabama, USA), March 1986 (J. Lewis and I. Knowles, Eds.).

RESONANCES IN SEMI-CLASSICAL ANALYSIS

Bernard Helffer
(after B. Helffer. J. Sjöstrand)

Département de Mathématiques - Université de NANTES
2, Rue de la Houssinière, 44072 NANTES Cédex
FRANCE
UA CNRS 758

I. Introduction

In the recent years, the study of the spectral properties of Schrödinger Operators : $-h^2\Delta + V$ on $L^2(\mathbf{R}^n)$ as $h \to 0$ has been intensively developped. Particularly, the precise behavior of the spectrum in some Intervall I(h) tending to 0 when $\lim\limits_{|x|\to\infty} V>0$ has been studied in great detail in $[\text{HE-SJ}]_{1-3}$, $[\text{SI}]_{3,4}$ [CO-DU-SE] and the tunneling effect is studied very precisely (not only in the case where the dimension is 1) with the help of the Agmon Metric $[\text{HE-SJ}]_1$, $[\text{SI}]_4$.

We are interested in this talk to the spectrum near 0 in the case when $\lim\limits_{|x|\to\infty} V<0$.

In this case, the problem can be very complicate. We don't have necessarily a natural selfadjoint extension of P and we are interested with eventually non real " eigenvalues " located near 0 (when $h \to 0$) called resonances.

Let us give 3 examples which enter in the theory we will give after

Ex 1 $n=2$, $P(h) = -h^2\,\Delta_{x_1,x_2} + (x_1^2 + x_2^2)(1-P_2\,(x_1,x_2))$

where P_2 is an elliptic positive homogeneous polynomial of degree 2.

This example appears in the Proof of the Bender. Wu Formula ($[\text{SI}]_2$).

Ex 2 $n=4$, $P(h) = -h^2\,\Delta_{x_1,x_2} + (|x_1|^2 + |x_2|^2)\,(1-|x_1|^2\,|x_2|^2)$

with $x_1 \in \mathbf{R}^2$, $x_2 \in \mathbf{R}^2$.

This example appears in the study of the Zeeman effect

Ex 3 $P(h) = -h^2\,\Delta + V$ where V is analytic outside a compact and s.t.

$V(x) = -1 + 0(\frac{1}{|x|})$ is holomorphic for $|\text{Im } x| < C^{-1}\,(1+|\text{Re } x|)$

The starting point for the study of resonances was the technic of " complex scaling " initiated in [AG-CO] and [BA-CO] and the theory was developped after by many authors (See for references before 1978 $[\text{SI}]_1$ and for the application to the Bender-Wu formula $[\text{SI}]_2$). I want in this talk to present a common work with J. Sjöstrand, announced in $[\text{SJ}]_2$, summarized in $[\text{HE-SJ}]_4$ and to appear in $[\text{HE-SJ}]_5$, on resonances and to mention more recent results obtained by other authors.

As says B. Simon in $[SI]_1$, complex scaling is an analytic continuation of a real canonical transformation.

In all the known examples, we have a family of canonical complex transformations from $\mathbf{C}^{2n} \to \mathbf{C}^{2n}$

Ex 4 $(x,\xi) \to (e^{i\theta} x, e^{-i\theta} \xi)$ $\qquad \theta \in \mathbf{R},\ t = \mathrm{tg}\,\theta$

ex 5 $(x,\xi) \to (x, \xi - it \nabla f)$ $\qquad t \in \mathbf{R}$

which sends $\mathbf{R}^n \times \mathbf{R}^n$ ($\sim T^* \mathbf{R}^n$) onto a Lagrangian submanifold (for $\mathrm{Im}\, d\xi \wedge dx$) of $\mathbf{C}^{2n}$: Λ_t.

When t is small, Λ_t can be parametrized by its projection onto $\mathbf{R}^{2n}$ with the help of a generating function G_t (Re x, Re ξ)

$$\mathrm{Im}\, x = \frac{\partial G_t}{\partial \,\mathrm{Re}\,\xi}(\mathrm{Re}\,x, \mathrm{Re}\,\xi),\ \mathrm{Im}\,\xi = -\frac{\partial G_t}{\partial \,\mathrm{Re}\,x}(\mathrm{Re}\,x, \mathrm{Re}\,\xi) \qquad (1.1)$$

In the two examples, $G_t = t\,G$ and $G(\mathrm{Re}\,x, \mathrm{Re}\,\xi) = \mathrm{Re}\,x.\mathrm{Re}\,\xi$ in ex. 4 and $G(\mathrm{Re}\,x, \mathrm{Re}\,\xi) = f(\mathrm{Re}\,x)$ in ex.5. G(or $\frac{\partial G_t}{\partial t}{}_{/t=0}$) is the principal symbol of the infinitesimal generator of the analytic dilation.

Our microlocal version of resonances corresponds to a generalization of this point of view. To study resonances for an operator P of principal symbol $p(x,\xi)$ ($P(h) = p(x,hD)$), we are looking for a function G (the escape function) s.t, if Λ_{tG} is the submanifold defined by (1), $p_{/\Lambda_{tG}}$ is elliptic at ∞ which permits us to consider some Fredholm theory in the spirit of [HE-RO]. We will introduce families of Hilbert-spaces attached to $H(\Lambda_{tG}, m)$ s.t the action of P on $H(\Lambda_{tG})$ is essentially equivalent to the action of a h-pseudodifferential operator of symbol $p_{/\Lambda_{tG}}$ on $L^2(\mathbf{R}^n)$.

2. A general theory of resonances

As explained in the introduction, the idea is to replace the phase space $\mathbf{R}^{2n}_{x,\xi}$ by a I-Lagrangian submanifold (i.e Lagrangian for the symplectic form $-\mathrm{Im}(dx \wedge d\xi)$) $\Lambda \in \mathbf{C}^{2n}$ near $\mathbf{R}^{2n}$ s.t $p_{/\Lambda}$ becomes elliptic at ∞ where $p = \xi^2 + V(x)$ is the principal symbol of P.

In the spirit of the modern theory of pseudodifferential operators (See [BE-FE])we first introduce the weights :

Let $r \geq 1$, $R > 0$ C^∞ functions on $\mathbf{R}^n_x$ s.t

$$r\,R \geq 1,\ \partial^\alpha R = O(R^{1-|\alpha|}),\ \partial^\alpha r = O(r\,R^{-|\alpha|}) \quad \forall \alpha \in \mathbf{N}^n \qquad (2.1)$$

Let $\tilde{r}(x,\xi) = (\xi^2 + r(x)^2)^{1/2}$. If $m(x,\xi,h)$ and $a(x,\xi,h)$ are C^∞, in x,ξ, we shall say that $a \in S(m)$ if

$$\partial_x^\alpha \partial_\xi^\beta a = O(m\,\tilde{r}^{-|\beta|} R^{-|\alpha|}) \text{ for all } \alpha,\beta \in \mathbf{N}^n \quad \text{(uniformly in h)} \qquad (2.2)$$

Let $V \in C^{\infty}(\mathbf{R}^n,\mathbf{R})$ s.t there exists a compact $K \subset \mathbf{R}^n$ and $C > 0$ s.t

(HI) V is analytic on $\mathbf{R}^n \setminus K$ and has an holomorphic extension to $\{x \in \mathbf{C}^n,\ \mathrm{Re}\, x \in \mathbf{R}^n \setminus K,\ |\mathrm{Im}\, x| < C^{-1} R(\mathrm{Re}\, x)\}$ verifying $|V(x)| \leqslant C\, r(\mathrm{Re}\, x)^2$

This hypothesis implies that $p \in S(\tilde{r}^{\,2})$.

(For example 1, take $R = (1+|x|^2)^{1/2}$, $r = (1+|x|^2)$

2, take $R = (1+|x|^2)^{-1/2}\,(1+|x_1|^2|x_2|^2)^{1/2}$

$r = (1+|x|^2)^{1/2}\,(1+|x_1|^2\,|x_2|^2)^{1/2}$

3, take $R = (1+|x|^2)^{1/2}$, $r = 1$)

Let us introduce the microlocal boxes of Beals-Fefferman defined for x,ξ real by : $B((x,\xi),\varepsilon)=\{(y,\eta) \in \mathbf{C}^{2n}$ (or $\mathbf{R}^{2n}$), $|\xi-\eta| \leqslant \varepsilon\tilde{r}(x,\xi)$, $|x-y| \leqslant \varepsilon\, R(x)\}$

We need a general hypothesis of ellipticity far from the real characteristic (setting $\Sigma_p = \{(x,\xi) \in \mathbf{R}^{2n},\ p(x,\xi) = 0\}$) :

(H2) For all $\varepsilon_0 > 0$, $\exists\, C_0 > 0$ s.t if $(x,\xi) \in \mathbf{R}^{2n}$ and $B((x,\xi),\varepsilon_0) \cap \Sigma_p = \emptyset$ then

$$|p(x,\xi)| \geqslant C_0^{-1}\ \tilde{r}(x,\xi)^2$$

Definition of the escape function

We shall say that G is a good escape function if

$\partial_{\xi_i} G \in S(R)$, $\partial_{x_j} G \in S(\tilde{r})$ (2.3)

and

$\exists$ a compact set $K \subset \Sigma_p$ and a constant $C > 0$ s.t (2.4)

$$H_p\, G \geqslant r^2/C \quad \text{in } \Sigma_p \setminus K$$

where $H_p = \left(\frac{\partial p}{\partial \xi},\ -\frac{\partial p}{\partial x}\right)$

Remarks

* (2.4) implies that there are no trapping rays starting of a point outside of K.

* (2.4) is deeply related to the Lavine-Mourre condition [LA], [MO].

* For examples (1) and (3), taking $G(x,\xi) = x.\xi$, the condition (2.4) is simply :

$$-2V - x\,.\,\frac{\partial V}{\partial x} \geqslant r^2/C \qquad \text{for } x \text{ outside a compact}$$

* For example (2), we can take $G(x,\xi) = \nabla f.\,\xi$ with f defined outside a nhd of 0 by

$$f(x) = \frac{|x_1|^2.|x_2|^2}{|x|^2} + \delta_1\,\chi\left(\frac{|x_1|-|x_2|}{\varepsilon|x|}\right).\,[|x_1|^2+|x_2|^2]$$

with $\varepsilon > 0$, $\delta_1 > 0$, $\chi(0) = 1$ with support in a nhd of 0.

* In fact we have to modify the natural escape function outside Σ_p by a function of the form : $G(x,\xi)\, \chi(\frac{|\xi|}{C_0\, r(x)}) + G(x,0)(1-\chi(\frac{|\xi|}{C_0\, r(x)})$

*
If $p(x,\xi)$ is not analytic in some compact set $\tilde{K}$ in x, we need also to work with an escape function which is independent of ξ for $x \in \tilde{K}$.

#

Under the hypotheses (1.3) and (1.4) we get that, for t small enough, $p_{/\Lambda_{tG}} \in S(\tilde{r}^2)$ and $|p/_{\Lambda_{tG}}| \geq t\, \tilde{r}^2/C$ outside a compact. (2.5)

It is difficult to say more about the pseudodifferential calculus. The spaces $H(\Lambda,m)$ where m is an order function defined on $\Lambda(\Lambda = \Lambda_{tG})$ are defined using the F.B.I transform of J. Sjöstrand $[SJ]_1$:

$$u \to Tu(\alpha,h) = \int_{\mathbf{R}^n} e^{i\,\varphi(\alpha,y)/h}\, t(\alpha,y,h)\, \chi(\frac{y-\mathrm{Re}\,\alpha_x}{R(\mathrm{Re}\,\alpha_x)})\, u(y)\, dy$$

$$\text{for } \alpha \in \Lambda, \text{ with } \varphi(\alpha,y) = (\alpha_x - y)\,\alpha_\xi + i\, C\frac{\tilde{r}(\mathrm{Re}\,\alpha)}{R(\mathrm{Re}\,\alpha)}\,(\alpha_x - y)^2$$

where t is an " elliptic symbol "

If H is a function on Λ s.t $-\mathrm{Im}(\alpha_\xi\, d\alpha_x)_{/\Lambda} = dH$, then, roughly,

$$H(\Lambda,m) = \{u \in \mathcal{D}',\ \int_\Lambda |Tu|^2\, m(\alpha)^2\, e^{-2\,H(\alpha)/h}\, d\alpha\} < \infty\}$$

Note here that if $G(x,\xi) = g(x)$ then $H(\Lambda_g,1) = L^2(\mathbf{R}^n, e^{-g/h})$. We then have the following theorem :

Theorem 2.1 ($[HE\text{-}SJ]_4$)

Under the hypotheses (H1), (H2) and (1.4), $\exists\varepsilon_0$, $\exists t_1$ s.t, if $0 < t \leq t_1$ and m is an order function, then $\exists h_0$ s.t for $0 < h \leq h_0$ we have a discret set $\Gamma(h) = \Gamma(h,t) \subset \Omega_t$

$(\Omega_t = [-\varepsilon_0,\varepsilon_0] \times [-\varepsilon_0\, t, 1])$ s.t

$(P-z) : H(\Lambda_t,\tilde{r}^2) \to H(\Lambda_t,1)$

is bijective for $z \in \Omega_t \,|\, \Gamma(h)$ and Fredholm, non bijective, of index 0 for $z \in \Gamma(h)$.

Moreover $\Gamma(h)$ is " reasonably " independant of G,t for h small enough.

" Flavor " of the proof : Using a Fourier integral calculus associated to Λ_t, we are " essentially reduced " to the study of classical p.d.o on $\mathbf{R}^n$ whose principal symbol is $p_{/\Lambda_{tG}}$.

Remarks

* It is possible to prove that $\Gamma(h) \subset \{\text{Im } z \leqslant 0\}$ and that if $z \in \Gamma(h)$ (z is called resonance) is real then a resonant state (i.e a generalized eigenvector in $H(\Lambda_t, \tilde{r}^2)$) is in fact an eigenvector in L^2.

* J. Sjöstrand [SJ]$_4$ has given refined results on the counting function of resonances inside subdomains of Ω_t.

§3 - Applications

3.1. The well inside the Island.

Let $U \subset \ddot{O} \subset \mathbf{R}^n$ where U is compact connected, $\ddot{O}$ open and connected. We assume that V is analytic, that (H1) and (H2) are satisfied and that we have a good escape function G. We assure moreover that :

$$V \leqslant 0 \text{ on } U \text{ (= the well)}, \; V > 0 \text{ on } \ddot{O} \setminus U, \; V < 0 \text{ in } \complement \overline{\ddot{O}} \text{ (= the sea)} \tag{3.1.1}$$

The escape function satisfies to (1.4) on $\Sigma_p / \complement\ddot{O}$ (3.1.2)

Let d the Agmon distance on the Island $\ddot{O}$ associated to the metric Max(V,0). If $S_0 = d(U, \partial\ddot{O})$, let $M_0 = \overline{B_d(U, S_0 - \eta)}$ ($\eta > 0$). As in [HE-SJ]$_1$, we consider the Dirichlet realization of $-h^2 \Delta + V$ in $L^2(M_0)$. Let $I(h)$ be an interval which tends to zero (when $h \to 0$) and $a(h)$ a function s.t :
$a(h) \geqslant C_\varepsilon^{-1} e^{-\varepsilon/h}$ ($\forall \varepsilon > 0$).

Let us assume that :

$$\text{Spectrum } (P_{M_0}) \cap (I(h) + [-2a, 2a] \setminus I(h)) = \emptyset \tag{3.1.3}$$

Let $b(h) \geqslant a(h)$ and $\Omega(h) = (I(h) + [-a,a]) \times [-b(h), b(h)]$ then we have

Theorem 3.1. [HE-SJ]$_5$

For h small enough, $\exists$ a bijection $b : Sp(P_{M_0}) \cap I(h) \to \Gamma(h) \subset \Omega(h)$ s.t $b(\mu) - \mu = O(e^{-(2 S_0 - \varepsilon(\eta))/h})$ where $\varepsilon(\eta) \to 0$ when $\eta \to 0$.

Remark

In connection with this result, let us mention [CODUKLESE], [AS-HA] and more recent announcements of Hislop and Sigal. One of the problems to compare the results is to prove that the resonances are the same !

#

Let us now consider the case where we can use B.K.W technics. As an application of theorem 3.1 we see that there exists a unique resonance $z(h)$ near the first eigenvalue $\mu_0(h)$ of P_{M_0}. If v is the resonant state, we have by the Green-Formula

$$\text{Im } z(h) \, \|v\|^2_{L^2(W)} = -h^2 \, \text{Im} \int_{\partial W} \left(\frac{\partial v}{\partial n}\right) . \, \bar{v} \, dS$$

and if we choose for W a small nhd of $\overline{B_d(U,S_0)}$ with C^∞ boundary, it is possible to see that the principal contribution in the formula come from points of ∂W near $\partial \ddot{O} \cap \overline{B_d(U,S_0)}$ where it is possible to extend the B.K.W construction starting from U (See $[\text{HE-SJ}]_1$) through the caustics using Airy-type integrals (Here U is assumed to be a nondegenerate Well).

Under some general conditions, we get :

$$- \operatorname{Im} z(h) = h^\nu f(h) e^{-2 S_0/h} \tag{3.1.4}$$

where f(h) is the realization of an analytic symbol.

3.2. The Bender-Wu Formula for the Zeemann-effect

This result can be applied to the proof of the Bender-Wu formulas in dimension > 1 for anharmonic oscillators (example 1) (See $[\text{SI}]_2$ for references and discussion). In the case of the Zeemann effect, it is interesting to study the asymptotic expansion E(B) of :

$$Z_B = -\Delta - r^{-1} + B(x^2+y^2) \quad \text{for } B > 0 \ (B \to 0),\ r = (x^2+y^2+z^2)^{1/2}$$

$$E(B) \sim -\frac{1}{4} + \sum_{n=1}^{\infty} \gamma_n B^n$$

Then using a trick suggested by S. Graffi, which reduces the study to the study of the Imaginary part of resonances of some anharmonic oscillator in $\mathbf{R}^4$ (example 2) we get the following result conjectured by Avron [AV].

Theorem $[\text{HE-SJ}]_5$

$$\boxed{\gamma_n = (-1)^n\, 1/2 \left(\frac{4}{\Pi}\right)^{5/2} \left(\frac{16}{\Pi^2}\right)^n (2n + 1/2)\,!\, \left(1 + O\left(\frac{1}{n}\right)\right)}$$

3.3. Recent results

Quite recently J. Sjöstrand $[\text{SJ}]_3$ and Briet-Combes-Duclos [BRICODU] have studied the localization of resonances generated by non-degenerate critical points. In a quite interesting paper, C. Gerard and J. Sjöstrand [GE-SJ] study in the same spirit semi-classical resonances generated by a closed trajectory of Hyperbolic type.

REFERENCES

[AG-CO] J. Aguilar - J.M. Combes
Comm. in Math. Physics, 22 (1971) p. 269-279

[AS-HA] M.S. Asbaugh - E.M. Harrell
Comm. in Math. Physics. 83 p. 151-170 (1982)

[AV] J. Avron
Annals of Physics 131 (1981) p. 73-94

[BA-CO] E. Baslev - J.M. Combes
Comm. in Math. Physics, 22 (1971) p. 280-294

[BE-FE] R. Beals - C. Feffermann
Comm. in Pure App. Math., 27 (1974) p. 1-24

[BRI-CO-DU] Ph. Briet - J.M. Combes - P. Duclos
On the location of resonances for Schrödinger operators in the semi-classical limit II Barrier top resonances. Preprint April 1986

[CO-THO] J.M. Combes - I. Thomas
Comm. Math. Phys. 34 (1973) p. 251

[CO-DU-SE] J.M. Combes - P. Duclos - R. Seiler
J. of Functional Analysis 52, p. 257 (1983)

[CO-DU-KLE-SE] J.M. Combes - P. Duclos - Klein - R. Seiler
The Shape resonance. Preprint Oct. 1985

[GE-SJ] G. Gérard - J. Sjöstrand
Semi-classical resonances generated by a closed trajectory of hyperbolic type. Preprint March 1986

[HA-SI] E. Harrell - B. Simon
Duke Math. J. 47, n° 4, Déc. 1980

[HE-RO] B. Helffer - D. Robert
J. of Functional Analysis, 53, n° 3 (1983) p. 246-268

[HE-SJ] B. Helffer - J. Sjöstrand
(1) Comm. in P.D.E, 9 (4) (1984), p. 337-408
(2) Ann. Inst. I.H.P. Vol. 42, n° 2, (1985) p. 127-212
(3) Publications of Polytechnique. July 1985
(4) Nato Series. Math. and Phys. Sciences. Vol. 168 p. 291-322 (1985)
(5) Resonances en limite semi-classique. to appear Bull de la S.M.F. (1986)

[LA] R. Lavine
Scattering theory in Math. Physics. Nato Series Reidel Vol. 9 p. 141-157

[MO] E. Mourre
Comm. in Math. Phys. 78 p. 391-408 (1981)

[SI] B. Simon
(1) Resonances and Complex scaling : A rigorous overview Int. J. of Quantum chemistry (1978)
(2) Int. J. of Quantum chemistry. Vol. XXI, p. 3-25 (1982)
(3) Ann. I.H.P. 38, p. 295-307 (1983)
(4) Ann. of Math. 120 p. 89-118 (1984)

[SJ] J. Sjöstrand
(1) Asterisque n° 95 (1982)
(2) Tunnel effects for semi-classical Schrödinger operators
Contribution to the workshop and Symposium an hyperbolic equations and related topics Katuda and Kyoto Sept. 1984 (to appear)

(3) Semi-classical resonances generated by non degenerate critical points. Preprint March 1986

(4) Estimations et calcul de résonances. Talk in Polytechnique June 1986

THE COMPETITION BETWEEN CURVILINEAR GEOMETRY AND CHEMISTRY IN REACTIVE FLUID FLOW.

J. Jones [1, 2, 3]

B. Bukiet [1, 2, 3]

Courant Institute of Mathematical Sciences
New York University
New York, N. Y. 10012

ABSTRACT

The methods of bifurcation theory and asymptotic analysis are used to study curvilinear perturbations of a plane wave in a reactive fluid. An asymptotic model is derived for the small curvature limit. The nonlinear competition between curvature and chemistry is manifested by a highly degenerate critical point and resulting nonuniform asymptotic behavior. Computer simulations will be presented showing agreement of this model with the reactive Euler equations in the small curvature limit.

1. INTRODUCTION

The strong degree of coupling between fluid dynamics and chemistry present in reacting fluid systems produces a variety of interesting phenomena, including nonlinear waves and instabilities. One example of a nonlinear wave is the detonation wave, which is created when the temperature rise across a shock wave in a combustable gas is sufficient to maintain a region of combustion behind the shock. In this report we perform an asymptotic analysis, in the limit of small curvature and large time, to study the effects of wavefront curvature on the structure and speed of a detonation wave. This analysis leads to a pair of autonomous ordinary differential equations which are formally equivalent to the full set of partial differential equations for the problem, to first order in the curvature. The system possesses a highly degenerate bifurcation point at zero curvature. Both eigenvalues of the linear part of the system are zero at the bifurcation

1. Supported in part by the Applied Mathematical Sciences subprogram of the Office of Energy Research, U. S. Department of Energy, under contract DE-AC02-76ER03077

2. Supported in part by the Army Research Office, grant DAAG29-85-0188

3. Supported in part by the National Science Foundation, grant DMS-83-1229

point, but more significant is the fact that resonant terms are missing from the Poincaré series for the bifurcation to all orders, so that the unfolding of the bifurcation is not determined by any finite degree polynomial vector field. It is hoped that further study of this bifurcation point will lend insight into the nonlinear instabilities characteristic of detonations.

The curvature of a diverging detonation wave generates expansion waves which cool the fluid in the reaction zone, consequently slowing down the reaction and lowering the wave speed. The main conclusion of this report, based on a combination of theoretical and numerical evidence, is that the wave speed depends linearly on the curvature to leading order, thus demonstrating the consistency of the model with the regularity assumptions employed in its derivation.

2. DERIVATION OF THE MODEL

The equations for spherically symmetric reactive Euler flow are

$$\mathbf{w}_t + \mathbf{f}(\mathbf{w})_r = \mathbf{C} - \mathbf{G}, \tag{1}$$

where

$$\mathbf{w} = \begin{pmatrix} \rho \\ m \\ e \\ \lambda \end{pmatrix} \quad \mathbf{f}(\mathbf{w}) = \begin{pmatrix} m \\ \dfrac{m^2}{\rho} + P \\ \dfrac{m}{\rho}(e+P) \\ m\dfrac{\lambda}{\rho} \end{pmatrix} \quad \mathbf{C} = \begin{pmatrix} 0 \\ 0 \\ 0 \\ R(\lambda,T) \end{pmatrix} \quad \mathbf{G} = \begin{pmatrix} \dfrac{m}{r} \\ \dfrac{m^2}{\rho r} \\ \dfrac{m}{\rho r}(e+P) \\ 0 \end{pmatrix}.$$

Here ρ is the density, $m = \rho u$ is the momentum density, P is the pressure and λ is the mass fraction of burned gas. We assume in the following a polytropic equation of state and Arrhenius kinetics, in which case the energy per unit volume is

$$e = \frac{P}{\gamma - 1} + (1 - \lambda)\rho q + \frac{\rho u^2}{2}.$$

The heat released during combustion is q and with $T = \dfrac{P}{\rho}$ as temperature, the reaction rate is

$$R(\lambda,T) = k(1 - \lambda)\exp\left(-\frac{E}{T}\right)$$

for $T \geq T_c$, where k is the rate multiplier, E is the activation energy and T_c is a critical temperature below which the reaction rate is taken to be identically zero. The quasilinear operator $\mathbf{f}_\mathbf{w}(\mathbf{w})$ has eigenvalues u, $u \pm c$ where the sound speed c is equal to $(\gamma P/\rho)^{1/2}$. It is well known that the plane wave detonation terminates at a sonic point $|u| = c$ when $\lambda = 1$ (in the frame of the shock) [4, 5]. Because of the lower wave speed, the diverging detonation achieves this sonic transition for some $\lambda < 1$. At this point the quasilinear operator $\mathbf{f_w}(\mathbf{w})$ becomes singular.

We wish to investigate the asymptotic behavior of a diverging detonation in the limit of large radius and time, relative to a reaction zone width and a characteristic decay time for transients in the reaction zone. Straightforward attempts to obtain an asymptotic expansion encounter nonuniformities at the sonic transition. The traditional method of resolving such nonuniformities is to construct an auxiliary inner expansion at the sonic point, and matching this to an outer expansion away from the sonic point. In the present case this method fails. There is no transition layer at the sonic point, rather there is a unique regular solution, with all other solutions either lacking a sonic transition, or possessing unbounded gradients and consequent nonuniform asymptotics. These issues are brought more sharply into focus by replacing the linear, nonautonomous first order system in the asymptotic expansion of (1) with a nonlinear, autonomous system which is equivalent to first order. This is accomplished by applying two approximations to (1). First, solutions of (1) are assumed to be quasi-steady state in the shock frame; the time derivatives may be neglected to first order, time entering as a parameter implicitly through the shock radius $z(t)$ and the wave speed $\dot{z}(t)$. Second, the flow curvature is set equal to the shock curvature in the region of asymptotic validity; deviations from this are second order. After a reduction of order and a convenient change of independent variable we arrive at the system

$$v_y = q(\gamma - 1)k(1 - \lambda)v - \frac{(\dot{z} - v)v}{z}c^2\exp\left(\frac{E\gamma}{c^2}\right) \tag{2}$$

$$\lambda_y = k(1 - \lambda)(c^2 - v^2)$$

where

$$c^2 = c_a^2 + \frac{\gamma - 1}{2}(\dot{z}^2 - v^2) + q(\gamma - 1)\lambda$$

where v is the velocity relative to the shock, oriented toward the origin, and y is a deformed spatial variable, which is 0 at the shock and infinity at the sonic transition. The sound speed in the constant ambient state ahead of the shock is denoted c_a. When $z = \infty$ the plane wave solution is recovered, and the vector field possesses an infinite codimensional bifurcation point at which the plane wave solution terminates. The condition that the Cauchy problem for (1) possess a regular solution for large times now translates into the condition that a solution of (2) must pass through a critical point of the vector field. The following result is therefore significant.

Theorem. Let $z < \infty$, but sufficiently large and assume that $T \geq T_c$ throughout the region behind the initial shock. Then the plane wave bifurcation point bifurcates into a unique critical point in the physical domain $0 \leq \lambda \leq 1$. This critical point is a saddle point and its coordinates in the

ν, λ plane are analytic functions of $\frac{1}{z}$.

There is a unique separatrix solution connecting the ambient state upstream of the shock via the Rankine-Hugoniot jump conditions to the critical point. This shooting problem determines the wave speed $\dot{z}$ implicitly as a function of z.

3. NUMERICAL VALIDATION

The stiffness and instability of reacting fluid systems makes them intrinsically difficult to study numerically. We have applied two methods to the diverging detonation problem. In the first method, operator splitting is used to separate the contributions of the flux terms $\mathbf{f}(\mathbf{w})_r$ from the geometrical and chemical source terms $\mathbf{C} - \mathbf{G}$. The homogeneous equation $\mathbf{w}_t + \mathbf{f}(\mathbf{w})_r = 0$ may then be solved by the random choice method. This method is unconditionally stable and is taken to be exact. In the second method we employ the model equations (2). For each radius z the shooting problem is solved numerically to determine the wave speed $\dot{z}$. The model equations are then integrated numerically to determine the gas states in the reaction zone. In Fig. 1 we plot the dependence of wave speed on shock curvature for the random choice method, and compare this to the results of the asymptotic model. Note that the leading order of the asymptotic model actually gives better correlation for very large radii than the full model, presumedly due to the spurious contributions of higher order terms. Fig. 2 illustrates the time dependence of the pressure behind the shock for the two methods. The results of a plane wave run are also shown for comparison. The oscillations in the random choice solutions are due to excitation of physical instabilities by the small level of noise introduced by the random choice algorithm. It is clear from the numerical solutions that the asymptotic model does indeed capture the leading order corrections to the wave speed.

One of the authors has employed front tracking in two dimensions to solve curved detonation problems with the reaction modeled as an instantaneous release of heat at the shock [1]. Because the reaction is not fully resolved, this method misses the interaction between the chemistry and the shock curvature and consequently yields incorrect wave speeds. One promising application of the work presented here is to combine wave speed calculations from the asymptotic model with the front tracking method to obtain high quality, multidimensional computations of curved detonations. This step is planned for future work.

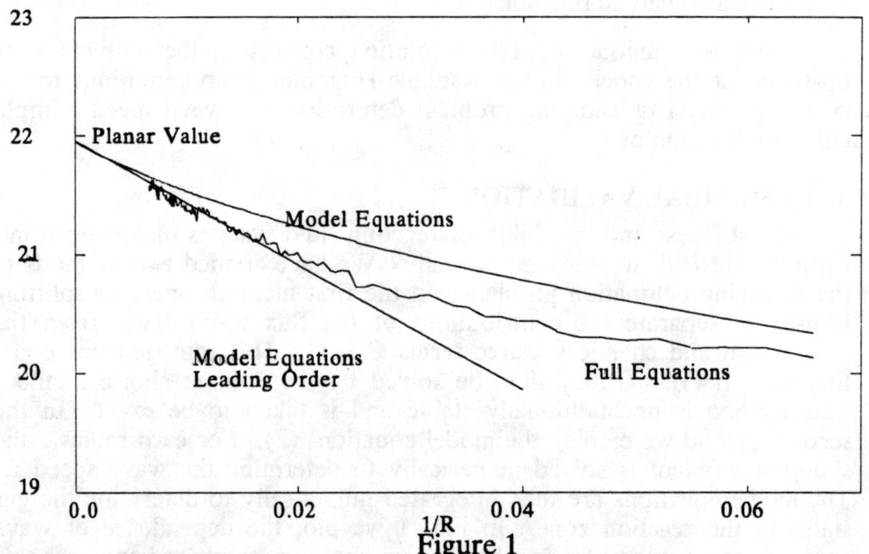

Figure 1

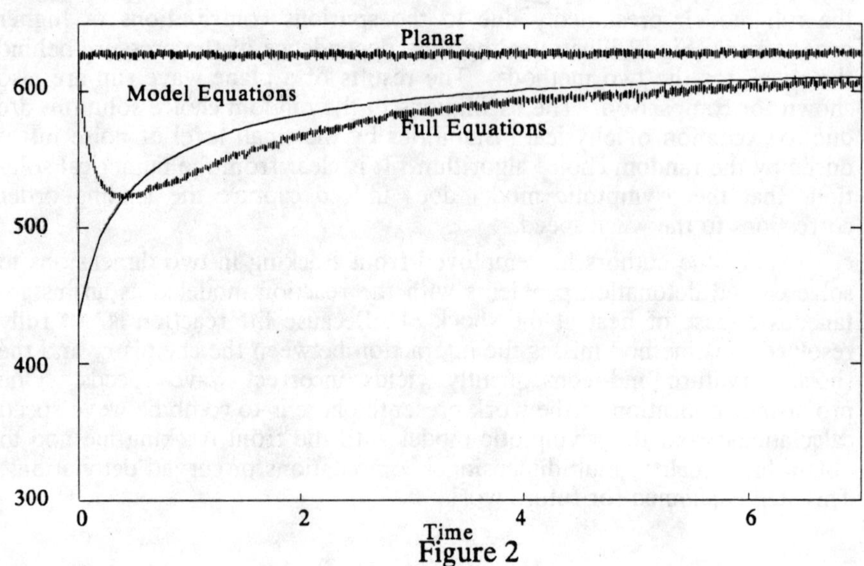

Figure 2

4. BIBLIOGRAPHY

1. B. Bukiet. *Application of Front Tracking to Two Dimensional Curved Detonation Fronts*. Submitted to SIAM J. Sci. Stat. Comp.
2. B. Bukiet. *The Effect of Curvature on Detonation Speed*. NYU Preprint.
3. B. Bukiet. *A Study of Some Numerical Methods for Two Dimensional Curved Detonation Problems*. NYU Thesis. (1986).
4. R. Courant and K. O. Friedrichs. *Supersonic Flow and Shock Waves*. Springer-Verlag, New York, (1976).
5. W. Fickett and W. C. Davis. *Detonation*. University of California Press, Berkeley, (1979).
6. W. Fickett. *Introduction to Detonation Theory*. University of California Press, Berkeley, (1985).
7. J. Jones. *Asymptotic Analysis of an Expanding Detonation*. To appear.
8. C. L. Mader. *Numerical Modeling of Detonations*. University of California Press, Berkeley, (1979).

RIGOROUS COMPUTER-ASSISTED RENORMALIZATION GROUP ANALYSIS

Hans Koch* and Peter Wittwer**
Department of Mathematics, Rutgers University
New Brunswick, NJ 08903, USA

ABSTRACT

Proofs involving the use of computers have played a major role in establishing the validity of renormalization group concepts in the field of nonlinear dynamical systems. We hope that in the not too distant future a similar statement can be made with reference to the theory of critical phenomena in statistical mechanics. Our aim is to develop the tools for a computer-assisted renormalization group analysis of Phi4 or the Ising model in three dimensions. The results presented here are a first step in this direction. We outline the ideas and techniques used in the construction of a non-Gaussian renormalization group fixed point for hierarchical scalar lattice field theories.

1. THE HIERARCHICAL MODEL

Renormalization is a scheme for determining the singularities of observables in systems with many degrees of freedom. We focus here on a Wilson-Kadanoff renormalization group (RG) transformation, tailored to an analysis of the singularities (critical indices) associated with the transition to non-Gaussian long distance behavior for scalar lattice field theories in three dimensions. A scalar lattice field

* Supported in part by the NSF under Grant No. DMS-8540879.
** Supported in part by the NSF under Grant No. DMR-8114726.

theory is a family of translation invariant probability measures on the space of field configurations $\phi : \Lambda \to \mathbb{R}$ on the three dimensional hypercubic lattice Λ. Let us consider measures of the form $d\mu(\phi) = d\mu_G(\phi)F(\phi)$, where $d\mu_G$ is the Gaussian measure whose convariance G is some fixed positive convolution operator on $\ell_2(\Lambda)$ similar to the inverse lattice Laplacean. We specify a RG transformation

$$N : d\mu(\phi) = d\mu_G(\phi)F(\phi) \mapsto d\tilde{\mu}(\phi) = d\mu_G(\phi)\tilde{F}(\phi) \tag{1}$$

by choosing an "averaging operator" S such that the "fluctuation covariance" $\Gamma \equiv G-S^*GS$ is strictly positive. Then N is formally defined by the equation

$$\tilde{F}(\phi) = \exp\{\tfrac{1}{2}(1-\rho^2)(\phi,G^{-1}\phi)-E\}\cdot\int d\mu_\Gamma(\psi)F(\rho S^*\phi+\psi), \tag{2}$$

where ρ is a constant that remains to be determined (field strength renormalization), and where $E = E(\rho,\Lambda)$ is defined by the condition that $\int N(d\mu_G) = 1$ for every ρ. This RG transformation acts on measures in two steps by first integrating out irrelevant (short range) fluctuations and then rescaling the remaining degrees of freedom.

The equations (1-2) should be viewed as guidelines for a rigorous definition of N. If G and S are chosen properly, then the RG transformation N may be expected to have the following properties.

(P1) N preserves normalization: $\int d\tilde{\mu} = \int d\mu$.

(P2) There is a particular choice of ρ such that N has a non-Gaussian fixed point $d\mu^*$.

(P3) The linearization DN of N at the fixed point has exactly one eigenvalue ρ of modulus larger than or equal to one.

(P4) $\nu = \ln(2)/\ln(\delta) \approx 0.64$, $\eta = -2\ln(\rho)/\ln(2) \approx 0.04$.

For the last statement we fixed scales by choosing the averaging operator to be S=2A, where A is defined by the equation

$$(A^*\phi)(i_1,i_2,i_3) = \sqrt{2}/4\cdot\phi(i_1 \bmod 2, i_2 \bmod 2, i_3 \bmod 2) \tag{3}$$

Note that (P1) is formally satisfied. The properties (P2-3) are suggested by the experimental observation of critical temperatures and self-similarity, and (P5) is suggested by various authors on the basis of approximate calculations for the Ising model.

A proof of (P1-4) seems beyond the present state of the art in RG analysis. One of the main obstacles is the lack of small expansion parameters. We shall now isolate this particular problem by considering the transformation N in a different environment, where some of the other difficulties are absent. The measures $d\mu$ considered now are characterized by functions of a single variable,

$$d\mu(\phi) = d\mu_G(\phi)\prod_j f([(A^*A\phi)(j)]^2), \tag{4}$$

and S,G will be chosen in such a way that the RG transformation (2) leaves a space of such measures invariant. To be more precise, let $S = 2A^*A^2$, and define G by iterating the equation $G = \Gamma + S^*GS$ with $\Gamma = 4\cdot\mathrm{Id}$. Because of the simple fluctuation covariance Γ, the integral in (2) now factorizes. The price we pay for this convenience is that G is not translation invariant. However it mimicks the long distance behavior of the inverse Laplacean. It is now easy to verify that, if we set $\rho = 1$, the transformation N maps the measure (4) into a measure of the same form, but with f replaced by $\tilde{f} = T(f)$,

$$T(f)(t^2) = \text{const.} \cdot\int ds \exp(-s^2)[f((t/\sqrt{2} + s)^2)]^8. \tag{5}$$

This "hierarchical model" was studied first by Gallavotti, near the trivial fixed point (f constant). Other hierarchical models have been analyzed near nontrivial fixed points, but only after introducing a small parameter which moves the nontrivial fixed point close to the trivial one [1)2)3)]. Our main result is the following [9)].

Theorem 1. There is a function f* on $\mathbb{R}$ with the following properties.

(i) f* is a fixed point of the transformation $\mathcal{T}$.

(ii) $f^*(r)\exp(kr) \to 0$ as $r \to +\infty$ for every $k<3/8$.

(iii) f* extends to an entire analytic function.

(iv) $(f^*)'(0) > 0$, i.e. $f^*(t^2)$ is not a Gaussian.

2. A COMPUTER-ASSISTED PROOF

It turns out to be very useful that, in contrast to Dyson's original hierarchical model, the RG transformation (5) has a second Gaussian fixed point in addition to f = const. This "high temperature" fixed point corresponds to $f(r) = \text{const}\cdot\exp(-3r/8)$ and is locally attracting. In fact, it attracts "every" exponentially decreasing function, as far as the behavior for $r \to +\infty$ is concerned, including the non-Gaussian fixed point. This feature may be more than just useful: there are other indications that the large field behavior is irrelevant for the long distance properties of field theories.

Our proof of Theorem 1 is based on the contraction mapping principle. We start with a reformulation of the fixed point problem. since $\mathcal{T}$ is expected to have a one dimensional expanding direction. The main steps can be described as follows.

(S1) Convert $\mathcal{T}$ into a more manageable form by choosing suitable coordinates in function space.

(S2) Guess a function h pointing approximately in the unstable direction of the new transformation $\mathcal{R}$.

(S3) Determine a coordinate H such that H(g+ch) is a monotonic function of c for each g.

(S4) Define $\mathcal{M}(g) = \mathcal{R}(g+ch)-ch$, where $c = c(f)$ is defined by the equation $H(\mathcal{R}(g+ch)) = H(g+ch)$.

(S5) Find a good approximate fixed point for the transformation $\mathcal{M}$.

(S6) Show that $\mathcal{M}$ is a contraction in a suitable ball around the approximate fixed point.

The procedure outlined here requires an unhealthy number of numerical calculations. But this is no concern to us since, following the ideas of Lanford[4], we perform the last two steps on a computer. Similar strategies have been applied to several other problems [5)6)7)8)]. This method is highly constructive and provides detailed information about the desired fixed point. As a purely numerical by-product (which could be made rigorous with some additional effort) we also have an approximate value for the largest eigenvalue δ of DT at the fixed point. We get $\ln(2)/\ln(\delta) = 0.65016...$, which comes surprisingly close to the value 0.64 quoted in (P4).

Motivated by a preliminary analysis of the map T, we restrict our attention to analytic functions of the form

$$f(r) = \exp(-r/3)g(7r/2), \quad ||g|| \equiv \Sigma_n |((d/dr)^n g)(0)| < \infty. \tag{6}$$

The change of coordinates $g \mapsto f$ defines the new RG transformation from which the would-be contraction M is constructed, following the steps (S2-4). The approximate fixed point referred to in (S5) is then obtained by writing a computer program which manipulates formal power series, truncating everything at a given degree N. In this approximation M is a nonlinear map on a N+1 dimensional space of Taylor coefficients. Now we simply iterate this map a large number of times, starting with appropriate initial data. If N is sufficiently large (in our case N = 55), this procedure provides us with a polynomial g_o for which the norm $\varepsilon \equiv ||M(g_o)-g_o||$ can be expected to be very small.

As a preparation for the next step (S6) we extend the numerical program by including the calculation of approximate values for ε, and for all the other quantities entering the assumptions of the contraction mapping principle. The computer program which is used to verify these assumptions (all are inequalities !) follows closely its numerical counterpart, except that roundoff errors and truncation errors are taken into account. This means in particular that real numbers are represented not by approximate values, but by rigorous upper and lower bounds. Similarly, a function g is no longer approximated by a polynomial, but a fixed number of intervals (bounds on Taylor coefficients and on the norm of the remainder) is used to describe a set Y in function space which contains g. Instead of just truncating, the "set arithmetic" subroutines now use rigorous functional inequalities in order to represent their result in the desired form. For example, product(X,Y,Z) has to return a set Z containing all possible products f*g with f in X and g in Y. More details can be found in [9].

3. REFERENCES

1) P.M. Bleher and Ya.G. Sinai, "Critical Indices for Dyson's asymptotically hierarchical models", Commun. Math. Phys. 45, 247-278 (1975).

2) P. Collet and J.-P. Eckmann, "A renormalization group analysis of the hierarchical model in statistical physics", Lecture Notes in Physics 74, Springer-Verlag, Berlin, Heidelberg (1978).

3) K. Gawedzki and A. Kupianinen, "Non-Gaussian fixed points of the block spin transformation. Hierarchical model approximation", Commun. Math. Phys. 89, 191-220 (1983).

4) O.E. Lanford III, "A computer-assisted proof of Feigenbaum's conjectures", Bull. A.M.S. 6, 427-434 (1984).

5) J.-P. Eckmann, H. Koch, and P. Wittwer, "A computer-assisted proof of universality for area-preserving maps", Memoirs of the A.M.S. 47, 289 (1984).

6) J.-P. Eckmann and P. Wittwer, "Computer methods and Borel summability applied to Feigenbaum's equation", Lecture Notes in Physics 227, Springer-Verlag, Berlin, Heidelberg (1985).

7) O.E. Lanford III and R. de la Llave, "Existence of a Fixed Point for Golden Mean Renormalization Group Equations for Maps of the Circle", to be published.

8) B. Mestel, Warwick University Thesis, Coventry, England, unpublished.

9) H. Koch and P. Wittwer, "A non-Gaussian renormalization group fixed point for hierarchical scalar lattice field theories", submitted to Commun. Math. Phys.

RIEMANN PROBLEMS AND THEIR APPLICATION TO ULTRA-RELATIVISTIC HEAVY ION COLLISIONS *

Bradley J. Plohr † ‡ and David H. Sharp†

Introduction

Future experiments may make possible the study of heavy ion collisions at sufficiently high energies that a quark-gluon plasma is formed, which subsequently expands, cools, and undergoes a phase transition to hadrons. It is thought that this process passes through a regime that can be modeled as a hydrodynamical flow.[2,1,7] Analysis of the nonlinear waves that are the predominant features of this flow leads to the study of Riemann problems for a relativistic gas undergoing a phase transition. The solutions of Riemann problems are also essential ingredients in effective methods for numerical modeling of such flows. This paper outlines the solution of the Riemann problem and presents the results of preliminary numerical computations of the flow.

In these preliminary calculations we have made two simplifications: baryon number density is neglected, and the flow is taken to consist of a cylindrically symmetric radial expansion together with a longitudinal Lorentz stretching. In extending these calculations, baryons are easily incorporated, and the circle of ideas discussed here can be applied to compute two-dimensional effects in the radial expansion.[6]

The equations governing the radial expansion of a relativistic fluid with vanishing baryon number density are

$$h(u)_t + f(u)_r = g(u,t,r) \ , \tag{1}$$

where $u = (e, \alpha)$, e being the energy density and α the radial rapidity,

$$h(u) = \left(\ (e+p)\cosh^2\alpha - p, \ (e+p)\cosh\alpha\sinh\alpha \ \right) \ , \tag{2}$$

$$f(u) = \left(\ (e+p)\cosh\alpha\sinh\alpha, \ (e+p)\sinh^2\alpha + p \ \right) \ , \tag{3}$$

and

$$g(u) = -(e+p)\left(\frac{\cosh\alpha}{t} + \frac{\sinh\alpha}{r}\right)\cdot(\cosh\alpha, \ \sinh\alpha) \ . \tag{4}$$

Following previous authors we have modeled both the hadronic and quark phases by a Stefan-Boltzmann equation of state, $p = p(e)$, with a sharp phase transition at a critical temperature of 200 MeV.

---

* Supported by the U. S. Department of Energy and the U. S. Army Research Office.

† Theoretical Division, Los Alamos National Laboratory, Los Alamos, NM 87545

‡ Permanent address: Computer Sciences Dept., University of Wisconsin–Madison, Madison, WI 53706

Solution of the Riemann Problem

The principal step in the numerical method we employ to solve Eq. 1 is the solution of its associated Riemann problem, *i.e.*, the initial-value problem for Eq. 1 with $g \equiv 0$ and with scale-invariant initial data: $u(t_0, r) = u_L$ for $r < r_0$ and $u(t_0, r) = u_R$ for $r > r_0$. Because Eq. 1 with $g \equiv 0$ is scale-invariant, we seek a (weak) solution of the Riemann problem that may be written in the form $u(t, r) = \hat{u}((r - r_0)/(t - t_0))$ for $t > t_0$.

The general solution of the Riemann problem is composed of smooth and discontinuous waves. A smooth wave, called a rarefaction, must satisfy

$$\left[-\lambda h'(\hat{u}(\lambda)) + f'(\hat{u}(\lambda))\right] \hat{u}'(\lambda) = 0 , \tag{5}$$

and is therefore associated with one of the characteristic speeds

$$\lambda_{\pm} = \tanh(\alpha \pm \tanh^{-1} c) \tag{6}$$

of Eq. 1 (c being the speed of sound $(dp/de)^{1/2}$). A discontinuous solution for which $\lim_{\lambda \nearrow s} \hat{u}(\lambda) = u_l$ and $\lim_{\lambda \searrow s} \hat{u}(\lambda) = u_r$ must satisfy the Rankine-Hugoniot conditions

$$-s\left[h(u_r) - h(u_l)\right] + f(u_r) - f(u_l) = 0 . \tag{7}$$

Not all such discontinuities are physical, however. Since a more complete model of gas flow would entail viscous terms in Eq. 1, we require that an admissible discontinuity arise as the limit, as the viscosity vanishes, of a traveling-wave solution of Eq. 1, augmented with viscous terms. Such admissible discontinuities, called shock waves, satisfy the Lax[8] criterion of being compressive with respect to one of the characteristic families, and they satisfy the thermodynamic requirement that entropy increase across a shock; surprisingly, however, we find discontinuous solutions in gas dynamics with phase transitions that satisfy these latter two conditions but do not admit viscous profiles.

The Riemann solution is constructed as follows. Let $W_-(u_L)$ be the locus of states u for which there is a sequence of λ_- rarefaction and shock waves connecting u_L to u. This locus, a continuous curve through u_L, is shown for two representative choices of u_L in Fig. 1. Points on $W_-(u_L)$ are depicted differently according to whether the last wave in the sequence is a rarefaction, shock, or composite wave (*viz.*, a shock moving at the characteristic speed of the fluid on its left). Similarly, define the locus $W_+(u_R)$ of states u for which a sequence of λ_+ waves connects u to u_R; $W_+(u_R)$ is the reflection of $W_-(u_R)$ through the line $\alpha = \alpha_R$. With these definitions, the solution of the Riemann

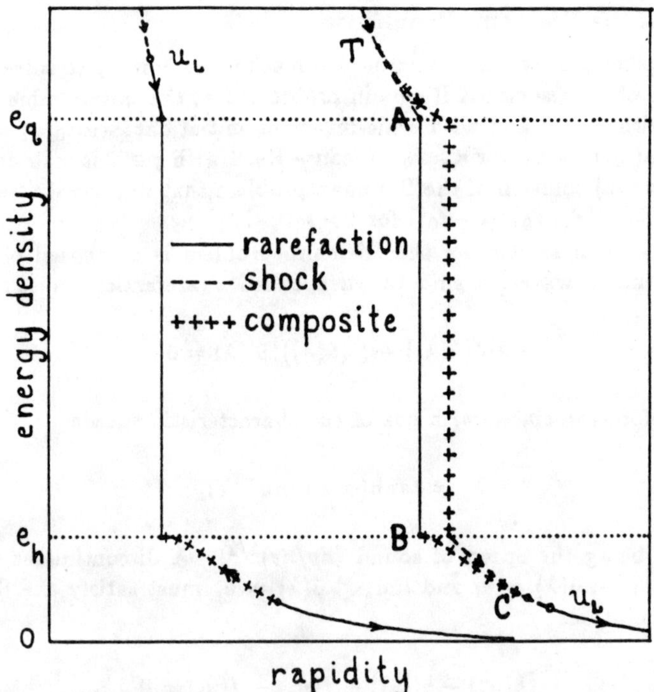

Fig. 1. The wave locus $W_{-}(u_L)$ for two choices of u_l. Points along T-A-B-C are not admissible.

problem is obtained by finding the intersection point u_M of the loci $W_{-}(u_L)$ and $W_{+}(u_R)$.

In Fig. 1 we have also drawn, along T-A-B-C, those solutions that obey the Lax criterion and the entropy condition but are not limits of viscous profiles. Were these solutions to be included, the solution of the Riemann problem in general would not be unique; for example, a discontinuity from u_L to a point between T and A will split into two λ_{-} shocks and a λ_{+} shock.

Tracked Random Choice Method

The random choice method is a numerical method for solving the general initial-value problem for a system of conservation laws. It is based on the constructive existence proof for solutions of conservation laws of Glimm,[4] and was adapted for numerical computation by Chorin.[3] In this method the solution is approximated at each time step by a piecewise-constant function. The solution is advanced to the next time step by solving a sequence of Riemann problems and sampling their solutions at random points to obtain an updated

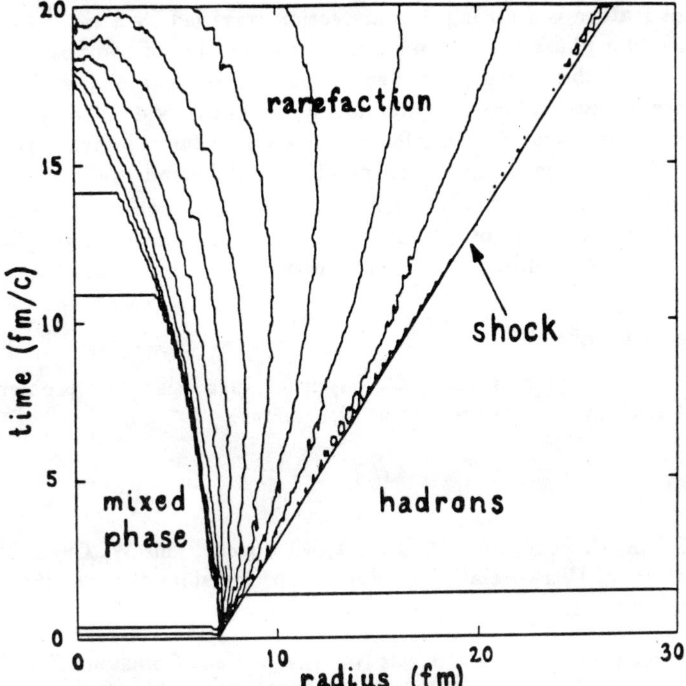

Fig. 2. Space-time contours of temperature for the expansion of a 230 MeV quark-gluon plasma.

piecewise-constant approximation. In order to account for the source term, the time-splitting method of Sod[10] is used. Also, we have incorporated front tracking[5] and mesh refinement[9] to better resolve the flow.

The solution given by the random choice method has the advantage, when compared to standard finite difference methods, of preserving the structure of discontinuous waves in the solution: shock waves remain perfectly sharp. The method is therefore able to detect formation of shock waves within regions extending over just a few mesh cells.

Computational Results

An example of the application of the tracked random choice method to ultra-relativistic heavy ion collisions is shown in Fig. 2. In this computation, quark-gluon plasma at a temperature of 230 MeV is initially confined to a sphere with radius 7 fm. The plasma is cooled by the longitudinal Lorentz stretching and expands into a sea of hadrons at a temperature of 20 MeV.

Fig. 2 depicts the space-time contours of temperature for the flow. The predominant feature is a strong λ_- rarefaction wave led by a weak λ_+ shock. The presence of a phase transition causes the formation of composite wave at the outer edge of the mixed-phase region; such a discontinuity within a rarefaction wave is unusual for gas dynamical flows, but is evident in our solution of the Riemann problem. Another feature, caused by the cylindrical geometry, is the formation of a weak λ_- secondary shock just behind the lead shock.

A full description of the solution of the Riemann problem, together with a more extensive analysis of the hydrodynamical regime occuring in ultra-relativistic heavy ion collisions, is in preparation.

Acknowledgements

The authors are grateful to G. Baym for suggesting this problem, and to him and J. Glimm for many helpful discussions.

References

1. G. Baym, B. Friman, J.-P. Blaizot, M. Soyeur, and W. Czyż, "Hydrodynamics of Ultra-Relativistic Heavy Ion Collisions," *Nucl. Phys.* A 407 (1983), pp. 541–570.

2. J. Bjorken, "Highly Relativistic Nucleus-Nucleus Collisions: The Central Rapidity Region," *Phys. Rev.* D 27 (1983), pp. 140–151.

3. A. Chorin, "Random Choice Solutions of Hyperbolic Systems," *J. Comp. Phys.* 22 (1976), pp. 517–533.

4. J. Glimm, "Solutions in the Large for Nonlinear Hyperbolic Systems of Equations," *Comm. Pure Appl. Math.* XVIII (1965), pp. 697–715.

5. J. Glimm, D. Marchesin, and O. McBryan, "Subgrid Resolution of Fluid Discontinuities II," *J. Comp. Phys.* 37 (1980), pp. 336–354.

6. J. Glimm, C. Klingenberg, O. McBryan, B. Plohr, D. Sharp, and S. Yaniv, "Front Tracking and Two Dimensional Riemann Problems," *Adv. Appl. Math.* 6 (1985), pp. 259–290.

7. K. Kajantie and L. McLerran, *Nucl. Phys.* B 214 (1983), p. 261.

8. P. Lax, "Hyperbolic Systems of Conservation Laws II," *Comm. Pure Appl. Math.* X (1957), p. 537.

9. B. Plohr, "Modeling of Shockless Acceleration of Thin Plates using a Tracked Random Choice Method," in preparation, 1986.

10. G. Sod, "A Numerical Study of a Converging Cylindrical Shock," *J. Fluid Mech.* 83 (1977), pp. 785–794.

New Monte Carlo Algorithms for Quantum Field Theory and Critical Phenomena, or How to Beat Critical Slowing-Down

Alan D. Sokal

Department of Physics
New York University
4 Washington Place
New York, NY 10003 USA

ABSTRACT

Monte Carlo computations in quantum field theory and statistical mechanics have been greatly hampered by critical slowing-down. We review recent progress in devising new Monte Carlo algorithms having radically reduced critical slowing-down. Emphasis is on "collective-mode" algorithms for spin systems and lattice field theories, and algorithms for the self-avoiding walk. We also discuss the rigorous analysis of these algorithms, which leads to interesting problems in the theory of Markov chains and dynamic critical phenomena.

1. Introduction

Mathematical physicists like to prove theorems; theoretical physicists like to derive formulae; but many important problems in physics are so strongly nonlinear that they seem to resist, at least at present, any kind of analytic attack. Quantum field theory and the statistical mechanics of critical phenomena offer numerous examples: Is the ϕ^4_4 (or QED_4) quantum field theory trivial? Does hyperscaling hold in the three-dimensional Ising model? These questions have been hotly debated since roughly 1950 and 1965, respectively; but despite the contributions of some of the most eminent physicists of our time (Landau, Källén, Kadanoff, Wilson ...), we are still very far from an analytic, first-principles answer. In despair, many theoretical physicists, and even some mathematical physicists, have turned to the computer. After all, before trying to prove theorems, it is helpful to have some idea what is *true*.

I should admit at the outset, therefore, that this talk will be largely propaganda: my aim is to advertise to mathematical physicists some of the fascinating problems which arise in the design and analysis of Monte Carlo algorithms for quantum field theory and critical phenomena. I hope to convince some of you that you can contribute to this field *without* ever touching a computer. There are two main kinds of contributions you can make:

• *Invent new algorithms.* This is urgent: in elementary-particle physics, for example, we have a candidate theory for the strong interactions, quantum chromodynamics (QCD), but we are unable to compare its predictions with experiment because we are unable to compute what its predictions are — its strongly nonlinear equations defy analytic solution. We can study QCD numerically by Monte Carlo, but current calculations are very crude: a really adequate computation of hadron masses in QCD with dynamical fermions, using conventional algorithms, would take, in my opinion, around 10^{14} seconds of CPU time on a

Cray X-MP supercomputer — that is, a few million years.* (On the other hand, if the exciting new algorithms that I'll describe in Section 3 turn out to be applicable to QCD — and it is far from clear that they will be — then this estimate could be reduced by a factor of 10 or even 100.) Obviously, improvements in computer technology will play an important role, but brute force is not enough: there is an urgent need for new and more efficient Monte Carlo algorithms. This is not primarily a question of clever programming, but of fundamental physics: for as I shall argue, the efficiency of an algorithm depends primarily on the *insight into the physical behavior of the model* which is implicit in the algorithm.

- *Rigorous analysis of algorithms.* These problems, though less urgent for physics, are very close to traditional mathematical physics, and may be ripe for solution. Mathematically, the problem is to prove good upper or lower bounds on the autocorrelation times of certain Markov chains. From a physical point of view, the goal is to erect a rigorous theory of dynamic critical phenomena analogous to the well-established rigorous study of static critical phenomena. Surprisingly, this field is largely virgin territory. The results are of practical as well as theoretical value: bounds on the autocorrelation time are valuable in planning Monte Carlo studies, and in comparing alternative algorithms.

My goal in this talk, therefore, is to give an overview of some of the key difficulties arising in Monte Carlo studies of quantum field theory and critical phenomena — notably the problem of *critical slowing-down* — and of some recent progress in devising new Monte Carlo algorithms which (partly) alleviate these difficulties. Section 2 contains a brief introduction to dynamic Monte Carlo methods. Section 3 describes Monte Carlo algorithms, old and new, for spin models and lattice field theories (*e.g.* Ising, ϕ^4, lattice gauge theories, etc.). Section 4 describes Monte Carlo algorithms for the self-avoiding walk. Section 5 introduces some tools for the rigorous analysis of Monte Carlo algorithms, and poses some open questions.

Some important topics will *not* be discussed here: purely computer improvements (*e.g.* vectorization, multi-spin coding); fermions (whether quenched or dynamic); and the choice of observables to be measured (*e.g.* Monte Carlo renormalization group).

2. Dynamic Monte Carlo Methods

All Monte Carlo work has the same general structure: given some probability measure μ (on some configuration space S), we wish to generate many random** samples from μ. How is this to be done?

Static Monte Carlo methods are those which generate a sequence of *independent* samples from the distribution μ. These techniques are widely used in Monte Carlo numerical integration [3]. But they are unfeasible for most applications in statistical physics and quantum field theory, where μ is the Gibbs measure of some rather complicated (high-dimensional) system.

*Be warned that this estimate is highly controversial. Other physicists' estimates may be a factor of 10^6 higher or lower.

**Of course, on the computer one does not use true random numbers but rather pseudo-random numbers, *i.e.* deterministic sequences which *seem* "random" *for most practical purposes*. The analysis of pseudo-random-number generators [1,2] is of great theoretical and practical importance, but it is not the subject of this lecture; here I shall pretend that all processes are truly random ones.

The idea of *dynamic* Monte Carlo methods is to invent an irreducible *Markov chain* with state space S having μ as its stationary distribution, and then to simulate this Markov chain on the computer. In other words, we invent (somehow or other) an irreducible transition probability kernel $P(x,dx')$ which leaves μ invariant, *i.e.*

$$\int \mu(dx)\, P(x,dx') \quad = \quad \mu(dx') \,. \tag{2.1}$$

The general principles of Markov-chain theory [4-7] then ensure that the Markov chain is irreducible and positive-recurrent, that μ is the *unique* stationary distribution, and that time averages converge (with probability 1) to μ-averages (irrespective of the initial state X_0). Thus, with sufficient computer time, one can in principle measure expectation values $<A>_\mu$ to arbitrary precision.

The trouble in practice, of course, is that the samples $X_0, X_1, \ldots$ from the Markov chain are not independent, but are in general strongly correlated. Crudely speaking, the statistical error in Monte Carlo work behaves as $\sim N^{-1/2}$, where N is the number of "effectively independent" samples. Thus, the statistical error in dynamic Monte Carlo is of order $\sim(\tau/T)^{1/2}$, where T is the total number of samples (run length) and τ is an "autocorrelation time" of the Markov chain. As we shall see, τ can get very large!

More precisely, let A be an observable in $L^2(\mu)$, and let

$$\rho_{AA}(t) \quad = \quad \frac{<A(s)\,A(s+t)> \; - \; <A(s)><A(s+t)>}{<A(s)^2> \; - \; <A(s)>^2} \tag{2.2}$$

be its normalized time-autocorrelation function in the *stationary* Markov process (*i.e.* "in equilibrium"). Typically $\rho_{AA}(t)$ decays exponentially ($\sim e^{-t/\tau}$) for large t; we define the *exponential autocorrelation time*

$$\tau_{exp,A} \quad = \quad \limsup_{t\to\infty} \; \frac{t}{-\log|\rho_{AA}(t)|} \tag{2.3}$$

and

$$\tau_{exp} \quad = \quad \sup_{A\in L^2(\mu)} \; \tau_{exp,A} \;. \tag{2.4}$$

Thus, τ_{exp} is the relaxation time of the slowest mode in the system. (It might be $+\infty$!)

An equivalent definition, which is useful for rigorous analysis, involves considering the transition probability P as an operator on $L^2(\mu)$. P is a contraction (in many but not all cases it is self-adjoint) and has a nondegenerate eigenvalue 1 with eigenvector equal to the constant function. Let R be the spectral radius of the remainder of P, *i.e.*

$$R \;\equiv\; \inf\left\{ r\colon \operatorname{spec} P \subset \{\lambda\colon |\lambda|\le r\} \bigcup \{1\} \right\}. \tag{2.5}$$

Then it is not difficult to show, using a slight generalization of the spectral radius formula [8], that $R = e^{-1/\tau_{exp}}$.

On the other hand, the statistical errors in Monte Carlo estimates of $<A>_\mu$ are controlled by the *integrated autocorrelation time*

$$\tau_{int,A} \quad = \quad \frac{1}{2}\sum_{t=-\infty}^{\infty} \rho_{AA}(|t|) \,. \tag{2.6}$$

If $\rho_{AA}(t) \approx e^{-t/\tau}$, then $\tau_{exp,A} \approx \tau_{int,A}$. This occurs in most but not all cases.

Now let μ be the Gibbs measure of some statistical-mechanical system at inverse temperature β. What typically occurs is that as β approaches the critical temperature β_c, the autocorrelation time τ diverges to *infinity* — so that the computational efficiency tends to *zero*! This behavior is called *critical slowing-down*; it plagues Monte Carlo studies of critical phenomena in statistical mechanics and of the continuum limit in quantum field theory. My goal in the remainder of this lecture is to explain: (1) the heuristic reason for critical slowing-down in conventional Monte Carlo algorithms; (2) some recent progress in devising new Monte Carlo algorithms with radically reduced critical slowing-down; and (3) some techniques and open problems in the rigorous analysis of critical slowing-down (dynamic critical phenomena).

3. Monte Carlo Methods for Spin Models and Lattice Field Theories

Consider, for purposes of illustration, a ϕ^4 model on a finite lattice $\Lambda \subset \mathbf{Z}^d$ of linear size L. The Hamiltonian is

$$H \;=\; -\frac{J}{2} \sum_{|i-j|=1} \phi_i \phi_j \;+\; \sum_i (A\phi_i^2 + \lambda\phi_i^4) \tag{3.1}$$

and the Gibbs measure is

$$d\mu(\phi) \;=\; Z^{-1}\, e^{-H(\phi)}\, d\phi\,. \tag{3.2}$$

A typical Monte Carlo method for this problem is the *single-site heat-bath algorithm*, defined as follows: For each site $i \in \Lambda$, define the probability kernel P_i to be the regular conditional probability for μ given $\{\phi_j\}_{j\neq i}$: in detail,

$$P_i(\phi, d\phi') \;=\; \mathrm{const}(\phi) \times e^{J\phi_i' \sum_{j:|j-i|=1} \phi_j - A\phi_i'^2 - \lambda\phi_i'^4}\, d\phi_i' \times \prod_{j\neq i} \delta(\phi_j' - \phi_j)\, d\phi_j'\,. \tag{3.3}$$

Algorithmically, what P_i does is to erase the old spin value ϕ_i and replace it by a new value ϕ_i' taken from the conditional distribution $E_\mu(\,\cdot\,|\{\phi_j\}_{j\neq i})$; all other spins are left unchanged. Then the heat-bath algorithm is defined either as

$$P \;=\; \frac{1}{|\Lambda|} \sum_{i\in\Lambda} P_i \tag{3.4}$$

(random choice of site) or as

$$P \;=\; P_1 P_2 \ldots P_{|\Lambda|} \tag{3.5}$$

(systematic sweep of sites). The essential behavior is the same in either case.

Why does the single-site heat-bath algorithm have critical slowing-down? The key fact is that the updates are *local*: in a single step of the algorithm, "information" is transmitted only to nearest neighbors. One might guess that this "information" executes a random walk around the lattice. In order for the system to reach an "effectively new" configuration, the "information" has to travel a distance of order ξ, the (static) correlation length. One would guess, therefore, that $\tau \sim \xi^2$ near criticality. This guess is correct for the Gaussian model (ϕ^4 model with $\lambda = 0$). It is only roughly correct for other models: in general, $\tau \sim \xi^z$ where z is a dynamic critical exponent. For ϕ^4 and Ising models, we have $z \approx 2$ [9]. Note that the same heuristic argument applies to *all* Monte Carlo algorithms (*e.g.* Metropolis, Langevin) with purely *local* updates.

It follows that the computational work needed to get *one* "independent" sample behaves as $\sim L^d \xi^z \geq \xi^{d-z}$. For accurate statistics one might want 10^6 "independent" samples. The reader is invited to plug in $L = 100$, $d = 4$ and get depressed. Note that the factor ξ^d is inherent in *all* Monte Carlo algorithms for spin models (but not for self-avoiding walks, see Section 4). The factor ξ^z could, however, conceivably be reduced or eliminated by a more clever algorithm.

What is to be done? Our knowledge of the *physics* of critical slowing-down tells us that the slow modes are the long-wavelength modes, if the updating is purely local. The natural solution is therefore to speed up those modes by *collective-mode* (nonlocal) updating [10, 11]. Of course, this is easier said than done: one must decide exactly *which* collective modes to speed up and *how*; and the algorithm must not be so nonlocal that its computational cost outweighs the reduction in critical slowing-down. Specific implementations of the collective-mode idea are thus highly model-dependent. At least three such algorithms have been invented so far:

- Fourier-accelerated Langevin (FAL) [11-13]
- Multi-grid Monte Carlo (MGMC) [14-18]
- Swendsen's algorithm for Potts models (including Ising) [19]

FAL and MGMC are very similar in spirit (though quite different technically); their performance is probably roughly similar. I'll describe MGMC very briefly.

One way of looking at the single-site heat-bath algorithm is that it proposes to add an unspecified number t to the spin ϕ_i, and then chooses t according to the appropriate conditional probability distribution. MGMC generalizes this by proposing updates in which a single unspecified number t is to be added simultaneously to *all* the spins in a subset $A \subset \Lambda$; again, t is chosen according to the appropriate conditional distribution. In MGMC, the subset A is taken successively to run through single-element subsets, cubes of side 2, cubes of side 4, etc. in a suitable order. Thus, MGMC updates *collective modes* (indicator functions of the blocks A) on *all length scales*. (In truth, this is not how the computation is performed, but it is mathematically equivalent and probably easier to understand physically.) I realize that this description of MGMC is so sketchy as to be probably incomprehensible. A somewhat different incomprehensible description can be found in [15]; a more detailed and hopefully comprehensible treatment will soon be available [16].

Variants on this algorithm can be devised for nonlinear σ-models and for $U(1)$ lattice gauge theories with or without *bosonic* matter fields [15, 16]. We don't yet know how to do non-Abelian gauge theories, but we're working on it!

Note that MGMC is philosophically very similar to the block-spin renormalization group (RG), but technically very different: RG *integrates out* the short-wavelength degrees of freedom, while MGMC *conditions* on them.

How well does MGMC perform? The answer is highly model-dependent:

- For the *Gaussian model*, it can be proven rigorously [15, 18] that τ is *bounded* as criticality is approached (empirically $\tau \approx 1-2$); the gain in efficiency over traditional algorithms thus grows to *infinity* near the critical point. This behavior is to be expected: the correct collective modes in the Gaussian model are sine waves, and the upshot of the MGMC convergence proof is that piecewise-constant waves are nearly as good.

- For the ϕ^4 *model*, numerical experiments [15, 16] show that τ diverges with the *same* dynamic critical exponent as in the heat-bath algorithm; the gain in efficiency thus approaches a *constant* factor, empirically ≈ 10. This behavior can be understood [16] as due to the double-well nature of the ϕ^4 potential, which makes MGMC ineffective on large blocks. Thus, the correct collective modes at long length scales are nonlinear excitations *not* well modelled by $\phi \to \phi + t\chi_A$; understanding what these excitations *are* would be the key to inventing a better collective-mode algorithm for the ϕ^4 model.

• For the $d = 2$ *XY model*, very preliminary data [17] show a more complicated behavior: As the critical temperature is approached from above, τ seems to diverge, perhaps with the same dynamic critical exponent as in the heat-bath algorithm; if so, the gain in efficiency would again approach a constant factor, of order $\approx 10 - 100$. On the other hand, below the critical temperature, τ seems to be very small ($\approx 2 - 3$); since for the heat-bath algorithm τ is infinite in this regime, the gain in efficiency is infinite as well. This behavior can also be understood physically: in the low-temperature phase the main excitations are spin waves, which are well handled by MGMC (as in the Gaussian model); but near the critical temperature the important excitations are widely separated vortex-antivortex pairs, which are apparently not easily created by the MGMC updates.

Collective-mode Monte Carlo is thus a general philosophy, not a cut-and-dried recipe. For each model, the challenge is to find the correct collective modes, and then to use this physical knowledge (and some ingenuity) to invent an efficient Monte Carlo algorithm. The field is wide open!

4. Monte Carlo Methods for the Self-Avoiding Walk

An *N-step self-avoiding walk* ω on the lattice $\mathbf{Z}^d$ is a sequence of *distinct* points $\omega_0, \omega_1, \ldots, \omega_N \in \mathbf{Z}^d$ such that each point is a nearest neighbor of its predecessor. Let c_N [resp. $c_N(x)$] be the number of N-step SAWs starting at the origin and ending anywhere [resp. ending at x]. Let $\langle\omega_N^2\rangle$ be the mean-square end-to-end distance of an N-step SAW. These quantities are believed to have the asymptotic behavior

$$c_N \sim \mu^N N^{\gamma-1} \tag{4.1}$$

$$c_N(x) \sim \mu^N N^{\alpha_{\text{sing}}-2} \qquad (x \text{ fixed} \neq 0) \tag{4.2}$$

$$\langle\omega_N^2\rangle \sim N^{2\nu} \tag{4.3}$$

as $N \to \infty$; here γ, α_{sing} and ν are critical exponents, while μ (the connective constant of the lattice) is the analogue of a critical temperature. The SAW has direct application in polymer physics [20], and is indirectly relevant to ferromagnetism and quantum field theory by virtue of its equivalence with the $n \to 0$ limit of the n-vector model [21].

The SAW has some advantages over spin systems for Monte Carlo work: Firstly, one can work directly with SAWs on an infinite lattice; there are no systematic errors due to finite-volume corrections. Secondly, there is no L^d (or ξ^d) factor in the computational work, so one can go closer to criticality. Thus, the SAW is an exceptionally advantageous "laboratory" for the numerical study of critical phenomena.

Different aspects of the SAW can be probed in three different ensembles:

- Free-endpoint grand canonical ensemble (variable N, variable x)
- Fixed-endpoint grand canonical ensemble (variable N, fixed x)
- Canonical ensemble (fixed N, variable x)

In the remainder of this section we survey some typical Monte Carlo algorithms for these ensembles.

• *Free-endpoint grand canonical ensemble*. Here the configuration space S is the set of all SAWs, of arbitrary length, starting at the origin and ending anywhere. The grand partition function is

$$\Xi(\beta) = \sum_{N=0}^{\infty} \beta^N c_N \tag{4.4}$$

and the Gibbs measure is

$$\mu(\omega) = \Xi(\beta)^{-1} \times \beta^{\omega} . \tag{4.5}$$

The "monomer activity" β is a user-chosen parameter satisfying $0 \le \beta < \beta_c \equiv \mu^{-1}$. As β approaches the critical activity β_c, the average walk length $<N>$ tends to infinity.

A dynamic Monte Carlo algorithm for this ensemble was proposed by Berretti and Sokal [22]. Its elementary moves are to delete the last link of the chain ($\Delta N = -1$) or to append one link to the chain ($\Delta N = +1$); the relative probabilities are chosen so as to leave the Gibbs measure μ invariant. This algorithm does have critical slowing-down: heuristic arguments and numerical evidence suggest strongly that $\tau \sim <N>^2$ [22], and this has very recently been proven [23] (see Section 5). It is worth comparing the computational work required for SAW versus Ising simulations: $<N>^2 \sim \xi^{2/\nu} = \xi^{\approx 3.4}$ for the $d=3$ SAW, versus $\xi^{d-z} = \xi^{\approx 5.0}$ for the $d=3$ Ising model. This vindicates our assertion that the SAW is an advantageous model for Monte Carlo studies of critical phenomena.

• *Fixed-endpoint grand canonical ensemble.* The configuration space S is the set of all SAWs, of arbitrary length, starting at the origin and ending at the fixed site x ($\neq 0$). The ensemble is as in the free-endpoint case, with c_N replaced by $Nc_N(x)$.

A dynamic Monte Carlo algorithm for this ensemble was proposed by Berg and Foerster [24] and Aragão de Carvalho, Caracciolo and Fröhlich [21, 25] (BFACF). The elementary moves are local deformations of the chain, with $\Delta N = 0, \pm 2$. The critical slowing-down in the BFACF algorithm is quite subtle. A heuristic argument [26] suggests that $\tau \sim <N>^{2-2\nu}$, but this is very likely wrong! Indeed, Sokal and Thomas [23] have recently proven the surprising result that $\tau_{exp} = +\infty$ for *all* $\beta \neq 0$. On the other hand, numerical experiments [27] show that $\tau_{int,N} \sim <N>^{3.0 \pm 0.4}$ (in $d=2$). Clearly, the BFACF dynamics is not well understood at present: further work, both theoretical and numerical, is needed.

• *Canonical ensemble.* Algorithms for this ensemble, based on local deformations of the chain, have been used by polymer physicists for nearly 25 years [28, 29]. So the recent proof [30] that all such algorithms are nonergodic (*i.e.* the transition matrix P is reducible) comes as a slight embarrassment. Fortunately, there does exist a *non-local* algorithm which is ergodic: the "pivot" algorithm, invented by Lal [31] and independently reinvented by MacDonald *et al.* [32] and by Madras [33]. The elementary move is as follows: choose at random a pivot point k along the walk ($1 \le k \le N-1$); choose at random a non-identity element of the symmetry group of the lattice (rotation or reflection); then apply the symmetry-group element to $\omega_{k-1}, \ldots, \omega_N$ using ω_k as a pivot. The resulting walk is accepted if it is self-avoiding; otherwise it is rejected and the walk ω is counted once more in the sample. It can be proven [33] that this algorithm is ergodic and preserves the equal-weight probability distribution.

At first thought the pivot algorithm sounds terrible (at least it did to me): for N large, nearly all the proposed moves will get rejected. This is in fact true: the acceptance fraction behaves $\sim N^{-p}$ as $N \to \infty$, where $p \approx 0.19$ in $d=2$ [33]. On the other hand, the pivot moves are very radical: after very few (5 or 10) accepted moves the SAW will have reached an "essentially new" conformation. One conjectures, therefore, that $\tau \sim N^{-p}$. Actually it is necessary to be a bit more careful: for *global* observables A (such as the end-to-end distance ω_N^2) one expects $\tau_{int,A} \sim N^{-p}$; but *local* observables (such as the angle between the 17th and 18th bonds of the walk) are expected to evolve a factor of N more slowly: $\tau_{int,A} \sim N^{1-p}$. Thus, the *slowest* mode is expected to behave as $\tau_{exp} \sim N^{1-p}$. (This is one of the unusual cases in which $\tau_{int,A}$ and τ_{exp} have different dynamic critical exponents.) For the pivot algorithm applied to *ordinary* random walk one can calculate the dynamical behavior exactly [33]: for *global* observables A the autocorrelation function behaves roughly like

$$\rho_{AA}(t) \sim \sum_{i=1}^{n} (1 - \frac{i}{n})^t , \tag{4.6}$$

from which it follows that

$$\tau_{int,A} \sim \log N \tag{4.7}$$

$$\tau_{exp,A} \sim N \tag{4.8}$$

— in agreement with our heuristic argument modulo logarithms. For the SAW, it is found numerically [33] that $\tau_{int,A} \sim N^{\approx 0.21}$ in $d=2$, also in close agreement with the heuristic argument. But rigorous analysis of the SAW case is completely lacking!

A careful analysis of the computational complexity of the pivot algorithm [33] shows that one "effectively independent" sample (at least as regards *global* observables) can be produced in a computer time of order N. This is a factor $\sim N$ more efficient than the Berretti-Sokal algorithm, a fact which opens up exciting prospects for high-precision Monte Carlo studies of critical phenomena in the SAW.

5. Rigorous Analysis of Dynamic Monte Carlo Methods

Mathematically, the problem is: given a probability kernel P, prove good upper or lower bounds on the autocorrelation time τ_{exp} — or equivalently, on the L^2 spectral gap $1-R$ — of the corresponding Markov chain. Physically, the issue is the rate of convergence to equilibrium for the stochastic dynamics defined by P. What makes this field difficult is that we are dealing with rather complicated statistical-mechanical systems, and we seek estimates which are reasonably sharp even near a critical point.

Nearly all known results assume that the Markov chain is *reversible*, *i.e.* that it satisfies the "detailed-balance condition"

$$\mu(dx)\, P(x,dx') = \mu(dx')\, P(x',dx) . \tag{5.1}$$

[This implies (2.1) but not conversely.] This condition is equivalent to the *self-adjointness* of P as an operator on $L^2(\mu)$. Many (but not all) of the Monte Carlo algorithms discussed above have this property.

The most straightforward and general results concern *lower* bounds on the autocorrelation time (*i.e.* upper bounds on the spectral gap). Here are some techniques:

- *Variational (Rayleigh-Ritz) method.* The variational method for proving bounds on the eigenvalues of self-adjoint operators goes back at least 80 years; its application to dynamic critical phenomena goes back at least 20 years [34]. The idea is to choose a trial vector which is a reasonable approximation to the next-to-leading eigenvector of P (the "slowest mode" of the system). For spin systems with local (*e.g.* single-site heat-bath) dynamics, a good trial vector is the total magnetization $\mathcal{M} \equiv \sum_i \phi_i$; it follows [34-36, 23] that $\tau_{exp} \geq \text{const} \times \chi$, where χ is the (static) susceptibility. This implies the critical-exponent inequality $z \geq \gamma/\nu$, which is reasonably sharp (for $d=3$ Ising, $z \approx 2$ while $\gamma/\nu \approx 1.97$). Likewise, for the Berretti-Sokal SAW dynamics, a variational calculation [23] shows that $\tau_{exp} \geq \text{const} \times \langle N \rangle^2$. Here even the constant may be sharp (or nearly so)!

• *Minimum-hitting-time method.* Let A and B be subsets of the state space S, and let T_{AB} be the m inimum time needed to get from A to B with nonzero probability. Then Sokal and Thom [23] prove, for any reversible Markov chain, the remarkable inequality

$$\tau_{exp} \geq \sup_{A,B} \frac{2(T_{AB} - 1)}{\log\left(\mu(A)\mu(B)\right)} \tag{5.2}$$

(an ome improvements thereof). Consider now the BFACF dynamics: it is not hard to fi. configurations ω, ω' with $\mu(\omega)\mu(\omega') \sim \beta^n$ but $T_{\omega\omega'} \sim n^2$, so it follows from (5.2) that $\tau_{exp} = +\infty$, *i.e.* there is *no* spectral gap. It would be useful to have a better understanding of the spectral properties of the BFACF dynamics.

• *Other hitting-time methods.* It is also possible to prove lower bounds on τ_{exp} in terms of mean [23] or exponential [37, 23] hitting times. These results are, however, of mainly theoretical interest: they are difficult to apply in concrete examples.

Upper bounds on the autocorrelation time (*i.e.* lower bounds on the spectral gap) are much more difficult to prove; indeed, it is far from trivial to prove (except at high temperature) even that $\tau_{exp} < \infty$! The proofs are thus highly model-dependent, and very little is known in general. For *reversible* Markov chains, it suffices [7, 8] to prove that hitting-time distributions decay exponentially ("geometric ergodicity"). By this method, Sokal and Thomas [23] have proven the bound $\tau_{exp} \leq \text{const} \times \langle N \rangle^2$ for the Berretti-Sokal dynamics, but the proof exploits very specific features of the model and is not easily generalized. For the single-site heat-bath dynamics, it is easy to show that $\tau_{exp} < \infty$ (uniform in the volume Λ) above the Dobrushin uniqueness temperature; indeed, this is precisely what the standard proof of the Dobrushin uniqueness theorem [38] does. One expects the same result to hold for *all* temperatures above critical, but this remains an open problem, despite recent progress by Aizenman and Holley [39]. For multi-grid Monte Carlo, rigorous convergence results are known only in the Gaussian case, where Fock-space arguments reduce the problem to a convergence proof for the corresponding deterministic multi-grid method [15, 16, 18], a problem which is quite interesting in its own right [40, 18].

This research was supported in part by NSF grant DMS-8400955.

REFERENCES

1. D.E. Knuth, *The Art of Computer Programming*, vol. 2, 2nd ed. (Addison-Wesley, Reading MA, 1981), chap. 3.
2. H. Niederreiter, Bull. Amer. Math. Soc. **84**, 957 (1978).
3. J.M. Hammersley and D.C. Handscomb, *Monte Carlo Methods* (Methuen, London, 1964), chap. 5.
4. J.G. Kemeny and J.L. Snell, *Finite Markov Chains* (Springer, New York, 1976).
5. M. Iosifescu, *Finite Markov Processes and Their Applications* (Wiley, Chichester, 1980).
6. K.L. Chung, *Markov Chains with Stationary Transition Probabilities*, 2nd ed. (Springer, New York, 1967).
7. E. Nummelin, *General Irreducible Markov Chains and Non-Negative Operators* (Cambridge Univ. Press, Cambridge, 1984).
8. A.D. Sokal, in preparation.

9. G.F. Mazenko and O.T. Valls, Phys. Rev. **B24**, 1419 (1981); J.K. Williams, J. Phys. **A18**, 49 (1985); R.B. Pearson, J.L. Richardson and D. Toussaint, Phys. Rev. **B31**, 4472 (1985).
10. M. Kalos, in Proceedings of the Brookhaven Conference on Monte Carlo Methods and Future Computer Architectures, May 1983 (unpublished).
11. G. Parisi, in *Progress in Gauge Field Theory* (1983 Cargèse lectures), ed. G. 't Hooft *et al*. (Plenum, New York, 1984).
12. G.G. Batrouni *et al.*, Phys. Rev. **D32**, 2736 (1985).
13. J.B. Kogut, Nucl. Phys. **B275 [FS17]**, 1 (1986).
14. A. Brandt, D. Ron and D.J. Amit, paper presented to the Second European Conference on Multigrid Methods (Cologne, October 1985).
15. J. Goodman and A.D. Sokal, Phys. Rev. Lett. **56**, 1015 (1986).
16. J. Goodman, A.D. Sokal and D. Zwanziger, in preparation.
17. R. Edwards, J. Goodman, A.D. Sokal and D. Zwanziger, in preparation.
18. J. Goodman and A.D. Sokal, in preparation.
19. R.H. Swendsen, private communication.
20. P.G. DeGennes, *Scaling Concepts in Polymer Physics* (Cornell Univ. Press, Ithaca NY, 1979).
21. C. Aragão de Carvalho, S. Caracciolo and J. Fröhlich, Nucl. Phys. **B215 [FS7]**, 209 (1983).
22. A. Berretti and A.D. Sokal, J. Stat. Phys. **40**, 483 (1985).
23. A.D. Sokal and L.E. Thomas, in preparation.
24. B. Berg and D. Foerster, Phys. Lett. **106B**, 323 (1981).
25. C. Aragão de Carvalho and S. Caracciolo, J. Physique **44**, 323 (1983).
26. A.D. Sokal, Comparative analysis of Monte Carlo methods for the self-avoiding walk, preliminary draft (1984).
27. S. Caracciolo and A.D. Sokal, J. Phys. **A19**, L797 (1986).
28. M. Delbrück, in *Mathematical Problems in the Biological Sciences* (Proc. Symp. Appl. Math., vol. 14), ed. R.E. Bellman (American Math. Soc., Providence RI, 1962).
29. P.H. Verdier and W.H. Stockmayer, J. Chem. Phys. **36**, 227 (1962).
30. N. Madras and A.D. Sokal, submitted to J. Stat. Phys.
31. M. Lal, Molec. Phys. **17**, 57 (1969).
32. B. MacDonald *et al.*, J. Phys. **A18**, 2627 (1985).
33. N. Madras and A.D. Sokal, in preparation.
34. K. Kawasaki, Phys. Rev. **148**, 375 (1966).
35. B.I. Halperin, Phys. Rev. **B8**, 4437 (1973).
36. R. Alicki, M. Fannes and A. Verbeure, J. Stat. Phys. **41**, 263 (1985).
37. R. Carmona and A. Klein, Ann. Probab. **11**, 648 (1983).
38. L.N. Vasershtein, Prob. Inform. Transm. **5**, 64 (1969); O.E. Lanford III, in *Statistical Mechanics and Mathematical Problems* (Lecture Notes in Physics #20), ed. A. Lenard (Springer, Berlin, 1973).
39. M. Aizenman and R. Holley, in Proceedings of the IMA Workshop on Percolation and Infinite Particle Systems (Springer, to appear).
40. W. Hackbusch, *Multi-Grid Methods and Applications* (Springer, Berlin, 1985).

MODELS OF QUASIPERIODIC STRUCTURES

Michel DUNEAU and André KATZ

Centre de Physique Théorique
Ecole Polytechnique, 91128 Palaiseau Cedex
FRANCE

ABSTRACT

The current attempts for structural models of quasicrystals are provided by different but connected methods which are discussed : 1) the grid method which is closely related to the theory of tilings, 2) the projection method which is more general and which gives an explicit Fourier transform of the structures and 3) the section method related to the theory of modulated crystals.

1. INTRODUCTION

In 1984 D.Shechtman et al.[1] announced the discovery of a new type of order in rapidly quenched alloys of Al-Mn, the diffraction patterns of which consisted of sharp Bragg peaks, like periodic crystals, and had a perfect icosahedral symmetry, which is inconsistent with any lattice translation group. Such patterns are now observed in many binary and ternary alloys [16]. Besides, other structures with 10-fold [2] and 12-fold [3] symmetry have recently been observed. These surprising properties of the diffraction patterns were reminiscent of the 2-dimensional tilings of R. Penrose [4,5], which present a 5-fold symmetry in a weak sense and which give rise to discrete optical Fourier transforms [6,7]. After the first pionnering works of R. Penrose, J. Conway [see 5], R. Robinson [see 5] and N. G. de Bruijn [8], the possible implications of these structures in physics were mentioned by A. L. Mackay [6] and R. Mosseri and J. F. Sadoc [7]. A 3-dimensional generalization of the Penrose tilings was described by P. Kramer and R. Neri [9]. The grid method, devised by de Bruijn to provide an algebraic construction of the Penrose tilings, allows to build tilings in any dimension. The projection method, which is now widely used in connection with structural models of quasicrystals [10,11,12,13,16], yields more general quasiperiodic patterns, the Fourier transforms of which can be explicitely computed. More general structures, such as modulated crystals, can be obtained by a related construction due to de Wolf, Janner and Janssen [14], called hereafter the section method .

2. THE GRID METHOD

Let E be the p-dimensional Euclidean space to be tiled and let $\{e_1,...,e_n\}$ be a set of n>p vectors in E of rank p. For each i=1,...,n, the grid G_i is defined as the set of parallel hyperplanes $x.e_i+\alpha_i \in \mathbb{Z}$, where α_i is a given real constant. The union $G=\cup G_i$ yields a partition of E-G into meshes and it is assumed that any intersection of p+1 grids is empty. Each mesh M can be labelled by $k=\{k_1,...,k_n\}\in \mathbb{Z}^n$ with $M=\cap\{x|\ k_i-1< x.e_i+\alpha_i<k_i\}$. Then, a dualization procedure transforms this multi-grid G into a tiling of E : to each mesh M, labelled by $\{k_1,...,k_n\}$, is associated a point $x=\sum k_i e_i$ in the $\mathbb{Z}$-module L spanned by $\{e_1,...,e_n\}$. This method has recently been extended by de Bruijn [16] to more general situations.

3. THE PROJECTION METHOD

The p-dimensional physical space E to be tiled is embedded in an n-dimensional space $\mathbb{R}^n$ (n>p) endowed with the cubic lattice $\mathbb{Z}^n$. A strip S is defined by shifting the unit cube γ_n of the lattice along E : $S=\{\xi+\zeta \mid \xi\in E, \zeta\in\gamma_n\}$. Then, for almost all positions of E, S contains a p-dimensional manifold (made of p-facets of the lattice), joining all the vertices of $\mathbb{Z}^n$ falling in the strip. The projection of this manifold on E yields a tiling by means of the projections of p-facets of $\mathbb{Z}^n$.

This construction depends on the three following points (see Fig.1a) :

1) The choice of the unit cube controls the profile of the strip in such a way that the p-manifold is unique and connected.

2) The tiling is periodic if and only if the orientation of E is rational i.e. if and only if E is parallel to a lattice plane of $\mathbb{Z}^n$. Aperiodic tilings can be thought of as interpolations between periodic ones, or as limits of periodic tilings with larger and larger unit cell.

3) For any given irrational orientation, the strip can be translated in $\mathbb{R}^n$. Generically, each time a lattice point ξ leaves the strip, a new point ζ enters the strip in such a way that (ξ,ζ) is the diagonal of some (p+1)-facet of $\mathbb{Z}^n$. In the $2\rightarrow1$ case the broken line "jumps" from one side of a unit square to the other. For any non trivial translation of the strip, such jumps occurs infinitely many times, in such a way that this construction yields a uncountable set of different tilings for each irrational slope.

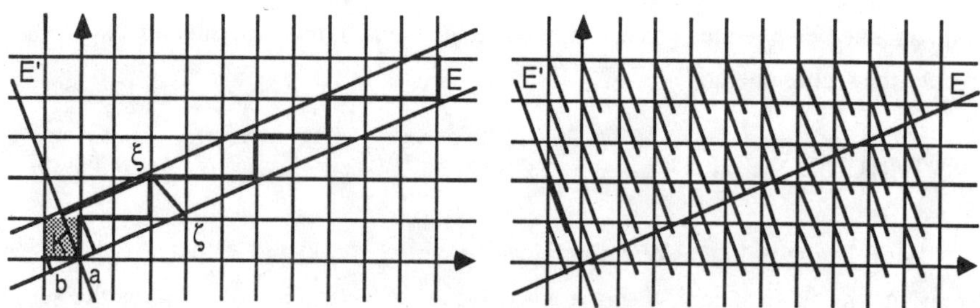

Fig 1a : The projection method. E is tiled by the projection of edges of $\mathbb{Z}^2$ falling in the strip S=E+K.

Fig 1b : The section method. The vertices of the tiling are given by the intersection of E with the set $\{-K+\xi\}$, $\xi\in\mathbb{Z}^2$.

This contruction is equivalent to the grid method, as proven in [15]. Actually, the meshes of the grid method correspond to the intersections of E with the hypercubes of $\mathbb{Z}^n$. The dualization amounts to project on E distinguished points of these hypercubes, which are exactly the lattice points which belong to the strip.

This procedure can be performed in an equivalent way : let π and π' be the orthogonal projectors on E and $E'=E^{\perp}$; then $\pi'(S)=\pi'(\gamma_n)$ is a convex polyhedron K of E'. A lattice point ξ falls in S iff its projection $x'=\pi'(\xi)$ falls in K. If the orientation of E and E' are both irrational, π and π' have a 1-1 restriction to $\mathbb{Z}^n$; then there is a 1-1 mapping between the vertices of the structure $X=\pi(S\cap\mathbb{Z}^n)$ and $\pi'(S\cap \mathbb{Z}^n)=K\cap\pi'(\mathbb{Z}^n)$.

More general patterns are obtained when the strip S=E+K is defined by means of any other bounded subset K of E'. The shape of K and its size determine the geometry of the pattern $\pi(S\cap\mathbb{Z}^n)$ and in particular its density and the nearest-neighbour distance. The structural problem raised by the Al-Mn quasicrystals requires to find a K which fits with the experimental data.

A measure m, associated to each pattern, is defined as the sum of Dirac deltas at each vertex. Let χ_S be the characteristic function of the strip and let $\mu=\chi_S.\Sigma_{\xi\in Z^n}\delta_\xi$. Then m can be deduced from μ by partial integration along E' : $m=\int_{E'}\chi_S.\Sigma_{\xi\in Z^n}\delta_\xi$. The Fourier transform μ^* is given by the convolution $\mu^*=(\chi_S^*)_*\,\Sigma_{\kappa\in Z^n}\delta_\kappa$ so that the Fourier transform m* is a section of μ^* i.e. $m^*=(\chi_S{}^*{}_*\,\Sigma_{\kappa\in Z^n}\delta_\kappa)\,|_{E^*}$. The characteristic function χ_S can be factorized in E×E' : $\chi_S(x,x')=\chi_K(x')$ where χ_K is the characteristic function of K in E'and its Fourier transform $\chi_S^*(k,k')=\delta(k)\,\chi_K^*(k')$ is carried by the dual E'* of E'. Then $\mu^*(\kappa)=\Sigma_{\lambda\in Z^n}\delta(k-l)\,\chi_S^*(k'-l')$ where $\kappa=(k,k')$ and

$\lambda=(l,l')$ are the decompositions in $E^*\times E'^*$. Finally, $m^*(k)=\Sigma_{\lambda\in Z^n}\,\delta(k-l)\,\chi_K^*(-l')$. As a consequence, m* is carried by the $\mathbb{Z}$-module L*, projection of the reciprocal lattice on E*. The dimension of this module is n-k where k is the dimension of the lattice $E'\cap\mathbb{Z}^n$. If $E'\cap\mathbb{Z}^n=\{0\}$ the amplitude at l is simply $\chi_K^*(-l')$ where $\lambda=(l,l')\in\mathbb{Z}^n$. The quasiperiodicity of the structure is thus a direct consequence of the construction.

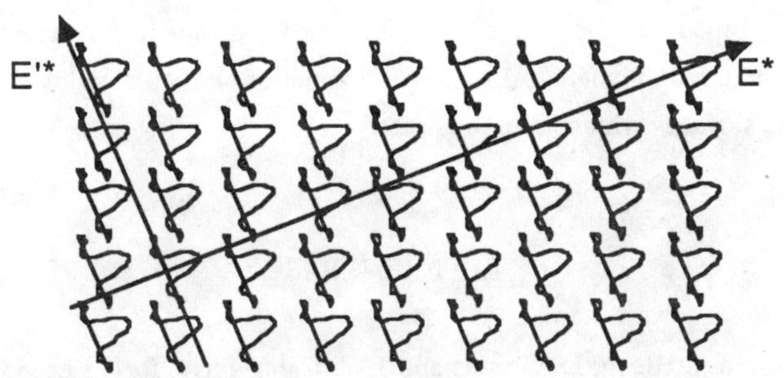

Fig 2 : The Fourier transform m* is the section (k'=0) of the convolution $\mu^*=(\chi_S^*)_*\,\Sigma_{\kappa\in Z^n}\,\delta_\kappa$.

The above construction implies a connection between the local geometry of the patterns and their Fourier spectrum. This connection can be weakened by the following generalization : let E_1 be a p-dimensional subspace of $\mathbb{R}^n$ transverse to E'. The strip is defined as above by S=E+K but the pattern is now given by the skew projection of $S\cap\mathbb{Z}^n$ on E, perpendicularly to E_1. In such a situation, the local geometry (for instance the tiles of a tiling) is fixed by E_1 whereas the periodicity properties depend on E.

4. THE SECTION METHOD

The structures described above can also be obtained in the following way : let K be the projection of the strip on $E'=E^\perp$. A lattice point ξ is in the strip S=E+K iff $\xi=x+x'$ for some x in E and x' in K so that x is the intersection of E with $-K+\xi$. The pattern is thus given by the intersection of E with the set $\{-K+\xi|\ \xi\in\mathbb{Z}^n\}$ (see Fig.1b). The associated measure m, defined as above, is now given by the restriction to E of the measure $\chi_{A*}\Sigma_{\xi\in Z^n}\,\delta_\xi$ associated to $\{-K+\xi|\ \xi\in\mathbb{Z}^n\}$, where $\chi_A(x,x')=\delta(x).\chi_K(-x')$ is carried by -K and where χ_K is the characteristic function of K. So $m=\chi_{A*}\Sigma_{\xi\in Z^n}\,\delta_\xi|_E$ and its Fourier transform m* can be deduced from the Fourier transform $\chi_A^*.\Sigma_{\kappa\in Z^n}\,\delta_\kappa$ by a partial integration on $E'^*=(E^*)^{\perp}$, i.e. $m^*=\int_{E'^*} \chi_A^*.\Sigma_{\kappa\in Z^n} \delta_\kappa$. Since $\chi_A^*(k,k')=\chi_K^*(-k')$ we have $\chi_A^*.\Sigma_{\kappa\in Z^n} \delta_\kappa= \Sigma_{\kappa\in Z^n} \chi_K^*(-k').\delta_\kappa$ and we find the same result for m^*.

This method is closely related to the construction of modulated crystals in the framework of de Wolf, Janner and Janssen [14], where the vertices are defined by the intersection of the physical p-dimensional space $E\subset\mathbb{R}^{p+1}$ with a set of lines (transverse to E) which is invariant with respect to a (p+1)-dimensional lattice. The above structures correspond to compact atomic surfaces which are related to incommensurate structures with a density modulation whereas displacive modulations are obtained with topologicaly different atomic surfaces which are unbounded.

REFERENCES

[1] D. Shechtman, I. Blech, D. Gratias and J.W. Cahn, Phys. Rev. Let 53 (1984) 195

[2] L. Bendersky, Phys. Rev. Lett. 55 (1985) 1461.

[3] T. Ishimaza, H. U. Nissen and Y. Fukano, Phys. Rev. Lett. 55 (1985) 511.

[4] R. Penrose, Math. Intelligencer 2 (1979) 32.

[5] M. Gardner, Sci. Am. 236 (1977) 110.

[6] A.L. Mackay, Physica 114 A (1982) 609, Kristallografiya **26** (1981) 909. Sov. Phys. Crystallogr. 26 (1981) 517.

[7] R. Mosseri and J. F. Sadoc, "Structure of non-crystalline materials 1982", ed. Gaskell et al. (London, Taylor and Francis, 1983).

[8] N.G. de Bruijn, Nederl. Akad. Wetensch. Proc. A 43 (1981) 39-66.

[9] P. Kramer and R. Neri, Acta Crystallogr. A 40 (1984) 580-587 .
P. Kramer, Acta Crystallogr. A 38 (1982) 257-264 .

[10] P.A. Kalugin, A.Yu. Kitayev and L.S. Levitov, J. Physique Lett. 46 (1985) 601, JETP Lett. 41 (1985) 145.

[11] M. Duneau and A. Katz, Phys. Rev. Let. 54 (1985) 2688.

[12] A. Katz and M. Duneau, J. de Physique 47 (1986) 181.

[13] V. Elser, Acta Cryst. A42 (1986) 36.

[14] A. Janner and J. Janssen, Physica B and C 99B-C (1980) 334.

[15] F.Gähler and J. Rhyner, J. Phys. A19 (1986) 267.

[16] Les Houches, Workshop on Aperiodic Crystals 11-22 March 1986, ed. L. Michel and D. Gratias, to be published in Journal de Physique (Colloque).

THE ISING MODEL IN A RANDOM MAGNETIC FIELD

John Z. Imbrie*

Departments of Mathematics and Physics
Harvard University
Cambridge, MA 02138
USA

ABSTRACT

In three or more dimensions, the random-field Ising model exhibits spontaneous magnetization at zero temperature. This behavior is explained in terms of a "contour gas" picture of the ground state. The proof settles a physics controversy regarding the value of the lower critical dimension of the model.

We consider the usual Ising model Hamiltonian

$$H = \sum_{\langle i,j\rangle} \frac{1}{2}\left(1 - \sigma_i\sigma_j\right) - \sum_i h_i\sigma_i ,$$

where $\sigma_i = \pm 1$ are Ising spins at sites $i \in \mathbb{Z}^d$, and where $\langle i,j\rangle$ denotes a nearest neighbor pair. The ground state, σ_{min}, is the spin configuration minimizing H. It is a function of the site-varying magnetic field h_i. We take each h_i to be an independent Gaussian random variable with mean zero and variance $\varepsilon^2 \ll 1$, and we take $\sigma_i = \pm 1$ for i outside a finite region $\Lambda \subset \mathbb{Z}^d$. Our main result is that for $d > 2$,

$$\text{Prob}\left(\sigma_0^{min} = -1\right) < \exp\left(-0\left(\varepsilon^{-2}\right)\right).$$

In other words, the spin at the origin almost always agrees with the

---

*Alfred P. Sloan Research Fellow.
Supported in part by the National Science Foundation under grants PHY-85-13554 and PHY 84-13285.

boundary conditions. The "contour gas" picture explains the long-range order (LRO) of the ground state using defect random variables (indexed by contours or domain walls) and their statistical properties. This approach avoids some of the shortcomings of the heuristic arguments of Imry and Ma, who first predicted LRO for $d > 2$ (i.e., lower critical dimension $d_\ell = 2$). Furthermore, the proof of LRO in $d = 3$ refuted the dimensional reduction prediction, $d_\ell = 3$.

The complete proof[1)], a shorter, less detailed version[2)], and a more pedagogical account[3)] are all given elsewhere. These papers and another[4)] may all be consulted for references to other literature in the random-field Ising model.

REFERENCES

1. Imbrie, J.Z., "The Ground State of the Three-Dimensional Random-Field Ising Model, Commun. Math. Phys. 98, 145-176 (1985).

2. Imbrie, J.Z., "The Ising Model in a Random Field: Long-Range Order in Three Dimensions," in: *Critical Phenomena, Random Systems, Gauge Theories*, Les Houches 1984. K. Osterwalder and R. Stora (eds.): North Holland, to appear.

3. Imbrie, J.Z., "Low Temperature Behavior in Random Ising Models," in: *Proceedings of the 1985 Workshop on Disordered Media*, the Institute for Mathematics and its Applications, Minneapolis, Minnesota, to appear.

4. Imbrie, J.Z., "Lower Critical Dimension of the Random-Field Ising Model", Phys. Rev. Lett. 53, 1747-1750 (1984).

RIGOROUS RESULTS ON ANDERSON LOCALIZATION

F.MARTINELLI[1] E.SCOPPOLA[2]

[1]Dipartimento di Matematica, Universita'"La Sapienza",Pz.A.Moro 2, 00185 Roma,Italy.

[2]Dipartimento di Fisica, Universita'"La Sapienza",Pz.A.Moro 2, 00185 Roma,Italy.

Abstract

We review some rigorous results concerning the Anderson localization.

§1

The Anderson model is a tight binding hamiltonian widely used in solid state physics in order to describe the motion of a single electron in a disordered crystal.The hamiltonian is given by

$$H(v) = -\Delta + v \quad \text{on} \quad \ell^2(Z^d) \qquad (1)$$

where $\{v_x\}_{x\in Z^d}$ are i.i.d.random variables with distribution dP(v) and Δ is the usual discrete Laplacian.

Let us consider for concreteness the typical case of a random potential uniformly distributed in the interval [0,1]. In the large disorder case $\lambda>>1$ or at very low energies the only possibility for a wave packet concentrated at time t=0 around the origin to travel a long distance is by means of <u>quantum tunneling through potential barriers</u>. Thus in order to understand the localization phenomena in this range of the parameters one has to understand why the <u>tunneling is not effective</u> over large distances.The key point is the following:

<u>the delocalization of a wave function among different wells due to tunneling is extremely unstable under perturbations of the potential and therefore it occurs only in very special situations.</u>

This important effect was discovered by G.Jona-Lasinio, F.Martinelli and E.Scoppola [2] in their analysis of the semiclassical limit of the one dimensional Schroedinger equation with a potential having a finite number of absolute minima. In order to apply the above ideas to the localization problem it is however crucial to understand how the does the analysis of tunneling extend to cases with an infinite number of wells. For this purpose in [3] it was constructed a hierarchical potential consisting of an infinite number of equal wells separated by barriers with the same height and arranged in such a way to give rise to a selfsimilar structure over a sequence of rapidly increasing length scales The result, in the case of symmetrically distributed and identical wells,was that quantum tunneling is actually taking place leading to a delocalization of the wave functions.In particular the succesive splittings of the eigenvalues of a single well due to the tunneling with the other wells led to a Cantor structure for the energy levels and the traveling of the particle, by jumping from one well to the next one, led to a logarithmic increase in time of the mean square distance $r^2(t)$.

However as soon as an arbitrarily small stochastic perturbation is added to each site of the lattice then tunneling over long distances disappears completely, the eigenfunctions become exponentially localized and the mean square distance $r^2(t)$ stays bounded uniformly in time.

The main tool in order to obtain such results was a multiscale analysis of the problem using the estimates on the decay of Green's functions in random potentials obtained by J.Frohlich and T.Spencer [4] in their detailed proof of the vanishing of the diffusion constant D(E) for the Anderson model at high disorder or low energy.

The proof of localization in the Anderson model, as given in [1] in collaboration with J.Frohlich and T.Spencer , follows closely the pattern of the proof in the randomly perturbed hierarchical case given in [3]. In particular in [1] it is shown that for large disorder or low energy the structure of the typical configurations of the random potential are, as far as tunnneling is concerned, of the same type of the hierarchical random case. This part of the proof is highly non trivial but it is also rewarding since the understanding of localization that emerges out of it is quite detailed.

§2 Description of the main results.

Let us assume that the probability distribution is Holder continuous.Then there exists a $\lambda_c < \infty$ such that:

A) Localization at high disorder [1]

i) The spectrum of H(v) is pure point with exponentially decaying eigenfunctions with probability one.

ii) The spread of a wave packet φ_0 under the dynamics generated by $e^{itH(v)}$ is uniformly bounded in time. More precisely for any φ_0 of compact support and any n >0 there exists a constant $C(n,\varphi_0)$ such that:

$$r^{(n)}(t,\varphi_0) \equiv \int dP(v) \sum |x|^n | e^{itH(v)}\varphi_0(x)|^2 < C(n,\varphi_0) \quad t > 0.$$

iii) Let E'(v) be one of the eigenvalues of H in I with eigenfunction φ' and let ξ' be its localization length (see [5] for the precise definition).Then:

$$\text{Prob}\{ \ E'(v)\in I \ ; \ |\varphi'| \text{ has an absolute maximum at } x=0 \text{ and } \xi' > L\} < \\ < L^{-p} \text{ for all L large enough.}$$

B) Localization for low energy.

i) ii) iii) hold for an arbitrary $\lambda>0$ and all energies sufficiently close to the edges of the spectrum.

C) Stability of localization [5]

Let H_1 be a bounded operator on $\ell^2(Z^d)$ with :

$H_1(x,y)=0 \qquad$ if $\qquad |x-y| >1.$

Then for all δ sufficiently small the results A) hold for the

perturbed Hamiltonian $H^p = H + \delta H_1$.
In particular this result allows to include the spin-orbit interaction or the interaction with a magnetic field [6].
It also gives localization at all energies in a bidimensional anysotropic Anderson model with small coupling in one direction [5] .In one dimension it is possible to relax considerably the assumptions and one obtains localization for <u>any</u> value of the coupling constant λ for <u>any</u> distribution $dP(v)$ having an <u>arbitrarily small moment</u> [7].

iii) Absence of two slow time evolution [5]
Suppose that the distribution $dP(v)$ is Holder continuous and assume that in the energy interval I the following estimate holds:

$$r^{(n)}(t,\varphi_0,I) < \text{const.}t \qquad \text{for any } t>0$$

Then if n is large enough localization holds for any energy $E \in I$.

This interesting result shows that localization different from exponential is very unlikely.
The above results also apply to the <u>continuous Schrodinger equation</u> with a random potential [8],[5],[9].Let us consider the following Hamiltonian:

$$H = -\Delta + \sum q_i V_0(x-x_i) \qquad (2)$$

where $\{x_i\}$ is a Poisson process on $\mathbf{R}^d$ with density ρ ,the q_i's are i.i.d. random variables uniformly distributed in [0,1] and V_0 is a non positive bounded function with support in the unit sphere.

D) <u>Localization in the continuous case</u>
For $\rho \ll 1$ the negative part of the spectrum of H consists of a dense set of eigenvalues with exponentially decaying eigenfunctions [5].

The techniques used in [1] can also be extended to other disordered systems with more complicate disorder like:
1) <u>Harmonic vibrations of a crystal lattice [1]</u>
2) <u>Wave propagation in wave guides with random boundaries [10]</u>

References

1) J.Frohlich,F.Martinelli,E.Scoppola,T.Spencer: "Constructive proof of localization in the Anderson tight binding model" Comm.Math.Phys.101 (1985)

2) G.Jona-Lasinio,F.Martinelli,E.Scoppola: "New approach to the semiclassical limit of quantum mechanics:I multiple tunnelings in one dimension" Comm.Math.Phys. 80 (1981).
(see also B.Helffer,J.Sjostrand "Multiple wells in the Semiclassica Limit I" Comm.P.D.E. 9 (1984) and "Puits multiples en limite semiclassique II" Ann.Inst.H.Poincare' 42 (1985))

3) G.Jona-Lasinio,F.Martinelli,E.Scoppola: "Multiple tunnelings in d-dimensions:a quantum particle in a hierarchical potential" Ann.Inst.H.Poincare' 42 N1 (1985)

4) J.Frohlich,T.Spencer: " Absence of diffusion in the Anderson tight binding model for large disorder or low energy" Comm. Math.Phys.

5) F.Martinelli,E.Scoppola: "Introduction to the mathematical theory of Anderson localization" (to appear in La Rivista del Nuovo Cimento).

6) J.Bellisard,D.R.Grempel,F.Martinelli,E.Scoppola: "Localization of electrons with spin orbit or magnetic interactions in a 2-D disordered crystal" Marseille preprint (to appear in Phys.Rev.Lett.)

7) R.Carmona,A.Klein,F.Martinelli: "Anderson localization for Bernoulli and other singular potentials" to appear in Comm.Math.Phys.

8) H.Holden,F.Martinelli: "On absence of diffusion for low energy for a random Schrodinger operator on $L^2(\mathbf{R}^d)$" Comm.Math.Phys. 93 (1984)

9) F.Martinelli,E.Scoppola: "Remark on the absence of absolutely continuous spectrum in the Anderson model for large disorder or low energy" Comm.Math.Phys. 97 (1985)

10) F.Martinelli: "The wave equation in random domains: localization of the normal modes in the small frequency region" Ann.Inst.H.Poincare' 43 N2 (1985)

PHYSICS OF DISORDERED MEDIA : SOME NEW RECENT RESULTS

R. Rammal
Centre de Recherches sur les Très Basses Températures,
C.N.R.S., BP 166 X, 38042 Grenoble-Cédex, FRANCE

ABSTRACT

The importance of the study of probability distributions is illustrated in three different problems. The first one is relevant for the transport properties (conductance, noise, ...) in percolation systems. The second one is spin dynamics in dilute ferromagnets. The third is wave transmission in non linear random media. New features recently discovered such as : infinite hierarchies of exponents, fluctuations from sample to sample, etc ... are also discussed.

1. INTRODUCTION

The main purpose of this talk is to illustrate, through three examples a basic feature of disordered media : The probability distributions that we need in order to have a deep knowledge of the physical behavior. In some cases scale invariant distributions can appear where an infinite set of exponents is necessary to fully describe the properties of the underlying system. The recent progress done on disordered materials tells us something about the "general principles" described in Ref. 1, but there is still a long way to go. Some illustrative examples for new concepts such as : non self-averaging quantities, fluctuation from sample to sample, universal fluctuations, hierarchies of exponents, ergodicity breaking, order parameter function, etc ..., will be the principal object of the next sections. Because of place limitation, we direct the reader to relevant references, such as Ref. 2 for spin glasses and Ref. 3 for localization problems, which are not covered in this talk.

2. CURRENT DISTRIBUTIONS IN RANDOM RESISTOR NETWORKS

- For lattice percolation networks, see :

Rammal, R., Tannous, C. and Tremblay, A.M.S., Phys. Rev. A 31, 2662 (1985).

Rammal, R., J. Physique-Lett. (Paris) 46, L129 (1985).

Rammal, R., Tannous, C., Breton, P. and Tremblay, A.M.S., Phys. Rev. Lett. 54, 1718 (1985).

- For continuum percolation systems, see :

Rammal, R., Phys. Rev. Lett. 55, 1428 (1985).

- For non linear random resistor networks, see :

Rammal, R. and Tremblay, A.M.S., Phys. Rev. Lett. 57, (1986).

- For an experimental realization, see :

Garfunkel, G. and Weissman, M.B., Phys. Rev. Lett. 55, 296 (1985).

3. ISING SPIN DYNAMICS NEAR THE PERCOLATION THRESHOLD

Rammal, R., J. Physique (Paris) 46, 1837 (1985).

Rammal, R. and Benoit, A., J. Physique-Lett. (Paris) 46, L667 (1985).

Rammal, R. and Benoit, A., Phys. Rev. Lett. 55, 649 (1985).

Henly, C., Phys. Rev. Lett. 54, 2030 (1985).

4. TRANSMISSION OF WAVES ACROSS RANDOM MEDIA

- For linear random media, see :

Abrikosov, A.A., Solid State Commun. 37, 997 (1981).

Melnikov, V.I., Sov. Phys. Sol. State 23, 444 (1981).

- Calculations of transmission in the presence of a magnetic field will be found in :

Douçot, B., PhD Thesis Grenoble (1986) unpublished.

- For non linear random media, see :

Devillard, P. and Souillard, B., J. Stat. Phys. 43, 423 (1986).

Douçot, B. and Rammal, R., Phys. Rev. Lett. 57, (1986).

Douçot, B. and Rammal, R., J. Physique (Paris) submitted (1986).

5. REFERENCES

1. Anderson, P.W., "Ill-Condensed Matter", Les Houches Summer School (North Holland, Amsterdam) p. 162 (1979).
2. Mézard, M., Parisi, G. and Virasoro, M.A., J. Physique-Lett. 46, L217 (1985).
3. Altshuler, B.L. and Khmelnitskii, D.E., J.E.T.P. Lett. 42, 359 (1985).

WAVES IN NON-LINEAR AND NON-HOMOGENEOUS MEDIA

Bernard SOUILLARD

Centre de Physique Théorique
Ecole Polytechnique
F-91128 Palaiseau
FRANCE

ABSTRACT

We review results on non-linear waves in inhomogeneous media. We discuss the stability or instability with respect to non-linearities of *Anderson localisation* in a random potential. We also describe new types of *bistability* phenomenon in the presence of a periodic modulation of the medium.

1 - NON-LINEAR WAVES IN RANDOM MEDIA

Linear waves in random media have been widely studied in the recent years, particularly in view of the Anderson localization phenomenon. In the present paragraph we discuss the stability or instability of Anderson localization with respect to non-linear perturbations in the medium.

But let us start with a brief summary of localization for linear waves. Consider for example a Schrödinger equation with a static random potential which occurs for example in the study of electron propagation in a solid in the presence of impurities located at random. We ask about the time behaviour of an electron initially located in some finite region of space. We also ask the nature of the proper modes of the equation, a related question. The remarquable phenomenon, called *Anderson localization* (for a discussion and a review on this problem, see e.g. Ref. [1]), is the following : if the disorder is sufficiently strong (i.e. the fluctuations of the potential are sufficiently wild) the proper modes of the equation, for a typical realization of the potential (i.e. with probability one) are exponentially localized and the spectrum of our self-adjoint operator is pure point with a dense set of eigenvalues and exponentially decaying wave functions. In dimensions 1 and probably 2 this occurs for any (arbitrarily small) disorder. Considering now time evolution, Anderson localization implies that, for an initial state starting in a finite region of space, the probability to find the particle outside some finite region of space will remain small *uniformly in time, for a large enough region* . Such a phenomenon is also naturally believed to imply that our electronic system is an insulator, since electrons are "trapped" by disorder and cannot anymore transport the electric current. Similar phenomena arise for other continuous and discrete linear wave equations when considering propagation in a disordered medium and apply for example to electromagnetic propagation in turbulent plasmas, to hydrodynamical waves on rough bottoms, to acoustic waves in random media ...[1]. Anderson localization is induced by phase interferences of the waves which are partially reflected by the fluctuations of the medium and interfere together with random phases.

Let us assume that we now add some non-linearity. This occurs for example in the discussion of polarons in condensed matter or in the propagation of electromagnetic waves in plasmas etc ... It is natural to ask whether localization is stable or unstable with respect to non-linearities and if there are new phenomenons appearing.

1.1 - The Case of Small Non-Linearities : Some Stability of Localization.

It is shown for a class of equations [2],[3] that if one introduces a small non-linearity in a situation for which the modes of the linear equation are exponentially localized, the perturbed non-linear equation still possesses some (and in fact infinitely many) exponentially decaying proper modes which thus correspond to localized solutions of the associated time evolution equation.

It does not follow from these results, in contrast to the linear case, that an initial condition, superposition of such modes, remains localized uniformly in time since for a non-linear equation there is no direct correspondance between the properties of the stationary solutions and the time evolution of an initial state. Actually such an extrapolation of the above results is not expected in most cases. In contrast one expects that most initial states will go to infinity in contrast to the exceptional solutions mentionned above. However, due to the localization of these stationary solutions, the delocalization should be very slow in analogy with Nichorochev theorem valid for a class of Hamiltonian sytems with a finite number of degrees of freedom. Such a result is not yet available but the papers quoted above [2],[3] represent the first step of the program devoted to prove some sort of stability of localization for some classes of small non-linear perturbations.

1.2 - General Non-Linearities : Apparition of New Phenomena.

For general non-linearities only one-dimensional results are available and we can think for example of a one-dimensional non-linear Schrödinger equation with a random potential. We again consider stationary solutions and more precisely we study [4] the transmission coefficient for a piece of length L of non-linear and disordered material and we want to know its behavior for L going to infinity. Because of the non-linearity the transmited intensity is not a linear function of the incident intensity and one must distinguish between the situation at fixed output and the one at fixed input.

The interesting point is the following [4] : the methods used in order to study the transmission coefficient for linear random equations, which are based on Furstenberg types of theorems, and yield an exponential decay of the transmission as a function of L, do not work any more. Actually there is a good reason for that : new phenomenon in the competition between non-linearities and disorder arise. It turns out that whereas for small non-linearities the transmission (i.e. the ratio of the transmited intensity to the incident one) still decays exponentially with the size of the system, we note that this is no more true for stronger non-linearities for which the transmission can be shown not to decay faster than a weak power law (the exact behavior depends on the type of non-linearity and on the regularity properties of the random potential). And there is a cross-over between these two regimes [4]. In any case the transmission can be proven to tend to zero. In this last result a crucial use was made of a method developped by Frish and Sulem [5] for the linear case but which does not yield the optimal result for it.

The precise behavior of the transmission as a function of L is related to the behavior at large energies of the Lyapunov exponent for the (linear) random Schrödinger equation which in turn depends on the differentiability properties of the potential. Very nice results on this last topic have been obtained recently by Delyon and Foulon [6] (related results have also been obtained by Carmona and Martinelli [7]).

Finally, recently, in the specific case of a white noise potential, Douçot and Rammal could study the same question in more details with other technics. They recover in particular the power law decay of Ref. [4], but they also discuss the behavior of the probability distribution of the transmission [8].

2 - BISTABILITIES IN AN INHOMOGENEOUS MEDIUM

As we have already mentionned, the transmited intensity through a piece of non-linear material is a non-linear function of the incident intensity, in contrast to the linear case. If we now keep the size L of the system constant and vary the incident intensity it may happen that for some ranges of incident energies, there are several possible transmited energies for the same incident energy. This phenomenon is called *bistability* and is widely used in particular in technological devices based on non-linear optics.

Recently the influence of disorder on the bistability phenomenon has been discussed by Baylis, Papanicolaou and White [9].

We have been interested in another problem namely the influence of a *periodic* modulation of the index of the medium [10]. We have discovered in this context a new type of bistability. It presents specific properties with respect to the previously known types of bistability :

- it is non-perturbative with respect to the non-linearity ;
- the transmitted energy exhibits plateaus as a function of the input intensity ;
- the frequency versus intensity propagation diagram is fractal.

This new type of bistability is connected with the stable or the chaotic properties of a family of related non-linear dynamical systems. Some possible applications of this new type of bistability have been discussed to polarons in condensed matter, and to optical, acoustical and electromagnetical devices.

3 - CONCLUSION

Let us first emphasize that the results described above hold for various classes of equations, but no doubt that some other classes of non-linearities can yield very different phenomenons. Although we may have not covered all the works concerning non-linear waves in inhomogeneous media, it appears readily from the above discussion that on one hand there are still few results : it is a difficult field. On the other hand it is an extremely interesting domain both from the theoretical point of view and for its connections with many applications. Finally it is a very exciting one in view of the fact that, as we have seen above, one obtains in some cases qualitatively new phenomenon arising. We have little doubt that this field will expand in the coming years !

4 - REFERENCES

[1] B. Souillard, *Spectral Properties of Discrete and Continuous Schrödinger Operators : a Review*, preprint to appear in the Proceedings of the meeting on "Random Media" held at the IMA Minnesotta (1985).

B. Souillard, *Waves and Electrons in Non-Homogeneous Media*, preprint to appear in the Proceedings of the Les Houches Summer School "Chance and Matter".

[2] J. Fröhlich, T. Spencer and J. Wayne, *Localization in Disordered Nonlinear Dynamical Systems*, J. Stat. Phys. **42** (1986) 247.

[3] J. Bellissard and M. Vitot, *Invariant Tori for an Infinite Lattice of Coupled Classical Rotators*, preprint.

[4] P. Devillard and B. Souillard, *Polynomially Decaying Transmission for the Non-Linear Schrödinger Equation in a Random Medium*, J. Stat. Phys. **43**, 423 (1986).

[5] U. Frisch and P.L. Sulem, *Total Reflection of a Plane Wave by a Semi-Infinite Random Medium*, J. Plasma. Phys. **8**, 217 (1972).

[6] F. Delyon and P. Foulon, *Adiabatic Invariants and Asymptotic Behavior of Lyapunov exponents of the Schrödinger Equation*, J. Stat. Phys. October 1986.

[7] R. Carmona and F. Martinelli, to be published.

[8] B. Douçot and R. Rammal, *On Anderson Localization in Non-Linear Media* , preprint to appear.

B. Douçot, Thesis (Grenoble 1986).

[9] A. Baylis, G. Papanicolaou and B. White, *Scattering by a Non-Linear Random Medium*, to appear.

[10] F. Delyon, Y.E. Lévy and B. Souillard, *Non-Perturbative Bistability in Periodic Non-Linear Media*, Phys. Rev. Lett.October 1986.

LOCAL BRS SYMMETRY AND HIGHER ORDER COCYCLES

Laurent Baulieu

Laboratoire de Physique Theorique et hautes energies

Paris, France

Two topics of much current research in mathematical physics are the BRST quantization of gauge theories and the understanding of anomalies in gauge theories. While a large part of the stimulus for this research is the revival of interest in string theory, much remains to be learned about gauge theories. Here, we shall present a new extension of BRST symmetry in the form of a local symmetry and demonstrate that this leads to an understanding, in the context of a field theory, of the higher cocycles that have been mathematically associated with chiral anomalies.[1] We obtain, furthermore, a unification of many approaches to the chiral anomaly. Namely, once the BRST symmetry is gauged, we can examine BRST current algebra. Anomalies and the cocyles associated with them are consequences of this current algebra. Moreover, the BRST charge derived from the BRST current is not nilpotent.

In the usual case of Yang-Mills theory, with Lie-algebra valued 1-form potential A, ghost c, antighost c, auxiliary field b, and matter Ψ with ghost numbers 0,1, -1,0,0 respectively, we have the following transformation under the BRST operator S

$$
\begin{aligned}
SA &= -Dc \\
Sc &= -\frac{1}{2}[c,c] \\
S\bar{c} &= 0 \\
Sb &= 0 \\
S\Psi &= -c\Psi
\end{aligned}
$$

As a consequence of closure of the algebra and the Jacobi identity, we have $S^2=0$. In addition for differential $d=dx^\mu \partial_\mu$, we have $d^2=0$ and $Sd+dS=0$. S acts as a differential operator in a direction perpendicular to d and graded by the sum of the ghost and Lorentz degree of forms.

If we define curvatures

and

$$F = (d+A)^2$$

$$F = (d+S+A+c)^2$$

the horizontality condition on the curvature F=F is easily seen to be equivalent to the BRST transformations.

The above BRST operator S defines a global symmetry with anti-commuting constant ξ

$$\delta_\xi \Phi = \xi S \Phi$$

We would like to make ξ local. This requires the introduction of a ghost number -1 gauge field α_μ, to compensate for $\partial_\mu \xi$, along with its ghost number 0 ghost field λ.

Defining the local BRST S (which reduces to the previous S for $\alpha=0$, $\lambda=1$), we have the following transformation laws:

$$\begin{aligned}
SA' &= dc' + [A',c'] \\
Sc' &= -\frac{1}{2}[c',c'] \\
Sc &= \lambda b \\
Sb &= 0 \\
S\alpha &= d\lambda \\
S\lambda &= 0
\end{aligned}$$

where $A'=A-\alpha c$, $c'=\lambda c$. Once again, the above BRST transformations are easily seen to be nilpotent and to be equivalent to the horizontality condition for the curvature

$$F' \equiv (d+S)(A'+c')+(A'+c')(A'+c') = dA'+A'A' \equiv F'$$

Moreover the invariant polynomials in the curvature that determine the Chern classes are also horizontal. This fact has important consequences for the current algebra.

Given the above local BRST transformation, one can construct very easily an invariant Lagrangian, L, with or without matter fields. Differentiating this Lagrangian with respect to α, the source of the BRST current, one obtains the form of the BRST current

$$J_\mu = \left.\frac{\delta L}{\delta \alpha_\mu}\right|_{\alpha=0}$$

Integration of J_0 over space yields the BRST charge Q, commutation with which determines the BRST transformations.

In gauge theories with anomalies, the BRST current algebra has interesting consequences. The anomaly can be calculated by n-point functions in the external gauge current or in the external BRST current. Both are related to the Chern-Simons polynomial of degree d+1, T_{d+1}, that is obtained from the invariant Chern polynomial, P^{inv}_{d+2}, of degree d+2 (d=dimension of space-time). Because of the horizontality condition

$$P^{inv}_{d+2}(F') = P^{inv}_{d+2}(F')$$

we have

$$(d+S)\ T_{d+1}(A'+c',F)=dT_{d+1}(A',F)$$

By the standard decent procedure

$$S\ T_{d+1}= -\ d\Delta_d^{\ 1}$$

where $\Delta_d^{\ 1}$ is the standard anomaly. However, $\Delta_d^{\ 1}$ is now dependent on α. By expanding in a Taylor series, one obtains not only the above anomaly, but also all the higher cocycles. Moreover, since these are obtained by differentiating with respect to α, one easily obtains that the presence of such gauge anomalies is equivalent to the violation of BRST Noether current conservation. By the standard technique of removing a derivative inside a time order product, one obtains the anomalous commutation relations involving the BRST current as well as the failure of the BRST charge to have square zero. For example, in two dimensions, the above procedure relates

$$\langle T(\partial_\mu A^\mu,\ A_\nu)\rangle$$

to

$$\langle T(\partial_\mu J^\mu_{BRST},\ J^{BRST}_\nu,\ \bar{c},\ \bar{c})\rangle$$

Removing the derivative from inside the time-order product gives the anomalous equal-time commutation relation

$$\{J_0,\ J_\mu\}_{ETC} \neq 0$$

Setting μ=0 and integrating over space yields $Q^2 \neq 0$.

References

1. The local BRST for gauge theories, gravity and string theories was first derived by L. Baulieu, C. Becchi, and R. Stora (in preparation). The derivation of the higher cocycles appears in L. Baulieu, B. Grossman and R. Stora LAPP preprint TH167 (to be published in Physics Letters B). See these papers for further references.

SUPERSTRINGS FROM THE BOSONIC STRING ON THE GROUP MANIFOLD

M.J. Duff
Theory Division, CERN, 1211 Geneva 23, Switzerland

ABSTRACT

We explore the possibility that superstrings have their origin in the purely bosonic string compactified on the group manifold, and whether this explains why physical space-time has four dimensions.

1. FERMIONS FROM BOSONS

The appearance of the rank 16, dimension 496, gauge groups $E_8 \times E_8$ and spin $32/Z_2$ as the only available candidates for anomaly free ten-dimensional superstrings[1)] prompted Freund[2)] to conjecture that the fundamental theory might be the 26-dimensional bosonic string, and that the ten-dimensional theories emerge after compactification on the torus $T^{16} = R^{16}/\Gamma$ where Γ is the even self-dual Euclidean lattice of $E_8 \times E_8$ or spin $32/Z_2$. In this picture, the fermions would appear as solitons of the bosonic theory. In addition to the 16 $[U(1)]^{16}$ elementary Kaluza-Klein gauge bosons, a further 480 gauge bosons would also emergy as Frenkel-Kac[3)] solitons since the string can wrap around the torus. The subsequent discovery of the $E_8 \times E_8$ and spin $32/Z_2$ heterotic strings[4)] brought the total number consistent superstrings to 5 as in Table 1, and only increased the desire for one underlying theory.

Table 1

Consistent superstrings

TYPE	SPINOR	STRING	LOW ENERGY THEORY
I [SO(32)]	Weyl + Majorana	open/closed	N = 1 supergravity + SO(32) Yang-Mills
IIA	Majorana	closed	N = 2 non-chiral supergravity
IIB	Weyl	closed	N = 2 chiral supergravity
Heterotic [SO(32)]	Weyl + Majorana	closed	N = 1 supergravity + SO(32) Yang-Mills
Heterotic [$E_8 \times E_8$]	Weyl + Majorana	closed	N = 1 supergravity + $E_8 \times E_8$ Yang-Mills

Noting that the bosonic string can also undergo spontaneous compactification from D to d dimensions on the simply-laced non-Abelian group manifold G of radius R provided[5] $R = \sqrt{\alpha'}$ and

$$D - 26 = \dim G \cdot \frac{C_A}{2 + C_A} = \dim G - \operatorname{rank} G \qquad (1)$$

where C_A is the second order Casimir in the adjoint representation, Nilsson, Pope and myself[6] proposed obtaining the d = 10 heterotic strings by choosing $G = E_8 \times E_8$ or spin $32/Z_2$ for which $C_A = 60$ and hence D = 506, d = 10. The origin of fermions was (and still is) less obvious than in the T^{16} compactification since G is simply connected. However, the appearance of the $E_8 \times E_8$ or spin $32/Z_2$ gauge bosons is easier to understand than the T^{16} case, since they are all just <u>elementary</u> Kaluza-Klein[7] fields, i.e., the gauge groups are just subgroups of the D = 506 general co-ordinate group. One nice feature of this group manifold approach is thus to maintain the Kaluza-Klein ideal of getting internal symmetries from space-time symmetries in a higher dimension; a feature which the superstrings with their primary Yang-Mills fields seemed to lack. The main reason why this traditional Kaluza-Klein idea fell out of favour was its inability to explain chiral fermions. This problem is now avoided by the bosonic string; we simply cut the Gordian knot and dispense with fermions altogether! Moreover, the old Kaluza-Klein trick of adding a cosmological constant in the higher dimensional theory with sign and magnitude so designed to cancel the one arising from compactification is now more respectable, since the required D-dimensional cosmological term $(D-26)/\alpha'$ is enforced by conformal invariance of the string. The Kaluza-Klein relation $R^2g^2 = \kappa^2$ between the Yang-Mills coupling g and the gravitational constant κ meant that the heterotic strings for which $\alpha' g^2 = \kappa^2$ might indeed admit of such an interpretation, but the Type I string for which $g^2 = \kappa\alpha'$ would not.

Independently, at about the same time, Casher et al.[8] took the idea one step further and showed how all closed superstrings; Type IIA, Type IIB, heterotic $E_8\times E_8$ and heterotic spin $32/Z_2$ could emerge from T^{16} compactification of the D = 26 bosonic string. The key idea was the identification of the transverse space-time SO(8) of the superstring with the diagonal subgroup of the transverse space-time SO(8) of the compactified bosonic string and SO(8) [internal]:

$$SO(8)\,[\text{space-time, super}] \subset SO(8)\,[\text{space-time, bosonic}] \times SO(8)\,[\text{internal}] \tag{2}$$

where SO(8) [internal] is a subgroup of G_R in the heterotic case and $G_L\times G_R$ in the case of Type II. In this way states transforming as spinor representations of SO(8) [internal] now transform as fermion representations of SO(8) [space-time,super].

Since the non-linear σ-model on the group manifold is equivalent to free bosons on the torus[5], it follows that our Kaluza-Klein approach and the Frenkel-Kac approach of Casher et al. are in fact equivalent.

Just like the wave-particle duality of quantum mechanics which picture one chooses is merely a matter of convenience. So far, the torus approach has proved more powerful for formal "stringy" results, whereas the elementary nature of the gauge fields in the Kaluza-Klein approach lends itself more readily to low-energy field theory considerations. A striking example of this is the derivation of the d = 10 Lorentz and Yang-Mills Chern-Simons terms summarized in Table 2. The identification of spin-connections with gauge potentials[9] is, as discussed by Nilsson, Pope, Warner and myself[10] just the field theoretic realization of the diagonal choice of space-time SO(8) discussed above. The identification of A with ω_- in going from the heterotic string to the Type II string had already been employed in the literature[11]. What is not generally appreciated, however, is that going from heterotic (one gravitino) to Type II (two gravitinos) requires exactly the same "fermions from bosons" phenomenon as going from bosonic (zero gravitinos) to heterotic (one gravitino). It is strange, therefore, that physicists who feel at ease with the first still remain sceptical about the second. These ideas are discussed in greater length in Refs. 12) and 13).

Table 2

Kaluza-Klein origin of d = 10 Chern-Simons terms. $\omega_\pm$ are the spin-connections with torsion $\pm\frac{1}{2}H$ and $(A,\tilde{A})$ are the Yang-Mills gauge potentials of (G_L,G_R).

STRING	DIMENSION	CONNECTIONS	$(\text{CURVATURE})^2$ TERMS	CHERN-SIMONS TERMS
bosonic	506	$\hat{\omega}_+,\hat{\omega}_-$	$\hat{R}_+^2 + \hat{R}_-^2$	$d\hat{H} = 0$
bosonic on G	10	$\omega_+,\tilde{A},\omega_-,A$	$R_+^2 - \tilde{F}^2 + R_-^2 - F^2$	$dH = \alpha'\mathrm{Tr}(\tilde{F}\wedge\tilde{F}-F\wedge F)$
heterotic	10	$\omega_+ = \tilde{A},\omega_-,A$	$R_-^2 - F^2$	$dH = \alpha'\mathrm{Tr}(R_+\wedge R_+-F\wedge F)$
Type II	10	$\omega_+ = \tilde{A},\ \omega_- = A$	0	$dH = \alpha'\mathrm{Tr}(R_+\wedge R_+-R_-\wedge R_-$

Of course, if the bosonic string really is the fundamental theory perhaps we should consider compactifications not from D to 10 dimensions but from D to four dimensions, i.e., on a group manifold for which from Eq. (1)

$$\text{rank } G = 22$$

But which G should we choose and why should the string prefer rank 22 to some other rank <26?

2. FOUR DIMENSIONAL SPACE-TIME FROM THE K3 LATTICE

The major unresolved problem of string theories is that of vacuum degeneracy. For although the higher-dimensional string equations have almost no parameters, all predictive power seems to be lost by the apparent multitude of different compactifications. In particular, it remains a mystery why the dimension of the uncompactified space-time should be just four. This section, based on a paper by Nilsson and myself[14)], is an attempt to resolve some of these questions.

We begin by recalling the recent work of Narain[15)] who considers compactifying into tori (10-d) and (26-d) dimensions of the right moving superstring and left moving bosonic string sectors respectively. Since the k-torus is given by R^k factored by a discrete lattice $\Gamma(T^k = R^k/\Gamma)$, the question devolves upon which Γ to choose. Narain points out that the condition for modular invariance is equivalent to self-duality for even Lorentzian lattices with (10-d) timelike and (26-d) spacelike directions. Let us denote such even self-dual lattices $\Gamma_{(p,q)}$, where (p,q) is the signature. Then all such Lorentzian lattices are isomorphic to the lattice

$$n\, E_8 \oplus q\, P_2 \qquad (3)$$

where

$$p - q = 8n \tag{4}$$

for some integer n, where E_8 is the root lattice of E_8 with Euclidean signature (8,0)

$$E_8 = \begin{bmatrix} 2 & -1 & 0 & 0 & 0 & 0 & 0 & 0 \\ -1 & 2 & -1 & 0 & 0 & 0 & 0 & 0 \\ 0 & -1 & 2 & -1 & 0 & 0 & 0 & 0 \\ 0 & 0 & -1 & 2 & -1 & 0 & 0 & 0 \\ 0 & 0 & 0 & -1 & 2 & -1 & 0 & -1 \\ 0 & 0 & 0 & 0 & -1 & 2 & -1 & 0 \\ 0 & 0 & 0 & 0 & 0 & -1 & 2 & 0 \\ 0 & 0 & 0 & 0 & -1 & 0 & 0 & 2 \end{bmatrix} \tag{5}$$

and where P_2 is a two-dimensional lattice with signature (1,1)

$$\begin{bmatrix} 0 & 1 \\ 1 & 0 \end{bmatrix} \tag{6}$$

corresponding to the group SU(2)×SU(2). This means that any even self-dual lattice $\Gamma_{(q+8n,q)}$ with $q > 0$ can be obtained from $nE_8 \oplus qP_2$ by means of an SO(8n+q,q) transformation. Distinct compactifications are then characterized by points in the coset

$$SO(p,q)/SO(p) \times SO(q)$$

There is a recent theorem due to Freedman[16)] that states that all even self-dual Lorentzian lattices are given by the "intersection form" of a simply-connected topological four-manifold M^4. See the article by Stern[17)] for a readable introduction to this branch of mathematics. Such a manifold will have Euler number

$$\chi = 2 + b_2 \tag{7}$$

where the second Betti number b_2 counts the number of harmonic two-forms. If we denote these two-forms by α_i $(i = 1,\ldots,b_2)$ then the intersection form is defined by

$$\Gamma_{ij} = \int_{M^4} \alpha_i \wedge \alpha_j \qquad (8)$$

and obviously has rank b_2. Its signature (p,q) is given by (b_2^+, b_2^-), and the Hirzebruch signature by

$$\tau = b_2^+ - b_2^- \qquad (9)$$

where b_2^+ count the number of self-dual two-forms and b_2^- the number of antiselfdual. (Hence τ must be a multiple of 8.) So the question of which is the right vacuum has been replaced by which is the right four-manifold.

Now Freedman's theorem involves <u>topological</u> four-manifolds but not every four-manifold is <u>differentiable</u>. For example, a necessary condition due to Rochlin[18] is that τ must be a multiple of 16. Suppose, just for fun, we use the criterion of <u>differentiability</u> of the four-manifold to narrow down the choice of vacuum.

Unfortunately, the question of which four-manifolds are smoothable (i.e., differentiable) is an outstanding problem in mathematics, but some very interesting results are known. For example

$$E_8 \oplus E_8$$

is not, even though $\tau = 16$! (This result, due to Donaldson[19], has created quite a stir in mathematical circles because it lies at the heart of the proof that R^4 has more than one differentiable structure.) P_2 on the other hand corresponds to the intersection form on $S^2 \times S^2$ which is differentiable. The problem then is to determine whether

$$2E_8 + q\,P_2$$

is smoothable for some $q > 0$.

Note that from the string point of view, our criterion of differentiability then means that we cannot go from $D = 26$ to $d = 10$ but may perhaps go to $d < 10$! The amazing fact is that the case $q = 3$ is differentiable and simply corresponds to the four-manifold K3 for which $b_2^+ = 19$, $b_2^- = 3$ and $b_2 = 22$.

K3 is defined as a quartic surface in complex projective three-space CP^3 by

$$K3 = \{ [z_0, z_1, z_2, z_3] \in \mathbb{C}P^3 : z_0^4 + z_1^4 + z_2^4 + z_3^4 = 0 \}$$

Applying Narain's techniques using this very special lattice leads to a heterotic string theory with unique space-time dimension $d = 7$. The low-energy limit corresponds to $d = 7$, $N = 2$ supergravity coupled to super-Yang-Mills with rank 19 gauge group $E_8 \times E_8 \times SU(2) \times SU(2) \times SU(2)$. [The remaining rank 3 gauge group simply corresponds to the three U(1)'s of $N = 2$ supergravity.] Corresponding theories in $d = 10-q < 7$ could also be obtained by taking the topological sum of K3 and (q-3) copies of $S^2 \times S^2$. Thus a four-dimensional theory could be obtained from

$$\Gamma_{(22,6)} = \Gamma [K3] \oplus 3 P_2$$

whose low-energy limit is $N = 4$ supergravity coupled to the rank 22 gauge group $E_8 \times E_8 \times [SU(2)]^6$. [Once again, the remaining rank six group simply corresponds to the six U(1)'s of $N = 4$ supergravity.] Unfortunately, the "minimal" theory is in $d = 7$ and there seems no compelling reason for adding three $S^2 \times S^2$ manifolds to K3.

So far we have followed Narain and considered only the heterotic string, but the situation becomes much more interesting if we adopt the point of view that the fundamental theory is the <u>bosonic</u> string. Once again we must compactify on a torus given by an even self-dual Lorentzian lattice but now with signature (26-d,26-d). The "minimal" theory in the sense described above is now given by

$$\Gamma_{(22,22)} = \Gamma[K3] \oplus \Gamma[\overline{K3}]$$

where $\overline{K3}$ corresponds to the four-manifold obtained from K3 by reversing the orientation and has $b_2^+ = 3$, $b_2^- = 19$ and $\tau = -16$. Hence

$$d = 26 - 22 = 4 \tag{10}$$

and we obtain a four-dimensional bosonic string with gauge group G×G, where G is the rank 22 group $E_8 \times E_8 \times SU(2)^6$.

Thus our objective is now to repeat the derivation of superstrings from bosonic strings discussed in Section 1 but now compactifying from D = 26 to d = 4 on the torus T^{22} defined by the intersection form of the four-manifold $K3 \oplus \overline{K3}$ [or, bearing in mind the previously discussed equivalence, from D = 518 to four on the group manifold $E_8 \times E_8 \times SU(2)^6$]. However, the outcome is no longer clear. In particular, it is unclear whether a chiral N = 1 theory would result. If a chiral theory does not emerge directly in this way, it may be necessary to go one stage further and compactify not merely on the torus T^{22} defined by $K3 \oplus \overline{K3}$ but on $T^{22}/\mathcal{G}$ where $\mathcal{G}$ is a discrete group. Factorings of T^6 by $\mathcal{G}$ have been considered by Dixon et al[20] but to obtain chiral fermions, it was necessary that $\mathcal{G}$ had fixed points thus leading to singularities, i.e., "orbifolds". From our K3 point of view, a more attractive possibility is that advocated by Lam and Li[21] who consider direct compactification from 26 to 4 via $T^{22}/\mathcal{G}$ where $\mathcal{G}$ acts on T^{22} without fixed points, so that $T^{22}/\mathcal{G}$ is a genuine manifold without singularities. These authors claim to obtain chiral N = 1 theories in this way while still preserving modular invariance. (They consider $E_8 \times E_8 \times SU(3)^2$ rather than $E_8 \times E_8 \times SU(2)^3$.]

The vital question remaining is why string theory should select intersection forms of differentiable manifolds, but if it does it would explain why we cannot remain in d = 10. And in answer to the question why space-time has four dimensions we would reply: because the second Betti number of K3 equals 22!

ACKNOWLEDGEMENTS

I am grateful for conversations with A. Chamseddine, B. Nilsson, C. Pope, D. Ross and N. Warner.

REFERENCES

1) Green, M.B. and Schwarz, J.H., Phys. Lett. B149, 117 (1984).

2) Freund, P.G.O., Phys. Lett. B151, 387 (1985).

3) Frenkel, I. and Kac, V.G., Inv. Math. 62, 23 (1980); Goddard, P. and Olive, D., in "Workshop on Vertex Operators in Mathematics and Physics", Berkeley (1983).

4) Gross, D., Harvey, J., Martinec, E. and Rohm, R., Phys. Rev. Lett. 54, 502 (1985); Nucl. Phys. B256, 253 (1985).

5) Witten, E., Comm. Math. Phys. 92, 455 (1984); Nemeschensky, D. and Yankielowicz, S., Phys. Rev. Lett. 54, 620 (1984); Altschüler, D. and Nilles, H.P., Phys. Lett. 154B, 135 (1985); Goddard, P. and Olive, D., Nucl. Phys. B257, 226 (1985); Jain, S., Shankar, R. and Wadia, S.R., Phys. Rev. D32, 2713 (1985); Bergshoeff, E. Randjbar-Daemi, S., Salam, A., Sarmadi, H. and Sezgin, E., Nucl. Phys. B269, 77 (1986).

6) Duff, M.J., Nilsson, B.E.W. and Pope, C.N., Phys. Lett. B163, 343 (1985), also published in Proc. Cambridge Workshop on Supersymmetry and its applications (June-July 1985), (Eds. Gibbons, Hawking and Townsend, C.U.P. 1986).

7) Duff, M.J., Nilsson, B.E.W. and Pope, C.N., Physics Reports 130, 1 (1986).

8) Casher, A., Englert F., Nicolai, H. and Taormina, A., Phys. Lett. B162, 121 (1985); see also Englert, F., Nicolai, H. and Schellekens, A., CERN preprint TH.4360/86 (1986).

9) Charap, J.M. and Duff, M.J., Phys. Lett. B69, 445 (1977).

10) Duff, M.J., Nilsson, B.E.W., Pope, C.N. and Warner, N.P., Phys. Lett. 171B, 170.

11) Hull, C.M., Nucl. Phys. B267, 266 (1986).

12) Duff, M.J., in Proceedings of the GR11 Conference, Stockholm, July 1986, CERN preprint TH.4568/86.

13) Duff, M.J., in Proceedings of the Fourth Capri Symposium, May 1986.

14) Duff, M.J. and Nilsson, B.E.W., Phys. Lett. 175B, 417 (1986).

15) Narain, K.S., Phys. Lett. B169, 41 (1986).

16) Freedman, M., Diff. J. Geom. 17, 357 (1983).

17) Stern, R.J., The Mathematical Intelligencer 5, 39 (1985).

18) Rochlin, V.A., Dokl. Akad. Nauk SSR 84,221 (1952).

19) Donaldson, S.K., Bull. Amer. Math. Soc. 8, 81 (1983).

20) Dixon, L., Harvey, J.A., Vafa, C. and Witten, E., Nucl. Phys. B261, 678 (1985).

21) Lam, C.S. and Da-Xi Li, McGill University preprints (1985).

DIFFERENTIAL ALGEBRAS IN FIELD THEORY AND THEIR ANOMALIES : TWO EXAMPLES

Raymond Stora

LAPP, BP. 909, 74019 Annecy-le-Vieux Cedex
FRANCE

Lecture given at the Luminy C.I.R.M. meeting on Harmonic Maps, 9-13 June 1986, and the VIIIth International Congress on Mathematical Physics, Marseille-Luminy, 16-25 July 1986.

ABSTRACT

The expression of gauge symmetries in local field theory proceeds via the construction of some differential algebras as was remarked some ten years ago. The construction relevant to Yang Mills theories is recalled. As another popular example, we have chosen to describe the covariant quantization of the free bosonic string in the metric background gauge.

The description of symmetries within most theoretical frameworks goes via groups, Lie algebras, and their representations. The description of gauge symmetries of local field theories, on the other hand, has required the construction of differential algebras which extend the "local" cohomology algebra of the gauge Lie algebra. This construction is commonly nicknamed the BRST construction[1)]. It has slowly crept into the bag of tricks of the gauge field theorists. It has been subjected to a number of variations and it has applied to a variety of possibly graded ("super") gauge Lie algebras. A more general construction, which I believe, is less well adapted to local field theory, has been performed by the Lebedev group, and applies to canonical dynamical systems with constraints[2)].

Within the Lagrangian framework, which is appropriate for the discussion of a number of symmetries of a geometrical nature, the general question is to present degenerate actions (i.e. whose hessian is degenerate) in a way appropriate for quantization i.e. remove degeneracy[3] of the action functional in a way which respects such principles as locality and geometrical invariances. In the simplest example of gauge theories to which the Faddeev Popov construction applies, the BRST constructions provide the correct differential algebras. In general, however the construction of a differential algebra requires the introduction of a system of auxiliary fields, an art which at the moment has not been turned into a routine. Because of the diversity of these constructions[4], it is far beyong the scope of this talk to review the work which has been done over the past few years during which the subject has come back to fashion, a fashion which has now fainted away to the benefit of string theory. To summarize that period[4] let us say that both the algebraic theory of anomalies, i.e. the study of the first local cohomology of gauge Lie algebras with values in spaces of local functionals of fields - among which, gauge fields - , and the topological study which relates anomalies with K theory via the Index of families of Dirac operators[5], have been pushed quite far. There remains one puzzle, though, namely that the topological study identifies anomalies with de Rham cohomology classes of the gauge group, whereas local cohomology is required, so that the U(1) anomaly, and, presumably also, the gravitational anomaly remain left out of this de Rham landscape. The corresponding cyclic cohomology construction[6] recently proposed does not seem to solve this locality problem either.

As already mentioned, the gauge Lie algebra cohomology is a piece of the BRST cohomology. The latter, which defines perturbative gauge theories[1] also seems to be of importance in the construction of string field theories[7]. Since there exist in the published literature a variety of "BRST" quantization schemes for the free string it has appeared appropriate to spell out what the orthodoxical BRST construction[8] is in this case, an exercise which, for some reason, has not been done so far.

We shall now recall the ten years old situation met in quantizing Yang Mills theories. At the tree level one starts with the classical action

$$S = S_{inv} + S_{gf} + S_{source} \tag{1}$$

S_{inv}, the invariant part of the action, depends on the Yang Mills field $\underline{a}$ and matter fields φ transforming in the known way under the gauge group $\mathcal{G}$ - and its Lie algebra Lie $\mathcal{G}$ - based on the structure group G, a compact Lie group:

$$\begin{aligned} \delta_{\underline{\omega}} \underline{a} &= d\underline{\omega} + [\underline{a},\underline{\omega}] \\ \delta_{\underline{\omega}} \varphi &= -t(\underline{\omega})\, \varphi \end{aligned} \qquad \underline{\omega} \in \text{Lie}\ \mathcal{G} \tag{2}$$

$$S_{gf} = \int dx\, [b_\alpha(x)\, g^\alpha(x) + \bar{\omega}_\alpha(x)\, s\, g^\alpha(x)] + Q(b)(x) \tag{3}$$

is the Faddeev Popov ($\phi\pi$) gauge fixing term. b is the Lagrange multiplier field, Q(b) a local quadratic form in b, $\bar{\omega}$ is the $\phi\pi$ antighost. The s operation is induced by :

$$\begin{aligned} s\, \underline{a} &= -d\omega - [\underline{a},\omega] \\ s\, \varphi &= -t(\omega)\varphi \\ s\, \omega &= -\frac{1}{2}[\omega,\omega] \\ s\, \bar{\omega} &= -b \\ s\, b &= 0 \end{aligned} \tag{4}$$

where the components of ω provide a set of generators of the cohomology algebra of Lie $\mathcal{G}$. g is a gauge function

$$g^\alpha(x) = \partial_\mu\, a^{\mu\alpha}(x) + \ldots \tag{5}$$

where a_μ^α are the components of $\underline{a}$.

The operation s is nilpotent and annihilates both S_{inv} and S_{gf}

$$S_{source} = \int dx\ [A^{\mu}_{\alpha} s\ a^{\alpha}_{\mu} + \phi\ s\ \varphi + \Omega\ s\ \omega]\ . \tag{6}$$

The s operation is extended by

$$sA^{\mu}_{\alpha} = s\phi = s\Omega = 0 \tag{7}$$

so that s also annihilates S_{source}.
There follows the Slavnov identity

$$\mathcal{S}(S) \equiv \int dx \left[\frac{\delta S}{\delta A^{\mu}_{\alpha}} \frac{\delta S}{\delta a^{\alpha}_{\mu}} + \frac{\delta S}{\delta \phi} \frac{\delta S}{\delta \varphi} + \frac{\delta S}{\delta \Omega} \frac{\delta S}{\delta \omega} - b \frac{\delta S}{\delta \varpi}\right](x) = 0 \tag{8}$$

The Slavnov operator

$$\mathcal{S}_S = \int dx \left[\frac{\delta S}{\delta A^{\alpha}_{\mu}} \frac{\delta}{\delta a^{\alpha}_{\mu}} + \frac{\delta S}{\delta a^{\alpha}_{\mu}} \frac{\delta}{\delta A^{\alpha}_{\mu}} + \frac{\delta S}{\delta \phi} \frac{\delta}{\delta \varphi} + \frac{\delta S}{\delta \varphi} \frac{\delta}{\delta \phi} + \frac{\delta S}{\delta \Omega} \frac{\delta}{\delta \omega} + \frac{\delta S}{\delta \omega} \frac{\delta}{\delta \Omega} - b \frac{\delta}{\delta \varpi}\right] \tag{9}$$

is nilpotent by virtue of the Slavnov idendity. The vertex functional which is the quantum extension of S :

$$\Gamma = S + \sum_{n>o} \hbar^n\ \Gamma^{(n)} \tag{10}$$

can be constructed perturbatively in such a way that

$$\mathcal{S}(\Gamma) = 0 \tag{11}$$

unless anomalies show up on the right hand side. These are classified as cohomology classes of $\mathcal{S}$.

A variant is obtained by eliminating the b field through its algebraic equations of motion and replacing S_{gf} by

$$S_{gf} = \int dx \left[\frac{(g-c)^2}{2} + \bar{\omega}\, s\, g\right] \tag{12}$$

where c is now a classical field and s is defined by

$$\begin{aligned} s\, c &= 0 \\ s\, \bar{\omega} &= -(g-c) \end{aligned} \tag{13}$$

or by

$$\begin{aligned} s\, g &= s\, c \\ s\, \bar{\omega} &= 0. \end{aligned} \tag{14}$$

The ensuring Slavnov identity which is now homogeneous quadratic, reads

$$\mathcal{S}(S) \equiv \int \frac{\delta S}{\delta A^{\mu}_{\alpha}} \frac{\delta S}{\delta a^{\alpha}_{\mu}} + \frac{\delta S}{\delta \phi} \frac{\delta S}{\delta \varphi} + \frac{\delta S}{\delta \Omega} \frac{\delta S}{\delta \omega} + \frac{\delta S}{\delta \bar{\omega}} \frac{\delta S}{\delta c} = 0 \tag{15}$$

Of the two possible choices of the s operation the first is non nilpotent whereas the second is nilpotent. The relevant nilpotent operator $\mathcal{S}_s$ obtained by polarization is itself nilpotent.

The situation for the free bosonic string is fairly similar[8]. We have

$$S_{inv} = \int d^2x\, g^{\alpha\beta} \sqrt{|g|}\, \partial_\alpha \vec{X} . \partial_\beta \vec{X} \tag{16}$$

where $x = (x^1, x^2)$ is a set of coordinates on the world sheet, $\{g^{\alpha\beta}\}_{\alpha,\beta=1,2}$ a metric $\vec{X} \in R^D$ $|g| = \det \|g_{\alpha\beta}\|$, . denotes a metric in R^D.

S_{inv} is invariant under the s operation corresponding to Diff $\ltimes$ Weyl :

$$\begin{aligned} s\, \vec{X} &\equiv L_\xi X = (\xi . \partial)\, \vec{X} \\ s\, g^{\alpha\beta} &\equiv L_\xi g^{\alpha\beta} = \partial . (\xi g^{\alpha\beta}) - \partial_\gamma \xi^\alpha g^{\gamma\beta} - \partial_\gamma \xi^\beta g^{\gamma\alpha} - \Omega g^{\alpha\beta} \\ s\, \Omega &\equiv L_\xi \Omega = (\xi . \partial) \Omega \end{aligned} \tag{17}$$

$$s\,\xi \equiv \frac{1}{2} L_\xi \xi = (\xi.\partial)\xi$$

ξ and Ω are the $\phi\pi$ ghosts corresponding respectively to Lie Diff $\pm$ Lie Weyl. In a Landau type background metric gauge $\overset{\circ}{g}$,

$$S_{gf} = \int d^2x\, \bar{b}_{\alpha\beta}(g^{\alpha\beta} - \overset{\circ}{g}{}^{\alpha\beta}) + \bar{\xi}_{\alpha\beta}\; s\, g^{\alpha\beta} \tag{18}$$

The s operation is extended by

$$s\,\bar{\xi}_{\alpha\beta} = -b_{\alpha\beta} \qquad s\,b_{\alpha\beta} = 0 \tag{19}$$

In this Landau type gauge we have to eliminate the b field, using the algebraic equation of motion

$$g = \overset{\circ}{g} \tag{20}$$

obtained by varying $\bar{b}$.

Omitting the upperscript $\circ$ and remembering that g represents now an external classical metric, the s operation is merely changed into

$$s\,\bar{\xi}_{\alpha\beta} = 0 \tag{21}$$

and S_{gf} into

$$S_{gf} = \int d^2x\, \bar{\xi}_{\alpha\beta}\; s\, g^{\alpha\beta}\,. \tag{22}$$

With

$$S_{source} = \int \vec{\mathfrak{X}}\; s\,\vec{X} + \Xi\; s\,\xi + \Theta\; s\,\Omega \tag{23}$$

we have the Slavnov identity

$$\int \frac{\delta S}{\delta \mathfrak{X}}\frac{\delta S}{\delta X} + \frac{\delta S}{\delta \Xi}\frac{\delta S}{\delta \xi} + \frac{\delta S}{\delta \Theta}\frac{\delta S}{\delta \Omega} + \frac{\delta S}{\delta \bar{\xi}_{\alpha\beta}}\frac{\delta S}{\delta g^{\alpha\beta}} = 0 \tag{24}$$

where

$$S = S_{inv} + S_{gf} + S_{source} \ . \tag{25}$$

We may now eliminate the Weyl ghost using the equation of motion

$$\Omega = \frac{1}{2} g_{\alpha\beta} L_\xi g^{\alpha\beta} \ . \tag{26}$$

This substitution may be performed both within the action and the s operation because

$$s(\Omega - \frac{1}{2} g_{\alpha\beta} L_\xi g^{\alpha\beta}) = 0 \ . \tag{27}$$

The s operation is thus changed into s_c :

$$s_c g^{\alpha\beta} = L_\xi g^{\alpha\beta} - g^{\alpha\beta} \frac{1}{2} g_{\gamma\delta} L_\xi g^{\gamma\delta} \tag{28}$$

$$s X = s_c X \qquad s \xi = s_c \xi \qquad s \bar{\xi}_{\alpha\beta} = s_c \bar{\xi}_{\alpha\beta} = 0.$$

Since

$$s_c |g| = 0 \tag{29}$$

one may eliminate one degree of freedom from $g_{\alpha\beta}$. Defining

$$\hat{g}_{\alpha\beta} = g_{\alpha\beta} \sqrt{|g|} \qquad \hat{\bar{\xi}}_{\alpha\beta} = \bar{\xi}_{\alpha\beta}/\sqrt{|g|} \tag{30}$$

we have

$$s_c \hat{g}_{\alpha\beta} = L_\xi \hat{g}_{\alpha\beta} \qquad s_c \hat{\bar{\xi}}_{\alpha\beta} = 0 \tag{31}$$

$$S_{gf} = \int \hat{\bar{\xi}}_{\alpha\beta} L_\xi \hat{g}_{\alpha\beta} \tag{32}$$

S_{gf} is now degenerate because of the zero modes

$$\hat{\bar{\xi}}_{\alpha\beta} \propto g_{\alpha\beta} \tag{33}$$

These can be eliminated without the introduction of extra ghosts by imposing the gauge condition

$$\hat{\bar{\xi}}_{\alpha\beta}\, \overset{\circ}{g}{}^{\alpha\beta} = 0 \qquad (s_c\, \hat{\bar{\xi}}_{\alpha\beta}\, \overset{\circ}{g}{}^{\alpha\beta} = 0\ !) \tag{34}$$

for some fixed backgroung metric $\overset{\circ}{g}$

$$s_c\, \overset{\circ}{g} = 0 \tag{35}$$

The unimodular $\hat{g}$ can then be parametrized in terms of two independent components $\hat{g}^{zz}$, $\hat{g}^{\bar{z}\bar{z}}$ if one uses the holomorphic coordinates determined by $\overset{\circ}{g}$ and orientation. Then we have

$$\hat{g}^{z\bar{z}} = \sqrt{1 + \hat{g}^{zz}\, \hat{g}^{\bar{z}\bar{z}}} \tag{36}$$

$$\begin{aligned} S_{inv} + S_{gf} = \int dz\, d\bar{z}\, \frac{1}{2} \Big[& \hat{g}^{zz} \partial_z X . \partial_{\bar{z}} X + 2\hat{g}^{z\bar{z}} \partial_z X . \partial_{\bar{z}} X + \hat{g}^{\bar{z}\bar{z}} \partial_{\bar{z}} \vec{X} . \partial_{\bar{z}} \vec{X} \\ & + \hat{\bar{\xi}}_{zz} L_\xi \hat{g}^{zz} + \hat{\bar{\xi}}_{\bar{z}\bar{z}} L_\xi \hat{g}^{\bar{z}\bar{z}} \Big] \end{aligned} \tag{37}$$

$$S_{source} = \int dz\, d\bar{z}\ \vec{\mathfrak{X}} . s_c \vec{X} + \Xi\, s_c \xi \tag{38}$$

and the Slavnov identity reads

$$0 = \int dz\, d\bar{z} \left[\frac{\delta S}{\delta \vec{X}} \frac{\delta S}{\delta \vec{\mathfrak{X}}} + \frac{\delta S}{\delta \Xi} \frac{\delta S}{\delta \xi} + \frac{\delta S}{\delta \hat{g}^{zz}} \frac{\delta S}{\delta \hat{\bar{\xi}}_{zz}} + \frac{\delta S}{\delta \hat{g}^{\bar{z}\bar{z}}} \frac{\delta S}{\delta \hat{\bar{\xi}}_{\bar{z}\bar{z}}} \right] \tag{39}$$

Whereas this is based on s_c :

$$s_c\, \hat{\bar{\xi}}_{zz} = s_c\, \hat{\bar{\xi}}_{\bar{z}\bar{z}} = 0 \qquad s_c\, \hat{g}^{\cdot\cdot} = L_\xi\, \hat{g}^{\cdot\cdot} \tag{40}$$

which is nilpotent, it can also be based on the non-nilpotent $\tilde{s}_c$:

$$s_c \hat{\xi}_{zz} = \frac{\delta S}{\delta \hat{g}^{zz}} = \Theta_{zz} \qquad \tilde{s}_c \hat{\xi}^{\bar{z}\bar{z}} = \frac{\delta S}{\delta \hat{g}^{\bar{z}\bar{z}}} = \Theta_{\bar{z}\bar{z}}$$

$$\tilde{s}_c \hat{g}^{\cdot\cdot} = 0 \tag{41}$$

The usual action is recovered by putting $\hat{g}^{zz} = \hat{g}^{\bar{z}\bar{z}} = 0$. Of course the remaining zero modes also have to be gauge fixed by the same method, but since these depend on the topology of the world sheet and/or boundary conditions we shall not go into details here.

The quantum theory is obtained by replacing S by the vertex functional

$$\Gamma = S + \hbar \Gamma^1 \tag{42}$$

The anomaly which may spoil the Slavnov identity reads, in presence of the Weyl ghost

$$\int d^2x \, \Omega \, (R(g) \sqrt{|g|} + \mu^2 \sqrt{|g|}) \tag{43}$$

where $R(g)$ is the scalar curvature of g (the last term can be eliminated by a local counterterm in Γ).

After elimination of the Weyl ghost, the anomaly reads

$$\int d^2x \, \mathring{\nabla}_\alpha \xi^\alpha R(\hat{g} \sqrt{|\mathring{g}|}) \sqrt{|\mathring{g}|} \tag{44}$$

They are both proportional to (D-26) ! The remarkable thing is that after elimination of the Weyl ghost and suitable field redefinitions the theory is conformal[9] i.e. expectation values of products of Θ_{zz}, $\Theta_{\bar{z}\bar{z}}$ factors factorize into expectation values of products of Θ_{zz} factors and expectation values of products of $\Theta_{\bar{z}\bar{z}}$ factors. This property is due to the contact terms produced by the nonlinearity of s in $\hat{g}^{zz}$, $\hat{g}^{\bar{z}\bar{z}}$ through the expression of $\hat{g}^{z\bar{z}}$. (cf. the introduction of the c^2 term in the Yang Mills case).

The treatment of the Slavnov symmetry we have presented here is a natural complement to various constructions presented in the literature[8]. It has the virtues to respect both geometry and nilpotency without the help of the equations of motion. Thus it ought to be a possible starting point to formulate second quantization. Whereas the conformality of the theory clearly results from the choice of gauge, the infinitesimal diffeomorphisms of the world sheet do belong to any construction of this type.

REFERENCES

1) Becchi C., Rouet A., Stora R., Phys. Lett. 52B (1975), Ann. Phys. 98, 287 (1976), in "Renormalization Theory", Erice 1975, G. Velo, A.S. Wightman eds, NATO ASI Series Vol. C23, Reidel (1976), in "Field Theory, Quantization and Statistical Physics", E. Tirapegui ed., Reidel (1981);
Zinn Justin J., in Trends of Elementary Particle Theory, Bonn, 1974, A. Rollnik, K. Dietz eds, Lecture Notes in Physics, Vol.37, Springer Verlag 1975, Nucl. Phys. B246, 246 (1984).

2) The bibliography can be traced back in M. Henneaux, Phys. Rep. 126, 1-66 (1985).

3) Schwarz A.S., Comm. Math. Phys. 67, 1-16 (1979).

4) A good sample of the literature can be traced back in:
Alvarez-Gaumé L., Ginsparg P., Ann. Phys. 161, 423 (1985);
Alvarez-Gaumé L., Erice Lectures 1985 (G. Velo, A.S. Wightman eds), Plenum Press, to appear; "Geometry, Anomalies, Topology", Argonne Symposium 1984, W.A. Bardeen, A.R. White eds, World Scientific Pub. (1985);
Stora R., in Gift Seminar, Jaca, Spain, 1985, World Scientific Pub. (1986).

5) Atiyah M.F., Singer I.M., P.N.A.S. 81, 2597 (1984);
Singer I.M., in "Elie Cartan et les Mathématiques d'aujourd'hui", Lyon 1984, Astérisque 1985, Société Mathématique de France, Paris, and references therein.

6) Connes A., Moscovici H., "Transgression of the Chern Character for Families and Cyclic Cohomology", IHES/M/86/33.

7) Siegel W., Phys. Lett. 151B, 391 (1985);
Witten E., Nucl. Phys. B268, 253 (1986);
Neveu A., this Conference and references therein;
See also J.M. Bismut's talk, this conference.

8) Baulieu L., Becchi C., Stora R., "On the Covariant Quantization of the Free Bosonic String", LAPP-TH-167, submitted for publication in Phys. Lett. B., and references therein.

9) Belavin A.A., Polyakov A.M., Zamolodchikov A.B., Nucl. Phys. B241, 383 (1984);
Friedan D., Qiu Z., Shenker S., Phys. Rev. Lett. 52, 1575 (1984), Phys. Lett. 151B, 37 (1985).

Conformal and holomorphic anomalies on Riemann surfaces and determinant line bundles

J.-B. Bost
Ecole Normale Superieure
45 Rue d'Ulm
75005 Paris, France

ABSTRACT. *We discuss a generalization of a formula of Quillen, which computes the curvature of the determinant line-bundle of a holomorphic vector bundle on a holomorphic family of Riemann surfaces. This formula has many applications to 2d-field theory (conformal anomalies; holomorphy properties of the string integrand; boson-fermion equivalence). Those are closely related to the work of Arakelov and Faltings on arithmetic surfaces.*

The main theme of this talk is the study of *regularized determinants.* Recall that a positive self-adjoint operator L on sections of a vector bundle over a compact manifold has pure point spectrum $\{\lambda_j\}$. One notes that formally

$$\frac{d}{ds}\Big(\sum_{\lambda_j \neq 0} \lambda_j^{-s}\Big)_{|s=0} = - \sum_{\lambda_j \neq 0} \log \lambda_j .$$

Let

$$\varsigma_L(s) = \sum_{\lambda_j \neq 0} \lambda_j^{-s} .$$

This function of s is well defined and holomorphic if $Re s \gg 0$, has a meromorphic continuation to $\mathbf{C}$, and 0 is not a pole of ς_L ([23]).Thus, it is natural to define the regularized determinant of L ([22]):

$$\det{}' L \stackrel{\text{def}}{=} e^{-\varsigma_L'(0)} .$$

Regularized determinants have an important role both in mathematics and in physics. In mathematics, they occur for example in differential geometry (torsion invariants) and in arithmetic geometry. In physics, they occur as gaussian functional integrals,partition functions... This talk is devoted to some recent interaction of the mathematical and physical viewpoints on regularized determinants of (modulus squared of) $\overline{\partial}$ operators on Riemann surfaces.

1 The Riemann-Roch-Grothendieck-Quillen formula

1.1. *Determinant line bundle.*

If Q is an elliptic differential operator on a compact manifold, one defines the (one dimensional) vector space

$$DET\,Q = (\Lambda^{max} Ker\,Q)^* \otimes (\Lambda^{max} Coker\,Q).$$

Formally $DET\,Q$ is the dual of the "maximal exterior power" of the "index" of Q, $Ker\,Q - Coker\,Q$.

Let $\pi : X \to S$ be a differentiable fiber bundle with compact fibers, E and F differentiable vector bundles on X, and $P = (P_s)_{s \in S}$ a smooth family of elliptic operators over the fibers $\pi^{-1}(s)$ of π, $P_s : C^\infty(\pi^{-1}(s), E) \to C^\infty(\pi^{-1}(s), F)$. Then there is a canonical structure of differentiable line bundle on the family of one dimensional vector spaces $(DET\,P_s)_{s \in S}$ ([21,11]).We denote by $DET\,P$ this line bundle over S.

Below, we consider holomorphic families of $\overline{\partial}$ operators on Riemann surfaces: then X and S are complex manifolds and $\pi : X \to S$ is a proper holomorphic submersion, the fibers of which have complex dimension one (such a $\pi : X \to S$ is called *a holomorphic family of Riemann surfaces*); E is a holomorphic vector bundle on X, and F is $E \otimes \overline{\omega}_{X|S}$, where $\omega_{X|S}$ is the line bundle of vertical (1,0)-forms on X; P is the $\overline{\partial}$ operator along the fibers of π, acting on E. We denote by $\overline{\partial}_E$ this operator. Then the determinant line bundle of $\overline{\partial}_E$ has a canonical holomorphic structure ([17,12]).

This notion of determinant line bundle has been introduced first within the framework of algebraic geometry ([24,17]). The construction for differentiable families of elliptic operators appears more recently in the literature ([8,19,21,11]), particularly in relation with the study of anomalies.

1.2. *The Quillen norm on the determinant line bundle.*

Let P be a family of elliptic differential operators as in the preceding section. Besides, let us choose a smooth family of Riemaniann metrics on the fibers $\pi^{-1}(s)$ (i.e. a smooth metric on the vertical tangent bundle $T_{X|S}$), and smooth hermitian metrics on E and F. Then, for any $s \in S$, P_s^* is well defined, and $Ker\,P_s$ and $Coker\,P_s$ have natural L^2 metrics, which define a metric $\|.\|_{L^2}$ on $DET\,P_s$. Generally, because of the jumps of the dimension of $Ker\,P_s$, $\|.\|_{L^2}$ is not a smooth metric on $DET\,P$. However, *the Quillen norm*

$$\|.\|_Q = (\det{}' P_s^* P_s)^{1/2} \|.\|_{L^2}$$

is always a smooth metric on $DET\,P$ (Quillen;[21,11]).

1.3. *The Riemann-Roch-Grothendieck-Quillen formula for smooth families of holomorphic curves.*

Let us recall that for any hermitian metric on a holomorphic vector bundle E, there is a unique unitary connection on E compatible with its holomorphic structure. Using this connection and the Chern-Weil formulae for the characteristic classes, we can associate to $(E, \|.\|)$ its *Chern forms* $c_i(E, \|.\|)$, its *Chern character form* $Ch(E, \|.\|)$, and its *Todd form* $Td(E, \|.\|)$. If σ is any differential form, we denote by $\sigma^{(k)}$ its component of degree k.

THEOREM. *Let $\pi : X \to S$ be a holomorphic family of Riemann surfaces, and E a holomorpic vector bundle over X. Let $\|.\|_T$ be a smooth hermitian metric on the complex line bundle $T_{X|S}$ and $\|.\|_E$ a smooth hermitian metric on E. Let $\|.\|_Q$ be the Quillen metric they define on $DET\,\overline{\partial}_E$. Then, one has the* Riemann-Roch-Grothendieck-Quillen formula:

$$c_1(DET\,\overline{\partial}_E, \|.\|_Q) = -\int_{X|S} \{Ch(E, \|.\|_E)Td(T, \|.\|_T)\}^{(4)} \quad (RRGQ).$$

(In this formula, $\int_{X|S}$ denotes the integration of differential forms along the fibers of π.)

The cohomological version of this formula (*i.e.*the equality RRGQ up to exact differential forms) is a direct consequence of the Atiyah-Singer index theorem for families, or of the Riemann-Roch-Grothendieck theorem (which gives a more precise equality, true in the rational Chow group of S).

The RRGQ formula has been proved by Quillen ([21]) when $X = X_0 \times S$ and the metrics are fixed, by Belavin and Knizhnik ([9,10]) when $E = T^n_{X|S}$. Bismut and Freed have proved an analogous statement for families of Dirac operators ([11]), from which one can deduce the general RRGQ formula ([12]).

The RRGQ formula can be viewed as giving the "holomorphic anomaly" of a holomorphic family of $\overline{\partial}$ operators.

2 Some consequences of the RRGQ formula.

In this section, we explain how to deduce formulae which connect together regularized determinants of various operators associated to Riemann surfaces, by applying the RRGQ formula to well chosen families.

2.1. First, let us explain how to recover from RRGQ the classical formulae giving the conformal anomaly (the "Liouville action"; cf. [20,1]). This derivation should make clear the link between the various kind of anomalies in 2d-field theory (cf. [2]).

Let M be a compact Riemann surface. Any conformal metric on M can be viewed as a hermitian metric on the holomorphic tangent bundle T, and defines a hermitian metric on the line bundle $\omega^n = T^{-n}$, for any n. From these metrics, we get a Quillen norm on the space $DET\,\overline{\partial}_{\omega^n}$. The conformal anomaly formula gives the ratio of the Quillen norms $\|.\|_{Q,0}$ and $\|.\|_{Q,1}$ obtained from two conformal metrics $\|.\|_0$ and $\|.\|_1$ on M: *if* $\|.\|_0 = e^f \|.\|_1$, *then*

$$\frac{\|.\|_{Q,1}}{\|.\|_{Q,0}} = \exp\left[\frac{6n(n-1)+1}{12\pi i}\int_M (fR + \partial f \wedge \overline{\partial f})\right].$$

In this formula, R denotes the curvature 2-form of the holomorphic line bundle T equipped with the metric $\|.\|_0$.

To prove this formula, let us choose a function $F \in C^\infty(M \times \mathbf{C}, \mathbf{R})$ such that the functions $f_s = F(.,s)$ depend only on $|s|$, $f_0 = 1$ and $f_1 = f$. If we apply RRGQ to the "trivial" family $S = \mathbf{C}$, $X = M \times \mathbf{C}$, $\pi(m,s) = s$, $E = \omega^n = \omega^n_{X|S} = T^{-n}_{X|S}$, to the metric $\|.\| = e^F \|.\|_0$ on $T_{X|S}$ and to the "product" metric on $E = T^{-n}_{X|S}$, we obtain after an easy computation:

$$\partial_s \overline{\partial}_s \log \|.\|_{Q,s} = \partial_s \overline{\partial}_s \left\{\frac{6n(n-1)+1}{12\pi i}\int_M (f_s R + \partial f_s \wedge \overline{\partial f_s})\right\}$$

where $\|.\|_s$ is the Quillen metric on $DET\,\overline{\partial}_{\omega^n}$ obtained from the metric $e^{f_s}\|.\|_0$ on M. As f_s and $\|.\|_{Q,s}$ depend only on $|s|$, that implies the announced anomaly formula[1].

Other general anomaly formulae can be proved in the same way.

2.2. Applied to suitable families, the RRGQ formula has been used to relate the regularized determinants of the twisted Dirac operator on a Riemann surface to theta functions ([5]), and to prove the holomorphy property of the string integrand (cf.[9,13,10] and G.Moore's contribution to these proceedings).

These results are statements about regularized determinants, the proofs of which follow roughly the same lines: they can be viewed as asserting that some (products of) DET line bundles equipped with Quillen metrics are isometric; these isometries are deduced from equalities of curvatures (computed by the RRGQ formula) combined with a compactness or symmetry argument.

2.3. Another example of that kind of proof is given by the proof of the "spin 1/2 -bosonization formula" of [14] (see also [10]).

Let M be a compact connected Riemann surface, of genus g, equipped with a conformal metric. Let us choose a canonical basis of $H_1(M, \mathbf{Z})$. To that choice is canonically associated a period matrix ω and an even spin-bundle $\mathcal{L}$ (the "vector of Riemann constants"). Then we have the following "bosonization formula" relating the regularized determinants of the (squared) Dirac operator $\overline{\partial}^*_{\mathcal{L}}\overline{\partial}_{\mathcal{L}}$ and of the scalar laplacian Δ on M:

$$\det \overline{\partial}^*_{\mathcal{L}}\overline{\partial}_{\mathcal{L}} = C_g \left(\frac{\det' \Delta}{A.\det Im\,\Omega}\right)^{-\frac{1}{2}} |\theta(0,\Omega)|^2$$

where C_g is a constant which only depends on g, A the area of M, and θ the Riemann theta function: $\theta(0,\Omega) = \sum_{m \in \mathbf{Z}^g} e^{2\pi i(\frac{1}{2}{}^t m\Omega m)}$.

[1] That proof has been worked out jointly with Th. Jolicœur.

The proof of this formula goes roughly as follows. [2]

Let S be the moduli space of curves of genus g equipped with a canonical basis of $H_1(M, \mathbf{Z}_2)$, and let $\pi : \mathcal{X} \to S$ the "universal curve" on S, and $\mathcal{L}$ the "universal spin-bundle". There is an explicit isomorphism between $(DET\,\overline{\partial}_{\mathcal{L}})^2$ and $(DET\,\overline{\partial}_{\mathcal{O}})^{-1}$, given by the multiplication by θ^2; RRGQ shows that, for any choice of metrics on the universal curve, this isomorphism preserves the curvatures of the Quillen metrics, hence is an isometry, up to a constant (because any pluriharmonic function on S is constant if $g \geq 3$).

We refer to [14,3,4] for details and for the physical derivation of this formula.

3 Arakelov-Faltings geometry (at infinite places) and 2d-field theory

3.1. Let us recall some definitions introduced by Arakelov in his work on intersection on arithmetic surfaces [6].

Let M be a connected compact Riemann surface of genus g and let μ be the Bergman 2-form of M, defined by $\mu = \frac{i}{2g}\sum_{j=1}^{g}\sigma_j \wedge \overline{\sigma_j}$ where (σ_j) is any basis of $\Omega^1(M)$ such that $\frac{i}{2}\int_M \sigma_j \wedge \overline{\sigma_k} = \delta_{jk}$. A metric on a holomorphic line bundle is said to be *admissible* if its curvature is μ up to a multiplicative constant.

Arakelov's Green function is the function $\log G$ on $M \times M$ (singular on the diagonal) defined by the conditions :

$$\partial_P \overline{\partial}_P \log G(P,Q) = i\pi\mu(P) - i\pi\delta(P,Q)$$

and

$$\int_M \mu(P) \log G(P,Q) = 0.$$

The *canonical metric* on the line bundle $\mathcal{O}(-P)$ is defined by $\|\mathbf{1}\|(Q) = G(P,Q)^{-1}$ where $\mathbf{1}$ is the canonical meromorphic section of $\mathcal{O}(-P)$.

The *Arakelov metric* on M is the conformal metric on M such that the metric it defines on (1,0)-forms satisfies the following "metric adjunction formula":

$$\|dz(P)\| = \lim_{Q \to P} \frac{|z(P) - z(Q)|}{G(P,Q)}.$$

This equality can be viewed as a condition on the "regularized" value of G on the diagonal ([3,4]):

$$: \log G(P,P) := 0.$$

Those metrics are admissible.

3.2. Let M be a compact Riemann surface, equipped with its Arakelov metric, and let ξ be a holomorphic line bundle on M, equipped with an admissible

[2] Strictly speaking, we can deduce this formula from RRGQ only, only if $g > 2$. This remark applies to the results of section 3 too.

metric. For any point P of M, we get an admissible metric on $\xi \otimes \mathcal{O}(-P)$ by multiplying the given metric on ξ by the canonical metric on $\mathcal{O}(-P)$. From these data, we obtain Quillen metrics on the (one dimensional) spaces $DET\,\overline{\partial}_\xi$ and $DET\,\overline{\partial}_{\xi\otimes\mathcal{O}(-P)}$.

On the other hand, the exact sequence of sheaves

$$0 \longrightarrow \xi \otimes \mathcal{O}(-P) \longrightarrow \xi \longrightarrow \xi_{|P} \longrightarrow 0$$

gives rise to a canonical isomorphism of one dimensional vector spaces:

$$I : (DET\,\overline{\partial}_{\xi\otimes\mathcal{O}(-P)})^{-1} \simeq (DET\,\overline{\partial}_\xi)^{-1} \otimes \xi_P.$$

(ξ_P is the fiber of ξ at P).

The main technical point of this section is that *this isomorphism is an isometry* when ξ_P is equipped with the given metric on ξ and the DET's are equipped with the Quillen metrics (up to a multiplicative constant, depending only on the genus g of M and of the degree d of ξ).

The proof of this statement is another example of the general method explained in 2.2 ([4,12]).

3.3. Thanks to that isometry, we can relate to regularized determinants various transcendental notions introduced for the purpose of "calculus on arithmetic surfaces" ([6,7,15,18]; such relations have been hinted at by Quillen, cf. [15]).

First, one deduce from it the following relation between the metric $\|.\|_F$ on $DET\,\overline{\partial}_\xi$ defined by Faltings in [15], for any ξ equipped with an admissible metric, and the Quillen metric $\|.\|_Q$ considered in the preceding section:

$$\|.\|_F = C(g,d)\left(\frac{det'\Delta}{A}\right)^{-1/2} \|.\|_Q.$$

In this formula, Δ is the scalar laplacian on M equipped with the Arakelov metric, A is the area of M, and $C(g,d)$ is a constant, which only depends on $g > 2$ and d.

Then, combining this formula with the "spin 1/2 -bosonization formula" above, one gets the following expression of the invariant $\delta(M)$ of Arakelov-Faltings (which occurs in Noether formula for arithmetic surfaces, see [2,15]):

$$\delta(M) = C(g) - 6\log\left(\frac{det'\Delta}{A}\right).$$

($C(g)$ is a constant which only depends on g).

We refer to [4,12] for detailed proofs.

3.4. The isometry I can be used to prove much more general bosonization formulae than the simple bosonization formula of section 2.3 : it allows us to handle arbitrary line bundles, and correlation functions involving any number of insertion points. We give here one of these formulae. We refer to [3,4] for details, and particularly for the physical derivation of these formulae.

We use the same notations as in section 2.3 .

Let $\lambda \geq 3$ be an integer, $\mathcal{L}_\lambda = \mathcal{L}^{2\lambda}$ the spin-λ bundle, and $\overline{\partial}_\lambda = \overline{\partial}_{\mathcal{L}_\lambda}$. Let $\mathcal{G}(P,Q)$ be a fermionic Green function for $\mathcal{L}_\lambda$, i.e. the kernel of a right inverse of $\overline{\partial}_\lambda$. Let us choose a basis $\psi_1, \ldots, \psi_k$ of $Ker \overline{\partial}_\lambda$ ($k = (2\lambda - 1)(g - 1)$ by Riemann-Roch). For any $q \geq 0$, let us choose $p = k + q$ insertion points $P_1, \ldots, P_k, Q_1, \ldots, Q_q$. Finally, let I be the jacobian map, $\overline{\Delta}$ the "vector of Riemann constants", and $\mathcal{N}$ the function on $\mathbf{C}^g/\mathbf{Z}^g + \Omega \mathbf{Z}^g$ defined by $\mathcal{N}(z) = e^{-2\pi {}^t(Im z)(Im \Omega)^{-1}(Im z)} |\theta(z, \Omega)|^2$ where θ is the Riemann theta function ($\theta(z,\Omega) = \sum_{m \in \mathbf{Z}^g} e^{2\pi i (\frac{1}{2} {}^t m \Omega m + {}^t m z)}$).

The announced formula reads:

$$\frac{\det \overline{\partial}_\lambda^* \overline{\partial}_\lambda}{\det(<\psi_i, \psi_j>)} \left\| \begin{vmatrix} \psi_1(P_1) & \ldots & \psi_k(P_1) & \mathcal{G}(Q_1, P_1) & \ldots & \mathcal{G}(Q_q, P_1) \\ \vdots & & \vdots & \vdots & & \vdots \\ \vdots & & \vdots & \vdots & & \vdots \\ \psi_1(P_p) & \ldots & \psi_k(P_p) & \mathcal{G}(Q_1, P_p) & \ldots & \mathcal{G}(Q_q, P_p) \end{vmatrix} \right\|^2$$

$$= C(g, \lambda, q) \left(\frac{\det{}' \Delta}{A. \det Im\, \Omega} \right)^{-\frac{1}{2}} \frac{\prod_{i<j} G(P_i, P_j)^2 \prod_{l<m} G(Q_l, Q_m)^2}{\prod_{r,s} G(P_r, Q_s)^2} \mathcal{N}(J)$$

where

$$J = I\left(\sum_i P_i - \sum_l Q_l \right) - (2\lambda - 1)\overline{\Delta}.$$

The left-(resp. right-)hand side of this formula is a fermionic (resp. bosonic) correlation function.

ACKNOWLEDGEMENTS. I would like to thank L. Alvarez-Gaumé, Th. Jolicœur, G. Moore, P. Nelson, I. Singer, C. Soulé and C. Vafa for discussions and collaborations. I am grateful to the MIT department of mathematics for its hospitality during April and May 1986. This work was supported in part by DOE contract DE-FG02-84-ER-40164-A001.

Bibliography

[1] O. Alvarez, Nucl. Phys. **B216**, 125 (1983).

[2] O. Alvarez, *Conformal anomalies and the index theorem*, UCB-preprint Ph-T185-39.

[3] L. Alvarez-Gaumé, J.-B. Bost, G.Moore, P. Nelson and C. Vafa, Harvard preprint HUTP-86/A039, to appear in Phys. Lett.B.

[4] L. Alvarez-Gaumé, J.-B. Bost, G.Moore, P. Nelson and C. Vafa, in preparation.

[5] L. Alvarez-Gaumé, G. Moore and C. Vafa, Comm. Math. Phys. **106** (1986).

[6] S.J. Arakelov, Izv. Akad. Nauk. SSSR Ser. Mat. **38** (1974) [=Math. USSR Izv. **8**, 1167 (1974)].

[7] S.J. Arakelov, Proc. Int. Cong. Vancouver 1974, vol.1, 405.

[8] M.F. Atiyah and I.M. Singer, Proc. Nat. Acad. Sc. USA, **81**, 2597 (1984).

[9] A.A. Belavin and V.G. Knizhnik, Phys. Lett. **B168**, 201 (1986).

[10] A.A. Belavin and V.G. Knizhnik, *Complex geometry and the theory of quantum strings.* Landau Institute Preprint, submitted to ZEPTF.

[11] J.-M. Bismut and D. Freed, *Geometry of elliptic families I,II.* Orsay preprint 85T47.

[12] J.-B. Bost, in preparation.

[13] J.-B. Bost and Th. Jolicœur, Phys. Lett. **B174**, 273 (1986).

[14] J.-B. Bost and P. Nelson, Phys. Rev. Lett. **57**, 795 (1986).

[15] G. Faltings, Ann. of Math. **119**, 387 (1984).

[16] D. Friedan, in *Recent advances in field theory and statistical mechanics, Les Houches, 1982* (eds. J.-B. Zuber and R. Stora; North-Holland 1984).

[17] F.F. Knudsen and D. Mumford, Math. Scand. **39**, 19 (1976).

[18] Yu.I. Manin, in *Arbeitstagung, Bonn 1984*, Springer Lecture Notes in Mathematics **1111** (1985).

[19] G. Moore and P. Nelson, Comm. Math. Phys. **100**, 83 (1985).

[20] A.M. Polyakov, Phys. Lett. B103, 207 (1981).

[21] D. Quillen, Funk. Anal. i Prilozen **19**, 37 (1985) [=Funct. Anal. Appl. **19**, 31 (1986)].

[22] D. Ray and I.M. Singer, Adv. in Math. **7**, 145 (1971).

[23] R.T. Seeley, Proc. Symp. Pure Math., AMS, vol. X, 288 (1967).

[24] *Séminaire de Géométrie Algèbrique 6*, Springer Lecture Notes in Mathematics **225**(1971).

HUTP-86/A074

Modular Forms and Multiloop String Physics [1]

Gregory Moore

Harvard University
Cambridge, MA 02138

and

Institute for Advanced Study
Princeton, NJ 08540

The theory of modular forms can be used to derive explicit formulæ for the bosonic string amplitudes at two and three loops, and to show that the vacuum amplitude of the heterotic string vanishes to all orders of perturbation theory.

[1] Talk presented at the VIIIth International Congress on Mathematical Physics at Marseille.

1. The Polyakov Path Integral

In the past year it was discovered that the integrand of the Polyakov string [1] satisfies beautiful holomorphic factorization properties [2][3][4][5][6][7][8]. We will review here two applications of this discovery to string theory: one can deduce explicit formulæ for the string integrand at two and three loops , [2][9][10][11][12] and one can show that the cosmological constant of the heterotic string vanishes to all orders of string perturbation theory [13][14].

The Polyakov ansatz for the vacuum amplitude of the string at g-loops is of the form

$$Z = \int \frac{\text{d(two dimensional geometry)d(string coordinates)}}{\text{volume(symmetry group)}} e^{-S} \tag{1}$$

where S is an action for the string coordinates in a two-dimensional "background." This action is assumed to be invariant under some (large) symmetry group.

For example, in the 26-dimensional closed bosonic string, 2-geometries consist of Riemannian metrics on a genus g parameter space Σ, and string coordinates X comprise an embedding of Σ into $\mathbf{R}^{26}$. The action is then just

$$S = \int_{\Sigma} \partial X^m \bar{\partial} X^m \tag{2}$$

where ∂ is the Cauchy-Riemann operator. In particular, S only depends on the complex structure on Σ induced by the two-metric.

The symmetry group is therefore { conformal rescalings } x {diffeomorphisms} and Z is therefore a finite dimensional integral of a density on the moduli space of genus g Riemann surfaces, $\mathcal{M}_g$.

The key discovery of last year was that the lift of this density to Teichmüller space T_g is of the form

$$\varsigma \wedge \bar{\varsigma} (\det Im\tau)^{-13} \tag{3}$$

where ς is a nonvanishing holomorphic $3g-3$ form on T_g.

2. Explicit Formulæ for g = 2,3

The explicit formulæ for $g = 2,3$ rely on the near coincidence of the Siegel upper half plane $\mathcal{H}_g$ with the quotient T'_g of Teichmüller space by the Torelli group for these genera.

Recall that if $\{a_i, b_i\}$ is a canonical homology basis then there is a unique basis of holomorphic 1-forms defined by

$$\int_{a_i} \omega_i = \delta_{ij} \, , \tag{4}$$

and then the period matrix

$$\tau_{ij} = \int_{b_i} \omega_j \tag{5}$$

is a nontrivial function of the Riemann surface and its marking. The matrix τ_{ij} is a $g \times g$ complex symmetric matrix with $Im\tau > 0$, and therefore lies in the Siegel upper half plane defined by

$$\mathcal{H}_g = \{\Omega | \Omega^{tr} = \Omega, Im\Omega > 0\}$$

By Torelli's theorem the period matrix uniquely determines the complex structure, so that T'_g injects into $\mathcal{H}_g$. We may then think of moduli space as the quotient of the image of T'_g by the action of Sp(2g, **Z**):

$$\Omega \to \tilde{\Omega} = (A\Omega + B)\,(C\Omega + D)^{-1} \tag{6}$$

corresponding to a change of marking with $\begin{pmatrix} A & B \\ C & D \end{pmatrix} \epsilon \; Sp\,(2g, \mathbf{Z})$.

For the familiar case of the torus, $T_1 = T'_1 = \mathcal{H}_1$. But for g> 1 dim $T'_g = 3g - 3$ while $dim\mathcal{H}_g = \frac{1}{2}g(g+1)$ and the dimensions only coincide for $g = 2,3$. In these two cases the spaces agree up to subvarieties of codimension one or more. In fact, at genus two the relation is

$$T'_{g=2} = \mathcal{H}_2 - \mathcal{D} \tag{7}$$

where

$$\mathcal{D} = \left\{ \begin{pmatrix} \tau & n \\ n & \tau_2 \end{pmatrix} | n \in \mathbf{Z}, \quad \tau_1, \tau_2 \right\} \tag{8}$$

is the space of period matrices which are diagonal mod **Z**. For $g = 3$ the relation is more complicated because of the hyperelliptic subvariety .

Thus, by lifting the Polyakov measure to T_g' for $g = 2, 3$ we obtain a holomorphic form ς defined on almost all of $\mathcal{H}_g$. If we write

$$\varsigma(\tau) = f(\tau) \prod_{i \leq j} d\tau_{ij} \tag{9}$$

on T_2' then, from the path integral form of the string measure we know that $f \neq 0$ on T_g while $f \to \infty$ on $\mathcal{D}$. Thus, f^{-1} is a holomorphic function on all of $\mathcal{H}_2$, vanishing on $\mathcal{D}$. On the other hand, for any g we have

$$det Im\Omega \to det Im\Omega |det C\Omega + D|^{-2} \tag{10}$$

$$\prod_{i \leq j} d\Omega_{ij} \to det(C\Omega + D)^{(g+1)} \prod_{i \leq j} d\Omega_{ij} \tag{11}$$

under a symplectic modular transformation.

By the modular invariance of the string integrand we know that f^{-1} is a modular form of weight ten.[1] Recall that a (Siegel) modular form ϕ of weight k is a holomorphic function on $\mathcal{H}_g$ such that under an $Sp(2g, \mathbf{Z})$ transformation

$$\phi(\tilde{\Omega}) = (det(C\Omega + D))^k \phi(\Omega) \tag{12}$$

By a remarkable theorem of Igusa [15] there is a unique modular form of weight 10 vanishing on $\mathcal{D}$. (In fact, Igusa worked out the entire *ring* of genus two Siegel modular forms.) This form can be written in terms of theta functions, more precisely, in terms of the thetanullwerte:

$$\vartheta \begin{bmatrix} \alpha \\ \beta \end{bmatrix} (0|\Omega) = \sum_{n \in \mathbf{Z}^g} e^{i\pi(n+\alpha)\cdot\Omega\cdot(n+\alpha) + 2\pi i(n+\alpha)\cdot\beta} \tag{13}$$

for half integral characteristics α, β. The formula is

$$f^{-1}(\Omega) = \prod_{\alpha,\beta} \vartheta^2 \begin{bmatrix} \alpha \\ \beta \end{bmatrix} \equiv (\Delta_{(2)})^2 \tag{14}$$

where the product is only taken over the *even* theta characteristics for which $4\alpha \cdot \beta \equiv 0$ mod(2). Thus the two loop vacuum energy is

$$Z_2 = \frac{\pi^3}{2} g^2 \left(\frac{T}{\pi}\right)^{25} \int_{\mathcal{F}_0} |d\Omega \Delta_2^{-2}|^2 (det\ Im\tau)^{-13} \tag{15}$$

[1] One can argue that there is no phase in this case, see [9] .

where g is the string coupling, T is the string tension, and $\mathcal{F}_2^0$ is a fundamental domain for the action of $Sp(4,\mathbf{Z})$ on $\mathcal{H}_2$ (with the diagonal period matrices removed).

We conclude this section with some remarks:

1) Igusa's theorem can also be used to fill in an annoying loophole in the argument for the uniqueness of the string measure. For $g \geq 3$ $\mathcal{M}_g$ admits a Satake compactification. This implies that the string integrand satisfies an important uniqueness property: Any two measures which do not vanish and have a lift to T_g of the form (3) must agree. In particular, the Polyakov form must coincide with the canonical density discovered by Mumford [16][17][2][3] . The argument fails for $g = 2$, since then there is no Satake compactification. In this case, however, the absence of weight zero Siegel modular forms implied by Igusa's result again guarantees uniqueness.

2) A similar but more intricate argument applies at genus 3. One finds that

$$f^{-1}(\Omega) = \prod_{\alpha,\beta} \vartheta^{\frac{1}{2}} \begin{bmatrix} \alpha \\ \beta \end{bmatrix} \tag{16}$$

where the product is again taken over even theta characteristics. This result was guessed in [9] and proved in [10][11] . In this case f^{-1} is a modular form of weight 9 and so has a minus sign ambiguity under a modular transformation, which suggests that the left movers of the bosonic string have a global anomaly at three loops. One should exercise caution with this interpretation since the string integrand and the appropriate regularized determinants differ by important factors (such as zero-mode determinants).

3. Physical Factorization

One amusing application of the explicit formula (15) is that one can verify the (intuitively obvious) factorization behavior of the string integrand on the boundary of moduli space. The boundary of moduli space (in the Deligne-Mumford compactification [18]) describes, roughly speaking, Riemann surfaces in which a closed curve has been shrunk to zero. If the curve is a nontrivial homology cycle the component is called Δ_0 and if the curve is homologically trivial but separates the surface into two surfaces of genus i and $g-i$, the component is called Δ_i. All the boundaries can be modelled by welding in the plumbing fixture defined by $zw = t$. See [2][7][8][9] and references therein for details. Here t is a parameter, $|t| < 1$, and the above family of Riemann surfaces describes an annulus degenerating into the one-point union of two disks. In terms of t the boundary of moduli space is described by $t = 0$. Using (15) one may check that, in the region of moduli space describing surfaces in which the handles move off to infinity (the component Δ_1)

$$\varsigma = \frac{dt\ d\tau_1 d\tau_2}{t^2}\,(a_0 + ta_1 + t^2 a_2 + \cdots) \tag{17}$$

verifying the second order pole observed by Belavin and Knizhnik [2]. Here τ_1, τ_2 are the modular parameters of the remaining tori. Furthermore one checks that

$$a_k(\tau_1,\tau_2) = a_k^{(1)}(\tau_1)a_k^{(2)}(\tau_2) \tag{18}$$

so that the full density behaves as

$$\sum_{k,\ell=0}^{\infty} d^2\tau_1\,\frac{a_k^{(1)}\overline{a_\ell^{(1)}}}{(Im\tau_1)^{13}}\left(\frac{dt d\bar{t}}{|t^2|^2}\,t^k\bar{t}^\ell\right)d^2\tau_2\,\frac{a_k^{(2)}\overline{a_\ell^{(2)}}}{(Im\tau_2)^{13}} \tag{19}$$

This expansion has the natural physical interpretation in terms of an exchange of particles between the handles with $L_0 = k, \bar{L}_0 = \ell$. When integrating over $\mathcal{M}$ the phase integral of t corresponds to the $L_0 - \bar{L}_0$ projection [2][7][9][12][19][20]. Indeed, in terms of the constant curvature metric the "proper distance" between the tori is $s = (2\pi)^{-1}log|t|^{-1}$ where

$$\int_{t\to 0}\frac{dt d\bar{t}}{|t^2|^2}\,t^k\bar{t}^\ell \to \delta_{k\ell}\int_{s\to\infty} ds\ e^{-4\pi(k-1)s} \tag{20}$$

In particular we see that the second order pole is due to the existence of tachyons [2][21]. Hence, in bosonic string theory, divergences arise both from tachyons *and* dilatons [22][23]. One can check that the first few coefficients a_k are indeed those expected from one point functions of vertex operators at zero momentum [9] .

Finally, we note that similar remarks apply for the boundary component Δ_0 and for two-loop scattering amplitudes. Factorization at three loops seems to rely on some (unproven?) identities for genus two ϑ-functions. Nevertheless, one can give a general argument for factorization on arbitrary surfaces, together with a formula for the complete vertex operator expansion [20][24].

4. The Heterotic String

Another useful application of modular forms involves an argument that the vacuum amplitude of the heterotic string vanishes in flat space.

Starting from the 2D supergravity form of the heterotic string [25][26][27][28][29] one may show that after gauge-fixing the symmetries (diffeomorphism, Weyl, super-Weyl, and supergravity) the amplitude takes the form

$$Z_t^{het} = \int_{\mathcal{F}_g} \frac{\varsigma \wedge \bar{\varsigma}}{(det\ Im\tau)^5}\ \psi_8 \bar{\xi} \tag{21}$$

where $\mathcal{F}_g$ is a fundamental domain for the action of the mapping class group on T_g, ς is the same differential form as in (3) and ψ_8, which arises from the 32 left-moving gauge fermions, is

$$\psi_8 = \begin{cases} (\sum \vartheta^8)^2 & \text{for } E_8 \times E_8 \\ \sum \vartheta^{16} & \text{for } Spin\ 32/\mathbf{Z}_2 \end{cases}$$

where the sum is over the even theta characteristics. ξ is, roughly speaking, the contribution of the right-movers. A formula for ξ in terms of functional determinants and supermoduli is given in [5][13] . One can show that $\bar{\xi}$ is antiholomorphic on T_g, so that absence of global anomalies [30] implies that ξ is a Teichmüller modular form of weight 8.

Again we may turn to Igusa in the case $g = 2, 3$ [2][9][11] . In those cases ψ_8 is the unique modular form of weight eight (in particular , it doesn't matter if we choose the form for $E_8 \times E_8$ or for Spin $32/\mathbf{Z}_2$) so $\xi = k\psi_8$ for some constant k. If $k \neq 0$ we obtain the partition function of a *bosonic* theory, and certainly not the heterotic string so that the vacuum amplitude must vanish at two [2][9][11] and three [2][11] loops.

We can go much further if we assume the physical factorization properties continue to hold which are known to hold for the bosonic string.[2] . Recall that the heterotic string is comprised of leftmovers from the bosonic string and rightmovers from the superstring. Thus the sum analogous to eqn. (19) must start out with a 2nd order pole in t but a *zeroth* order pole in $\bar{t}$. Comparing with Eq. (21) and recalling that ς has a 2nd order pole (see (17)) we learn that ξ must have a second order zero on all the boundary components Δ_i for $0 \le i \le [g/2]$.

It was shown in [13] that no Teichmüller modular form of weight eight with 2nd order zeroes on Δ_i exists for $g \le 19$.[3] In [14] it was then shown that no such modular form exists for $g \ge 17$. Thus the vacuum amplitude vanishes to all orders of perturbation theory.

In superstring theory, the only expected divergences in scattering amplitudes come from dilaton one-point functions at zero momentum. Some of these we have just shown are

[2] The methods of [20] can probably be extended to the heterotic string.

[3] It was claimed in [13] that $g \le 20$. A more careful evaluation of the relevant lower bound on the slope of the effective divisor showed that the bound is only good for $g \le 19$.

zero. Thus, the result gives strong evidence for finiteness, although that is hardly a proof. If we had a clear understanding of an algebro-geometric interpretation of the scattering amplitudes in superstring theory, then perhaps the theory of modular forms could be used to prove more powerful finiteness results.

The most serious limitation of the above result is that it only applies to flat space. In fact, we would like to know how the cosmological constant can be zero *after* supersymmetry breaking, and for applications to this problem we have proved too much: if the amplitude vanishes point-by-point on moduli space then all mass levels are degenerate (which is practically the same as supersymmetry, and certainly not observed in nature) [31]. If modular forms are relevant to the cosmological constant problem we must find a good reason for the integral of a *nonzero* density to be zero. In fact, such things are known to be true in the theory of modular forms as a result of the Hecke theory. If two modular (cusp) forms of the same weight have different eigenvalues under a Hecke operator then their Petersson inner product must vanish. Note that it is precisely the Petersson inner product which arises in the string expressions for one-loop diagrams, although the modular forms which do appear are typically not cusp forms (i.e. there are tachyons and massless particles) and have negative weight (since we can compactify at most 10 or 24 dimensions). Nevertheless, perhaps an argument along these lines can be used to show that a string in a nonsupersymmetric background can nonetheless have a vanishing cosmological constant.

Acknowledgments

I would like to thank L. Alvarez-Gaumé, J.-B Bost, A. Cohen, P. Ginspang, J. Harris, J. Harvey, P. Nelson, I. Singer, A. Strominger, J. Polchinski, and C. Vafa for much help and many fruitful collaborations. I would also like to thank E. Witten for many useful and clarifying remarks on Hecke operators. Finally, I would like to thank the organizers of the Marseille conference for their invitation to speak. This work was supported in part by the Harvard Society of Fellows, by NSF-PHY 85-15249, and by Dept. of Energy Grant No. DE-AC02-76ERO2220.

References

[1] A. Polyakov, Phys. Lett. **103B** (1981) 207; Phys. Lett **103B** (1981) 211

[2] A. A. Belavin and V. G. Knizhnik, Phys. Lett. **168B**(1986)201; "Complex Geometry and the Theory of Quantum Strings" Landau Inst. preprint, submitted to ZETF

[3] J. Bost and J. Jolicoer, Phys. Lett. **174B**(1986)273

[4] R. Catenacci, M. Cornalba, M. Martinelli, and C. Reina, Phys. Lett. **172B** (1986) 328.

[5] L. Alvarez Gaumé, G. Moore, and C. Vafa, Comm. Math. Phys. **106** (1986) 1

[6] D. Freed, "Determinants, Torsion, and Strings," to appear in Comm. Math. Phys

[7] D. Friedan and S. Shenker, Phys. Lett. **175B**(1986)287; Chicago preprint EFI-86-18B

[8] P. Nelson, "Lectures on Strings and Moduli Space," HUTP-86/ A047

[9] G. Moore, Phys. Lett. **176B**(1986) 369

[10] A. Belavin, V. Knizhnik, A. Morozov, A. Perelonov, "Two-and three-loop amplitudes in the bosonic string theory," preprint ITEP-59 (1986).

[11] A. Morozov, "Explicit formulae for one, two, three, and four- loop string amplitudes," preprint.

[12] A. Kato, Y. Matsuo, and S. Odake, Tokyo University preprint UT-489.

[13] G. Moore, J. Harris, P. Nelson, and I. Singer, "Modular forms and the cosmological constant," Phys. Lett. **178B**(1986)167.

[14] M. C. Chang and Z. Ran, "Divisors on $\overline{M}_g$ and the cosmological constant," preprint.

[15] J.-I. Igusa, Amer. J. Math. **84** (1962) 175; Amer. J. Math. **86** (1964) 392.

[16] D. Mumford, Ens. Math. **23** (1977) 39.

[17] A. A. Beilinson and Yu. I. Manin, "The Mumford Form and the Polyakov Measure in String Theory," submitted to CMP.

[18] P. Deligne and D. Mumford, Publ. IHES **36**(1969)75.

[19] A. Cohen, G. Moore, P. Nelson, and J. Polchinski, "Semi-Off-Shell Amplitudes," HUTP-86/A028, Nucl. Phys. B, to appear

[20] A. Cohen, G. Moore, and J. Polchinski, in preparation.

[21] S. Wolpert, Maryland preprint MD86-10-SAW,1986

[22] R. Rohm, Nucl. Phys. **B237**(1984)553.

[23] J. Polchinski, Comm. Math. Phys. **104**(1986)37.

[24] The methods of [20] rely on functional integrals and are therefore not completely rigorous. Part of the factorization theorem for the bosonic string has been given a rigorous proof from the aglebro-geometric viewpoint by J.-B. Bost.

[25] C. Hull and E. Witten, Phys. Lett. **160B** (1985) 398.

[26] D. Friedan, E. Martinec and S. Shenker, Phys. Lett. **160B** (1985) 55

[27] G. Moore, P. Nelson and J. Polchinski, Phys. Lett. **169B** (1986) 47

[28] M. Evans and B. Ovrut, Phys. Lett. **171B**(1986)177; Penn. preprint UPR-0294-T;J. Gates and Nishino, preprint.

[29] P. Nelson and G. Moore, Nucl. Phys. **B274**(1986)509.

[30] E. Witten, in W. A. Bardeen and A. R. White, eds. *Geometry Anomalies Topology* (World Scientific, Singapore, 1985).

[31] I would like to thank J. Harvey, J. Polchinski and A. Strominger for remarks on this point

THE THOULESS EFFECT IN THE HIERARCHICAL MODEL

BLEHER P.M.

Keldysh Institute of Applied Mathematics, Moscow A47

We study phase transitions in a one-dimensional chain of spins $\sigma(i) = \pm 1$, $i \in \mathbb{N} = \{1,2,3,\ldots\}$, with the hierarchical Hamiltonian

$$H(\sigma) = - \sum_{i<j} \frac{\varphi(d(i,j))}{d^2(i,j)} \sigma(i)\,\sigma(j),$$

where $d(i,j)$ is a "hierarchical distance" on $\mathbb{N}$ and $\varphi(x) > 0$ is a slowly changing function. To define the hierarchical distance consider finite sequences $V_{kn} = \{(k-1)2^n+1, \ (k-1)2^n+2, \ldots, \ k2^n\}$, $k \geq 1$; $n \geq 0$, and set $n(i,j) = \min\{n \mid \exists k : \{i,j\} \subset V_{kn}\}$. Then $d(i,j) = 2^{n(i,j)}$, $i \neq j$, and $d(i,i) = 0$. Note that $d(i,j)$, $i \neq j$, takes only values of the form 2^n, $n \geq 1$. Put

$b_n = \varphi(2^n)$. Assume that $\lim\limits_{n\to\infty} \frac{b_{n+1}}{b_n} = 1$, $b_n > 0$.

Dyson proved in [1] the following fine result.

<u>Theorem</u> : If $\lim\limits_{n\to\infty} (b_n/\log n) = 0$ then the spontaneous magnetization $m(\beta) = 0$ for all β and there is no phase transition. If $b_n/\log n \geq 1$ then $m(\beta) > 0$ for large β and there is a phase transition. If $b_n = \log n$ then there is a Thouless effect at the transition, i.e. $m(\beta)$ has a jump at the critical point β_c.

We study the case $b_n = \log(n+n_o)$, where $n_o \gg 1$ and prove limit theorems at $\beta \geq \beta_c$. Note that the existence of the critical point $\beta_c = \beta_c(n_o)$ and the Thouless effect at $\beta=\beta_c$ follows, for any $n_o \geq 0$, from Dyson's arguments.

Consider spins $\sigma(1),\dots,\sigma(2^n)$ distributed according to a finite Gibbs distribution $\mu_n(\sigma;\beta) = \Xi_n^{-1}(\beta)\exp(\beta \sum_{1\le i<j\le 2^n} \frac{\varphi(d(i,j))}{d^2(i,j)}\sigma(i)\sigma(j))$.

To describe limit distributions of appropriately normalized sums of the spins $\sigma(1),\dots,\sigma(2^n)$ introduce a real random variable ξ whose characteristic function is

$$\chi(\tau) = \exp\left\{\sum_{m=-\infty}^{0} 2^m\left[\exp\left(\frac{i\tau}{2^m}\right)-1\right] + \sum_{m=1}^{\infty} 2^m\left[\exp\left(\frac{i\tau}{2^m}\right) - 1 - \frac{i\tau}{2^m}\right]\right\}$$

Note that ξ satisfies the equation

$$\frac{\xi_1 + \xi_2}{2} = \xi + C_o ,$$

where ξ_1,ξ_2 are independent and ξ-distributed and the equality of random variables is understood as the one of their distributions. Such type random variables were considered by Levy and we call ξ the Levy random variable. Note that ξ has a purely point distribution with atoms at points $\{\frac{j}{2^l},\ j\in\mathbb{Z},\ l\ge 0\}$.

<u>Theorem 1</u> : Let $b_n = \log(n+n_o)$. There exists $N_o > 0$ such that $\forall\ n_o \ge N_o$ $\forall\ \beta\ge\beta_c=\beta_c(n_o)\ \exists\ \gamma(\beta)$:

$$\lim_{n\to\infty}\left[\left|\frac{\sigma(1)+\dots+\sigma(2^n)}{2^n}\right| - m_n(\beta)\right] n^{\gamma(\beta)} = C\xi$$

where $$m_n(\beta) = E\left|\frac{\sigma(1)+\dots+\sigma(2^n)}{2^n}\right| = \sum_{\sigma}\left|\frac{\sigma(1)+\dots+\sigma(2^n)}{2^n}\right|\mu_n(\sigma;\beta),$$

$$\lim_{n\to\infty} m_n(\beta) = m(\beta) > 0,$$

$C > 0$, ξ is the Levy random variable.

Theorem 2 : Under the conditions of Theorem 1 limit Gibbs states exist and the number of extremal Gibbs states is 1 for $\beta < \beta_c$ and 2 for $\beta \geq \beta_c$.

This result supports a conjecture of Simon concerning the Thouless effect.

To formulate our last result, let us give the following definition. A sequence $b_n > 0$ is called ε-slowly changing if $\left|\frac{b_{n+1}}{b_n} - 1\right| < \varepsilon$ for all $n \geq 1$ and $\lim \frac{b_{n+1}}{b_n} = 1$.

Theorem 3 : $\exists\ \varepsilon > 0$ such that $\forall\ \varepsilon$-slowly changing sequence b_n such that

$$\lim_{n\to\infty} \frac{b_n}{\log n} = \infty, \qquad \lim_{\beta\to\beta_c} m(\beta) = 0.$$

That means that the phase transition at β_c is of continuous type. This confirms a conjecture of Dyson in [1].

REFERENCES

[1] Dyson, F.J., An Ising Ferromagnet with Discontinuous Long-Range Order; Commun. Math. Phys., 21, 269-283 (1971).

SYMMETRY BREAKING IN THE CLASSICAL N-VECTOR HIERARCHICAL MODEL

P.M. BLEHER

Keldysh Institute of Applied Mathematics, Moscow A47

This work was done by the author in collaboration with P. Major. We study symmetry breaking in the classical N-vector hierarchical model (HM). The Hamiltonian of the model is $H(\sigma) = - \sum_{i\neq j} \frac{(\sigma(i),\sigma(j))}{d^a(i,j)}$ where $\sigma(i) \in R^N$, $i=1,2,\ldots,$ are classical N-vector spins, $N>1$; $a>1$ is a parameter of the model and $d(i,j)$ is the "hierarchical distance" being defined in the following way : let $V_{kn} = \{(k-1)2^n+1,(k-1)2^n+2,\ldots,k2^n\}$, $k=1,2,3,\ldots;$ $n\geq 0$, and $n(i,j) = \min\{n | \exists k : \{i,j\} \subset V_{kn}\}$. Then $d(i,j) = 2^{n(i,j)-1}$.

A finite Gibbs distribution of the HM is

$$\mu_n(d\sigma;\beta) = \Xi_n^{-1}(\beta) \exp\Big(\frac{\beta}{2} \sum_{\substack{i,j=1\\ i\neq j}}^{2^n} \frac{(\sigma(i),\sigma(j))}{d^a(i,j)}\Big) \prod_{i=1}^{2^n} \nu(d\sigma(i)).$$

We consider the case when the free measure $\nu(dx)$ is of the form

$$\nu(dx) = L^{-1} \exp\Big(-\frac{u}{4} |x|^4 - \frac{1}{2} |x|^2 - r(x)\Big)\, dx ,$$

where the "remainder term" $r(x)$ has the following properties :

1) $r(x)$ depends only on $|x|$, $r(x) = r_o(|x|)$;

2) $\|r(x)\|_{C^4(\mathbb{R}^N)} \le u^{1+\delta}$, $\delta>0$. In that case we say that $\nu(dx)$ belongs to the class S_u. We shall assume that $u>0$ is small enough (a condition on the smallness of u is included in the formulations of the theorems). That means that $\nu(dx)$ is close to a Gaussian distribution. Note that the Gibbs distribution $\mu_n(d\sigma;\beta)$ is invariant with respect to simultaneous rotations of spins $\sigma(j) \to U\sigma(j)$, $U \in O(N)$.

We prove limit theorems for sums $\sigma(1)+\ldots+\sigma(2^n)$ of μ_n-distributed random variabes $\sigma(1),\ldots,\sigma(2^n)$ as $n\to\infty$. The situation is different for different values of the parameter a. Five cases should be differentiated : 1) $a\geq 2$; 2) $a=2$ with slowly increasing corrections ; 3) $2>a>3/2$; 4) $a=3/2$; 5) $3/2>a>1$. We consider these five cases in the sequence 1), 5), 4), 3), 2). Denote ξ_N the standard Gaussian N-dimensional random variable with zero mean and unit variance.

<u>Theorem 1</u> : $\forall\ a\geq 2\ \ \exists\ u_o>0$ such that $\forall\ \nu\in S_u$, for $0<u<u_o$, and $\forall\ \beta>0$

$$\lim_{n\to\infty} \frac{\sigma(1)+\ldots+\sigma(2^n)}{(2^n)^{1/2}} = \sqrt{\chi(\beta)}\ \xi_N$$

Here and later the convergence of random variables is understood as the one of their distribution densities in the norm $\|\cdot\| = \|\cdot\|_{C(\mathbb{R}^N)} + \|\cdot\|_{L^1(\mathbb{R}^N)}$. Theorem 1 means that for $a\geq 2$ the local central limit theorem (CLT) holds for <u>any</u> $\beta>0$, so there is no symmetry breaking in that case. The value $\chi(\beta)>0$ is the variance of the limit distribution.

<u>Theorem 2</u> : (see [1]). $\forall\ 3/2 > a > 1\ \exists\ u_o > 0$ such that $\forall\ \nu\in S_u$, for $0<u<u_o$, $\exists$ a critical point $\beta_c = \beta_c(a,\nu) > 0$ and

$$\lim_{n\to\infty} \frac{\sigma(1)+\ldots+\sigma(2^n)}{(2^n)^{1/2}} = \sqrt{\chi(\beta)}\ \xi_N \quad , \quad \text{if}\ \ \beta < \beta_c$$

$$\lim_{n\to\infty} \frac{\sigma(1)+\ldots+\sigma(2^n)}{(2^n)^{a/2}} = \frac{\xi_N}{\sqrt{a_o\beta}} \quad \text{if} \quad \beta = \beta_c$$

$$\lim_{n\to\infty} \frac{|\sigma(1)+\ldots+\sigma(2^n)|-2^n M_n(\beta)}{(2^n)^{1/2}} = \sqrt{\chi(\beta)}\ \xi_1 \ , \ \text{if } \beta>\beta_c$$

where $a_o = \dfrac{2}{1-2^{1-a}}$, and $\lim_{n\to\infty} M_n(\beta) = M(\beta) > 0$ ($M(\beta)$ is the spontaneous magnetization). Moreover $\chi(\beta) \sim C_{\pm}|\beta-\beta_c|^{-1}$ as $\beta \to \beta_c \pm 0$, $M(\beta) \sim C\sqrt{\beta-\beta_c}$ as $\beta \to \beta_c + 0$.

We use the notation $F(\beta) \sim G(\beta)$ as $\beta \to \beta_c$ in the sense that $\lim F(\beta)/G(\beta) = 1$. The notation $\beta \to \beta_c+0$ (resp. $\beta \to \beta_c-0$) means $\beta \downarrow \beta_c$ (resp. $\beta \uparrow \beta_c$). Theorem 2 shows that for $3/2 > a > 1$ the limit distribution at $\beta = \beta_c$ is Gaussian and the critical exponents are classical.

<u>Theorem 3</u> : For $a=3/2$ $\exists u_o>0$ such that $\forall v\in S_u$, for $0<u<u_o$, $\exists$ a critical point $\beta_c = \beta_c(v) > 0$ with

$$\lim_{n\to\infty} \frac{\sigma(1)+\ldots+\sigma(2^n)}{(2^n)^{1/2}} = \sqrt{\chi(\beta)}\ \xi_N \ , \quad \text{if} \quad \beta < \beta_c$$

$$\lim_{n\to\infty} \frac{\sigma(1)+\ldots+\sigma(2^n)}{(2^n)^{a/2}} = \frac{\xi_N}{\sqrt{a_o\beta}} \quad \text{if} \quad \beta = \beta_c$$

$$\lim_{n\to\infty} \frac{|\sigma(1)+\ldots+\sigma(2^n)|-2^n M_n(\beta)}{(2^n)^{1/2}\, n^{1/2}} = \chi(\beta)\ \xi_1 \ , \ \text{if } \beta>\beta_c$$

where $\lim_{n\to\infty} M_n(\beta) = M(\beta) > 0$. Moreover $\chi(\beta) \sim C_{\pm} |\beta-\beta_c|^{-1}$, as $\beta \to \beta_c \pm 0$, and $M(\beta) \sim C\sqrt{(\beta-\beta_c)|\ln(\beta-\beta_c)|}$ as $\beta \to \beta_c+0$.

Thus logarithmic corrections arise when $a = 3/2$.

<u>Theorem 4</u> (see [2]) : $\exists\, \varepsilon_o = \varepsilon_o(N)$ and $u_o>0$ such that $\forall$ $a=3/2+\varepsilon$, where $0<\varepsilon<\varepsilon_o$, and $\forall$ $\nu\in S_u$, where $0<u<u_o$, $\exists$ a critical point $\beta_c = \beta_c(a,\nu) > 0$ with

$$(1) \qquad \lim_{n\to\infty} \frac{\sigma(1)+\dots+\sigma(2^n)}{(2^n)^{1/2}} = \sqrt{\chi(\beta)}\, \xi_N \quad , \quad \text{if} \quad \beta < \beta_c$$

$$(2) \qquad \lim_{n\to\infty} \frac{\sigma(1)+\dots+\sigma(2^n)}{(2^n)^{a/2}} = \frac{\xi_N^*}{\sqrt{\beta}} \quad \text{if} \quad \beta = \beta_c$$

$$(3) \qquad \lim_{n\to\infty} \frac{|\sigma(1)+\dots+\sigma(2^n)|-2^n M_n(\beta)}{(2^n)^{a-1}} = \frac{\xi_1^-}{M(\beta)\sqrt{\beta}} \quad \text{if } \beta>\beta_c$$

where ξ_N^* is a non-Gaussian isotropic N-dimensional random variable, ξ_1^- is a non-Gaussian 1-dimensional random variable (both ξ_N^* and ξ_1^- depend on a) and $\lim_{n\to\infty} M_n(\beta) = M(\beta) > 0$. Moreover $\chi(\beta) \sim C_- |\beta-\beta_c|^{-\gamma}$ as $\beta \to \beta_c-0$, $M(\beta) \sim C\, |\beta-\beta_c|^{\omega}$ as $\beta \to \beta_c+0$, $\gamma = \dfrac{a-1}{\log_2\lambda_1}$, $\omega = \dfrac{1-a/2}{\log_2\lambda_1}$, where $\lambda_1>1$ is an eigenvalue of an integral linear operator (of the linearized renormalization group transformation).

<u>Note</u>. The distribution densities of ξ_N^* and ξ_1^- are analytic fast decreasing functions. Moreover $\xi_1^- = \dfrac{1}{2}\sum_{k=o}^{\infty} \dfrac{\sum_{j=1}^{2^k}(\eta_{kj}-E\eta_{kj})}{2^{k(a-1)}}$ where the random variables η_{kj} are independent and χ^2-distributed with (N-1)

degrees of freedom.

In paper [3] Gawedzki and Kupiainen proved the limit theorem (2) at $\beta=\beta_c$ for any $3/2 < a < 2$ if N is large enough. For any $1 < a < 2$ existence of long-range order at low temperatures (β is large) was proved in the original paper [4] of Dyson. At low temperatures we have in addition to Theorem 4 the following result.

<u>Theorem 4'</u> : $\forall\ 3/2 < a < 2 \quad \exists\ u_o > 0$ and $\beta_1 > \beta_o > 0$ such that $\forall\ \nu\in S_u$, where $0 < u < u_o$, the limit (1) exists if $\beta < \beta_o$, and the limit (3) exists if $\beta > \beta_1$.

In paper [5] Dyson formulated a general principle for the marginal case a=2. Let $H(\sigma) = -\sum_{i\neq j} \frac{(\sigma(i),\sigma(j))}{d^2(i,j)} \varphi(d(i,j))$. Denote $b_n = \varphi(2^n) > 0$. Then the phase transition exists iff the series $\sum_{n=0}^{\infty} \frac{1}{b_n}$ converges and in that case it is of continuous type (i.e. $M(\beta)$ tends to zero when $\beta \to \beta_c+0$). In support of this principle Dyson proved in [5] the existence of long range-order at large β under the condition of the convergence of the series $\sum_{n=0} \frac{1}{b_n}$ in the classical Heisenberg model i.e. when N=3 and $\nu(dx) = (4\pi)^{-1}\ \delta(|x|-1)dx$.

We proved the Dyson principle for $\nu\in S_u$, $0 < u << 1$. Namely the following results hold. We say that the sequence $b_n > 0$ is slowly changing if $\left|\frac{b_{n+1}}{b_n} - 1\right| \leq \frac{1}{2}$ and $\lim_{n\to\infty} \frac{b_{n+1}}{b_n} = 1$.

<u>Theorem 5</u> : If $b_n = \varphi(2^n) > 0$ is a slowly changing sequence and the series $\sum_{n=0} \frac{1}{b_n}$ converges then $u_o > 0$ exists such that $\forall\ \nu\in S_u$, where $0 < u < u_o$, the CLT holds at any $\beta > 0$.

__Theorem 6__ : If $b_n = \varphi(2^n) > 0$ is a slowly changing sequence and the series $\sum_{n=0}^{\infty} \frac{1}{b_n}$ diverges then $u_o > 0$ exists such that $\forall\ \nu \in S_u$, where $0 < u < u_o$, a critical point $\beta_c > 0$ exists with

$$\lim_{n\to\infty} \frac{\sigma(1)+\ldots+\sigma(2^n)}{(2^n)^{1/2}} = \sqrt{\chi(\beta)}\ \xi_N \quad , \quad \text{if} \quad \beta < \beta_c$$

$$\lim_{n\to\infty} \frac{|\sigma(1)+\ldots+\sigma(2^n)|-2^n M_n}{(2^n/(b_n M_n))} = \frac{\bar{\xi}_1}{\sqrt{\beta}} \quad , \text{ if} \quad \beta \geq B_c$$

where $\lim_{n\to\infty} M_n(\beta) = M(\beta) > 0$ if $\beta > \beta_c$;

$\lim_{n\to\infty} M_n(\beta) = 0$ and $\lim_{n\to\infty} M_n/(\frac{1}{2} \sum_{l=n}^{\infty} \frac{1}{b_\ell})^{1/2} = 1$, if $\beta = \beta_c$,

and $\lim_{\beta\to\beta_c+0} M(\beta) = 0$.

Note. Here $\bar{\xi}_1 = \frac{1}{2} \sum_{k=0} \frac{\sum_{j=1}^{2^k} (\eta_{kj} - E\eta_{kj})}{2^k}$.

REFERENCES

[1] Bleher, P.M., Major, P., Renormalization of Dyson's hierarchical vector-valued φ^4-model at low temperatures. Commun. Math. Phys., 95, 487-532 (1984).

[2] Bleher, P.M., Major, P., Critical phenomena and universal exponents in statistical physics, submitted to Annals of Probability.

[3] Gawedzki, K., Kupiainen, A., Non-Gaussian fixed points of the block spin transformation. Hierarchical model approximation.

Commun. Math. Phys., 89, 191-220 (1983).

[4] Dyson, F.J., Existence of a phase transition in a one-dimensional Ising ferromagnet. Commun. Math. Phys., 12, 91-107 (1969).

[5] Dyson, F.J., An Ising ferromagnet with discontinuous long-range order. Commun. Math. Phys., 21, 269-283 (1971).

LOW-TEMPERATURE PHASE TRANSITIONS IN ANNNI MODEL

E.I. Dinaburg, A.E. Mazel

Schmidt Institute of Earth Physics
USSR Academy of Sciences

The first-order phase transitions of classical lattice models defined by a translation-invariant Hamiltonian with a finite interaction radius and with a finite number of periodic ground states and the phase diagrams of such models are described by the Pirogov-Sinaï theory (PS-theory) [1,2,3,]. In the frame of this theory it is assumed that the Hamiltonians of these models satisfy some condition of nondegeneracy, known as the Gerzik-Pirogov-Sinaï condition (GPS-condition) see /2/, /4/. Roughly the condition consists in the fact that the surface energy of phase-separation surface is proportional to the area of this surface with a large constant of the proportionality (~ the inverse temperature). In this theory the ground state means a single configuration of physical variables. The latter fact does not allow to use the theory for models with continuous spin and for some other models in which an ensemble of configurations is naturally meant under ground state. In particular the Potts model, the discrete Widow-Robinson model belong to such models. For such models the success can be achieved by an extension of the PS-theory [5, 6, 7, 8].

It is known that the PS-theory reduces the consideration of a physical model to the study of an appropriate contour model in which the interaction is of a "hard-core" type : any configuration of contours is admissible if their supports do not intersect and the Hamiltonian of the contour model is free and is represented as a sum of single-contour Hamiltonians. The studies of phase transitions in the models considered in [5, 6, 7] is reduced to the investigations of contour models with interaction. The Hamiltonian of such models is the sum of a free Hamiltonian and of a small perturbation term represented as a sum of two-contour Hamiltonian, three-contour Hamiltonian, etc. In all cases it is assumed that the GPS-condition is true for a single-contour Hamiltonian.

However this condition is not true for the ANNNI model and there appeared the necessity of a further extension of the contour models with interaction connected with the replacement of GPS-condition by a condition of a more general type. Such extension is contained in the survey [8] where it is used for the analysis of phase coexistence curves in the ANNNI model.

Now we give some rigorous definitions and statements.

Let $\mathbb{Z}^d$ be a d-dimensional integer-valued lattice in $\mathbb{R}^d$, $d \geqslant 2$, τ be some set which elements γ are called contours, $F : \tau \to R$ be a real-valued function called a contour functional, $P : \tau \to \Omega_c(\mathbb{Z}^d)$ be a map of the set τ into the set $\Omega_c(\mathbb{Z}^d)$ of finite connected subsets of $\mathbb{Z}^d$. The set $P(\gamma)$ is called the support of the contour γ and is noted Supp γ.

We assume also that for any unit shift S of $\mathbb{Z}^d$ the map $S^\rho : \tau \to \tau$ is defined. This map satisfies the following conditions : the functional F is invariant under S^ρ and the diagram

$$\begin{array}{ccc} \tau & \xrightarrow{S^\rho} & \tau \\ P \Big\downarrow & & \Big\downarrow P \\ \Omega_c(\mathbb{Z}^d) & \xrightarrow{S^*} & \Omega_c(\mathbb{Z}^d) \end{array}$$

is commutative ($S^* A = S^{-1} A$ for any $A \in \Omega_c(\mathbb{Z}^d)$). Fix some integer $r_o > 0$. Any given finite collection of contours $\Gamma = \{\gamma_i\}_{i\in I}$ is called the configuration of contours if dist (Supp γ_i, Supp γ_j) $\geq r_o$ for $i \neq j$, $i, j \in I$. Here dist $(x,y) = \max_{1\leq i\leq d} |x^i - y^i|$, $x = (x^1, \ldots, x^d)$, $y = (y^1, \ldots, y^d) \in \mathbb{Z}^d$.

We assume that the contour model is defined by a Hamiltonian H which on every configuration Γ has the value

$$H(\Gamma) = \Sigma_{\gamma\in\Gamma} \; F(\gamma) + G(\Gamma)$$

where $G(\Gamma)$ is the interaction energy of contours of Γ. If $G(\Gamma) \equiv 0$ and for $F(\gamma)$ the GPS-condition is true then the PS-theory can be applied. If $F(\gamma)$ satisfies the GPS-condition and $G(\Gamma)$ possesses some property of clusterness which will be described later then we are in the limits of the contour models considered in [5, 6, 7].

We change the GPS-condition : $F(\gamma) \geqslant C|\text{Supp}|$ with a sufficiently large C, by a weaker condition :

there exist constants $\mu > 0$, $\lambda > 0$ such that

$$\sum_{\gamma : \text{Supp}\, \gamma \ni z} \exp\{-F(\gamma) + \mu\, |\text{Supp}\, \gamma|\} < \lambda$$

for each point $z \subset \mathbb{Z}^d$.

Now we state the property of clusterness of the interaction $G(\Gamma)$. The subconfiguration Γ' of the configuration Γ is called a maximal admissible subconfiguration of Γ if

a) $\Gamma' \subset \Gamma' \Rightarrow \gamma' \in \Gamma$;

b) for any pair $\gamma', \gamma'' \in \Gamma$ it does not exist a contour $\gamma \in \Gamma$ separating this pair ;

c) Γ' is a maximal configuration with respect to a) and b).

In other words Γ' consists of all outer contours of Γ or of all contours contained in the interior of some contour of Γ and being mutually outer. We should remember that the contour of Γ is called outer if it is not enclosed by any other contour of Γ.

We assume that the following conditions are true for $G(\Gamma)$:

1. $G(\Gamma) = \Sigma_{\Gamma' \subset \Gamma}\ G(\Gamma'|\Gamma)$ where the sumation goes over all maximal admissible subconfigurations Γ' of Γ ;

2. $G(\Gamma'|\Gamma) = \Sigma_{A_{\Gamma'}}\ G_{\Gamma}(A_{\Gamma'})$, $A_{\Gamma}' \in \mathcal{A}(\Gamma')$ where $\mathcal{A}(\Gamma')$ is the set of subsets $A_{\Gamma'}$ of the lattice $\mathbb{Z}^d$ such that there exist at least two contours $\gamma', \gamma'' \in \Gamma'$ for which $\text{Supp}\, \Gamma' \cup \text{Supp}\, \Gamma'' \cup A_{\Gamma'}$, is a ω-connected subset of $\mathbb{Z}^d$ (ω is a fixed integer ; $\omega < r_o$) ;

3. $|G_\Gamma(A_{\Gamma'})| < \rho^{|A_{\Gamma'}|}$ with a positive constant ρ ;

4. $G_\Gamma(A_{\Gamma'})$ is invariant under translations of $\mathbb{Z}^d$. The condition 2) means that the interaction is of a Markovian type. As usually we define for any finite volume $V \in \mathbb{Z}^d$ the diluted contour partition function

$$Z(V|F,G) = \Sigma_{\Gamma\subset V} \exp(-H(\Gamma))$$

Here $\Gamma \subset V$ means that supp $\gamma \subset V$ and supp $\gamma \cap \partial V = \emptyset$ for any $\gamma \in \Gamma$. By analogy the crystal contour partition function of the contour γ is by definition

$$Z(\gamma|F,G) = \Sigma_{\Gamma\subset \mathrm{Int}\gamma} \exp(-H(\gamma\cup\Gamma)),$$

where Int γ is the interior of γ and the sumation goes over all configurations $\Gamma \subset$ Int γ which are giving with γ an admissible configuration.

<u>Theorem 1</u>. There exist constants $\lambda_o > 0$, $\rho_o > 0$ such that for all $\lambda < \lambda_o$, $\rho < \rho_o$ there exists a constant μ_o : $\mu_o(\lambda,\rho)$ such that for any $\mu > \mu_o$.

$$\ln Z(V|F,G) = a(F,G)|V| + \Delta(V|F,G),$$

$$\ln Z()(\gamma|F,G) = -F(\gamma) + a(F,G) \mid \mathrm{Int}\ \gamma| + \Delta(\mathcal{A}) (\gamma|F,G)$$

and

$$|\Delta(V|F,G)| < b(F,G), |\partial V|,$$

$$|\Delta^{()}(\gamma|F,G) < b^{(a)}(F,G) . |\partial \mathrm{Int}\gamma|,$$

where a (F,G) is the free energy of the contour model and b(F,G), $b^{(a)}(F,G) \to 0$ when λ, $\rho \to 0$.

Since the GPS-condition is not true this theorem cannot be

directly proved with the standard technique of polymer expansions. The proof of the theorem is based on some modification of Minlos-Sinaï correlation equations[11].

Now we shall consider the ANNNI model. We would remind you that the ANNNI model is defined by the formal Hamiltonian

$$H(\varphi) = -J_0 \Sigma_{(x,y)\in \mathbb{Z}^{(1)}_{hor}} \varphi(x).\varphi(y) - J_1 \Sigma_{(x,y)\in \mathbb{Z}^{(1)}_{ver}} \varphi(x)\varphi(y) + J_2 \Sigma_{(x,y)\in \mathbb{Z}^{(2)}_{ver}} \varphi(y)\,\varphi(x)$$

where $x,y \in \mathbb{Z}^3$, the spin $\varphi(x)$ can take the values ± 1, $\mathbb{Z}^{(1)}_{hor}$ is the set of horizontal bonds of $\mathbb{Z}^3$ having length 1, $\mathbb{Z}^{(1)}_{ver}$ ($\mathbb{Z}^{(2)}_{ver}$) is the set of vertical bonds of $\mathbb{Z}^3$ having length 1(2). This model was introduced more than twenty years ago by Domb [12] and Elliott [13] and was recently attracting much attention in connection with experimental results concerning compounds of rare-earth elements [14]. The deep analysis of the phase diagram of the ANNNI model was performed by Fisher and Selke (see [15]). Using a formal perturbation theory the authors have shown that for low temperatures T there are infinitely many separation-phase lines on the plane (T,δ) starting at the point (0,0) where $\delta = J_1 - 2J_2$. Our analysis [8], [16], [17], gives rigorous results about phase coexistence curves.

First of all we describe the ground states (g,s) of the ANNNI model. The configuration $\varphi(\mathbb{Z}^3)$ is called the layered one (l.c.) and is marked in $\varphi^1(\mathbb{Z}^3)$ if it has a constant value on every horizontal plane $x^3 = \text{const}$ $(x=x^1, x^2, x^3) \in \mathbb{Z}^3)$. In l.c. $\varphi^1(\mathbb{Z}^3)$ the sequence of k neighbouring horizontal planes with the same value of spin equal a bounded above and below by horizontal planes on which the spin has the value $-a$ is called the layer with thickness k. For periodic l.c. we shall use the notation $(k_1{}^{S_1}; k_2{}^{S_2}, \ldots, k_m{}^{S_m})$ introduced in [15]. This notation shows that the period of $\varphi^1(\mathbb{Z}^3)$ considered as the sequence of layers (independent of a sign of spin) contains S_2 layers with thickness k_1, S_2 layers with thickness k_2 and so on. With the aim of unification of notations we note the configurations (3) by $(3,2^\circ)$ and the configurations $\varphi = \text{const}$ by (F) or $(3,2^{-1})$. Besides we denote $(3,2^{\geq n})$, the infinite family of l.c. consisting of the layers with thickness 3, separated one from the other by no less than $(n-2)$ layers with thickness 2.

As proved in [15] the g.s. of the ANNNI model are the l.c. and besides for $\sigma>0$ the g.s. is the l.c. (F) and for $\delta > 0$ the g.s. is the l.c. (2). For $\delta = 0$ any l.c. not containing the layers with thickness 1 is the g.s. It can be shown that for $\delta > 0$ and for $\delta<0$ the PS-theory can be applied when the inverse temperature is sufficiently large. On the contrary if δ is close to zero the PS-theory cannot be used in view of infinite degeneracy. Just this case is interesting for us and we consider the following values of parameters : $\beta > \beta_o$, $|\delta| < e^{-J_o\beta}$, $J_o > 100J_1$.

In the neighbourhood of the point (0,0) in the plane $(1/\beta,\delta)$ the supposed phase diagram has the following structure : the infinite number curves $\delta = \delta_n(\beta)$ ($n = 1, 2, \ldots$) pass through the point (0,0) such that each curve $\delta n(\beta)$ is the co-existence curve for phases corresponding to l.c. $(3,2^{n-2})$, $(3,2^{n-1})$.

The proof of existence of curve $\delta_1(\beta)$ is obtained in (16) and the curves $\delta_n(\beta)$ ($n>1$) in [17] with the help of different extensions of a contour method. The application of the contour model with the interaction described above permits, as we believe, to give a simpler and clearer proof of the existence of the curves $\delta_n(\beta)$. The detailed account is given in survey [8]. Here we give the summary of the main ideas of the proof.

To remove the infinite degeneracy of g.s. we use the following method. For all $n = 0, 1, 2, \ldots$ we define the functions $f_n(\beta) = e^{-2\beta J_0 4n}$ which are called main orders. The functions $O(e^{-2\beta J_0 4n})$ for $\beta \to \infty$ are called n-small. Further we successively construct the phase diagram in the neighbourhood of the point (0,0) with the exactness of n-small functions. It is naturally to name such diagram the phase diagram to the nth order of the perturbation theory. If we replace n by n+1 then the phase diagram is defined more precisely. For $n = 0$ the phase diagram consists of a single curve $\delta'_0(\beta) \equiv 0$ of infinite degeneracy. In first order of perturbation theory it splits into the co-existence curve $\delta_1(\beta)$ of finite number of g.s. and into the curve $\delta'_1(\beta)$ of infinite degeneracy.

Similarly in the nth order of perturbation theory the curve $\delta_{n-1}(\beta)$ splits into the co-existence curve $\delta_n(\beta)$ of finite number of g.s. and into the curve $\delta'_1(\beta)$ of infinite degeneracy and so on. Let us give the exact definitions.

Let C(x) be the unit cube with center at the point $x \in \mathbb{Z}^3$. Further we shall identify any set $A \subset \mathbb{Z}^3$ with $\cup_{x\in A} C(x)$ without reserve.

Therefore, ∂A – the set of boundary faces of A, $\partial A(ver)$ – the set of vertical faces belonging to ∂A, $\partial A(hor)$ – the set of horizontal faces of ∂A are correctly defined. Naturally $\partial A(ext)$ is the external boundary of A ; $\partial A(int)$ is the internal boundary of A ; $\partial A(ext, ver)$ is the external vertical boundary of A ; $\partial A(int, ver)$ is the internal vertical boundary of A ; $\partial A(ext,hor)$ is the external horizontal boundary of A. For arbitrary $A \subset \mathbb{Z}^3$ we denote the union of the vertices of single bonds and double vertical bonds with at least one vertex belonging to A as $R_{1,2}(A)$. The set $A \subset \mathbb{Z}^3$ is called (1,2)-connected if for any a_1, $a_k \in A$ there is a chain $(a_1, a_2), (a_2, a_3), \ldots, (a_{k-1}, a_k) \in A$ consisting of single bonds and double vertical bonds. Let us call elementary excitation (e.e.) a pair $\varepsilon = (t, \varphi(R_{1,2}(t))$ where $t \in \mathbb{Z}^3$ (denoted by $t(\varepsilon)$ or Supp ε) is the (1,2)-connected set called the support of ε and where $\varphi(R_{1,2}(t))$ is the configuration denoted by $\varphi(\varepsilon)$ or φ_ε) for which $\tilde{\varphi}_\varepsilon(x) = \{\varphi_\varepsilon(x)$ with $x \in R_{1,2}(t)\backslash t$; $-\varphi_\varepsilon(x)$ with $x\in t\}$ is the layered configuration in the volume $R_{1,2}(t)$. The arbitrary configuration φ and the e.e. ε are said to be compatible $(\varepsilon \leftrightarrow \varphi)$ if $\varphi(R_{1,2}(t)) = \tilde{\varphi}_\varepsilon(R_{1,2}(t))$. The statistical weight $W(\varepsilon) = \exp\{-\beta E(\varepsilon)\}$ where $E(\varepsilon) = H(\varphi_\varepsilon(R_{1,2}(t))) - H(\tilde{\varphi}_\varepsilon(R_{1,2}(t)))$ is naturally attributed to each e.e.ε. We say that the e.e. ε′ and ε″ are matched if $t'(\varepsilon') \cup t''(\varepsilon'')$ is not the (1,2)-connected subset of $\mathbb{Z}^3$ and if $\varphi_{\varepsilon'}(R_{1,2}(t') \cap R_{1,2}(t'')) = \varphi_{\varepsilon''}(R_{1,2}(t') \cap R_{1,2}(t''))$. The collection $\{\varepsilon_i\}$ of e.e. is matched if all of its e.e. are mutually matched. The notion of congruence of e.e. ($\varepsilon' \cong \varepsilon''$) is defined naturally.

Now we inductively define classes of e.e. finite with respect to congruence : $\varepsilon_0 = \emptyset$; $\varepsilon_1 = \varepsilon_0 \cup \{\varepsilon : |\partial t(ver,ext)| \leqslant 4,\ \tilde{\varphi}_\varepsilon$ has no layers with the thickness $1\}$; $\varepsilon_2 = \varepsilon_0 \cup \varepsilon_1 \cup \{\varepsilon : |\partial t(ver,ext)| \leqslant 8, \tilde{\varphi}_\varepsilon$ consist only of the layers with thickness 2 or 3 ; ... ; $\varepsilon_n = \cup\{\varepsilon : |\partial t(ver,ext-| \leqslant 4n,\ \tilde{\varphi}_\varepsilon$ consists only of the layers with thickness 3 separated by no less than n−2 layers with thickness $2\}$.

Let φ^1 be an arbitrary l.c. and ε −a class of e.e. The set of configurations differing from φ^1 by the arbitrary matched collection of e.e. belonging to ε and compatible with φ^1 is called the e.e. gas of the configuration φ^1 corresponding to the class ε.

The expression

$$\Xi(V|\varphi^1,\varepsilon) = \sum_{\{\varepsilon_i\} \in V,\ \varepsilon_i \in \varepsilon\ ,\ \varepsilon_i \leftrightarrow \varphi^1} \prod_i W(\varepsilon_i)$$

is named the partition function in volume V for the configuration φ^1 on the e.e. gas from class ε.

For $\Xi(V|\varphi^1,\varepsilon_n)$ we use the notation $\Xi_n(V|\varphi^1)$. By definition the free energy of this gas is

$a(\varphi^1,\varepsilon) = \lim_{V\to\infty} \ 1/|V| \ \Xi \ (V|\varphi^1,\varepsilon)$.

In particular $a_n(\varphi^1) = a(\varphi^1,\varepsilon_n) = \lim_{V\to\infty} \ 1/|V| \ \Xi_n(V|\varphi^1)$.

The proof of the existence of the free energy and its computation is based on a standard technique of polymer expansions that gives the following :

$$\ln\Xi_n(V|\varphi^1)= \{\varepsilon_i{}^{\alpha}{}_i\}^c\!: \varepsilon_i \in V\Sigma, \ \ \varepsilon_i \in \varepsilon_n, \ \varepsilon_i \leftrightarrow \varphi^1 \ r^{(\{\varepsilon_i{}^{\alpha}\}^c)} \ \ i \ \ w(\varepsilon_i)^{\alpha_i}$$

where $\{\varepsilon_i{}^{\alpha}{}_i\}^c$ is the connected e.e. collection, i.e. a collection in which each ε_i is included α_i times and any two e.e. from this collection are connected by the chain with mutually non matched e.e. from the above mentioned collection ; $r(\{\varepsilon_i{}^{\alpha}{}_i\}^c)$ is a coefficient defined by mutual position of ε_i and by the numbers α_i (see [9]). It is sufficient for the convergence of series (1) as $V\to\infty$ that $w(\varepsilon)<e^{-k|t(\varepsilon)|}$ with the absolute constant k, and we have that $w(\varepsilon_i)\leqslant \exp\{-2\beta J_o|\partial t(ver)|\}$. Therefore for the existence of $a_n(\varphi^1)$ there appears a limitation on $\beta > \beta_n = kn/8J_o$ since $\max_{\varepsilon\in\varepsilon_n}\{|t|.|\partial t(ver)|^{-1}\}=n/4$. It is the main limitation on β that our method gives. The other limitations deteriorate this estimate only in absolute constant times. Let us also define the truncated gas free energy

$$a_n{}^T(\varphi^1) = \lim_{V\to\infty} 1/|V| \ \Sigma^T$$

where Σ^T is the truncated sum in the right side of (1) in which only terms with $\sum_i \alpha_i . |\partial t_i(\text{ext,ver})| \leqslant 4n$ are left. (We mean that $\varepsilon_i = (t_i, \varphi_i)$. Let $h(\varphi^1) = \lim_{V \to \infty} \beta(H(\varphi^1(V)))/|V|$ be the specific energy of the l.c. φ^1 and $e_n(\varphi^1) = a_n T(\varphi^1) - h(\varphi^1)$. L.c. φ^1 is called g.s. with the exactness n-small functions if $e_n(\varphi^1) = \sup\{e_n(\varphi^1_1)$ where supremum is taken all over the l.c. φ^1_1. Now we can formulate the result on a phase diagram in the nth order of the perturbation theory.

<u>Lemma 1</u>. For any n = 1,2, ... there exist β_n such that at $\beta > \beta_n$ on the plane (β^{-1}, δ) there are curves $\delta = \delta'_n(\beta)$ and $\delta = \delta_n(\beta) = \delta_{n-1}(\beta) + O(f_{n-1}(\beta))$, for which the following statements are valid: in the n-small neighbourhood of the curve $\delta_n(\beta)$ g.s. with the exactness of n-small functions are configurations of the type $(3,2^{n-2})$, $(3,2^{n-1})$ and for the rest of l.c. φ^1

$$e_n(\varphi^1) = e_n((3,2^{n-2})) - \alpha_n(\varphi^1) f_n(\beta)$$

where $\alpha_n(\varphi^1) > \bar{\alpha}_n > 0$ and $\bar{\alpha}_n$ does not depend on β ; in the n-small neighbourhood of the curve $\delta = \delta'_n(\beta)$ g.s. with the exactness of n-small functions are configurations of the type $(3,2^{\geqslant n-1})$, (2) and for the rest of l.c. φ^1

$$e_n(\varphi^1) = e_n((2)) - \alpha'_n(\varphi^1) f_n(\beta)$$

where $\alpha'_n(\varphi^1) > \bar{\alpha}'_n > 0$ and $\bar{\alpha}'_n$ does not depend on β.

Now we are to explain how are done the model "factorization" with respect to e.e. gas and the reduction to the contour models with interaction and with n-small contours statistical weight .

We fix n and study the neighbourhood of the curve

$\delta = \delta_n(\beta)$.

Let $P_n = 2(2n = 1-(2n-1)$ be the common period of the type $(3,2^{n-2})$, $(3,2^{n-1})$ configurations, $\mathbb{Z}^3{}_{P_n} = P_n \mathbb{Z}^3$ be a sublattice of $\mathbb{Z}^3$,C_{P_n} be an arbitrary cube of $\mathbb{Z}^3$ being an elementary cube of $\mathbb{Z}^3{}_{P_n}$. $C_{P_n}(x)$ be such a cube containing $x \in \mathbb{Z}^3$. Further we consider only volumes V, made of C_{P_n} cubes.

Let us take an arbitrary configuration $\varphi(\mathbb{Z}^3)$ coinciding out of a finite V with one of the configurations of the set $Ph_n = \{(3,2^{n-2}),(3,2^{n-1})\}$. The restriction of this configuration to V is called an ideal configuration if it does not contain e.e. from ε_n. For each $\varphi(V)$ one can construct a unique ideal configuration $\varphi^{id}(V)$ so that $\varphi(V)$ is produced from $\varphi^{id}(V)$ by adding certain e.e. belonging to ε_n.

If $\varphi^{id}(C_{P_n})(x))$ coincides with $\mathscr{R}(C_{P_n}(x))$, then the configuration $\varphi(V)$ at the point $x \in V$ is by definition in the phase $\mathscr{R} \in Ph_n$. The set $B(\varphi)$ of points x where $\varphi(V)$ is not in any of the phases $\mathscr{R} \in Ph_n$ is called the configuration boundary. Suppose $\bar{B}(\varphi) = \{x \in \mathbb{Z}^3 | \operatorname{dist}(B(\varphi), x) \leqslant P_n\}$. Connected components $\bar{B}_i(\varphi)$ of the set $\bar{B}(\varphi)$ are called precontours of the configuration φ. Let $\partial\bar{B}_i(\varphi)$ be the boundary of the $\bar{B}_i(\varphi)$. In the band with width P_n belonging to $\bar{B}_i(\varphi)$ and adjacent to $\partial\bar{B}_i(\varphi)$ the configuration $\varphi^{id}(B_i(\varphi))$ coincides with one of the phases $\mathscr{R} \in Ph_n$. Hence there is a uniquely defined function ph on $\partial\bar{B}_i(\varphi)$ with values in Ph_n equal on each connected component of $\partial\overline{B}_i(\varphi)$ to the value of the phase $\mathcal{R}$, phase in which there are points adjacent to $\partial\overline{B}_i(\varphi)$.

The contours of the configuration φ is a triple $\gamma = (b, ph, \{\varepsilon_i\})$ where $b = \overline{B}_i(\varphi)$ is a precontour of φ ; ph is the function defined above with values in Ph_n ; $\{\varepsilon_i\}$ is the φ e.e. collection belonging to $\overline{B}_i(\varphi)$ with supports of ε_i belonging to preboundary band $\overline{B}_i(\varphi)\cap(\overline{B}(\varphi)\setminus B(\varphi))$.

For the set b called support of γ (Suppγ) we naturally define : Ext γ is the exterior of the contour ; Int γ is the interior of the contour spliting into connected sets O_i ; the boundary ∂b of the set b is equal to $\partial b(ext)\cup(\cup\partial b^{(i)}(int))$ where ∂b (ext) is the external boundary of the contour and $\partial b^{(i)}(int)$ is a connected component of the internal boundary of the contour. Suppose $\mathcal{R}_i = ph(\mathcal{R}^{(i)}(int))$, $\overline{\mathcal{R}} = ph(\partial b(ext))$ and let us define the contour statistical weight $w(\gamma)$ in the following way :

$$w(\gamma) = \lim_{V\to\infty} \frac{\Xi_V(\gamma)}{\Xi_n(V|\overline{\mathcal{R}}) \exp\{-\beta H(\overline{\mathcal{R}}_V)\}}$$

$$_{i}\ \frac{\exp\{a_n(\overline{\mathcal{R}}) - h(\overline{\mathcal{R}}))\,|O_i|\}}{\exp\{a_n(\mathcal{R}_i) - h(\mathcal{R}_i))|O_i|\}},$$

where $\Xi_V(\gamma)$ is a partition function with the boundary condition $\overline{\mathcal{R}}$ on the ensemble of configurations $\varphi(V)$ for which $\overline{B}(\varphi) = \gamma$.

The following estimates can be obtained for $w(\gamma)$.

Theorem 2. Let $\tilde{\gamma} = (b,ph)$ be the set of all contours with fixed support and function ph ; $W_\Sigma(\tilde{\gamma}) = \Sigma_{\gamma\in\tilde{\gamma}} w(\gamma)$. Then there exists $\beta^*_n > \beta_n$ such that in the region of parameter values : $\beta > \beta^*_n$, $\delta\epsilon\delta_n(\beta) + O(f_n(\beta)$ for any $\tilde{\gamma}$

$$-\ln W_\Sigma(\tilde{\gamma}) = \Sigma_i \Psi_1(\partial b^{(i)}(ver)) + \Psi_2(b)$$

where $\partial d^{(i)}$ (ver) is an arbitrary connected component of ∂b(ver), summation goes over all these components, Ψ_1, Ψ_2 are functionals satisfying the estimates

$\Psi_1(\partial b^{(i)}(ver)) \geqslant \max\{\beta K(n)|\partial b^{(i)}(ver)|,\ 2\beta(4J_0 n + 0{,}1J_1)\}$,

$\Psi_2(b) \geqslant d(n)\ |\ b\ |\ e^{-2\beta 4J_0 n}$,

k(n) and d(n) being positive constants. Similar estimates are valid as well for $w(\gamma)$.

Let us consider the matched collection $\Gamma'=(\gamma_1,\gamma_2, \ldots, \gamma_m)$ of mutually outer contours. Let

$$\bar{G}^{\bar{\mathcal{H}}}(\Gamma') = \lim_{V\to\infty} \Xi^1_V(\Gamma')\, \Xi_n(V|\bar{\mathcal{H}})^{-1}\, e^{\beta H(\bar{\mathcal{H}}_V)} \times \tag{2}$$

$$\times \Big[\prod_{j=1}^{m} \frac{\Xi_V{}^1(\gamma_j)}{\Xi_n(V|\bar{\mathcal{H}}) e^{-\beta H(\mathcal{H}_V)}} \Big]^{-1}$$

where $\Xi^1_V(\Gamma')$ ($\Xi^1_V(\gamma_j)$) is the partition function with boundary condition $\bar{\mathcal{H}}$ over configurations $\gamma(V)$ for which the collection of external contours of the set $\bar{B}(\gamma)$ coincides with the collection Γ' (with the contour γ_j).

One can show that the limit (2) exists for a sufficiently large β. For each $\mathscr{H}\in Ph_n$ let us consider the set $P^{\mathscr{H}}$ consisting of all contours $\gamma^{\bar{\mathscr{H}}} = (b, ph, \{\varepsilon_1\})$ for which $\bar{\mathscr{H}}(\gamma) = \mathscr{H}$. For $\mathscr{H} \in Ph_n$ let us define the contour model with the contour functional $F^{\mathscr{H}}(\gamma)$ and with the interaction $G^{\mathscr{H}}(\Gamma)$ described as follows. For any collection Γ suppose $G^{\mathscr{H}}(\Gamma) = \sum_{\Gamma'} G^{\mathscr{H}}(\Gamma'|\Gamma)$ where $G^{\mathscr{H}}(\Gamma'|\Gamma) = \bar{G}^{\mathscr{H}}(\Gamma') = \bar{G}^{\mathscr{H}}(\Gamma'|\beta,\delta)$.

We are looking for the contour functional $F^{\mathscr{H}}(\gamma)$ proceeding from the following relations :

1) $\quad -h(\mathscr{H}) + a_n(\mathscr{H}) + a(F^{\mathscr{H}}, G^{\mathscr{H}})$

does not depend on $\mathscr{H}\in Ph_n$;

2) $$\lim_{V\to\infty} \frac{\Xi^1_V(\gamma)}{\Xi_n(V|\mathscr{H})\,.\,e^{-\beta H(\mathscr{H}_V)}} \equiv Z^{(a)}\ (\gamma \mid F^{\mathscr{H}}, G^{\mathscr{H}})$$

for any $\gamma\in P^{\mathscr{H}}$. For β and δ belonging to n-small neighbourhood of the curve $\delta = \delta_n(\beta)$ one can obtain contour functionals $F^{\mathscr{H}}$ such that $F^{\mathscr{H}}$ and $G^{\mathscr{H}}$ give contour models with the interaction of the type described before. From the condition 1) and with $\beta > \beta^*_n$ one can find the curve $\delta = \delta^*_n(\beta)$.

<u>Main theorem</u>. Suppose that $J_o > 100\, J_1$. For any n = 1,2, ... there is inverse temperature β^*_n such that for $\beta > \beta^*_n$ in the region $\delta = \delta_n(\beta) + o(f_n(\beta))$ there exists the curve $\delta^*_n(\beta)$ on which there are 8n different limit Gibbs distribution corresponding to

the boundary conditions $(3,2^{n-2})$ or $(3,2^{n-1})$.

One technique can be applied to the antiferromagnet Ising Model on the face centered cubic lattice L with nearest neighbour interaction as well. This model is described by the formal Hamiltonian

$$H = J \sum_{(x,\, y) \in L} \varphi(x)\varphi(y) - h \sum_{x \in L} \varphi(x)$$

where $J > 0$ is a coupling constant ; h is an external magnetic field; the spin $\varphi(x)$ can take the value ± 1 ; (x,y) is the nearest neighbour pair. As demonstrated in (18) the whole region of values of the parameter h is divided into three domains :

1) $S_1 = \{ h < -12J \ , \ h > 12J \}$ where the g.s. is unique ;

2) $S_2 \neq \{h \neq \pm 4J, -12J < h < 12J\}$, where the number of g.s. is infinite but the specific entropy per spin is equal to zero ;

3) $S_3 = \{\pm 4J, \pm 12J \}$ are the so called superdegenerated points where this entropy is different from zero.

Our technique can be applied to the investigation of the domain S_2. Here the situation is even simpler than in the ANNNI model because under suitable definition of e.e. class the infinite degeneracy disappears at once all over the domain S_2 and no sequence of curves arises on the phase diagram as in case of the ANNNI model. On the contrary in the whole neighbourhood of the superdegenerated points the situation is much more complex and at present is not found out [3, 18].

Obviously due to symmetry it is sufficient to consider

$h > 0$. The main parameters for us is $\delta = \{4J - h$ when $0 \leqslant h \leqslant 4J$; $h - 4J$ when $4J \leqslant h \leqslant 8J$; $12J - h$ when $8J \leqslant h < 12J$ (evidently $0 < \delta < 4J$). The function $f_o(\beta) = \exp\{-\min(8\delta\beta, 16J\beta)\}$ will be called the main order, the necessary e.e. class including all e.e. (i.e. spin flips in the connected set) with the statistical weight $w(\varepsilon) \geqslant \exp\{-1.01 \min(8\delta\beta, 16J\beta)\}$. Among all g.s. in the domain S_2 this class will point out 6 configurations in the region $0 \leqslant h < 4J$ (we shall mark them φ^{A2B_2}) and 4 configurations in the region $4J<h<12J$ (we shall mark them $\varphi^{A_3,B}$). Configurations $\varphi^{A_2B_2}$ and φ^{A_3B} are defined as follows.

Face centered cubic lattice contains just 4 cubic sublattices. Configurations of the type $\varphi^{A_2B_2}$ take the value +1 on those two sublattices and take the value -1 on the two sublattices left. For the configurations of the type φ^{A_3B} only one of this sublattices has the sign "−", the others having the sign "+". Also for this model similarity with the ANNNI model lies in the fact that corresponding contour models with interaction are of the same type.

<u>Theorem 3</u>. There is $c > 0$ such that in the region of parameter values : $J > 0$, $0 \leqslant h < 4J$, $\beta > c(4J - h)^{-1}$ there exist limit Gibbs distributions for each of the 6 configurations of type $\varphi^{A_2B_2}$; in the region of parameter values $J > 0$, $4J < h < 12J$, $\beta \geqslant c(h-4J)^{-1}$, $\beta \geqslant c(12J - h)^{-1}$ there exist limit Gibbs distributions for each of the 4 configurations of type φ^{A_3B}.

REFERENCES

1. Pirogov S.A., Sinai Ya.G. Teoret.Mat.Fiz. 25 (1975),26(1976)

2. Sinai Ya. G. : Theory of phase transitions. Budapest : Acad. Kiado 1982.

3. J. Slawny. Preprint. Blacksburg, Virginia (1985)

4. Dobrushin R.L., Shlosman S.B. Problemy ustoichivosti osnovnyh sostoyanii. Preprint. Institut Probl em Peredachi Informacii AN SSSR (1984).

5. Bricmont J., Kuroda K., Lebowitz J.J. Stat. Phys. 33(1983)

6. Dinaburg E.I., Sinai Ya. G. Trudy Konferencii po Mat. Fiz. Dubna (1984).

7. Bricmont J., Kuroda K., Lebowitz J. Preprint. (1985)

8. Dinaburg E.I., Mazel A.E., Sinai Ya. G. SSR Math. Phys. 6(1986)

9. Seiler E., Gauge Theories as a Problem in Constructive Quantum Field Theory and Statistical Mechanics. Lect. Notes in Phys., 159, Springer-Verlag, Berlin-Heidelberg-New York (1982)

10. Malyshev V.A. Minlos R.A. Sluchainye Gibbsovskie Polya. M., Nauka (1985).

11. Minlos R.A., Sinai Ya. G. Trudy Mosc. Mat. Obsch. 17(1967), 19 (1968).

12. Domb C., Adv. Phys. 9, N° 34 (1960).

13. Elliott R.J., Phys. Rev. 124 (1961).

14. Bak P. , von Boehm J., Phys. Rev. B 21,5297 (1980)

15. Fisher M.E., Selke W., Phil.Trans.Royal. Soc., 302 (1981)

16. Dinaburg E.I., Sinai Ya. G., Comm. Math. Phys., 98 (1985)

17. Mazel A.E. Teoret. Mat. Fiz., 62 (1986)

18. Lebowitz J.L., Phani M.K. , Steyer D.F., J. Stat. Phys., 38 (1985)

NONANALYTIC FEATURES OF THE FIRST ORDER PHASE TRANSITION IN CLASSICAL LATTICE GAS MODELS

S.N. ISAKOV

Urals Polytechnical Inst. Sverdlovska - USSR

Numerous examinations show that the lattice gas model possesses all main features of the classical continuous gas and at the same time is more convenient for the theoretical analysis of the phase transitions [1-4]. This correspondence becomes better when the ratio of the interaction radius to the lattice period grows, although nearest-neighbour interactions are already good enough to study phase transitions.

This work is the continuation of [5-6] and concerns the first order phase transitions study. It is proved that the pressure is an analytic function up to the phase boundary for the lattice gas model with finite pair interactions. Besides, the existence of the limits of all derivatives with respect to the chemical potential μ on the phase boundary and their asymptotic behaviour $(k!)^{d/(d-1)}$ for a large number of derivatives k (d is the dimension, $d \geqslant 2$) is proved. Thus, the first order phase transition point is an essential singularity. The relation of this result with the metastable state theory is discussed in § 4. The special properties of the two-dimensional Ising model with nearest-neighbour interaction make it possible to improve the estimates on the k-th derivative and yield the critical index equal to 15/8 for the width of the metastable region near the critical point.

1. Contour representation of the model

We consider the d-dimensional cubic lattice consisting of unit hypercubes. Two states "+" and "−" are possible for each cube corresponding to the presence or the absence of a particle in the cube. The energy of a state ν in the volume Λ (ν is a fixed set of signs in Λ) is equal to

$$E_\nu = \sum_{1 \leqslant i < j \leqslant N_\nu} u(r_i - r_j) - \mu N_\nu, \tag{1}$$

where subscripts i,j enumerate the particles, r_i, r_j are d integer numbers, N_ν is the number of plus, $u(r) = u(-r)$, $u(r) = 0$ if $|r| > r_R$.

The particles are φ-ordered in a certain state if the signs of the state have d periods along the vectors $r_1^\varphi, \ldots, r_d^\varphi$. The elementary volume $|\varphi|$ is equal to the determinant $|\varphi| = |r_1^\varphi \ldots r_d^\varphi| > 0$. We suppose that the vectors $r_1^\varphi, \ldots, r_2^\varphi$ are defined so that $|\varphi|$ takes on the minimal possible value. Let $|\varphi|^+$ be the number of particles in the elementary volume ($0 \leqslant |\varphi|^+ < |\varphi|$). $|\varphi|$ modifications $\varphi_1, \varphi_2, \ldots$ of the φ-order exist so that φ_i coincide with φ_1 after a certain lattice translation. We also introduce the ψ-order ($0 < |\psi|^+ \leqslant |\psi|$, $\kappa = |\psi|^+/|\psi| - |\varphi|^+/|\varphi| > 0$) with the aim to examine the $\varphi \to \psi$ transition. An index γ will be used to enumerate all possible modifications ($1 \leqslant \gamma \leqslant |\varphi| + |\psi|$).

We define a one-to-one correspondence between the unit cubes $e(r_i)$ and cubes Q_i of size $2q+1$($q = 1,2,3,\ldots$) with $e(r_i)$ as central cube. Here q is the minimal possible number with the conditions : 1) $u(r_i - r_j) = 0$ if $e(r_j) \notin Q_i$, 2) each of the $|\varphi| + |\psi|$ modifications has different sets of signs in the volume $q \times (2q+1)^{d-1}$, when an arbitrary axis is chosen to reduce the cube Q_i.

Let ν^γ denote a state on the infinite lattice which differs from the γ-order in a finite region. This state determines the set $\{e(r_i)\}_\nu$ so that each Q_i does not belong to any of $|\varphi| + |\psi|$ modifications. We define the set of signs for $\{e(r_i)\}_\nu$ as a contour representation Ω^γ_ν of the state ν^γ. It is possible that $\{e(r_i)\}_\gamma$ divides the lattice into an exterior part and several finite volumes $\Lambda^{\gamma_1}_1$, $\Lambda^{\gamma_2}_2$, ... with γ_1, γ_2, ... modifications. The set of signs for $\{e(r_i)\}_\nu$ determines the modifications γ_1, γ_2, ... in $\Lambda^{\gamma_1}_1$, $\Lambda^{\gamma_2}_2$, ... (see condition 2 for Q_i), so Ω^γ_ν determines the state ν^γ.

The representation Ω^γ_ν can in general be divided into several sets $\Omega^\gamma_\nu = \Omega^{\gamma_1}_{\nu_1} \cup \Omega^{\gamma_2}_{\nu_2} \cup$ ... such that $\Omega^{\gamma_k}_{\nu_k}$ are representations of certain states $\nu_k{}^{\gamma_k}$. It is possible that $\gamma_k \neq \gamma$ because one representation can be inside another. A contour Γ^γ is defined as a representation Ω^γ which cannot be divided any more. The subscript α will be used to enumerate contours. A contour Γ^γ_α contains the volume $\Lambda_\alpha = \Lambda^{\gamma_1}_{\alpha_1} \cup \Lambda^{\gamma_2}_{\alpha_2} \cup$ ... (particularly, $\Lambda_\alpha = \varnothing$). The distance between $\Lambda^{\gamma_k}_{\alpha_k}$ and $\Lambda^{\gamma_m}_{\alpha_m}$ ($\gamma_k \neq \gamma_m$) is not less than $q + 1$ because of the condition 2. Let $\Lambda^{\varphi_1}_\alpha$ be the part of Λ_α with φ_1-order modification. We have $\Lambda_\alpha = \Lambda^\varphi_\alpha \cup \Lambda^\psi_\alpha$, $\Lambda^\varphi_\alpha = \Lambda^{\varphi 1}_\alpha \cup \Lambda^{\varphi 2}_\alpha \cup$... The number of cubes in the contour $\Gamma^\gamma{}_\alpha$ (in the set $\{e(r_j)\}_\alpha$) will be denoted as $|\Gamma^\gamma_\alpha|$ and the number of cubes in Λ_α will be denoted as $|\Lambda_\alpha|$.

Several contours $\Gamma^{\gamma_1}_1, \Gamma^{\gamma_2}_2, \ldots$ represent a conformed set, if no pair has common cubes and each contour $\Gamma^{\gamma_\alpha}_\alpha$ is placed in the volume with the γ_α-order modification. A one-to-one correspondence exist between contour representations Ω^γ_ν and conformed sets of contours $\{\Gamma^{\gamma_\alpha}_\alpha\}_\nu$.

We shall examine the partition functions only for volumes Λ_α which are created by some contour Γ^γ_α. It means that we have γ_κ boundary conditions in every connected part Λ^{γ_κ} of the volume Λ_α. Only those cubes $e(r_j) \in \Lambda_\alpha$, for which the whole cube Q_j belong to Λ_α will be taken in account. Then a one-to-one correspondence exists between all possible states (sets of signs for $\{e(r_j)\}_\alpha$) and conformed sets of contours in Λ_α.

The main restriction for the potential u(r) is as follows. The energy of each state ν^γ, for a certain value $\mu = \mu_0$, represented by the only contour Γ^γ_α(compared with the energy of the pure γ-order) satisfies $E_\alpha \geqslant |\Gamma^\gamma_\alpha|$ (the Peierls condition). Thus, the φ-order is a $|\varphi|$-times degenerated ground state for $\mu_0 - |\varphi|^+/|\varphi| < \mu < \mu_0$ and the ψ-order is the ground state for $\mu_0 < \mu < \mu_0 + |\psi|^+/|\psi|$. The addition of energy $U_\alpha(\mu)$ for the only contour Γ^φ_α and arbitrary μ is equal to

$$U_\alpha = E_\alpha - \kappa\,(\mu - \mu_0) V^\psi_\alpha, \qquad (2)$$

where $V^\psi{}_\alpha = \Delta N_\alpha/\kappa$, ΔN_α – the additional number of plus for the state ν^φ_α in comparison with the pure φ-order.

The constant V_α^ψ is of order $|\Lambda_\alpha^+|$:

$$|V_\alpha^\psi - |\Lambda_\alpha^\psi|| < |\Gamma_\alpha^\varphi|/\kappa \tag{3}$$

The permutation $\varphi \leftrightarrows \psi$ changes the sign in front of κ in (2).
The additional energy U_ν^γ for an arbitrary state ν^γ is equal to the sum $\sum U_\alpha^\gamma \alpha$ over the contours of ν^γ (see condition 1 for Q_i). We admit here that one contour may contain an another one.

The partition function for a fixed volume Λ_α^γ is defined as $Z(\Lambda_0^\gamma) = \sum \exp(-\beta E_\nu^\gamma)$ with summation over all sets of signs for cubes taking part in the statistics. Thus, $Z(\Lambda_0^\gamma)$ has a normalization such that the weight of the nonperturbed state is equal to 1.

We have $Z(\Lambda_0) = \prod_\gamma Z(\Lambda_0^\gamma)$ as it follows from the condition 1 for Q_i and

$$Z(\Lambda_0^\gamma) = 1 + \sum \exp(-\beta U_\alpha), \tag{4}$$

where the summation is taken over all not empty conformed sets of contours in Λ_0^γ. In fact we need a somewhat different formulation for $Z(\Lambda_0^\gamma)$:

$$Z(\Lambda_0^\gamma) = 1 + \sum \prod_\alpha W_\alpha^\gamma \tag{5}$$

with summation over all sets of non-connected contours (when no pair has a common cube $e(r_i)$) of one and the same modification γ. The weights W_α^γ are the following :

$$W^{\gamma}_{\alpha} = \exp(-\beta E_{\alpha} \pm h\, V^{\gamma}_{\alpha})\, Z\,(\Lambda_{\alpha}) /\, Z^{\gamma}(\Lambda_{\alpha}), \tag{6}$$

where $h = \beta\, \kappa\, (\mu - \mu_0)$, "+" corresponds to $\gamma = \varphi_k$ and on the contrary, $Z^{\gamma}(\Lambda_{\alpha})$ is the partition function for the γ boundary conditions in the volume Λ_{α} in all its parts. Thus, the representation (5), (6) is the cluster expansion of the partition function $Z(\Lambda^{\gamma}_0)$.

We shall write some estimates which may refer both to the restricted partition functions containing not all possible contours and to the complete partition functions $Z(\Lambda_0)$. Such restricted partition functions will be denoted as $Z^R(\Lambda_0)$. The partition function $Z\,(\Lambda^{\gamma}_0 - e(r_0))$ contains all the contours except $\Gamma^{\gamma}_{\alpha} : e(r_0) \in \Gamma^{\gamma}_{\alpha}$, $e(r_0)$ a fixed cube in Λ_0. Thus, $Z(\Lambda^{\gamma}_0 - e(r_0))$ is an example of $Z^R(\Lambda^{\gamma}_0)$. Another example of restricted partition function is $Z_v(\Lambda^{\gamma}_0)$. It does not contain large contours Γ^{γ}_{α} which satisfy $|\Lambda_{\alpha}| > v$ ($v=1,2,3,\ldots$). The partition functions $Z(\Lambda^{\gamma}_0 - e(r_0))$, $Z_v(\Lambda^{\gamma}_0)$ may be also somehow restricted and so we consider $Z^R(\Lambda^{\gamma}_0 - e(r_0))$, $Z^R_v(\Lambda^{\gamma}_0)$. Notice that $Z_v(\Lambda_0) = Z(\Lambda_0)$ if $v > |\Lambda_0| - 5^d$.

We define a diagram D^{γ} as a connected set of contours $\Gamma^{\gamma}_1, \Gamma^{\gamma}_2, \ldots$ taken $K_1, K_2, \ldots$ times. A subscript δ will be used to enumerate diagrams. The order of D^{γ}_{δ} is $|D^{\gamma}_{\delta}| = K_1|\Gamma^{\gamma}_1| + K_2|\Gamma^{\gamma}_2| + \ldots$ Let Z^{γ}_{δ} be the partition function (5) based on the contours $\Gamma^{\gamma}_1, \Gamma^{\gamma}_2, \ldots$ We derive the power expansion for $\ln Z^{\gamma}_{\delta}$ when $W^{\gamma}_1, W^{\gamma}_2, \ldots$ are taken as independent variables. Then the weight W^{γ}_{δ} of the diagram D^{γ}_{δ} is equal to the term corresponding to the set $D^{\gamma}_{\delta}: W^{\gamma}_{\delta} = \mu_{\delta} (W^{\gamma}_1)^{k_1} (W^{\gamma}_2)^{k_2} \ldots$, where μ_{δ} depends only on the graph of the contour bonds in D^{γ}_{δ}. This definition leads to the formula

$$\ln Z^R(\Lambda_o^\gamma) = \sum_\delta W_\delta^\gamma, \qquad (7)$$

where the summation is taken over all possible diagrams based on the contours of $Z^R(\Lambda_o^\gamma)$. We suppose here that $|W_\alpha^\gamma|$ are small enough such that the series (7) converges.

The number of contours with the fixed order $|\Gamma_\alpha^\gamma| = m$ and a cube $e(r_o) \in \Gamma_\alpha^\gamma$ is bounded by $C^m{}_1$ ($C_1 = C_1(d) > 2$). The number of diagrams translationally invariant along the vectors $r_1^\gamma, \dots r_d^\gamma$ is also bounded by C_2^m ($C_2 = C_2(d) > 2$).

2. Regions of analyticity and singularity

<u>Lemma 1</u>. Let the complex weights W_α^γ of the contours Γ_α^γ contained in $Z^R(\Lambda_o^\gamma)$ satisfy the condition

$$|W_\alpha^\gamma| < (t/C_1)^{|\Gamma_\alpha^\gamma|} \qquad (8)$$

($o < t < (3C_2)^{-1}$). Then, the series (7) converges exponentially and one has the following estimates :

$$|\ln (Z^R(\Lambda_o^\gamma) / Z^R(\Lambda_o^\gamma - e))| < t, \qquad (9)$$

$$|\ln Z^R(\Lambda_o^\gamma)| < t\, |\Lambda_o^\gamma|, \qquad (10)$$

$$|W^\gamma{}_\delta| < |D^\gamma{}_\delta|(3t)^{|D^\gamma{}_\delta|} \qquad (11)$$

for an arbitrary restricted partition function $Z^R(\Lambda_o^\gamma)$ and a cube $e \in \Lambda_o^\gamma$.

The inequality (9) is proved with the help of a mathematical induction on $e(r_i) \in \Lambda^\gamma_o$ applied to

$$Z^R(\Lambda_o^\gamma) = Z^R(\Lambda_o^\gamma - e) + \sum W_\alpha^\gamma Z^R(\Lambda_o^\gamma - \Gamma_\alpha^\gamma)\,, \tag{12}$$

where the summation is taken over all contours Γ_α^γ containing e. The estimate (10) is the corollary of (9), and the Cauchy formula for $\ln Z_\delta^\gamma$ together with (10) gives (11).

<u>Lemma 2</u>. Let the conditions of lemma 1 be valid and let the weights W_α^γ be analytic functions of h in G satisfying

$$\left|\frac{d}{dh}\ln W_\alpha^\gamma\right| < 2^{|\Gamma_\alpha^\gamma|}. \tag{13}$$

Then,
$$\left|\frac{d}{dh}\ln(Z^R(\Lambda_o^\gamma)/Z^R(\Lambda_o^\gamma - e))\right| < t. \quad (0 < t < (9C_2)^{-1}). \tag{14}$$

The proof is realized with the help of equality (12) and its derivative.

We define the functions

$$p_\nu^\gamma = \frac{1}{|\gamma|}\sum W_\delta^\gamma \quad (\gamma = \varphi, \psi) \tag{15}$$

with the summation over all translationally invariant diagrams D_δ^γ on the infinite lattice with restricted contours Γ_α^γ : $|\Lambda_\alpha| \leqslant \nu$. The functions p_∞^γ contain diagrams with all possible contours.

Lemma 3. Let the conditions of lemma 1 remain valid for $0 < t < (9C_2)^{-1}$ and $Z^R(\Lambda_0^\gamma)$ be $Z_\nu(\Lambda_0^\gamma)$. Then, the series (15) converges exponentially and

$$|\ln Z_\nu(\Lambda_0^\gamma) - |\Lambda_0^\gamma| p_\nu^\gamma| < t\,|\Gamma_0|. \tag{16}$$

The existence of the thermodynamic limit $p_\nu^\gamma = \lim |\Lambda_0^\gamma|^{-1} \ln Z_\nu(\Lambda_0^\gamma)$ when $|\Lambda_0^\gamma| / |\Gamma_0| \to \infty$ follows from (16). In particular,

$$|\ln Z(\Lambda_0^\gamma) - |\Lambda_0^\gamma| p_\infty^\gamma| < t\,|\Gamma_0| \tag{17}$$

hold true for not restricted partition function $Z(\Lambda_0^\gamma)$.

In this case the limit $p_\infty^\gamma = \lim |\Lambda_0^\gamma|^{-1} \ln Z(\Lambda_0^\gamma)$ exists and is equal to the pressure $p = p_\infty^\gamma + p_c^\gamma$ with the term $p_c^\gamma = -\beta E^\gamma$, where E^γ is the energy the nonperturbed state γ for one cube ($E^\varphi(\mu_0) = E^\psi(\mu_0)$, $E^\gamma(\mu) = E^\gamma(\mu_0) - (\mu-\mu_0)|\gamma|^+/|\gamma|$, $p_c^\psi - p_c^\varphi = h$).

We consider β, h as independent complex thermodynamic variables $\beta = \beta' + i\beta''$, $h = h' + ih''$ and suppose that β' is sufficiently large : $\beta' > \beta_0$, i.e. $W_\alpha^\gamma = W_\alpha^\gamma(\beta,h)$, $p_\nu^\gamma = p_\nu^\gamma(\beta,h)$.

The definition of $\tau^\varphi(\nu)(\nu = 1,2,3,\ldots)$ is $\tau^\varphi(\nu) = \max_\alpha V_\alpha^\psi / E_\alpha$, where the maximum is taken over all restricted contours Γ_α^φ : $|\Lambda_\alpha^\psi| \leqslant \nu$. The functions $\tau^\varphi(\nu)$, $\tau^\psi(\nu)$ are positive and do not decrease.

For large ν the following asymptotics behavior is evident :

$$\tau_{-} < \tau^{\varphi,\psi}(\nu)/\nu^{1/d} < \tau_{+}\,, \tag{18}$$

where τ_{-}, τ_{+} are positive constants. The function $\tau(\nu)$: $\tau(1)=\max(\tau^{\varphi}(1),\tau^{\psi}(1))$, $\tau(\nu) = \max\,[\tau^{\varphi}(\nu),\ \tau^{\psi}(\nu),\ \tau(\nu-1)(\nu/\nu-1)^{1/2d}]$ (ν = 2,3,...) increases and has the same asymptotics behavior (18).

Theorem 1. The infinite succession of regions $G_1 \supset G_2 \supset$.. exist on the complex plane h with the following properties. The estimates (8)-(11), (13), (14), (16) are true in G_ν for $t=C_1\exp(1-\sqrt{\beta'})$ and arbitrary Λ_0^γ, $e(r_0)$ after the substitution $Z^R \to Z_\nu^R$ in (9), (10), (14). Besides, $p_\nu^\varphi(h)$, $p_\nu^\psi(h)$ are analytic functions in G_ν with bounds $|p_\nu^\gamma| < t$, $|\frac{d}{dh}\,p_\nu^\gamma| < t$.

The equation

$$\mathrm{Re}\,(p_\nu^\varphi - p_\nu^\psi) = h' \tag{19}$$

has the unique solution $h' = h'_\nu(h")$ for arbitrary fixed $h"(-\infty < h" < \infty)$ in G_ν and

$$|h'_y(h") - h'_\nu(h")| < 2t^{\sqrt{\nu}} \tag{20}$$

for all $y : y > \nu$. The region G_ν is the union of circles $|h-h'_{\nu-1}(h")| \leqslant r_\nu$, $r_\nu = (\beta' - \sqrt{\beta'})/\tau(\nu)$ with centers on the infinite line $h' = h'_{\nu-1}(h")$ $(h'_0(h") \equiv 0)$.

The proof is realized by mathematical induction on ν. Inequality (13) follows from (6) and (14). The estimate (8) is proved with the help of the inequality $|f(h)-f(h_0)| \leqslant \max|f'(z)|\,|h-h_0|$ and besides the definitions of G_ν, r_ν, $\tau(\nu)$ should be taken into consideration. The order of the addition $p^\gamma_{\nu+1} - p^\gamma_\nu$ determines the estimate (20).

A certain line $h' = h'_\infty(h'')$ is the limit of the successive G_ν, $\nu \to \infty$. We define the regions G^φ : $-\beta'\kappa/2 \leqslant h' \leqslant h'_\infty(h'')$, G^ψ : $h'_\infty(h'') \leqslant h' \leqslant \beta'\kappa/2$ $(-\infty<h''<\infty)$ joining the line $h' = h'_\infty(h'')$ on the both sides.

<u>Theorem 2</u>. The estimates (8)-(11), (13), (14), (16) remain valid in G^γ for arbitrary Λ^γ_0 and $t = C_1\exp(1-\sqrt{\beta'})$ without the exchange $Z^R \to Z^R_\nu$. The functions p^γ_ν, p^γ_∞ are analytic functions in $G^\gamma \cup G_\nu$, G^γ and the finite limits for all the derivatives $p^\gamma_{\infty,k}(h'') = \frac{d^k}{dh^k} p^\gamma_\infty(h)|h \to h'_\infty(h'')$, $h \in G^\gamma$ exist ($k = 1,2,\ldots$, h'' is arbitrary fixed). The series (15) has all its derivatives and converges exponentially on the line $h = h'_\infty(h'')$.

The idea of the proof may be found in [6] where a similar statement is proved for the d-dimentional Ising model with the nearest-neighbour interactions (φ-order consists of minus and ψ-order has only plus).

The real part of the pressure $\mathrm{Re}(p^\gamma_\infty + p^\gamma_c)$ continuously changes when h crosses the line $h'_\infty(h'')$, in particular, the pressure $p = p^\gamma_\infty + p^\gamma_c$ is a continuous function for real h,β. At the same time the density $\rho^\gamma(\beta,h) = \kappa\frac{\partial p^\gamma_\infty}{\partial h} + |\gamma|^+/|\gamma|$ has a jump of order $\kappa(1-O(t))$, i.e. $h = h'_\infty(h'')$ is the first order phase transition line.

<u>Lemma 4</u>. The pressure $p_\infty^\gamma(h)$ is an analytic function for all h", h' < $h'_\infty(h")$ when φ-order contains only minus.

The same statement is true for $p_\infty^\psi(h)$, $h' > h'_\infty(h")$ if the ψ-order consists of plus. In particular, the free energy of the ferromagnetic model ($\varphi \to \psi$ is the minus → plus transition) is an analytic function everywhere on the complex plane h but the line $h' = h'_\infty(h")$. Note that $h'_\infty(-h") = -h'_\infty(h")$ and $h'_\infty(h") = 0$ for real β.

<u>Lemma 5</u>. Let η, φ, ψ be 3 orders and the Peierls condition hold true both for $\eta \to \varphi$ and $\varphi \to \psi$ transitions ($|\eta|^+/|\eta| < |\varphi|^+/|\varphi| < |\psi|^+/|\psi|$). Then $p_\infty^\varphi(h)$ is an analytic function in the region between the two phase transition lines.

Besides that, each transition $\varphi \to \psi$ has its sign inversion $\tilde{\psi} \to \tilde{\varphi}$ after permutation of plus and minus in the orders. The potentials E_α^φ, $E_\alpha^{\tilde{\varphi}}$ in φ and $\tilde{\varphi}$ orders coincides because of the symmetry of the model. Let η, φ, ψ,... ($0 = |\eta|^+/|\eta| < |\varphi|^+/|\varphi| < ... < |\varphi|^+/|\varphi| < |\eta|^+/|\eta| = 1$) be some orders and let the Peierls condition be fulfilled for all transitions $\eta \to \varphi$, $\varphi \to \psi$,... . Then the pressure is an analytic function on the whole complex plane μ except the phase transition lines. Thus, the complete investigation of the regions of analycity on the plane μ is possible for a wide class of potentials u(r).

The problem of existence of the analytic continuation for $p_\infty^\gamma(h)$ through the line $h' = h'_\infty(h")$ arises naturally as soon as the limits of all the derivatives exist on this line. We cannot answer to this question for arbitrary

β,h", but for the most important case of real β,h ($\beta'' = 0$, $h'' = 0$) the following theorem gives a negative answer.

Theorem 3. Let $\beta = \beta' > \beta_0$ and let the following limits

$$\lim_{\nu\to\infty} \tau^{\varphi}(\nu)\,\nu^{-1/d} = \lim_{\nu\to\infty} \tau^{\psi}(\nu)\,\nu^{-1/d} = \tau_\infty \tag{21}$$

exist and coincide. Then,

$$p^{\gamma}_{\infty k}(0) \simeq (K\,!)^{\frac{d}{d-1}}\left[\frac{\tau_\infty d}{(d-1)(\beta+C_k)}\right]^{\frac{dk}{d-1}} \tag{22}$$

where $|C_k| < 3 \ln C_1 + 5$, for arbitrary κ, $\kappa > K_0(\beta)$. The estimate (22) remain valid for $(-1)^k \frac{d^k p^{\psi}_\infty}{dh^k}|h\to h'(o)$, $h\in G^{\psi}$. Thus, the point $h=h'_\infty(o)$ is an essential singularity for $p^{\varphi}_\infty(h)$, $p^{\psi}_\infty(h)$.

The main features of the proof may be found in [6] where the Ising model was considered.

We require two main conditions for the potential u(r) according to theorem 3 : the Peierls condition for $\mu=\mu_0$ and the existence of the limit (21). The latter is a stronger proposition than the estimate (18) which is the corollary of the Peierls condition. In fact the condition (21) is always fulfilled together with (18) for numerous examples of u(r). Thus, the requirement (21) does not essentially restrict the class of available solutions. In particular, $\tau_\infty=(2d)^{-1}$ for the Ising model with nearest-neighbour interaction.

The continuous gas model with finite pair interaction ($u(r) = o$, if $|r| > r_R$) is the limit of the lattice gas model when $t \to 0$, t the period of the lattice. The formula (22) is valid only for arbitrary large, but fixed ratios r_R/t. However the continuous limit $t \to 0$ for (22) deserves attention, although this result is not a rigorous one. We require $u(r) = u(|r|)$. In this way we expect the optimal contour (21) to be a d-sphere with energy $E_\alpha = \alpha S$, where S is the area of the sphere and α the coefficient of the surface stress. The density of the liquid phase $\rho = |\psi|^+ / |\psi|\, t^d$ has a certain positive limit when $t \to 0$. The nonperturbed order for the gaseous phase consists only of minus . We introduce the new variable $\chi = \beta(\mu - \mu_0)$ instead of h, because $K_{|t \to 0} \to 0$. Then,

$$\frac{d^k p^\varphi_\infty}{d\chi^k}\Bigg|_{\chi \to \chi'_\infty(o), \chi \in G^\varphi} \sim (k!)^{\frac{d}{d-1}} \left[\frac{(\Gamma(\frac{d}{2}+1))^{\frac{1}{d-1}} \rho}{(\sqrt{\pi}(d-1)\beta\alpha)^{\frac{d}{d-1}}} \right]^k \tag{23}$$

where $V^{d-1}/S^d = \Gamma(d/2+1) / (\sqrt{\pi} d)^d$ (V is the volume of the d-ball) is taken into consideration.

3. Two-dimentional Ising model

The uncertainty on the asymptotic behavior (22) is of exponential order $\sim \exp(CK/\beta)$. The exact values are found only for the coefficient $(K!)^{d/(d-1)}$ and the part of the exponential factor corresponding to infinite β. More detailed information about the contours than the simple estimate $C^m{}_1$ is necessary if we want to obtain more exact asymptotics. Such calculations are available for the two-dimentional Ising model with the nearest-neighbour interaction $I < 0$.

In this case φ-order contains only minus. Each contour $\Gamma^{\varphi}_{\alpha}$ consists of two stratums : exterior (minus) and interior (plus). Now we shall use simpler contour representation [6]. In these terms, a contour Γ_{α} is the joint system of unit edges which separate plus and minus in a fixed state ν. In fact contours $\Gamma^{\varphi}_{\alpha}$ and Γ^{ψ}_{α} coincide. Only the boundary conditions determine the index φ or ψ for Γ_{α}. Two contours are joint if they have a common vertex. The weight of $\Gamma^{\varphi}_{\alpha}$ is

$$W^{\varphi}_{\alpha} = \omega^{l_{\alpha}} e^{hs_{\alpha}} Z^{\varphi}(\Lambda^{\psi}_{\alpha}) / Z(\Lambda^{\psi}_{\alpha})$$

where $\omega = e^{2\beta I}$, l_{α} is the length of the contour (the number of edges in $\Gamma^{\varphi}_{\alpha}$), $h = 2\beta H$, H is the magnetic field, $\Lambda_{\alpha} = \Lambda^{\psi}_{\alpha}$ $(\Lambda^{\varphi}_{\alpha} = \varnothing)$, $S_{\alpha} = V^{\psi}_{\alpha}$ is the number of plus for the only contour $\Gamma^{\varphi}_{\alpha}$. Besides we have $Z^{\varphi}(\Lambda^{\psi}_{\alpha}, h) = Z(\Lambda^{\psi}_{\alpha}, -h)$, $W^{\varphi}_{\alpha}(h) = W^{\psi}_{\alpha}(-h)$, $W^{\varphi}_{\alpha}\big|_{h=o} = \omega^{l_{\alpha}}$.

The polygonal way $\Gamma_{\alpha nm}$ is defined as a joint set of edges with 2 or 4 edges beginning at each vertex on the way except the vertices (o,o) and (n,m) where only 1 or 3 edges begin. We denote the restricted partition function $Z^{R}(\Lambda)$ which does not contain the contours joint with the vertices of Γ_{α} $(\Gamma_{\alpha nm})$ as $Z(\Lambda - \Gamma_{\alpha})$ $(Z(\Lambda - \Gamma_{\alpha nm}))$.

<u>Lemma 6</u>. The thermodynamic limit

$$R_{\alpha}(\omega) = \lim_{D(\Gamma\alpha\ ,\ \Gamma) \to \infty} Z(\Lambda - \Gamma_{\alpha}) / Z(\Lambda) \tag{24}$$

exist in the circle $|\omega| < \omega_o$, ω_o small enough, h=o for arbitrary fixed contour Γ_{α} (or way $\Gamma_{\alpha nm}$). Here $D(\Gamma_{\alpha},\Gamma)$ is the distance between Γ_{α} and the boundary of Λ. Besides, $R_{\alpha}(\omega) = 1 + O(\omega^{4})$.

Let $Z_\alpha(\Lambda^\varphi)$ be a restricted partition function $Z^R(\Lambda^\varphi)$, containing Γ^φ_α, all contours $\Gamma^\varphi_{\alpha'}$, $(S_{\alpha'} < S_\alpha)$ and perhaps some contours with $S_{\alpha'} = S_\alpha$. Besides that $D(\Gamma_\alpha, \Gamma) \geqslant l_\alpha$ is required. The partition function $Z'_\alpha(\Lambda^\varphi)$ contains the same contours but Γ^φ_α.

Lemma 7. Let the contour Γ^φ_α satisfies inequalities $S_\alpha > 1/8 (k/\ln_\omega)^2$, $l_\alpha < 1/3\sqrt{S_\alpha}$. Then,

$$d^k/dh^k \ \ln(Z_\alpha(\Lambda^\varphi)/Z'_\alpha(\Lambda^\varphi))\Big|_{h=0} = \lambda_k S^k_\alpha M^k \omega^{l_\alpha} R_\alpha(\omega), \tag{25}$$

where $M = (1-(2\omega/(1-\omega^2))^4)^{1/8}$ is the spontaneous magnetization, $k > k_o$, $0 < \lambda^{-1} < \lambda_k < \lambda$, k_o and λ depending only on ω and sufficiently large.

The proof is based on some intermediate estimates of theorem 2 [6].

We introduce the sums

$$< \sigma_{oo}\, \sigma_{nm} > = \Sigma\, R_{\alpha nm}(\omega)\omega^{l_{\alpha nm}} \tag{26}$$

over all possible ways from (o,o) to (n,m). These sums $< \sigma_{oo}\, \sigma_{nm} >$ coincide with the high-temperature pair correlation functions of the model for w=-thβI. Besides we consider the generating function

$$F(t_1,t_2) = \sum_{n,m=-\infty}^{\infty} < \sigma_{oo}\sigma_{nm} > t_1^n\, t_2^m \tag{27}$$

Lemma 8. The function $F(e^x,e^y)$ $(o < \omega < \omega_o)$ is analytic for real x,y in the square $\ln\omega < x < -\ln\omega$, $\ln\omega < y < -\ln\omega$ except the convex analytic line of poles given by the equation $1/F(e^x, e^y) = o$. This line has the following axes of symmetry : OX, OY, $Y = \pm X$ and it approaches the boundary of the region (the square $(-2\ln\omega) \times (-2\ln\omega)$) when $\omega \to o$.

The function $\varphi(z)$ $(-\infty< z<\infty)$ is defined as $\varphi(dy/dx)=y-dy/dx\ x$, where $y=y_p(x) > o$ is the line of poles.
Note that $\varphi(z) = \varphi(-z) > o$, $\varphi(z) = z\varphi\ (1/z)$.

Lemma 9. The estimate

$$<\sigma_{oo}\sigma_{nm}> = \lambda_n e^{-n\varphi(m/n)} \qquad (28)$$

hold for $-n \leqslant m \leqslant n$, where $1/n < \lambda_n < n$, $n > n_o$, $n_o(\omega)$ is sufficiently large.

The line $1/F(e^x, e^y) = o$ possesses the following property. Let the weight of the element dx be $dP = \varphi\ (dy/dx)\ dx$ for an arbitrary line $y = y(x)$. Then the line (contour) $y = \pm Cy_p(x)$ has the minimal weight $CP_p(\omega)$ for the given area S. Here $C = \sqrt{S/S_p(\omega)}$, $S_p(\omega)$ is the area of the contour $1/F(e^x,e^y)=o$. and $P_p(\omega)$ its weight. Besides, $P_p(\omega) = 2S_p(\omega)$ and the calculations give

$$S_p(\omega) = (2\ln\omega)^2 - 2/3\ \pi^2 - (8\omega^2 + O(\omega^4))\ln\omega. \qquad (29)$$

Lemma 10.

$$\sum_\alpha R_\alpha \omega^{l_\alpha} = \exp(-\sqrt{SS_p(\omega)} + \lambda_s), \qquad (30)$$

where the summation is taken over all translationally invariant contours Γ_α^φ with $S_\alpha = S$, $S > S_o$, $S_o(\omega)$ sufficiently large, $l_\alpha < 1/3\sqrt{S}$, $|\lambda_s| < S^{1/4}\ln S$.

Theorem 4. The improved asymptotic behavior (22) for the two-dimentional Ising model is

$$\left.\frac{d^k p_\infty^\varphi}{dh^k}\right|_{h\to -o} = (k!)^2\left(\frac{M(\omega)}{S_p(\omega)}\right)^k e^{o(k)k} \qquad (31)$$

where $|o(k)| < k^{-1/2}\ln k$, $k > k_o(\omega)$, $o < \omega < \omega_o$.

So we obtain the exact value for the exponential dependence in k of the asymptotics.

4. Metastable state

The thermodynamic functions in the metastable state cannot be calculated directly with the help of the thermodynamic limit for partition function. It is well known that there is no irregular behaviour of the system when it passes the transition point to the metastable state. Therefore it is naturally to define the pressure for the metastable state $p_\infty^\varphi(\Delta h)$, $\Delta h = h'-h'_\infty(o) > o$ with the help of the series

$$p_\infty^\varphi(\Delta h) \sim p_\infty^\varphi(o) + p_{\infty,1}^\varphi\cdot\Delta h + p_{\infty,2}^\varphi\frac{(\Delta h)^2}{2!} + \dots, \qquad (32)$$

taking into consideration the existence of all the limits $\left.\frac{d^k p_\infty^\varphi}{dh^k}\right|_{h\to h'(o),\ h\in G^\varphi} = p_{\infty,k}^\varphi$. The convergence of the series (32) for any

$\Delta h \neq 0$ is equivalent to the existence of the analytic continuation near the first order phase transition point $\Delta h = 0$. In this case we obtain the determined values of $p^{\varphi}_{\infty}(\Delta h)$ in the metastable state ($\Delta h > 0$). In fact the series (32) does not converge for any $\Delta h \neq 0$, and so it is an asymptotic one. The terms of (32) first decrease up to the m-th number ($m \sim [(d-1)\beta/\tau d]^{d}/(\Delta h)^{d-1}$) and then increase to infinity. Therefore it is impossible to calculate more exact values of $p^{\varphi}_{\circ}$ (Δh) (32) than $p^{\varphi}_{\infty m}(\Delta h)^{m}/m! \sim \varepsilon$,

$$\varepsilon = \exp[-(\beta/\tau_{\infty} d)^{d}((d-1)/\Delta h)^{d-1}].$$

At first the uncertainty ε (33) seems to be some kind of defect of the theory which prevents us to have the exact values of p^{φ}_{∞} for $\Delta h > 0$. It should be noticed that the metastable state has a bounded time of life and therefore essentially differs from the stable one. More detailed analysis shows the impossibility to obtain the exact values of the thermodynamic functions in the metastable state. We fix Δh, $0 < \Delta h \ll 1$ and consider a volume Λ^{φ}. Then ψ-ordered states with one large contour $\Gamma^{\varphi}_{\alpha}$ containing almost all the volume Λ^{φ} have the maximum probability, if the volume Λ^{φ} is sufficiently large for given β, Δh. However the continuous evolution to the ψ-order state (stable state) from the φ-order one (metastable state) is possible only through states ν with sufficiently small probabilities. We examine the weight $W^{\varphi}_{\alpha} \sim \exp(-\beta E_{\alpha} + V^{\psi}_{\alpha}\ \Delta h)$ in order to estimate these probabilities. At first W^{φ}_{α} decreases and then increases unboundedly, when the size of $\Gamma^{\varphi}_{\alpha}$ grows to infinity. Thus, a critical size for $\Gamma^{\varphi}_{\alpha}$ (a droplet) exist so that smaller contours prefer to decay and larger contours prefer to grow.

The probability of transition from the metastable state to the stable one strongly depends on the form of the critical contour. This probability ~ W^{φ}_{α} takes on its maximum $W^{\varphi}_{o} \sim \exp(-\beta V_o^{(d-1)/d}/\tau_{\infty} + V_o \Delta h)$ for the optimal contour (21) where the minimum E_{α} for a given $V^{\varphi}_{\alpha} = V_o$ is realized. The minimum of $W^{\varphi}_{o}(V_o)$ corresponds to $V_o = V_m$,

$$V_m = ((d-1)\beta/\tau_{\infty} d\Delta h)^d \qquad (34)$$

(the critical probability), i.e. V_m is the volume of the critical droplet.

It is natural to define the partition function and the pressure for the metastable state as $Z_v(\Lambda^{\varphi})$, p^{φ}_v, where v is the integer part of V_m. So we neglect the large contours which overflop the system. Notice that $h' \in G^{\varphi} \cup G_v$ according to theorem 2 and thus the thermodynamic limit $p^{\varphi}_v = \lim |\Lambda^{\varphi}|^{-1} \ln Z_v(\Lambda^{\varphi})$ exist. However we cannot exactly distinguish those near-critical contours which should be excluded. Really, there are two possibilities for the system when such contour appears : to decay into the stable state or to remain in the metastable one. In such a way the uncertainty reveals itself. It is remarkable that $\exp(-\beta V_m^{(d-1)/d}/\tau_{\infty} + V_m \Delta h) = \varepsilon$, $W^{\varphi}_{o}(V_m) \sim \varepsilon$ and we obtain the same value for the uncertainty which was found for the asymptotic expansion (32).

The attempts to define exact values for the thermodynamic functions give no use. For example, we cannot define pressure to be exactly equal to $p^{\varphi}_v(\Delta h)$ because $v = v(\Delta h)$ and therefore the function $p^{\varphi}_v(\Delta h)$ has a lot of jumps of order $\varepsilon(\Delta h)$.

It is known that the metastable region boundary approaches the first order phase transition line when $\Delta T = T_c - T \to 0$ (T_c is the critical temperature).

Now it should be taken into consideration that there is no any determined boundary. We can only fix $\varepsilon_0 > 0$ and examine the solution $\Delta h = f_0(\Delta T)$ of the equation $\varepsilon(T, \Delta h) = \varepsilon_0$ on the plane (T,h). This line $\Delta h = f_0(\Delta T)$ defines the metastable region boundary for the level of uncertainty $\varepsilon=\varepsilon_0$. We can admit that the functions $\Delta h=f_1(\Delta T), \Delta h=f_2(\Delta T), \ldots$ for the levels $\varepsilon_1, \varepsilon_2, \ldots$ have one and the same law of decreasing $\Delta h \sim (\Delta T)^{\mu}$. The index μ can be exactly calculated for the two-dimentional Ising model thanks to the asymptotic formula (31). Though we proved (31) only for sufficiently small ω ($\omega < \omega_0$), it is natural to suppose that the asymptotics (31) remain valid up to the critical temperature $\omega < \omega_c$. Then, $M \sim (\Delta T)^{1/8}$. Besides, $\varphi(z,T) \sim \Delta T.\ \varphi_c(z)$ [1].
Therefore $S_p \sim (\Delta T)^2$ and

$$p^{\varphi}_{\infty k} \sim (k!)^2 (\Delta T)^{-15/8\,k} \quad (k > k_0). \tag{35}$$

Simple calculations give the law $\Delta h \sim (\Delta T)^{15/8}$ for the metastable region boundary.

Formula (35) shows that the critical index of the functions $p^{\varphi}_{\infty,k}(\Delta T)$ decreases with the step 15/8 for sufficiently large k. It should be noticed that the free energy $p^{\varphi}_{\infty,0}$, the spontaneous magnetization $p^{\varphi}_{\infty,1}$ and the initial successibility $p^{\varphi}_{\infty,2}$ have the following critical indices : 2, 1/8, -7/4 [1], i.e. have the same difference 15/8.

Therefore we can admit

$$p^{\varphi}_{\infty,k} \sim (k!)^2\,(\Delta T)^{2-15/8\,k}$$

for all k = 1,2,3,...

REFERENCES

[1] M.E. FISHER, The nature of critical points. Colorado, Boulder, 1965.

[2] D. RUELLE, Statistical mechanics, Rigorous results. New-York, Amsterdam, W.A. Benjamin, inc. 1969.

[3] Ya.G. SINAI, Phase transitions theory (in Russian), Moskow, Nauka 1980.

[4] H.E. STANLEY, Introduction to phase transitions and critical phenomena, Oxford, Clarendon press 1971.

[5] S.A. PIROGOV, Ya. G. SINAI, Theor. Math. Phys. (in Russian), 25, N3, 358-369,1975, 26, N1, 61-76, 1976.

[6] S.N. ISAKOV, Commun Math. Phys., 95, 427-443, 1984.

[7] G. BAKER, D. KIM, J. Phys. A., Math. Gen., 13, 103-106, 1980.

[8] D. CAPOCACCIA, M. CASSANDRO, E. OLIVIERI, Commun. Math. Phys., 39, 185, 1974.

[9] A.G. BASSUYEV, Theor. Math. Phys. (in Russian), 58, N1, 121-136, N2, 261-278.

GRAPH COLOURING : A WAY TO VARIETY OF NEW CORRELATION INEQUALITIES

S.B. SHLOSMAN

Institute of Information Transmission Problems,
Academy of Sciences, Moscow, USSR

In this paper we describe an algorithm for obtaining infinitely many correlation inequalities for even Ising ferromagnets. This series contains, together with some well-known, several conjectured earlier as well as new inequalities.

Let $\sigma_1,\ldots,\sigma_N$ be Ising spins with the distribution given by the density

$$p(\sigma_1,\ldots,\sigma_N) = Z^{-1} \exp\left\{\sum_{1\le i<j\le N} J_{ij}\,\sigma_i\sigma_j\right\}$$

with respect to the measure $d\bar{\sigma} = \prod_i d\nu(\sigma_i)$, where

$$J_{ij} \ge 0\ ,$$

Z is the partition function,

ν is any measure from the Simon-Griffiths class S[1] (which includes, together with the classical case $\sigma = \pm 1$, also other discrete spins, uniform distribution on $[-a,a]$, $\nu = \frac{1}{2n+1}(\delta(.-n)+\ldots+\delta(.+n))$, Gaussian symmetric distribution, φ^4-distribution and some other measures).

Let k be an integer, $2k < N$. Suppose that for any partition

$\mathbb{P} = \{P_1,\dots,P_s\}$ of the set $K = \{1,\dots,2k\}$ into even nonempty subsets some real number $a(\mathbb{P})$ is fixed. We would like to know whether the inequality

$$\sum_{\mathbb{P}} a(\mathbb{P}) \prod_i \langle \prod_{j\in P_i} \sigma_j \rangle \geq 0 \qquad (1)$$

holds for any N , $\{J_{ij}\}$ and $\nu \in S$.

The following statement is a sufficient condition.

Let V be any partition of K into k disjoint pairs, $V = \{v_1, \dots, v_k\}$, and G be any graph with the set V as the set of sites. By a colouring of G we mean any partition of V, such that any two sites which are joined by a bond in G belong to different elements of it (≡ have different colours). For example,

1) all sites have different colours,
2) all sites have the same colour, provided there are no bonds in G.

Let T be some colouring of (G,V). Then we can define the corresponding partition $\mathbb{P}(T)$ of K : the subset $P_i \subset K$, $P_i \in \mathbb{P}(T)$ contains all those pairs of points of K, which have the same colour according to T.

Theorem 1. Suppose that for all partitions V of K into pairs and for all graph structures G on V the inequality

$$\sum_T a(\mathbb{P}(T)) \geq 0 \qquad (2)$$

holds (Here the summation goes over all colourings T of G). Then the inequality (1) also holds.

The proof goes essentially in the same way, as in [2], [3]. We are going on to examples

1) The simplest inequality of type (1) - the first GKS inequality ([4]) :

$$\langle\sigma_K\rangle \geq 0 \qquad (\sigma_K = \sigma_1,\dots,\sigma_{2k})$$

In this case

$$a(P) = \begin{cases} 1, \; P = \{P_1\}, \quad P_1 = \{1,\dots,2k\} \\ 0 \quad \text{otherwise.} \end{cases}$$

Hence, $a(P) \geq 0$, and so condition (2) holds.

2) Unfortunately, the second GKS inequality ([4]) :

$$\langle\sigma_1\dots\sigma_{2k}\rangle \geq \langle\sigma_1\dots\sigma_{2\ell}\rangle\,\langle\sigma_{2\ell+1}\dots\sigma_{2k}\rangle \;, \quad 0 < l < k \quad ,$$

does not follow from our theorem. Indeed, in that case

$$a(P) = \begin{cases} 1, \; P = \{P_1\}, \; P_1 = \{1,\dots,2k\}, \\ -1, \; \mathbb{P} = \{P_1,P_2\}, \; P_1 = \{1,\dots,2l\}, \; P_2 = \{2l+1,\dots,2k\}, \\ 0 \quad \text{otherwise} \end{cases}$$

Consider now the pair partition $V = \{(1,2),(3,4),\dots,(2k-1,2k)\}$, and let G be a graph structure on V, such that two sites $(2p-1,2p)$, $(2q-1,2q)$ are connected iff $p \leq l < q$. Clearly, in that case $\sum_T a(\mathbb{P}(T)) < 0$.

3) Let $\mathcal{L}$ be a family of partitions of K, $\mathcal{L} = \{P_i\}$,

$P_i = \{P_1^{(i)},\dots,P_t^{(i)}\}$, $t = t(i)$, where the sets $P_1^{(i)},\dots,P_t^{(i)}$ are disjoint and even, i.e. $|P_j^{(i)}| = 2l_j^{(i)}$. Following Newman [5], we call the family $\mathcal{L}$ admissible, if every pair-partition of K is a refinement of some partition $\mathbb{P}$ in $\mathcal{L}$. Let

$$a(P) = \begin{cases} -1, & |\mathbb{P}| = 1 \\ +1, & \mathbb{P}\in\mathcal{L}\ ,\ |P| > 1 \\ 0 & \text{otherwise.} \end{cases}$$

Now, let V be some pair-partition of K, and consider the graph G on V with no bonds at all. By definition, there is a colouring T of G, such that $P(T) \in \mathcal{L}$. Hence, for this G the sum $\sum_T a(P(T)) \geq 0$. The same is true for other G-s, which amounts to the inequality :

$$\langle\sigma_K\rangle \leq \sum_{P\in\mathcal{L}} \prod_{P_j\in\mathbb{P}} \langle\sigma_{P_j}\rangle . \tag{3}$$

It was conjectured by Newman [5], who proved it for the case when all $t(i) = 2$. The general case was proven later by Brydges, Fröhlich and Sokal [6].

4) Our method permits us to bound the quantity $\sum_P \prod_i \langle\sigma_{P_i}\rangle - \langle\sigma_K\rangle$ also from above. We present here the simplest case. For any pair partition $\mathbb{D} = \{\mathbb{D}_1,\dots,\mathbb{D}_k\}$ of K, let $\mathbb{R}(\mathbb{D}) = \{\mathbb{P}\in\mathcal{L} : \mathbb{D}$ is a refinement of $P\}$. We say that the partition $\mathbb{P}$ separates two points $x,y \in K$ (notation : $x\mathbb{P}y$), if they belong to different elements of it. Let

$$N_1(\mathbb{D},\mathcal{L}) = \#\,\mathbb{R}(\mathbb{D}),$$

$$N_2(\mathbb{D},\mathcal{L}) = \max_{x,y\in K\ ,\ x\mathbb{D}y} \#\,\{\mathbb{P}\in\mathbb{R}(\mathbb{D}) : x\mathbb{P}y\}$$

Evidently, $N_2(\mathbb{D},\mathcal{L}) \leq N_1(\mathbb{D},\mathcal{L})$. Let

$$N(\mathbb{D},\mathcal{L}) = \max \{N_1(\mathbb{D},\mathcal{L}) - 1, \quad N_2(\mathbb{D},\mathcal{L})\}.$$

Then one has the inequality :

$$\langle\sigma_K\rangle - \sum_{\mathbb{P}\in\mathcal{L}} \prod_{P_j\in P} \langle\sigma_{P_j}\rangle + \sum_{\mathbb{D}} N(\mathbb{D},\mathcal{L}) \prod_{i=1}^{k} \langle\sigma_{D_i}\rangle \geq 0 . \tag{4}$$

This inequality seems to be new.

Let us consider now the case $|K| = 6$, with containing all 15 partitions of K into 2+4. Evidently is admissible. It is easy to see that for all $\mathbb{D}$, $\mathbb{N}_1(\mathbb{D},\mathcal{L}) = 3$, $N_2(\mathbb{D},\mathcal{L}) = 2$, hence $N(\mathbb{D},\mathcal{L}) = 2$. The inequality (4) in that case is nothing else than Cartier - Percus - Sylvester inequality ([7], [8], [9]) : $u_6 \geq 0$.

5) Our condition for the correlation inequality to hold acquires more attractive form in the case when the left-hand side of (1) can be expressed as a symmetric combination of Ursell functions u_n , which are defined by

$$u_n(\sigma_{i_1},\ldots,\sigma_{i_n}) \equiv u_n(i_1,\ldots,i_n) \equiv \frac{\partial^n}{\partial h_1\ldots\partial h_n} \ln \langle\exp \sum_{j=1}^{n} h_j \sigma_{i_j}\rangle\Big|_{\bar{h}\equiv 0} .$$

Let H be any graph (without loops and multiple bonds). Let b be its bond connecting the sites v_1, v_2 and let $v_3,\ldots,v_k$ be the rest of them. The contracted graph H(b) result from the identification of v_1 and v_2 : it has (k-1) sites - w, $v_3,\ldots,v_k$, with v_i being connected with v_j, $i,j\geq 3$, in H(b) iff they were connected in H, and with w being connected with v_i, $i \geq 3$, in H(b) iff v_i was connected with at least one of v_1, v_2 in H. It follows that H(b) also has no loops or multiple bonds.

Now, let us define the function R(.) on the set of all graphs by the following recursion :

1) R (tree) = 1

2) R (disconnected graph) = 0

3) R(H) = R(H(b)) + R(H\b),

where H\b is the result of deleting the bond b from H. It can be shown that we obtain thus some well-defined function.

For s, $p_1 < \ldots < p_s$, $q_1, \ldots, q_s$ given integers,

$$p_1 q_1 + \ldots p_s q_s = k \ , \tag{6}$$

let a $(s,\overline{p},\overline{q})$-splitting of a graph H be any family

$$H_1^{(1)}, \ldots, H_{q_1}^{(1)}, H_1^{(2)}, \ldots, H_{q_2}^{(2)}, \ldots H_{q_s}^{(s)}$$

of its subgraphs, such that any site of H belongs exactly to one $H_i^{(j)}$: the number of sites in $H_i^{(j)}$ is p_j and v',v" $\in H_i^{(j)}$ are connected in $H_i^{(j)}$ iff they are connected in H.

<u>Theorem 2</u> [10] : suppose one is given a function $a(s,\overline{p},\overline{q})$, satisfying (6), such that for any graph H with k sites

$$\sum_{(s,\overline{p},\overline{q})} a(s,\overline{p},\overline{q}) \sum_{H_1^{(1)}, \ldots, H_{q_s}^{(s)}} R(H_1^{(1)}) \ldots R(H_{q_s}^{(s)}) \geq 0, \tag{7}$$

where the inner summation goes over all $(s,\overline{p},\overline{q})$-splittings of H. Then for all even Ising ferromagnets

$$\sum_{(s,\bar{p},\bar{q})} a(s,\bar{p},\bar{q})\,(-1)^{k-q_1-\ldots-q_s} \times$$

$$\times \Sigma \left[u_{2p_1}(\Lambda_1^{(1)})\ldots u_{2p_1}(\Lambda_{q_1}^{(1)})u_{2p_2}(\Lambda_1^{(2)})\ldots u_{2p_s}(\Lambda_{q_s}^{(s)})\right] \geq 0 ,$$

where the inner summation goes over all partitions of K into $q_1+\ldots+q_s$ disjoint subsets, $\Lambda_1^{(1)},\ldots,\Lambda_{q_s}^{(s)}$, such that exactly q_i of them contains $2p_i$ elements, $i=1,\ldots,s$.

6) By definition of the function R, $R(H) \geq 0$ for all H ; hence by theorem 2

$$(-1)^{k-1}\, u_{2k}(\sigma_1,\ldots,\sigma_{2k}) \geq 0 ,$$

(for details, see [2]). This proves the old conjecture of Newman [11] and Feldman [12].

7) It can be shown (see [3]), that for any graph H with k sites

$$k\,R(H) \leq (k-1) \sum_{\alpha=1}^{k} R(H_\alpha) ,$$

where H_α is obtained from H by eliminating the α-th site together with all the ajacent bonds.

Hence

$$(-1)^{k-1}u_{2k}(1,\ldots,2k) \leq \frac{k-1}{k}\,(-1)^{k-2}\, u_{2k-2}(1,\ldots,\hat{i},\ldots,\hat{j},\ldots,2k)\, u_2(i,j),$$

where the sum is over all choices of pairs (i,j) from (1,...2k). We can iterate the last inequality. The final iteration has the form

$$(-1)^{k-1}\, u_{2k}(\sigma_1,\dots,\sigma_{2k}) \le (k-1)!\, \Sigma\langle\sigma_{i_1}\sigma_{j_1}\rangle\dots\langle\sigma_{i_k}\sigma_{j_k}\rangle ,$$

where the sum goes over all pair partitions of K. This series of inequalities is among those conjectured by Newman [13].

8) In some cases the resulting inequality includes an arbitrary real parameter. For example,

$$u_6(1,\dots,6) \ge a\, \Sigma'\, u_4(i_1,\dots,i_4)\, u_2(i_5,i_6) +$$

$$+ \frac{1}{3}\min(0,a,1+2a,2+3a)\, \Sigma''\, u_2(i_1,i_2)u_2(i_3,i_4)u_2(i_5,i_6), \tag{8}$$

where the sum Σ' runs through all fifteen 2+4 partitions, and the sum Σ" - through all fifteen 2+2+2 partitions. Here $a \in \mathbb{R}^1$ is arbitrary. In a weaker form the inequality (8) was obtained in [14] (though for a larger class of free measures).

REFERENCES

[1] Simon, B., Griffiths, R.B., Comm. Math. Phys. 33, 145, (1973).

[2] Shlosman, S.B., Comm. Math. Phys. 102, 679 (1986).

[3] Shlosman, S.B., Theor. Ver. Pril., to appear.

[4] Griffiths, R.B., J. Math. Phys. 8, 478-484 (1967).
Kelly, D.G., Sherman, S., J. Math. Phys. 9, 466 (1969).

[5] Newman, C., Z. fur Wahrsch. 33, 75 (1975).

[6] Brydges, D., Fröhlich, J., Sokal, A., Comm. Math. Phys. 91, 117 (1983).

[7] Cartier, P., in Lect. Notes Math., N°383, 242 (1974).

[8] Percus, J.K., Comm. Math. Phys. 40, 283 (1975).

[9] Sylvester, G., Comm. Math. Phys. 42, 209 (1975).

[10] Shlosman, S.B., Reports of Acad. of Sci. USSR, to appear.

[11] Newman, C., in Lect. Notes Phys. N°25, (G. Velo, A.S. Wightman, Eds.) Springer (1973).

[12] Feldman, J., Canad. J. Phys. 52, 1583 (1984).

[13] Newman, C., Comm. Math. Phys. 41, 1 (1975).

[14] Kondo, K., Otofuji, T., Sugiyama, Y., J. Stat. Phys. 40, 563 (1985).

POSTER SESSIONS

organized by Ph. COMBE and R. RODRIGUEZ

EQUILIBRIUM STATISTICAL MECHANICS, INTEGRABLE SYSTEMS, KAC MOODY ALGEBRAS, FIELDS, QUANTA, ENTROPY, STOCHASTIC METHODS

R. ALICKI, M. FANNES
Dilatation of quantum dynamical semigroups with classical Brownian motion.

A. AMANN — Integrable Ergodic W* Systems.

J. AMBJORN, Ph. de FORCRAND, F. KOUKIOU, D. PETRITIS
Monte Carlo simulation of triangulated random surfaces. Self avoiding random walks in 3 dimensions.

A. BINESH — A simple calculation method in random walk problems.

G. CAGINALP — The phase field approach to understanding supercooling, surface tension and anisotropy.

C. CARVALHO — The second virial coefficient of the quantum Lorentz gas.

H. CHEN, C. CHIN — Exact solution of the periodic Liouville equation.

D. D ANTONI — Universal structure of local algebras.

J. DE CONINCK — Les lois de probabilité indéfiniment divisibles en Mécanique Statistique.

A. EL MELLOUKI — Random surfaces, wetting and layering.

G. EPIFANIO, C. TRAPANI
Quasi *algebras valued quantized fields.

M. FORGER, M. JACQUES
Higher conservation laws for 10 dim supersymmetric Yang Mills theories.

T. HARA, H. TASAKI
Rigorous control of logarithmic corrections in 4.dim spin systems.

H. HASEGAWA — Self contained framework Stochastic Mechanics for reconstructing the Onsager Machlup theory.

M. JACQUES — Infinite dimensional Lie Algebras acting on the solutions space of various σ models.

L.F. KO, B. Mc COY
Some exact results for the electric correlation functions of the 8-vertex model.

G. KOLEROV Quantization and projective geometry.

R. KOTECKY, H. LAANAIT, A. MESSAGER, J. RUIZ
Potts model in the standard Pirogov Sinaï theory.

R. KUIK Clustering, global Markov properties and the infinite volume transfer operator.

A. KUNDU Gauge equivalence of non linear integrable models.

C. MAES, J.L. LEBOWITZ
The harmonic Ising interface.

F. MATHOT Integral decomposition of partial *algebras of closed operators.

J. MIEKISZ Equilibrium states of ferromagnetic systems.

J. NAUDTS The generalized Lee-Yang theorem.

P. PEARCE Magnetic hard squares : exact solutions.

D. PETZ On the sufficiency of quantum communication channels.

M. REQUARDT Some rigorous results from the Statistical Mechanics of phase boundaries and fluid near a plane wall.

S. SHLOSMAN Graph colouring ; a way to variety of new correlation inequalities.

K. SINHA Quantum Stochastic processes, Boson Fermion relations and stop time operators in Fock space.

Y. SOULET Particular representations of a locally convex *algebra. Application to infinite number of particles systems.

S. STAMENKOVIC, N. TONCHEV, V. ZAGREBNOV
Exactly solvable model for structural phase transition with a gaussian type anharmonicity.

S.T. TEMKO, S.K. KUZ'MIN
On the statistical theory of space clusters.

W. TIMMERMAN On the states of topological algebras of unbounded operators.

B. TORRESANI Unitary positive energy representations of the Gauge Group.

J. VIGFUSON Upper bounds on the critical temperature for 1D long range Ising spin models.

CLASSICAL FIELD THEORY, NONEQUILIBRIUM STATISTICAL MECANICS, GENERAL Q.F.T.

J. AVAN, H. DE VEGA, J.M. MAILLET
Self dual Yang Mills fields, conformal invariance and Backlund transformation.

R. BENGURIA, L. JEANNERET
Equations of the Korteweg de Vries type with non trivial conserved quantities.

Y. ELSKENS — Non unitary deformations in the hydrodynamic limit.

A. FORSTER — Spacial and spatio temporal autorotation of elementary particles in Relativity.

W. GARCZYNSKI — Senjanovic measure in a path integral quantization of constrained systems.

J.P. GAZEAU, M. HANS
Quantization of Q.E.D. in De Sitter space.

P. HORVATHY, J.H. RAWNSLEY
On the stability of monopoles.

K. KONDO — Charged states and Higgs mechanism in the lattice U(1) Higgs model.
Correlation inequalities and triviality of scalar field theory.

C.N. KTORIDES, L. PAPALOUCAS
A geometric realization of Van Hove's quantization prescription.

C. MAES — The harmonic Ising interface.

M. MANOLESSOU — The $\phi^4{}_4$ equations of motion. A fixed point method for the construction of a non trivial Wightman Quantum Field Theory.

C. MARCHIORO — An example of absence of turbulence for any Reynolds number.

M. O'CARROL — A convergent perturbation theory for particle masses in Euclidean lattice Quantum Field Theory.

A. PALANQUES-MESTRE
Maximal substraction and anomalous dimension in 1/N Q.E.D.

D. RADULESCU — About the mathematical foundations of Classical Mechanics and Special Relativity.

M. REQUARDT — An investigation of (non additive) scattering invariants in (non)relativistic Classical Mechanics and Quantum(Field)Theory.

A. SENGUPTA On a discretised spectral approximation in neutron transport theory.

F. SMITH Jr. Spin (8) Gauge field theory.

F. STROCCHI Mass/energy gap associated to symmetry breaking. A generalized Goldstone theorem for long range interactions.

D. TCHRAKIAN Generalized Yang Mills systems in 4p dimensions.

CONTRUCTIVE FIELD THEORY, DYNAMICAL SYSTEMS, QUANTUM GRAVITY, SUPERSYMMETRIES

I. AREF'EVA, I.V. VOLOVICH
Hyperbolic manifolds as vacuum solutions in the Kaluza Klein theory.

J. BECKERS N=2 supersymmetric Quantum Mechanics and the harmonic oscillator.

M.R. BROWN, A. OTTEWILL
Exact and approximate renormalized stress tensors in curved space time.

M. CARFORA, M. MARTINELLI, A. MARZUOLI, R. SILVOTTI
Conformal anomaly and spacetime curvature in string theory.

G. CLEMENT Kaluza-Klein conformastationary solutions.

M. CORNALBA, R. KATENACCI, M. MARTINELLI, C. REINA
Algebraic geometry and path integrals for closed strings.

J. FOURNIER, G. LEVINE, M. TABOR
Singularity clustering in the Duffing oscillator.

A. GIELERAK Some Gibbsian aspects of the Euclidian Field Theory.

J. GROENEVELD Construction of complete solution classes of the Cvitanovic-Feigenbaum equation.

H. GUO The Chern-Simons cohomology in Quantum Theories.

J. HARNAD, M. JACQUES
Formal power series solutions of supersymmetric (N=3) Yang Mills equations.

P. HJORTH Classical dynamics in a stongly magnetized one species plasma.

S. ISOLA, R. LIVI A single particle approach to broken ergodicity.

R. LIMA, R. VILELA MENDES
Stability of invariant circles in a class of dissipative maps.

V. MAKHANKOV Exact solutions of the vector non linear Schrödinger equation.

F. NISHOFF, D.J. SMIT
Hierarchies of integrable non linear evolution equations and Wess Zumino Lagrangians.

H. PEDERSEN Einstein metrics, spinning top motion and monopoles.

B. PIETTE, L. VINET
Morse and Pöschl-Teller supersymmetric Schrödinger equations.

V. PROTOPOPESCU Duals and propagators : a canonical approach to non linear equations.

G. QUISPEL Universal functional equations for period doubling in constant Jacobian maps.

A. RIECKERS Three operator algebraic models of the Josephson effect.

B. ROSSETTO Singular approximation of chaotic slow-fast dynamical systems.

N. SANCHEZ Semi classical Quantum Gravity : recent results in 2 and 4 dimensions.

G. SLADE The effective potential : Legendre transformation and energy density.
The diffusion of self avoiding random walk in high dimensions.

W. WARCHALL Explicit set of spherical standing waves for the non linear Klein Gordon equation.

W. ZACHARY Existence and finite dimensionality of attractors for the Landau Lifschitz equation.

SCHRODINGER OPERATORS, SEMICLASSICAL METHODS, GENERAL QUANTIZATION

A. ANDERSON — Exact propagator for the N sphere in terms of classical paths.

J.P. ANTOINE, F. GESZTESY, J. SHABANI
Exactly solvable models of surface interactions in non relativistic Quantum Mechanics.

J.P. ANTOINE, F. MATHOT
Unbounded (bi)commutants of partial OP *algebras.

M. ASHBAUGH, R.BENGURIA
On ratios of the eigenvalues of Schrödinger operators with positive potentials.

J. AVAN, H. DE VEGA
Large orders as perturbative expansion by inverse scattering transform in angular momentum.

D. BESSIS — Exact projection methods for the ground state of the Schrödinger equation.

J. DEREZINSKI — Exponential bounds in cones for eigenfunctions of N body Schrödinger operators.

P. EXNER — On mathematical models for quantum point - contact spectroscopy.

C. GERARD, A. GRIGIS
Precise estimates of tunneling for semi-classical Schrödinger operators.

P. GHOSH — Transmission in a system of potential barriers.

A. GIBSON — Low energy scattering with Coulomb plus polarization potentials.

G. HAGEDORN — Multiple scales and the time independant Born Oppenheimer approximation.

M. HOFFMANN-OSTENHOF, T. HOFFMANN-OSTENHOF, J. SWETINA
Continuity properties near infinity and asymptotics of solutions of Schrödinger equations in exterior domains.

A. JENSEN — Commutator methods applied to Schrödinger operators.

D. LAMBERT, A. RONVEAUX
Elliptic and ultra hyperbolic differential operators connected by Hurwitz transformation.

A. MARTINEZ — Estimates of tunneling for symmetric double well.

H. SIEDENTOP — Density functionals for angular momentum channels.

M. SPERA — Quantization of Abelian varieties.

F. STEINER — Spectral sum rules for the circular Aharonov Bohm quantum billard.

H.P. WANG — Time decay of scattering solutions of semiclassical Schrödinger operators.

DISORDERED SYSTEMS, COMPUTATIONAL PHYSICS, ANOMALIES, STRINGS, SUPERSTRINGS, GENERAL RELATIVITY, OTHERS

S.T. ALI, P. BUSCH, R. GAGNON, F.E. SCHROECK Jr.
Current conservation as a geometric property of space time.

J. BASTO GONCALVES
Phase reduction : a control theoretical approach.

J. BECKER, V. HUSSIN, P. WINTERNITZ
Nonlinear differential equations with superposition principals and the exceptional Lie Group G_2.

J.P. BOUCHAUD, A. COMTET, A. GEORGES, P. LEDOUSSAL
An exactly solvable disorder in 1D correlated random potential and diffusion in random medium.

P. BUSCH, F.E. SCHROECK Jr.
Some open problems in measurement theory in light of recent results.

W. CEGLA — Remarks on the causal logic.

P. CHRUSCIEL — Remarks on energy in General Relativity.

A. DAGAZIAN, B. OBI, R. PARIS
The eignevalues of the simplified ideal magnetohydrodynamic ballooning equation.

A. DEOUD, R. PARIS
Asymptotic of a class of high order linear ordinary differential equations and applications.

J.P. GAZEAU — SO(3,2) de Sitter recurrence theorems and homogeneous propagators.

A. HONG TUAN Zero critial dimension for ϕ^3 Field Theory.

M. KATANAYEV, I.V. VOLONOVICH
Bosonic string model with dynamical geometry.

S. KHEIFETS Electromagnetic field of a charge going through a hole in metallic screen.

W. KLINK, P.E.T. JORGENSEN
Quantum Mechanics and Nilpotent Groups.

E. MASCHKE, B. SARAMITO
Bifurcation of non linear Tearing modes in magnetohydrodynamics.

M. MEHTA Basic sets of invariant polynomials for finite reflection groups.

A. NENCKA-FICEK Uncertainty relations and fractals.

I. PAVLOTSKY Vlasov's and Wigner's equations in the postnewtonian approximation.

G. PRON'KO Quantum theory of string in the 4 dim. space time.

O. PENROSE Phase transitions on fractal lattices with long range interactions.

W. SCHNEIDER Generalized Levy distributions as limit laws.

L. ZSIDO Robbins Monro type estimations for the solution of Ax=B.

Talks presented at the Congress but not published

Main Talks

M. BERRY : "Classical chaos and quantum Eigenvalues"

J.T. LEWIS : "Dynamical systems and quantum optics"

A. LIBCHABER : "Experiment on quasi periodicity"

B. SIMON : "Recent developments in the analysis of Random Hamiltonians"

Topical talks

Equilibrium statistical mechanics

M. AIZENMAN : "Non perturbative analysis of critical behaviour"

Fields, Quanta, Entropy, Understanding Heritage and Perspectives

S. WORONOWICZ : "An example of a noncommutative differential calculus"

Integrable systems and Kac-Moody Algebras

M. SATO : "$\mathcal{D}$-modules and integrable systems"

Dynamical systems

M. FEIGENBAUM : "......................."

Quantum gravity

B. WHITING, J. YORK : "A discussion of the Hamiltonian form of classical and quantum gravity"

Supersymmetry

I. AREF'EVA : "Field theory of gauge invariant interaction of superstrings"

Disordered systems

J. FRÖHLICH : "Disordered magnets and spin glasses"

J. CHAYES : "A mean field spin glass with short-range Interactions"

L. CHAYES : "Percolation and percolation methods for disordered systems"

Strings and superstrings

J. HARVEY : "String compactifications in orbifolds recent developments"

Authors Index

		Pages
S.J.	AL'BER	447
S.	ALBEVERIO	409
H.	ARAKI	354
V.I.	ARNOLD	1
V.B.	BARIAKHTAR	512
L.	BAULIEU	366-743
J.M.	BISMUT	17
P.M.	BLEHER	786-789
A.	BOHM	638
J. B.	BOST	768
D.	BUCHHOLZ	381
G.	CASATI	463
P.	COLLET	467
T.	DAMOUR	57
I.	DAUBECHIES	675
R.	DE LA LLAVE	516
S.	DESER	557
E.I.	DINABURG	796
R.L.	DOBRUSHIN	73
S.	DOPLICHER	489
M.	DUFF	746
M.	DUNEAU	724
H.	EPSTEIN	517
P.	FEDERBUSH	597
G.	FELDER	605
M.	FORGER	641
E.S.	FRADKIN	92-548
K.	FREDENHAGEN	499
M.	FREIDLIN	470
K.	GAWEDZKI	318
E.	GETZLER	431
P.	GODDARD	390
S.	GOLDSTEIN	482
R.	GRIMM	651
H.	GROSSE	667

		Pages
V.	GUILLEMIN	106
J.	HALLIWELL	549
J.	HARNAD	644
J.	HARTLE	566
M.	HAZEWINKEL	120
B.	HELFFER	687
M. R.	HERMAN	138
D.	IAGOLNITZER	622
J.Z.	IMBRIE	729
S.N.	ISAKOV	816
J.	JONES	696
T.	KENNEDY	311
H.	KOCH	702
A.	KUPIAINEN	628
O.E.	LANFORD	532
E.	LIEB	185
F.	MARTINELLI	731
V.P.	MASLOV	197
S.	MIRACLE-SOLE	331
G.	MOORE	776
M.	MÜLLER	658
A.	NEVEU	217
S.P.	NOVIKOV	226
D.	OLIVE	242
T.	PAUL	675
CH.E.	PFISTER	338
B.	PLOHR	708
E.	PRESUTTI	486
R.	RAMMAL	736
D.	RAND	537
V.	RIVASSEAU	257
H.	ROEMER	662
L.	ROSEN	614
D.	RUELLE	273
H.	RUMPF	582

S.	SHLOSMAN	839
K.	SIBOLT	505
YA.G.	SINAI	283
M.	SIRUGUE-COLLIN	421
E.K.	SKLYANIN	402
A.	SOFFER	292
A.	SOKAL	714
B.	SOUILLARD	738
R.	STORA	757
C. E.	WAYNE	531
S.-T.	YAU	305
M.	YOR	439
M.	ZAHRADNIK	347

List of Participants

P	AIZENMAN	Rutgers Univ. - USA
G.	AKDENIZ	Univ.of Istanbul- TR
S.J.	AL'BER	Moscow - USSR
C.	ALBANESE	ETH-Zurich -CH
S.	ALBEVERIO	Bochum Univ - D
L.	ALVAREZ-GAUME	Harvard Univ - USA
A.	AMANN	ETH-Zurich - CH
A.	ANDERSON	Univ.of Maryland - USA
G.	ANGERAND	Paris - F
J.P.	ANTOINE	U.C. Louvain la Neuve-B
H.	ARAKI	Kyoto Univ. - Japan
I.Y.	AREF'EVA	Steckhlov Math.Inst., Moscow- USSR
V.I.	ARNOLD	Moscow Univ.- USSR
R.	ARTUSO	Dipart.fisica-Milano- I
J.	ASCH	Tech.Univ.Berlin - D
M.	ASHBAUGH	Univ.of Missouri -USA
M.	ATIYAH	Univ.of Oxford - GB
J.	AVAN	LPTHE - Paris -F
H.	BACRY	Univ.Aix-Marseille - CPT-F
E.	BALSLEV	Math.Inst. Aarhus - DK
V.B.	BARIAKHTAR	Inst.Metals Phys. - Kiev -USSR
D.	BARSKY	Rutgers Univ. - USA
J.	BASTO GONCALVES	Univ.of Porto - P

G.	BATTLE	Texas A & M Univ. - USA
L.	BAULIEU	LPTHE-Paris - F
B.	BAUMGARTNER	Univ.of Wien - A
J.	BECKERS	Univ.of Liège - B
H.	BEHNCKE	Univ.of Osnabruck - RFA
W. D.	BEIGLBOCK	Heidelberg Univ.. - RFA
J.	BELLISSARD	CPT- Marseille - F
F.	BENATTI	SISSA, Trieste- I
J.A.	BENAVIDES	Leiden, Univ. - NL
G.	BENFATTO	Univ of Roma - I
R.D.	BENGURIA	Univ.of Santiago -CHILE
F.	BENTOSELA	CPT- Marseille - F
M.	BERGVELT	ITP - Amsterdam- NL
A.	BERKOVICH	USA
A.	BERNUI	SISSA - Trieste - I
M.	BERRY	H.H.Will Phys.Lab.-Bristol - GB
M. L.	BERTOTTI	SISSA - Trieste - I
D.	BESSIS	CEA-Saclay - F
C.	BILLIONNET	Ec.Polytechnique - Palaiseau -F
A. R.	BINESH	Birjand Univ.- IRAN
J.M.	BISMUT	Univ.paris-Sud -F
PH.	BLANCHARD	Bielefeld Univ. - D
P.M.	BLEHER	Keldysh Institute - Moscow - USSR
F.	BLUM	Techn.Hochs. Aachen - D
N.N.	BOGOLIUBOV JR	Moscow Univ. - USSR
A.	BOHM	Univ.of Texas-Austin-USA
G. F.	BOLZ	Univ.of Bielefeld - D
P.J.M.	BONGAARTS	Univ.Leiden - NL
H. J.	BORCHERS	Univ.of Gottingen - D
J. B.	BOST	E N S - Paris- F
C.	BOURRELY	CPT- Marseille - F
A.	BOVIER	ETH-Zurich -CH
J. M.	BRAGA	Lausanne - CH
H.	BREZIS	Université Paris VI - F
J.	BRICMONT	U.C. Louvain la Neuve - B
P.	BRIET	CPT - Marseille - F
J. A.	BROOKE	Univ.of Saskatchewan - CANADA
E.	BRUNING	Univ. Bielefeld -D
D.	BUCHHOLZ	Univ.Hamburg - D
P.	BUDINICH	SISSA, Trieste - I

G.	BURDET	CPT- Marseille - F
P.	BUSCH	Univ.Koln - D
G.	BYRNES	Sydney Univ.. - AUSTRALIA
F.	CALHEIROS	Fac.Engenharia-Porto - P
E.	CALICETI	Modena - I
D.	CANGEMI	Univ. Lausanne - CH
M.	CARFORA	Pavia Univ. - I
E.	CARLEN	MIT, Cambridge - USA
M. R.	CARTER	Stanford Univ.- USA
M. C.V	CARVALHO	CFMC, Lisbonne - P
G.	CASATI	Univ. Milan - I
W.	CEGLA	Univ. of Wroclaw -PL
A.	CELLETTI	Roma - I
K.	CHADAN	LPTHE-Univ. Paris Sud - F
G.	CHATELET	Paris - F
J.	CHAYES	Cornell Univ. - USA
L.	CHAYES	Cornell Univ. - USA
P.	CHIAPPETTA	CPT - Marseille - F
C.H.	CHIN	Univ.Hsin Cho - Taiwan
J.S.R.	CHISHOLM	Univ. of Kent - GB
P.	CHRUSCIEL	Polish Ac. of Sci.,Varsawa - PL
M.	CIBILS	Univ. de Genève - CH
G.	CLEMENT	Univ.Constantine -Algérie
A.	COHEN	Univ. de Paris VIII - F
O.	COHENDET	CPT - Marseille - F
P.	COLLET	Ecole Polytech., Palaiseau - F
Ph.	COMBE	CPT- Marseille -F
J.M.	COMBES	Univ. Toulon and CPT - F
M.	COMBESCURE	Univ. de Paris Sud - F
A.	COMTET	Univ. de Paris Sud - F
F.	CONSTANTINESCU	Frankfurt Univ. - D
R.	COQUEREAUX	CPT- Marseille - F
J.	CORBETT	Macquarie Univ. - Australia
A.	COSTE	CPT- Marseille - F
M.	COURBAGE	UNIV.PARIS VI - F
J.	CUNTZ	Univ. Aix-Marseille II and CPT - F
C.	D'ANTONI	Univ. dell'Aquila-Roma- I
F.	D'ISEP	Torino - I
B.	D'ONOFRIO	Univ. dell'Aquila-Roma - I

T.	DAMOUR	Observatoire de Meudon - F
I.	DAUBECHIES	Courant Institute - USA
G. F.	DE ANGELIS	Univ. di Salerno - I
A.	DE BIVAR WEINHOLTZ	CMAF, Lisboa - P
C.	DE CALAN	Ecole Polytechn., Palaiseau - F
J.	DE CONINCK	Univ. de l'Etat, Mons - B
D.	de FALCO	Univ. di Salerno - I
R.	DE LA LLAVE	Princeton Univ. - USA
E.	de RAFAEL	CPT, Marseille - F
G.F.	DELL'ANTONIO	Univ. di Roma - I
P.	DELORME	Luminy -Marseille - F
J.	DEREZINSKI	Univ. Varsawa - PL
S.	DESER	Brandeis Univ.- USA
C.	DEWITT-MORETTE	Univ. of Texas at Austin - USA
G.	DI GENOVA	SISSA, Trieste - I
E.I.	DINABURG	Schmidt Inst.-Moscow - USSR
R.L.	DOBRUSHIN	USSR Ac. Sci., Moscow - USSR
S.	DOPLICHER	Univ. di Roma - I
T.	DORLAS	Gröningen Univ. - NL
P.	DRAZEN	Belgrade - Yu
D.A.	DUBIN	Open Univ.,Milton Keynes-GB
P.	DUCLOS	Univ. Toulon and CPT - F
B.	DUCOMET	S.P.T.N., CEA - F
M.	DUFF	CERN - CH
M.	DUNEAU	Ecole Polytechn.,Palaiseau - F
F.	DUNLOP	Ecole Polytechn.,Palaiseau - F
CH.	DUVAL	Univ.Aix-Marseille II-CPT - F
A.	EL MELLOUKI	Univ. Louvain la Neuve - B
J.	ELHADAD	UNIV.Aix-Marseille I and CPT - F
Y.	ELSKENS	U.L. Bruxelles - B
G.	EPIFANIO	Univ. Palerme - I
H.	EPSTEIN	IHES-Bures /Yvette - F
P.	EXNER	JINR - Dubna - USSR
L.	FADDEEV	Steklov Math. Inst.,Leningrad - USSR
F.	FALCETO	Univ. Zaragoza - E
C.	FALCOLINI	Univ. di Roma - I
M.	FANNES	Univ. C. Louvain la Neuve - B
P. A.	FARIA DA VEIGA	Ecole Polytechn.,Palaiseau - F
P.	FEDERBUSH	Univ. of Michigan - USA

M.	FEIGENBAUM	Cornell Univ. - USA
G.	FELDER	ETH-Zurich - CH
J. S.	FELDMAN	Univ. of British Columbia - Canada
P.	FERRERO	CPT, Marseille - F
R.	FIGARI	Univ. di Napoli - I
F.	FIGLIOLINI	Roma - I
M.	FLATO	Univ.Dijon - F
M.	FORGER	CERN - CH
A.	FORSTER	Univ. Toulon - F
J.D.	FOURNIER	Observ. de Nice - F
E.S.	FRADKIN	Lebedev Inst., Moscow - USSR
K.	FREDENHAGEN	Hamburg Univ. - D
M.	FREIDLIN	Moscow - USSR
R.	FROESE	Univ. of British Columbia - Canada
J.	FROHLICH	ETH-Zurich - CH
A.	GALINDO TIXAIRE	Complutense Univ., Madrid - E
G.	GALLAVOTTI	Univ.Roma - I
W.	GANS	Univ. Berlin - D
A	GARBAZEWSKI	Univ.Wroclaw - PL
X.	GARBET	CEN Cadarache - F
O.	GARCIA PRADA	Valladolid - E
W.	GARCZYNSKI	Wroclaw Univ. - PL
A.	GAUDEFROY	L.M.A - Marseille - F
K.	GAWEDZKI	IHES - Bures /Yvette - F
J. P.	GAZEAU	Univ. Paris VII - F
H.O.	GEORGII	Munich Univ. - D
C.	GERARD	Univ. Paris Sud - F
F.	GESZTESY	Graz Univ. - A
E.	GETZLER	E.N.S., Paris - F
J. M.	GHEZ	Univ. Toulon and CPT - F
R.	GIACHETTI	Univ. di Firenze - I
A. G.	GIBSON	Univ. of New Mexico - USA
R.	GIELERAK	Wroclaw Univ. - PL
F.	GIERES	LAPP, Annecy - F
J.	GINIBRE	Univ. Paris Sud - F
M.	GINOCCHIO	Univ. Paris VII - F
J.	GLIMM	Courant Inst.NY - USA
E.	GLINER	Washington Univ. - USA
P.	GODDARD	Cambridge Univ. - GB

J. L.	GOITY	Univ. de Chile - Chile
K. M.	GOLDEN	Rutgers Univ. - USA
G. A.	GOLDIN	Rutgers Univ. - USA
S.	GOLDSTEIN	Rutgers Univ. - USA
S.	GOLIN	Univ. Bielefeld - D
V.	GORINI	Univ. Milano - I
J. M.	GRAF	ETH-Zurich - CH
V.	GRECCHI	Univ. Modena - I
W.	GREENBERG	Virginia Polyt., Blacksburg - USA
R.	GRIMM	LAPP, Annecy - F
J.	GROENEVELD	Utrecht Univ. - NL
H.	GROSSE	Univ. of Vienna - A
A.	GROSSMANN	CPT, Marseille - F
C.	GRUBER	Ecole Polytechn.,Lausanne - CH
D.	GUIDO	Univ. di Roma - I
V.	GUILLEMIN	MIT, Cambridge - USA
L.	GUILLOPE	Inst. Fourier, Grenoble - F
H.Y.	GUO	Beijing Univ. - Rep of China
R.	HAAG	Hamburg Univ. - D
P.	HAAGENSEN	New York Univ. - USA
Z.	HABA	Univ. of Wroclaw - PL
G. A.	HAGEDORN	Virginia Polyt.Inst,Blacksburg - USA
J.	HALLIWELL	Univ. of Cambridge - GB
J.	HAMBERG	Lund Univ. - S
M.	HANS	Univ. Paris VII - F
T.	HARA	Univ. of Tokyo - Japan
J.	HARNAD	Ecole Polytechn., Montréal - Canada
E.	HARRELL	Georgia Inst., Atlanta - USA
J.	HARTLE	UCLA - USA
J.	HARVEY	Princeton Univ. - USA
H.	HASEGAWA	Kyoto Univ. - Japan
M.	HASHIZUME	Hiroshima Univ. - Japan
M.	HAZEWINKEL	Amsterdam - NL
G. C.	HEGERFELDT	Univ. Göttingen - RFA
B.	HELFFER	Univ. de Nantes- F
M. R.	HERMAN	Ecole Polytechn., Palaiseau - F
A.	HESLOT	Paris - F
A.	HILBERT	Univ. Bielefeld - RFA
P.	HISLOP	Univ.Toronto - Canada

P.G.	HJORTH	UCLA - USA
R.	HOEGH KROHN	Univ. of Oslo - N
T.	HOFFMANN-OSTENHOF	Univ. of Wien - A
M.	HOFFMANN-OSTENHOF	Univ. of Wien - A
H.	HOLDEN	Univ.oslo - N
M.	HOLSCHMEIDER	CPT,Marseille - F
R.	HONG TUAN	Univ. Paris Sud - F
P.	HORVATHY	Inst.f. Adv. Studies, Dublin - IRL
A.	HUBACHER	ETH-Zurich - CH
A.	HUBER	Univ. Kiel - D
N. M.	HUGENHOLTZ	Univ. Groningen - NL
W.	HUNZIKER	Inst. Phys. Théor., Zurich - CH
V.	HUSSIN	Univ. de Liège - B
D.	IAGOLNITZER	CEN-Saclay - F
H. J.	IMBENS	Utrecht - NL
J.	IMBRIE	Harvard Univ. - USA
S.N.	ISAKOV	Urals Polyth.Inst. - USSR
M.	ISRAEL	Irvine Univ. _ USA
W.	ISRAEL	Univ. of Edmonton - Canada
K.	ITO	Kyoto Univ. - Japan
M.	JACQUES	Univ. de Louvain la Neuve - B
A.	JADCZYK	Wroclaw Univ. - PL
A.	JAFFE	Harvard Univ. - USA
R.	JANCEL	Univ. Paris VII - F
B.	JAWERTH	Washington Univ. in St-Louis -USA
A.	JENSEN	Aarhus Univ. - DK
M.	JIMBO	Kyoto Univ. - Japan
G.	JONA-LASINIO	Univ. di Roma - I
M.	JONES	Boulder Univ. - USA
J.	JONES	Courant Inst. - USA
G.	KAISER	Univ. of Lowell - USA
R.	KALLSTROM	Royal Inst. of Techn.,Stockholm -S
S.	KAMPHORST	Univ. de Genève - CH
G.	KARNER	Univ. Bielefeld - D
W.	KARWOWSKI	Wroclaw Univ.- PL
D.	KASTLER	Univ. Aix-Marseille II et CPT - F
T.	KENNEDY	IHES,Bures /Yvette - F
T. W.	KEPHART	Van der Built Univ. - USA
S.	KHEIFETS	Stanford Univ. USA

N. N.	KHURI	Rockfeller Univ. - USA
C.	KILI	
C.	KING	Princeton Univ. - USA
J.R.	KLAUDER	Bell Lab. -Murray-Hill - USA
W. H.	KLINK	Univ. of Iowa - USA
S.	KNABE	T U Berlin - D
H.	KOCH	Rutgers Univ. - USA
G.Y.	KOLEROV	JINR - DUBNA - USSR
K.-I.	KONDO	Univ. of Tokyo - Japan
C.	KORTHALS ALTES	CPT Marseille - F
R.	KOTECKY	Charles Univ., Praha - CS
F.	KOUKIOU	Univ. Lausanne - CH
T.	KRUEGER	Univ.Bielefedl - D
R.	KUIK	Univ.Groningen - NL
A.	KUPIAINEN	Helsinki Univ. - SF
S.T.	KURODA	Gakushuin Univ.-Tokyo - Japan
L.	LAANAIT	CPT, Marseille - F
G.	LACHAUD	CIRM-Marseille - F
A.	LAMBERT	Univ. Aix-Marseille II,CPT, Marseille -
O.E.	LANFORD III	IHES, Bures /yvette - F
D.J.	LARNER	Reidel Publishing Company-NL
J.	LASCOUX	Ecole Polytechn.,Palaiseau - F
G.	LASSNER	Karl-Marx Univ., Leipzig - DDR
J.L.	LEBOWITZ	Rutgers Univ. - USA
O.	LEGRAND	Observatoire de Nice - F
H.	LEHMANN	II Inst. Theor. Phys.,Hamburg - D
J.T.	LEWIS	Inst. of Adv.Studies, Dublin - IRL
A.	LIBCHABER	Univ. of Chicago - USA
E.	LIEB	Princeton (USA) and IHES - F
G.	LINDBLAD	Royal Inst. of Techn.,Stockholm - S
M. E.	LOEWE	Univ. of Texas, Austin - USA
J.-F.	LOISEAU	Univ. de Pau - F
M.	LOSS	F.U. Berlin - D
C.H.	LUDERS	Univ. of Göttingen - D
G.	MACK	I.T P, Hamburg - D
K.	MADDALY	Caltech - USA
C.	MAES	Louvain - B
M.	MAIOLI	Univ. di Modena - I
V.G.	MAKHANKOV	J I N R,Dubna - USSR

M.	MAKOWKA	ITP, Lausanne - CH
M.	MANOLESSOU	IHES,Bures/Yvette - F
P.	MARCHETTI	Univ. di Padova - I
C.	MARCHIORO	Univ.Roma - I
A.	MARINI	
J.	MARION	Univ. d'Aix-Marseille II - F
F.	MARTINELLI	Univ. di Roma - I
A.	MARTINEZ	Univ.Paris-Sud - F
R.	MARTINI	Twente Univ., Enschede - NL
A.	MARZUOLI	Univ. di Pavia I
V.P.	MASLOV	USSR Acad. of Sci., Moscow - USSR
V.	MASTRANGELO	CNAM, Paris - F
F.	MATHOT- DEBACKER	U. Louvain la Neuve - B
C.	MAUDUIT	Univ.Aix-Marseille II - F
R. D.	MAWHINNEY	Harvard Univ. - USA
B.	McCOY	SUNY at Stonybrook - USA
M.	MEBKHOUT	Univ.Aix-Marseille II and CPT - F
M.L.	MEHTA	SPhT-CEN Saclay - F
E.	MENOSSI	Ferrara - I
P.	MERY	Univ. Aix-Marseille II and CPT - F
A.	MESSAGER	CPT Marseille - F
L.	MICHEL	IHES,Bures /Yvette - F
J.	MIEKISZ	Univ. of Texas at Austin - USA
P. M.	MILHEIRO DE OLIVEIRA	Facult. de Engenharia, Porto - P
S.	MIRACLE-SOLE	CPT, Marseille - F
J.	MISGUICH	CEA Cadarache - F
V.	MOAURO	Univ. di Trento - I
J.	MONTALDI	Univ. of Warwick - GB
A.	MONTORSI	Torino - I
G.	MOORE	Harvard Univ. - USA
G.	MORROS	CEN Cadarache - F
F.	MOURGUES	CEN Cadarache - F
E.	MOURRE	CPT, Marseille - F
P.	MOUSSA	CEN Saclay - F
M.	MÜLLER	Univ. Karlsruhe - D
S.	MÜLLER	Univ.Saarbrucken - D
V. F.	MULLER	Univ. Kaiserslautern - RFA
R.	MURENZI	Univ. Louvain la Neuve - B
R.	MURRAY	Austin - USA

S. A.	MUTANGADURA	Univ. of Zimbabwe - Zimbabwe
B.	NAGEL	Royal Inst. of Techn.,Stockholm - S
V.	NARODITZKY	Univ. California - USA
J.	NAUDTS	Univ. Antwerpen - B
H.	NENCKA FICEK	Polish Ac. of Sci., Poznan - PL
C.	NESSMANN	Univ. of Pennsylvania - USA
A.	NEVEU	CERN - CH
F.W.	NIJHOFF	Univ. of Utrecht - NL
F.	NILL	Max-Planck Inst., Munchen -RFA
S.P.	NOVIKOV	Steklov Math. Inst., Moscow - USSR
M. L.	O'CARROLL	Univ. Fed. de Minas Gerais, Brasil
C.	OGUEY	Ecole Polytech., Palaiseau- F
D.	OLIVE	Imperial College, London - GB
E.	OMERTI	Univ. di Trento - I
E.	ORLANDI	Univ. di Roma - I
K.	OSTERWALDER	ETH Zentrum - Zurich - CH
A. C.	OTTEWILL	Oxford Univ. - GB
R.	OUZILOU	Univ. de St Etienne - F
A.	PALANQUES-MESTRE	Univ. de Barcelona - E
R.	PARIS	CEN Cadarache - F
S.	PASQUIER	Univ. de Genève - CH
S.N.	PATNAIK	Constantine - Algérie
T.	PAUL	CPT, Marseille - F
I.P.	PAVLOCKIJ	Keldysh Inst.,Moscow - USSR
S.	PAYCHA	Bochum Univ. - D
P. A.	PEARCE	Melbourne Univ. - Australie
H.	PEDERSEN	Odense Univ. - DK
A.	PELLEGRINOTTI	Univ.Camerino - I
S.	PENATI	Univ. Milano - I
V.	PENNA	Turino - I
O.	PENROSE	Heriot-Watt Univ.,Edingurgh-GB
J.K.	PERCUS	Courant Inst. - USA
M.	PERRIN	CPT, Marseille - F
M.	PERROUD	Univ. Montréal - Canada
G.	PETERS	Royal Inst.,Stockholm - S
D.	PETRITIS	Univ. de Lausanne - CH
D.	PETZ	Math.Inst. ,Budapest - H
CH.E.	PFISTER	Ecole Polytechn., Lausanne - CH
P.	PICCO	CPT,Marseille - F

P.	PICCOLI	SISSA, Trieste - I
B.	PIETTE	Univ. Louvain la Neuve - B
C.-A.	PILLET	ETH Zurich - CH
B.	PLOHR	Univ. of Wisconsin - USA
M.C.	POLIVANOV	Steklov Inst.,Moscow - USSR
A.	PRELLER	Univ. d'Aix-Marseille II - F
E.	PRESUTTI	Univ. di Roma - I
T.	PRICHETT	CPT, Marseille - F
G.	PRON'KO	Inst.High Energ.Phys,Moscow - USSR
V.	PROTOPOPESCU	Oak Ridge Natl. Lab. - USA
R.	QUISPEL	Univ. Utrecht - NL
D.C.	RADULESCU	Rutgers Univ. - USA
G. A.	RAGGIO	ETH-Zurich - CH
G.	RAKAVY	Hebrew Univ., Jerusalem - Israël
R.	RAMMAL	CRTBT, Grenoble - F
D.	RAND	Univ. of Warwick - GB
G.	RAUZY	Univ.Aix Marseille II - F
H.	REEH	Univ. Göttingen - D
A. C.	REIS DE PAIVA	Univ. Frankfurt - D
M.	REQUARDT	Univ.Gottingen - D
J.	REZENDE	Facult. de Ciencias, Lisboa - P
M.	RIBEIRO DE FARIA	Univ.Braga - P
J. L.	RICHARD	CPT, Marseille - F
L. M.	RICHARDSON	Inst. of Phys., Bristol - GB
G.	RIDEAU	Univ. Paris VII - F
A.	RIECKERS	Univ. Göttingen - D
G.	RIELA	Univ. Palermo - I
V.	RIVASSEAU	Ecole Polytechn., Palaiseau - F
S.	ROBINSON	North Carolina Univ.-USA
F.	ROCCA	Univ. de Nice - F
R.	RODRIGUEZ	CPT, Marseille - F
H.	ROEMER	Univ. Freiburg - D
A.	RONVEAUX	Univ. N.D. de la Paix, Namur - B
H.	ROOS	Univ. Göttingen - D
L.	ROSEN	Univ. of British Columbia -Canada
B.	ROSSETTO	Univ. de Toulon - F
M.	ROTHSTEIN	I.A.S - Princeton - USA
A.	ROUET	CPT, Marseille - F
M.	ROULEUX	Chatellerault - F

W.	RUEHL	Univ. Kaiserslautern - D
D.	RUELLE	IHES,Bures/Yvette - F
S.	RUIJSENAARS	Tübingen Univ. - D
J.	RUIZ	CPT, Marseille - F
H.	RUMPF	Wien Univ. - A
A.	RUTHERFORD	Univ. of British Columbia - Canada
C.	SABBAH	Ecole Polytechn., Palaiseau - F
P.	SADOWSKI	Warsawa - PL
A.	SAENZ	Naval Res.-Washington - USA
Y.	SAINT-AUBIN	Univ. de Montréal - Canada
N.	SANCHEZ	Observatoire de Meudon - F
B.	SARAMITO	CEA Fontenay aux Roses - F
M.	SATO	Kyoto Univ. - Japan
S.	SCARLATTI	Bochum Univ. - D
W. R.	SCHNEIDER	Brown Boveri R.C., Baden - CH
R.S.	SCHOR	Belo Horizonte - Brazil
H.	SCHULZE	Univ. Göttingen - D
E.	SCOPPOLA	Univ.Roma - I
E.	SEILER	Max-Planck Inst., Munchen - D
R.	SEILER	Tech. Univ. Berlin - D
R.	SENEOR	Ecole Polytechn., Palaiseau - F
A.	SENGUPTA	Indian Inst. Tech., Kanpur - India
G.L.	SEWELL	Queen Mary College, London - GB
J.	SHABANI	Univ. du Burundi - Burundi
S.	SHLOSMAN	Inst.Problem Inform.,Moscow - USSR
S.	SHTRIKMAN	Weizmann Inst.-Rehovot-Israël
K.	SIBOLT	Max-Planck Inst., Munchen - D
H.	SIEDENTOP	Univ. Braunschweig - D
I.E.	SIGAL	Univ. of California, Irvine - USA
R.	SILVOTTI	Univ. Pavia - I
B.	SIMON	Caltech - USA
A.	SIMONI	Univ. di Napoli - I
YA.G.	SINAI	Landau Inst., Moscow - USSR
K. B.	SINHA	Indian Stat. Inst., New Delhi - India
M.	SIRUGUE-COLLIN	Univ.Aix Marseille I et CPT - F
E.	SKIBSTED	Aarhus Univ. - DK
E.K.	SKLYANIN	Steklov Math.Inst.,Leningrad-USSR
G.	SLADE	Mc Master Univ., Hamilton-Canada
F.D.	SMITH ,JR.	Cartersville - USA

A.	SOFFER	Caltech - USA
A.	SOKAL	Courant Institute - USA
J. P.	SOLOVEJ	Princeton Univ. - USA
G.	SOMMER	Univ. Bielefeld - D
B.	SOUILLARD	Ecole Polytechn., Palaiseau - F
Y.	SOULET	Univ. de Toulouse - F
J.M.	SOURIAU	Univ.d'Aix-Marseille I et CPT - F
M.	SPERA	Univ. di Roma - I
F.	STEINER	Univ. of Hamburg - D
A.	STEINMANN	ETH-Zurich - CH
O.	STEINMANN	Univ. Bielefeld - D
D.	STERNHEIMER	Univ.Dijon - F
R.	STORA	LAPP, Annecy - F
L.	STREIT	Univ. Bielefeld - D
F.	STROCCHI	SISSA, Trieste - I
R.	STROFFOLINI	Univ.Napoli - I
J.	STUBBE	Univ. Bielefeld - D
Y.M.	SUKHOV	Inst. Prob. Inform., Moscow - USSR
C.	SUNYACH	Paris - F
A.	SUTO	Univ. Lausanne - CH
V.	TAPIA	SISSA, Trieste - I
P.	TAXIL	CPT, Marseille - F
P.	TCHAMITCHIAN	Univ. Aix-Marseille III and CPT - F
T.	TCHRAKIAN	St Patrick's College, Kildare - IRL
V.	TENNA	Turino - I
D.	TESTARD	CPT., Marseille - F
A.	TETA	SISSA ,Trieste - I
B.	THALLER	F.U. Berlin - D
W.	TIMMERMANN	JINR, Dubna - USSR
G.	TOMBERGER	Univ. Osnabruck -D
N.	TONCHEV	JINR Moscow - USSR
B.	TORRESANI	CPT, Marseille - F
C.	TRAPANI	Univ. Palermo - I
R.	TRIAY	Univ. Aix-Marseille and CPT - F
R.G.	TROSS	Ottawa Univ. - Canada
G.	TURCHETTI	Univ. di Bologna I
A.	UHLMANN	Karl Marx Univ., Leipzig - DDR
G.	VALLEE	Univ.Nice - F
A.	VAN ENTER	Tchnion,Haifa - Israël

L.	VAZQUEZ	Univ. Complutense, Madrid - E
G.	VELO	Univ. di Bologna - I
A.	VERBEURE	Univ. Leuven - B
D.	VERSTEGEN	CERN - CH
J. O.	VIGFUSSON	Univ. Zurich - CH
G.	VILASI	Univ. di Salerno - I
M.	VILLANI	Univ. di Bari - I
M.	VITTOT	CPT, Marseille - F
E. J.	VLACHYNSKY	Univ. of Sydney - Australia
X.-P.	WANG	Univ. de Nantes - F
H.	WARCHALL	North Tx State Univ. Denton-USA
C. E.	WAYNE	Pennsylvania State Univ. - USA
R.	WERNER	Univ. Osnabruck - D
J.	WESS	Univ. Karlsruhe - D
B.	WHITING	Univ. of North Carolina - USA
G.L.	WIERSMA	Inst. Phys. Theor., Amsterdam - NL
A.S.	WIGHTMAN	Caltech - USA
N.	WILDBERGER	Univ. of Toronto - Canada
C.	WILLIAMSON	Univ. of Missouri - USA
M.	WINNINK	Univ.Groningen - NL
W.	WOLLNY	Techn. Hoschule Darmstadt - RFA
S.	WORONOWICZ	Univ. Warsaw - PL
W.	WRESZINSKI	Univ. de Sao Paulo - Brazil
J.	WRIGHT	Univ. Virginia - USA
W.	WYSS	Univ.Colorado,Boulder - USA
S.-T.	YAU	Univ. of California, San Diego - USA
J.	YNGVASON	Univ. of Iceland - Iceland
M.	YOR	Univ.P.M.Curie, Paris - F
F.	ZACCARIA	Univ. di Napoli - I
W.	ZACHARY	U.S. Naval Res. Lab., Washington - USA
V.A.	ZAGREBNOV	JINR,Dubna - USSR
M.	ZAHRADNIK	Charles Univ,Praha - CS
N.	ZANGHI	Univ.Bielefeld - D
R.	ZEKRI	CPT,Marseille - F
L.	ZSIDO	Math. Inst., Stuttgart - D

RANDALL LIBRARY-UNCW
3 0490 0090349 +